Biomimetic Organic Synthesis

Edited by Erwan Poupon and Bastien Nay

Related Titles

Nicolaou, K. C., Chen, J. S.

Classics in Total Synthesis III

Further Targets, Strategies, Methods

2011
ISBN: 978-3-527-32958-8

Dewick, P. M.

Medicinal Natural Products

A Biosynthetic Approach

Third Edition
2009
ISBN: 978-0-470-74168-9

Dalko, P. I. (ed.)

Enantioselective Organocatalysis

Reactions and Experimental Procedures

2007
ISBN: 978-3-527-31522-2

Breslow, R. (ed.)

Artificial Enzymes

2005
ISBN: 978-3-527-31165-1

Berkessel, A., Gröger, H.

Asymmetric Organocatalysis

From Biomimetic Concepts to Applications in Asymmetric Synthesis

2005
ISBN: 978-3-527-30517-9

Nicolaou, K. C., Snyder, S. A.

Classics in Total Synthesis II

More Targets, Strategies, Methods

2003
ISBN: 978-3-527-30684-8

Biomimetic Organic Synthesis

Volume 2
Terpenoids, Polyketides, Polyphenols, Frontiers in Biomimetic Chemistry

Edited by Erwan Poupon and Bastien Nay

WILEY-VCH Verlag GmbH & Co. KGaA

The Editors

Prof. Dr. Erwan Poupon
Université Paris-Sud
Faculté du Pharmacie
5, rue Jean-Baptiste Clément
92260 Châtenay-Malabry
France

Dr. Bastien Nay
Museum National d'Histoire
Naturelle, CNRS
57, rue Cuvier
75005 Paris
France

All books published by **Wiley-VCH** are carefully produced. Nevertheless, authors, editors, and publisher do not warrant the information contained in these books, including this book, to be free of errors. Readers are advised to keep in mind that statements, data, illustrations, procedural details or other items may inadvertently be inaccurate.

Library of Congress Card No.: applied for

British Library Cataloguing-in-Publication Data
A catalogue record for this book is available from the British Library.

Bibliographic information published by the Deutsche Nationalbibliothek
The Deutsche Nationalbibliothek lists this publication in the Deutsche Nationalbibliografie; detailed bibliographic data are available on the Internet at <http://dnb.d-nb.de>.

© 2011 Wiley-VCH Verlag & Co. KGaA, Boschstr. 12, 69469 Weinheim, Germany

All rights reserved (including those of translation into other languages). No part of this book may be reproduced in any form – by photoprinting, microfilm, or any other means – nor transmitted or translated into a machine language without written permission from the publishers. Registered names, trademarks, etc. used in this book, even when not specifically marked as such, are not to be considered unprotected by law.

Composition   Laserwords Private Ltd., Chennai
Printing and Binding   Strauss GmbH, Mörlenbach
Cover Design   Schulz Grafik-Design, Fußgönheim

Printed in the Federal Republic of Germany
Printed on acid-free paper

ISBN: 978-3-527-32580-1
ePDF ISBN: 978-3-527-63477-4
ePub ISBN: 978-3-527-63476-7
Mobi ISBN: 978-3-527-63478-1

Foreword

The beauty and diversity of the biochemical pathways developed by Nature to produce complex molecules is a good source of inspiration for chemists who want to guided in their synthetic approach by biomimetic strategies. The first biomimetic syntheses were reported at the beginning of the 20th century, with the famous examples of Collie's and Robinson's related to the synthesis of phenolics (orcinol) and alkaloids (tropinone). Since then, the number of reported biomimetic syntheses, especially in the last twenty years, has increased, demonstrating the power of these approaches in contemporary organic and bioorganic chemistry. Biomimetic strategies allow the construction of complex natural products in a minimum of steps which is in accordance with the "atom economy" principle of green chemistry and, in addition, simple reagents can be used to access the targets. Furthermore, the bioorganic consequences of such successful syntheses allow the comprehension of the biosynthetic origin of natural compounds and these processes can produce sufficient quantities of pure products to achieve biological investigations.

The biomimetic synthesis field came to maturity thanks to interconnexions between biosynthetic studies and organic synthesis, especially in the total synthesis of complex molecules. Biomimetic syntheses could even be considered as the latest stage of biosynthetic studies, confirming or invalidating the intimate steps leading to natural product skeletons. For example, the Johnson's polycyclization of squalene precursors is one of the most impressive achievements in this field. This is still organic synthesis as the reactions are taking place in the chemist's flask under chemically controlled experimental conditions, while biosynthetic steps can involve enzymatic catalysis, at least to a certain extent. However, concerning complex biochemical transformations, the exact role of enzymes has not always been clear, and has even been questionned by synthetic chemists.

The two book volumes *"Biomimetic Organic Synthesis"* fill the gap in the organic chemistry literature on complex natural products. These books gather 25 chapters from outstanding authors, not only dealing with the most important families of natural products (alkaloids, terpenoids, polyketides, polyphenols...), but also with biologically inspired reactions and concepts which are truly taking part in biomimetic processes. By assembling these books, the editors **E. Poupon** and **B. Nay** succeeded in gathering specialists in complex natural product chemistry for the benefit of the synthetic chemist community. With an educational effort in discussions and schemes, and in comparing both the biosynthetic routes and the biomimetic achievements, the demonstration of the power of the biomimetic strategies will become obvious to the readers in both research and teaching areas. These books will be a great source of inspiration for organic chemists and will ensure the continued development in this exciting field.

ESPCI-ParisTech Paris, France   *Janine Cossy*

Contents to Volume 1

Part I   Biomimetic Total Synthesis of Alkaloids   *1*

1   Biomimetic Synthesis of Ornithine/Arginine and Lysine-Derived Alkaloids: Selected Examples   *3*
Erwan Poupon, Rim Salame, and Lok-Hang Yan

2   Biomimetic Synthesis of Alkaloids Derived from Tyrosine: The Case of FR-901483 and TAN-1251 Compounds   *61*
Huan Liang and Marco A. Ciufolini

3   Biomimetic Synthesis of Alkaloids Derived from Tryptophan: Indolemonoterpene Alkaloids   *91*
Sylvie Michel and François Tillequin

4   Biomimetic Synthesis of Alkaloids Derived from Tryptophan: Dioxopiperazine Alkaloids   *117*
Timothy R. Welch and Robert M. Williams

5   Biomimetic Synthesis of Alkaloids with a Modified Indole Nucleus   *149*
Tanja Gaich and Johann Mulzer

6   Biomimetic Synthesis of Manzamine Alkaloids   *181*
Romain Duval and Erwan Poupon

7   Biomimetic Synthesis of Marine Pyrrole-2-Aminoimidazole and Guanidinium Alkaloids   *225*
Jérôme Appenzeller and Ali Al-Mourabit

8   Biomimetic Syntheses of Alkaloids with a Non-Amino Acid Origin   *271*
Edmond Gravel

9	**Biomimetic Synthesis of Azole- and Aryl-Peptide Alkaloids** *317*
	Hans-Dieter Arndt, Roman Lichtenecker, Patrick Loos, and Lech-Gustav Milroy

10	**Biomimetic Synthesis of Indole-Oxidized and Complex Peptide Alkaloids** *357*
	Hans-Dieter Arndt, Lech-Gustav Milroy, and Stefano Rizzo

Contents to Volume 2

Preface *XVII*
List of Contributors *XIX*
Biomimetic Organic Synthesis: an Introduction *XXIII*
Bastien Nay and Erwan Poupon

Part II Biomimetic Synthesis of Terpenoids and Polyprenylated Natural Compounds *395*

11	**Biomimetic Rearrangements of Complex Terpenoids** *397*
	Bastien Nay and Laurent Evanno
11.1	Introduction *397*
11.2	Beginning with Monoterpene Rearrangements *397*
11.2.1	Historical Overview of Monoterpene Rearrangements: A Century since Wagner's Structure of Camphene *397*
11.2.2	Kinetics of the Monoterpene Rearrangement and Relation with the Catalytic Landscape in Terpene Biosynthesis *399*
11.3	Biomimetic Rearrangements of Sesquiterpenes *401*
11.3.1	Caryophyllenes in Sesquiterpene Biosyntheses *401*
11.3.2	Biomimetic Studies in the Caryolane and Clovane Series *402*
11.3.3	Biomimetic Studies in the Triquinane Series *404*
11.3.4	Oxidative Rearrangements in the Silphinane Series: the Penifulvins *405*
11.3.5	Miscellaneous Sesquiterpene Rearrangements *406*
11.4	Diterpene Rearrangements *408*
11.4.1	Dead End Products in the Biomimetic Synthesis of Antheridic Acid from Gibberellins *408*
11.4.2	Biomimetic Synthesis of Marine Diterpenes from *Pseudopterogorgia elisabethae* *410*
11.4.3	Biomimetic Relationships among Furanocembranoids *414*
11.4.4	Miscellaneous Diterpenes *417*
11.5	Triterpene Rearrangements *420*

11.6	Some Examples of the Biomimetic Synthesis of Meroterpenoids *424*	
11.7	Conclusion *425*	
	References *428*	

12	**Polyprenylated Phloroglucinols and Xanthones** *433*	
	Marianna Dakanali and Emmanuel A. Theodorakis	
12.1	Introduction *433*	
12.2	Polycyclic Polyprenylated Phloroglucinols *433*	
12.2.1	Introduction and Chemical Classification *433*	
12.2.2	Biosynthesis of PPAPs *434*	
12.2.3	Biomimetic Synthesis of PPAPs *436*	
12.2.3.1	Biomimetic Total Synthesis of (±)-Clusianone *438*	
12.2.3.2	Biomimetic Approach to the Bicyclic Framework of Type A PPAPs *439*	
12.2.3.3	Biomimetic Synthesis of (±)-Ialibinone A and B and (±)-Hyperguinone B *440*	
12.2.4	Non-biomimetic Synthesis of PPAPs *441*	
12.2.4.1	Total Synthesis of Garsubellin A *441*	
12.2.4.2	Total Synthesis of Nemorosone and Clusianone through Differentiation of "Carbanions" *443*	
12.2.4.3	Total Synthesis of (−)-Hyperforin *445*	
12.2.4.4	Total Synthesis of Clusianone *448*	
12.2.5	Concluding Remarks *451*	
12.3	Polyprenylated Xanthones *452*	
12.3.1	Introduction and Chemical Classification *452*	
12.3.2	Biosynthesis of Polyprenylated Xanthones *454*	
12.3.3	Biomimetic Synthesis of Caged *Garcinia* Xanthones *455*	
12.3.3.1	Nicolaou Approach to Forbesione and Gambogin *458*	
12.3.3.2	Theodorakis' Unified Approach to Caged *Garcinia* Xanthones *459*	
12.3.3.3	Synthesis of Methyllateriflorone *459*	
12.3.3.4	Non-biomimetic Synthesis of the Caged *Garcinia* Xanthones *460*	
12.3.3.5	Concluding Remarks *463*	
	References *464*	

	Part III **Biomimetic Synthesis of Polyketides** *469*	
13	**Polyketide Assembly Mimics and Biomimetic Access to Aromatic Rings** *471*	
	Grégory Genta-Jouve, Sylvain Antoniotti, and Olivier P. Thomas	
13.1	Introduction *471*	
13.2	Polyketide Assembly Mimics *472*	
13.2.1	Type-a Mimics *475*	

13.2.1.1	Malonyl Activation 475
13.2.1.2	Without Malonyl Activation 477
13.2.2	Type-b Mimics 478
13.2.2.1	Malonyl Activation 479
13.2.2.2	Without Malonyl Activation 482
13.2.3	Type-c Mimics 483
13.3	Biomimetic Access to Aromatic Rings 485
13.3.1	Biomimetic Access to Benzenoid Derivatives 487
13.3.2	Biomimetic Access to Naphthalenoid Derivatives 492
13.3.3	Biomimetic Access to Anthracenoid Derivatives 494
13.3.4	Biomimetic Access to Tetracyclic Derivatives 495
13.3.4.1	Biomimetic Access to Tetracenoid Derivatives 495
13.3.4.2	Biomimetic Access to Tetraphenoid Derivatives 496
13.3.4.3	Biomimetic Access to Benzo[a]tetracenoid Derivatives 498
13.4	Conclusion 499
	References 499

14 Biomimetic Synthesis of Non-Aromatic Polycyclic Polyketides 503
Bastien Nay and Nassima Riache

14.1	Introduction 503
14.2	Biomimetic Studies in the Nonadride Series 504
14.2.1	Dimerization Process towards Isoglaucanic Acid 504
14.2.2	The Unresolved Case of CP-225917 and CP-263114 505
14.3	Biomimetic Syntheses Involving the Diels–Alder Reaction 506
14.3.1	Biomimetic Diels–Alder Reactions Affording Decalin Systems 506
14.3.2	Biomimetic Diels–Alder Reactions Affording Tetrahydroindane Systems 509
14.3.3	Biomimetic Diels–Alder Reactions Affording Spiro Systems 512
14.3.4	Biomimetic TADA Reactions toward FR182877, Hexacyclinic Acid and the Parent Cochleamycin A and Macquarimicin A 514
14.3.5	Biomimetic TADA Reactions toward Spinosyns 521
14.4	Biomimetic Cascade Reactions 524
14.4.1	A Metalated Ionophore Template for the Biomimetic Synthesis of Tetronasin 524
14.4.2	The 6,5,6-Fused System and Macrocycle of Hirsutellones: Work Yet to Be Done? 525
14.5	Conclusion 530
	References 530

15 Biomimetic Synthesis of Polyether Natural Products via Polyepoxide Opening 537
Ivan Vilotijevic and Timothy F. Jamison

| 15.1 | Introduction 537 |
| 15.2 | Synthetic Considerations: Baldwin's Rules 538 |

15.2.1	Control of Regioselectivity in Intramolecular Epoxide-Opening Reactions *539*	
15.3	Polycyclic Polyethers: Structure and Biosynthesis *539*	
15.3.1	Polyether Ionophores *539*	
15.3.2	Polyethers Derived from Squalene *542*	
15.3.3	Ladder Polyethers *545*	
15.4	Epoxide-Opening Cascades in the Synthesis of Polycyclic Polyethers *550*	
15.4.1	Epoxide-Opening Cascades in the Synthesis of Polyether Ionophores *550*	
15.4.2	Applications of Epoxide-Opening Cascades in the Synthesis of Ionophores *554*	
15.4.3	Epoxide-Opening Cascades in the Synthesis of Squalene-Derived Polyethers *558*	
15.4.4	Epoxide-Opening Cascades in the Synthesis of Ladder Polyethers *565*	
15.4.4.1	Iterative Approaches *565*	
15.4.4.2	Epoxide-Opening Cascades Leading to Fused Polyether Systems *567*	
15.4.4.3	Applications of Epoxide-Opening Cascades in the Synthesis of Ladder Polyethers *580*	
15.5	Summary and Outlook *583*	
	References *584*	
16	**Biomimetic Electrocyclization Reactions toward Polyketide-Derived Natural Products** *591*	
	James Burnley, Michael Ralph, Pallavi Sharma, and John E. Moses	
16.1	Introduction *591*	
16.2	Electrocyclic Reactions *592*	
16.3	Polyketides *593*	
16.4	Fatty Acid Biosynthesis *594*	
16.5	Biomimetic Analysis *597*	
16.6	6π Electrocyclizations *598*	
16.6.1	Tridachiahydropyrones *599*	
16.6.2	Tridachione Family *603*	
16.6.3	Pseudorubrenoic Acid A *608*	
16.6.4	Torreyanic Acid *610*	
16.7	8π Systems and the Black 8π–6π Electrocyclic Cascade *612*	
16.7.1	Endiandric Acids *612*	
16.7.2	Nitrophenyl Pyrones: SNF4435 C and D *618*	
16.7.3	Ocellapyrones *621*	
16.7.4	Elysiapyrones *624*	
16.7.5	Shimalactones *625*	
16.8	Biological Electrocyclizations and Enzyme Catalysis *628*	
16.9	Conclusion *631*	
	Acknowledgments *632*	
	References *632*	

Part IV Biomimetic Synthesis of Polyphenols *637*

17 Biomimetic Synthesis and Related Reactions of Ellagitannins *639*
Takashi Tanaka, Isao Kouno, and Gen-ichiro Nonaka
17.1 Introduction *639*
17.2 Biosynthesis of Ellagitannins *641*
17.3 Biomimetic Total Synthesis of Ellagitannins *642*
17.3.1 Chemical Synthesis of Ellagitannins by Biaryl Coupling of Galloyl Esters *642*
17.3.2 Ellagitannins with $^1C_4$ Glucopyranose Cores *645*
17.3.3 Synthesis of an Allagitannin with 3,6-(R)-HHDP Group *651*
17.3.4 Synthesis of Ellagitannins by Double Esterification of Hexahydroxydiphenic Acid *651*
17.3.5 Biomimetic Synthesis of Dimeric Ellagitannin *658*
17.4 Conversion of Dehydroellagitannins into Related Ellagitannins *659*
17.4.1 Reduction of DHHDP Esters *659*
17.4.2 Reaction with Thiol Compounds and the Biomimetic Synthesis of Chebulagic Acid *662*
17.4.3 Other Reactions of DHHDP Esters *663*
17.5 Reactions of *C*-Glycosidic Ellagitannins *663*
17.5.1 Conversion between Pyranose-Type Ellagitannins and *C*-Glycosidic Ellagitannins *665*
17.5.2 Reaction at the C1 Positions of *C*-Glycosidic Ellagitannins *665*
17.5.3 Oxidation of *C*-Glycosidic Ellagitannins *669*
17.6 Conclusions and Perspectives *670*
References *672*

18 Biomimetic Synthesis of Lignans *677*
Craig W. Lindsley, Corey R. Hopkins, and Gary A. Sulikowski
18.1 Introduction to Lignans *677*
18.1.1 Biomimetic Synthesis of Lignans *681*
18.1.1.1 Biomimetic Synthesis of Podophyllotoxin-Like Lignans *681*
18.1.1.2 Biomimetic Synthesis of Furofuran Lignans *681*
18.1.1.3 Biomimetic Synthesis of Benzoxanthenone Lignans *683*
18.1.1.4 Biomimetic Synthesis of Benzo[*kl*]xanthene Lignans *686*
18.2 Conclusion *688*
References *691*

19 Synthetic Approaches to the Resveratrol-Based Family of Oligomeric Natural Products *695*
Scott A. Snyder
19.1 Introduction *695*
19.2 Biosynthetic Approaches *697*
19.3 Stepwise Synthetic Approaches *705*
19.3.1 Work toward Single Targets within the Resveratrol Family *705*

19.3.2	Towards a Universal, Controlled Synthesis Approach	709
19.4	Conclusions	717
	Acknowledgments	717
	References	718

20 Sequential Reactions Initiated by Oxidative Dearomatization. Biomimicry or Artifact? *723*
Stephen K. Jackson, Kun-Liang Wu, and Thomas R.R. Pettus

20.1	Overview	723
20.2	Oxidative Dearomatization Sequences and the Initial Intermediate	723
20.3	Intermolecular Dimerizations	724
20.4	Successive Intermolecular Reactions	727
20.5	Intramolecular Cycloadditions	729
20.6	Other Successive Intramolecular Cascade Sequences	731
20.7	Successive Tautomerizations and Rearrangements	733
20.8	Sequential Ring Rupture and Contraction	737
20.9	Sequential Ring Rupture and Expansion	739
20.10	Successive Intramolecular and Intermolecular Reactions	741
20.11	Natural Products Hypothesized to Conclude Phenol Oxidative Cascades	741
20.12	Conclusion	747
	References	747

Part V Frontiers in Biomimetic Chemistry: From Biological to Bio-inspired Processes *751*

21 The Diels–Alderase Never Ending Story *753*
Atsushi Minami and Hideaki Oikawa

21.1	Introduction	753
21.2	Diels–Alderases Found in Nature	754
21.2.1	Lovastatin Nonaketide Synthase	755
21.2.2	Macrophomate Synthase	756
21.2.3	Solanapyrone Synthase	758
21.3	Intramolecular Diels–Alder Reactions Possibly Catalyzed by Dehydratase or DH-Red-Domain of PKS or Hybrid PKS-NRPS	760
21.3.1	Equisetin and Chaetoglobosin (Compactin, Lovastatin, Solanapyrone)	761
21.3.2	Kijanimicin, Chlorothricin, and Tetrocarcin A	763
21.3.3	Indanomycin	764
21.3.4	Spinosyn	766
21.4	Diels–Alder Reactions after Formation of Reactive Substrates by Oxidation Enzymes	767
21.4.1	Oxidation of Phenol and Catechol to Reactive Dienone and Orthoquinone	768

21.4.2	Conjugated Diene Derived from Dehydrogenation of Prenyl Side Chain 775
21.4.3	Cyclopentadiene Formation Derived from Dehydrogenation 779
21.5	Summary 779
	References 782

22 Bio-Inspired Transfer Hydrogenations 787
Magnus Rueping, Fenja R. Schoepke, Iuliana Atodiresei, and Erli Sugiono

22.1	Introduction 787
22.2	Nature's Reductions: Dehydrogenases as a Role Model 787
22.3	Brønsted Acid Catalyzed Transfer Hydrogenation of Imines, Imino Esters, and Enamines 788
22.4	Asymmetric Organocatalytic Reduction of N-Heterocycles 800
22.4.1	Asymmetric Organocatalytic Reduction of Quinolines 800
22.4.2	Asymmetric Brønsted Acid Catalyzed Hydrogenation of Indoles 805
22.4.3	Asymmetric Brønsted Acid Catalyzed Hydrogenation of Benzoxazines, Benzothiazines, Benzoxazinones, Quinoxalines, Quinoxalinones, Diazepines, and Benzodiazepinones 806
22.4.4	Asymmetric Organocatalytic Reduction of Pyridines 813
22.5	Asymmetric Organocatalytic Reductions in Cascade Sequences 814
22.6	Conclusion 817
	References 818

23 Life's Single Chirality: Origin of Symmetry Breaking in Biomolecules 823
Michael Mauksch and Svetlana B. Tsogoeva

23.1	Introduction 823
23.2	Autocatalytic Enantioselective Reactions 825
23.3	Autocatalysis and Self-replication 833
23.4	Polymerization and Aggregation Models of Enantioenrichment 834
23.5	Phase Equilibria 835
23.6	Adsorption on Chiral Surfaces 837
23.7	Spontaneous Symmetry Breaking in Conglomerate Crystallizations 837
23.8	Symmetry Breaking in Reaction–Diffusion Models, Collision Kinetics, and Membrane Diffusion 840
23.9	Concluding Remarks and Outlook 840
	References 841

Part VI Conclusion: From Natural Facts to Chemical Fictions 847

24 Artifacts and Natural Substances Formed Spontaneously 849
Pierre Champy

24.1	Introduction 849
24.2	Glucosidases as Triggers for Formation of By-products 852

24.3	Oxidation Processes 853	
24.3.1	Thiol Oxidation 853	
24.3.2	Oxidation Processes of Oxygenated Functions 857	
24.3.3	Newly Oxygenated Products 859	
24.3.4	Oxidative Coupling 864	
24.3.5	The *N*-Oxide and Oxoalkaloid Cases 864	
24.4	Exposure to Light 870	
24.4.1	Isomerization and Epimerization 870	
24.4.2	Rearrangements 872	
24.4.3	Photocycloaddition and Photodimerization 876	
24.5	Heat and Pressure 878	
24.5.1	Epimerization and Isomerization In or Out of Solutions 880	
24.5.2	Hydrodistillation 880	
24.5.3	Decarboxylation Processes 885	
24.5.4	Supercritical CO_2 885	
24.6	Alkaline Media 888	
24.6.1	Amination Processes 888	
24.6.2	Other Base-Catalyzed Reactions 892	
24.7	Acidic Conditions during Purifications 895	
24.7.1	Epimerization 895	
24.7.2	Hydrolysis 897	
24.7.3	Other Acid-Catalyzed Reactions 900	
24.8	Protic Solvents 903	
24.8.1	Lactonic Compounds: Epimerization, Transesterification 905	
24.8.2	Esterification, Transesterification 908	
24.8.3	"Apparent Alkylations" 910	
24.8.4	Formation of Acetals 913	
24.9	Acetone-Derived Artifacts 916	
24.10	Halogenated Solvents 919	
24.11	Protoberberines, a *"Cabinet de Curiosités"* 921	
24.12	Conclusion 925	
	References 930	

Index 935

Preface

When we decided to start this project, at the end of 2008, we were perfectly aware that the amount of work to provide on it, the Biomimetic Organic Synthesis saga, would be very important. In fact, we were far from reality since the field not only concerns the huge universe of natural product chemistry, but also tends to embrace many fields beyond. We tried to design this book according to natural product chemistry principles, mainly by compound classes, and hope that few of them slipped our notice. Hopefully, the contributors who were asked to write a chapter in their respective field have welcomed this project with a great enthusiasm and worked hard to finish their chapter on time. Our editing adventure is now ending and we want now to warmly thank all of them for their outstanding contribution to this lengthy book. We also want to pay tribute to Professor François Tillequin, so happy with natural product chemistry, who recently passed away. Special thanks are also due to the staff of Wiley-VCH especially to Dr Gudrun Walter and Lesley Belfit for excellent collaboration.

Biomimetic synthesis is the construction of natural products by chemical means using Nature's hypothetical or established strategies, *i.e.* starting from synthetic mimicry of Nature's biosynthetic precursors, ideally by way of biologically compatible reactions. In theory, this principle can be applied to all natural product classes, from the simplest to the most complex compounds. Yet the activation methods in the laboratory can be far from Nature's enzymatic environment, and the biomimetic step can then be more difficult than expected at first glance. The way may therefore be tricky, even for a skilled chemist. We hope this book will delight readers by materializing most of organic synthesis concepts built from biochemical (biosynthetic) inspirations. Fortunately, readers may find solutions to synthetic problems or, at least, find a new way to improve their knowledge, as we did.

Enjoy reading.

March 2011

Erwan Poupon
Université Paris-Sud, Châtenay-Malabry, France
Bastien Nay
Muséum National d'Histoire Naturelle, Paris, France

List of Contributors

Sylvain Antoniotti
Université de
Nice-Sophia Antipolis
Faculté des Sciences
Départment de Chimie
28 AvenueValrose
06108 Nice Cedex 2
France

Iuliana Atodiresei
RWTH Aachen University
Institute of Organic Chemistry
Landoltweg 1
52074 Aachen
Germany

James Burnley
University of Nottingham
Faculty of Science
School of Chemistry
University Park
Nottingham NG7 2RD
United Kingdom

Pierre Champy
Université Paris-Sud 11
Chimie des Substances
Naturelles CNRS
UMR 8076 BioCIS
Faculté de Pharmacie
5 rue Jean-Baptiste Clément
92296 Châtenay-Malabry
France

Marianna Dakanali
University of California
San Diego
Department of Chemistry and
Biochemistry
9500 Gilman Drive
La Jolla
San Diego, CA 92093-0358
USA

Laurent Evanno
Muséum National
d'Histoire Naturelle
Unité Molécules de
Communication et Adaptation
des Micro-organismes associée au
CNRS (UMR 7245)
57 rue Cuvier
75005 Paris
France

List of Contributors

Grégory Genta-Jouve
Université de
Nice-Sophia Antipolis
Faculté des Sciences
Départment de Chimie
28 AvenueValrose
06108 Nice Cedex 2
France

Corey R. Hopkins
Vanderbilt University
Medical Center
Department of Chemistry
Department of Pharmacology
Vanderbilt Program in Drug
Discovery
Nashville, TN 37272-6600
USA

Stephen K. Jackson
University of California
Department of Chemistry
and Biochemistry
Santa Barbara, CA 93106-9510
USA

Timothy F. Jamison
Massachusetts
Institute of Technology
Department of Chemistry
77 Massachusetts Avenue
Cambridge, MA 02139
USA

Isao Kouno
Nagasaki University
Graduate School of
Biomedical Sciences
Department of Molecular
Medicinal Sciences
1-14 Bunkyo-machi
Nagasaki 852-8521
Japan

Craig W. Lindsley
Vanderbilt University
Medical Center
Department of Chemistry
Department of Pharmacology
Vanderbilt Program in
Drug Discovery
Nashville, TN 37272-6600
USA

Michael Mauksch
University of
Erlangen- Nuremberg
Department of Chemistry
and Pharmacy
Henkestrasse 42
91054 Erlangen
Germany

Atsushi Minami
Hokkaido University
Graduate School of Science
Division of Chemistry
Sapporo 060-0810
Japan

John E. Moses
University of Nottingham
Faculty of Science
School of Chemistry
University Park
Nottingham NG7 2RD
United Kingdom

Bastien Nay
Muséum National
d'Histoire Naturelle
Unité Molécules de
Communication et Adaptation
des Micro-organismes associée au
CNRS (UMR 7245)
57 rue Cuvier
75005 Paris
France

Gen-ichiro Nonaka
Usaien Pharmaceutical Company
Ltd. 1-4-6 Zaimoku
Saga 840-0055
Japan

Hideaki Oikawa
Hokkaido University
Graduate School of Science
Division of Chemistry
Sapporo 060-0810
Japan

Thomas R.R. Pettus
University of California
Department of Chemistry
and Biochemistry
Santa Barbara, CA 93106-9510
USA

Michael Ralph
University of Nottingham
Faculty of Science
School of Chemistry
University Park
Nottingham NG7 2RD
United Kingdom

Nassima Riache
Muséum National
d'Histoire Naturelle
Unité Molécules de
Communication et Adaptation
des Micro-organismes associée au
CNRS (UMR 7245)
57 rue Cuvier
75005 Paris
France

Magnus Rueping
RWTH Aachen University
Institute of Organic Chemistry
Landoltweg 1
52074 Aachen
Germany

Fenja R. Schoepke
RWTH Aachen University
Institute of Organic Chemistry
Landoltweg 1
52074 Aachen
Germany

Pallavi Sharma
University of Nottingham
Faculty of Science
School of Chemistry
University Park
Nottingham NG7 2RD
United Kingdom

Scott A. Snyder
Columbia University
Department of Chemistry
Havemeyer Hall
3000 Broadway
New York, NY 10027
USA

Erli Sugiono
RWTH Aachen University
Institute of Organic Chemistry
Landoltweg 1
52074 Aachen
Germany

Gary A. Sulikowski
Vanderbilt University
Medical Center
Department of Chemistry
Department of Pharmacology
Vanderbilt Program
in Drug Discovery
Nashville, TN 37272-6600
USA

Takashi Tanaka
Nagasaki University
Graduate School of
Biomedical Sciences
Department of Molecular
Medicinal Sciences
1-14 Bunkyo-machi
Nagasaki 852-8521
Japan

Emmanuel A. Theodorakis
University of California
San Diego
Department of Chemistry and
Biochemistry
9500 Gilman Drive
La Jolla
San Diego, CA 92093-0358
USA

Olivier P. Thomas
Université de
Nice-Sophia Antipolis
Faculté des Sciences
Départment de Chimie
28 AvenueValrose
06108 Nice Cedex 2
France

Svetlana B. Tsogoeva
University of
Erlangen-Nuremberg
Department of Chemistry
and Pharmacy
Henkestrasse 42
91054 Erlangen
Germany

Ivan Vilotijevic
Massachusetts
Institute of Technology
Department of Chemistry
77 Massachusetts Avenue
Cambridge, MA 02139
USA

Kun-Liang Wu
University of California
Department of Chemistry
and Biochemistry
Santa Barbara, CA 93106-9510
USA

Biomimetic Organic Synthesis: an Introduction

Bastien Nay and Erwan Poupon

> *Nature always makes the best of possible things*
> Aristotle

1
General remarks

"Biomimetic", "biomimicry" and "biologically inspired" are terms that can be used whenever Nature symphonic processes inspire human creation. This will encompass science, arts, architecture and so on. In this book, we will focus our attention on organic chemistry. To assemble these two volumes, we were spoiled for choice. A selection of topics was made to give a wide perspective on biomimetic synthesis, especially when dedicated to natural product chemistry and total synthesis which will constitute the major part and a guiding principle all along this book. We are of course conscious that entire fields are left aside such as material chemistry or supramolecular chemistry. Yet we wish that this book will convince most readers that applying biomimetic strategies to organic synthesis can provide a shortcut toward efficiency, beauty and originality, in a wide scope of fields (Figure 1).

2
Natural products as a vital lead

For many decades, the question why living organisms of all kingdoms produce secondary metabolites ("natural products") has been the subject of many debates. As soon as the first structures were determined, chemists also started thinking about the possible origin of the molecules [1].

Natural products are at the center of chemical ecology and have been forged in the crucible of Darwinian evolution. Many theories have tried to explain the incredible diversity of natural substances, including appealing views concluding that the living organisms that may be selected by evolution are the ones that favor chemical diversity [2], which may be the product of biochemical combinatorial processes. It is needless to remind here the importance of secondary metabolites

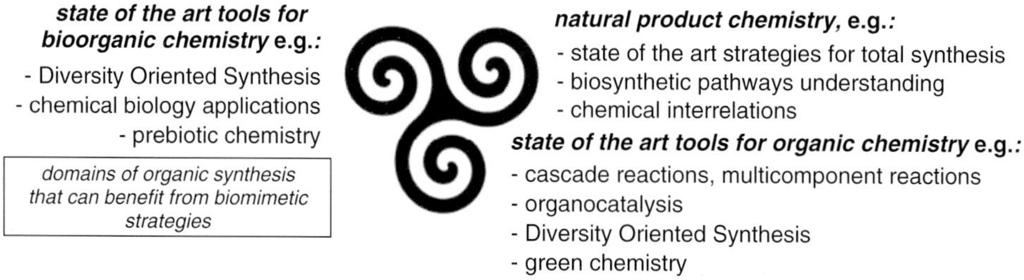

state of the art tools for bioorganic chemistry e.g.:
- Diversity Oriented Synthesis
- chemical biology applications
- prebiotic chemistry

domains of organic synthesis that can benefit from biomimetic strategies

natural product chemistry, e.g.:
- state of the art strategies for total synthesis
- biosynthetic pathways understanding
- chemical interrelations

state of the art tools for organic chemistry e.g.:
- cascade reactions, multicomponent reactions
- organocatalysis
- Diversity Oriented Synthesis
- green chemistry

Figure 1 Tentacular influence of biomimetic strategies.

for Humanity notably as a source of drug candidates, pharmaceuticals, flavours, fragrances, food supplements. This aspect has been widely covered over the years.

Back to the biological functions, activities of natural substances may be explain because they interact with and modulate almost all type of biological targets including proteins (enzymes, receptors, and cytoskeleton), membranes, or nucleic acids. Here again, important notions such as the conservation of protein domains in living organisms or the selection of privileged scaffolds have been discussed and should not be ignored by chemists interested in natural substances [3].

3
Biomimetic synthesis

Biomimetic synthesis is the construction of natural products by chemical means using Nature's hypothetical or established strategy. It therefore stands in close relation with biosynthetic studies. Engaged in biomimetic strategies, the chemists

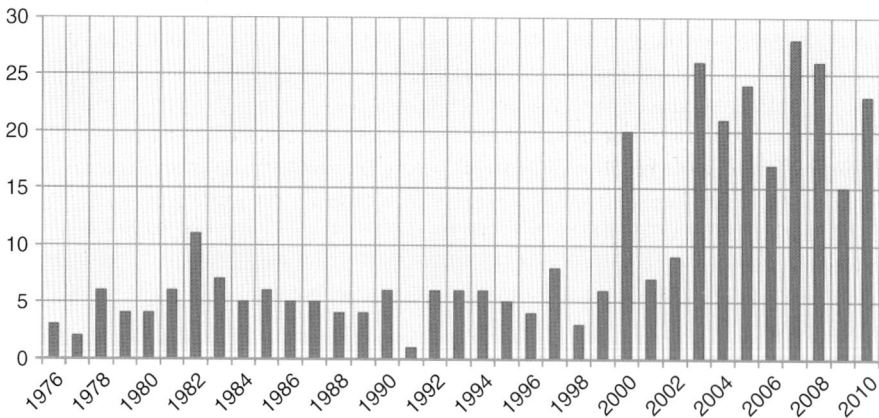

Figure 2 Analysis of bibliographical search in SciFinder with the terms "biomimetic total synthesis" from 1960 to 2010, leading to 339 references (30/10/2010).

will face pragmatic issues for planning their synthesis but will undoubtedly wonder about the exact role of enzymes in nature's way to construct sometimes highly complex structures. From highly evolved biosynthetic pathways involving enzymes with very high selectivity to less evolved routes or less specific enzymic catalysis, secondary metabolism pathways embrace a wide range of chemical efficiency. Biomimetic strategies will often, usually unintentionally, point out this aspect.

An increasing number of total syntheses have been termed "biomimetic" or "biosynthetically inspired" and so on, especially during the last decade. Basically a quick search in *SciFinder*, using the term "biomimetic total synthesis" over the period 1960–2010, afforded 339 occurrences, beginning in 1976. As illustrated in Figure 2, the last ten years have shown an increasing number of publications in this field.

Different situations and "degree of biomimicry" can then be instinctively distinguished when closely analyzing the final total synthesis including:

- a total synthesis featuring a biomimetic crucial step after a multistep total synthesis of the natural product precursor;
- a total synthesis featuring a biomimetic cascade reaction from more or less simple precursors.

Many examples in both situations will be found in this book. Simple parameters can at first sight help defining the relevance of the biomimetic step especially in terms of complexity generation. These include among others: the number of new carbon-carbon bonds and cycles formed, the number of changes in hybridization state of carbon atoms, the global oxidation state and also the stereochemical changes. The chemical reactions borrowed from Nature tool box for building carbon-carbon bonds that will emerge will particularly stand out century-old reactions such as aldolization, Claisen condensation, Mannich reaction and Diels-Alder and other cycloadditions. Situations where self-assembly relies merely on inherent reactivity of the precursors are probably situations that will be most likely mimicked successfully in the laboratory [4]. Beautiful examples will be presented in this book. Since simplicity should be the hallmark of total syntheses approaching the perfect or ideal total synthesis [5], the use of biomimetic strategies can advantageously bring solutions to intricate synthetic problems [6]. Let us finally add that by many aspects we will not debate here, biomimetic strategies may fulfill the criteria of "green chemistry" and "atom economy" when exploiting for example multicomponent strategies [7].

4
On the organization of the book, in close relation with secondary metabolism biochemistry

The chief purpose of this book is not to give a full coverage of the main biosynthetic pathways of secondary metabolites. Yet a particular care has been brought by authors in providing basic key elements of biosynthesis in the different chapters, to make them as comprehensive as possible to readers. If more has to be known about biosynthetic elements, we suggest referring to excellent books that have already covered the subject [8].

4.1
Alkaloids

An alkaloid is a cyclic organic compound containing nitrogen in a negative oxidation state which is of limited distribution among living organisms. This is a modern definition for a heterogeneous class of natural substances given by S. W. Pelletier in the first volume of the series of famous periodical books *Alkaloids* [9].

In our *Biomimetic Organic Synthesis*, all chapters related to alkaloids have been gathered in the first volume of this edition. Many classifications were proposed for this class of compounds. They could be based on the biogenesis, structure, biological origin, spectroscopic properties or also biological properties. The great lack of general principles towards a unified classification is obvious and the borderline between alkaloids *sensu stricto* and other natural nitrogen-containing secondary metabolites (such as peptides or nucleosidic compounds) is often unclear. A classification based on the nitrogen source of the alkaloid will guide

Chapter 2.1 by R. Salame, L.-Y. Yan and E. Poupon

Biomimetic Synthesis of Ornithine/Arginine and Lysine-Derived Alkaloids: Selected Examples

Chapter 2.2 by H. Liang and M. A. Ciufolini

Biomimetic synthesis of alkaloids derived from tyrosine: the case of FR-901483 and TAN 1251 compounds

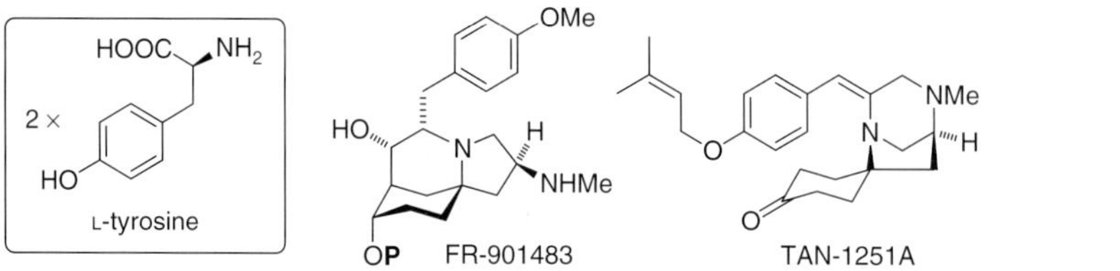

Scheme 1 Alkaloids derived from ornithine/arginine, lysine and tyrosine.

our choice of topics tackled in this book. This has the advantage of linking the biosynthetic origin (and thereby the biomimetic approach) and the chemical structure of the secondary metabolites. Accordingly, chapters will be devoted to alkaloids primarily deriving from ornithine, arginine, lysine, tyrosine (Scheme 1) and of course tryptophan (which highly diverse chemistry will be envisaged in three chapters, Scheme 2). Particularly, we thought that the important class of indolomonoterpenic alkaloids, despite largely discussed along the years, deserved an overview chapter putting forward crucial ideas and challenges when approaching their chemistry. A large array of natural substances isolated from microorganisms displays a diketopiperazine ring system in more or less rearranged form. Because of constant efforts towards the comprehension of their biosynthesis and their total synthesis, a chapter is dedicated to these alkaloids. In fact, they are probably among the secondary metabolites that have largely benefited from biomimetic strategies

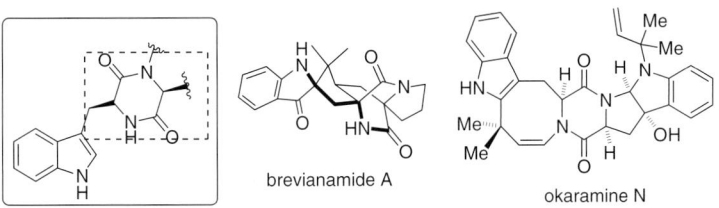

Scheme 2 Alkaloids derived from tryptophan.

XXVIII | *Biomimetic Organic Synthesis: an Introduction*

Chapter 2.6 by R. Duval and E. Poupon

Biomimetic synthesis of manzamine alkaloids

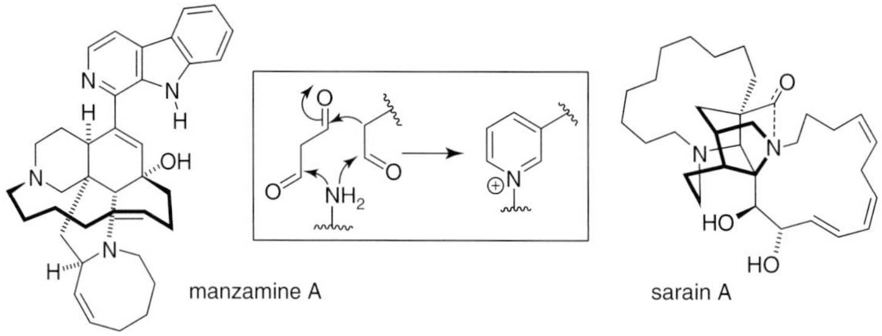

Chapter 2.7 by J. Appenzeller and A. Al Mourabit

Biomimetic synthesis of marine pyrrole-2-aminoimidazole and guanidium alkaloids

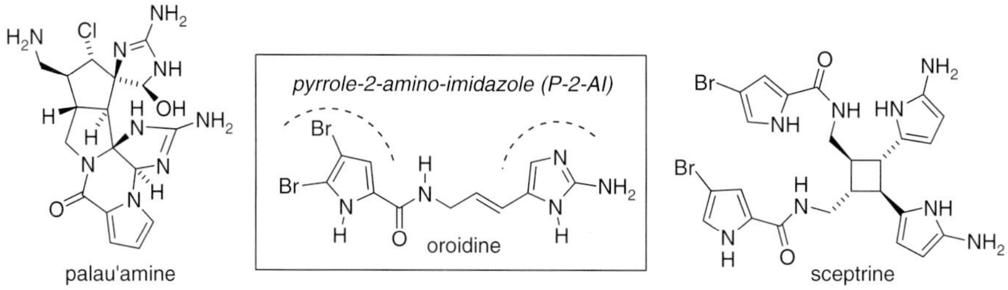

Scheme 3   Focus on two classes of complex marine alkaloids.

with undeniable success. Also of special interest in biomimetic chemistry, several alkaloids are derived in nature by profound modifications of the indole nucleus itself giving rise to secondary metabolites for which the biosynthetic origin is not obvious at first glance. The examples of quinine and camptothecin were among the first structures where such phenomena were suspected. Up to now, such biomimetic syntheses imply an initial oxidation step of the indole nucleus, which selected examples are disclosed in a proper chapter.

Two chapters will also cover the biomimetic synthesis of two classes of important marine alkaloids: the manzamine type alkaloids and the pyrrole-2-aminoimidazole alkaloids (Scheme 3). Alkaloids encompass also secondary metabolites obviously deriving from terpenes/steroids or polyketides, they will be considered as well in an individualized chapter (Scheme 4). Despite more related to polyketides in terms of biosynthetic machinery (see below), peptides alkaloids will be covered by two chapters in this section (Scheme 5).

4.2
Terpenes and terpenoids

Terpenes and terpenoids will be covered by chapters of the second volume of *Biomimetic Organic Synthesis*. They are made by terpene cyclases which catalyze

Chapter 2.8 by E. Gravel

Biomimetic synthesis of alkaloids with a non amino-acid origin

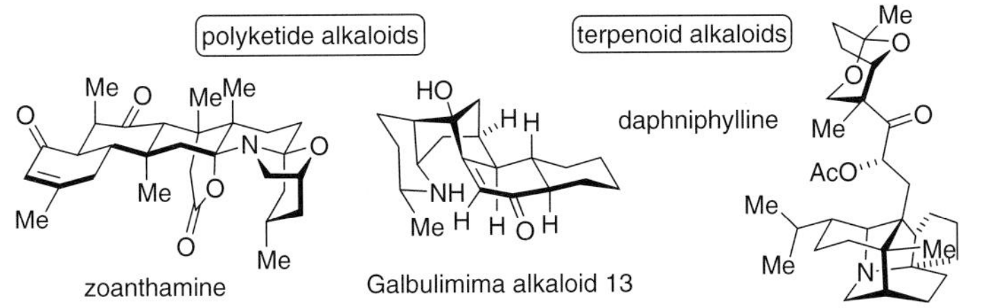

Scheme 4 Polyketide and terpenoid alkaloids.

Chapter 2.9 by H.-D. Arndt, R. Lichtenecker, P. Loos and L.-G. Milroy

Biomimetic synthesis of azole- and aryl-peptide alkaloids

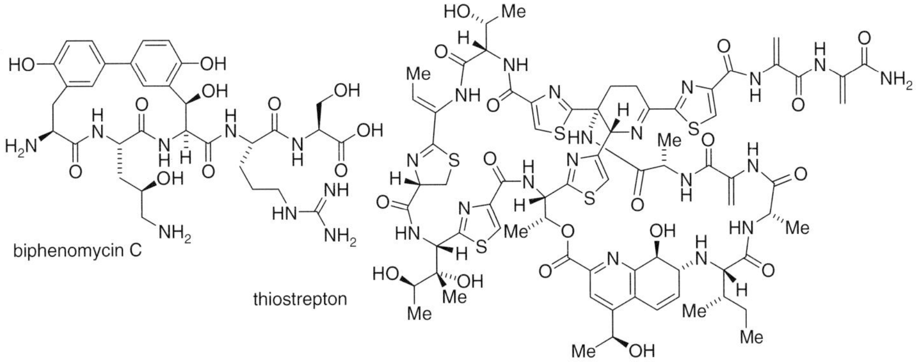

Chapter 2.10 by H.-D. Arndt, L.-G. Milroy, S. Rizzo

Biomimetic synthesis of indole-oxidized and complex peptide alkaloids

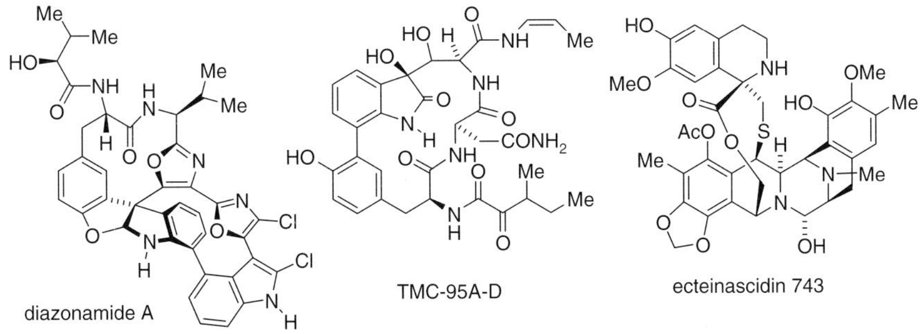

Scheme 5 Complex peptide alkaloids.

highly efficient reactions at the origin of such a rich chemistry. The cationic cascade aspect of terpene biosynthesis from oligomers of activated forms of isoprene is very appealing for many biomimetic endeavors. Current aspects of terpene biosynthesis include the interesting notion of accuracy of terpene cyclases, an issue that is closely

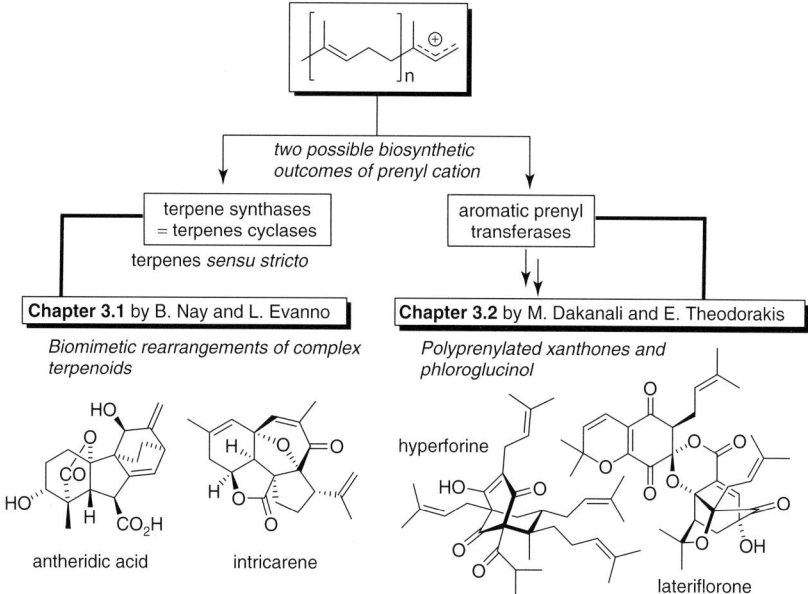

Scheme 6 Biomimetic synthesis of terpenes and terpenoids.

related to the quest for selectivity in biomimetic synthesis of such compounds. Leading review articles have already been published elsewhere about cationic cascade cyclizations [10]. A chapter will focus on the post-polycyclization events, dealing with biomimetic rearrangements of already complex terpene structures. Polyprenylated secondary metabolites resulting primarily from the transfer of prenyl units to aromatic rings by aromatic prenyl transferases, and sometimes followed by rearrangements, will be covered by another chapter in the second volume (Scheme 6). Other aspects of terpene alkaloids have been developed in the first volume of *Biomimetic Organic Synthesis* (especially, the reader can refer to chapters 2.3 and 2.8).

4.3
Polyketides

Manipulations of polyketide gene clusters have contributed to a revolution in the comprehension of polyketides (PK), and also of non-ribosomal peptides (NRP), biosynthesis. The genome sequencing of numerous PK and NRP producing microorganisms has revealed a large number of cryptic metabolites mostly unknown. Current challenges include the discovery of such natural compounds by allowing the expression of the corresponding genes ("turn them on") and the programming/reprogramming of fungal PKS. Not to be forgotten is the implication of the PK pathways in aromagenesis in nature *via* the biosynthesis of phenols. We therefore

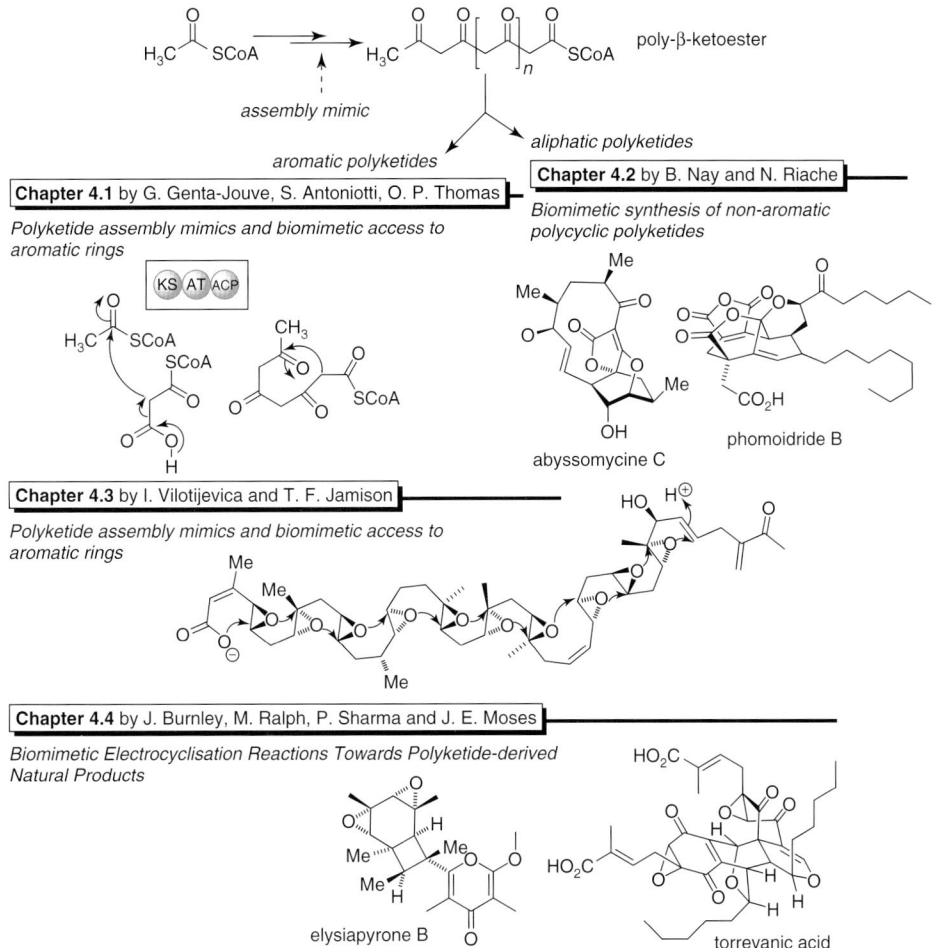

Scheme 7 Biomimetic synthesis of polyketides.

decided to ask for a contribution on biomimetic mimics of the fundamental steps of PK assembly and phenol ring formation. Turning our attention to more complex structures, beautiful examples of biomimetic synthesis of complex non aromatic polycyclic-PKs will be presented. The two following chapters will deal with two specific classes of natural substances characterized by their seminal mechanism of formation: *i.e.* polyepoxide ring opening and electrocyclization (Scheme 7).

4.4
Polyphenolic compounds

Another important biosynthetic route to aromatic rings in nature is provided by the shikimate/chorismate pathway. Simple phenolic acids enter the biosynthesis

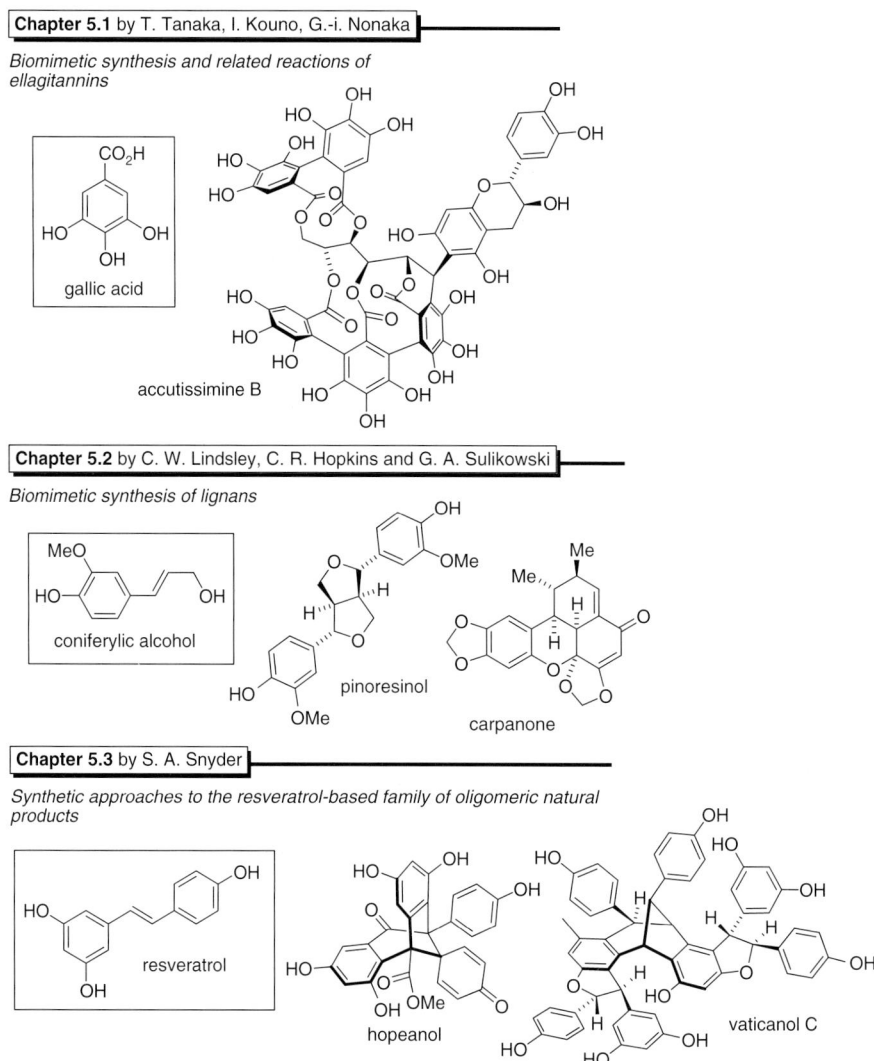

Scheme 8 Biomimetic synthesis of polyphenolic natural substances.

of sometimes highly complex ellagitannins, a class of hydrolysable tannins widely studied for their health benefits (Scheme 8). Among phenylpropanoids natural substances directly derived from chorismate are the lignans that are discussed in the following chapter. Typical extended phenylpropanoids include compounds such as flavonoids and stilbenes. The chemistry of flavonoids has been widely studied and reviewed over the years [11]. This is not the case for natural substances deriving from resveratrol which hold center stage in the last few years because of the growing importance of resveratrol itself in human health, and because of new developments in the total synthesis of this very interesting class of polycyclic molecules. For these last three classes of molecules (ellagitannins, lignans, resveratrol derived), radical phenolic couplings plays a center role as the main source of carbon-carbon bonds.

4.5
Frontiers in biomimetic synthesis

At the cross-roads of methodology and total synthesis, a few topics will show how nature observation, especially enzymic mechanisms, can lead to new discoveries in organic chemistry (Scheme 9). A discussion on the engaging issue of occurrence of the Diels-Alder reaction in nature will be conducted in a chapter. The exponential impact of organocatalysis in organic chemistry will be illustrated by the challenging problem of transfer hydrogenations in a bio-inspired manner. Once again a plethora of review articles and books deals with the other aspects of organocatalysis [12]. Finally, by many aspects, biomimetic organic chemistry may be closely linked

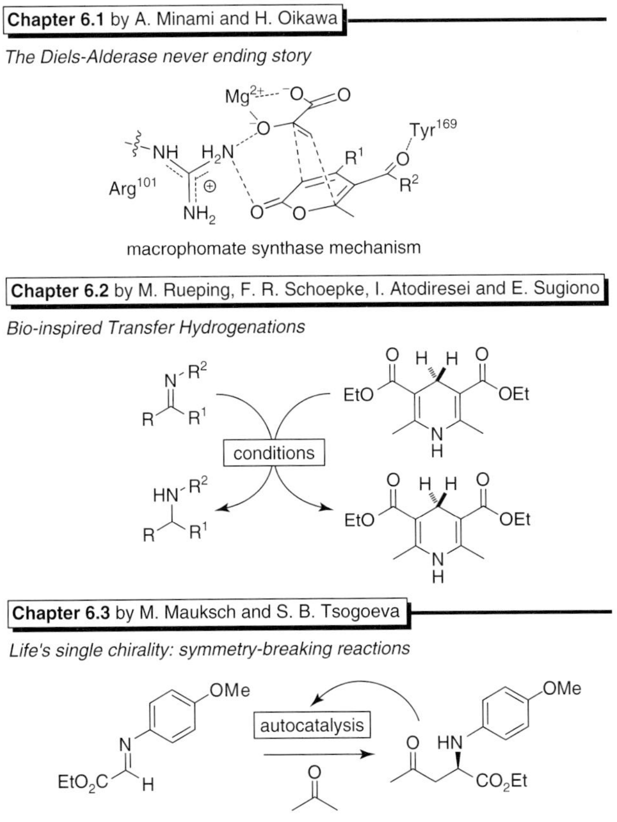

Scheme 9 Frontiers in biomimetic organic synthesis.

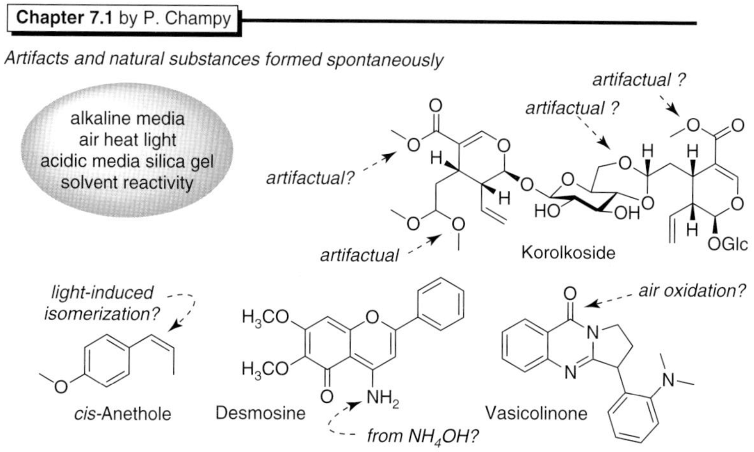

Scheme 10 Artifacts as a matter of debate for the conclusion.

to prebiotic chemistry. Key-words such as spontaneous evolution, molecular and supramolecular self-organization of organic molecules can indeed refer to both domains. A chapter will be devoted to the emergence of life single chirality on earth in a manner, once again, understandable to a broad readership.

Eventually, we thought that a chapter about artifacts in natural product chemistry might provide the matter of debate for an open conclusion, just to spin out the discussion (Scheme 10). May the readers enjoy their trip in the fascinating science of *Biomimetic Organic Synthesis*.

References

1. See, this article of great interest: Thomas, R. (2004) *Nat. Prod. Rep.*, **21**, 224–248.
2. See among others: (a) Firn, R.D. and Jones, C.G. (2009) *J. Exp. Bot.*, **60**, 719–726 and references cited therein; (b) Jenke-Kodama, H. and Dittmann, E. (2009) *Phytochemistry*, **70**, 1858–1866.
3. See among others: (a) Breinbauer, R., Vetter, I.R., and Waldmann, H. (2002) *Angew. Chem. Int. Ed.*, **41**, 2878–2890; (b) Bon, R.S. and Waldmann H. (2010) *Acc. Chem. Res.*, **43**, 1103–1114 and references cited therein; (c) Dobson, C.M. (2004) *Nature*, **432**, 824–828 and references cited therein; (d) Welsch, M.E., Snyder, S.A., and Stockwell, B.R. (2010) *Curr. Opin. Chem. Biol.*, **14**, 347–361.
4. (a) Gravel, E. and Poupon, E. (2008) *Eur. J. Org. Chem.*, 27–42; (b) E.J. Sorensen (2003) *Bioorg. Med. Chem.*, **11**, 3225–3228.
5. (a) Wender, P.A., Handy, S.T., and Wright, D.L. (1997) *Chemistry & Industry*, 765; (b) Wender, P.A. and Miller, B.L. (2009) *Nature*, **460**, 197–20; (c) Gaich, T. and Baran, P.S. (2010) *J. Org. Chem.*, **75**, 4657–4673.
6. Among other review articles, interesting thoughts and historical perspectives are discussed in: (a) Scholz, U. and Winterfeldt, E. (2000) *Nat. Prod. Rep.*, **17**, 349–366; (b) de la Torre, M.C. and Sierra, M.A. (2004) *Angew. Chem. Int. Ed.*, **43**, 160–181; (c) Heathcock, C.H. (1996) *Proc. Natl. Acad. Sci. USA*, **93**, 14323–14327.
7. Touré, B.B. and Hall, D.G. (2009) *Chem. Rev.*, **109**, 4439–4486.

8. (a) Dewick, P.M. (2009) *Medicinal natural products: a biosynthetic approach*, 3rd Edition, Wiley, Chichester (UK); (b) Bruneton, J. (2009) *Pharmacognosie, phytochimie et plantes médicinales*, 4th Edition, Tec et Doc, Paris; (c) see also the book series: Barton, D., Nakanishi, K., Meth-Cohn, O. (Eds) (1999) *Comprehensive Natural Products Chemistry, 1–9*, Elsevier Science Ltd, Oxford; (d) Mander, L. and Liu, H.-W. (Eds) (2010) *Comprehensive Natural Products Chemistry II, 1–10*, Elsevier Science Ltd, Oxford; (d) See also the monthly issues of *Nat. Prod. Rep.*
9. Pelletier, S.W. (1983) The nature and definition of an alkaloid in *Alkaloids: Chemical and Biological Perspectives*, Vol. 1 (ed. Pelletier, S.W.) Wiley-Interscience, New York, pp. 1–32.
10. For example, see the following early and late reviews: (a) Johnson, W.S. (1976) *Bioorg. Chem.*, **5**, 51–98; (b) Yoder, R.A. and Johnston, J.N. (2005) *Chem. Rev.*, **105**, 4730–4756.
11. Andersen, Ø. M. and Markham, K.R. (Eds) (2006) *Flavonoids: chemistry, biochemistry, and applications*, CRC Taylor and Francis, Boca Raton.
12. (a) Berkessel, A. and Groger, H. (2005) *Asymmetric organocatalysis: from Biomimetic Concepts To Applications In Asymmetric Synthesis*, Wiley-VCH, Weinheim; (b) Reetz, M.T., List, B., Jaroch, S., and Weinmann, H. (eds) (2008) *Organocatalysis*, Springer Verlag, Berlin; (c) with specific applications in total synthesis, see for example: Marquéz-López, E., Herrera, R.P., and Christmann, M. (2010) *Nat. Prod. Rep.*, **27**, 1138–1167.

Part II
Biomimetic Synthesis of Terpenoids and Polyprenylated Natural Compounds

11
Biomimetic Rearrangements of Complex Terpenoids

Bastien Nay and Laurent Evanno

11.1
Introduction

"Rearrangements form an integral part of terpene chemistry and sometimes take a truly spectacular course." So was introduced the Simonsen lecture delivered by Ourisson in February 1964 at the Imperial College of Science and Technology, London [1]. This lecture gave an interesting overview dealing with some impressive terpene rearrangements, the most spectacular course of which, according to the author, had truly biomimetic likeness. Besides biomimetic polyolefin cyclizations to give (poly)terpenes, which have been extensively reviewed by others [2],[1] complex rearrangements offer a second aspect of terpene biomimetic chemistry. Both aspects have been intensively studied since the nineteenth century, beginning in 1802 with Kindt's experiments on turpentine oil transformation into *artificial camphor* [3]. The amount of work carried out is huge and obviously will not be treated exhaustively here. Recent examples have rather been privileged although we have tried to keep "classics" close to the discussion.

11.2
Beginning with Monoterpene Rearrangements

11.2.1
Historical Overview of Monoterpene Rearrangements: A Century since Wagner's Structure of Camphene

In 1899, Georg Wagner described the pinacol-related rearrangement of borneol (**1a**) into camphene (**2**) [4] and of α-pinene (**5**) into bornyl chloride (**4a**) [5]. The

---

1) An example of polyolefin cyclization is used to introduce Section 11.5 dealing with triterpene rearrangements. It will be the only one described in this chapter, which deals with rearrangements of advanced biomimetic structures.

Biomimetic Organic Synthesis, First Edition. Edited by Erwan Poupon and Bastien Nay.
© 2011 Wiley-VCH Verlag GmbH & Co. KGaA. Published 2011 by Wiley-VCH Verlag GmbH & Co. KGaA.

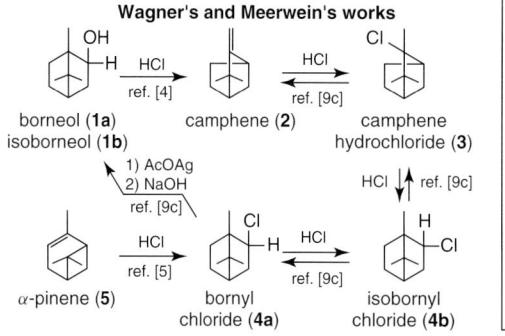

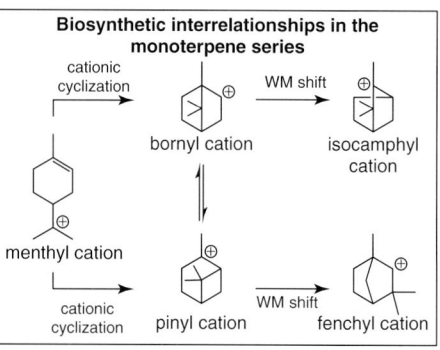

Scheme 11.1 Wagner's and Meerwein's work on the rearrangement of borneols (**1**), α-pinene (**5**), and camphene (**2**) (stereochemistry not specified) [4, 5, 9c], and relation to the biosynthetic routes of the corresponding cation (in the box) [6, 7].

transformations were performed in the presence of a strong acid and if we consider the general biosynthetic pathways for monoterpenes [6, 7] they are probably the earliest biomimetic transformations in the terpene series (Scheme 11.1). Not only did this work put an end to the eighteenth century controversy over the structure of camphene [8] but it opened the way to useful chemical transformations in organic synthesis. In 1914, Hans Meerwein started a series of discussions on the mechanism of this reaction [9].[2] He finally suggested that the rearrangements were initiated by the ionization of the substrates into the corresponding carbocations [9c]. Meerwein's carbocation concept and ionic mechanisms stand as an important breakthrough in organic chemistry. Strictly, the Wagner–Meerwein rearrangement can thus be defined as the carbocation-promoted 1,2-carbon shift of an alkyl or a hydrogen. Many related processes have been described involving longer, especially transannular, migration events. Wagner–Meerwein rearrangements can proceed in tandem as a succession of hydrogen or methyl (or alkyl) shifts. As shown in the following, the reaction is classical in terpene synthesis and biosynthesis [7].

Considering the economic importance of terpenes in the chemical, pharmaceutical, and fragrance industries, the rearrangement of α-pinene (**5**) into camphene (**2**) is a pivotal reaction in terpene manufacture, being the basis of the synthesis of camphor. The process has been improved and now relies on heterogeneous catalysis (Scheme 11.2). The isomerization of α-pinene (**5**) can be carried out over TiO_2, giving large amounts of camphene (**2**) accompanied by monocyclic terpenes resulting from ring opening–elimination reactions (terpinenes, cymene, limonene) [11, 12]. Further acid-catalyzed isomerization gives isobornyl acetate (**6**), which yields camphor (**7**) after hydrolysis and oxidation.

---

2) This was complementary to Meerwein's own work on the mechanism of the pinacol rearrangement published a year before: see Reference [10].

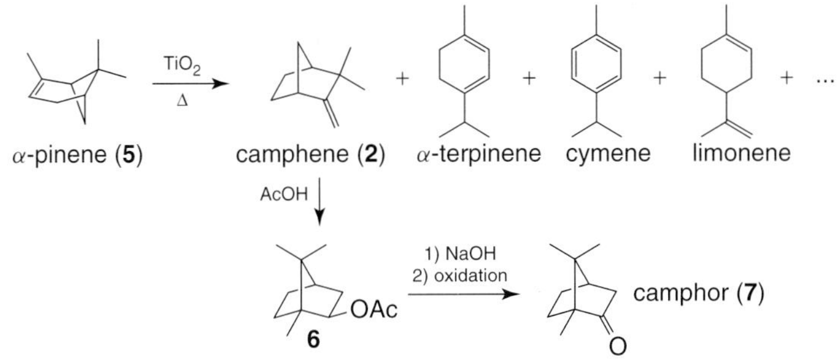

Scheme 11.2 Industrial process for the production of camphor (7) [11, 12].

11.2.2
Kinetics of the Monoterpene Rearrangement and Relation with the Catalytic Landscape in Terpene Biosynthesis

The Wagner–Meerwein rearrangement of a substrate R_1X results from a combination of thermodynamic and kinetic conformational events. It is conditioned by the formation of the carbocation R_1^+, which is the rate-determining step, after which structural changes occur ($R_1^+ \rightarrow R_2^+$). Formation of the carbocation is favored under acid catalysis and solvolytic conditions [13]. Depending on the complexity of R_1^+, the reaction should not be restricted to a single mode of rearrangement (Figure 11.1). Knowing the activation barrier $\Delta G^{\ddagger}$ of the carbocation rearrangement is important for understanding the overall reaction mechanism from R_1X. Especially, this barrier "dictates" the choice of experimental conditions used to carry out these reactions [13].

Sorensen has studied the kinetics of terpene rearrangements and the thermodynamic stability of various carbonium ions under acidic and superacidic conditions [13, 14]. For a given reaction, it was possible to determine which of these processes has the lowest barrier by noting the products formed under varying experimental conditions. In particular, Sorensen studied the rearrangement of

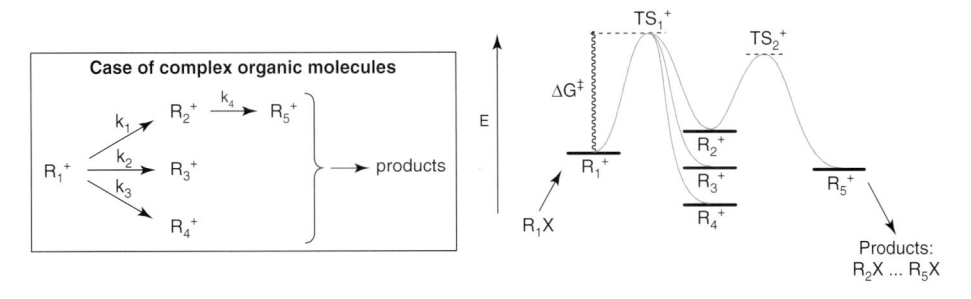

Figure 11.1 Reaction diagram of Wagner–Meerwein rearrangements involving complex intermediates: the reaction is not restricted to a single product.

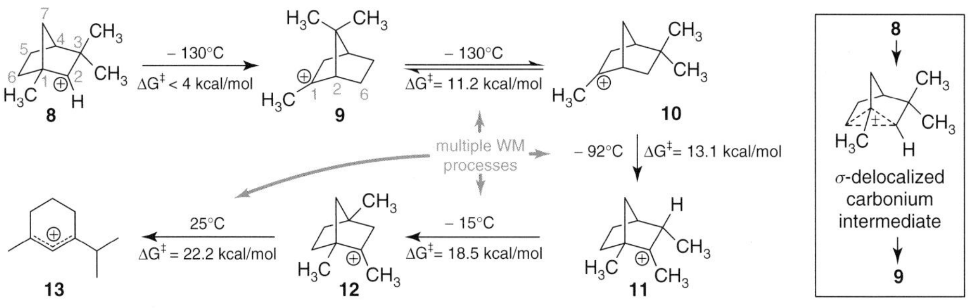

Scheme 11.3 Rearrangement of the fenchyl cation (**8**) according to Sorensen [13].

the fenchyl cation (**8**) generated from α-fenchene (Scheme 11.3). In this system, the cation to cation barriers were readily overcome in formic acid as the solvolysis medium. The 1,2-shifts could involve partially σ-delocalized carbonium ions (box in Scheme 11.3) [15].

It was not possible to slow down the formation of the cation **9**, and even at −130 °C the combination of complex Wagner–Meerwein processes leading to the β-fenchene cation **10** was too rapid to be measured by NMR. However at −92 °C the mixture evolved toward the ε-fenchene cation **11**. The final ions **12** and **13** were only observed on raising the temperature to −15 and 25 °C, respectively. These observations and related works allowed the activation energies to be inferred for each step (Scheme 11.3). Typically, the Wagner–Meerwein alkyl 1,2-shift of a tertiary cation into a parent tertiary cation has a low activation barrier <4 kcal mol^{-1} [13]. When the resulting cation is secondary instead, the activation energy increases by 5.5 kcal mol^{-1}.

We complete this discussion with the interesting computer-assisted work of Johnson and Collins, who analyzed the multiple-rearrangement products of the norbornyl and fenchyl (**8**) cations [16]. They elaborated road maps for determining sequences of Wagner–Meerwein rearrangements, 6,2-*endo*-hydride shifts, 3,2-*exo*-hydride shifts, 3,2-*exo*-methyl shifts, and the "double Wagner–Meerwein" rearrangement which interconverts, given the rearranged cations. Important mechanistic information resulted from the use of graph models and computer programs. Applying this to Sorensen's work on the rearrangement of the fenchyl cation (**8**, Scheme 11.3), there would be 330 theoretically distinguishable cationic intermediates, reduced to a minimum of 165 racemic "nodes" [16b]. The system appeared to seek out those structures that are thermodynamically most stable at a given temperature.

We can bring these observations together with more recent works on terpene biosynthesis. The role of enzymes has been pointed out not only for the formation of a reactive carbocation but also in providing a stabilizing template for the cyclization reaction as well as controlling water access. In these terpene synthases, one cationic species only, or a few among the possible tens or hundreds, would be preferred and stabilized in an unreactive aprotic environment by way of interactions with the π-systems of aromatic residues [17]. By the controlled mutation of specific amino acids in the active site, it has been possible to define *catalytic landscapes*, in work used to understand the cyclization potential of linear C_{15}-precursors [18].

11.3
Biomimetic Rearrangements of Sesquiterpenes

With more important structural complexity, the sesquiterpenes offer larger rearrangement capabilities and higher chemical diversity than monoterpenes. Historical work on longifolene, as an analog of camphene, showed a very similar rearrangement pattern under acidic conditions, yet with additional reactivity or with limitations imposed by the shielding effect provided by the large transannular bridge [1, 19]. In this section, we will see how complex polycyclic and densely functionalized compounds can be obtained not only by cationic rearrangements but also by oxidative cleavage and cyclization cascades. Special attention will be devoted to caryophyllane polycyclic derivatives, which offer a rich and diversified chemistry.

11.3.1
Caryophyllenes in Sesquiterpene Biosyntheses

Caryophyllene (**14**) (trans isomer) and isocaryophyllene (**15**) (cis isomer) are the biosynthetic starting point for numerous polycyclic sesquiterpenes [20]. They are available on a large scale for use in the flavoring and fragrance industry. The chemistry of caryophyllenes has been well reviewed by Collado *et al.* [21]. We present herein a short description of relevant information about these compounds with regard to the biomimetic synthesis of sesquiterpenes. The caryolanyl (**16**) and clovanyl (**17**) cations are derived from the cationic cyclization and rearrangement of **14** (Scheme 11.4), whereas the presilphiperfolanyl cation (**18**) and its rearranged forms (silphinyl **19** and terrecyclanyl **20**) are derived from the isomer **15**. The co-occurrence of these different sesquiterpene skeletons in common living organisms shed light on their biogenetic relationships. ^{13}C NMR studies and MM1 calculations [22] have shown that there are four possible conformations of *trans*-caryophyllene (**14**) distinguished by the relative positions of the exocyclic methylene and olefinic methyl groups (Scheme 11.4). A 3 : 1 ratio has been measured between the main conformations, $\beta\alpha$ and $\beta\beta$, having a low inversion barrier (16.25 ± 0.11 kcal mol^{-1}). Treatment of caryophyllene (**14**) in acidic media (acetic, formic, sulfuric acids, and superacidic media) led to numerous rearranged compounds after 30 min, the relative yields of which reflect the relative population of the possible conformers.

For longer reaction times (several days), the major products were caryolan-1-ol (from cation **16**) and clov-2-ene (from cation **17**). These results suggested reversible interconversion between intermediates that eventually evolved toward the more stable isomers [21]. In superacidic conditions (SO_2FCl) at $-124\,°C$, the distribution of products changed [23]. At this temperature, the various conformers of caryophyllene

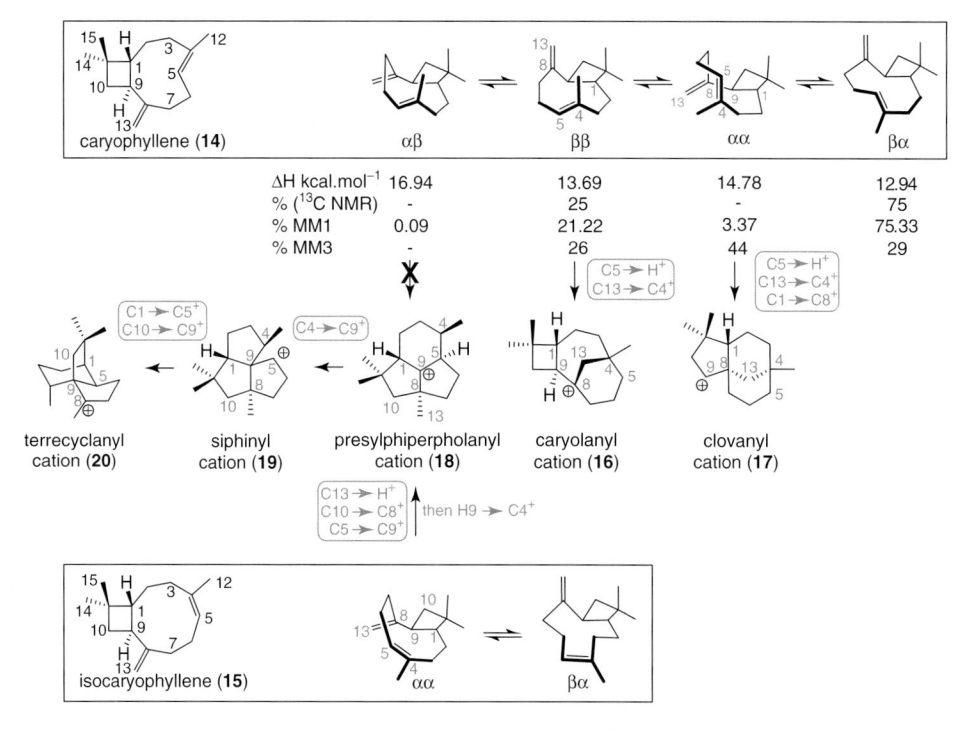

Scheme 11.4 Conformational equilibrium of caryophyllene (**14**) and isocaryophyllene (**15**) and general biosynthetic origin of polycyclic sesquiterpenes (numbering according to **14** and **15**).

(**14**) are not rapidly interconverting, providing evidence for a significant transition barrier between the $\alpha\alpha$ and $\beta\beta$ conformations. This gave experimental support for the results of population predictions based on MM3 calculations, suggesting proportions of 44, 29, and 26% for the $\alpha\alpha$, $\beta\alpha$, and $\beta\beta$ conformers, respectively [24]. The interconversion between caryophyllene (**14**) and isocaryophyllene (**15**) has been reported under thermal conditions (pyrolysis at >240 °C); it proceeds through a series of Cope [3,3]-sigmatropic rearrangements [25]. The presylphiperpholane (type **18**), silphinane (type **19**), and neoclovane skeletons (not shown) were only observed on treating isocaryophyllene (**15**) in acidic media, and were accompanied by other rearrangement products. To better control access to the different skeletons from caryophyllenes, the use of activated derivatives, for example, epoxides or tosylates, has been employed (see below) [26, 27].

11.3.2
Biomimetic Studies in the Caryolane and Clovane Series

As discussed above, under acidic and superacidic conditions, caryophyllene (**14**) can undergo non-controlled rearrangements, mainly leading to caryolane and clovane

11.3 Biomimetic Rearrangements of Sesquiterpenes

derivatives, yet with low selectivity. Transformations performed on the α- and β-oxides of **14** afforded the caryolane and clovane frameworks, respectively, with much higher selectivity. In the presence of hydrated bismuth triflate utilized as a proton donor, caryophyllene-β-oxide **21**, having an αα conformation, underwent cyclization on the α-face of the molecule, generating the bridgehead carbocation **22** (Scheme 11.5). The good overlap between the cation orbital and the transannular cyclobutane bond promoted Wagner–Meerwein rearrangement into **23**. After β-elimination of a proton, 2-clovene (**24**) was obtained [27].

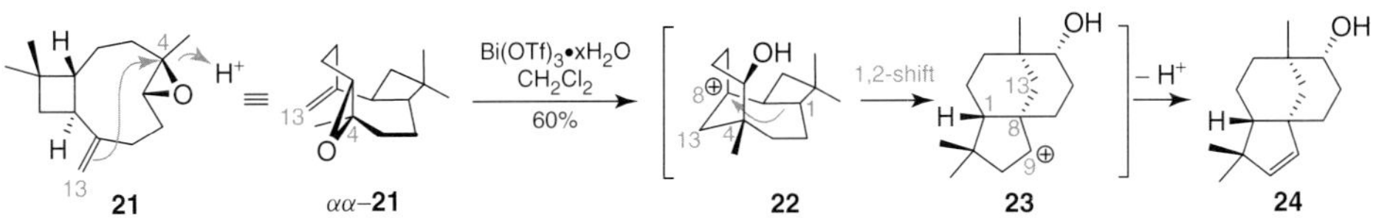

Scheme 11.5 Biomimetic synthesis of 2-clovene (**24**) through acid-catalyzed rearrangement of caryophyllene-β-oxide **21**.

The Brønsted acid $(PhO)_2P(O)OH$ promoted the same rearrangement of **21** towards the clovanyl cation **23**, which was readily trapped by magnolol (**25**) available in the reaction mixture (Scheme 11.6a). This afforded clovanemagnolol (**26**), a natural product with neuroregenerative properties isolated from *Magnolia obovata* [28]. A similar reaction was performed with caryophyllene α-oxide **27**, involving a ββ conformation. In this case, a non-matching alignment of the cation orbital with the transannular cyclobutyl C–C bond of cation **28** prevented the Wagner–Meerwein shift. The carbocation **28** was instead directly trapped by magnolol (**25**) to give caryolanemagnolol (**29**, Scheme 11.6b). The low yields

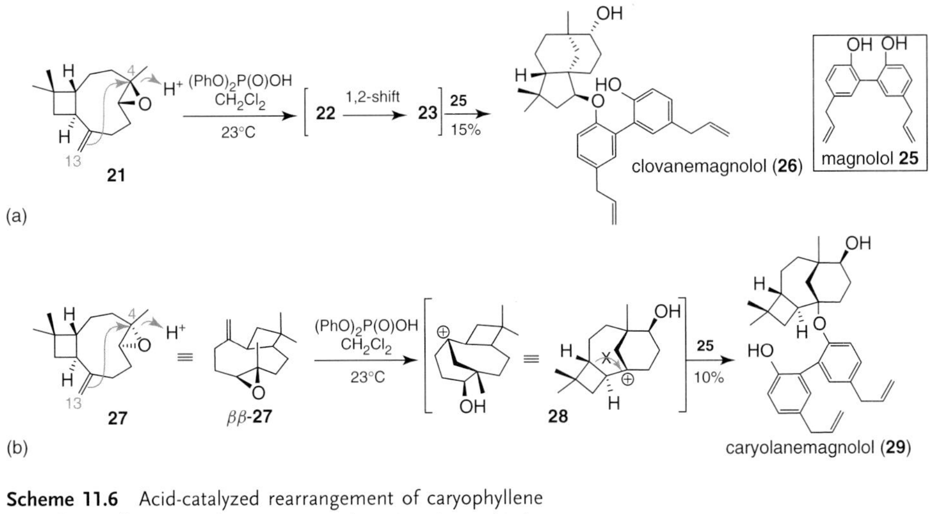

Scheme 11.6 Acid-catalyzed rearrangement of caryophyllene α- and β-oxides (**21** and **27**): two-step biomimetic synthesis of (a) clovanemagnolol (**26**) and (b) caryolanemagnolol (**29**).

(10–15%) of these transformations are balanced by the limited number of steps for this biomimetic route [two steps from caryophyllene (**14**)]. Better trapping yields were obtained when 4-bromophenol was used instead of **25**, releasing a synthetic intermediate, but the two-step biomimetic synthesis makes this route particularly competitive.

11.3.3
Biomimetic Studies in the Triquinane Series

Isocaryophyllene (**15**) holds the $\alpha\alpha$ conformation and underwent under acidic conditions (H_2SO_4, ether) a Wagner–Meerwein 1,2-shift of C10 from C9 onto C8 (Scheme 11.7a) [29]. Subsequent transannular electrophilic cyclization of **31** created the C5–C9 bond of the presilphiperfolane skeleton (**32**), which was neutralized as presilphiperfolene (**33**) in 35% global yield. Better control of the rearrangement in aqueous conditions was obtained when the process was performed on the tosylate **35**, which originated from the caryophyllene β-oxide **21** (Scheme 11.7b) [30]. In this route, the methyl at C8 was missing after ozonolysis of the exocyclic double bond of **21**. Furthermore, the endocyclic *trans*-olefin (**34**) was regenerated by the reduction of the epoxide. Stereochemical considerations of the rearrangement of **35** ruled out a concerted pathway involving departure of the tosylate group. Steric hindrance of the tosylate would favor a $\alpha\beta$-conformation and the solvolytic formation of a short-lived caryophyllenyl carbocation. Then, rapid rearrangement resulted in the formation of the norpresilphiperfolanol (**36**). According to the authors, although the methyl at C8 was missing in this route, the reaction provided a chemical precedent for a biogenetic connection between the *trans* caryophyllene series and the presilphiperfolanols.

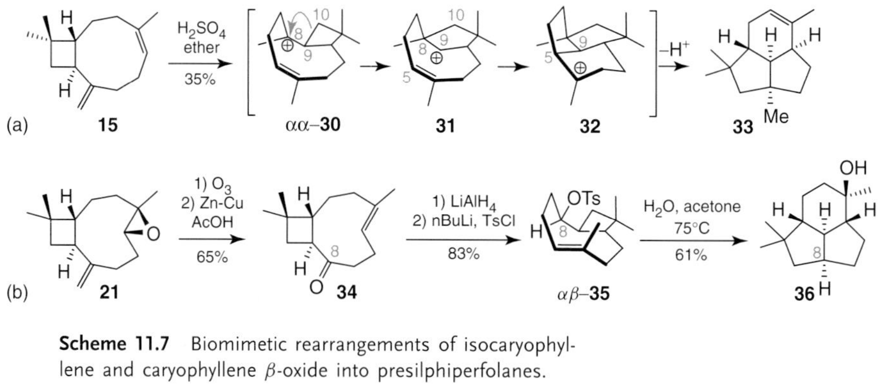

Scheme 11.7 Biomimetic rearrangements of isocaryophyllene and caryophyllene β-oxide into presilphiperfolanes.

Presilphiperfolanol (**37**) can easily undergo ring contraction in the presence of H_2SO_4–SiO_2 in benzene to generate silphiperfol-6-ene (**39**, Scheme 11.8) [31]. After the formation of the carbocation **18**, contraction of the cyclohexane ring into **38** was followed by a methyl 1,2-shift. Subsequent return to neutrality by β-elimination of a proton afforded silphiperfolene (**39**), accompanied by α-terrecyclene (**40**) as

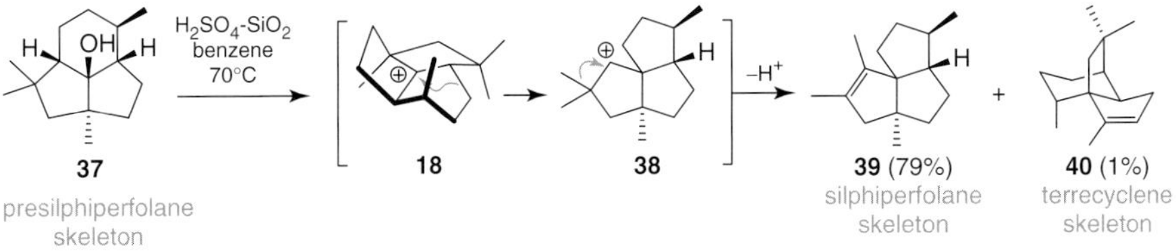

11.3 Biomimetic Rearrangements of Sesquiterpenes

Scheme 11.8 Biomimetic rearrangement of presilphiperfolanol (**37**) into silphiperfolene (**39**) and the secondary product terrecyclene (**40**).

a minor product (1%). It was presumed to be formed by the Wagner–Meerwein rearrangement of an intermediary silphinyl cation (not shown).

In another experiment, the formation of α-terrecyclene (**40**) was rationalized by the specific transformation of silphinyl mesylate **41** through double Wagner–Meerwein rearrangement (80% yield) (Scheme 11.9) [32]. Formolysis promoted the abstraction of the mesylate group from **41** to give cation **42**. The migration of methine C7 from C11 to C1 afforded the bicyclo[3.3.1]nonane system **43**. Then, migration of the methylene C5 from C4 to C11 gave the cation **44**, achieving at the same time the complex transformation of **41** and biomimetic synthesis of α-terrecyclene (**40**).

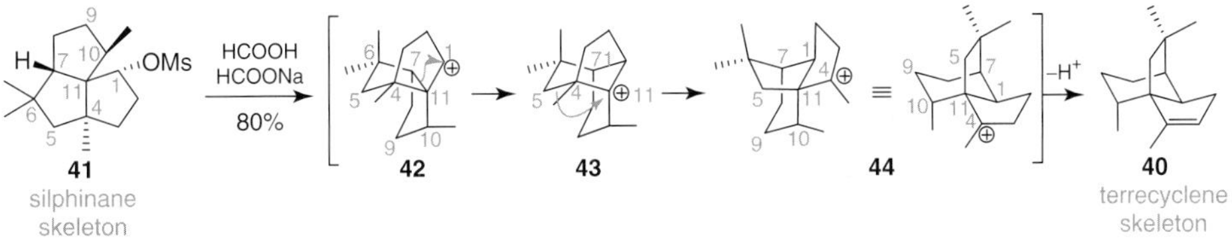

Scheme 11.9 Biomimetic rearrangement of the silphinyl mesylate (**41**) into terrecyclene (**40**).

11.3.4
Oxidative Rearrangements in the Silphinane Series: the Penifulvins

Penifulvins A–C (**45**–**47**) have been isolated from the fungus *Penicillium griseofulvum* [33]. Among these compounds, **45** showed significant insecticide activity against the fall armyworm *Spodoptera frugiperda*. The originality of penifulvins lies in a complex dioxafenestrane core in which four fused cycles share a common quaternary center (Scheme 11.10). The biosynthetic hypothesis has been secured by the co-occurrence of 12-hydroxy-silphinen-15-oic acid (**48**) in the fungus, which was extracted along with **45**–**47**. Compound **48** would be the precursor of **50**.

In 2009, Gaich and Mulzer reported the first total synthesis of **45** [34a], which was followed in 2010 by those of **46** and **47** [34b]. The work features a biomimetic step consisting of the oxidative rearrangement of a silphinene intermediate (**52**), derived from the photochemical rearrangement of the aromatic compound **51**, into the dioxafenestrane core. Ozonolysis of the olefin led to the dialdehyde intermediate **53**,

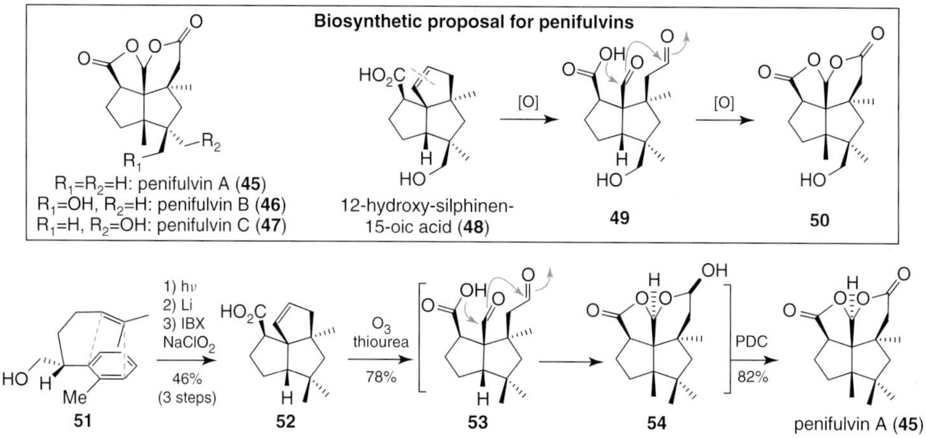

11 Biomimetic Rearrangements of Complex Terpenoids

Scheme 11.10 Biosynthetic proposal for penifulvins (box) and Mulzer's biomimetic synthesis of penifulvin A (**45**).

which underwent domino cyclization of both oxygenated cycles. The lactol **54** was not isolated, being oxidized in the presence of PDC (pyridinium dichromate) to give the natural product **45**. Using this key-transformation, penifulvin A was synthesized racemically in five steps from *o*-tolylacetic acid (14% overall yield), or enantioselectively in eight steps (8% overall yield).

11.3.5
Miscellaneous Sesquiterpene Rearrangements

Neopupukeananes (structure type **57**) and trachyopsanes (structure type **58**) are coexisting marine sesquiterpenes derived from amorphane (Scheme 11.11). According to Srikrishna's biosynthetic proposal and synthetic work, the trachyopsanes would be derived from neopupukeananes via a Wagner–Meerwein shift leading to

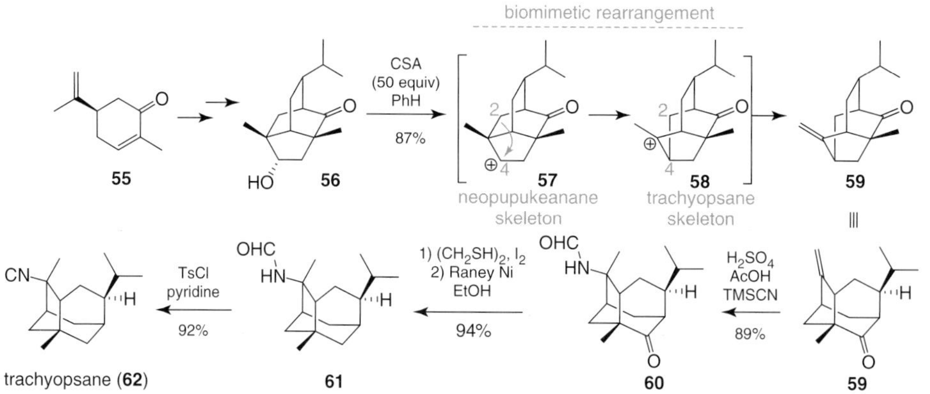

Scheme 11.11 Biomimetic synthesis of trachyopsane A (**62**).

C2–C4 bond formation from cation **57** [35]. The neopupukeanane substrate **56** was constructed in a very short and enantioselective sequence from *(R)*-carvone (**55**) [36]. Cation **57** was generated by treatment of **56** in the presence of 50 equiv of CSA (camphorsulfonic acid) in refluxing benzene. Rearrangement into the trachyopsane cation **58** was followed by β-elimination to give trachyops-2(14)-ene-7one (**59**) in 87% yield. Compared to other acidic systems [HCO$_2$H, TFA (trifluoroacetic acid), MsOH, p-toluenesulfonic acid (PTSA), and BF$_3$·OEt$_2$/TFA or MsOH], CSA gave the best results for this rearrangement. The synthesis of 2-isocyanotrachyopsane (**61**) was completed by reduction of the ketone and installment of the isonitrile group.

The panasinsane skeleton (**63**) is found in the roots of *Panax ginseng* (Scheme 11.12). From this plant were also isolated neoclovane and ginsenane (**64**) compounds, all showing substantial variations within the tricyclic structure. Collado studied the rearrangements of panasinsane derivatives prepared from caryophyllene oxide [37]. Novel structures were discovered and the 4,5,6-fused tricyclic skeleton of a panasinsane precursor was converted into ginsenol derivatives under tetracyanoethylene-catalyzed solvolysis.

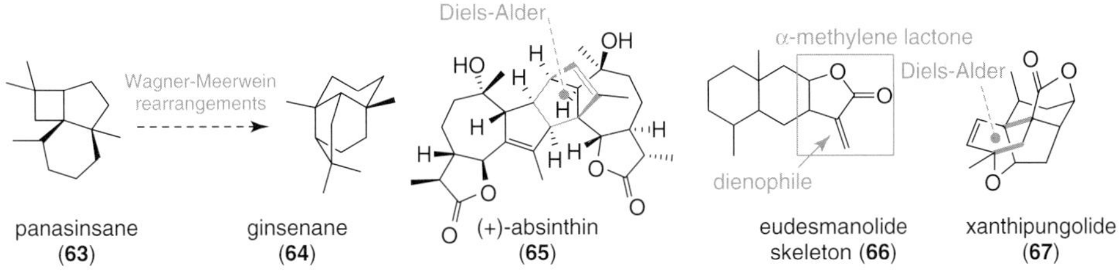

Scheme 11.12 Biomimetic synthesis of miscellaneous sesquiterpenes.

Diene-containing terpenoids can react with parent dienophile structures during biomimetic Diels–Alder reactions, affording dimers or heterodimers. The triterpene (+)-absinthin (**65**, Scheme 11.12) is actually a dimeric sesquiterpenoid isolated from *Artemisia absinthium*. It has been synthesized in nine steps and 19% overall yield from *O*-acetylisophotosantonic lactone, which is readily available from the photolysis of santonin and which holds a reactive cyclopentadiene moiety [38]. The biomimetic dimerization of the sesquiterpene precursor via regio- and stereospecific Diels–Alder reaction proceeded spontaneously in 72% yield upon leaving it neat at room temperature for ten days. Five additional steps were needed to obtain absinthin. Importantly, (+)-absinthin was previously found to afford the monomeric sesquiterpene artabsin through a retro-Diels–Alder reaction upon heating [39], suggesting a biosynthetic relationship between both monomeric and dimeric series.

Natural α-methylene lactones are particularly interesting dienophiles that have been used in biomimetic Diels–Alder reactions. Plagiospirolides A and B are C$_{35}$ terpenic natural products, the biosynthesis of which was supposed to arise from the Diels–Alder reaction between an eudesmanolide sesquiterpene (**66**) (containing an α-methylene lactone dienophile, Scheme 11.12) and a fusicoccane diterpene (containing a cyclopentadiene). The biomimetic Diels–Alder reaction has been reproduced experimentally by heating both precursors in benzene at 60 °C [40]. Similar transformations of eudesmanolides have been reported although the required conditions could sometimes hardly be regarded as physiological [41]. Finally, the sesquiterpene xanthipungolide (**67**) (Scheme 11.12) was synthesized by irradiating an EtOH solution of an α-methylene lactone precursor found in the same plant and bearing a dienone moiety [42]. The reaction involved a tandem process starting with the electrocyclization of the dienone to reveal a diene that spontaneously underwent intramolecular [4 + 2] cyclization with the proximate α-methylene lactone dienophile.

11.4
Diterpene Rearrangements

Polycyclic diterpenes arise from the electrophilic cyclization of the linear geranylgeranyl diphosphate precursor **68**. Important biosynthetic intermediates have been described (Scheme 11.13), among which are verticillene (**70**), casbene (**71**), and both enantiomers of copalyl diphosphate [(−)-**72** shown, and (+)-**72**] leading to kaurenes (**73**) [7]. All are precursors of various structures of great interest. Carbon skeleton reorganization and oxygenation lead to important chemical diversity and have inspired many synthetic chemists to use biomimetic strategies.

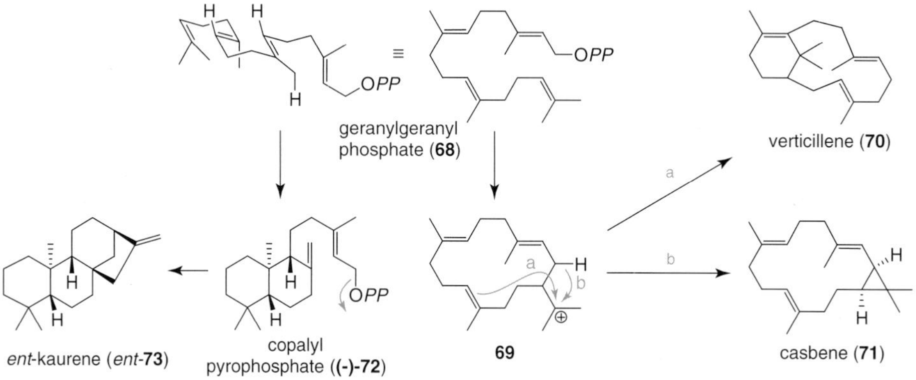

Scheme 11.13 Examples of early chemical diversity in diterpene biosynthesis.

11.4.1
Dead End Products in the Biomimetic Synthesis of Antheridic Acid from Gibberellins

Antheridic acid (**78**, Scheme 11.14) is an antheridium-inducing factor (antheridiogen) isolated from the fern *Anemia phyllitis* [43]. Close similarities between **78** and fungal gibberellins, for example, gibberellins A_7 (**74**) and A_9 (**75**), suggested

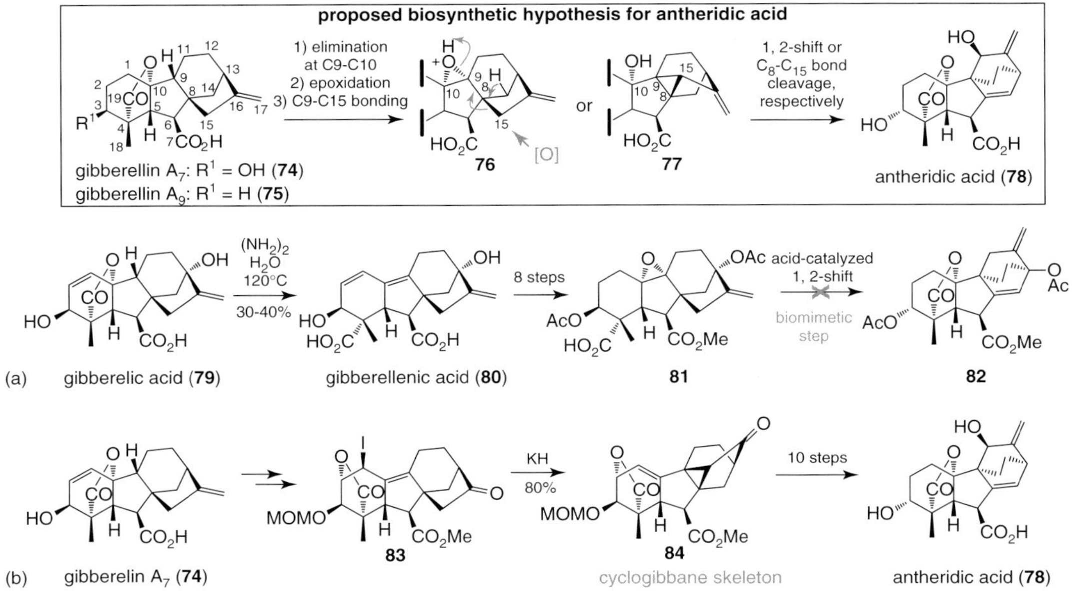

Scheme 11.14 Unsuccessful biomimetic strategy for synthesis towards antheridic acid (**78**) (a) and an alternative synthetic route (b).

a common biosynthetic relationship.[3] A Wagner–Meerwein 1,2-shift induced by acid-catalyzed opening of an epoxide (**76**) was initially proposed by Nakanishi [43a]. This hypothesis also raised the prospect of achieving the biomimetic synthesis of **78** from readily available precursors derived from gibberellins. Mander undertook this task and prepared epoxide **81** from gibberellenic acid **80**, which is accessible by treatment of gibberellic acid (**79**) in hot water (Scheme 11.14a) [45].

Unfortunately, the reactions of epoxide **81** with a range of Lewis acids did not produce any evidence of the desired rearrangement into the product **82** having the antheridic acid skeleton. In fact, the carbocyclic core remained unchanged and only products from lactonization or elimination were observed (not shown). Failure was attributed to the presence of the 17-methylene group in the D-ring, which would be responsible for the reduced migration aptitude of C15, and also to the proximity of the C19 carboxylic acid in the epoxide **81**. A more secure strategy was finally elected to perform the synthesis of antheridic acid (**78**) from gibberellin A_7 (**74**, Scheme 11.14b). The approach consisted in the construction of the cyclogibbane derivative **84** by means of intramolecular enolate alkylation. Then, a controlled fragmentation of the cyclopropane ring achieved construction of the tetracyclic core of **78**, the synthesis of which was completed after additional steps. Several antheridiogens based on the 9,15-cycloderivative **77** have been isolated from natural sources, which make this approach particularly attractive by revealing biogenetic likeness [46]. However, the synthetic intermediates in this route are not biomimetic.

11.4.2
Biomimetic Synthesis of Marine Diterpenes from *Pseudopterogorgia elisabethae*

The gorgonian octocoral *Pseudopterogorgia elisabethae* is the source of interesting rearranged diterpenes, including colombiasin A (**85**) [47], elisabethin A–D (A: **86**) [48, 49], elisabanolide (**87**) [48, 49], and elisapterosin A and B (B: **88**) [49] (Scheme 11.15), which are biogenetically linked to amphilectane diterpenes. Many of them have shown significant antituberculosis and/or antitumor activities, but they are famous to organic chemists because of the important work undertaken using biomimetic approaches. Although colombiasin A and elisapterosin B have been successfully worked out, the case of elisabethin A remains unresolved [50].

According to Rodríguez, the biosynthesis of the *Pseudopterogorgia* diterpenes would start with the formation of the serrulatane bicyclic skeleton (Scheme 11.15) [47–49]. The complex polycyclic systems of colombianes and elisapteranes would be formed by [4 + 2] and [5 + 2] cycloadditions from serrulatane, respectively, although electrophilic cyclizations and Wagner–Meerwein rearrangements have

3) Gibberellins are biogenetically derived from the ent-kaurane skeleton ent-73 (Scheme 11.13), the rearrangements of which have been well described in the literature; for example, see Reference [44]. The biogenetic-like interconversions of the tetracyclic diterpene in the beyerane–kaurane series (including atiserane, hibaane, trachylobane, and phyllocladane) were achieved using carbonium ion rearrangements very similar to those of bridged bicyclic monoterpenes.

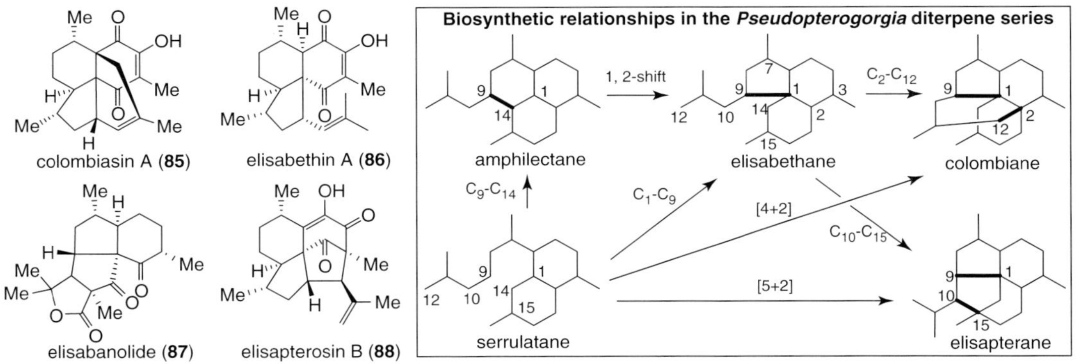

Scheme 11.15 Representative examples of marine diterpenes from *Pseudopterogorgia elisabethae* and biosynthetic hypotheses [49] (atom numbering according to the elisabethane core).

also been proposed. The elisabane skeleton would arise from oxidative cleavage of the quinonoid ring of elisapteranes.

The first synthesis of colombiasin A (**85**) was reported in 2001 by Nicolaou [51]. The strategy featured two Diels–Alder reactions, the first one to build the bicyclic moiety of a serrulatane-type intermediate and the second to complete the construction of the colombiane skeleton in a biomimetic manner (Scheme 11.16). The bicyclic serrulatane intermediate **93** was build from the diene **89** and the quinone precursor **90** through intermolecular Diels–Alder cyclization while a Claisen rearrangement promoted by Pd(PPh$_3$)$_4$ allowed for branching of the side chain on **92**. The intramolecular Diels–Alder reaction was performed on sulfone **94** following cheletropic extrusion of SO$_2$ to unmask the diene. Only the *endo* product was observed after 20 min of reaction at 180 °C in toluene (sealed tube), affording the colombiane skeleton **95** in 89% yield. The cycloaddition did not work when performed directly on the free diene, although it was successfully carried out on 7-epi-**93** for the synthesis of 7-epi-colombiasin A. After deoxygenation and deprotection, the total synthesis of colombiasin A (**85**) was completed. Although the work was initially done racemically, Nicolaou was able to perform an asymmetric Diels–Alder reaction between substrates **89** and **90**, using the chiral Makami catalyst [(R)-BINOL-TiCl$_2$]. This allowed the absolute configuration of the natural product to be determined.

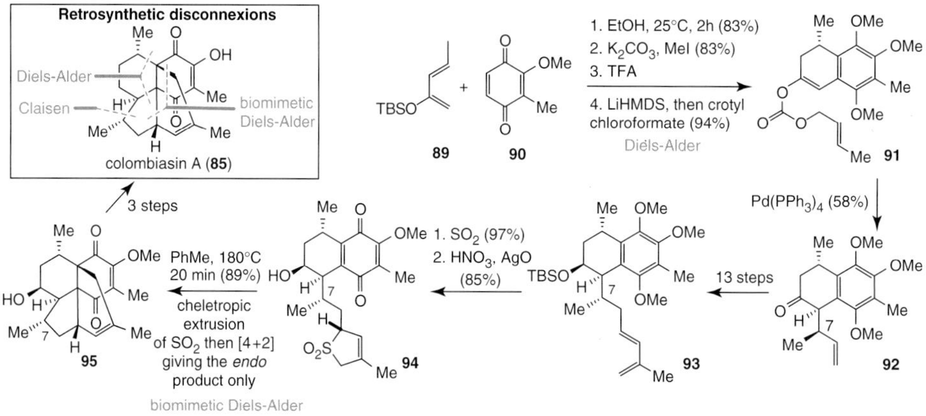

Scheme 11.16 Nicolaou's synthesis of colombiasin A (**85**).

Kim and Rychnovsky developed a unified strategy for the asymmetric synthesis of (−)-colombiasin A (**85**) and elisapterosin B (**88**) [52], which supported Rodríguez's biosynthetic proposal. Here again, the *cis*-decalin serrulatane skeleton was synthesized by an intermolecular Diels–Alder reaction between the quinone **90** and the enantiomerically pure diene **97** made from the asymmetric alkylation of the propionamide (**96**) of Myer's pseudoephedrine auxiliary (Scheme 11.17). The Diels–Alder reaction was carried out in the presence of LiClO$_4$ in Et$_2$O, which provided the only successful conditions, giving the *endo* product **98** as the major stereoisomer. Finally, functionalization of the side chain led to the

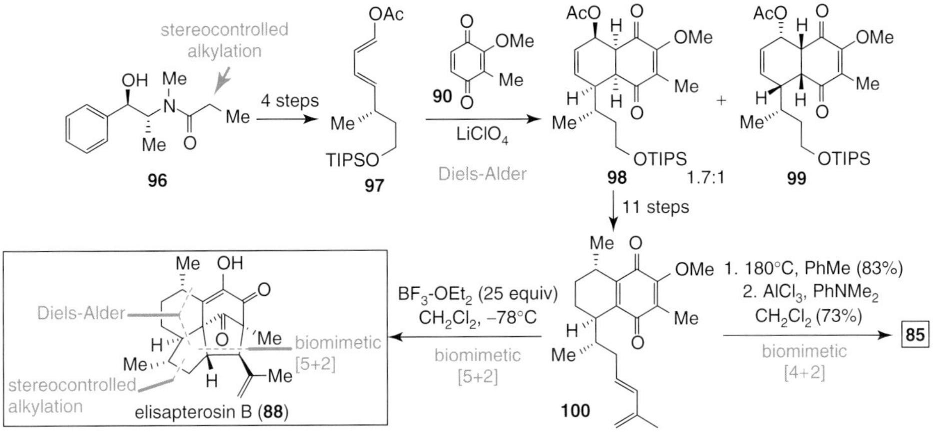

Scheme 11.17 Rychnovsky's synthesis of colombiasin A (**85**) and elisapterosin B (**88**).

diene moiety on the serrulatane skeleton (**100**) for the biomimetic intramolecular cycloaddition.

When the serrulatane substrate **100** was submitted to thermal [4 + 2] cycloaddition (toluene, 180 °C), and after an additional step of demethylation, (−)-colombiasin A (**85**) was obtained in 83% yield. The natural product was prepared in 17 steps and 3.9% overall yield from **96**. When submitted to a large excess of BF$_3$·OEt$_2$ (25 equiv.) in CH$_2$Cl$_2$ at −78 °C, **100** underwent [5 + 2] cycloaddition to yield (−)-elisapterosin B (**88**) in 41% yield as a mixture of separable diastereoisomers. The proposed [5 + 2] cycloaddition had good precedent from Joseph-Nathan's work on BF$_3$·OEt$_2$-catalyzed [5 + 2] cycloadditions of naturally occurring sesquiterpenes [53].[4)] Elisapterosin B (**88**) was obtained in 16 steps and 2.6% overall yield from **96**.

Other studies on the biomimetic synthesis of *Pseudopterogorgia* diterpenes have been reported by others, differing mainly in the strategy used to synthesize the serrulatane precursor. Harrowven used the Moore rearrangement to make a dihydroquinone advanced intermediate, synthesizing **85** and **88** in 12 and 11 steps, respectively, from (−)-dihydrocarvone [55]. Jacobsen used chiral chromium complexes as catalysts to promote the intermolecular Diels–Alder reaction leading to the serrulatane core from quinone **90** [56]. Davies used the combined C–H

---

4) In his work published in 1987, Joseph Nathan described the [5 + 2] cycloaddition of the sesquiterpene perezone (i) into α- and β-pipitzol (ii and iii). The reaction was highly stereoselective in the presence of BF$_3$·OEt$_2$, yielding a 9:1 mixture of both isomers. It was also strongly influenced by electronic factors. This reaction had been described in 1885 under thermal conditions [54].

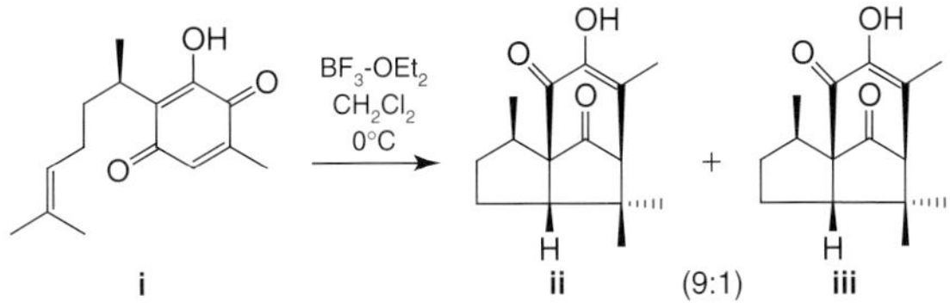

activation/Cope rearrangement between a 2-diazoester and dihydronaphthalene derivatives, using a chiral dirhodium complex catalyst, to set up the side chain [57].

11.4.3 Biomimetic Relationships among Furanocembranoids

The furanocembranoids and their rearranged derivatives are diterpenes isolated from marine sources, most of them from gorgonian corals [58]. Remarkable bioactivities have been associated with these structurally interesting compounds and, thus, they have been of great interest to the synthetic chemist community. Furanocembranoids feature a 14-membered carbocycle usually encompassed by furan and butenolide heterocycles associated with a high oxidation level at variable positions. Rearrangements of the furanocembrane skeleton and oxidative cleavages further increase the chemical diversity, as shown by the photochemical transformation of bipinnatin J (**101**) into the related pseudopterane (**102**) and gersolane (**103**, **104**) skeletons by Rodríguez (Scheme 11.18), respectively, formed from the photochemically allowed $[_{\sigma}2_s + _{\pi}2_s]$ (or 1,3-sigmatropic rearrangement) and $[_{\sigma}2_a + _{\pi}2_a]$ cycloadditions [59]. Both skeletons were formed by irradiation of **101** in acetonitrile. This experiment provided evidence for the biosynthetic connection of pseudopteranes and gersolanes to furanocembranes. More complex polycyclic skeletons have also been encountered, resulting from [4 + 2], [5 + 2], [2 + 2], or [4 + 3] cycloadditions, as exemplified by intricarene, bielschowskysin, and rameswaralide described below.

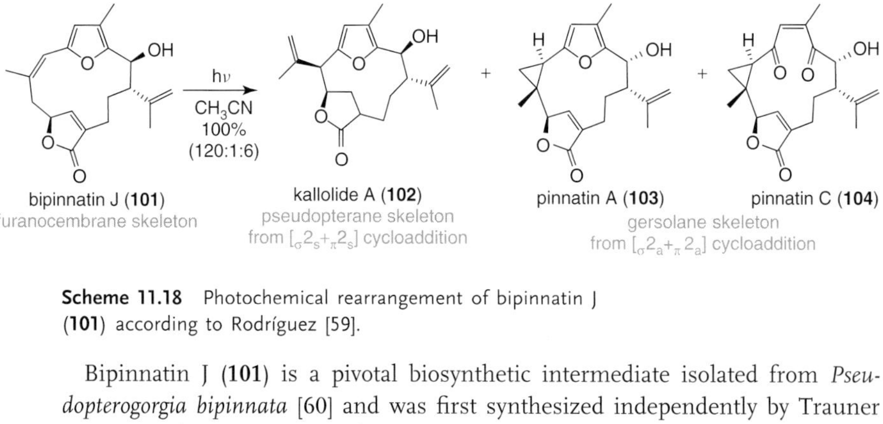

Scheme 11.18 Photochemical rearrangement of bipinnatin J (**101**) according to Rodríguez [59].

Bipinnatin J (**101**) is a pivotal biosynthetic intermediate isolated from *Pseudopterogorgia bipinnata* [60] and was first synthesized independently by Trauner [61a], Rawal [62], and soon after by Pattenden [63]. In their work, Trauner [61b] and Pattenden figured out the biosynthetic origin of intricarene (**111**), a polycyclic compound issued from a transannular 1,3-dipolar cycloaddition (Scheme 11.19). In all cases, bipinnatin J was obtained by the macrocyclization of the linear precursor **105** under Nozaki–Hiyama conditions ($CrCl_2/NiCl_2$ or $CrCl_2$). Trauner reduced **101** into rubifolide (**106**), a putative biosynthetic precursor of the series, which in turn was oxidized into isoepilophodione B (**107**) in the presence of *m*CPBA (*m*-chloroperbenzoic acid) or alternatively by the selective and biomimetic addition

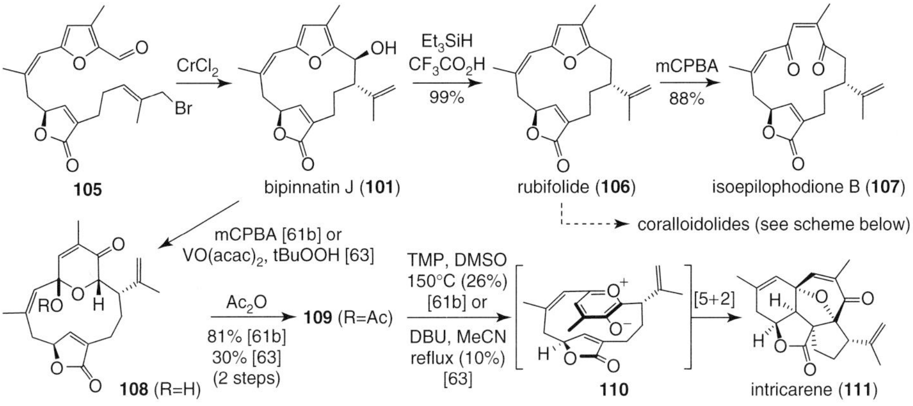

Scheme 11.19 Bipinnatin J (**101**) as a pivotal natural precursor of furanocembranoids.

of singlet oxygen to the furan moiety of **106** [61b]. In the same work, Trauner used *m*CPBA to oxidize **101** into the sensitive hydroxypyranone **108**, the acetylation of which furnished the direct precursor of intricarene **109**. Treatment of **109** with the hindered base 2,2,6,6-tetramethylpiperidine (TMP) in dimethyl sulfoxide (DMSO) at 150 °C afforded of the oxidopyrylium species **110**, which underwent transannular 1,3-dipolar cycloaddition to yield **111** in 26% yield [64]. According to the authors, temperatures in excess of 150 °C cannot be deemed biomimetic. The requirement of such conditions thus opens up the possibility that an enzyme (possibly a monooxygenase) could mediate the reaction. Similarly, Pattenden oxidized bipinnatin J (**101**) to the hydroxypyranone **108** in the presence of VO(acac)$_2$ and *t*BuOOH [63]. After acetylation to **109** and treatment under basic conditions [DBU (1,8-diazabicyclo[5.4.0]undec-7-ene), MeCN, reflux], the pyrone intermediate **110** was generated, the [5 + 2] cycloaddition of which gave intricarene **111** through an *endo* transition state, in 10% yield (not optimized).

Rubifolide (**106**), synthesized by reduction of bipinnatin J (**101**), was the starting material used by Trauner for the biomimetic synthesis of coralloidolides A, B, C, and E (**112**–**115**, Scheme 11.20) [65]. The work used Nature's strategy of *oxidative diversification* to obtain this family of diterpenes, which originated from an alcynocean coral. Chemoselective nucleophilic epoxidation at C11–C12 of **106** provided coralloidolide A (**112**) as a single diastereoisomer. Oxidative cleavage of the furan ring of **112** in the presence of *m*CPBA afforded coralloidolide E (**113**) with a similar degree of chemoselectivity. The reactivity of this compound under various reaction conditions was then explored. After extensive experimentation, the authors found that treatment of **113** with Sc(OTf)$_3$ hydrate in dioxane led to clean conversion into the rearranged product coralloidolide B (**114**) in 63% yield. This biomimetic conversion would proceed in a stepwise manner, with the catalyst taking part both in hydration of the enedione and in the subsequent transannular attack of the epoxide at C11. Alternatively, coralloidolide C (**115**) was obtained by treatment of **113** in the presence of DBU in excess. The base catalyzed

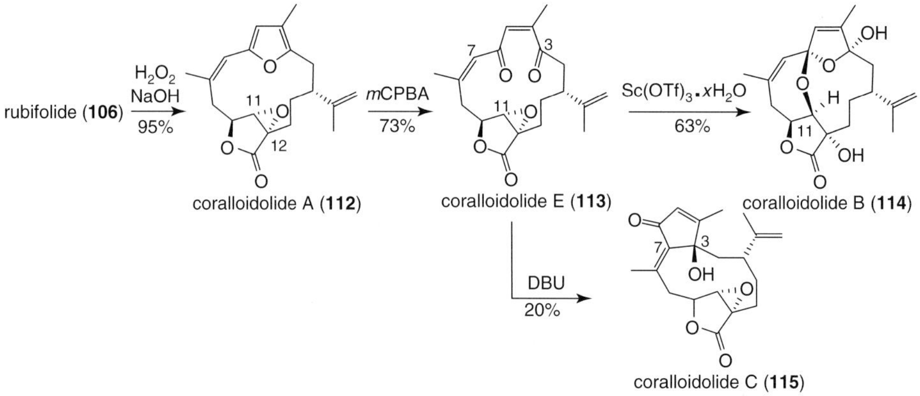

Scheme 11.20 Conversion of rubifolide (**106**) into coralloidolides A, B, C, and E (**112–115**).

the transannular aldol addition of C7 onto C3 to close the cyclopentenone ring. Remarkably, no protecting groups were used during these biomimetic syntheses.

Bielschowskysin (**117**) is a complex furanocembranoid isolated by Rodríguez from *Pseudopterogorgia kallos* [66], with a new skeleton that was initially supposed to originate from the verrillane core during biogenesis (Scheme 11.21). However, some authors have made the hypothesis that it could be formed photochemically by the [2 + 2] cycloaddition of a precursor derived from the oxygenation of bipinnatin

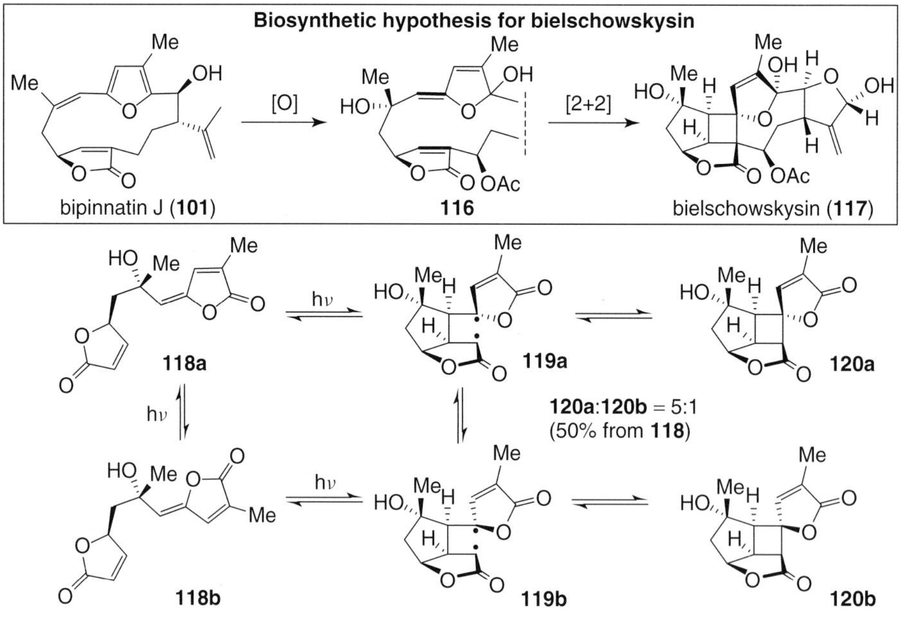

Scheme 11.21 Biomimetic studies toward bielschowskysin (**117**).

J (**101**) [61a, 67, 68]. Doroh and Sulikowski provided the bases toward the total synthesis of **117** [69]. The stereoselective intramolecular [2 + 2] photocycloaddition of the 5-alkylidene-2(5H)-furanone **118a** led to the tetracyclic intermediate **120a** as the major cycloadduct (**120a** : **120b** = 5 : 1). The substrate was irradiated in acetone, leading to isomerization into **118b** and to the 1,4-biradical intermediates **119a** and **119b**. This furnished a model study validating the [2 + 2] cycloaddition biomimetic route toward **117**. In a related work, Lear described the synthesis of the tricyclo[3.3.0]oxoheptane core of **117** by a [2 + 2] cycloaddition of an allene-butenolide substrate [68].

More recently, Pattenden and Winne proposed an intramolecular [4 + 3]-cycloaddition approach to rameswaralide (**122**, Scheme 11.22) inspired by biosynthetic speculations [70]. Rameswaralide is a polycyclic furanocembrane-related diterpene isolated from the soft coral *Sinularia dissecta* [71]. According to Pattenden and Winne's proposal, **122** would be derived from rubifolide (**106**) or a related compound. From this precursor, oxidative transformations would lead to furan opening to give an activated intermediate **121**, the [4 + 3]-cycloaddition of which would result in rameswaralide (**122**). In the same report, an alternative route from plumarellide (**124**), another furanocembranoid, would involve a 1,2-rearrangement. This speculation was tested on the synthetic seco-furanobutenolide **126**, which was submitted to acid-catalyzed rearrangement. Treatment of **126** in TFA, in the presence of small amounts of water at room temperature, soon afforded the rearranged polycyclic product **127** in quantitative yield. The exact sequence of this intriguing rearrangement could not be unraveled during this work and the authors postulated that either a [4 + 3]-cycloaddition or a [4 + 2]-cycloaddition followed by 6 → 7 ring expansion would occur. Yet another type of rearrangement was also discussed, involving a concerted [6 + 4]-cycloaddition.

11.4.4
Miscellaneous Diterpenes

Pallavicinolide A (**128**) is a rearranged seco-labdane diterpene isolated from a Japanese liverwort (Scheme 11.23) [72], and was recently synthesized in 32 steps from 2-methyl-1,3-cyclohexanedione [73]. The synthesis featured a Grob fragmentation performed on intermediate **129**, the singlet oxygen oxidation of a furan ring, and an intramolecular Diels–Alder cycloaddition on substrate **130** as the biomimetic step. All retrosynthetic disconnections in this synthesis are reminiscent of the hypothetical biogenetic route for the natural product.

In the proposal for the biosynthesis of the icetexane-based diterpene perovskone (**136**), it was suggested that the skeleton may arise from the addition of geranyl diphosphate to an icetexone precursor (Scheme 11.24). Majetich used this hypothesis for his total synthesis of this intricate natural product [74]. The tricyclic precursor **132** (rings A–C) was synthesized from 1-bromo-2,3,5-trimethoxybenzene. It was engaged in a Diels–Alder reaction, catalyzed by Eu(fod)$_3$ at 45 °C, with *trans*-α-ocimene (**133**), making ring D of intermediate **134**. When pursuing the reaction at 110 °C, ring E was closed by an ene-reaction while ring F was formed by

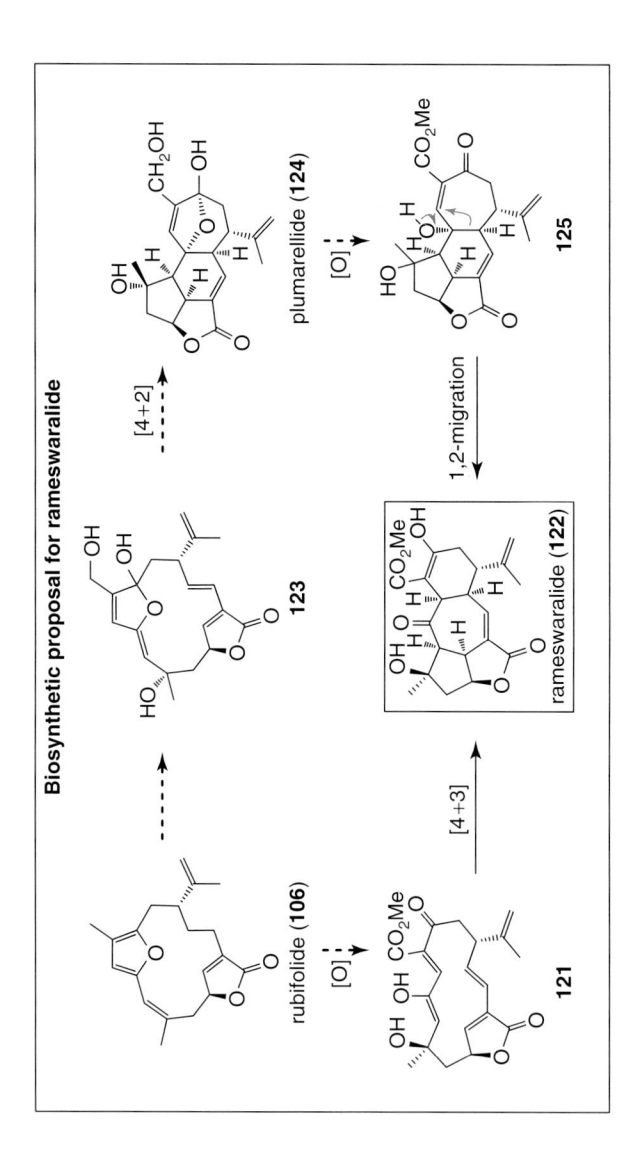

418 | 11 Biomimetic Rearrangements of Complex Terpenoids

Biosynthetic proposal for rameswaralide

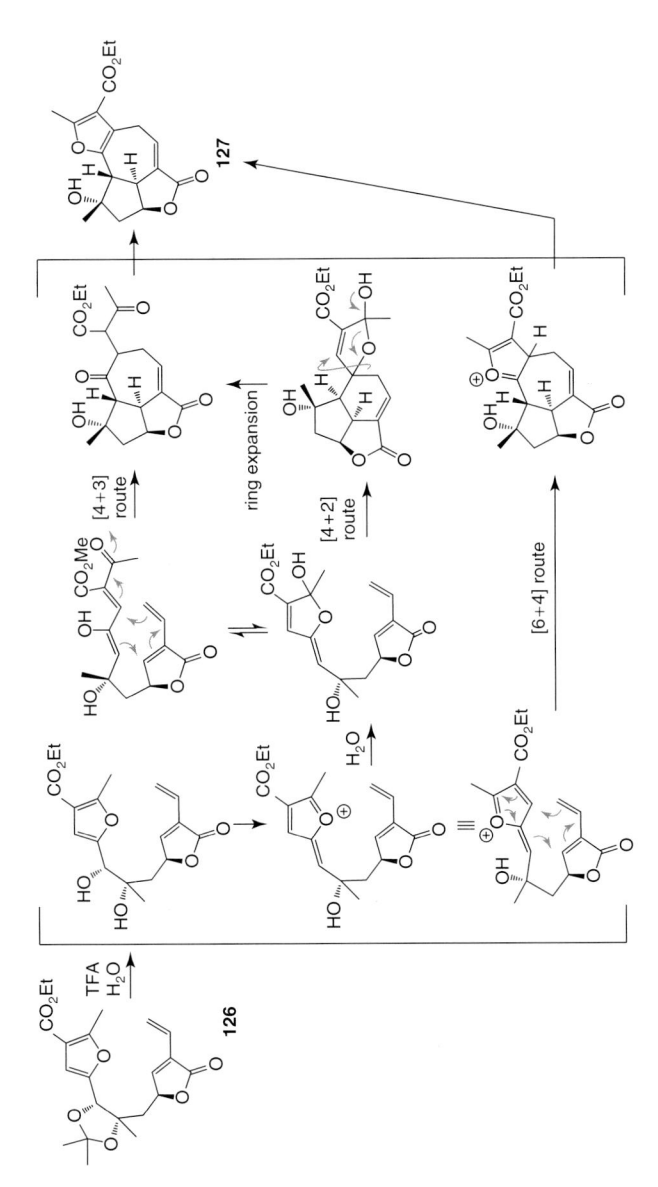

Scheme 11.22 Biomimetic studies toward rameswaralide (**122**).

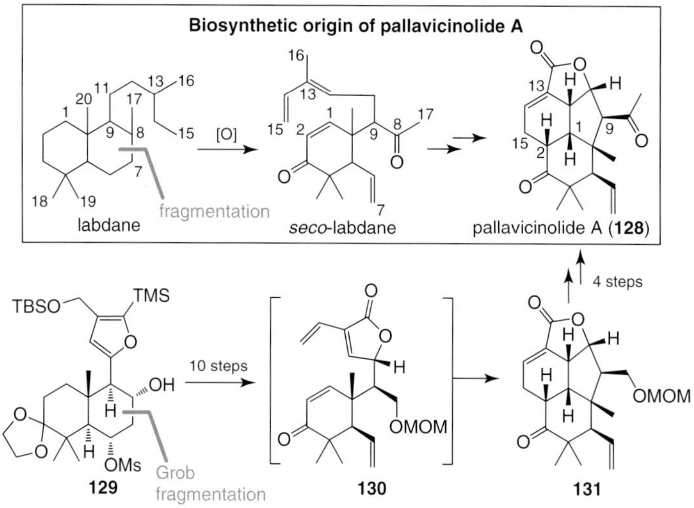

Scheme 11.23 Biomimetic synthesis of pallavicinolide A (**128**).

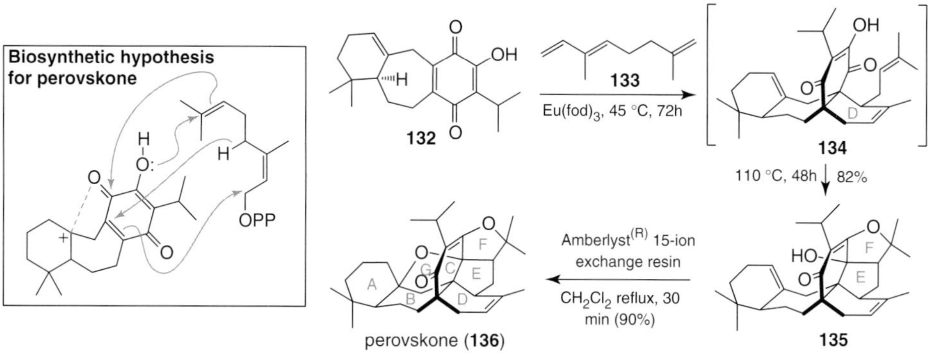

Scheme 11.24 Biomimetic synthesis of perovskone (**136**).

subsequent addition of the hydroxyl on the remaining olefin. Final acid treatment of **135** completed the synthesis of perovskone (**136**) by closing ring G.

11.5
Triterpene Rearrangements

Triterpenes are derived from the cationic cyclization of squalene-based precursors. One of the most beautiful biomimetic transformations of this type, the pentacyclization of the linear polyene **137** towards sophoradiol (**140**, Scheme 11.25), was published by Johnson in 1994 [75]. The outcome of this cyclization relied on the presence of a fluorine atom on C13, which was use to control the regioselectivity of the cyclization, while being easily and regioselectively removed in the presence of

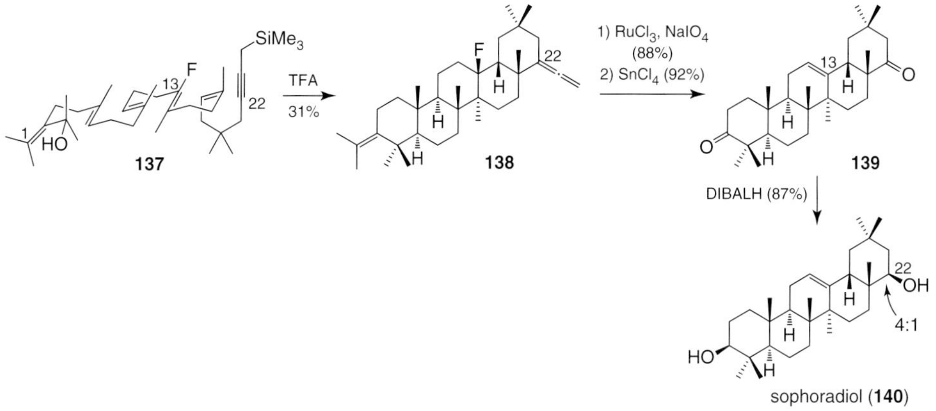

Scheme 11.25 Example of polyolefin cyclization: the biomimetic synthesis of sophoradiol (**140**).

tin tetrachloride. In the absence of fluorine on C13, a five-membered ring would have been formed. The biomimetic cyclization was performed in TFA, giving the pentacyclic intermediate **138** in 31% yield. After the cyclization process, return to neutrality was provided by TMS-elimination from the terminal propargylsilane. The polycyclization could also be performed in 50% yield in the presence of the Lewis acid SnCl$_4$, but was accompanied by fluorine elimination at C13, a side reaction yet incompatible with the planned oxidative cleavage of the exocyclic double bonds. Sophoradiol (**140**) was then obtained in three high-yielding steps. This impressive achievement, representing the first controlled polycyclization towards a pentacyclic system, stands as a magnificent conclusion to Johnson's career.

Important biologically relevant compounds bear, most frequently, the tetracyclic or the pentacyclic C$_{30}$ framework of this series, which can even be truncated with possible loss of several carbons. There are numerous examples of polycyclizations of linear intermediates toward such structures, performed with synthetic or methodological purposes that cannot be described in this chapter. The reader is invited to consult excellent reviews published by others on this field [2].

Further biosynthetic rearrangements of the polycyclic C$_{30}$ framework can lead to modified skeletons. For example, cyclopamine (**148**, Scheme 11.26) is a teratogenic steroidal alkaloid isolated from the lily corn *Veratrum californicum* [76]. It strongly inhibits hedgehog signaling during embryogenesis. The name of the compound originates from the Cyclops-like malformations of lambs, observed when pregnant sheep ate the plant. Cyclopamine consists of a 14(13 → 12)*abeo*-cholestane core (or C-*nor* D-*homo*) arising from a transposition in the cholestane skeleton [77].

The total synthesis of **148** was accomplished in 2009 by Giannis from a commercially available steroid (Scheme 11.26) [78]. It was presumed that the cationic rearrangement of 12β-hydroxy steroids into their C-*nor*-D-*homo* counterparts would be a reliable approach, as demonstrated by others [79]. Since 12β-hydroxy steroids are rare and a properly functionalized one resembling the cyclopamine ABCD skeleton was not available, it was envisioned that the hydroxyl could be introduced

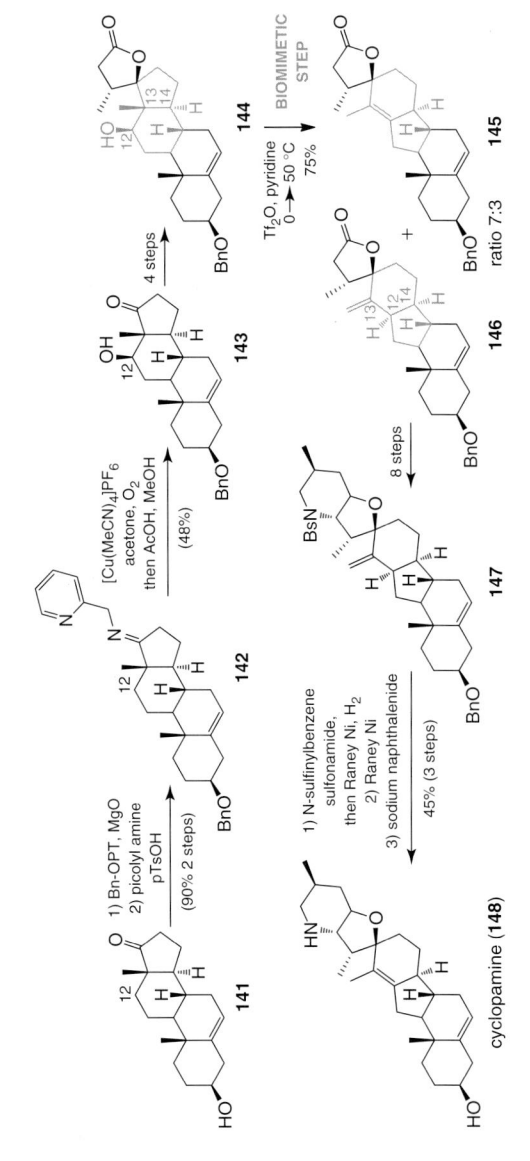

Scheme 11.26 Semi-synthetic approach to cyclopamine (**148**); Bn-OPT: 2-benzyloxy-1-methylpyridinium triflate.

at position 12 by selective C–H activation of dehydroepiandrosterone (**141**). This was undertaken by treating the 2-picolylimine **142** with tetrakis(acetonitrilo)copper(I) hexafluorophosphate and molecular oxygen in acetone, furnishing the 12β-hydroxy derivative **143**. The reaction was performed with complete regioselectivity and stereoselectivity.

After installment of the spiro lactone moiety of intermediate **144**, the biomimetic Wagner–Meerwein rearrangement of the CD ring system was performed by exposure of the alcohol to trifluoromethanesulfonic anhydride in pyridine. The 14(13 → 12)*abeo*-cholestane core was isolated in nearly quantitative yields as a 3 : 7 mixture of regioisomers **145** and **146**, respectively. Even though it would be necessary to convert the exocyclic double bond into the endocyclic isomer, the compound **146**, as the dominant isomer, was engaged in the next steps. The fused piperidine cycle was installed in eight steps before double bond isomerization using an Alder-ene reaction in the presence of a sulfurizing agent. Final deprotections led to cyclopamine (**148**). The synthesis was undertaken in 20 steps and 1% overall yield from the steroid **141**. At the same time several isomeric compounds were obtained for structure–activity studies.

Betulin (**149**, R = CH$_2$OH), an abundant triterpene isolated from the birch tree, shares the same lupane pentacyclic framework as betulinic acid (**150**, R =CO$_2$ H) and lupeol (**153**, R = CH$_3$) (Scheme 11.27). The expansion of ring E through Wagner–Meerwein rearrangement of **149** was first reported in 1922 by Schulze and Pieroh by use of formic acid [80]. This biomimetic transformation afforded allo-betulin (**151**) with the oleane framework, a compound that exhibits many biological

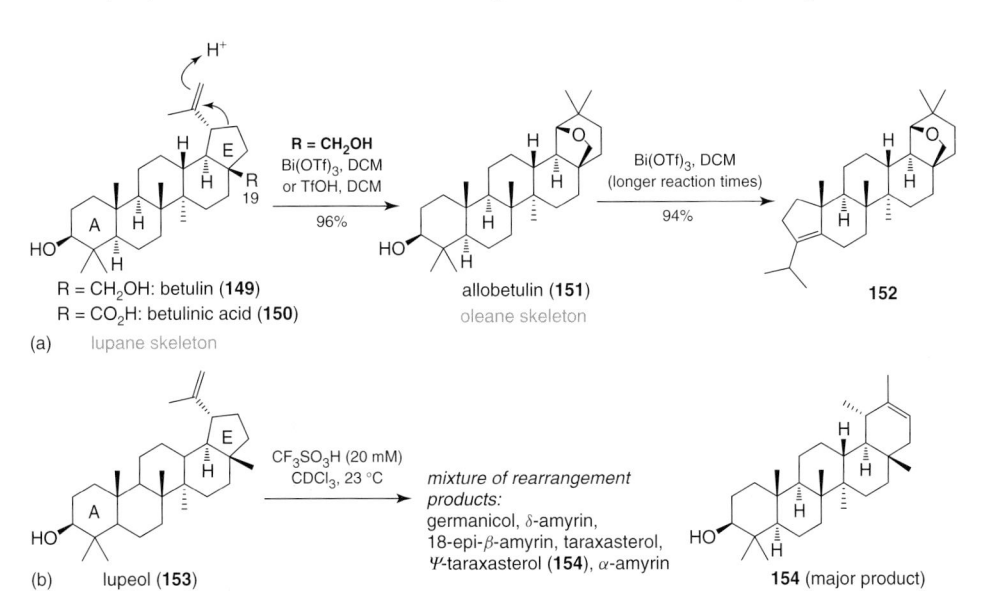

Scheme 11.27 Biomimetic conversion of lupane into oleane skeletons: (a) Bi(III) salt catalysis by Salvador [27]; (b) Corey acid-catalyzed rearrangement of lupeol [82].

activities (e.g., antifeedant, anti-inflammatory, cytotoxic). During the reaction, a Wagner–Meerwein 1,2-shift was accompanied by tandem formation of an ether bridge following a step-by-step carbocation mechanism. The reaction has been well studied and can be catalyzed by various acids. For example, montmorillonite K10 and kaolinite have been reported to convert quantitatively **149** into **151** [81]. More recently, the use of "eco-friendly" bismuth(III) salts as catalysts has been reported (Scheme 11.27a) [27]. In fact, the acid protons released from the hydrolysis of $Bi(OTf)_3 \cdot xH_2O$ were shown to be the true catalytic species during the formation of compounds **151** since no reaction occurred when a proton scavenger was added to the reaction mixture. For longer reaction times in the presence of the bismuth(III) salt, a second contraction at the A-ring of **151** provided the end product **152**.

Recently, Corey reported the total synthesis of lupeol (**153**) [82]. On this occasion, the one-step conversion of **153** into the naturally occurring pentacyclic triterpenes germanicol, δ-amyrin, 18-epi-β-amyrin, taraxasterol, ψ-taraxasterol (**154**), and α-amyrin via cationic intermediates was studied (Scheme 11.27b). Treating lupeol with a 20 mM solution of triflic acid in $CDCl_3$ at 23 °C resulted in a mixture of the six rearrangement products, monitored by ^1H NMR, the major product of the reaction being ψ-taraxasterol (**154**). In this rearrangement the main drawback was the non-controlled return to neutrality, generating a complex mixture of isomeric products.

11.6
Some Examples of the Biomimetic Synthesis of Meroterpenoids

Meroterpenoids are natural products that have their terpenic origin mixed in with another biosynthetic pathway. Compounds mixing terpene and polyketide features, including phenols, have provided many examples of biosynthetic rearrangements. In prenylated phenols it is common to observe covalent interactions between the terpene part and the phenol ring. This chemistry has been described elsewhere in this book.

The biomimetic conversion of a geranylated pyrone (**157**) towards the skeleton of transtaganolides (**155**) and basidiolides (**156**), biologically active compounds isolated from the Mediterranean plant *Thapsia garganica*, has been reported recently (Scheme 11.28) [83]. This involved a sequence of Ireland–Claisen rearrangement of **157** into **158** and an intramolecular Diels–Alder reaction into **159**, following the biosynthetic disconnections.

Erinacine E (**165**) is a rearranged diterpene glycoside, isolated from the fungus *Hericium erinaceum*, with neuritogenic activities [84]. Its biosynthesis would result from a double C–C linkage between the glycoside part and the cyathane diterpene nucleus of an intermediate derived from erinacine P. The cyathane glycoside derivative **160**, analogous to erinacine P, has been synthesized by Nakada (Scheme 11.29) [85]. After Swern oxidation into **161**, a conjugate addition–elimination occurred to make the C2'–C13 bond of compound **162**, the structure of which was close to the natural products striatals A–D. Subsequent treatment with DBU installed the

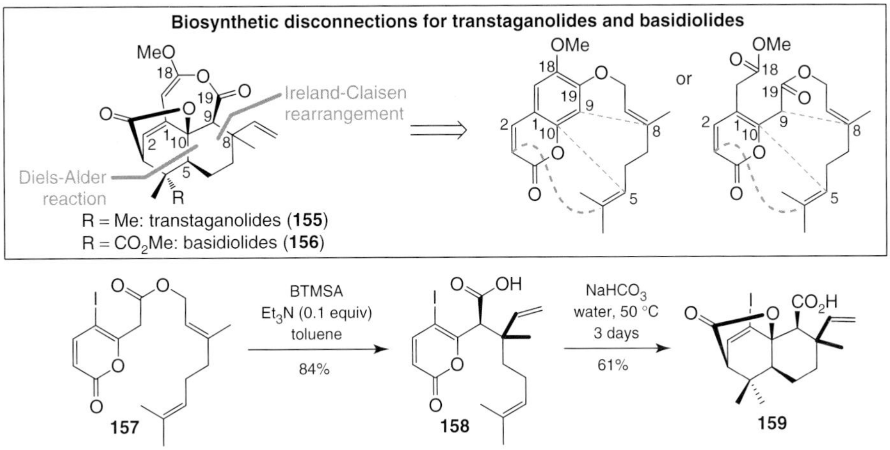

Scheme 11.28 Johansson's biomimetic study towards transtaganolides (**155**) and basidiolides (**156**) (BTMSA: N,O-bistrimethylsilylacetamide).

C4′–C15 bond in the strained hexacyclic intermediate **164**, and deprotection led to erinacine E (**165**).

The biomimetic Cope-rearrangement of globiferin (**167**), isolated from the antimalarial root extract of *Cordia globifera*, led to cordiachrome C (**166**) (Scheme 11.30) [86]. The co-occurrence of both metabolites in the same plant provided substantial information on the biogenetic relationship between the natural products. Although harsh conditions were employed for the rearrangement of **167**, the racemic character of cordiachrome C (**166**) has suggested a non-enzymatic biosynthetic pathway.

The asymmetric biomimetic synthesis of (−)-longithorone A (**172**), a cytotoxic marine natural product, featured a double Diels–Alder reaction of properly functionalized *ansa*-farnesylhydroquinone derivatives **168** and **169** (Scheme 11.31) [87]. The stereoselective intermolecular cycloaddition of both atropoisomeric substrates resulted in heterodimerization to give **170**. After activation as the bis(quinone) **171**, the intramolecular cycloaddition afforded the optically pure natural product **172**. This work provided a unique example of chirality transfer in complex molecule synthesis involving chiral atropoisomers.

11.7
Conclusion

This chapter has illustrated how complex terpenoid compounds can be obtained using Nature's biosynthetic strategy. Striking examples were discussed, showing the efficiency of this strategy, sometimes under conditions having physiological likeness. Some reactions occurred spontaneously, but most of them were undertaken with the use of a physical or chemical activation. The Wagner–Meerwein

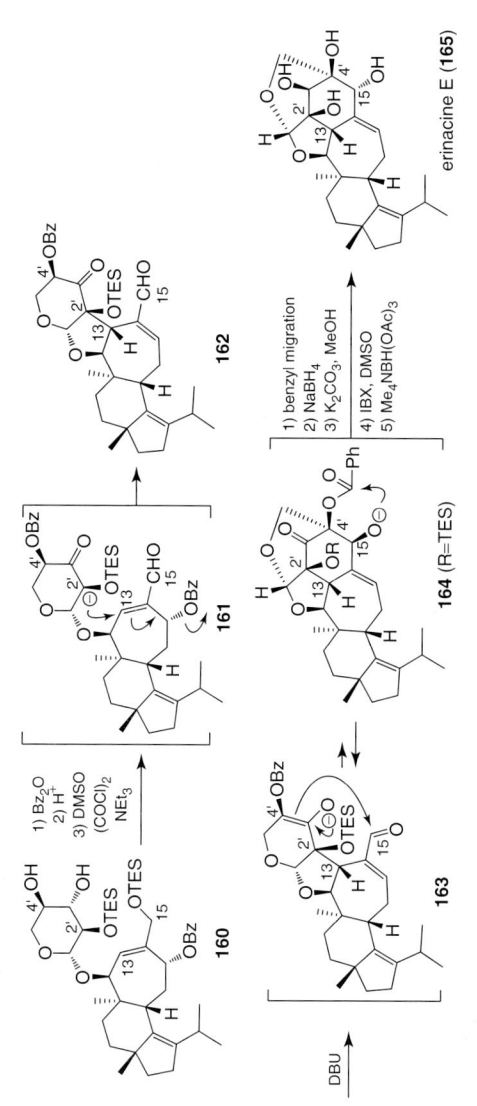

Scheme 11.29 Nakada's biomimetic synthesis of erinacine E (**165**).

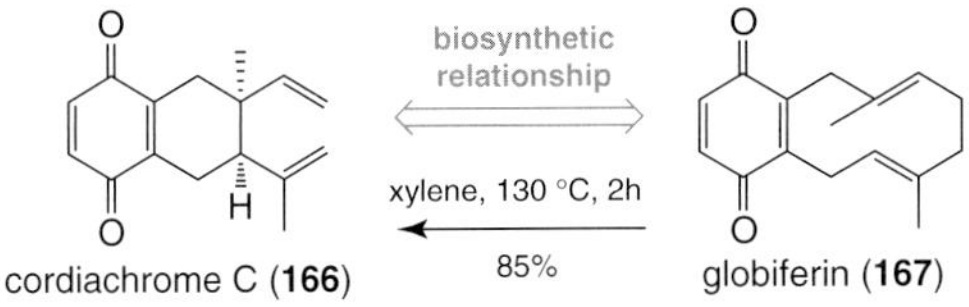

Scheme 11.30 Biomimetic conversion of globiferin (**167**) into cordiachrome C (**166**).

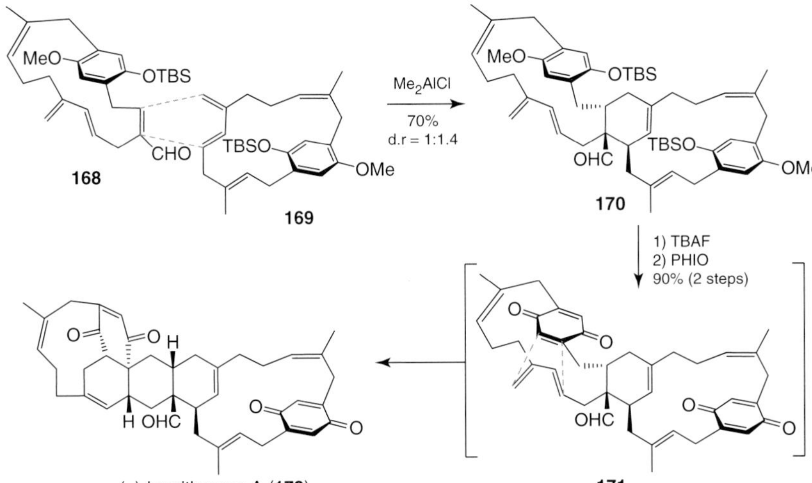

Scheme 11.31 Biomimetic synthesis of longithorone A (**172**).

rearrangement, which was discovered more than a century ago, has been used extensively in terpene biomimetic synthesis, using weak to strong acid catalysis. It is also the key reaction for diversification in terpene biogenesis. Furthermore, many biomimetic cycloadditions were undertaken by photochemical or thermal activation. The photochemical conditions are present as sunlight activation in Nature and could proceed without the involvement of an enzyme. The thermal conditions, which can be very harsh in the laboratory to reach the proper activation barrier for the cycloaddition or the rearrangement, usually do not occur in Nature. In such cases the activation barrier might be overcome by enzymes during biosynthesis, with somewhat better stereoselectivity for chiral natural products.

For the most strained reactant systems of biosynthetic hypotheses, quantum chemical calculations have provided theoretical catalysts called *theozymes* [88]. This work predicts the catalytic system to be used in a reaction, with important applications in biological systems. Salvileucalin B (**175**) is a polycyclic rearranged diterpene related to the neoclerodane structure isolated from *Salvia leucantha* (Scheme 11.32) [89]. The strained tricyclo[3.2.1.0^{2,7}]octane moiety of this natural product caused chemists to question its biosynthetic origin. Fortunately, it was isolated at the same time as the biogenetically correlated salvileucalin A (**173**), so

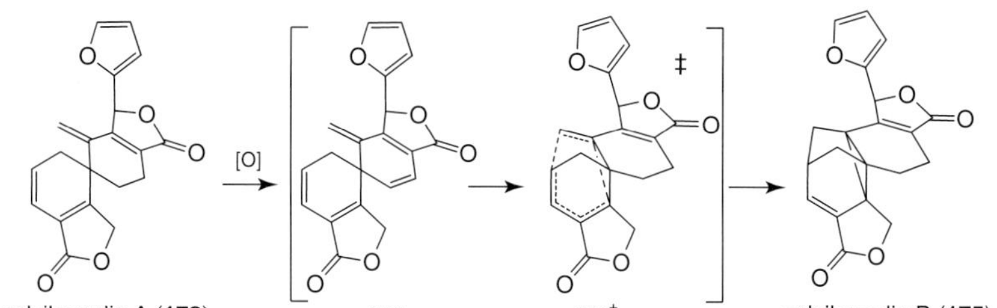

Scheme 11.32 Salvileucalin A (**173**), a biosynthetic precursor of salvileucalin B (**175**).

that a biosynthetic scheme could be postulated. Quantum chemical calculations have revealed how functional group arrays present in a known enzyme active site could accelerate the intramolecular Diels–Alder reaction proposed to occur during the biosynthesis of **175** [90]. Several theozymes were screened to find that, after the oxidation of **173** to obtain the activated compound **174**, the cycloaddition could be promoted by selective binding of the transition state **174**‡ to the active site of a catalytic antibody species. The theozyme would lower the activation barrier of the cycloaddition from circa 25 to 20 kcal mol^{-1}, a value that is at the high end of the range of typical barriers for enzyme-catalyzed reactions. The work also pointed out that a biomimetic total synthesis of **175** might benefit from performing the Diels–Alder reaction prior to installing the "bottom" lactone. It was shown that an organocatalyst (thiourea) or a Lewis acid (AlMe$_2$Cl) could promote this transformation. This synthesis has yet to be performed.

References

1. Ourisson, G. (1964) *Proc. Chem. Soc.*, 274–282.
2. For example, see the following early or late reviews: (a) Johnson, W.S. (1976) *Bioorg. Chem.*, **5**, 51–98; (b) Yoder, R.A. and Johnston, J.N. (2005) *Chem. Rev.*, **105**, 4730–4756.
3. Kindt, H.H. (1802) *Trommsdorff J. Pharm.*, **11**, 132.
4. Wagner, G. (1899) *J. Russ. Phys. Chem. Soc.*, **31**, 680.
5. (a) Wagner, G. and Brickner, W. (1899) *Chem. Ber.*, **32**, 2302–2325; (b) Wagner, G. and Slawinski, K. (1899) *Chem. Ber.*, **32**, 2064–2086.
6. Banthorpe, D.V., Charlwood, B.V., and Francis, M.J.O. (1972) *Chem. Rev.*, **72**, 115–155.
7. Dewick, P.M. (2002) *Medicinal Natural Products: A Biosynthetic Approach*, 2nd edn, John Wiley & Sons, Ltd, Chichester.
8. Birladeanu, L. (2000) *J. Chem. Ed.*, **77**, 858–863.
9. (a) Meerwein, H. (1914) *Liebigs Ann. Chem.*, **405**, 129–175; (b) Meerwein, H. and van Emster, K. (1920) *Chem. Ber.*, **53**, 1815–1829; (c) Meerwein, H. and van Emster, K. (1922) *Chem. Ber.*, **55**, 2500–2528.
10. Meerwein, H. (1913) *Liebigs Ann. Chem.*, **396**, 200–225.
11. Swift, K.A.D. (2001) in *Fine Chemicals Through Heterogeneous Catalysis* (eds R.A. Sheldon and H. van Bekkum), Wiley-VCH Verlag GmbH, Weinheim, pp. 242–246.

12. Gscheidmeier, M. and Häberlein, H. (1998) US Patent US005826202A.
13. Sorensen, T.S. (1976) *Acc. Chem. Res.*, **9**, 257–265.
14. (a) Huang, E., Ranganayakulu, K., and Sorensen, T.S. (1972) *J. Am. Chem. Soc.*, **94**, 1779–1780; (b) Huang, E., Ranganayakulu, K., and Sorensen, T.S. (1972) *J. Am. Chem. Soc.*, **94**, 1780–1782; (c) Haseltine, R., Huang, E., Ranganayakulu, K., Sorensen, T.S., and Wong, N. (1975) *Can. J. Chem.*, **53**, 1876–1890; (d) Haseltine, R., Ranganayakulu, K., Wong, N., and Sorensen, T.S. (1975) *Can. J. Chem.*, **53**, 1901–1914.
15. Olah, G.A., de Member, J.R., Lui, C.Y., and Porter, R.D. (1971) *J. Am. Chem. Soc.*, **93**, 1442–1446.
16. (a) Johnson, C.K. and Collins, C.J. (1974) *J. Am. Chem. Soc.*, **96**, 2514–2523; (b) Collins, C.J., Johnson, C.K., and Raaen, V.F. (1974) *J. Am. Chem. Soc.*, **96**, 2524–2531.
17. Allemann, R.K. (2008) *Pure Appl. Chem.*, **80**, 1791–1798.
18. O'Maille, P.E., Malone, A., Dellas, N., Hess, B.A., Smentek, L., Sheehan, I., Greenhagen, B.T., Chappell, J., Manning, G., and Noel, J.P. Jr. (2008) *Nat. Chem. Biol.*, **4**, 617–623.
19. For an acid-catalyzed ring contraction of longifolene into isolongifolene, see: Ranganathan, R., Nayak, U.R., Santhanakrishnan, T.S., and Dev, S. (1970) *Tetrahedron*, **26**, 621–630.
20. Bohlmann, F., Zdero, C., Jakupovic, J., Robinson, H., and King, R.M. (1981) *Phytochemistry*, **20**, 2239–2244.
21. Collado, I.G., Hanson, J.R., and Macias-Sanchez, J. (1998) *Nat. Prod. Rep.*, **15**, 187–204.
22. (a) Shirahama, H., Osawa, E., Chhabra, B.R., Shimokawa, T., Yokono, T., Kanaiwa, T., Amiya, T., and Matsumoto, T. (1981) *Tetrahedron Lett.*, **22**, 1527–1528; (b) Guella, G., Chiasera, G., N'Diaye, I., and Pietra, F. (1994) *Helv. Chim. Acta*, **77**, 1203–1221.
23. Gatilova, V.P., Korchagina, D.V., Rybalova, T.V., Gatilov, Yu.V., Dubovenko, Zh.V., and Barkhash, V.A. (1989) *Zh. Org. Khim.*, **25**, 320–332.
24. Fitjer, L., Malich, A., Paschke, C., Kluge, S., Gerke, R., Rissom, B., Weiser, J., and Noltemeyer, M. (1995) *J. Am. Chem. Soc.*, **117**, 9180–9189.
25. Olhoff, G., Uhde, G., and Schulte-Elte, K.H. (1967) *Helv. Chim. Acta*, **50**, 561–570.
26. Collado, I.J., Hanson, J.R., Hernandez-Galan, R., Hitchcock, P.B., Macias-Sanchez, A.J., and Racero, J.C. (1998) *Tetrahedron*, **54**, 1615–1626.
27. Salvador, J.A.R., Pinto, R.M.A., Santos, R.C., Le Roux, C., Matos Beja, A., and Paixão, J.A. (2009) *Org. Biomol. Chem.*, **7**, 508–517.
28. Cheng, X., Harzdorf, N.L., Shaw, T., and Siegel, D. (2010) *Org. Lett.*, **12**, 1304–1307.
29. Cameron, A.F., Hannaway, C., Roberts, J.S., and Robertson, J.M. (1970) *Chem. Commun.*, 248–249.
30. Shankar, S. and Coates, R.M. (1998) *J. Org. Chem.*, **63**, 9177–9182.
31. Coates, R.M., Ho, J.Z., Klobus, M., and Wilson, S.R. (1996) *J. Am. Chem. Soc.*, **118**, 9249–9254.
32. Coates, R.M., Ho, J.Z., Klobus, M., and Zhu, L. (1998) *J. Org. Chem.*, **63**, 9166–9176.
33. (a) Shim, H.S., Swenson, D.C., Gloer, J.B., Dowd, P.F., and Wicklow, D.T. (2006) *Org. Lett.*, **8**, 1225–1228; (b) Shim, H.S., Gloer, J.B., and Wicklow, D.T. (2006) *J. Nat. Prod.*, **69**, 1601–1605.
34. (a) Gaich, T. and Mulzer, J. (2009) *J. Am. Chem. Soc.*, **131**, 452–453; (b) Gaich, T. and Mulzer, J. (2010) *Org. Lett.*, **12**, 272–275.
35. Srikrishna, A., Ravi, G., and Venkata Subbaiah, D.R.C. (2009) *Synlett*, 32–34.
36. Srikrishna, A. and Gharpure, S.J. (1999) *Tetrahedron Lett.*, **40**, 1035–1038.
37. Amigo, C.F.D., Collado, I.G., Hanson, J.R., Hernández-Galán, R., Hitchcock, P.B., Macías-Sánchez, A.J., and Mobbs, D.J. (2001) *J. Org. Chem.*, **66**, 4327–4332.
38. Zhang, W., Luo, S., Fang, F., Chen, Q., Hu, H., Jia, X., and Zhai, H. (2005) *J. Am. Chem. Soc.*, **127**, 18–19.
39. Vokáč, K., Samek, Z., Herout, V., and Šorm, F. (1968) *Tetrahedron Lett.*, **9**, 3855–3857.

40. Kato, N., Wu, X., Nishikawa, H., Nakanishi, K., and Takeshita, H. (1994) *J. Chem. Soc. Perkin Trans. 1*, 1047–1053.
41. Matusch, R. and Häberlein, H. (1987) *Liebigs Ann. Chem.*, 455–457.
42. Ahmed, A.A., Jakupovic, J., Bohlmann, F., Regaila, H.A., and Ahmed, A.M. (1990) *Phytochemistry*, **29**, 2211–2215.
43. (a) Nakanishi, K., Endo, M., Näf, U., and Johnson, L.F. (1971) *J. Am. Chem. Soc.*, **93**, 5579–5581; (b) Zanno, P.R., Endo, M., Nakanishi, K., Näf, U., and Stein, C. (1972) *Naturwissenschaften*, **59**, 512.
44. Coates, R.M. and Bertram, E.F. (1969) *J. Chem. Soc. D*, 797–798; (b) Coates, R.M. and Bertram, E.F. (1971) *J. Org. Chem.*, **36**, 3722–3729; (c) Coates, R.M. and Kang, H.Y. (1987) *J. Chem. Soc., Chem. Commun.*, 232–233; (d) McGrindle, R. and Roy, R.G. (1971) *J. Chem. Soc. C*, 1018–1020.
45. (a) Furber, M. and Mander, L.N. (1987) *J. Am. Chem. Soc.*, **109**, 6389–6396; (b) Furber, M. and Mander, L.N. (1988) *Tetrahedron Lett.*, **29**, 3339–3342.
46. Yamauchi, T., Oyama, N., Yamane, H., Murofushi, N., Schraudolf, H., Pour, M., Furber, M., and Mander, L.N. (1996) *Plant Physiol.*, **111**, 741–745.
47. Rodríguez, A.D. and Ramírez, C. (2000) *Org. Lett.*, **2**, 507–510.
48. Rodríguez, A.D., González, E., and Huang, S.D. (1998) *J. Org. Chem.*, **63**, 7083–7091.
49. Rodríguez, A.D., Ramírez, C., Rodríguez, I.I., and Barnes, C.L. (2000) *J. Org. Chem.*, **65**, 1390–1398.
50. For a non-biomimetic total synthesis of elisabethin A, see: (a) Heckrodt, T.J. and Mulzer, J. (2003) *J. Am. Chem. Soc.*, **125**, 4680–4681; for a critical highlight on the total synthesis of elisabethin A, see: (b) Zanoni, G. and Franzini, M. (2004) *Angew. Chem. Int. Ed.*, **43**, 4837–4841.
51. (a) Nicolaou, K.C., Vassilikogiannakis, G., Mägerlein, W., and Kranich, R. (2001) *Angew. Chem. Int. Ed.*, **40**, 2482–2486; (b) Nicolaou, K.C., Vassilikogiannakis, G., Mägerlein, W., and Kranich, R. (2001) *Chem. Eur. J.*, **7**, 5359–5371.
52. Kim, A.I. and Rychnovsky, S.D. (2003) *Angew. Chem. Int. Ed.*, **42**, 1267–1270.
53. Joseph-Nathan, P., Garibay, M.E., and Santillan, R.L. (1987) *J. Org. Chem.*, **52**, 759–563.
54. Anschütz, R. and Leather, W. (1885) *Chem. Ber.*, **18**, 715–717.
55. Harrowven, D.C., Pascoe, D.D., Demurtas, D., and Bourne, H.O. (2005) *Angew. Chem. Int. Ed.*, **44**, 1221–1222.
56. Boezio, A.A., Jarvo, E.R., Lawrence, B.M., and Jacobsen, E.N. (2005) *Angew. Chem. Int. Ed.*, **44**, 6046–6050.
57. Davies, H.M.L., Dai, X., and Long, M.S. (2006) *J. Am. Chem. Soc.*, **128**, 2485–2490.
58. Roethle, P.A. and Trauner, D. (2008) *J. Nat. Prod.*, **25**, 298–317.
59. (a) Rodríguez, A.D. and Shi, J.-G. (1998) *J. Org. Chem.*, **63**, 420–421; (b) Rodríguez, A.D., Shi, J.-G., and Huang, S.D. (1998) *J. Org. Chem.*, **63**, 4425–4432.
60. Rodríguez, A.D. and Shi, J.-G. (1998) *J. Org. Chem.*, **63**, 420–421.
61. (a) Roethle, P.A. and Trauner, D. (2006) *Org. Lett.*, **8**, 345–347; (b) Roethle, P.A., Hernandez, P.T., and Trauner, D. (2006) *Org. Lett.*, **8**, 5901–5904.
62. Huang, Q. and Rawal, G.H. (2006) *Org. Lett.*, **8**, 543–545.
63. Tang, B., Bray, C.D., and Pattenden, G. (2006) *Tetrahedron Lett.*, **47**, 6401–6404.
64. For a recent review on cycloadditions involving oxidopyrylium species, see: Singh, V., Krishna Vikrant, U.M., and Trivedi, G.K. (2008) *Tetrahedron*, **64**, 3405–3428.
65. Kimbrough, T., Roethle, P.A., Mayer, P., and Trauner, D. (2010) *Angew. Chem. Int. Ed.*, **49**, 2919–2621.
66. Marrero, J., Rodríguez, A.D., Baran, P., Raptis, R.G., Sánchez, J.A., Ortega-Barria, E., and Capson, T.L. (2004) *Org. Lett.*, **6**, 1661–1664.
67. Bray, C.D. and Pattenden, G. (2006) *Tetrahedron Lett.*, **47**, 3937–3939.
68. Miao, R., Gramani, S., and Lear, M.J. (2009) *Tetrahedron Lett.*, **50**, 1761–1733.
69. Doroh, B. and Sulikowski, G.A. (2006) *Org. Lett.*, **8**, 903–906.
70. Pattenden, G. and Winne, J.M. (2009) *Tetrahedron Lett.*, **50**, 7310–7313.
71. Ramesh, P., Reddy, N.S., Venkateswarlu, Y., Reddy, M.V.R., and Faulkner, D.J. (1998) *Tetrahedron Lett.*, **39**, 8217–8220.

72. Toyota, M., Sata, T., and Asakawa, Y. (1998) *Chem. Pharm. Bull.*, **46**, 178–180.
73. (a) Dong, J.-Q. and Wong, H.N.C. (2009) *Angew. Chem. Int. Ed.*, **48**, 2351–2354; (b) Dong, J.-Q. and Wong, H.N.C. (2009) *Angew. Chem.*, **121**, 2387–2390.
74. Majetich, G. and Zhang, Y. (1994) *J. Am. Chem. Soc.*, **116**, 4979–4980.
75. Fish, P.V. and Johnson, W.S. (1994) *J. Org. Chem.*, **59**, 2324–2335.
76. Keeler, R.F. (1968) *Phytochemistry*, **7**, 303–306.
77. Heretsch, P., Tzagkraroulaki, L., and Giannis, A. (2010) *Angew. Chem. Int. Ed.*, **49**, 3418–3427.
78. Giannis, A., Heretsch, P., Sarli, V., and Stöβel, A. (2009) *Angew. Chem. Int. Ed.*, **48**, 7911–7914.
79. Hirschmann, R., Snoddy, C.S., and Wendler, N.L. (1952) *J. Am. Chem. Soc.*, **74**, 2693–2694.
80. Schulze, H. and Pieroh, K. (1922) *Chem. Ber.*, **55**, 2332–2346.
81. Li, T.-S., Wang, J.-X., and Zheng, X.-J. (1998) *J. Chem. Soc., Perkin Trans. 1*, 3957–3965.
82. Surendra, K. and Corey, E.J. (2009) *J. Am. Chem. Soc.*, **131**, 13928–13929.
83. Larsson, R., Sterner, O., and Johansson, M. (2009) *Org. Lett.*, **11**, 657–660.
84. Kawagishi, H., Shimada, A., Hosokawa, S., Mori, H., Sakamoto, H., Ishiguro, Y., Sakemi, S., Bordner, J., Kojima, N., and Furukawa, S. (1996) *Tetrahedron Lett.*, **37**, 7399–7402.
85. Watanabe, H. and Nakada, M. (2008) *J. Am. Chem. Soc.*, **130**, 1150–1151.
86. Dettrakul, S., Surerum, S., Rajviroongit, S., and Kittakoop, P. (2009) *J. Nat. Prod.*, **72**, 861–865.
87. Layton, M.E., Morales, C.A., and Shair, M.D. (2002) *J. Am. Chem. Soc.*, **124**, 773–775.
88. Tantillo, D.J., Chen, J., and Houk, K.N. (1998) *Curr. Opin. Chem. Biol.*, **2**, 743–750.
89. Aoyagi, Y., Yamazaki, A., Nakatsugawa, C., Fukaya, H., Takeya, K., Kawauchi, S., and Izumi, H. (2008) *Org. Lett.*, **10**, 4429–4432.
90. Tantillo, D.J. (2010) *Org. Lett.*, **12**, 1164–1167.

12
Polyprenylated Phloroglucinols and Xanthones

Marianna Dakanali and Emmanuel A. Theodorakis

12.1
Introduction

Guttiferae (Clusiaceae) is a family of plants that includes more than 37 genera and 1600 species [1]. Characteristic to this family is its large variation in plant morphology, which makes it an important group of plants for the study of floral diversification and evolutionary plasticity. Although mainly confined to the tropical areas, this family also includes the genus *Hypericum*, a plant that is found widely around the Mediterranean area. Several plants of the Guttiferae family have a rich history in ethnomedicine for their broad-spectrum antibacterial and healing properties. For instance, the antibacterial and antidepressant activities of *Hypericum perforatum* (St. John's wort) have been noted in traditional European medicine. In fact, *Hypericum* extracts have been tested in various clinical trials and are currently used in certain countries for the treatment of depressive, anxiety, and sleep disorders [2–11]. On the other hand, members of the *Garcinia* genus of tropical trees have considerable value as sources of medicines, pigments, foodstuffs, and lumber [12, 13].

Chemically, the Clusiaceae family of plants constitutes a rich source of polyprenylated acylphloroglucinols and xanthones. Both chemical classes have generated substantial interest due to their fascinating chemical structures and potent bioactivities [10]. This chapter summarizes the chemical classification, biosynthesis, and synthetic approaches toward these compounds.

12.2
Polycyclic Polyprenylated Phloroglucinols

12.2.1
Introduction and Chemical Classification

The chemical structures of all known polycyclic polyprenylated acylphloroglucinols (PPAPs) can be classified into three types that are related to the biosynthesis

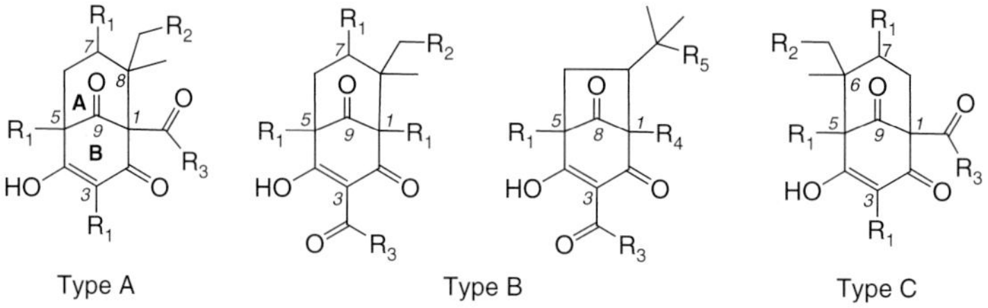

Figure 12.1 Classification of polycyclic polyprenylated acylphloroglucinols (PPAPs).

R_1 = Me, C_5H_9, or $C_{10}H_{17}$
R_2 = H or prenyl
R_3 = i-Pr, i-Bu, s-Bu,
 Ph, 3-(OH)C_6H_4, or 3,4-(OH)$_2C_6H_3$
R_4 = Me, R_5 = OH or R_4, R_5 = CH_2CHR_6
R_6 = H, C(CH$_3$) = CH$_2$, or C(CH$_3$)$_2$OH

proposal (Scheme 12.2 below). Types A, B(I), and C PPAPs are distinguished by the presence of a highly oxygenated bicyclo[3.3.1]nonane-2,4,9-trione motif, while type B(II) PPAPs contain a bicyclo[3.2.1]octane-2,4,8-trione carbon framework (Figure 12.1). In all types, the bicyclic motif is further decorated with prenyl, geranyl, and related side chains. Certain family members contain additional rings, formed by cyclizations between the β-diketone and an alkene, leading to adamantanes, pyrano-fused, or other cyclic substructures. Type A PPAPs have an acyl-substituent at C1 adjacent to a quaternary C8 carbon, while the type B compounds have the acyl-substituent at C3. The more rare type C PPAPs have the acyl group at C1 but the quaternary carbon is located at the distant C6 (Figure 12.1) [14]. For the purpose of this chapter, the carbon numbering of these molecules is based on the nemorosone numbering [14].

Figure 12.2 shows representative members of the PPAPs. Hyperforin (**1**), one of the bioactive ingredients of *Hypericum perforatum* [15, 16], belongs to the type A PPAPs and is also known for its antibacterial [17] and anticancer properties [9, 18]. Another type A phloroglucinol is garsubellin A (**2**), a natural product noted for its activities against neurodegenerative diseases. In fact, recent studies have shown that garsubellin A induces biosynthesis of acetylcholine, a neurotransmitter that at low concentrations can lead to Alzheimer's disease [19]. Nemorosone (**3**) exhibits antimicrobial [20, 21], cytotoxic, and antioxidant activities [20], while, clusianone (**4**), a type B(I) PPAP, is known for its anti-HIV activity [22]. The type B(II) enaimeone A (**5**) was isolated from *Hypericum papuanum*, the leaves of which are used in the traditional medicine of Papua New Guinea for treating sores [23]. Garcinielliptone M (**6**), a type C PPAP, isolated from *Garcinia subelliptica*, shows potential anti-inflammatory activity [24].

12.2.2
Biosynthesis of PPAPs

PPAPs derive biosynthetically from the less complex monocyclic polyprenylated acylphloroglucinols (MPAPs), a class of natural products isolated from plants of the

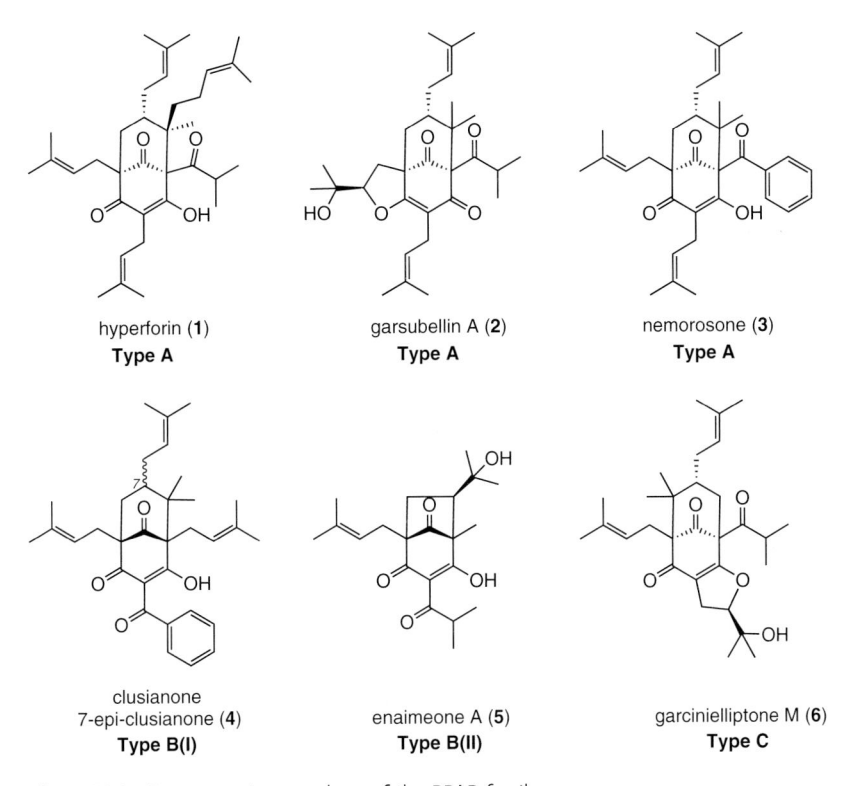

Figure 12.2 Representative members of the PPAP family.

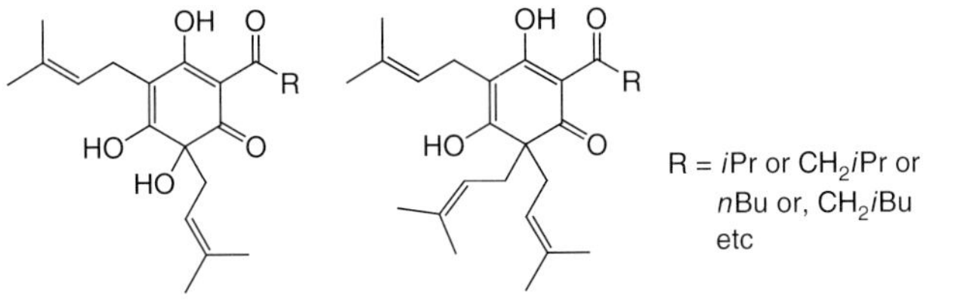

- hyperforin (1) — Type A
- garsubellin A (2) — Type A
- nemorosone (3) — Type A
- clusianone 7-epi-clusianone (4) — Type B(I)
- enaimeone A (5) — Type B(II)
- garcinielliptone M (6) — Type C

α-acids (7)

β-acids (8)

R = *i*Pr or CH$_2$*i*Pr or *n*Bu or, CH$_2$*i*Bu etc

Figure 12.3 Monocyclic polyprenylated acylphloroglucinols (MPAPs) from *Humulus lupulus*.

Myrtaceae and Cannabinaceae families [10]. There are two main classes of MPAPs: the diprenylated α-acids (7) and triprenylated β-acids (8) (Figure 12.3). α-Acids are responsible for the flavor and bitter taste of beer [25] while β-acids show, among other properties, free radical scavenger activity [7, 26].

Labeling and enzymological experiments have provided evidence that bitter acids are biosynthesized via condensation of three malonyl-CoA and one acyl-CoA, such as isobutyryl-CoA (9) (Scheme 12.1) [27–30]. The intermediate polyketide 10 can then cyclize via an intramolecular Dieckmann condensation to produce acylphloroglucinol 11 [31, 32]. Subsequent prenylations (or geranylations) occur via an enzymatic process that involves prenyltransferase-catalyzed reactions

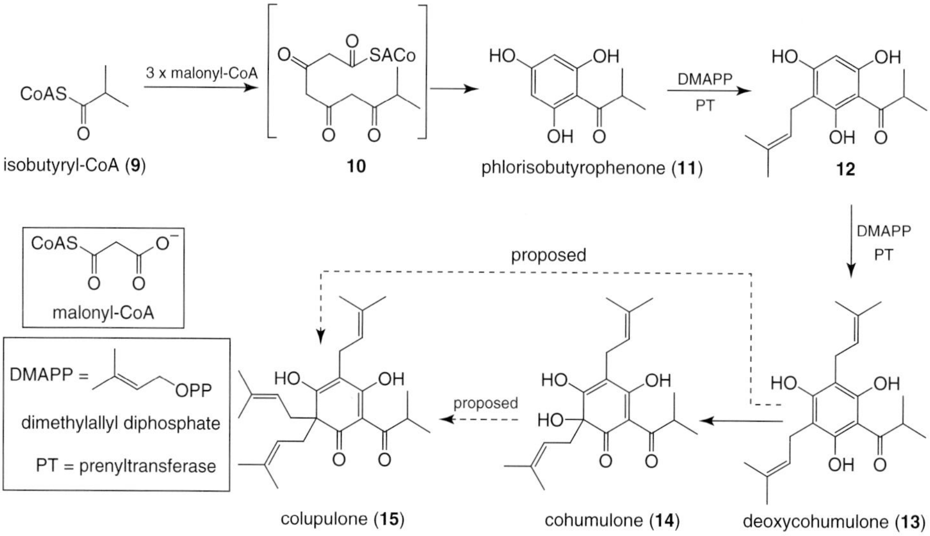

Scheme 12.1 Biosynthesis of acylphloroglucinol **11** and bitter acids **14** and **15**.

of the appropriate diphosphates with phloroglucinol [25, 33–37]. This stepwise prenylation process is illustrated in Scheme 12.1 for the construction of **12** and **13** (deoxycohumulone) from **11**. It has been shown that chemical and/or enzymatic oxidation of **13** can lead to cohumulone (**14**), a representative α-acid. It has also been proposed that additional prenylation of **13** can form colupulone (**15**), a typical β-acid [25].

Type A and type B PPAPs are proposed to arise via reaction of acylphloroglucinol **16** with prenyl diphosphate. The resulting carbocation **17** can be attacked by the C1' or C5' enol to produce compounds **18** or **19**, respectively, that represent type A and type B PPAPs (Scheme 12.2) [14]. Acylphloroglucinol **20**, containing a prenylated C1' center, is the proposed intermediate of type C PPAPs. Specifically, reaction of **20** with prenyl diphosphate can form carbocation **21** and, after cyclization, bicyclic **22**, a type C PPAP [10].

12.2.3
Biomimetic Synthesis of PPAPs

A 2006 review summarizes the synthetic efforts toward PPAPs [10]. These studies set the stage for the first total synthesis of garsubellin A, reported by the Shibasaki group [38], and were followed a few months later by a synthesis from the Danishefsky group [39]. Since then, additional total syntheses have been published of either type A [hyperforin (**1**), garsubellin A (**2**), nemorosone (**3**)] [40–42], or type B [clusianone (**4**)] [41, 43–45] PPAPs. Among all published approaches there are only two strategies built upon biomimetic considerations that involve: (i) formation of a fully prenylated B ring and (ii) construction of the A ring via a cation-based alkylative dearomatization. Both strategies departed from the

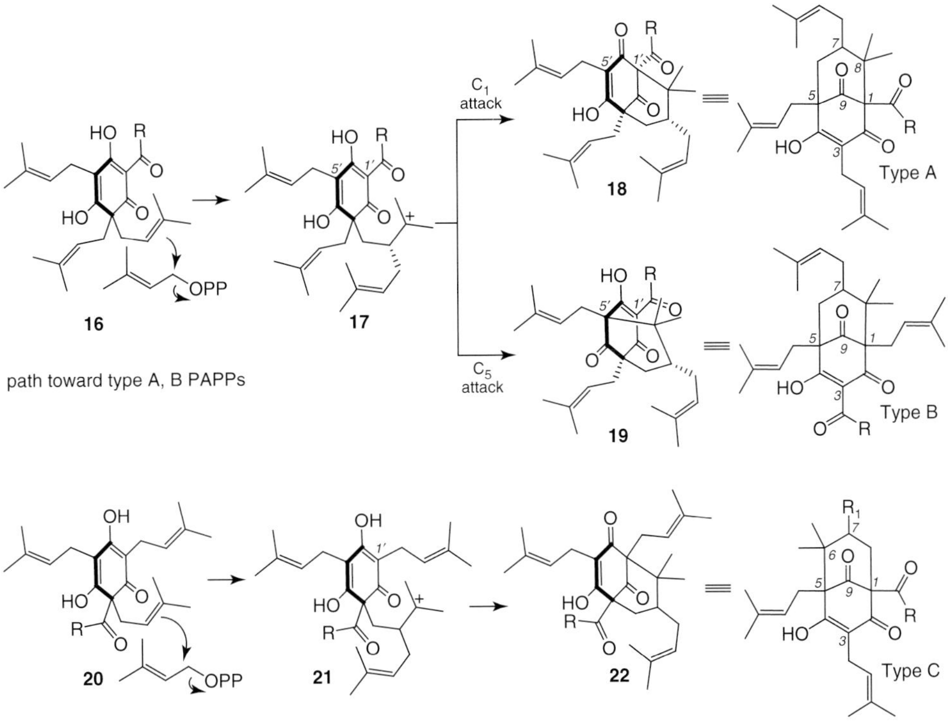

Scheme 12.2 Biosynthesis of type A, type B, and type C PPAPs from MPAPs.

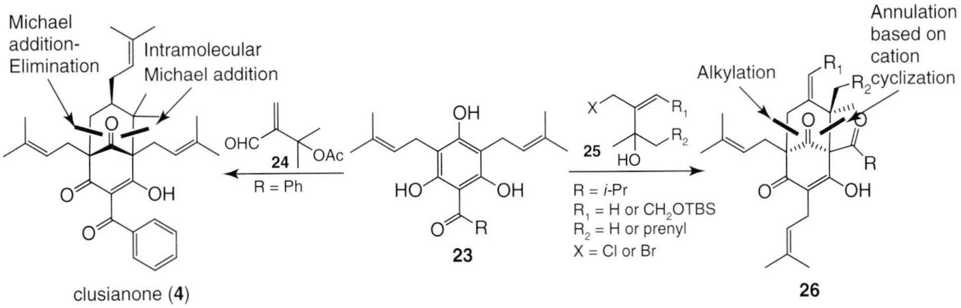

Scheme 12.3 Biomimetic approaches toward type A and type B PPAPs.

bis-prenylated acylphloroglucinol **23**, a key intermediate in the biosynthesis of these natural products (Scheme 12.3). The first one, reported by the Porco group, resulted in the total synthesis of clusianone (type B PPAP) [45] using a double Michael reaction as a key step. More recently, the Couladouros group produced compound **26**, representing the fully functionalized bicyclic core of the type A PPAPs, by a double alkylation of **23** with allyl electrophile **25** [46].

12.2.3.1 Biomimetic Total Synthesis of (±)-Clusianone

The Porco synthesis of clusianone (**4**) is summarized in Scheme 12.4 and features a double Michael addition of clusianophenone B (**27**) with α-acetoxy enal **24**. The bicyclic product **28** was methylated to form **29** as a mixture of regioisomers [45]. The employment of enal **24** proved to be a useful handle for the ensuing installation of the prenyl group at C7 (Scheme 12.4). Addition of vinyl magnesium bromide to aldehyde **29** and acetylation of the resulting alcohol gave access to acetate **30**. Palladium-mediated formate reduction followed by cross metathesis using Grubbs second-generation catalyst (**31**) yielded methylated clusianone **32** (81% over two steps). Finally, demethylation of **32** afforded (±)-clusianone (**4**) as a mixture of enol tautomers.

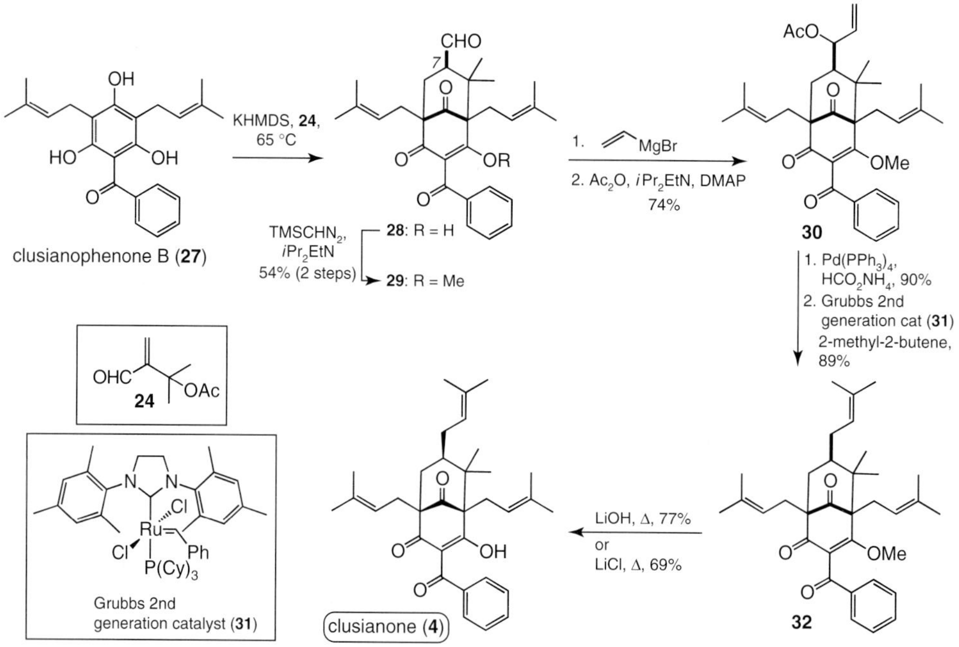

Scheme 12.4 Biomimetic synthesis of clusianone by Porco, Jr. et al.

The double Michael reaction, used for the conversion of **27** into **28**, deserves an additional comment. The authors reported that heating of this reaction at 65 °C led to desired compound **28** via epimerization of the C7 aldehyde stereocenter. Interestingly, performing this reaction at 0 °C formed the adamantane-like compound **34** via an intramolecular aldol reaction. When **34** was treated with potassium hexamethyldisilazane (KHMDS) at 65 °C compound **28** was synthesized, presumably via a retro-aldol epimerization process (Scheme 12.5). This observation allows the synthesis of adamantane-like compounds that are structurally related to hyperibone K (**36**).

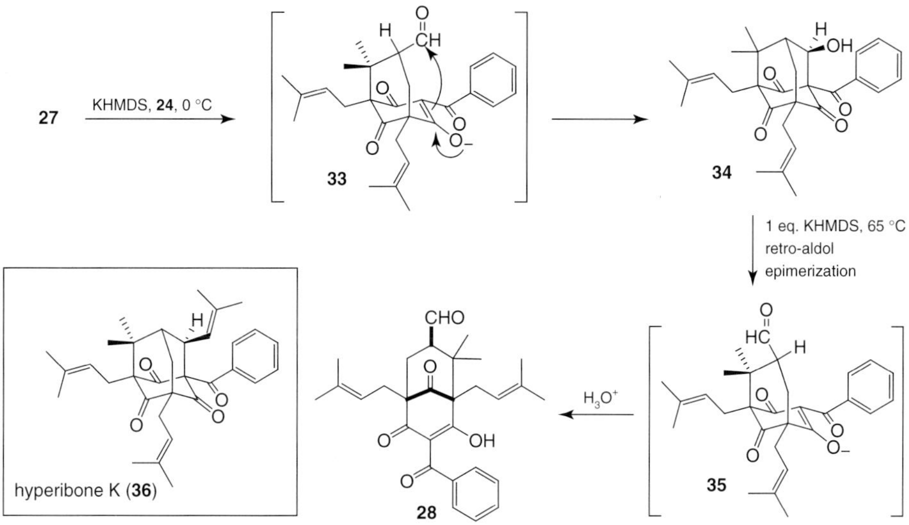

Scheme 12.5 Observations related to the double Michael reaction **27** → **28**.

12.2.3.2 Biomimetic Approach to the Bicyclic Framework of Type A PPAPs

The Couladouros approach towards the bicyclic core of type A PPAPs is highlighted in Scheme 12.6 [46]. C-alkylation of deoxycohumulone (**13**) with chloride **37**, under a two-phase solvent system at pH 14, produced compound **38** as a mixture of two diastereoisomers. Acetylation of the C4 hydroxyl group of **38** followed by mesylation

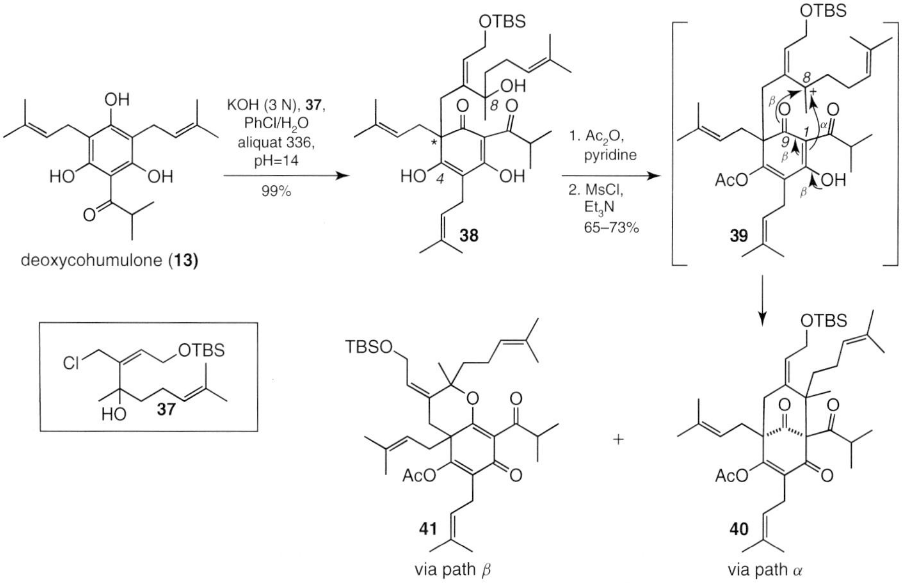

Scheme 12.6 Biomimetic formation of the type A carbon framework.

of the C8 alcohol produced bicyclic motifs **40** and **41** in 1.5 : 1 ratio, presumably via common intermediate **39**. Compound **41** is generated via O-alkylation of the C9 enol to the intermediate C8 carbocation formed during the reaction (path β). Notably, the presence of the double bond, allylic to the tertiary alcohol, is of importance for the stabilization of the cation **39**.

The authors have also evaluated a Michael addition for construction of the type A skeleton, in a similar manner as that presented above, in the synthesis of clusianone [45]. This approach was only successful for the synthesis of compounds non-substituted at C8 (Scheme 12.7). These results illustrate the difficulty in synthesizing the fully functionalized carbon framework of type A PPAPs, where the quaternary bridgehead C1 is located next to the fully substituted C8.

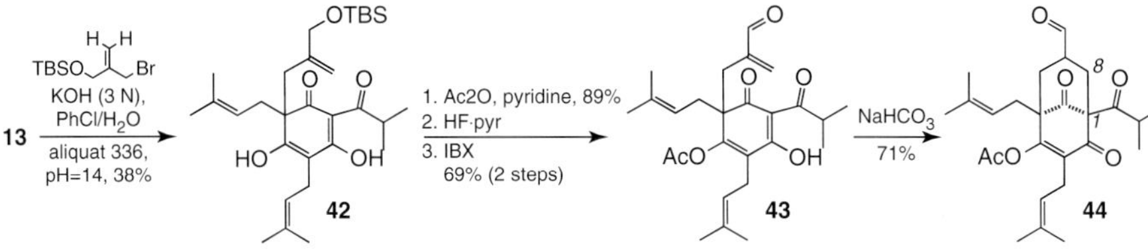

Scheme 12.7 Construction of a type A PPAP motif via an intramolecular Michael addition.

12.2.3.3 Biomimetic Synthesis of (±)-Ialibinone A and B and (±)-Hyperguinone B

Very recently a biomimetic synthesis of PPAPs isolated from *Hypericum papuanum* was reported. Scheme 12.8 depicts the total synthesis of racemic ialibinones A and

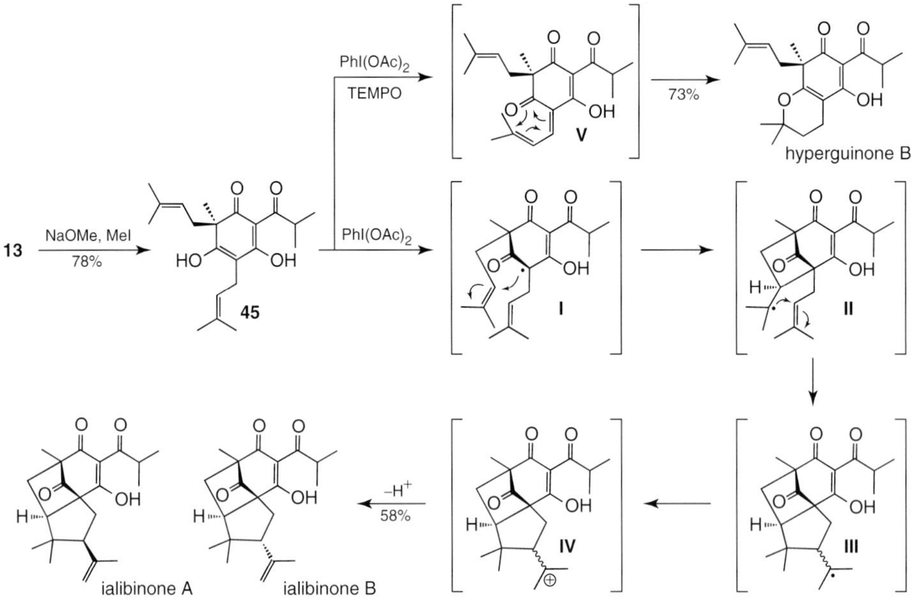

Scheme 12.8 Biomimetic synthesis of PPAPs via oxidative cyclization reactions.

B and hyperguinone B [47]. The synthesis of all three natural products started with compound **13**, which is the same starting material used in Couladouros' approach and similar to that used in Porco's synthesis of clusianone. The C-methylation of **13** was carried out by treatment with NaOMe–MeI to give **45** in very good yield. Reaction of **45** with PhI(OAc)$_2$ gave a 1 : 1 mixture of ($\pm$)-ialibinones A and B in 58% combined yield. The reaction most likely proceeds via an initial single-electron oxidation of **45** to give intermediate I. A stereoselective 5-*exo*-trig cyclization of this radical onto the pendant prenyl group would then give tertiary radical II, which can undergo a second cyclization onto the other prenyl group to give the tertiary radical III. Finally, an additional single-electron oxidation of III would lead to the tertiary carbocation IV and then to the final racemic natural products. In contrast, treatment of **45** with PhI(OAc)$_2$ in the presence of TEMPO (2,2,6,6-tetramethylpiperidine-1-oxyl) gave ($\pm$)-hyperguinone B in 73% yield. The reaction presumably proceeds via hydride abstraction by the *in situ* generated TEMPO cation to give intermediate V, which undergoes a 6π-electrocyclization leading to the pyran ring of hyperguinone B.

12.2.4
Non-biomimetic Synthesis of PPAPs

In addition to the biomimetic approaches toward PPAPs, presented above, there are also a few non-biomimetic total syntheses that have been reported in the literature in the past few years. In this section we present briefly the total synthesis of garsubellin A, nemorosone, and hyperforin – representative members of type A PPAPs – and the synthesis of clusianone, a type B PPAP.

12.2.4.1 Total Synthesis of Garsubellin A

The Shibasaki synthesis of garsubellin A (**2**) is summarized in Scheme 12.9 [38]. An interesting structural feature of this natural product is the additional ring (C-ring) that is formed by oxidative cyclization of a prenyl group. Key to the synthesis was the conversion of compound **47** into **51** via two steps: (i) a stereoselective Claisen rearrangement of **47** → **49** via intermediate **48**, which installs the alkene substituents at C1 and C5 syn to each other, and (ii) a ring-closing metathesis using the Hoveyda–Grubbs catalyst **50**. The fused tetrahydrofuran ring was constructed after allylic oxidation, hydrolysis of the carbonate, and Wacker oxidative cyclization. Finally, Stille coupling of **53** with tributyl(prenyl)tin completed the total synthesis of ($\pm$)-garsubellin A.

The Danishefsky synthesis of garsubellin A (**2**) is highlighted in Schemes 12.10 and 12.11 [39]. Compound **54**, representing the B ring of the target molecule, was converted into acetonide **55** in five steps in 43% overall yield (Scheme 12.10). Treatment of **55** with HClO$_4$ at 80 °C led to a mixture of bicyclic adducts **57** and **58** that, upon further heating, produced initially **59** and ultimately adduct **60** (71% isolated yield).

Iodocarbocyclization of **61** provided **62** using standard iodolactonization conditions (Scheme 12.11). Conceptually, this reaction is similar to the Se-mediated

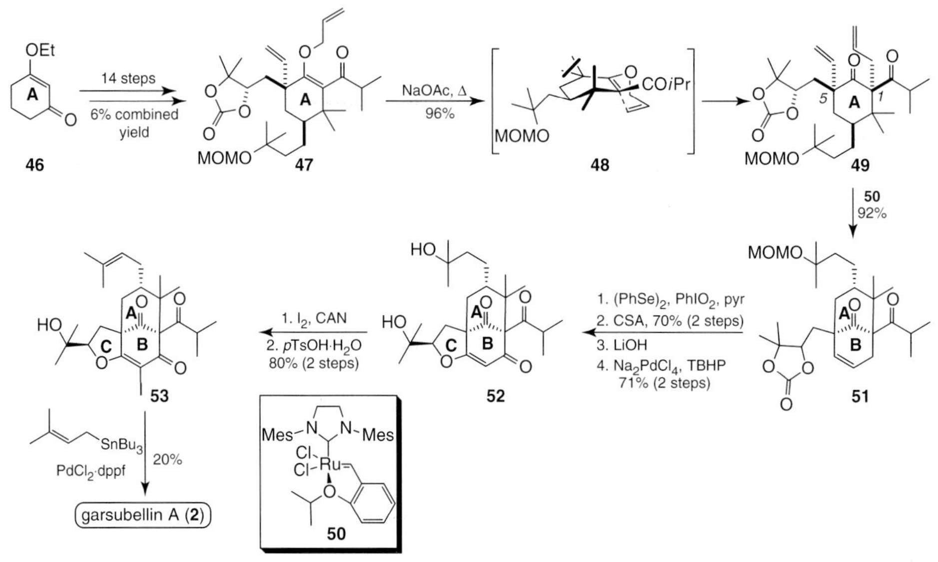

Scheme 12.9 Total synthesis of garsubellin A by the Shibasaki group.

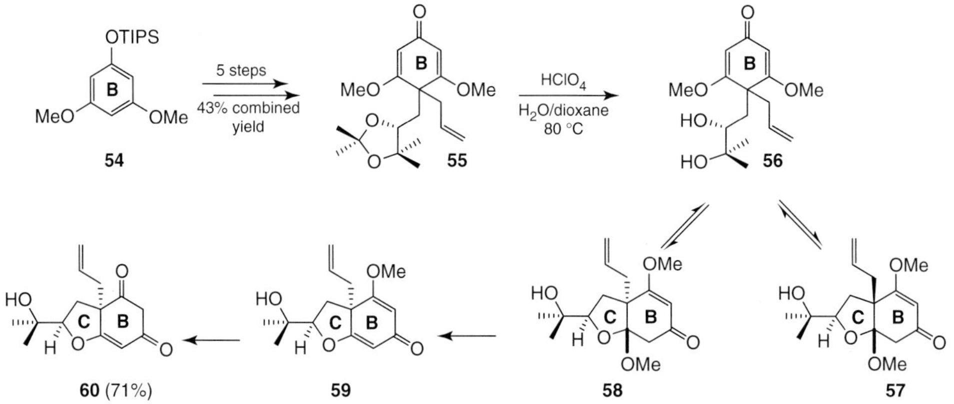

Scheme 12.10 Formation of the B-C ring system of garsubellin A by Danishefsky et al.

cyclization approach reported by Nicolaou during his efforts towards the synthesis of garsubellin [48, 49]. Compound **62** underwent additional iodination to form triiodide **63**. Treatment of **63** with excess isopropylmagnesium chloride led to compound **64** via an intramolecular Wurtz cyclopropanation and subsequent allylation. Reaction of **64** with TMSI (trimethylsilyl iodide) formed **65**, containing the tricyclic framework of garsubellin A. Finally, the synthesis of garsubellin A (**2**) was completed after decoration of **65** with the appropriate prenyl and acyl substituents.

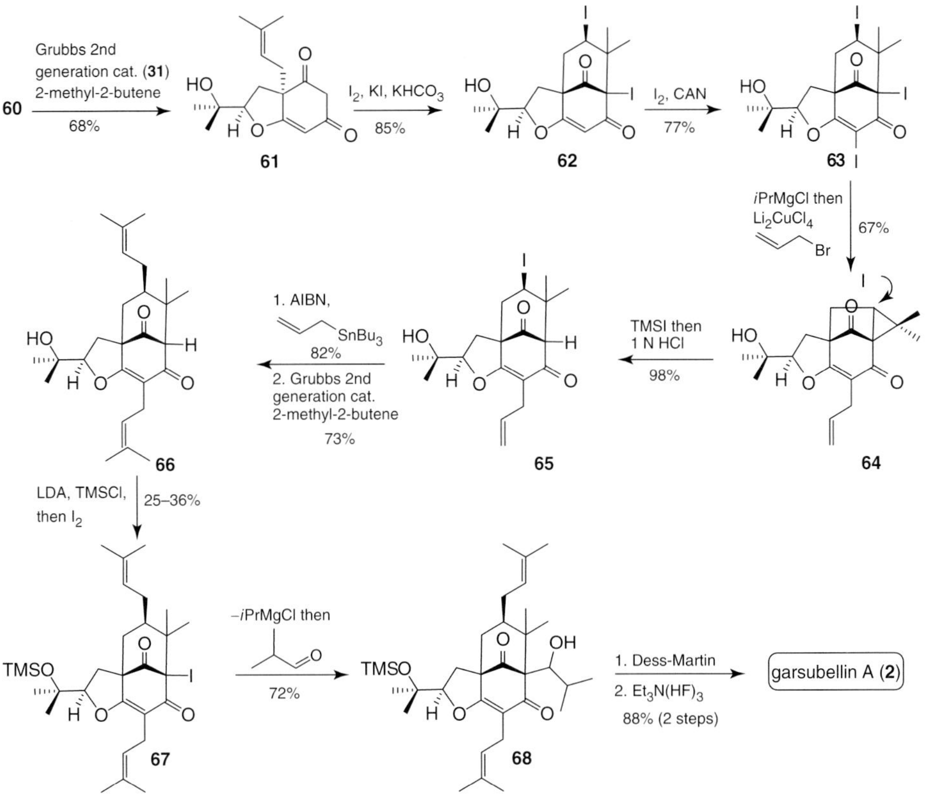

Scheme 12.11 Total synthesis of garsubellin A by the Danishefsky group.

12.2.4.2 Total Synthesis of Nemorosone and Clusianone through Differentiation of "Carbanions"

More recently, the Danishefsky group has extended the above strategy to the synthesis of nemorosone (**3**) and clusianone (**4**) [41]. These natural products proved to be more demanding targets than garsubellin A, due to the lack of the furano-fused ring that provides further stability to the molecule. A significant complication arose during the iodonium-induced carbocyclization of compound **70** (Scheme 12.12). In addition to the desired product **73**, the authors obtained substantial quantities of **71** and **72** (53% combined yield). Most likely, these side products were formed via O-alkylation of an iodonium intermediate. Gratifyingly, treatment of **71** and **72** with zinc in aqueous THF could regenerate **70** in high yield, allowing recycling of the starting material. Reaction of **73** with isopropylmagnesium chloride produced cyclopropane adduct **74** that, in turn, upon treatment with TMSI gave rise to bicyclic motif **75**. Radical allylation then produced compound **76**, a common intermediate in the synthesis of nemorosone and clusianone.

Completion of the synthesis of **3** and **4** required differentiation of carbons C1 and C3 of common intermediate **76**. To this end, treatment of **76** with an excess

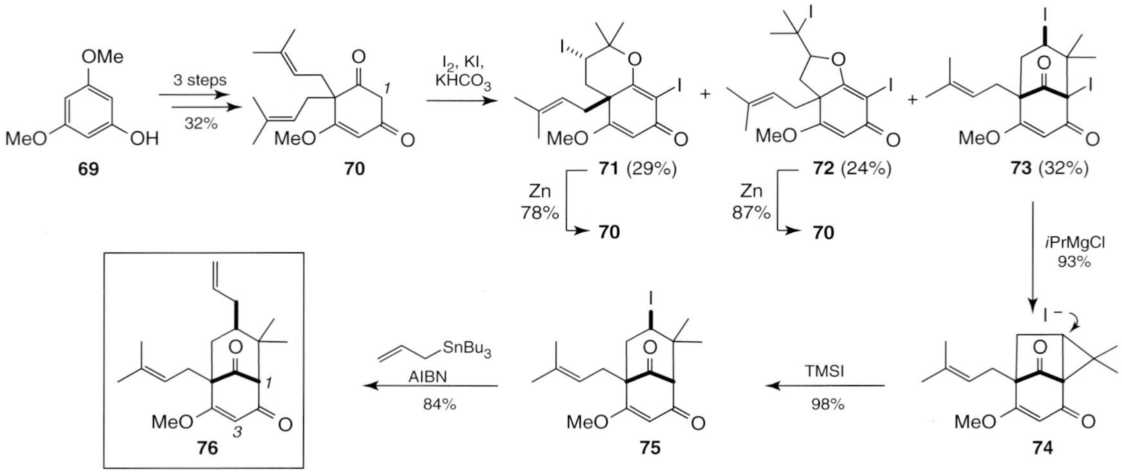

Scheme 12.12 Formation of the common intermediate **70** towards the synthesis of nemorosone and clusianone.

of LDA (lithium diisopropylamide) and TMSCl (trimethylsilyl chloride) followed by oxidative quenching with iodine provided **77** in moderate yield along with **78**, which can in turn be converted into **77** by applying the same reaction conditions (Scheme 12.13). The total synthesis of **3** was completed after, first, acylation of C1 and then lithium-mediated allylation at C3.

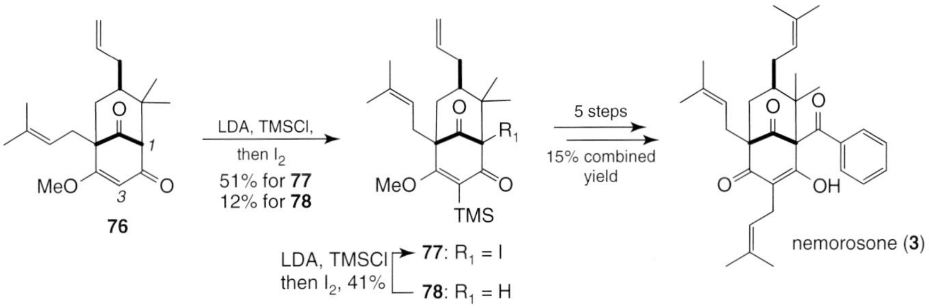

Scheme 12.13 Total synthesis of nemorosone.

On the other hand, treatment of **76** with LDA without the presence of TMSCl and quenching with benzaldehyde gave rise to **79**, after acylation at C3 (Scheme 12.14). Conversion of **79** into its C1 iodo derivative and transformation of the iodo group into an allyl functionality concluded the synthesis of clusianone (**4**).

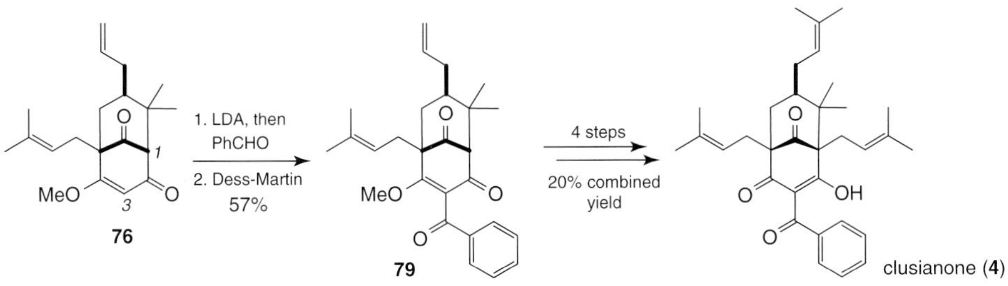

Scheme 12.14 Total synthesis of clusianone by the Danishefsky group.

12.2.4.3 Total Synthesis of (−)-Hyperforin

(+)-Hyperforin (**1**), a type A PPAP, contains an additional chiral quaternary center compared to the other herein mentioned acylphloroglucinols, thus presenting greater difficulties towards its total synthesis. In early 2010 a catalytic asymmetric total synthesis of (−)-hyperforin was published by Shibasaki and coworkers [42]. Notably, this is the first catalytic asymmetric synthesis of any PPAP, since the previously reported asymmetric synthesis of clusianone involved a late-stage kinetic resolution (Section 12.2.4.4) [50].

Compound **83**, having the desired stereochemistry at C7 and C8 carbons, was constructed via a catalytic asymmetric Diels–Alder reaction between **80** and **81** (Scheme 12.15). Compound **83** was then converted into allyl ether **84**, which underwent a stereoselective Claisen rearrangement to form ketone **86**. This rearrangement proceeded with high selectivity from the β face, most likely due to the

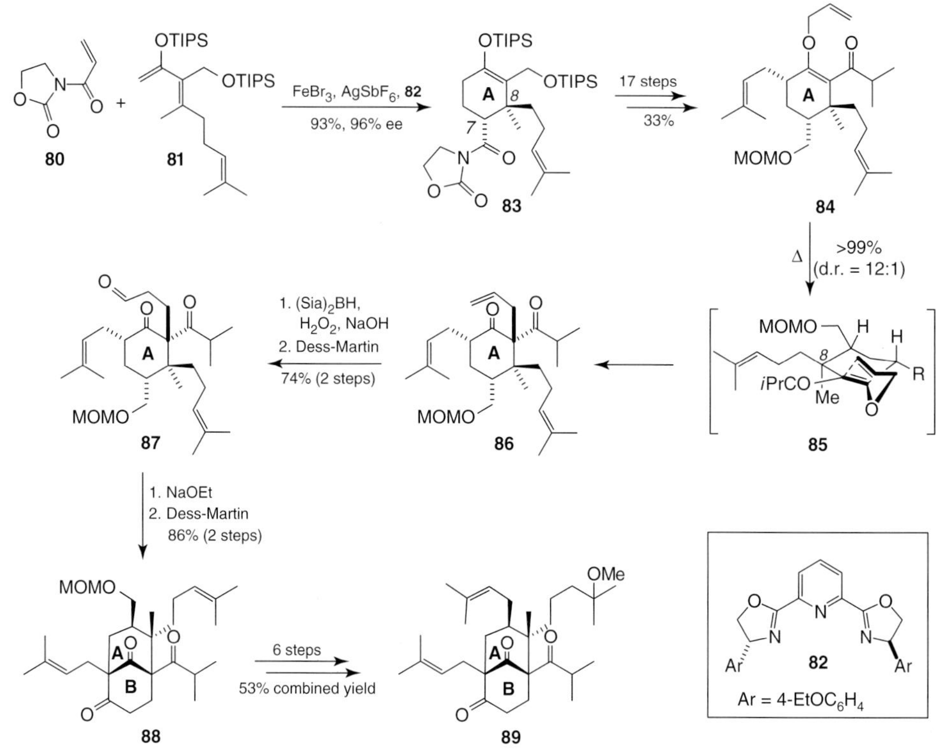

Scheme 12.15 Catalytic asymmetric synthesis of the bicyclic framework of hyperforin.

pseudo-axial orientation of the methyl group at C8, which blocks the α face, as shown in intermediate **85**. Initial attempts to form the B ring of hyperforin via an olefin metathesis approach proved unsuccessful, prompting the development of an alternative strategy using an intramolecular aldol reaction [51]. With this in mind, hydroxylation and oxidation of the terminal alkene of **86** produced aldehyde **87** that after intramolecular aldol reaction and oxidation gave rise to bicyclic adduct **88**. Peripheral decoration of **88** afforded ketone **89** in six steps (53% combined yield).

Scheme 12.16 shows the completion of the (−)-hyperforin synthesis from **89**. Oxidation of C2 proved to be more difficult than anticipated since any attempt at nucleophilic addition in **90** failed. Furthermore, efforts to induce a [3.3] sigmatropic rearrangement of xanthate **91** led to dithionate **92** after a [1.3] rearrangement. This finding, however, allowed the use of a vinylogous Pummerer rearrangement for the oxidation of C2. To this end, dithionate **92** was converted into methylsulfoxide **93** and, after Pummerer rearrangement, to the desired allylic alcohol **94** (three steps, 61% combined yield). Finally, the prenyl group at C3 was installed by intramolecular allyl transfer via a π-allyl-palladium intermediate and cross metathesis to give (−)-(**1**).

Scheme 12.16 Total synthesis of *ent*-hyperforin.

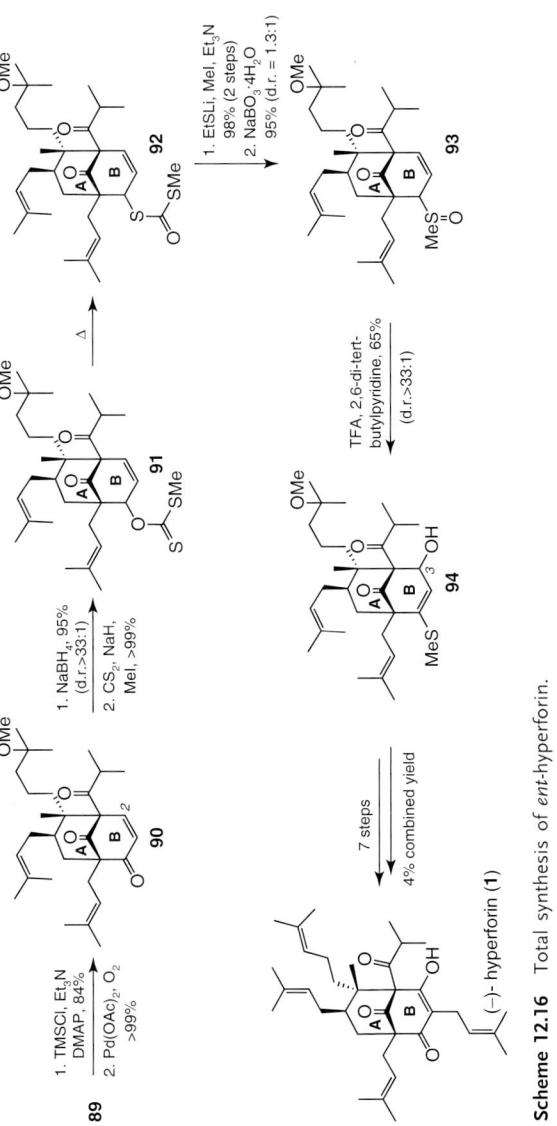

12.2.4.4 Total Synthesis of Clusianone

In 2006 the Simpkins laboratory reported the first total synthesis of (±)-clusianone using as a key step a regioselective lithiation of enol ether derivatives [43]. The construction of the bicyclic core was inspired by the approach described by Spessard and Stoltz (Scheme 12.17) [52]. These authors accomplished a diastereoselective conversion of enol ether **95** into bicyclic trione **96**, using dichloromalonate as the electrophile. Notably, construction of the bicyclo[3.3.1]nonane by the use of malonyl dichloride was reported initially by Effenburger [53].

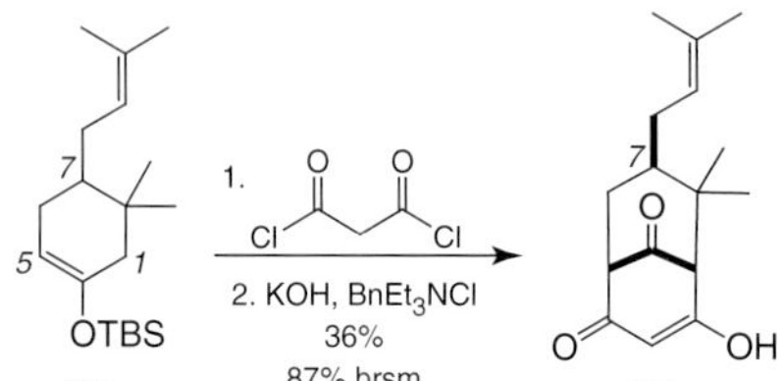

Scheme 12.17 Model studies toward the synthesis of bicyclo[3.3.1]nonane-2,4,9-triones.

Scheme 12.18 summarizes Simpkins' synthesis of clusianone [43]. Notably, enol ether **98** was pre-functionalized with a prenyl group at the C1 center to overcome problems deriving from the steric hindrance of this center after the formation of the bicycle. In fact, Danishefsky had already reported similar problems during the installation of such a substituent in the synthesis of garsubellin A (Section 12.2.4.1). Interestingly, prenylation of **100** at the bridgehead position afforded compound **101** in high yield. The same efficiency was observed for the acylation of C3 by the action of lithium tetramethyl piperidine (LTMP), leading, after hydrolysis, to racemic clusianone. Importantly, prenylation of racemic **100** using a chiral base, such as **103**, led to selective alkylation of the (−) isomer via a kinetic resolution process. This reaction allowed assignment of the absolute configuration of clusianone, isolated from *Clusia torresii*, as the (+) isomer. Interestingly, the (−) isomer of **4** matches the data of a compound reported by a Brazilian group [54]. Based on this, it is possible that clusianone exists in Nature in either enantiomeric form [50].

The above strategy was then extended to a formal synthesis of garsubellin A [40]. Specifically, application of the Effenburger-type cyclization to **105** afforded **110**, a compound isolated by the Danishefsky group *en route* to the synthesis of garsubellin A (Scheme 12.19) [39].

Marazano and coworkers have also published their synthesis of (±)-clusianone incorporating a similar approach [44]. In their synthesis the starting enol ether **114** contains all three prenyl groups, yielding, after reaction with dichloromalonate, compound **115**, which has both quaternary bridgehead centers (Scheme 12.20). Notably, formation of the desired product **115** was accompanied with side product **116** and desilylated compounds **112/113**. Under different Lewis acid catalysis, this reaction produced significant amounts of bicyclic adduct **117**. In principle the double alkylation of **114** to **115** has some biosynthetic relevance. However, the sequence of ring construction (formation of the B ring at the end) is not in accordance with the biosynthesis scenario in which the B ring is formed at the

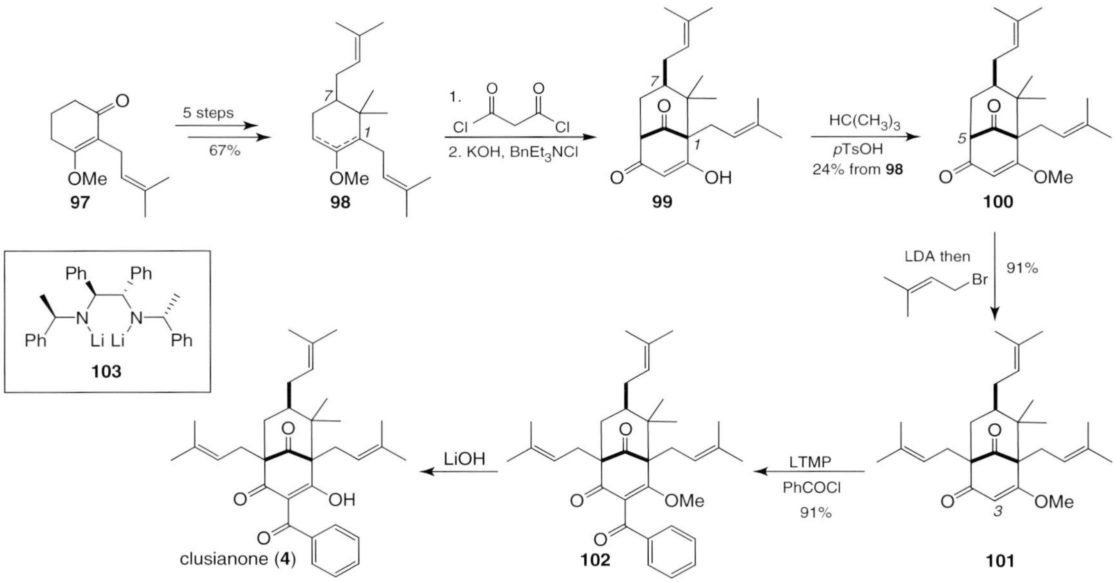

Scheme 12.18 Total synthesis of clusianone by the Simpkins group.

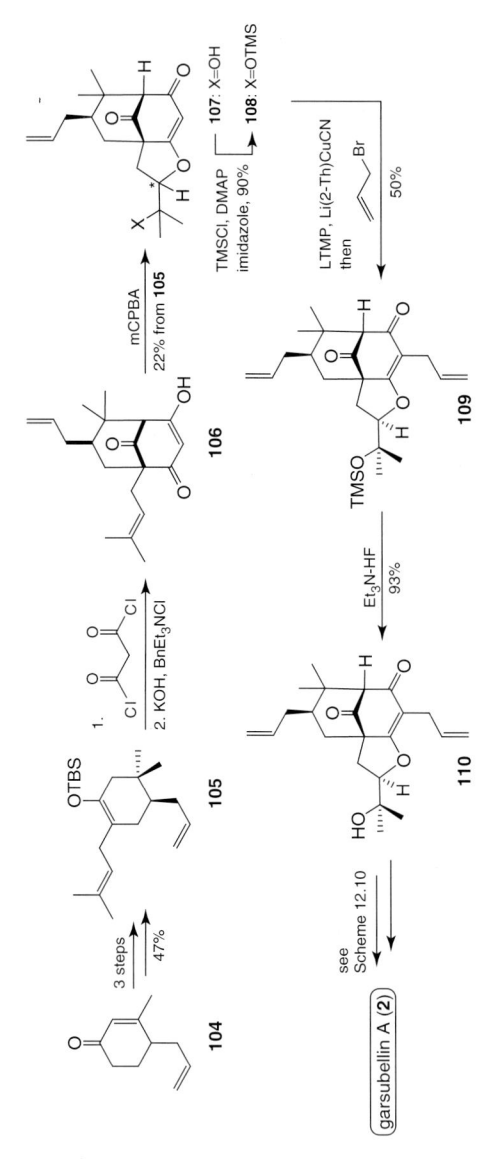

450 | *12 Polyprenylated Phloroglucinols and Xanthones*

Scheme 12.19 Formal synthesis of garsubellin A by the Simpkins group.

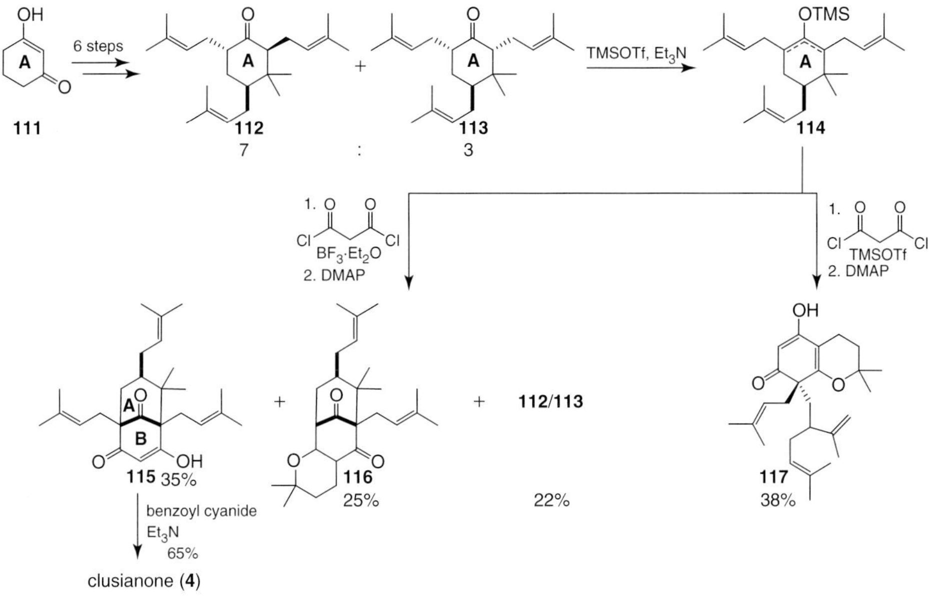

Scheme 12.20 Total synthesis of clusianone by Marazano and coworkers.

beginning. Based on this, such an approach cannot be considered as a biomimetic strategy.

12.2.5
Concluding Remarks

In short, biomimetic approaches toward PPAPs are based on double alkylation on a functionalized B ring with an electrophile. These strategies require fewer steps in a linear sense and can potentially produce the natural product target in higher yield. The total synthesis of racemic clusianone, by the Porco group [45], was completed in seven synthetic steps, starting from the biosynthetic intermediate **27**, with a total yield of 25%. The approach is based on biosynthetic hypotheses that involve alkylative dearomatization of phloroglucinols and carbocation-mediated cyclization. This method can be applied with minor changes for the synthesis of other type B PPAPs and can also be implemented to the synthesis of adamantane-like PPAPs. The strategy presented by Couladouros et al. [46] can give access in few steps and satisfactory yields to several compounds possessing the bicyclic framework of natural products for SAR (structure–activity relationship) studies. In contrast to other reported methods, the two quaternary centers (C1 and C8) are connected during the cyclization step, thereby eliminating steric hindrance problems encountered in other reported approaches. A similar approach, and the most recently published biomimetic synthesis of PPAPs [47], starts with alkylation of a functionalized B ring, yet formation of the second ring occurs after oxidative cyclization. On the other hand, the non-biomimetic strategies require many steps, as shown by the Danishefsky group syntheses of garsubellin A, nemorosone, and clusianone that were completed in 17, 14, and 12 steps respectively, or by the synthesis of (−)-hyperforin by the Shibasaki group, which required 44 steps. The advantage of the non-biomimetic approaches is that they are more flexible in terms of the synthetic route followed to the target and can lead to asymmetric final products. In such a way the asymmetric total synthesis of (−)-hyperforin was achieved.

In addition to the synthetic strategies described herein, there is still ongoing interest towards the synthesis of PPAPs and their analogs. New approaches have appeared in the literature, aimed at a facile method for the construction of the bicyclic core of PPAPs that will allow easier access to those compounds for their biological evaluation and SAR studies [55–61].

12.3
Polyprenylated Xanthones

12.3.1
Introduction and Chemical Classification

Polyprenylated xanthones constitute a subclass of a larger class of compounds known as xanthones, all bearing a dibenzo-γ-pyrone scaffold [62]. Polyprenylated xanthones can further be divided, according to their oxidation degree, into mono-, di-, tri-, and so on, oxygenated compounds. Recently, the classification, synthesis, and biological evaluation of simple xanthones have been reviewed extensively [62–71]. In this chapter we focus on the so-called caged *Garcinia* xanthones (CGXs), owing to their similarities (isolation, biosynthesis) to the aforementioned polyprenylated phloroglucinols.

Caged xanthones are natural products isolated from plants of the genus *Garcinia* (Guttiferae family) that are found in lowland rainforests of India, Indochina, Indonesia, West and Central Africa, and Brazil [1, 72]. The most studied member of this family is gambogic acid (**118**), a compound isolated from gamboge, the resin of *Garcinia hanburyi* (Figure 12.4). Common to the chemical structure of all CGXs is a xanthone backbone in which the C ring has been converted into an unusual 4-oxa-tricyclo[4.3.1.0^{3,7}]dec-8-en-2-one ring (caged) scaffold (see structure **120**) [73, 74]. This general motif can be further decorated with different substituents on the aromatic ring A and/or can be oxidized to yield a wide range of compounds, representative members of which are shown in Figure 12.4. This concept is exemplified by the structure of forbesione (**125**), a natural product isolated from *Garcinia forbesii* [75] and *Garcinia hanburyi* [76]. Specifically, prenylation at the C5 center of forbesione (gambogic acid numbering) gives access to the gaudichaudione scaffold [77], represented here by deoxygaudichaudione A (**126**) [78]. Alternatively, prenylation of **125** at C5 followed by cyclization with the pendant phenol gives access to the morellin scaffold, represented here by desoxymorellin (**123**) [79]. Progressive oxidations at the C29 center of **123** produce morellinol (**122**), morellin

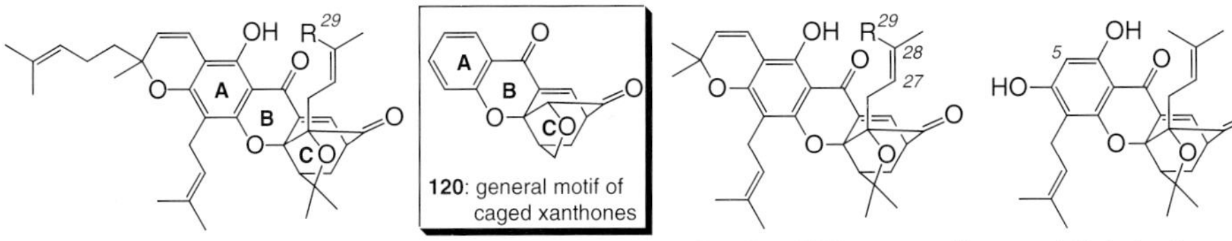

118: R = CO₂H: gambogic acid
119: R = CH₃: gambogin

120: general motif of caged xanthones

121: R = CHO: morellin
122: R = CH₂OH: morellinol
123: R = CH₃: desoxymorellin
124: R = CO₂H: morellic acid

125: forbesione

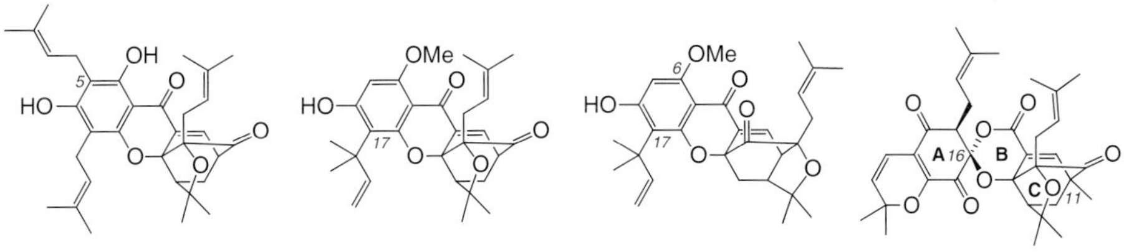

126: deoxygaudichaudione A **127:** 6-*O*-methylbractatin **128:** 6-*O*-methylneobractatin **129:** lateriflorone

Figure 12.4 Representative members of the caged *Garcinia* xanthone family.

(121), and morellic acid (124) [80]. Geranylation at the C5 center of forbesione affords, after formation of the pyran ring, gambogin (119) [81]. Further oxidation at C29 leads to the structure of gambogic acid (118) [76]. Compounds arising from isomerization around the C27=C28 double bond have also been isolated. Thus, morellin (121), having the cis configuration about the C27=C28 double bond, is known to isomerize to the trans isomer, isomorellin [82]. Similar observations have been reported for gambogic acid [83]. In contrast, the bractatin subfamily (127, 128) [84, 85] provides examples of forbesione-type natural products that contain a reverse prenyl group at the C17 center.

Although the vast majority of the CGXs contain the general motif **120**, there are a few examples of natural products with alternative cage structures or with additional oxidations of the xanthone motif. For instance, 6-*O*-methylneobractatin (128) is the only natural product known to contain a modified caged scaffold, referred to as the *neo*-motif [84, 85]. In addition, in the structure of lateriflorone (129) [86], the caged motif is attached to a spiroxalactone core, which is likely a product of oxidation of the xanthone B ring.

Biologically, CGXs are known for their antimicrobial and anticancer activities and are widely used as herbal medicines in traditional Eastern medicine [83, 87–89]. Initial biological studies with semi-purified gamboge extracts documented its antiprotozoal activities, thus lending support for its indigenous use in the treatment of enteric diseases [90–94]. It has also been shown that morellin (121) and gambogic acid (118) exhibit a high specific growth inhibitory effect on Gram-positive bacteria *in vitro* and a protective action against experimental staphylococcal infections in mice [95–98]. In addition to their antimicrobial activity, most CGXs have received a great deal of attention for their anticancer activity [99, 100]. An ever increasing body of evidence indicates that these compounds are cytotoxic against various cancer cell lines at low micromolar concentrations [79].

12.3.2
Biosynthesis of Polyprenylated Xanthones

Biosynthetically, the xanthone backbone of the caged compounds is assumed to derive from common benzophenone intermediates that are synthesized in a similar manner to that described for PPAPs. Their oxygenation patterns indicate a mixed shikimate (formation of C ring)–acetate (formation of A ring) pathway [101–106]. The proposed biosynthesis is exemplified with the synthesis of maclurin (**134**) and 1,3,5,6-tetrahydroxyxanthone (**135**) in Scheme 12.21 [107–110]. Shikimic acid, derived from the shikimic pathway, can be converted into protocatechuic acid (**131**) after oxidation, dehydration, and enolization. Reaction of **131** with coenzyme A (HSCoA) can produce activated ester **132** that can further react with three units of malonyl-coenzyme A to yield intermediate **133**. A Dieckmann condensation gives rise to benzophenones, such as maclurin (**134**). Depending upon the benzophenone produced, this is a branch point in the biogenesis of other benzophenone-type natural products. It is generally accepted that xanthones such as 1,3,5,6-tetrahydroxyxanthone (**135**) are formed by means of phenolic coupling of the benzophenone precursors [109, 110].

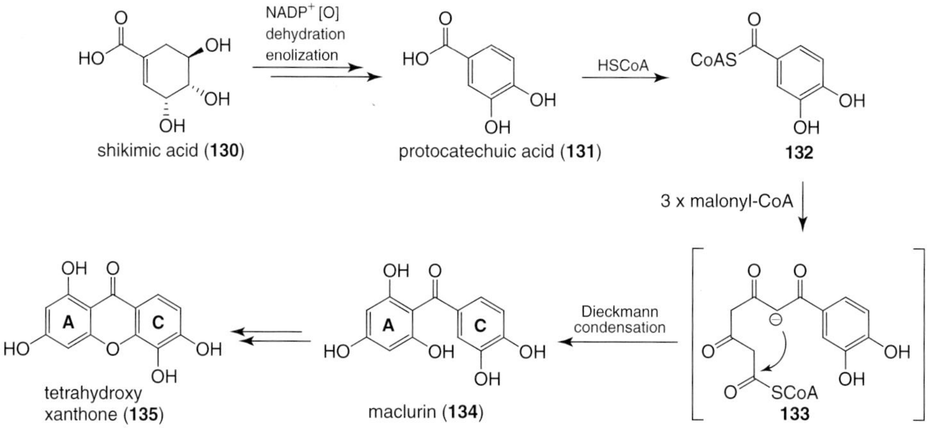

Scheme 12.21 Biosynthesis of the backbone of xanthones in higher plants.

Different hypotheses have been proposed for the biosynthetic conversion of simpler xanthones such as **135** into the more complex caged structures. The first proposal, illustrated in Scheme 12.22, requires at an early stage the prenylation of a xanthone, as **136**, at C11 and C13 positions to produce **137** [111]. An oxidation–reduction–oxidation sequence of reactions is then required to form the final caged structure **141**. Essential to this hypothesis is a presumed nucleophilic attack by the C13 tertiary alcohol of **138** on the pendant prenyl group that could initiate a cyclization cascade leading to the caged structure **140**. Nonetheless, as shown with structure **139**, neither the molecular geometry nor the reactivity required for this cascade is optimal, making this proposal unlikely.

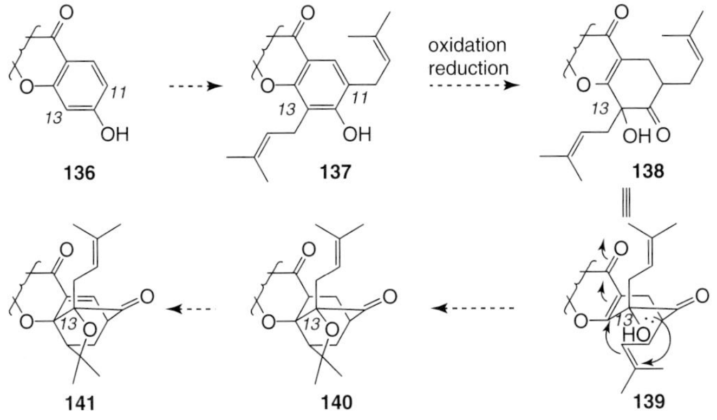

Scheme 12.22 Proposed biosynthesis of the CGX motif via a cascade of nucleophilic attacks.

A more plausible biosynthetic scenario stems from the pioneering work of Quillinan and Scheinmann [112]. In their work, they proposed that the caged motif can be formed via a Claisen rearrangement followed by a Diels–Alder reaction on the intermediate dienone (Scheme 12.23). The authors also provided experimental evidence in support of the Claisen/Diels–Alder reaction cascade: upon heating of compound **143**, prepared by allylation of mesuaxanthone B (**142**), at 190 °C for 14 h they observed products showing NMR signals characteristic of cage structure **145**. At present, there have been no labeling experiments testing the biosynthetic feasibility of the Claisen/Diels–Alder reaction cascade. Nonetheless, additional support for the validity of this proposal was obtained by recent studies in which retro-Diels–Alder fragments have been detected in mass spectroscopy studies of several CGXs [113, 114].

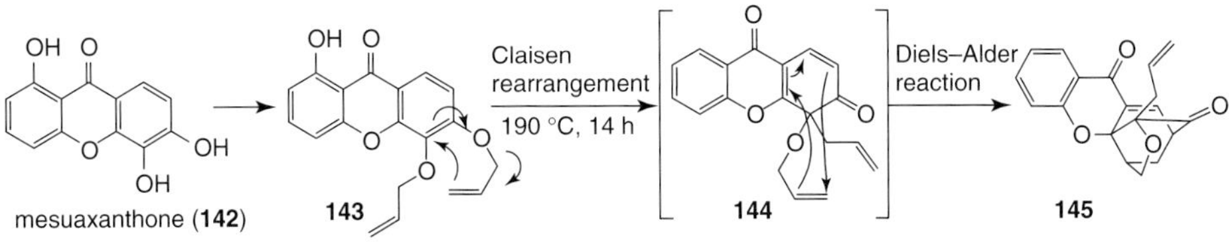

Scheme 12.23 Proposed biosynthesis of the CGX motif via a Claisen/Diels–Alder reaction cascade.

12.3.3
Biomimetic Synthesis of Caged *Garcinia* Xanthones

Inspired by Quillinan and Scheinmann's proposed biosynthesis, both the Nicolaou [115] and Theodorakis [116] groups evaluated the tandem Claisen/Diels–Alder sequence for the synthesis of representative members of the CGX family. Their work provided further support for the proposed biosynthetic hypothesis.

Starting from the tris-allylated xanthone **146** both groups investigated the possibility of synthesizing forbesione (**125**) in one pot. In principle, exposure of such motif to heat could produce four products arising from a combination of two competing C-ring Claisen/Diels–Alder reactions, leading to regular and neo-caged motifs, and two A-ring Claisen migrations, producing C17 and C5 prenylations. Working with methoxy xanthone **146c**, the Nicolaou group was the first to describe its conversion into methyl forbesione **147c** and methyl neoforbesione (**148c**) in a 2.4 : 1 ratio and 89% combined yield (Scheme 12.24) [115]. On the other hand, studies by the Theodorakis group showed that heating of xanthone **146a** led only to the isolation of forbesione **147a** and isoforbesione (**149a**). The neo-C-ring isomers

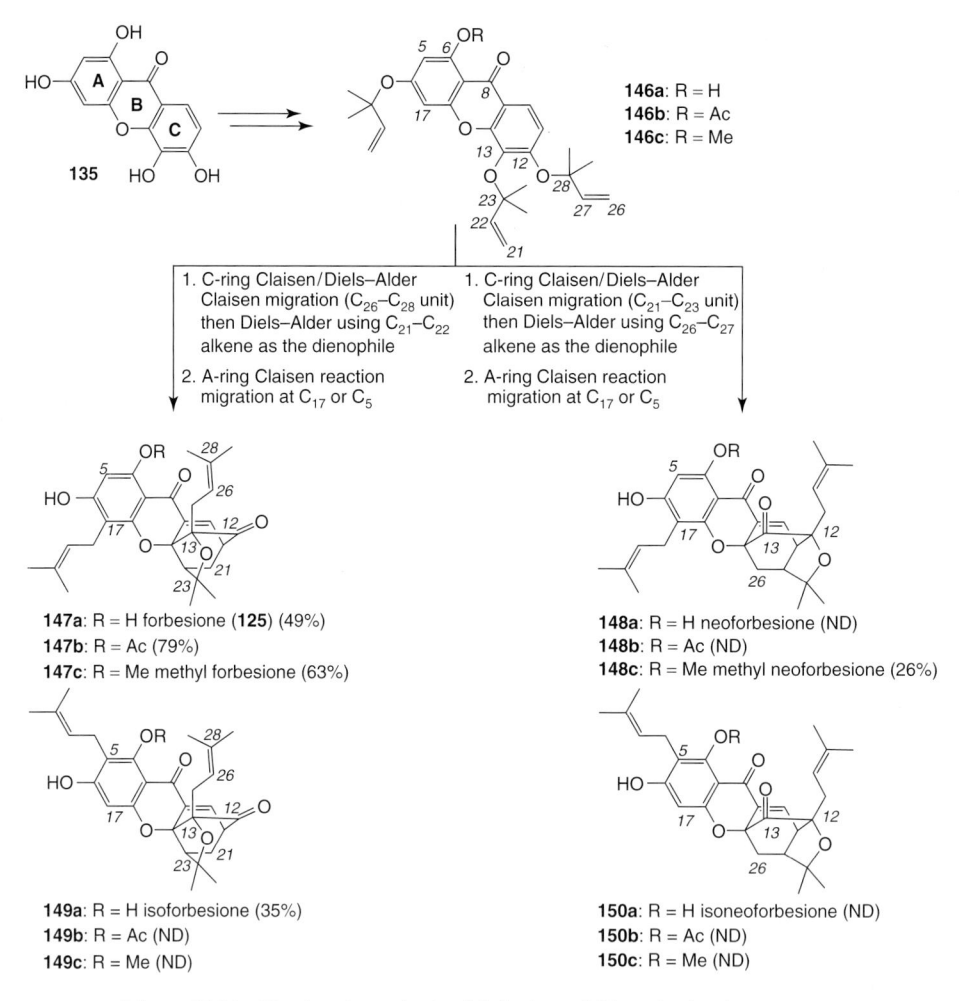

146a: R = H
146b: R = Ac
146c: R = Me

1. C-ring Claisen/Diels–Alder
Claisen migration (C_{26}–C_{28} unit)
then Diels–Alder using C_{21}–C_{22}
alkene as the dienophile

2. A-ring Claisen reaction
migration at C_{17} or C_5

1. C-ring Claisen/Diels–Alder
Claisen migration (C_{21}–C_{23} unit)
then Diels–Alder using C_{26}–C_{27}
alkene as the dienophile

2. A-ring Claisen reaction
migration at C_{17} or C_5

147a: R = H forbesione (**125**) (49%)
147b: R = Ac (79%)
147c: R = Me methyl forbesione (63%)

148a: R = H neoforbesione (ND)
148b: R = Ac (ND)
148c: R = Me methyl neoforbesione (26%)

149a: R = H isoforbesione (35%)
149b: R = Ac (ND)
149c: R = Me (ND)

150a: R = H isoneoforbesione (ND)
150b: R = Ac (ND)
150c: R = Me (ND)

Scheme 12.24 Biomimetic synthesis of forbesione (**125**) and related structures via a Claisen/Diels–Alder/Claisen reaction cascade.

148a and **150a** were not detected in this case. More impressively, the O6-acetylated xanthone **146b** afforded, upon heating, solely acetyl forbesione (**147b**) [116]. Similar observations have been reported more recently by other groups [117]. The above-described results towards the synthesis of forbesione along with the results from several model studies [118] can be summarized as follows:

- The C-ring Claisen/Diels–Alder rearrangement proceeds first and is followed by an A-ring Claisen reaction.
- The site-selectivity of the A-ring Claisen rearrangement (C17 versus C5 prenylation) is controlled by the steric and electronic effects of the C6 phenolic substituent.
- The site-selectivity of the C-ring Claisen/Diels–Alder reaction is attributed to and governed by the electronic density of the C8 carbonyl-group. Being para to the C12 allyloxy unit, the electron-deficient C8 carbonyl carbon polarizes selectively the O–C28 bond and facilitates its rupture. In turn, this leads to a site-selective Claisen rearrangement of the C12 allyloxy unit onto the C13 center, thereby producing exclusively the regular caged motif found in the structure of forbesione (**147a**).
- Substitution of the C6 phenol can regulate the electronic density of the C8 carbonyl group, thus affecting the site selectivity of the C-ring Claisen/Diels–Alder reaction.

The experimental findings on the tandem Claisen/Diels–Alder/Claisen reaction cascade provide useful insights regarding the biosynthesis of all known CGXs [118]. All these natural products (representative examples shown in Figure 12.4) share a common caged motif, exemplified by structure **120**, except for 6-O-methylneobractatin (**128**), which contains the neo-caged motif. The remote electronic effects of the seemingly innocuous 6-O-methyl group may explain the concomitant biosynthesis of both 6-O-methylbractatin (**127**) and 6-O-methylneobractatin (**128**).

Studies by the Nicolaou group have shown that the Claisen/Diels–Alder reaction can be accelerated in the presence of polar solvents [119]. For instance, as depicted for the synthesis of gambogin (Scheme 12.25), the conversion of allyl ether **159** into caged structure **160a** and the neo-isomer **160b** was dramatically accelerated upon changing the solvent from benzene to DMF to a MeOH–water (1:2) mixture. It has been proposed that polar aprotic solvents, such as DMF, and more impressively protic solvents, such as water, can accelerate the Claisen rearrangement by stabilizing its polar transition state [120–124]. The concurrent acceleration of the Diels–Alder component of this cascade may be due to the hydrophobic effect of water [125] rather than to a polarity or hydrogen-bonding phenomena [126–128]. Computational studies on the above-mentioned reaction have also concluded that the Claisen rearrangement is reversible and the energetics of the irreversible Diels–Alder cyclization can determine the product formation [129].

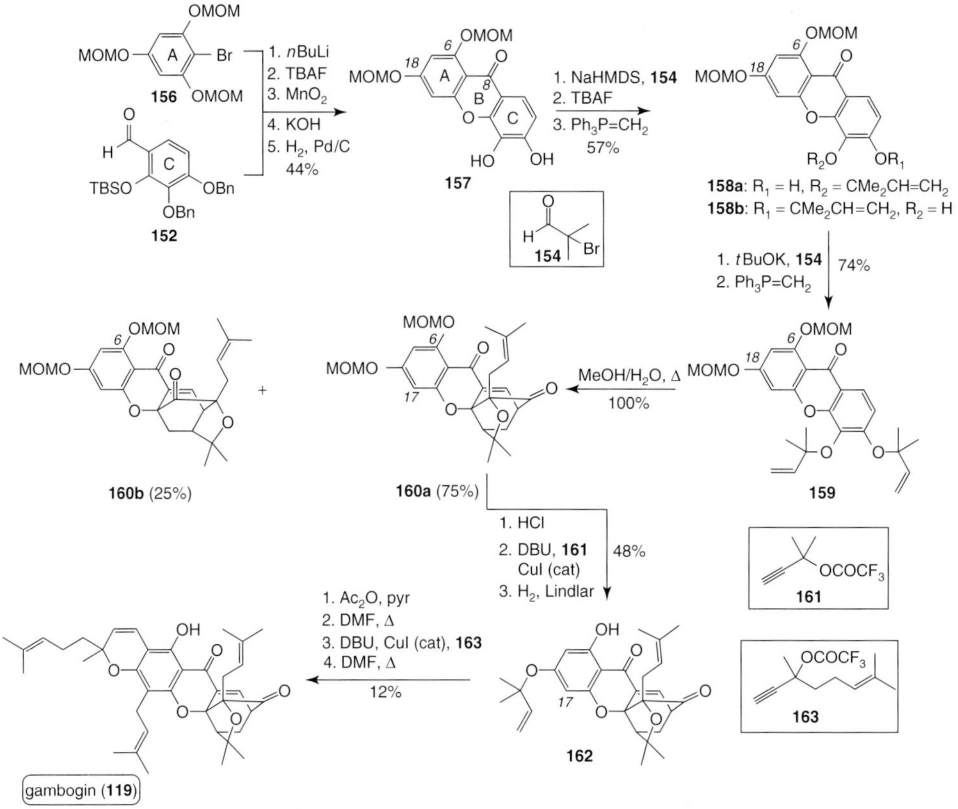

Scheme 12.25 Biomimetic synthesis of gambogin by the Nicolaou group.

12.3.3.1 Nicolaou Approach to Forbesione and Gambogin

Scheme 12.26 depicts the synthetic strategy developed by Nicolaou and Li [115] for the synthesis of 6-O-methylforbesione (**147c**). Xanthone **153** was generated in five steps and in 78% combined yield starting from the aryl bromide **151** and the benzaldehyde **152**. Treatment of **153** with α-bromoisobutyraldehyde (**154**) under basic conditions followed by Wittig olefination produced a mixture of the diallylated compounds **155a** and **155b** that, after reiteration of the alkylation/olefination reactions, yielded the triallylated xanthone **146c**. Heating of **146c** in DMF at 120 °C induced the Claisen/Diels–Alder/Claisen reaction cascade to produce compound **147c** along with its neo-isomer **148c** in 89% combined yield.

In a similar manner, the total synthesis of gambogin was achieved starting from the partially protected xanthone **157** (Scheme 12.25) [119]. This time the Claisen/Diels–Alder reaction proceeded quantitatively in refluxing MeOH–H$_2$O (1 : 2) to produce the regular caged motif **160a** along with its neo-isomer **160b** in a 3 : 1 ratio. Methoxymethyl (MOM) deprotection of **160a** followed by propargylation with alkyne **161** at C18 and partial reduction with Lindlar catalyst gave rise to compound **162**. Gambogin was then synthesized after a sequence of four reactions

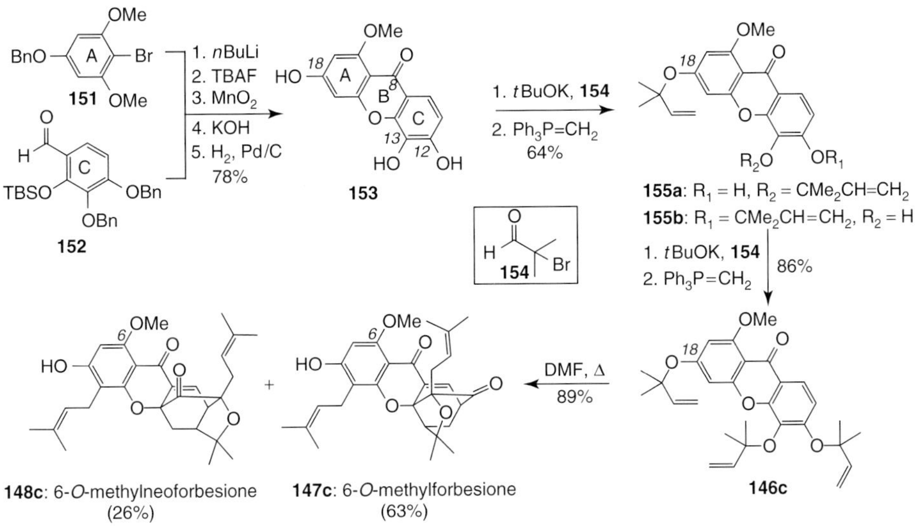

Scheme 12.26 Biomimetic synthesis of 6-O-methylforbesione (**147c**).

that included: (i) acetylation of C6 phenol; (ii) Claisen rearrangement to install the prenyl group at C17; (iii) propargylation of the resulting phenol with alkyne **163**; and (iv) Claisen rearrangement to form the dihydropyran ring of the natural product.

12.3.3.2 Theodorakis' Unified Approach to Caged *Garcinia* Xanthones

The common structural motif of most CGXs suggests that they can be synthesized by functionalizing the A ring of forbesione. Along these lines, the Theodorakis group developed a strategy that uses forbesione (**125**) to gain access to representative members of the gaudichaudiones, morellins, and gambogins [118].

As illustrated in Scheme 12.27, $ZnCl_2$-mediated condensation of phloroglucinol (**164**) with benzoic acid **165** produced xanthone **135**. Propargylation of **135** with the propargyl chloride **166** followed by partial reduction using Lindlar catalyst and acetylation of phenol at C6 gave rise to compound **146b**. Heating of **146b** (DMF, 1 h, 120 °C) set the stage for a site-selective Claisen/Diels–Alder/Claisen reaction cascade that produced, after deprotection of the C6 acetate, forbesione (**125**) in 72% combined yield. Further decoration of the A ring of forbesione gave access to more functionalized CGX family members. Specifically, propargylation of the C18 phenol of forbesione with chloride **166** afforded, after Lindlar reduction and Claisen rearrangement, deoxygaudichaudione A (**126**). On the other hand, propargylation of **125** and immediate Claisen rearrangement formed desoxymorellin (**123**). Finally, condensation of forbesione (**125**) with citral (**168**) in Et_3N produced gambogin (**119**).

12.3.3.3 Synthesis of Methyllateriflorone

It has been proposed that the unprecedented spiroxalactone motif of lateriflorone (**129**) could be formed by condensation of two fully functionalized fragments, **169** and **170** (Scheme 12.28) [86]. An alternative and likely more biosynthetically

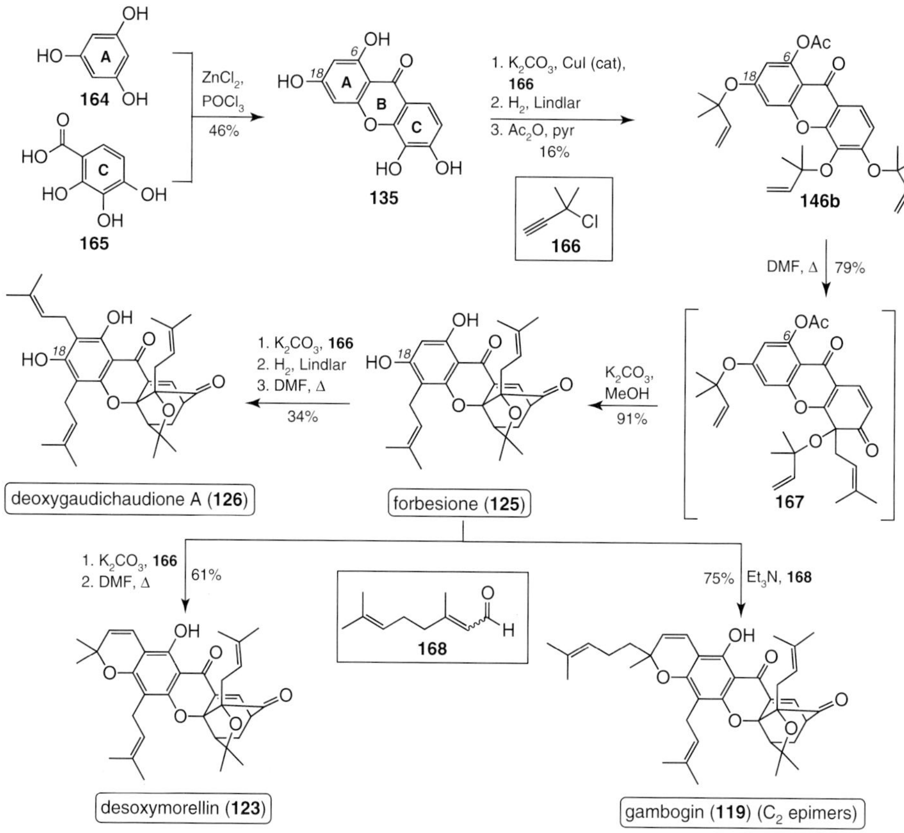

Scheme 12.27 Unified biomimetic synthesis of CGXs by the Theodorakis group.

relevant hypothesis could involve conversion of xanthone (**172**) into dioxepanone (**171**) that, upon hydrolysis and spirocyclization at the C16 center, could form the spiroxalactone ring system of lateriflorone.

Quite recently, the Nicolaou group has reported a synthesis of C11-methyllateriflorone (**178**) (Scheme 12.29) [130]. Key to the strategy was the coupling of orthogonally protected hydroquinone **173** with acid **174** that after selective deprotection of the C7 MOM ether produced compound **175** (61% combined yield). Oxidation of **175** in the presence of iodosobenzene bis(trifluoroacetate) in methanol, followed by heating under acidic conditions formed spiroxalactone **177**. Acid-catalyzed hydrolysis of **177** gave rise to C11-methyllateriflorone (**178**) in 66% yield.

12.3.3.4 Non-biomimetic Synthesis of the Caged *Garcinia* Xanthones

An alternative non-biomimetic synthesis of CGXs relies on a tandem Wessely oxidation/Diels–Alder reaction cascade. Yates and coworkers applied this strategy to the synthesis of caged structures reminiscent of the CGX motif. Thus, treatment

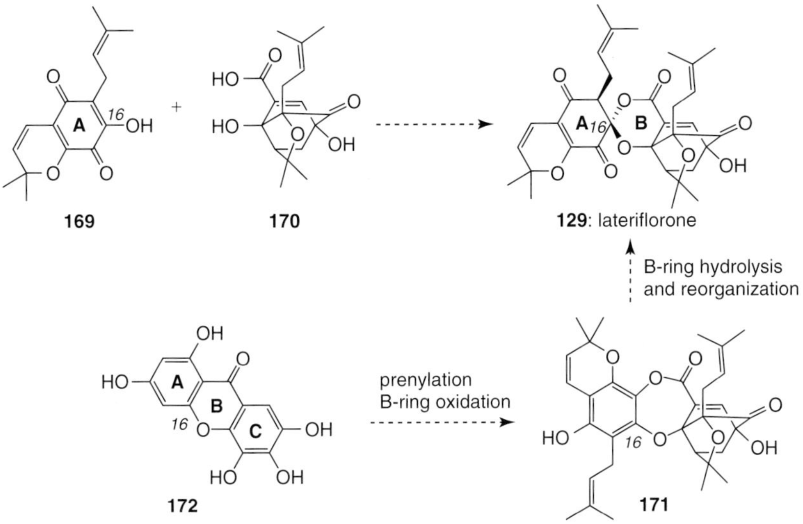

Scheme 12.28 Proposed biosynthesis of lateriflorone (**129**).

of phenol **179** with Pb(OAc)$_4$ in acetic acid produced 2,4-cyclohexadienone **180** that, upon heating at 140 °C, formed compound **181** (gambogic acid numbering) (Scheme 12.30) [131]. In a previous study, xanthene **182** was treated with lead tetraacrylate [formed *in situ* by Pb(OAc)$_4$ and acrylic acid] to produce dienone **183** and, after an intramolecular Diels–Alder reaction, caged compound **184** [132].

Theodorakis and coworkers [133] investigated the application of the Wessely/Diels–Alder strategy for the synthesis of a more hydroxylated caged motif related to the structure of lateriflorone (**129**) (Figure 12.4). Treatment of **185** with Pb(OAc)$_4$ in acrylic acid–dichloromethane produced, after heating in refluxing benzene (80 °C), tricyclic lactone **187** in 82% combined yield (Scheme 12.31). Crystallographic studies established that **187** is a constitutional isomer of the desired structure **190** and is reminiscent of the so-called neo-caged structure. The connectivity of compound **187** suggested that during the Wessely oxidation the acrylate unit was attached exclusively at the more electronically rich C11 center of **185**, instead of the desired C13 carbon. In turn, this produced dienone **186** that subsequently underwent an efficient Diels–Alder cycloaddition with the pendant acrylate dienophile. To alter the connectivity of the caged structure, one could have the acetoxy group preinstalled at the C13 center and promote the migration of the prenyl group. Along these lines, heating of allyl ether **188** in *m*-xylene (140 °C) gave rise exclusively to caged motif **190** via a Claisen rearrangement and Diels–Alder cycloaddition. The selectivity of the Claisen rearrangement at the C13 center can be explained by considering that intermediate **189** has the necessary geometry that allows it to be trapped as the Diels–Alder adduct.

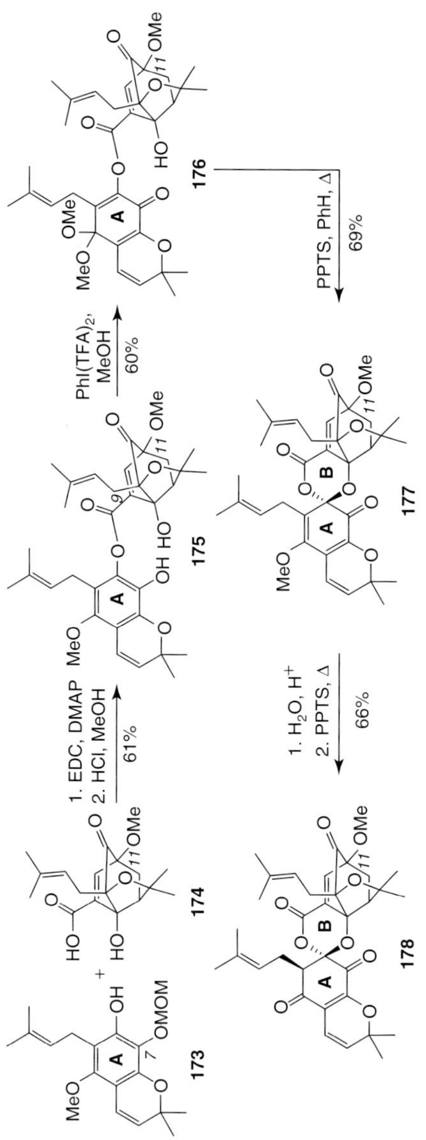

Scheme 12.29 Total synthesis of C11-methyllateriflorone (**178**).

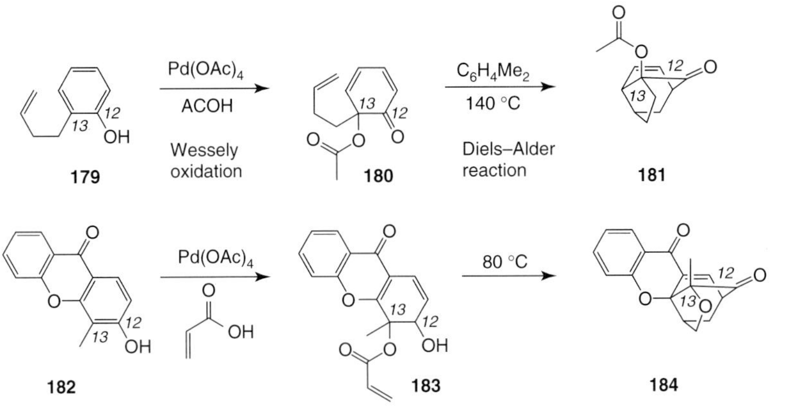

Scheme 12.30 Representative examples of caged structures, 181 and 184, formed via a Wessely oxidation/Diels–Alder reaction cascade.

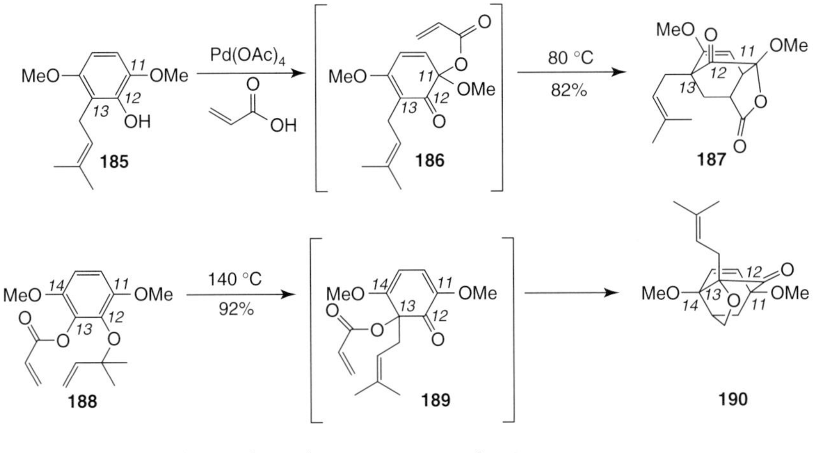

Scheme 12.31 Synthesis of caged structures 187 and 190.

12.3.3.5 Concluding Remarks

CGXs are a family of polyprenylated xanthones that have a remarkable chemical structure, inspiring biosynthesis, and significant medicinal potential. Their chemical structure is represented by an unusual xanthone backbone in which the C ring has been converted into a 4-oxa-tricyclo[4.3.1.0^{3,7}]dec-8-en-2-one (caged) scaffold. Their biosynthesis is proposed to involve a cascade of Claisen and Diels–Alder reactions and has provided the inspiration for the development of efficient laboratory syntheses of the parent molecules and designed analogs. Their medicinal value stems from their use in ethnomedicine and remains still largely unexplored [134]. The recent advances in the synthesis of these compounds have paved the way for the generation of analogs with the desired pharmacological and biological profile. In particular, the biosynthetically inspired Claisen/Diels–Alder reaction cascade can reliably produce the caged motif of CGXs in excellent yields. On the other hand, the non-biomimetic Wessely/Diels–Alder strategy can form analogs of the caged motif that cannot be made by fragmentation of the natural products. It is very likely that these strategies will be used for the development of more potent CGX analogs. One limitation of both strategies is that, at present, they both deliver racemic mixtures of the caged structures. Thus, the development of an enantioselective variant of the Claisen/Diels–Alder and Wessely/Diels–Alder reaction cascades still needs to be addressed.

References

1. Gustafsson, M.H.G., Bittrich, V., and Stevens, P.F. (2002) *Int. J. Plant Sci.*, **163**, 1045–1054.
2. Woelk, H. (2000) *Br. Med. J.*, **321**, 536–539.
3. Di Carlo, G., Borrelli, F., Ernst, E., and Izzo, A.A. (2001) *Trends Pharmacol. Sci.*, **22**, 292–297.
4. Miller, A.L. (1998) *Altern. Med. Rev.*, **3**, 18–26.
5. Barnes, J., Anderson, L.A., and Phillipson, J.D. (2001) *J. Pharm. Pharmacol.*, **53**, 583–600.
6. Müller, W.E. (2003) *Pharmacol. Res.*, **47**, 101–109.
7. Verotta, L. (2002) *Phytochem. Rev.*, **1**, 389–407.
8. Rodriguez-Landa, J.F. and Contreras, C.M. (2003) *Phytomedicine*, **10**, 688–699.
9. Medina, M.A., Marti'nez-Poveda, B., Amores-Sa'nchez, M.I., and Quesada, A.R. (2006) *Life Sci.*, **79**, 105–111.
10. Ciochina, R. and Grossman, R.B. (2006) *Chem. Rev.*, **106**, 3363–3386.
11. Linde, K., Berner, M.M., and Kriston, L. (2008) *Cochrane Database Syst. Rev.* (Art. No.: CD000448). doi: 10.1002/14651858.CD000448.pub3.
12. Mabberley, D.J. (1997) *The Plant-Book: A Portable Dictionary of the Vascular Plants*, 2nd edn, Cambridge University Press, New York.
13. Kumar, P. and Baslas, R.K. (1980) *Herba Hung.*, **19**, 81–91.
14. Cuesta-Rubio, O., Valez-Castro, H., Frontana-Uribe, B.A., and Cardenas, J. (2001) *Phytochemistry*, **57**, 279–283.
15. Müller, W.E., Singer, A., Wonnemann, M., Hafner, U., Rolli, M., and Schäfer, C. (1998) *Pharmacopsychiatry*, **31** (Suppl. 1), 16–21.
16. Mennini, T. and Gobbi, M. (2004) *Life Sci.*, **75**, 1021–1027.
17. Schempp, C.M., Pelz, K., Wittmer, A., Schöpf, E., and Simon, J.C. (1999) *Lancet*, **353**, 2129.
18. Quiney, C., Billard, C., Salanoubat, C., Fourneron, J.D., and Kolb, J.P. (2006) *Leukemia*, **20**, 1519–1525.
19. Fukuyama, Y., Kuwayama, A., and Minami, H. (1997) *Chem. Pharm. Bull.*, **45**, 947–949.
20. Cuesta-Rubio, O., Frontana-Uribe, B.A., Ramirez-Apan, T., and Cardenas, J. (2002) *Naturforsch. Teil C*, **57**, 372–378.
21. Lokvam, J., Braddock, J.F., Reichardt, P.B., and Clausen, T.P. (2000) *Phytochemistry*, **55**, 29–34.
22. Piccinelli, A.L., Cuesta-Rubio, O., China, M.B., Mahmood, N., Pagano, B., Pavone, M., Barone, V., and Rastrelli, L. (2005) *Tetrahedron*, **61**, 8206–8211.
23. Winkelmann, K., Heilmann, J., Zerbe, O., Rali, T., and Sticher, O. (2001) *Helv. Chim. Acta*, **84**, 3380–3392.
24. Weng, J.-R., Tsao, L.-T., Wang, J.-P., Wu, R.-R., and Lin, C.-N. (2004) *J. Nat. Prod.*, **67**, 1796–1799.
25. Zuurbier, K.W.M., Fung, S.-Y., Scheffer, J.C., and Verpoorte, R. (1998) *Phytochemistry*, **49**, 2315–2322.
26. Gerhäuser, C. (2005) *Eur. J. Cancer*, **41**, 1941–1954.
27. Adam, P., Arigoni, D., Bacher, A., and Eisenreich, W. (2002) *J. Med. Chem.*, **45**, 4786–4793.
28. Drawert, F. and Beier, J. (1974) *Phytochemistry*, **13**, 2149–2155.

29. Drawert, F. and Beier, J. (1976) *Phytochemistry*, **15**, 1693–1694.
30. Drawert, F. and Beier, J. (1976) *Phytochemistry*, **15**, 1695–1696.
31. Klingauf, P., Beuerle, T., Mellenthin, A., El-Moghazy, S.A.M., Boubakir, Z., and Beerhues, L. (2005) *Phytochemistry*, **66**, 139–145.
32. Liu, B., Falkenstein-Paul, H., Schmidt, W., and Beerhues, L. (2003) *Plant J.*, **34**, 847–855.
33. Zuurbier, K.W.M., Fung, S.-Y., Scheffer, J.C., and Verpoorte, R. (1995) *Phytochemistry*, **38**, 77–82.
34. Hecht, S., Wungsintaweekul, J., Rohdich, F., Kis, K., Radykewicz, T., Schuhr, C.A., Eisenreich, W., Richter, G., and Bacher, A. (2001) *J. Org. Chem.*, **66**, 7770–7775.
35. Eisenreich, W., Rohdich, F., and Bacher, A. (2001) *Trends Plant Sci.*, **6**, 78–84.
36. Gabrielsen, M., Rohdich, F., Eisenreich, W., Grawert, T., Hecht, S., Bacher, A., and Hunter, W.N. (2004) *Eur. J. Biochem.*, **271**, 3028–3035.
37. Boubakir, Z., Beuerle, T., Benye, L., and Beerhues, L. (2005) *Phytochemistry*, **66**, 51–57.
38. Kuramochi, A., Usuda, H., Yamatsugu, K., Kanai, M., and Shibasaki, M. (2005) *J. Am. Chem. Soc.*, **127**, 14200–14201.
39. Siegel, D.R. and Danishefsky, S.J. (2006) *J. Am. Chem. Soc.*, **128**, 1048–1049.
40. Ahmad, N.M., Rodeschini, V., Simpkins, N.S., Ward, S.E., and Blake, A.J. (2007) *J. Org. Chem.*, **72**, 4803–4815.
41. Tsukano, C., Siegel, D.R., and Danishefsky, S.J. (2007) *Angew. Chem. Int. Ed.*, **46**, 8840–8844.
42. Shimizu, Y., Shi, S.-L., Usuda, H., Kanai, M., and Shibasaki, M. (2010) *Angew. Chem. Int. Ed.*, **49**, 1103–1106.
43. Rodeschini, V., Ahmad, N.M., and Simpkins, N.S. (2006) *Org. Lett.*, **8**, 5283–5285.
44. Nuhant, P., David, M., Pouplin, T., Delpech, B., and Marazano, C. (2007) *Org. Lett.*, **9**, 287–289.
45. Qi, J. and Porco, J.A. Jr. (2007) *J. Am. Chem. Soc.*, **129**, 12682–12683.
46. Couladouros, E.A., Dakanali, M., Demadis, K.D., and Vidali, V.P. (2009) *Org. Lett.*, **11**, 4430–4433.
47. George, J.H., Hesse, M.D., Baldwin, J.E., and Adlington, R.M. (2010) *Org. Lett.*, **12**, 3532–3535.
48. Nicolaou, K.C., Pfefferkorn, J.A., Kim, S., and Wei, H.X. (1999) *J. Am. Chem. Soc.*, **121**, 4724–4725.
49. Nicolaou, K.C., Pfefferkorn, J.A., Cao, G.-Q., Sanghee, K., and Kessabi, J. (1999) *Org. Lett.*, **1**, 807–810.
50. Rodeschini, V., Simpkins, N.S., and Wilson, C. (2007) *J. Org. Chem.*, **72**, 4265–4267.
51. Shimizu, Y., Kuramochi, A., Usuda, H., Kanai, M., and Shibasaki, M. (2007) *Tetrahedron Lett.*, **48**, 4173–4177.
52. Spessard, S.J. and Stoltz, B.M. (2002) *Org. Lett.*, **4**, 1943–1946.
53. Schönwälder, K.-H., Kollatt, P., Stezowski, J.J., and Effenburger, F. (1984) *Chem. Ber.*, **117**, 3280–3296.
54. de Oliveira, C.M.A., Porto, A.M., Bittrich, V., Vencato, I., and Marsaioli, A.J. (1996) *Tetrahedron Lett.*, **37**, 6427–6430.
55. Abe, M. and Nakada, M. (2007) *Tetrahedron Lett.*, **48**, 4873–4877.
56. Abe, M., Saito, A., and Nakada, M. (2010) *Tetrahedron Lett.*, **51**, 1298–1302.
57. Kraus, G.A. and Jeon, I. (2008) *Tetrahedron Lett.*, **49**, 286–288.
58. Mehta, G. and Bera, M.K. (2009) *Tetrahedron Lett.*, **50**, 3519–3522.
59. Mitasev, B. and Porco, J.A. Jr. (2009) *Org. Lett.*, **11**, 2285–2288.
60. Takagi, R., Inoue, Y., and Ohkata, K. (2008) *J. Org. Chem.*, **73**, 9320–9325.
61. Pouplin, T., Tolon, B., Nuhant, P., Delpech, B., and Marazano, C. (2007) *Eur. J. Org. Chem.*, 5117–5125.
62. Sousa, M.E. and Pinto, M.M.M. (2005) *Curr. Med. Chem.*, **12**, 2447–2479.
63. Vieira, L.M.M. and Kijjoa, A. (2005) *Curr. Med. Chem.*, **12**, 2413–2446.
64. Silva, A.M.S. and Pinto, D.C.G.A. (2005) *Curr. Med. Chem.*, **12**, 2481–2497.
65. Gales, L. and Damas, A.M. (2005) *Curr. Med. Chem.*, **12**, 2499–2515.

66. Pinto, M.M.M., Sousa, M.E., and Nascimento, M.S.J. (2005) *Curr. Med. Chem.*, **12**, 2517–2538.
67. Riscoe, M., Kelly, J.X., and Winter, R. (2005) *Curr. Med. Chem.*, **12**, 2539–2549.
68. Pinto, M.M.M. and Castanheiro, R.A.P. (2009) *Curr. Org. Chem.*, **13**, 1215–1240.
69. Pouli, N. and Marakos, P. (2009) *Anticancer Agents Med. Chem.*, **9**, 77–98.
70. Han, Q.-B. and Xu, H.-X. (2009) *Curr. Med. Chem.*, **16**, 3775–3796.
71. El-Seedi, H.R., El-Ghorab, D.M.H., El-Barbary, M.A., Zayed, M.F., Goransson, U., Larsson, S., and Verpoorte, R. (2009) *Curr. Med. Chem.*, **16**, 2581–2626.
72. Sultanbawa, M.U.S. (1980) *Tetrahedron*, **36**, 1465–1506.
73. Ollis, W.D., Ramsay, M.V.J., Sutherland, I.O., and Mongkolsuk, S. (1965) *Tetrahedron*, **21**, 1453–1470.
74. Ahmad, S.A., Rigby, W., and Taylor, R.B. (1966) *J. Chem. Soc. (C)*, 772–779.
75. Leong, Y.-W., Harrison, L.J., Bennett, G.J., and Tan, H.T.-W. (1996) *J. Chem. Res. (S)*, 392–393.
76. Wang, L.L., Li, Z.L., Xu, Y.P., Liu, X.Q., Pei, Y.H., Jing, Y.K., and Hua, H.M. (2008) *Chin. Chem. Lett.*, **19**, 1221–1223.
77. Cao, S.-G., Sng, V.H.L., Wu, X.-H., Sim, K.-Y., Tan, B.H.K., Pereira, J.T., and Goh, S.H. (1998) *Tetrahedron*, **54**, 10915–10924.
78. Han, Q.-B., Wang, Y.-L., Yang, L., Tso, T.-F., Qiao, C.-F., Song, J.-Z., Xu, L.-J., Chen, S.-L., Yang, D.-J., and Xu, H.-X. (2006) *Chem. Pharm. Bull.*, **54**, 265–267.
79. Tao, S.-J., Guan, S.-H., Wang, W., Lu, Z.-Q., Chen, G.-T., Sha, N., Yue, Q.-X., Liu, X., and Guo, D.-A. (2009) *J. Nat. Prod.*, **72**, 117–124.
80. Sukpondma, Y., Rukachaisirikul, V., and Phongpaichit, S. (2005) *Chem. Pharm. Bull.*, **53**, 850–852.
81. Asano, J., Chiba, K., Tada, M., and Yoshii, T. (1996) *Phytochemistry*, **41**, 815–820.
82. Nair, P.M. and Venkataraman, K. (1964) *Indian J. Chem.*, **2**, 402–404.
83. Lin, L.-J., Lin, L.-Z., Pezzuto, J.M., Cordell, G.A., and Ruangrungsi, N. (1993) *Magn. Reson. Chem.*, **31**, 340–347.
84. Thoison, O., Fahy, J., Dumontet, V., Chiaroni, A., Riche, C., van Tri, M., and Sévenet, T. (2000) *J. Nat. Prod.*, **63**, 441–446.
85. Thoison, O., Cuong, D.D., Gramain, A., Chiaroni, A., Hung, N.V., and Sévenet, T. (2005) *Tetrahedron*, **61**, 8529–8535.
86. Kosela, S., Cao, S.-G., Wu, X.-H., Vittal, J.J., Sukri, T., Masdianto, M., Goh, S.-H., and Sim, K.-Y. (1999) *Tetrahedron Lett.*, **40**, 157–160.
87. Jiangsu New Medical College (1997) *Dictionary of Chinese Traditional Medicines*, Shanghai Scientific and Technical Publishers, Shanghai.
88. Jinxiang, Y. (1989) *Chin. J. Cancer Res.*, **1**, 75–78.
89. Panthong, A., Norkaew, P., Kanjanapothi, D., Taesotikul, T., Anantachoke, N., and Reutrakul, V. (2007) *J. Ethnopharmacol.*, **111**, 335–340.
90. Rao, P.L.N. and Verma, S.C.L. (1951) *J. Sci. Ind. Res. B*, **10**, 184–185.
91. Gupta, V.S., Rao, A.V.S.P., and Rao, P.L.N. (1963) *Indian J. Exp. Biol.*, **1**, 146–147.
92. Sani, B.P. and Rao, P.L.N. (1966) *Indian J. Exp. Biol.*, **4**, 27–28.
93. Santhanam, K. and Rao, P.L.N. (1968) *Indian J. Exp. Biol.*, **6**, 158–159.
94. Puttanna, C.R. and Rao, P.L.N. (1968) *Indian J. Exp. Biol.*, **6**, 150–152.
95. Santhanam, K. and Rao, P.L.N. (1969) *Indian J. Exp. Biol.*, **7**, 34–36.
96. Sani, B.P. and Rao, P.L.N. (1969) *Indian J. Chem.*, **7**, 680–684.
97. Verma, S.C.L. and Rao, P.L.N. (1967) *Indian J. Exp. Biol.*, **5**, 106–109.
98. Rao, D.R., Gupta, T.R., Gupta, V.S., Rao, K.V.N., and Rao, P.L.N. (1963) *Indian J. Chem.*, **1**, 276–277.
99. Batova, A., Lam, T., Wascholowski, V., Yu, A.L., Giannis, A., and Theodorakis, E.A. (2007) *Org. Biomol. Chem.*, **5**, 494–500.
100. Chantarasriwong, O., Cho, W.C., Batova, A., Chavasiri, W., Moore, C., Rheingold, A.L., and Theodorakis, E.A.

(2009) *Org. Biomol. Chem.*, **7**, 4886–4894.
101. Dewick, P.M. (1998) *Nat. Prod. Rep.*, **15**, 17–58.
102. Knaggs, A.R. (2003) *Nat. Prod. Rep.*, **20**, 119–136.
103. Herrmann, K.M. and Weaver, L.M. (1999) *Annu. Rev. Plant Biol.*, **50**, 473–503.
104. Beerhues, L. and Liu, B. (2009) *Phytochemistry*, **70**, 1719–1727.
105. Gottlieb, O.R. (1968) *Phytochemistry*, **7**, 411–421.
106. Dewick, P.M. (2009) *Medicinal Natural Products: A Biosynthetic Approach*, 3rd edn, John Wiley & Sons, Inc., Hoboken, New Jersey.
107. Locksley, H.D., Moore, I., and Scheinmann, F. (1967) *Tetrahedron*, **23**, 2229–2234.
108. Carpenter, I., Locksley, H.D., and Scheinmann, F. (1969) *Phytochemistry*, **8**, 2013–2025.
109. Bennett, G.J. and Lee, H.-H. (1988) *J. Chem. Soc., Chem. Commun.*, 619–620.
110. Bennett, G.J., Lee, H.-H., and Das, N.P. (1990) *J. Chem. Soc., Perkin Trans. I*, 2671–2676.
111. Kartha, G., Ramachandran, G.N., Bhat, H.B., Nair, P.M., Raghavan, V.K.V., and Venkataraman, K. (1963) *Tetrahedron Lett.*, **4**, 459–472.
112. Quillinan, A.J. and Scheinmann, F. (1971) *J. Chem. Soc., Chem. Commun.*, 966–967.
113. Yemul, S.S. and Rama Rao, A.V. (1974) *Org. Mass Spectrom.*, **9**, 1063–1072.
114. Han, Q., Yang, L., Liu, Y., Wang, Y., Qiao, C., Song, J., Xu, L., Yang, D., Chen, S., and Xu, H. (2006) *Planta Med.*, **72**, 281–284.
115. Nicolaou, K.C. and Li, J. (2001) *Angew. Chem. Int. Ed.*, **40**, 4264–4268.
116. Tisdale, E.J., Slobodov, I., and Theodorakis, E.A. (2003) *Org. Biomol. Chem.*, **1**, 4418–4422.
117. Li, N.-G., Wang, J.-X., Liu, X.-R., Lin, C.-J., You, Q.-D., and Guo, Q.-L. (2007) *Tetrahedron Lett.*, **48**, 6586–6589.
118. Tisdale, E.J., Slobodov, I., and Theodorakis, E.A. (2004) *Proc. Natl. Acad. Sci. U.S.A.*, **101**, 12030–12035.
119. Nicolaou, K.C., Xu, H., and Wartmann, M. (2005) *Angew. Chem. Int. Ed.*, **44**, 756–761.
120. Gajewski, J.J. (1980) *Acc. Chem. Res.*, **13**, 142–148.
121. Ganem, B. (1996) *Angew. Chem. Int. Ed.*, **35**, 936–945.
122. Ganem, B. (1996) *Angew. Chem.*, **108**, 1014–1023.
123. Severance, D.L. and Jorgensen, W.L. (1992) *J. Am. Chem. Soc.*, **114**, 10966–10968.
124. Gajewski, J.J. (1997) *Acc. Chem. Res.*, **30**, 219–225.
125. Tanford, C. (1980) *The Hydrophobic Effect*, 2nd edn, John Wiley & Sons, Inc., New York.
126. Breslow, R. (2004) *Acc. Chem. Res.*, **37**, 471–478.
127. Lindström, U.M. (2002) *Chem. Rev.*, **102**, 2751–2772.
128. Grieco, P.A. and Kaufman, M.D. (1999) *J. Org. Chem.*, **64**, 6041–6048.
129. Hayden, A.E., Xu, H., Nicolaou, K.C., and Houk, K.N. (2006) *Org. Lett.*, **8**, 2989–2992.
130. Nicolaou, K.C., Sasmal, P.K., and Xu, H. (2004) *J. Am. Chem. Soc.*, **126**, 5493–5501.
131. Bhamare, N.K., Granger, T., John, C.R., and Yates, P. (1991) *Tetrahedron Lett.*, **32**, 4439–4442.
132. Bichan, D.J. and Yates, P. (1972) *J. Am. Chem. Soc.*, **94**, 4773–4774.
133. Tisdale, E.J., Chowdhury, C., Vong, B.G., Li, H., and Theodorakis, E.A. (2002) *Org. Lett.*, **4**, 909–912.
134. Chantarasriwong, O., Batova, A., Chavasiri, W., and Theodorakis, E.A. (2010) *Chem. Eur. J.*, **16**, 9944–9962.

Part III
Biomimetic Synthesis of Polyketides

13
Polyketide Assembly Mimics and Biomimetic Access to Aromatic Rings

Grégory Genta-Jouve, Sylvain Antoniotti, and Olivier P. Thomas

> Out of intense complexities intense simplicities emerge.
> *Winston Churchill*

13.1
Introduction

Polyketides represent one of the most complex groups of natural products, exemplified by renowned families of compounds such as polyphenols, macrolides, polyethers, and polyenes. To date, around 10 000 polyketides have been characterized and they are recognized as useful templates for the discovery and the design of new drugs [1]. Despite the fact that highly complex structures are frequent, a striking characteristic of this family of compounds arises from the structural simplicity of their biosynthetic precursors and of their assembly. The understanding of polyketide biosynthesis during the last two decades was facilitated by the relative simplicity of these iterative biosynthetic steps. Because many polyketides are produced by microorganisms their biosynthetic enzymes and genes were among the first to be identified [2, 3]. Polyketide synthases (PKSs) belong nowadays to the best known and described polyenzymatic complexes and their huge potential to afford polyketides diversity was reviewed in 2009 by Hertweck [4]. A few years earlier, a comprehensive review by Staunton *et al.* related the history and key achievements in the discovery of the logic of polyketide biosynthesis [5] and also in 2001 Whiting reviewed the chemistry of natural phenolic compounds across the twentieth century [6]. In the first part of this chapter, we will therefore focus on biomimetic polyketide synthesis.

As we will see in the second part of this chapter, the original idea that some aromatic compounds may be derived from linear polyketone chains first came from Collie in the late nineteenth century [7]. Even if the main interest of Robinson's laboratory in Oxford focused on alkaloid and terpene biosynthesis, this idea was later supported by Robinson himself and particularly by a former fellow of his laboratory, Arthur Birch, in the 1950s. He was the first to postulate and demonstrate, using isotopically enriched precursor, that acetic acid was the sole

Biomimetic Organic Synthesis, First Edition. Edited by Erwan Poupon and Bastien Nay.
© 2011 Wiley-VCH Verlag GmbH & Co. KGaA. Published 2011 by Wiley-VCH Verlag GmbH & Co. KGaA.

molecule responsible for the construction of linear polyketone chains giving rise to aromatic natural products such as 6-methylsalicylic acid [8]. As a consequence, the Claisen condensation appeared as the key reaction leading to the creation of the C–C connections between acetic acid units. Birch's discovery triggered a high number of scientific reports that all showed that acetic acid not only lead to aromatic compounds but also to structurally more complex polyketides. Indeed, a succession of reductive, oxidative, and cyclization processes of the linear polyketone chain would be able to afford the high diversity of the polyketide family. These simple chemical processes inspired a flurry of synthetic organic methodologies trying to mimic the natural strategy to synthesize complex bioactive molecules.

The first work reported by Claisen in 1881 described the condensations of ketones with aldehydes under strongly basic conditions [9], but the Claisen condensation is now well recognized as a C–C bond formation between two esters leading to a β-keto ester. Because these "synthetic" conditions were not consistent with "natural" enzymatic conditions it was further discovered that two additional steps were selected during evolution to lower the pK_a at the α-position of the acid derivative to facilitate this condensation. First, acetic, propionic, or butyric acids are linked to coenzyme A (CoA) or to the thiol residue of a polyenzymatic complex through a thioester bond, which induces a decrease of approximately two pK_a units compared with an ester. In this chapter, we will detail in-depth the results of chemical studies using thioesters for the Claisen condensation. A second development of biological systems to facilitate this C–C bond formation was a key carboxylation at the α position of the acyl derivatives, leading to highly reactive malonyl derivatives. This activation was found to be catalyzed by a carboxylase and biotin appeared as the cofactor transferring carbon dioxide to this position [10]. Here also we will focus mainly on synthetic strategies using malonyl derivatives. In the infancy of polyketide biomimetic chemistry, pioneer chemists developed reactions mainly aimed at mimicking Nature and improving the understanding of metabolic pathways. Later on, the logic of biomimetic synthesis also inspired organic chemists in the design of new synthetic methodologies often applied to the total synthesis of complex bioactive products. The examples presented in this chapter illustrate these two aspects of biomimetic polyketides and polyaromatic natural products.

13.2
Polyketide Assembly Mimics

The mechanism of a polyketide assembly between two acetyl units has been studied extensively and is closely related to fatty acid assembly. The C–C connection originates from a decarboxylative Claisen condensation between one nucleophilic malonyl unit linked to the polyenzymatic complex by a thioester bridge and a second electrophilic acetyl derived thioester leading to a β-keto thioester (Scheme 13.1). Nevertheless, the exact sequence of the mechanistic steps is not fully accepted and

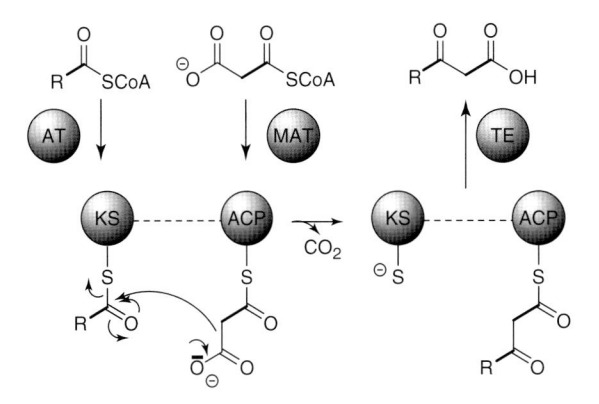

Scheme 13.1 Mechanism of the key C–C connection between two units of a polyketide.

diverse experimental proof supported three distinct mechanisms: concerted, stepwise addition–decarboxylation, and stepwise decarboxylation–addition. We will detail later in this chapter how biomimetic approaches can help in understanding the real biosynthetic events.

PKSs are complex polyenzymatic systems closely related to fatty acid synthases (FASs) [4]. The larger chemical diversity produced by PKS is due to a broader tolerance toward the precursors but also to the absence of some reductive steps always present for FASs. PKSs have been classified in several types depending on their global architecture and function. Iterative and non-iterative type I PKS mainly produce non-aromatic compounds, including a large family of bioactive macrolactones represented by erythromycin (**1**) (Figure 13.1). In type II PKS, a unique module composed of three domains is used in an iterative manner to yield most of the aromatic and polyaromatic polyketides exemplified by doxorubicin (**2**). Finally, type III PKS can metabolize a broad range of starter units followed by various iterative steps and cyclization, often leading to aromatic compounds like tetrahydrocannabinol (**3**).

For type I multimodular PKS, each module is responsible for the addition of an acetyl or propionyl unit. Several domains are constitutive of a module: a ketosynthase (KS), an acyl carrier protein (ACP), and optionally an acyl transferase (AT)

Erythromycin A (**1**)   Doxorubicin (**2**)   Tetrahydrocannabinol (**3**)

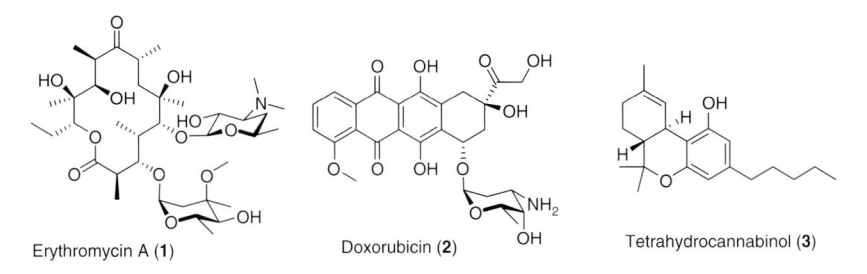

Figure 13.1 Structures of three polyketides produced by the three types of PKS.

and malonyl acyl transferase (MAT). Iteration of the process leads to polyketones and finally a *chain termination domain* (TE) allows the release of the product as a carboxylic acid or a lactone.

The high diversity of polyketides arises from the presence of additional domains in some modules of the PKS. The simplest modules with no reductive step lead to a β-keto acyl derivative (Scheme 13.2). To classify the biomimetic reactions further reported we will name this transformation a "type-a reaction." A "type-b reaction" will give directly the β-hydroxy acyl derivative, involving an additional reductive process due to the presence of a ketoreductase (KR) domain in the PKS module. Furthermore, the presence of a dehydratase (DH) domain will allow the biosynthesis of an α, β-unsaturated acyl derivative, classified as a "type-c reaction" in biomimetic approaches. Finally, an enoyl reductase module (ER) could be present to saturate the chain by a formal hydrogenation. This last process is mostly representative of the FAS and biomimetic access to this connection type is beyond the scope of this chapter.

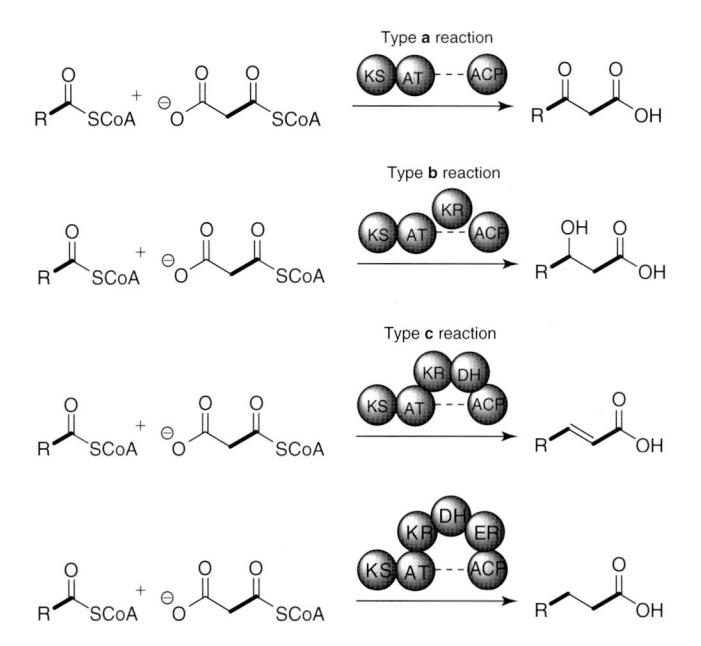

Scheme 13.2 The four types of product formation for a PKS module.

As a consequence, this section is divided according to the type of PKS assembly the synthetic chemist plans to mimic. In the case of a type-a mimic, the electrophile entity can be chosen among the acyl derivatives, whereas for type-b and -c mimics the acyl derivative should be replaced by an aldehyde or an equivalent to keep the corresponding oxidation state in a one-step procedure. In all these cases, several nucleophiles were tested; we will first focus on the most "biomimetic" reactions corresponding to a decarboxylative Claisen condensation using malonic acid half-thioesters (MAHTs).

13.2.1
Type-a Mimics

As early as 1967, Lynen proposed a biochemical process based on an acyl transfer from a MAHT to an enzyme bound thioester as the key step, explaining a polyketide unit assembly [11]. Nucleophilic activation of the acyl through biotin carboxylation was proven to occur in the biosynthetic assembly of a polyketide unit. We have thus decided to first detail biomimetic studies using malonyl derivatives as nucleophiles and, in a second part, condensations that do not require a decarboxylative step.

13.2.1.1 Malonyl Activation

The first experiments designed to mimic a type-a PKS module were reported in 1975 by Scott et al. The acyl transfer was tested on a chemical template to mimic the enzymatic machinery and to facilitate the transfer through an intramolecular process. To this end, Scott et al. used catechol derivative 4 with both hydroxyl groups acylated, one by an acetyl and one by a malonyl group [12]. Strongly basic conditions in the presence of isopropylmagnesium bromide allowed them to isolate the modified acyl-transferred catechol derivative 5 in 30% yield (Scheme 13.3).

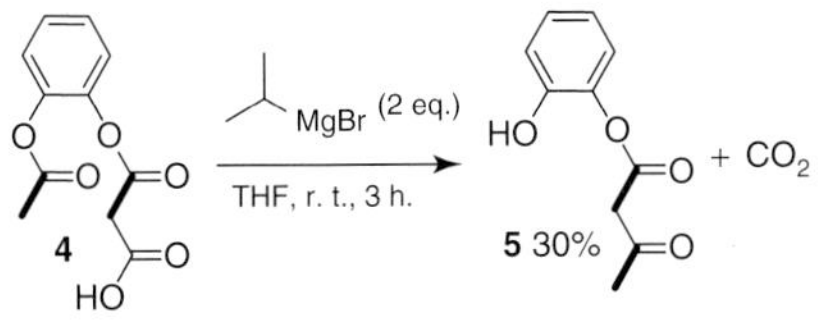

Scheme 13.3 Scott's conditions for the intramolecular acyl transfer.

The intramolecular nature of the acyl transfer was demonstrated and chelation of the magnesium cation with the malonate in 6 was assumed to be essential for control of the C-acylation over O-acylation (Scheme 13.4).

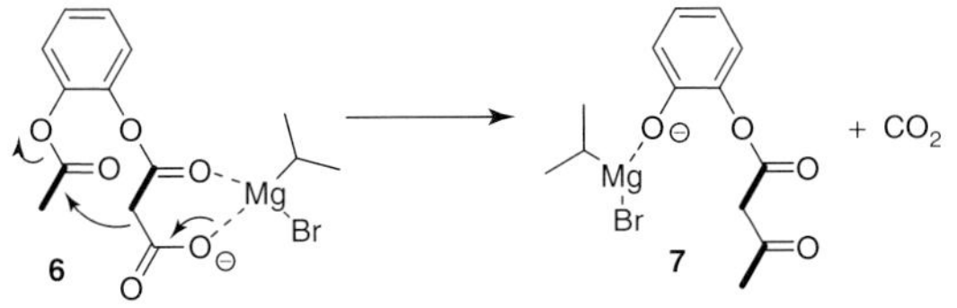

Scheme 13.4 Proposed mechanism for Scott's acyl transfer.

Three years later, Kobuke and Yoshida studied an analogous intermolecular reaction using the acyl and malonyl thioesters 8 and 9, respectively, which could be considered closer to AcylCoA and MalonyCoA and therefore more biomimetic than the catechol derivative 4 [13]. They were able to isolate and characterize the β-ketoacyl derivative 10 (60% yield), the product of a type-a condensation between both units, using catalytic Mg(II) salts and imidazole as a soft organic base (Scheme 13.5). This study evidenced the major role played by the thioester present in the polyenzymatic systems. By lowering the pK_a of the α hydrogen, this

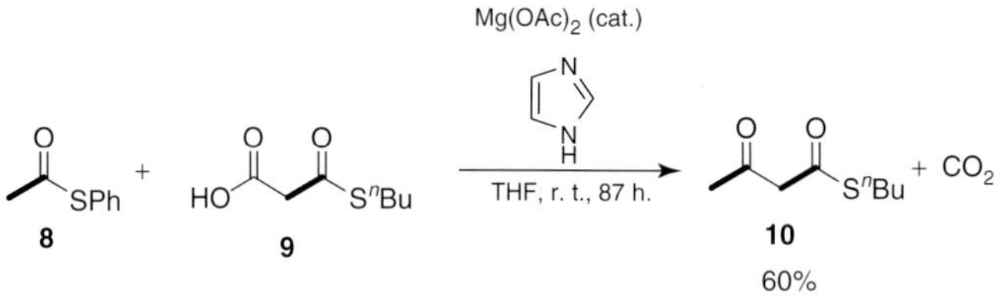

Scheme 13.5 Biomimetic conditions of Kobuke and Yoshida, using thioesters, imidazole, and catalytic Mg(II) salts.

acyl derivative allows the use of a soft organic base and a catalytic amount of metal, avoiding harsh conditions and getting closer to real biosynthetic mimics.

Worth noting is the use of imidazole as a base, which improves the biomimetic character of the process by emulation of histidine-rich PKS. Even if the catalytic sites of PKS were not described as containing metals, the presence of Mg(II) salts can help in the decarboxylation step through coordination with the malonyl group (Scheme 13.6).

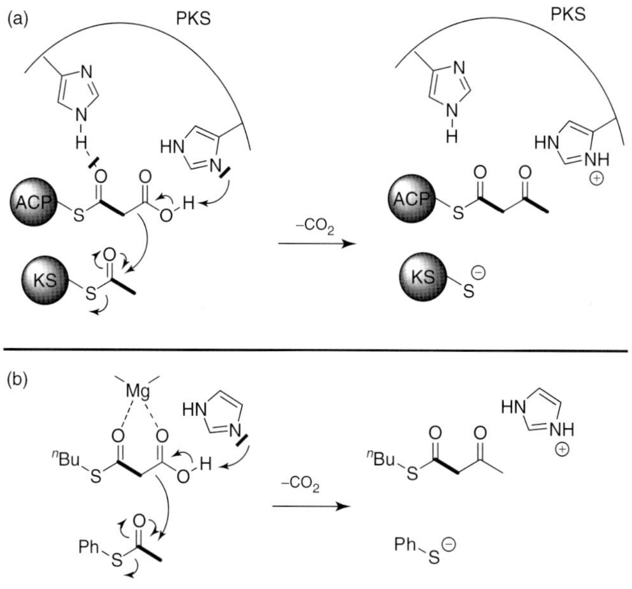

Scheme 13.6 Proposed mechanism for the condensation between thioesters in: (a) the active site of a PKS and (b) the biomimetic conditions of Kobuke and Yoshida.

At the same time, Masamune extended the scope of this reaction to other bi-substituted malonyl and imidazolyl acyl derivatives [14]. Subsequently, no significant advance in this field was made during the next 20 years until Matile and coworkers found that, under similar conditions, the self-condensation of a malonyl thioester was possible when derived from a *para*-methoxythiophenol such as **11**, increasing at the same time the acidity of the α-proton [15]. In this case, even if Mg(OAc)$_2$ remained the most potent metal catalyst, 5-nitrobenzimidazole was more efficient than imidazole as an organic base and could be used in catalytic amounts to afford the self-condensed product **12** in 34% yield, along with the non-decarboxylated precursor **13** (37%), both issuing from the Claisen self-condensation (Scheme 13.7).

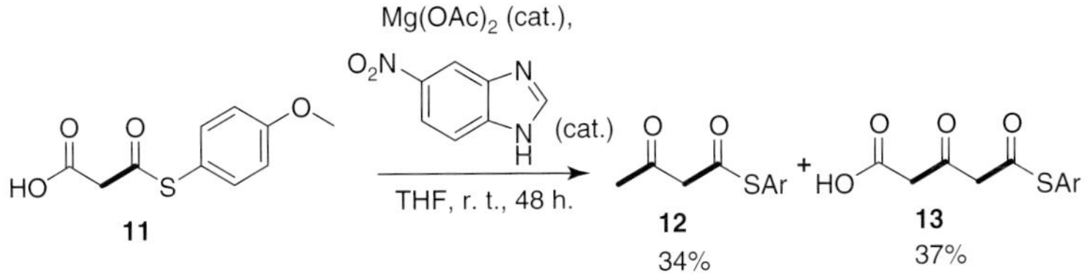

Scheme 13.7 Matile's biomimetic conditions for the self-condensation of malonyl thioesters.

Given the frequency and the amounts of some by-products of decarboxylation or hydrolysis of the thioesters isolated, the authors concluded that "Claisen self condensation toward polyketide was questionable" under similar conditions. Mimicking a long evolutionary polyenzymatic complex could not be an easy task!

Decarboxylative intramolecular Claisen condensations were also studied by changing the template. The group of Harisson published several reports on the use of glycoluril to promote the acyl transfer. One of the reports focused on the description of specific conditions for a malonyl condensation [16], but most of their work showed that activation by a malonyl was not necessary and we will therefore detail their results in the following section.

The use of MAHT as nucleophiles was particularly adequate for soft and biomimetic reaction conditions, nevertheless their preparations were sometimes troublesome and synthetic chemists mostly developed methodologies based on non-activated acyl derivatives.

13.2.1.2 Without Malonyl Activation

Because the presence of a carboxylic acid substituent increases the acidity of the α-proton of the malonyl by around two pK_a units, an acyl transfer with a simple acyl nucleophile requires strong basic conditions. Important advances in this field were obtained by the group of Harisson, who used glycoluril **14** as a template for an intramolecular acyl transfer [17]. The original properties of this template allowed them to repeat the acyl transfer in several subsequent iterative steps, culminating in the addition of four C$_2$ units in the example of adduct **15** (Scheme 13.8) [18].

This approach was applied to isotopically ^{2}H and ^{13}C labeled acyl derivatives, which allowed the synthesis of these important labeled precursors used for biosynthetic studies [19]. Kinetic and mechanistic studies were undertaken to understand the regioselectivity of the reaction during the acyl transfer [20].

All these approaches were dedicated to the construction of β-polyketones and they belong to the type-a mimic class. One of the conclusions that can be underlined

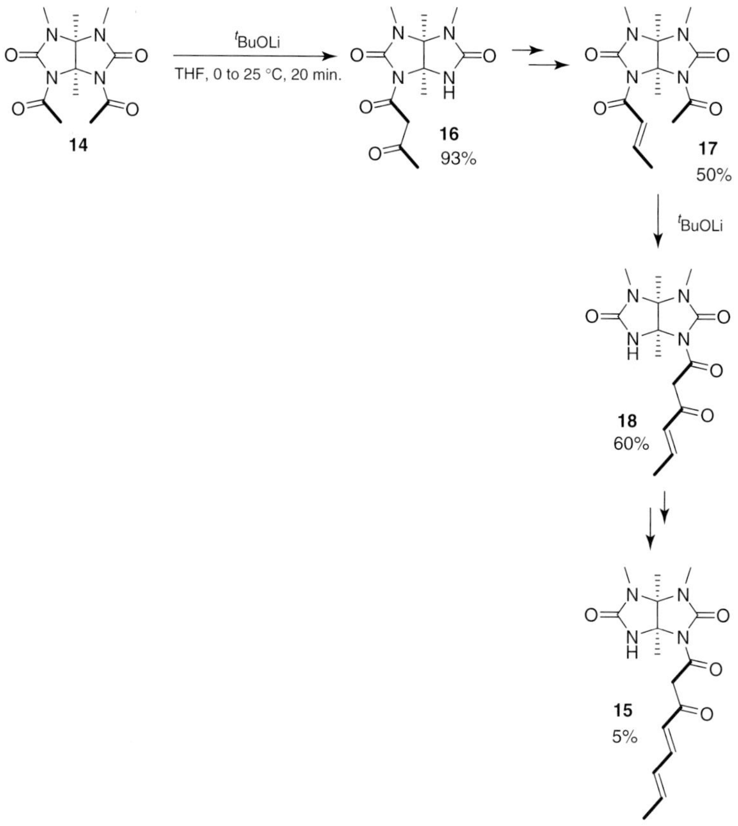

Scheme 13.8 Harisson's intramolecular iterative acyl transfer using glycoluril **14** as a template.

is the subtle balance of the conditions necessary to mimic the catalytic site of an enzyme. The detailed work of the group of Harisson in this field is worth noting, which also lead to labeled polyketones. Another challenging issue was to obtain β-hydroxyacyl derivatives in a unique one-pot procedure, thus mimicking a type-b PKS module.

13.2.2
Type-b Mimics

In this case the electrophile entity is an aldehyde or an equivalent that can undergo a nucleophilic addition under mild conditions to afford the desired β-hydroxyacyl product. In this case also, activation of the nucleophilic acyl moiety has been studied. A key point of these studies came from the control of the stereochemistry of the nucleophilic addition, an enzyme catalysis leading to a unique stereoisomer during the formation of the stereogenic center. Aldol-type reactions using acid derivatives have been greatly studied and reviewed [12]. In this chapter we will only detail biomimetic reactions using malonyl half thioesters (MAHT) or unactivated thioesters that were recognized as very soft nucleophile for an aldol addition.

13.2.2.1 Malonyl Activation

A first catalytic procedure that did not require *in situ* aldolate functionalization, or an excess of nucleophile, was developed by the group of Shair with MAHT in 2003 [22]. In this case, catalysis with the magnesium salts previously used for type-a mimics were not efficient and after a small screening Cu(II) salts were identified as the best catalysts for the addition of MAHT **19** on aldehyde **20**, affording the β-hydroxythioester **21** in good yield. Benzimidazole derivatives were found to be necessary as an additional soft organic base for this condensation (Scheme 13.9).

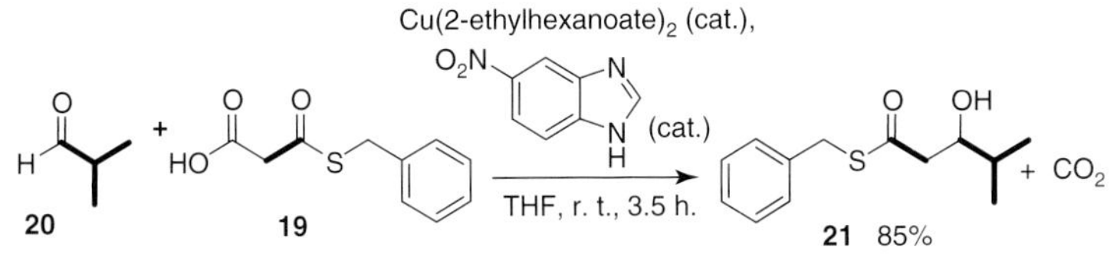

Scheme 13.9 Shair's catalytic aldol condensation of aldehydes with a MAHT.

Worth noting are the exceptionally mild conditions of this reaction, which was performed at room temperature and which did not require dry solvents and vessels usually necessary for a catalytic aldol reaction. Increasing the electrophilicity of the aldehyde with electron-withdrawing substituents even improved the yields of the desired products. Interestingly, the reaction was successfully applied to MAHT substituted at the α-position by a methyl, thus mimicking a propionate insertion in a polyketide chain. In this case, the diastereoselectivity was largely in favor of the *syn* products, which renders this reaction even more interesting.

Subsequent studies on this important reaction focused on the enantioselectivity that could be induced by the presence of a chiral ligand coordinating the metal catalyst. The first report of a small enantiomeric excess using Cu(II) triflate salts and a chiral bisbenzimidazole derived from tartric acid was reported by the group of Cozzi [23]. At the same time, the group of Shair greatly improved diastereoselective and enantioselective excesses of **22** with methyl MAHT **23** as the nucleophile and a bisoxazoline ligand associated with the copper(II) triflate catalyst (Scheme 13.10) [24].

The presence of the thioester proved to be critical for the nucleophilic addition to proceed under very mild condition and allow further modifications to access to more complex structures. Shair and coworkers underlined a significant compatibility of the reaction conditions with several substrates, including substrates bearing protic functional groups that are often troublesome for such type of additions. Mechanistic insights into polyketide biosynthesis were also obtained working in these biomimetic conditions [25]. It appeared indeed much easier to detail and interpret all the kinetics parameters of such a simplified reaction than to study these

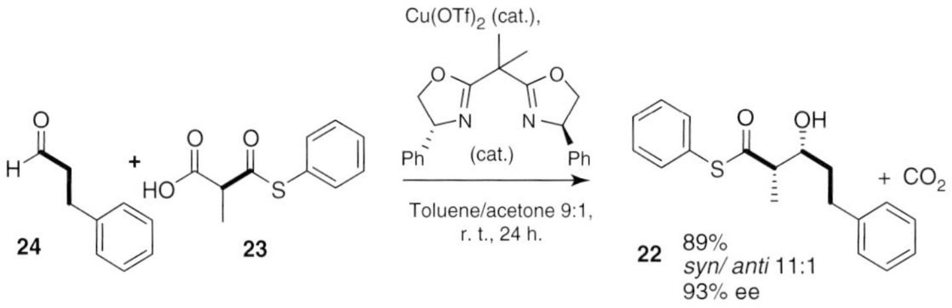

Scheme 13.10 Shair's asymmetric catalytic aldol condensation of aldehydes with methyl MAHT.

parameters in a complex polyenzymatic system. Kinetic experiments using labeled precursors suggested that decarboxylation of the malonyl moiety occurred after a preliminary addition and not before, as usually stated for FASs and polyketases (Scheme 13.11) [26].

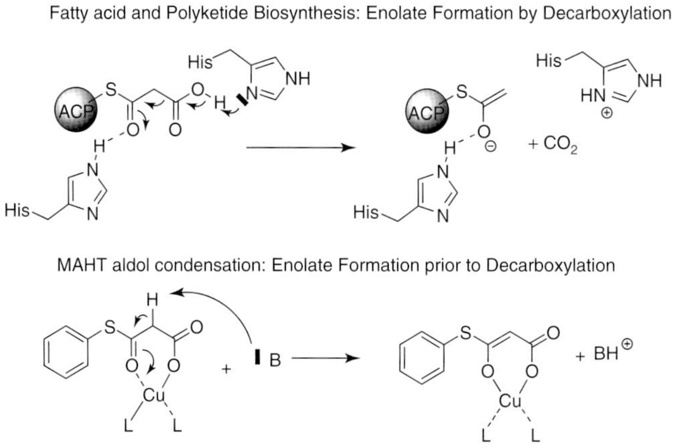

Scheme 13.11 Two distinct mechanisms of enolate formation before nucleophilic addition.

A metal-free decarboxylative addition of MAHT to activated carbonyl compounds was further developed by Fagnou and coworkers [27]. For the first time, they were able to extend a decarboxylative Claisen condensation to malonic acid half oxoesters (MAHOs) **25** without the use of a metal catalyst, and the ketol product **26** was formed in good yield in the presence of triethylamine starting from ethyl pyruvate (**27**) at room temperature (Scheme 13.12).

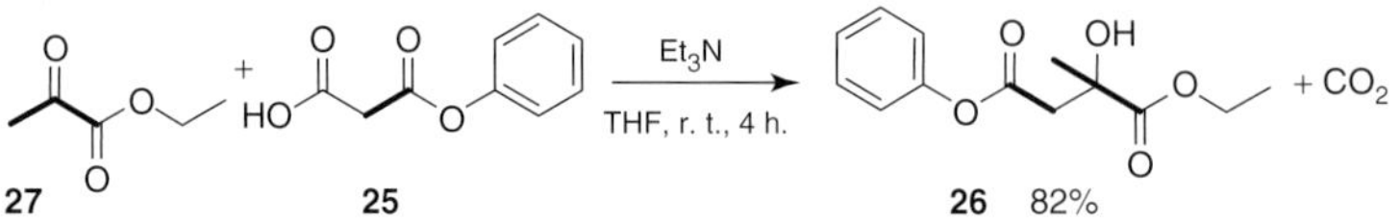

Scheme 13.12 Fagnou's metal-free decarboxylative condensation with a MAHO.

13.2 Polyketide Assembly Mimics | 481

These metal-free conditions also allowed a deeper understanding of the reaction mechanism in particular by diffusion ordered spectroscopy (DOSY) NMR experiments. Identifying the post-addition/pre-decarboxylation adduct **28** by NMR for both MAHT and MAHO, the authors undoubtedly demonstrated that the stepwise addition–decarboxylation was the unique mechanism possible for this reaction, leading to the same conclusions as the group of Shair. They were also able to demonstrate the reversibility of the first addition step and kinetic studies revealed the higher reactivity of MAHT over MAHO (Scheme 13.13). As a consequence, these results shed some light on the real biosynthetic mechanism.

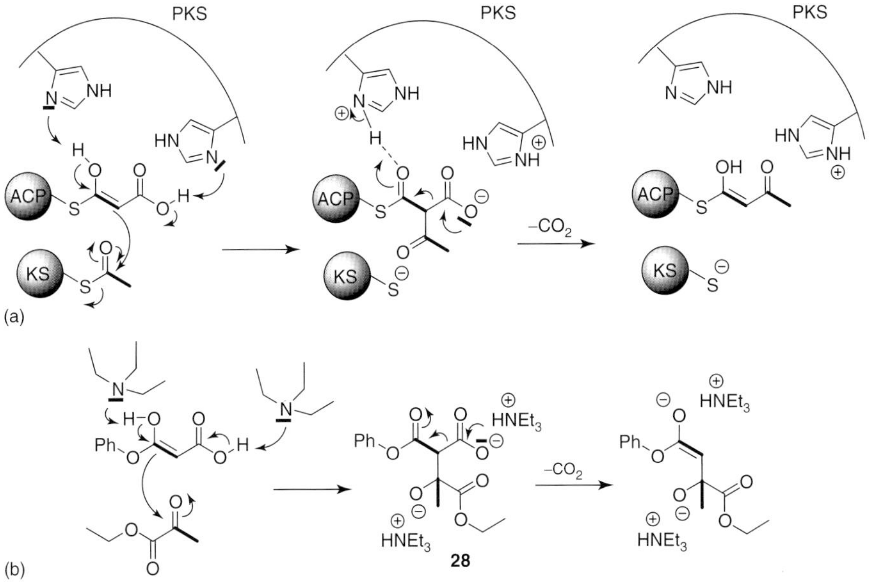

Scheme 13.13 Stepwise addition–decarboxylation for the C–C connection: (a) in polyketide biosynthesis; (b) under Fagnou's conditions.

This result is highly significant as it is well known that PKS enzymes do not require a metal ion to perform the C–C condensation (see, for example, Reference [28]). As extensions of these important catalytic decarboxylative Claisen condensations with MAHT to organic synthesis, the carbonyl group has been replaced by other unsaturated electrophiles that undergo, in the same manner, nucleophilic addition for the creation of important C–C bonds.

As striking examples, a catalytic decarboxylative condensation of MAHT to imines has been found to be catalyzed by chiral *Cinchona* alkaloids, leading to important chiral β-aminothioesters [29]. Other similar conditions were developed for the asymmetric 1,4-decarboxylative addition of MAHT to nitroolefins, also catalyzed by chiral *Cinchona* alkaloids [30]. Indeed, the group of Wennemers was inspired by the cysteine, histidine, and asparagine triad described in the catalytic active site of

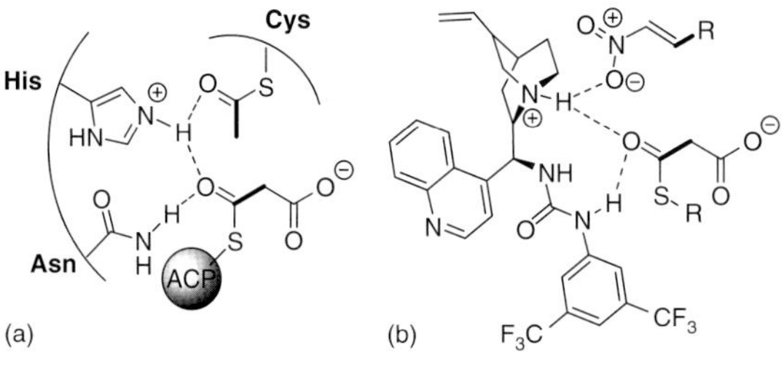

Scheme 13.14 Rational design behind an asymmetric and organocatalytic addition of MAHT to nitroolefins by Wennemers: (a) the cysteine, histidine, and asparagine triad catalytic active site of some PKS enzymes; (b) the designed organocatalyst.

some PKS enzymes to develop an organocatalytic and asymmetric version of the addition of MAHT to an appropriate electrophile (Scheme 13.14).

In ethyl vinyl ether as solvent the reaction is performed at room temperature to afford the addition product in around 60% yield and with an enantiomeric excess of ca. 80%.

13.2.2.2 Without Malonyl Activation

Thioesters alone have long been recognized as important enolizable substrates for nucleophilic addition to several functional groups, and malonyl activation is not always necessary if strong bases are used [31]. In a biomimetic context, the group of Coltart developed an aldol addition of non-activated thioesters **29** under very mild conditions, leading to β-hydroxythioester **30** [32]. Here also, choosing a thioester was justified by the increased acidity of the α-proton, which avoids the use of strong bases like lithium diisopropylamide (LDA) or t-BuOK. The method involved an organic amine base and Mg(II) salts (Scheme 13.15).

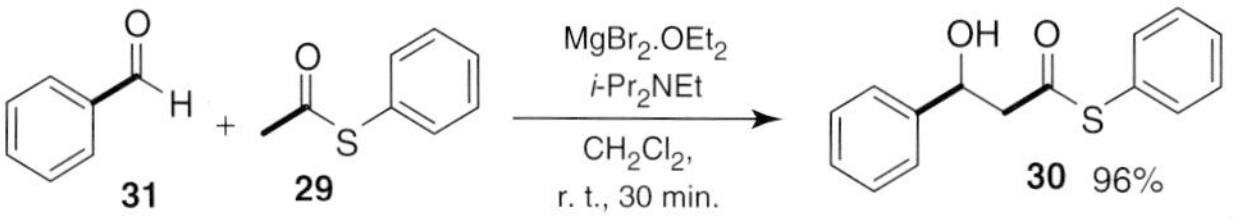

Scheme 13.15 Coltart's aldol addition with non-activated thioester.

Importantly, Mg(II) and Cu(II) salts emerged as the best Lewis acids to promote the condensation. Even if catalytic conditions were not developed at this stage, the wide scope of this reaction allowed several extensions to C–C bond forming reactions for organic chemists. Type-a mimics have also been studied under these conditions, but to a lesser extent, and to activate the nucleophile N-acylbenzotriazoles and pentafluorophenyl esters are necessary to give access to 1,3-diketones in good yields [33]. As another extension of this work, an asymmetric version of the Mannich condensation of thioesters was developed by this group using *Cinchona* alkaloids as chiral organocatalysts [34]. Also avoiding enolate formation through malonic activation, the group of Barbas III developed an asymmetric version of Michael and Mannich additions of thioesters to α,β-unsaturated aldehydes catalyzed by chiral proline derivatives [35, 36]. To increase the acidity of the α-proton they used the S-trifluoroethyl thioester **32**, which reacted with the Michael acceptor **33** to yield the addition product **34** with a high enantioselectivity (Scheme 13.16).

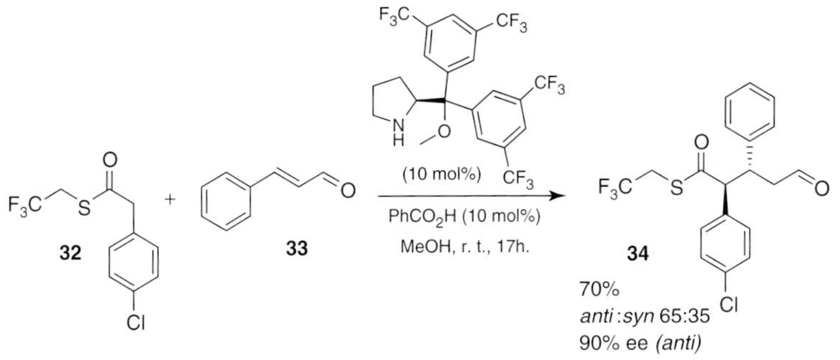

Scheme 13.16 Barbas III asymmetric and organocatalytic addition of thioesters to Michael acceptors.

The aldol reaction is not restricted to the use of thioester as nucleophiles and numerous other acyl derivatives, even in a chiral environment, have been developed. The Evans oxazolidinones belong to the most renowned examples in this field, but, in this case, the biomimetic nature of the addition is not so relevant [20–37].

13.2.3
Type-c Mimics

The last challenge in the biomimetic C–C assembly of two acyl units leading to polyketides was to obtain a dehydrated acyl product (type-c mimics). The condensation product can be obtained starting from an aldehyde and a subsequent dehydration affords the unsaturated acyl derivative. These studies are particularly challenging as, in a one-pot procedure, the synthetic chemist would mimic a complex module consisting of five domains. As above, we will mainly address reports on biomimetic conditions using thioesters as nucleophiles.

With a malonyl diester electrophile in the presence of piperidine under strong conditions the reaction is known as the *Knoevenagel condensation*, leading to an α-olefinic diester moiety [38]. The Doebner modification includes the use of pyridine with a mono- or dicarboxylic acid and a subsequent decarboxylation step yields an unsaturated acyl derivative [39]. Application of this strategy to MAHO often leads to a mixture of *(Z)* and *(E)* stereoisomers as well as α,β and β,γ unsaturated regioisomers. Control of the selectivity is troublesome and the harsh conditions required for this condensation prompted new studies for this reaction [40]. A green process using simultaneous microwave and ultrasonic irradiation was reported for such applications [41].

Very few studies focused on the development of mild biomimetic conditions to promote this condensation. A very interesting catalytic version using 4-dimethylaminopyridine (DMAP) as the sole catalyst was developed by List and coworkers in 2005, featuring high selectivity in favor of the α,β unsaturated **35** compound with an *(E)* configuration, starting with aldehyde **36** and MAHO **37** (Scheme 13.17) [42].

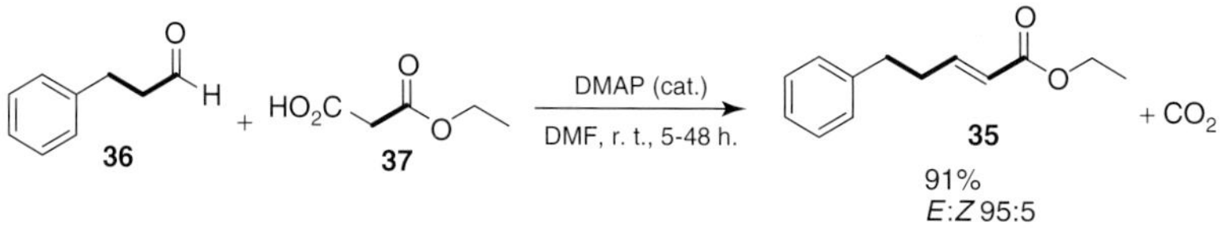

Scheme 13.17 List's condensation of MAHO in a type-c mimic.

To obtain very soft conditions for this condensation, a catalytic version of this procedure in wet THF was proposed with phenylacetaldehyde (**38**) as electrophile and MAHT **39** as nucleophile [43]. After a rapid screening of several metallic catalysts, Yb(OTf)$_3$ appeared as the most effective to obtain the desired dehydrated product, with Cu(II) salts giving mostly the aldol product as described by Shair. The use of 5-methoxybenzimidazole as the base afforded a very high regioselectivity in favor of the β,γ unsaturated thioester **40** with an *(E)* configuration. Interestingly, the regioselectivity was inverted in the case of aliphatic aldehyde **41**, suggesting thermodynamic control of the reaction (Scheme 13.18).

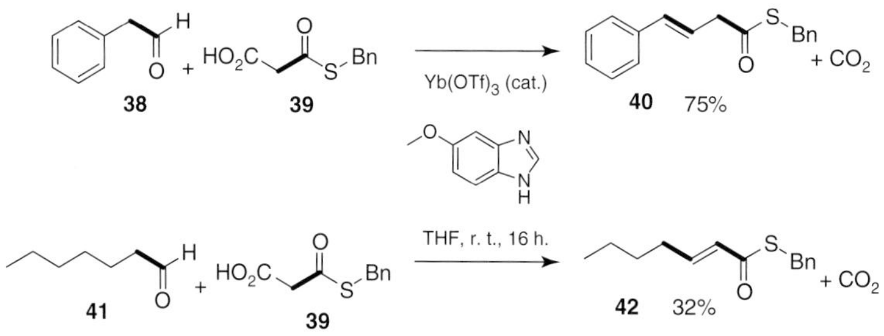

Scheme 13.18 A type-c reaction mimic with MAHT.

In this case also, the mechanism was suggested to proceed via a first addition step followed by decarboxylation of the reactive intermediate. To get further insight into the mechanism of this condensation, the role of the metal cation should be clearly assessed but as usually described the decarboxylation could be concomitant with the dehydration step. To date, no work has been reported on a condensation of type-c without malonyl activation.

This first part has focused on the biomimetic methods developed by synthetic chemists to mimic the assembly of acyl units for the construction of important natural products. In biological systems, the long alkyl chain is released after a cyclization or a simple hydrolysis. Historically, the first biosynthetic studies on polyketides were performed on aromatic compounds formed mainly by type II and III PKS of non-reduced β-polyketone chains. As a first consequence, a retrobiosynthetic approach would consider these aromatic end-products as masked 1,3-dicarbonyl precursors. Indeed, Birch reduction of a mono- or disubstituted benzene derivative 43 affords a 1,4-cyclohexadiene 44 in a regioselective manner, further leading to the 1,3-dicarbonyl compound 45 after subsequent ozonolysis of both double bonds (Scheme 13.19). Applications of this approach to the biomimetic syntheses of natural products were reviewed by Hilt et al. in 2009 [44].

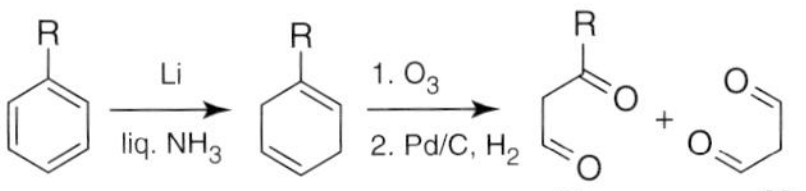

Scheme 13.19 Birch reduction–ozonolysis reaction sequence leading to 1,3-dicarbonyls.

We will detail in the second part of this chapter the direct biomimetic constructions of aromatic mono- and polycyclic polyketides starting from β-polyketones.

13.3
Biomimetic Access to Aromatic Rings

The first biomimetic reaction leading to an aromatic polyketide dates back to the late nineteenth century when James Collie obtained orcinol (47) by boiling 4-pyranone 48 in a barium hydroxide solution (Scheme 13.20) [7]. For the first time, linear polyketone 49 was proposed as an intermediate in the cyclization and aromatization into 47 but this result remained mostly ignored by the scientific community at the time of its disclosure.

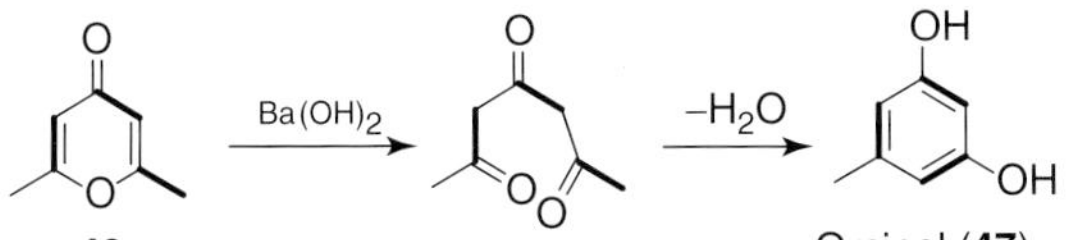

Scheme 13.20 First biomimetic access to an aromatic ring by Collie and coworkers.

More than 50 years after, the group of Birch suggested that acetic acid could be the unique precursor of the alkyl chains of β-polyketones. He was further able to prove the role of these alkyl chains in the biosynthesis of some aromatic compounds like 6-methylsalicylic acid using feeding experiments with ^{14}C labeled acetic acid [45]. In 1955, Robinson was able to present a general biosynthetic scheme for the construction of aromatic polyketides in his benchmark book *The Structural Relations of Natural Products* [46]. The biosynthesis of polyketides was further confirmed to be the result of iterative condensations of acetyl and malonyl units to form a polyketoacid that could undergo two types of cyclization: an aldol condensation (A) or a Claisen condensation (B) (Scheme 13.21) [47].

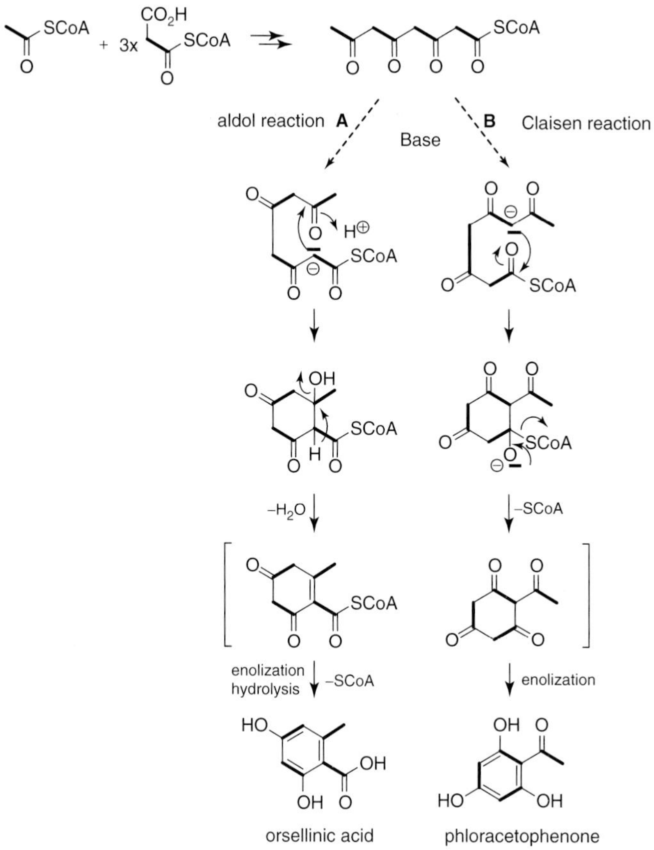

Scheme 13.21 Two biosynthetic pathways leading to aromatic polyketides.

In 2001, Thomas presented a biosynthetic classification of fused-ring aromatic polyketides based on their mode of cyclization [48]. Indeed, fungi and streptomycetes were found to utilize distinct polyketide-folding strategies for the construction of polycyclic systems. The mode F folding utilized by fungi leads, for example, to islandicin (**50**) produced by *Penicillium islandicum*, whereas the mode S folding yields actinorhodin (**51**) produced by *Streptomyces coelicolor* (Scheme 13.22).

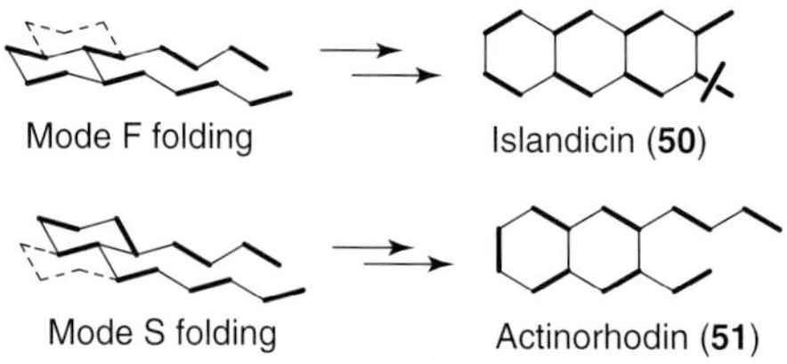

Scheme 13.22 Two modes of folding leading to polycyclic aromatic polyketides.

Differences appear in the number of acetate units present in the first aromatic ring (left-hand side): two units for the mode F folding and three units for the mode S.

No exception to this observation has been reported so far. Eukaryotes and prokaryotes are then characterized by distinct biosynthetic pathways and the mode F folding was more recently identified in plants [49]. These observations would have to be taken into consideration for biomimetic approaches. The group of Bringmann reported the first natural product proceeding from these two distinct modes of folding in two organisms. For the first time, convergence of these two distinct biosynthetic pathways was observed for chrysophanol (**52**) [50]. It was even shown that a third metabolic pathway was involved in the formation of this secondary metabolite. The anthracenoid **53**, previously synthesized by a biomimetic approach, was proved to be an intermediate towards compound **52** following the mode S' folding in a second strain of *Streptomyces*. These modes of cyclization appeared to be organism-specific, a fact of high evolutionary interest (Scheme 13.23) [51].

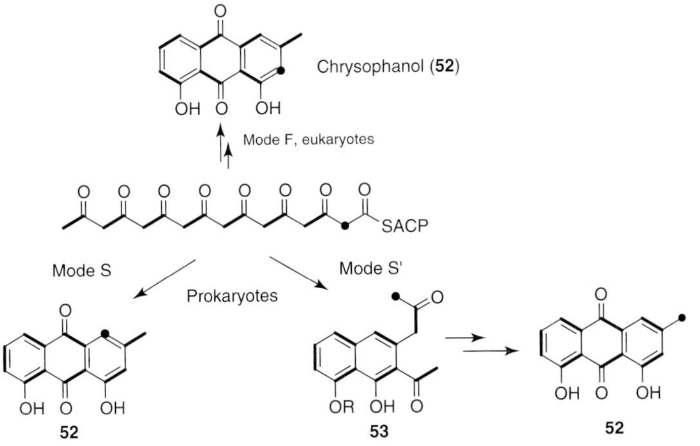

Scheme 13.23 A third mode of cyclization towards chrysophanol (**52**).

At the same time, the first studies on biomimetic access to simple aromatic polyketides were reported. The major difficulties were closely associated with the synthetic routes leading to the preparation of long β-polyketones. Figure 13.2 presents an overview of important findings in this field. Since the 1970s, the synthetic effort moved from tri- up to dodeca-β-ketoacids by improvement of the acylation methods (Figure 13.2).

13.3.1
Biomimetic Access to Benzenoid Derivatives

Even though the cleavage of pyrones appeared as a first access to tri-β-carbonyl entities, two major routes were designed to obtain these important precursors (Scheme 13.24) [52]:

488 | *13 Polyketide Assembly Mimics and Biomimetic Access to Aromatic Rings*

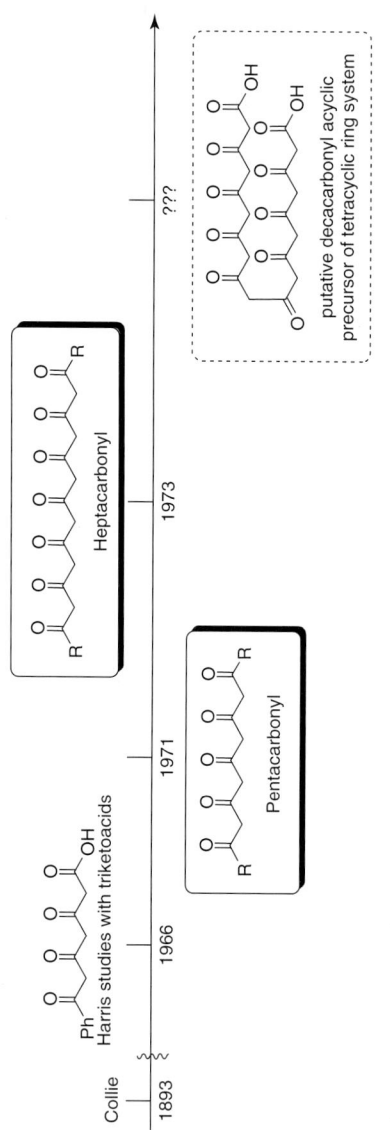

Figure 13.2 Historical achievements toward the synthesis of β-polyketones and biomimetic aromatic polyketides.

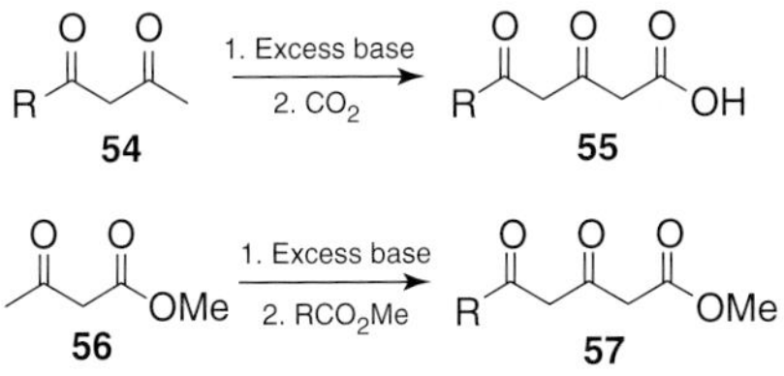

Scheme 13.24 Alternatives for the synthesis of tri-β-carbonyl entities.

1) Carboxylation of β-diketones such as **54** into diketoacids of type **55**;
2) acylation of the dianion of acetoacetic acid ester **56** into **57**.

One result of particular biogenetic significance was the cyclization of the thioacid derivative **58** into pyrone **59** obtained by the group of Harris (Scheme 13.25) [53]. Birch has been able to prove that thioacid **58** was much more prone to cyclize into resorcinol derivative **59** than the corresponding β-triketone [54].

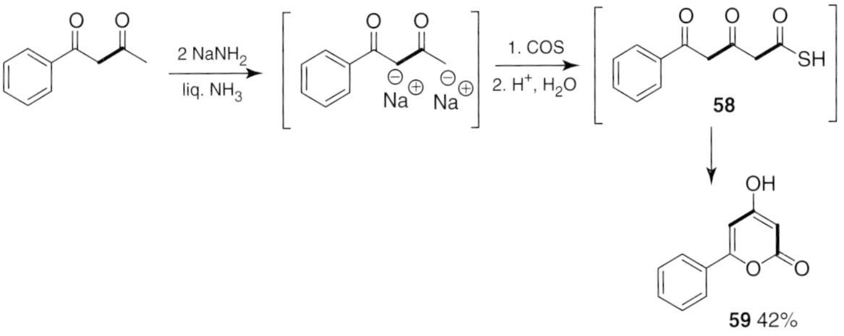

Scheme 13.25 Harris' biomimetic access to pyrones.

Tetra-β-carbonyl compounds have attracted much more attention as they are the smallest polycarbonyl systems that can provide reasonable models of biological systems, giving access to benzenoid natural products. The use of LDA proved to be much more efficient in yielding the reactive trianion intermediates [55]. As described in Scheme 13.21, these intermediates can lead to a resorcinol or a phloroglucinol compound in addition to two other pyrones. The group of Harris was able to observe the formation of 6-phenyl-β-resorcylic acid (**60**) by cyclization of 7-phenyl-3,5,7-trioxoheptanoic acid (**61**) under mild "physiological" conditions at an appropriate acidic pH (Scheme 13.26) [56].

Esterifying the carboxylic acid resulted in another mechanism of cyclization. Indeed, when treated in a pH 8.5 buffered aqueous solution, ester **62** led exclusively to resorcinol derivative **63**, while in an aqueous solution of potassium hydroxide at −5 °C the same ester **62** gave a mixture of phloroglucinol **64** and resorcylic derivatives **63**, **65**, and **66**. These results are in accordance with the fact that deprotonation of the acid derivative **61** can only lead to aldol products and that Claisen condensations require esterification of the carboxylic acid. Notably, in these

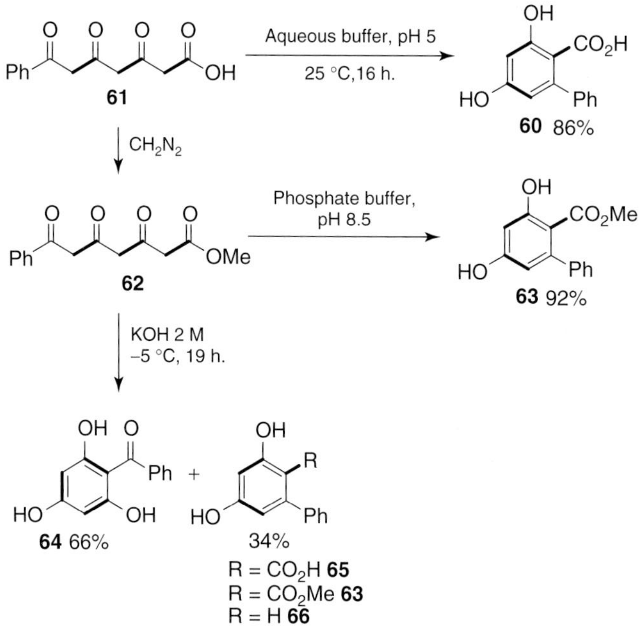

Scheme 13.26 Harris' biomimetic cyclization of β-triketoacids.

cases, Mg(II) salts were not efficient acid catalysts for promoting the cyclization of the polyketide intermediates.

An interesting study, developed by the groups of Scott and Money, and reviewed in 1970, used pyrones as masked tetraketides [57]. Under similar conditions, Crombie and James underlined the role of Mg(II) in performing the self-condensation of bispyrone **67** into phloroglucinol derivative **68**, without reporting the yield (Scheme 13.27) [58]. Chelate formation and geometry prevents aldol-type cyclization and Claisen condensation provides benzenoid derivative **68**.

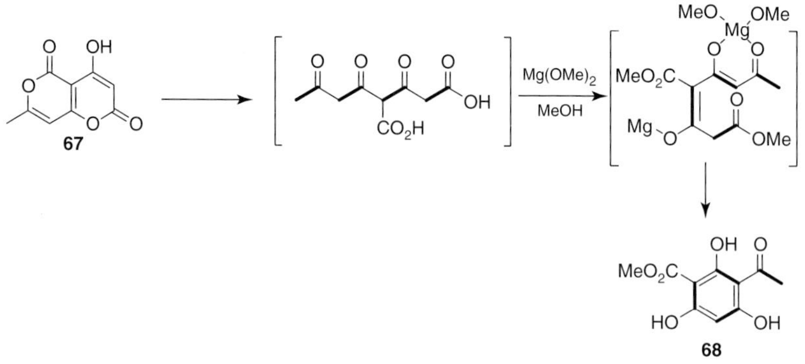

Scheme 13.27 Use of pyrones as masked tetraketide.

A similar synthetic strategy using 1,3-dioxin-4-one as masked tetra-β-carbonyl synthons was developed recently by the group of Barrett. They first reported the biomimetic syntheses of bioactive resorcylate lactones produced by the marine fungus *Hypoxylon oceanicum* LL-15G256 [59]. The strategy was based on a late-stage biomimetic aromatization into resorcylate derivatives. As an initial application, thermolysis of the 1,3-dioxin-4-one **69** and *in situ* trapping of the transient ketene **70** by the chiral alcohol **71** gave the tri-β-ketoester **72**, which was not isolated due to a high instability (Scheme 13.28). Because the use of previous Harris conditions or analogous conditions at pH 9 were unsuccessful [60], aromatization was performed using base-catalyzed aldol condensation and subsequent addition of a strong acid. Ring-closing metathesis was thus applied to cyclize ester **73** into the natural *(S)*-(−)-zearalenone (**74**). This approach was further slightly modified to give access to (+)-montagnetol and (+)-erythrin [61].

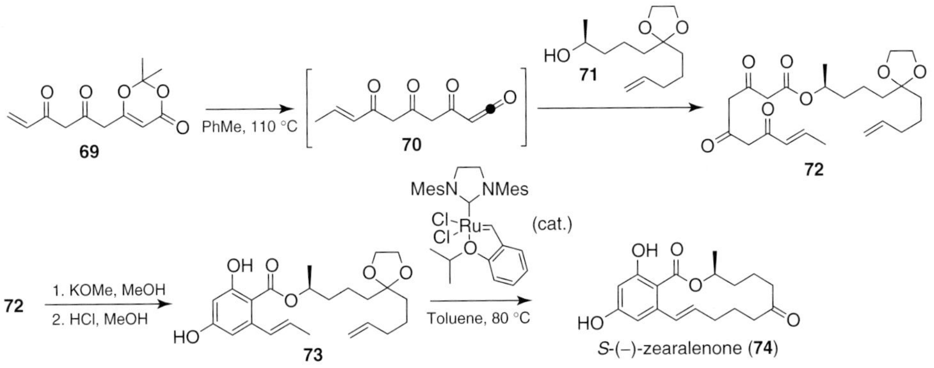

Scheme 13.28 Barrett's biomimetic total synthesis of zearalenone (**74**).

The outcome of the cyclization was even more difficult to control with non-protected pentacarbonyl intermediates, which were first synthesized by the group of Harris [62]. Coumarin derivative **75** was obtained in good yield starting from the unprotected compound **76** in slightly acidic or basic conditions. Upon using more basic conditions (KOH) this compound led to the other major aldol condensed product **77** (Scheme 13.29).

Notably, no Claisen condensation product was isolated under these conditions even with the methyl ester of **76**. At the same time, Scott and Money undertook similar studies with a three-rings fused pyranone [63]. Several other region-isomers were obtained but yields did not exceed 15%.

An alternative approach towards protected hexaketides was developed by the group of Schmidt, where the methyl ether of triacetic lactone **78** underwent an acylation by the trianion of heptane-2,4,6-trione **79** to yield the monoprotected hexaketide-derived intermediate **80** [64]. In these conditions only the substituted chromone **81** was obtained, albeit in low yield (Scheme 13.30).

Even though the transformation of the above-mentioned benzenoid compounds into naphthalenoids has been performed [65], no direct biomimetic transformations from penta-β-carbonyl intermediates could be performed, and other

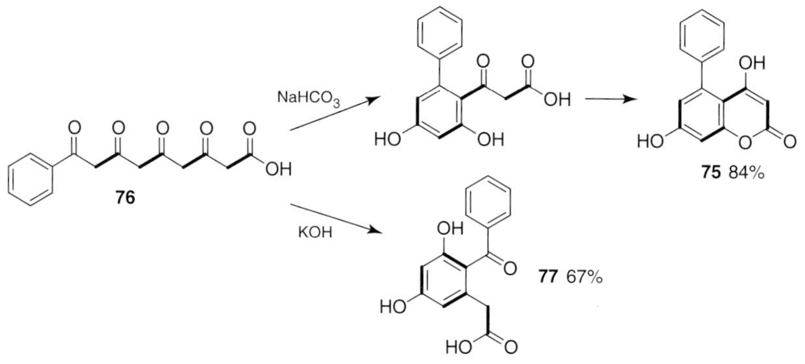

Scheme 13.29 Harris' biomimetic cyclizations of pentacarbonyl derivative **76**.

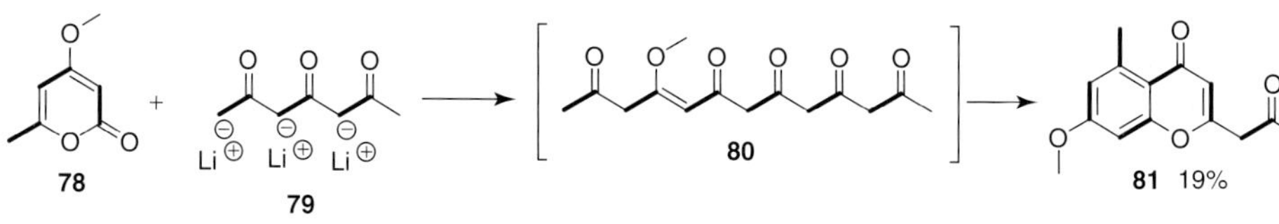

Scheme 13.30 Schmidt's condensation of masked hexaketides.

strategies using hexa-β-carbonyl derivatives were necessary to afford naphthalenoid derivatives.

13.3.2
Biomimetic Access to Naphthalenoid Derivatives

A tris(pyrone), analogous to the bispyrone **67**, was used by the group of Money as a masked hexacarbonyl derivative but neither benzenoid nor naphthalenoid products could be isolated, which definitely defined the limit of their approach.

At the same time, the group of Harris was able to synthesize β-hexaketone **82** by double acylation of acetylacetone with the aryl β-ketoester **83** (Scheme 13.31) [66].

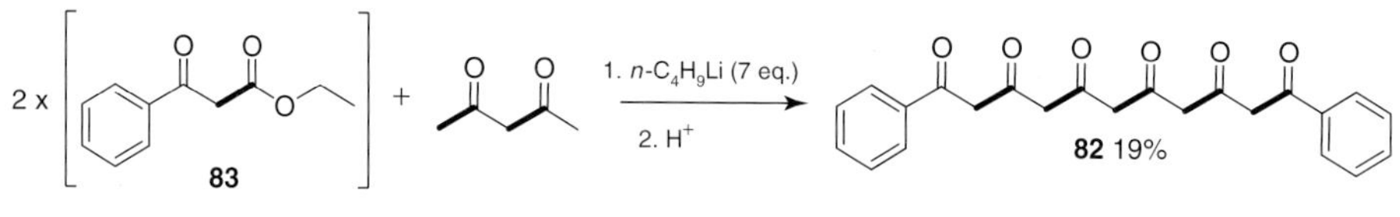

Scheme 13.31 Preparation of β-hexaketones by polyanion acylation with β-ketoester monoanions.

Cyclization of hexaketide **84** could give rise to four naphthyl derivatives, but the reaction appeared to be highly regioselective. Depending on the basic or acidic conditions, naphthalenoid region-isomers **85** or **86** were the major products (Scheme 13.32).

The approach of Harris culminated into the biomimetic total syntheses of 6-hydroxymusizin (**87**), as well as the related heterocyclic metabolites barakol (**88**)

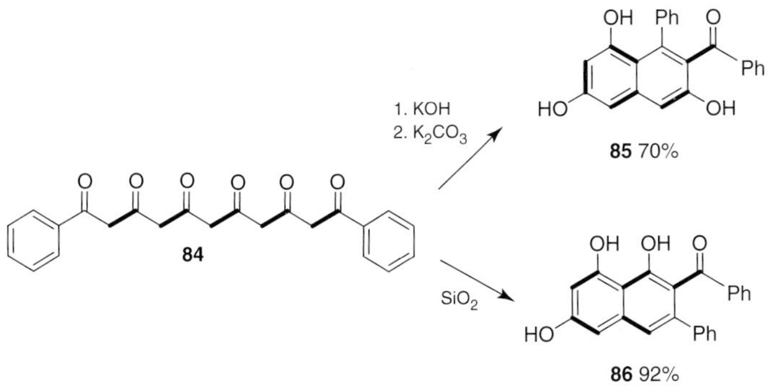

Scheme 13.32 Regioselectivity in the naphthyl cyclization of β-hexaketones.

and eleutherinol (**89**), using terminally protected hexaketone **90** and heptaketone **91** (Scheme 13.33) [67].

The mild conditions of all these reactions are of particular interest, confirming that biomimetic approaches could be useful methods by which to access natural products in good yields.

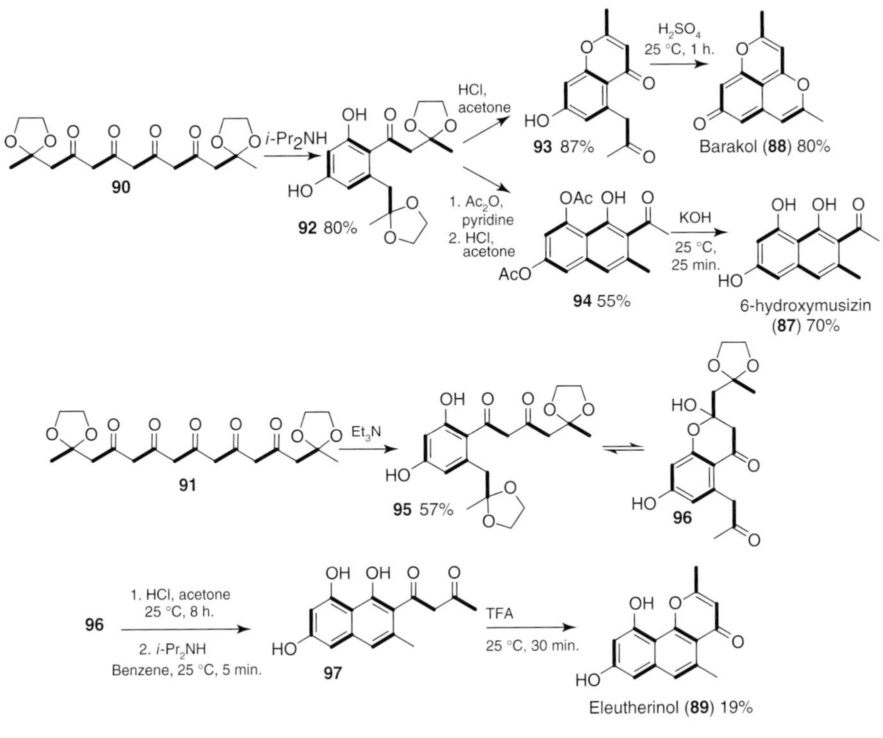

Scheme 13.33 Harris' biomimetic syntheses of 6-hydroxymusizin (**87**), barakol (**88**), and eleutherinol (**89**).

The group of Yamaguchi addressed this issue by a closely similar approach, accessing modified β-polyketide diester from the dianion of methyl acetoacetate (**98**) and hydroxyglutarate diester **99** (Weiler reaction) [68]. As an example, 1,8-naphthalenediol derivative **100** was obtained in good yield after cyclization of **101** in the presence of Ca(II) salts (Scheme 13.34) [69].

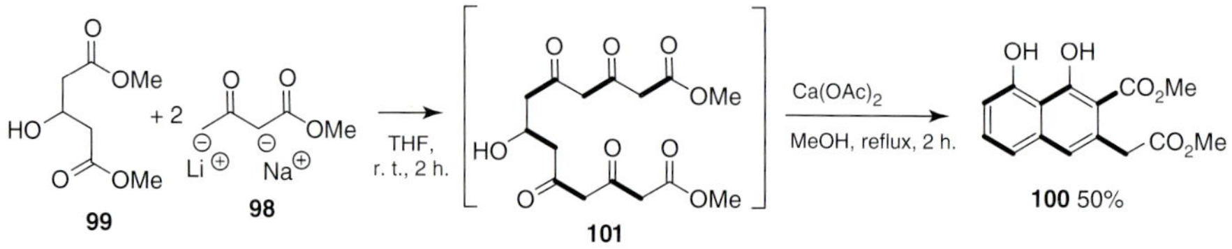

Scheme 13.34 Yamaguchi's aromatic cyclization of polyketides into naphthalenoids.

An interesting application of this methodology led to the biomimetic synthesis of the naphthopyran polyketide (±)-nanaomycin A (**102**) (Scheme 13.35) [70]. An additional acetyl was introduced by a Claisen condensation between **100** and AcO*t*Bu to yield **103**, which underwent a reduction–lactonization–protection sequence leading to **104**. Transformation of the lactone into the methyl ketone **105** was followed by deprotection and oxidation to give **102**.

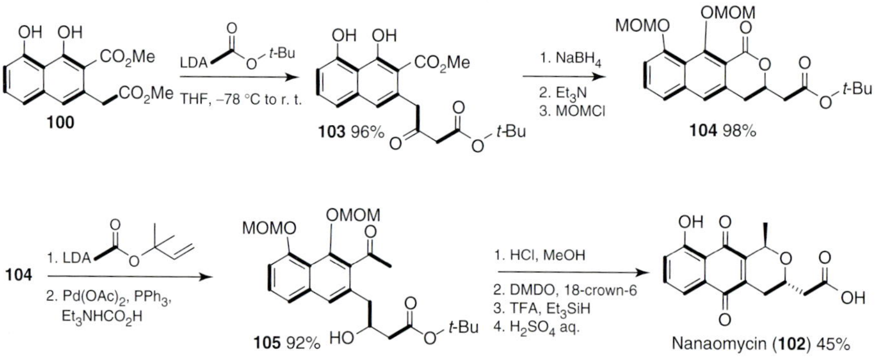

Scheme 13.35 Yamaguchi's biomimetic synthesis of nanaomycine (**102**).

These biomimetic approaches have been further extended to the synthesis of important polycyclic aromatic polyketides, some of which exhibit high potential as antimicrobial compounds.

13.3.3
Biomimetic Access to Anthracenoid Derivatives

For longer and regular β-polyketones, control of the regioselectivity of the cyclization proved to be highly challenging and modification of the protection or modification of some carbonyl moieties was found to be necessary. The group of Harris generalized their approach with terminally protected heptapolyketones such as **106** to perform the syntheses of the anthracenoids emodin (**107**) and chrysophanol (**52**) using an alternative approach for the synthesis of precursor **109** (Scheme 13.36) [71]. All these results could now been interpreted in the light of the biosynthetic considerations developed by the group of Thomas and Bringmann (Section 13.3).

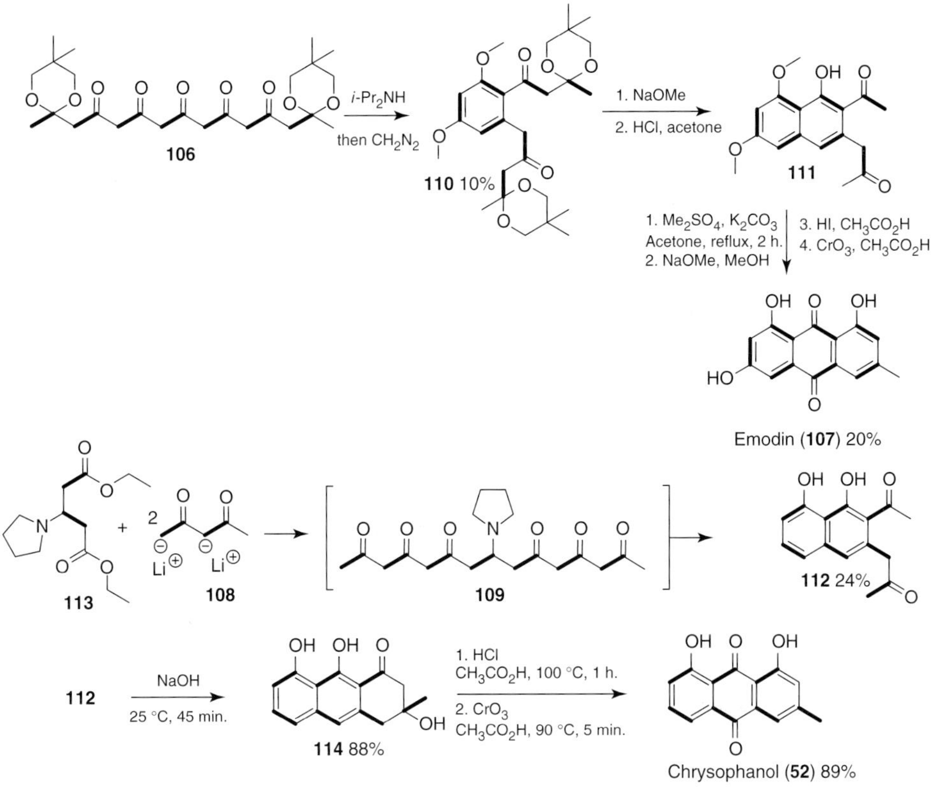

Scheme 13.36 Harris' biomimetic syntheses of emodin (**107**) and chrysophanol (**108**).

The biomimetic formation of the key aromatic intermediates **111** and **112** was followed by additional steps using conventional chemistry to achieve the biomimetic total syntheses. In both cases, the quinone functionality was introduced in the last step of the syntheses upon CrO_3 oxidation of the central phenol.

13.3.4
Biomimetic Access to Tetracyclic Derivatives

13.3.4.1 Biomimetic Access to Tetracenoid Derivatives

Tetracyclines have been recognized as very important antimicrobials and the group of Harris applied their methodology to the synthesis of pretetramide (**115**) and its methyl derivative, which are key biosynthetic intermediates toward

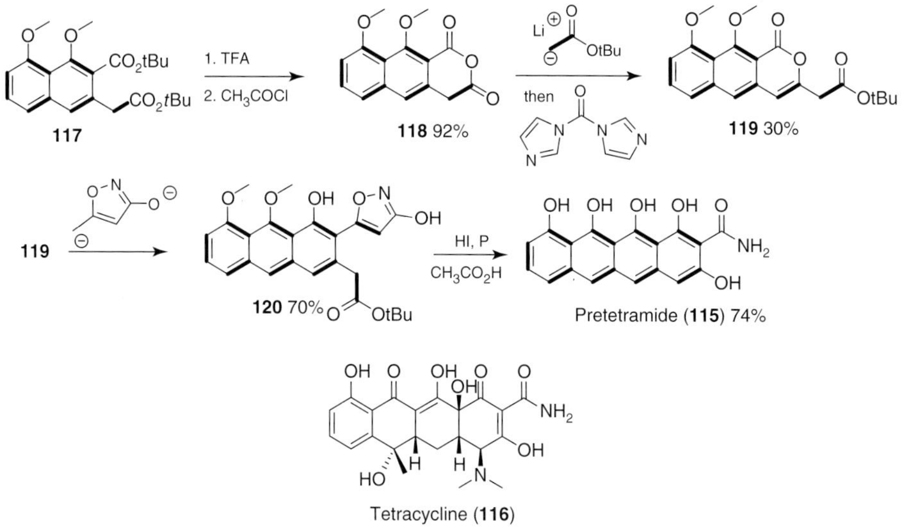

Scheme 13.37 Harris' biomimetic synthesis of pretetramide (**115**).

tetracycline (**116**) (Scheme 13.37) [72]. They started from naphthenoid **117**, a close analog of previously used glutarates whose reactivity was enhanced by using the anhydride form **118** obtained after dealkylative cyclization promoted by trifluoroacetic acid (TFA). A first addition of AcO*t*Bu anion and subsequent dehydration was followed by the addition of the dianion of hydroxymethylisoxazole as a β-ketoamide equivalent. Acidic cyclization in the presence of phosphorus led to the desired pretetramide (**115**). The biosynthesis of pretetramide, featuring an amide function, which is unusual in the polyketides realm, involves the initial co-condensation of nine malonylCoA units. The elongation is followed by the action of an amidotransferase, introducing the amide function, to yield an enzyme-bound polyoxoamide further sequentially cyclized by several enzymes [73].

13.3.4.2 Biomimetic Access to Tetraphenoid Derivatives

An extension of the previous work of the group of Yamaguchi led to the biomimetic synthesis of the anthracenoid intermediate **121** starting from the diester **122** (Scheme 13.38). The use of aromatic diesters appeared as a simple procedure, based on Ca-promoted cyclizations, to allow the synthesis of the functionalized anthracenoid **121** following acetylation of the hydroxyl groups to obtain a stable product.

After the partial synthesis of important naphthoquinones [74], one of the most prominent landmark accomplishments of this methodology led to the biomimetic synthesis of (−)-urdamycinone B (**123**), a potent antibiotic produced by *Streptomyces fradie* (Scheme 13.39) [75]. Access to the naphthol derivative **124** was performed following the same cyclization by Ca(II) salts. An additional acetyl unit was

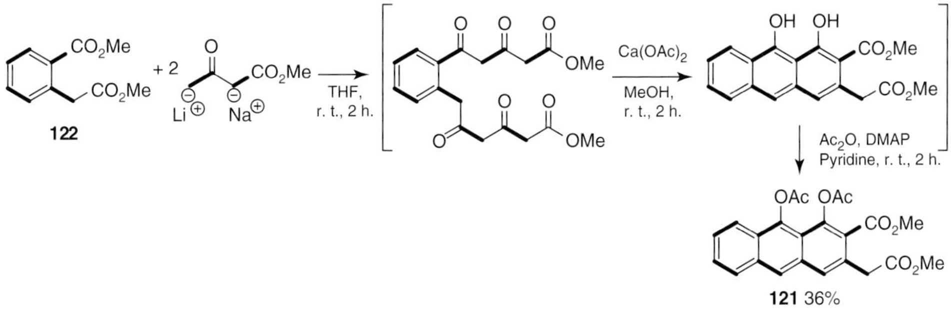

13.3 Biomimetic Access to Aromatic Rings | 497

Scheme 13.38 Yamaguchi's access to anthracenoids.

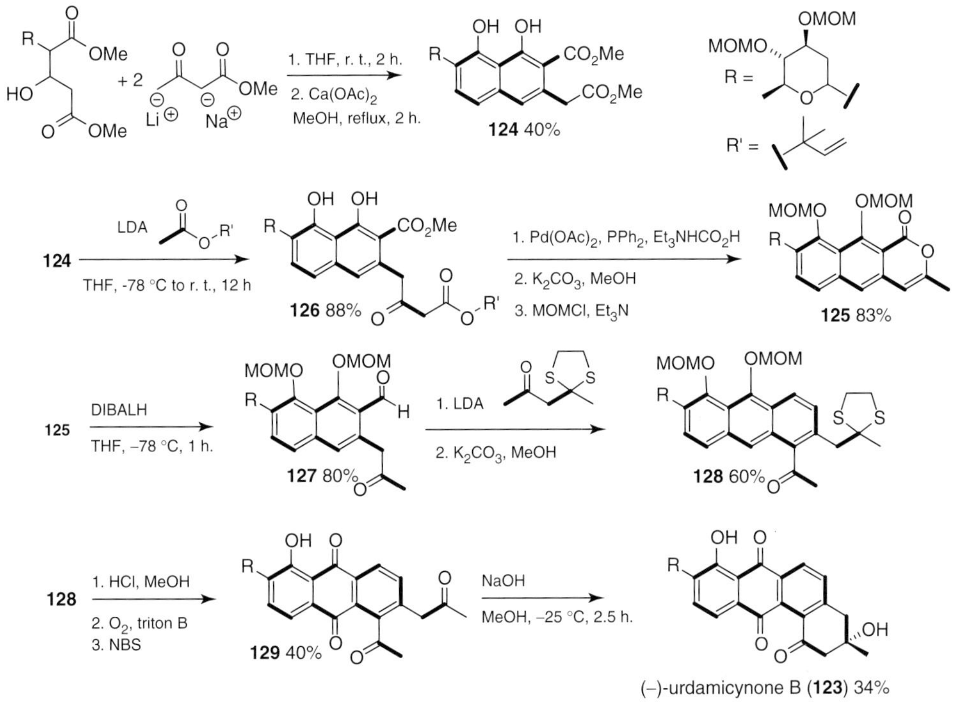

Scheme 13.39 Yamaguchi's biomimetic synthesis of (−)-urdamicynone (**123**).

then added by Claisen condensation of **124** with dimethylallyl acetate in the presence of LDA. Deallyloxycarbonylation catalyzed by a Pd(II) complex followed by lactonization and methoxymethyl (MOM)-protection of the hydroxyl groups yielded **125** in 83% from **126**. Controlled reduction of the enol lactone moiety of **125** by diisobutylaluminum hydride (DIBALH) led to the reactive aldehyde **127**, which was submitted to a double inter/intra aldolization reaction leading to the protected diketone **128**. Deprotection and oxidation by molecular oxygen in the presence of triton B afforded the quinone intermediate **129** in 40% yield, before a final base-induced aldol annelation yielded the target (−)-urdamicynone (**123**).

Angucycline antibiotics were also the main targets of Krohn and coworkers. An alternative biomimetic approach based on a late biomimetic cyclization was developed to synthesize these compounds in good yields [76]. For example, they used dibromonaphthoquinone **130** as the starting material to synthesize tetrangomycin (**131**) by a combination of biomimetic and conventional chemistries (Scheme 13.40) [77].

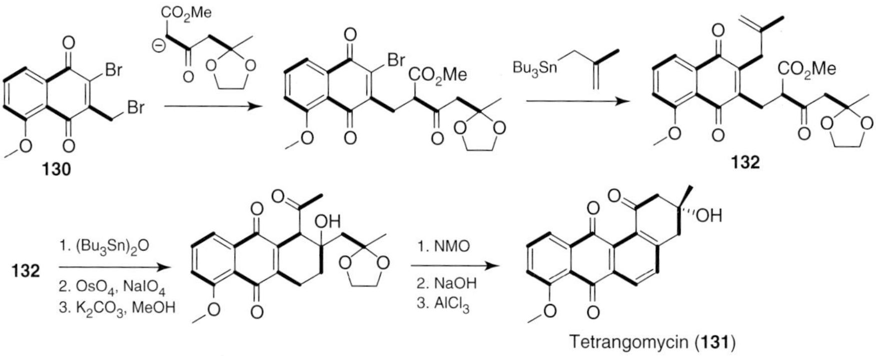

Scheme 13.40 Krohn's biomimetic synthesis of tetrangomycin (**131**).

The same group was able to perform the syntheses of the closely related angucyclinone, aquayamycin, and WP 3688-2 using a SmI_2-mediated cyclization of ketides [78, 79].

13.3.4.3 Biomimetic Access to Benzo[*a*]tetracenoid Derivatives

The last step towards complex aromatic polyketides was access to benzo[*a*]tetracenoid derivatives, which necessitated dodecaketide equivalents. The group of Krohn undertook synthetic studies to construct these very important antibiotics, which culminated in the biomimetic synthesis of 5,6-dideoxypradione (**133**) closely, which is related to pradimicin A (**134**) (Scheme 13.41) [80]. While the synthesis of the anthracenequinone **135** began with a Diels–Alder reaction between the appropriate naphthoquinone and diene, construction of the remaining bicycle was performed by a biomimetic approach using very mild conditions.

Beyond the scope of this chapter, but worth noting for those interested in such phenolic compounds, is the biomimetic synthesis of aquaticol from hydroxycuparene, a C15 phenolic derivative, through phenol dearomatization and dimerization by Diels–Alder cycloaddition by the group of Quideau [81].

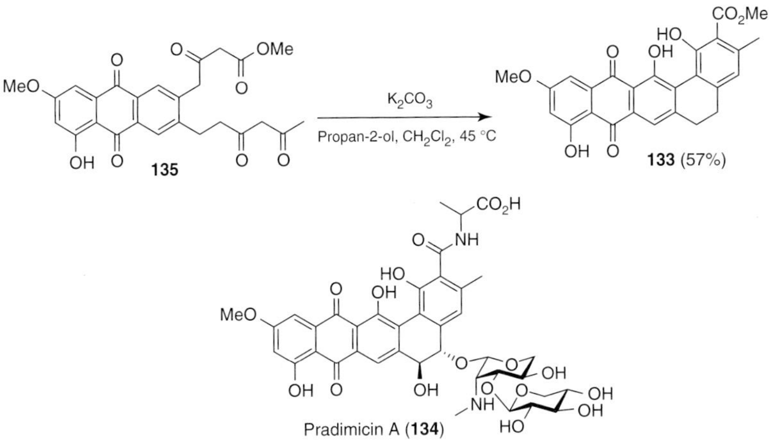

Scheme 13.41 Krohn's biomimetic synthesis of **133**.

13.4
Conclusion

Throughout this chapter we have demonstrated the parallel advances in the biochemical understanding of the polyketide biosynthetic pathways and the development of powerful chemical methods aimed at mimicking these pathways. We first noticed that pioneering work begun in the nineteenth century. But is this field still alive and what are the prospects of biomimetic synthesis in the field of polyketides?

Two distinct answers can be given to these questions. There is renewed interest in the search for polyketide assembly mimics, as evidenced by recent synthetic methodologies developed in this field. Our increasing knowledge of the enzymatic active sites is now opening the way for new powerful asymmetric organocatalytic methods for the construction of key C–C connections. In terms of the second question, the biological pathways inspired a flurry of biomimetic total syntheses of complex aromatic polyketides and the pioneering work is now used by several groups of chemists worldwide. Tetra- and pentacyclic highly bioactive substances are now accessible by straightforward routes mimicking the cyclization/aromatization biosynthetic strategies. Very powerful chemical routes, inspired by the functioning of biological systems optimized during millions of years of evolution, have today given rise to renewed interest in the context of a synthetic chemistry at the heart of sustainable development. Without doubt we have not yet reached the apogee of biomimetic chemistry but we may have just passed its infancy.

References

1. Rohr, J. (2000) *Angew. Chem. Int. Ed.*, **39**, 2847–2849.
2. Smith, S. and Tsai, S.-C. (2007) *Nat. Prod. Rep.*, **24**, 1041–1072.

3. Hill, A.M. (2006) *Nat. Prod. Rep.*, **23**, 256–320.
4. Hertweck, C. (2009) *Angew. Chem. Int. Ed.*, **48**, 4688–4716.
5. Staunton, J. and Weissman, K.J. (2001) *Nat. Prod. Rep.*, **18**, 380–416.
6. Whiting, D. (2001) *Nat. Prod. Rep.*, **18**, 583–606.
7. Collie, J.N. and Myers, W.S. (1893) *J. Chem. Soc.*, **63**, 122–128.
8. Birch, A.J., Massy-Westropp, P.A., and Moye, C.J. (1955) *Aust. J. Chem.*, **8**, 539–544.
9. Claisen, L. and Claparede, A. (1881) *Ber. Deut. Chem. Ges.*, **14**, 2460–2468.
10. Zempleni, J., Wijeratne, S.S., and Hassan, Y.I. (2009) *Biofactors*, **35**, 36–46.
11. Lynen, F. (1967) *Biochem. J.*, **102**, 381–400.
12. Scott, A.I., Wiesner, C.J., Yoo, S., and Chung, S.-K. (1975) *J. Am. Chem. Soc.*, **97**, 6277–6278.
13. Kobuke, Y. and Yoshida, J. (1978) *Tetrahedron Lett.*, **19**, 367–370.
14. Brooks, D.W., Lu, L.D.-L., and Masamune, S. (1979) *Angew. Chem. Int. Ed.*, **18**, 72–74.
15. Sakai, N., Sordé, N., and Matile, S. (2001) *Molecules*, **6**, 845–851.
16. Chen, H. and Harrison, P.H. (2002) *Can. J. Chem.*, **80**, 601–607.
17. Sun, S. and Harisson, P.H. (1992) *Tetrahedron Lett.*, **33**, 7715–7718.
18. Sun, S. and Harisson, P.H. (1994) *J. Chem. Soc., Chem. Commun.*, 2235–2236.
19. Sun, S., Edwards, L., and Harisson, P.H. (1998) *J. Chem. Soc., Perkin Trans. 1*, 437–448.
20. Chen, M., Won, K., McDonald, R.S., and Harisson, P.H. (2006) *Can. J. Chem.*, **84**, 1188–1196.
21. (a) Carreira, E.M. (1999) in *Comprehensive Asymmetric Catalysis*, vol. 3 (eds E.N. Jacobsen, A. Pfaltz, and H. Yamamoto), Springer, Heidelberg, pp. 997–1065; (b) Mahrwald, R. (2004) *Modern Aldol Reactions*, vol. 2, Wiley-VCH Verlag GmbH, Weinheim.
22. Lalic, G., Aloise, A.D., and Shair, M.D. (2003) *J. Am. Chem. Soc.*, **125**, 2852–2853.
23. Orlandi, S., Benaglia, M., and Cozzi, F. (2004) *Tetrahedron Lett.*, **45**, 1747–1749.
24. Magdziak, D., Lalic, G., Lee, H.M., Fortner, K.C., Aloise, A.D., and Shair, M.D. (2005) *J. Am. Chem. Soc.*, **127**, 7284–7285.
25. Fortner, K.C. and Shair, M.D. (2007) *J. Am. Chem. Soc.*, **129**, 1032–1033.
26. Davies, C., Heath, R.J., White, S.W., and Rock, C.O. (2000) *Structure*, **8**, 185–195.
27. Blaquiere, N., Shore, D.G., Rousseaux, S., and Fagnou, K. (2009) *J. Org. Chem.*, **74**, 6190–6198.
28. Zhang, Y.-M., Hurlbert, J., White, S.W., and Rock, C.O. (2006) *J. Biol. Chem.*, **281**, 17390–17399.
29. Ricci, A., Pettersen, D., Bernardi, L., Fini, F., Fochi, M., Perez Herrera, R., and Sgarzani, V. (2007) *Adv. Synth. Catal.*, **349**, 1037–1040.
30. Lubkoll, J. and Wennemers, H. (2007) *Angew. Chem. Int. Ed.*, **46**, 6841–6844.
31. Benaglia, M., Cinquini, M., and Cozzi, F. (2000) *Eur. J. Org. Chem.*, 563–572.
32. Yost, J.M., Zhou, G., and Coltart, D.M. (2006) *Org. Lett.*, **8**, 1503–1506.
33. Lim, D., Fang, F., Zhou, G., and Coltart, D.M. (2007) *Org. Lett.*, **9**, 4139–4142.
34. Kohler, M.C., Yost, J.M., Garnsey, M.R., and Coltart, D.M. (2010) *Org. Lett.*, **12**, 3376–3379.
35. Alonso, D.A., Kitagaki, S., Utsumi, N., and Barbas, C.F. III (2008) *Angew. Chem. Int. Ed.*, **47**, 4588–4591.
36. Utsumi, N., Kitagaki, S., and Barbas, C.F. III (2008) *Org. Lett.*, **10**, 3405–3408.
37. Le Sann, C., Munoz, D.M., Saunders, N., Simpson, T.J., Smith, D.I., Soulas, F., Watts, P., and Willis, C.L. (2005) *Org. Biomol. Chem.*, **3**, 1719–1728.
38. Jones, G. (1967) in *Organic Reactions*, vol. 15, John Wiley & Sons, Inc., New York, pp. 204–599.
39. Augustine, J.K., Naik, Y.A., Poojari, S., Chowdappa, N., Sherigara, B.S., and Areppa, K. (2009) *Synthesis*, 2349–2356.
40. Lopez Herrera, F.J. and Pino Gonzalez, M.S. (1986) *Carbohydr. Res.*, **152**, 283–291.
41. Peng, Y. and Song, G. (2003) *Green Chem.*, **5**, 704–706.

42. List, B., Doehring, A., Hechavarria Fonseca, M.T., Wobser, K., van Thienen, H., Rios Torres, R., and Llamas Galilea, P. (2005) *Adv. Synth. Cat.*, **347**, 1558–1560.
43. Berrue, F., Antoniotti, S., Thomas, O.P., and Amade, P. (2007) *Eur. J. Org. Chem.*, 1743–1748.
44. Hilt, G. and Weske, D.F. (2009) *Chem. Soc. Rev.*, **38**, 3082–3091.
45. Birch, A.J., Massy-Westropp, P.A., and Moye, C.J. (1955) *Aust. J. Chem.*, **8**, 539–544.
46. Robinson, R. (1955) *The Structural Relations of Natural Products*, Clarendon Press, Oxford.
47. Birch, A.J. and Donovan, F.W. (1953) *Aust. J. Chem.*, **6**, 360–368.
48. Thomas, R. (2001) *ChemBioChem*, **2**, 612–627.
49. Bringmann, G., Wohlfarth, M., Rischer, H., Rückert, M., and Schlauer, J. (2000) *Angew. Chem. Int. Ed.*, **39**, 1464–1466.
50. Bringmann, G., Noll, T.F., Gulder, T.A.M., Grüne, M., Dreyer, M., Wilde, C., Pankewitz, F., Hilker, M., Payne, G.D., Jones, A.L., Goodfellow, M., and Fiedler, H.-P. (2006) *Nat. Chem. Biol.*, **2**, 429–433.
51. Bringmann, G., Gulder, T.A.M., Hamm, A., Goodfellow, M., and Fiedler, H.-P. (2009) *Chem. Commun.*, 6810–6812.
52. Reviews: (a) Harris, T.M. and Harris, C.M. (1977) *Tetrahedron*, **33**, 2159–2185; (b) Harris, T.M. and Harris, C.M. (1986) *Pure Appl. Chem.*, **58**, 283–294.
53. Harris, T.M. and Harris, C.M. (1969) *Tetrahedron*, **25**, 2687–2691.
54. Birch, A.J., Cameron, D.W., and Richards, R.W. (1960) *J. Chem. Soc.*, 4395–4400.
55. Harris, T.M., Murphy, G.P., and Poje, A.J. (1976) *J. Am. Chem. Soc.*, **98**, 7733–7741.
56. (a) Harris, T.M. and Carnay, R.L. (1966) *J. Am. Chem. Soc.*, **88**, 5686–5687; (b) Harris, T.M. and Carney, R.L. (1967) *J. Am. Chem. Soc.*, **89**, 6734–6740.
57. Money, T. (1970) *Chem. Rev.*, **70**, 553–560.
58. Crombie, L. and James, A.W.G. (1966) *Chem. Commun.*, 357–359.
59. Navarro, I., Basset, J.-F., Hebbe, S., Major, S.M., Werner, T., Howsham, C., Bräckow, J., and Barrett, A.G.M. (2008) *J. Am. Chem. Soc.*, **130**, 10293–10298.
60. Barrett, A.G.M., Morris, T.M., and Barton, D.H.R.J. (1980) *J. Chem. Soc., Perkin Trans. 1*, **10**, 2272–2277.
61. Basset, J.-F., Leslie, C., Hamprecht, D., White, A.J.P., and Barrett, A.G.M. (2010) *Tetrahedron Lett.*, **51**, 783–785.
62. Harris, T.M. and Murphy, G.P. (1971) *J. Am. Chem. Soc.*, **93**, 6708–6709.
63. Comer, F.W., Money, T., and Scott, A.I. (1967) *Chem. Commun.*, 231–233.
64. Stockinger, H. and Schmidt, U. (1976) *Liebigs Ann. Chem.*, 1617–1625.
65. Baker, P.M. and Bycroft, B.W. (1968) *Chem. Commun.*, 71–72.
66. Wittek, P.J. and Harris, T.M. (1973) *J. Am. Chem. Soc.*, **95**, 6865–6867.
67. Harris, T.M. and Wittek, P.J. (1975) *J. Am. Chem. Soc.*, **97**, 3270–3271.
68. Review: Yamaguchi, M. (1992) in *Studies in Natural Products Chemistry*, vol. 11 (ed. A.U. Rahman), Elsevier Science BV, Amsterdam, pp. 113–149.
69. Yamaguchi, M., Hasebe, K., and Minami, T. (1986) *Tetrahedron Lett.*, **27**, 2401–2404.
70. Yamaguchi, M., Nakamura, S., Okuma, T., and Minami, T. (1990) *Tetrahedron Lett.*, **31**, 3913–3916.
71. Harris, T.M., Webb, A.D., Harris, C.M., Wittek, P.J., and Murray, T.P. (1976) *J. Am. Chem. Soc.*, **98**, 6065–6067.
72. (a) Gilbreath, S.G., Harris, C.M., and Harris, T.M. (1988) *J. Am. Chem. Soc.*, **110**, 6172–6179; (b) Harris, T.M., Harris, C.M., Oster, T.A., Brown L.E. Jr., and Lee, J.Y.-C. (1988) *J. Am. Chem. Soc.*, **110**, 6180–6186; (c) Harris, T.M., Harris, C.M., Kuzma, P.C., Lee, J.Y.-C., Mahalingam, S., and Gilbreath, S.G. (1988) *J. Am. Chem. Soc.*, **110**, 6186–6192.
73. Pickens, L.B., Kim, W., Wang, P., Zhou Watanabe, K., Gomi, S., and Tang, T. (2009) *J. Am. Chem. Soc.*, **131**, 17677–17689.
74. Yamaguchi, M., Hasebe, K., Higashi, H., Uchida, M., Irie, A., and Minami, T. (1990) *J. Org. Chem.*, **55**, 1611–1623.

75. Yamaguchi, M., Okuma, T., Horiguchi, A., Ikeura, C., and Minami, T. (1992) *J. Org. Chem.*, **57**, 1647–1649.
76. Review: Krohn, K. (2002) *Eur. J. Org. Chem.*, 1351–1362.
77. Krohn, K., Boker, N., Florke, U., and Freund, C. (1997) *J. Org. Chem.*, **62**, 2350–2356.
78. Krohn, K., Frese, P., and Florke, U. (2000) *Chem. Eur. J.*, **6**, 3887–3896.
79. Krohn, K., Vidal, A., Tran-Thien, H.T., Florke, U., Bechthold, A., Dujardin, G., and Green, I. (2010) *Eur. J. Org. Chem.*, 3080–3092.
80. (a) Krohn, K. and Bernhard, S. (1999) *Eur. J. Org. Chem.*, 3099–3103; (b) Krohn, K., Bernhardt, S., Florke, U., and Hayat, N. (2000) *J. Org. Chem.*, **65**, 3218–3222.
81. Gagnepain, J., Castet, F., and Quideau, S. (2007) *Angew. Chem. Int. Ed.*, **46**, 1533–1535.

14
Biomimetic Synthesis of Non-Aromatic Polycyclic Polyketides
Bastien Nay and Nassima Riache

14.1
Introduction

The biosynthesis of polyketides is extraordinarily orchestrated by proteins called polyketide synthases (PKSs) which are always highly organized modular or iterative enzymatic complexes [1]. They are analogous to the fatty acid synthases that catalyze the decarboxylation of the malonate precursor to drive the elongation of acyl chains. Depending on the functional attributes of the PKS enzymatic complex, the elongated carbon chain will be more or less functionalized, with a tuned reactivity for specific rearrangements, especially cyclizations, toward the natural product.

The "minimal PKS" is made of an acyl carrier protein (ACP), an acyl transferase (AT), and a ketosynthase (KS), which lead to highly reactive poly-β-ketoacyl chains. When the chain reaches the appropriate length (this can be controlled by a chain length factor) it undergoes Claisen condensations to furnish phenolic compounds such as actinorhodin, tetracycline, or doxorubicin [1c, 2]. Additional modules can be involved within the PKS, catalyzing ketone reduction, dehydration, and enoyl reduction. In such a complete PKS, the carbon chain is saturated, as in fatty acids. These additional modules are optional and can be missing, which means that the functionalization can stop at the β-hydroxy or the enoyl stage, thus increasing the variability of functional groups within the elongated chain. In many cases, the linear product of the PKS will be able to undergo complex rearrangement towards polycyclic structures. This is the biosynthetic path for numerous polyketide natural products, some of which are described in this chapter.

The science of synthesis was soon interested in polyketide complexity. It has been obvious to many authors that mimicking the biosynthetic route would provide efficient access to this complexity. The biomimetic synthesis of polyketides embraces a wide range of organic reactions. One of the first biomimetic syntheses of aromatic polyketides was reported by Collie and Myers in 1893 [3], consisting of the condensation of a trienone into the phenolic compound orcinol. Since then, this strategy has been used to synthesize numerous polyaromatic natural products [4]. To obtain the poly-β-ketoacyl precursors, some authors embarked

Biomimetic Organic Synthesis, First Edition. Edited by Erwan Poupon and Bastien Nay.
© 2011 Wiley-VCH Verlag GmbH & Co. KGaA. Published 2011 by Wiley-VCH Verlag GmbH & Co. KGaA.

on the biomimetic decarboxylative condensation of malonate onto acetate units. Especially successful attempts were realized by use of enzyme mimics [5].[1]

Studies of the biomimetic synthesis of non-aromatic polycyclic polyketides have grown impressively over the past 20 years. This has brought important keys to the comprehension of biosynthetic steps and of the reactivity of biosynthetic intermediates. Rapid and efficient synthetic methods have been collected for the construction of biologically interesting polyketides. In this chapter, we give a comprehensive overview of those works, describing significant examples with variable complexity and reactivity.

14.2
Biomimetic Studies in the Nonadride Series

14.2.1
Dimerization Process towards Isoglaucanic Acid

Glaucanic acid (**2**) is a bis(anhydride) derived from the formal $[6\pi+4\pi]$ cycloaddition of two C_9-units (**1**) (Scheme 14.1). It was first isolated in 1931 from *Penicillium glaucum* [6]. With regard of its C_9 structural features the name nonadride was suggested by Sutherland and coworkers [7, 8]. The biosynthesis of **2** was studied by Sutherland, who demonstrated the acetate origin of glaucanic acids [9]. When the synthetic [^{3}H]-labeled C_9-precursor **1** was fed to the growing mould *Penicillium purpurogenum*, another nonadride producer, 51% of the [^{3}H]-activity was incorporated into glauconic acid (**4**). The mould was thus capable of effecting the dimerization of **1** into **2** and **4**. There were numerous possible mechanisms for this condensation, but an attractive one was the $[6\pi+4\pi]$ cycloaddition of the anion of **1** with the cisoid form of the same anhydride to give **2**. The reaction via an *endo* transition state would give the correct relative stereochemistry. This dimerization was attempted *in vitro* by treating anhydride **1** with triethylamine in dimethylformamide (DMF) [9b, 10]. Yet it only gave 4% of the epimeric *exo* product, iso-glaucanic acid (**3**). No other bis(anhydrides) of this type were formed during the reaction. The difference between the biochemical and chemical routes leading to **2** and **3**, respectively, was

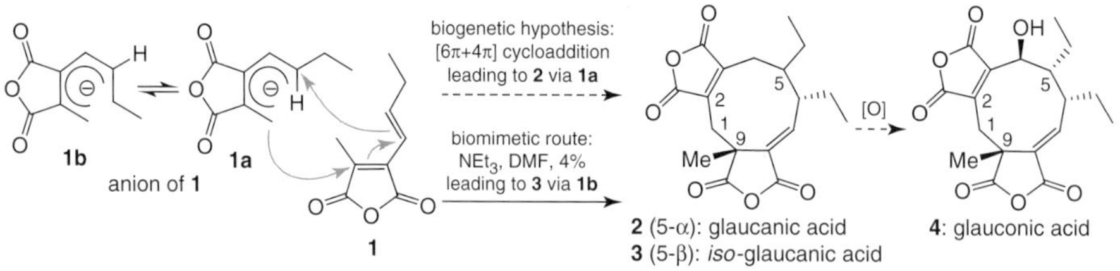

Scheme 14.1 Nonadride biogenetic hypothesis and Sutherland's biomimetic studies.

---

1) For a comprehensive discussion on polyketide enzyme mimics and on the biomimetic synthesis of aromatic polyketides, see Chapter 13.

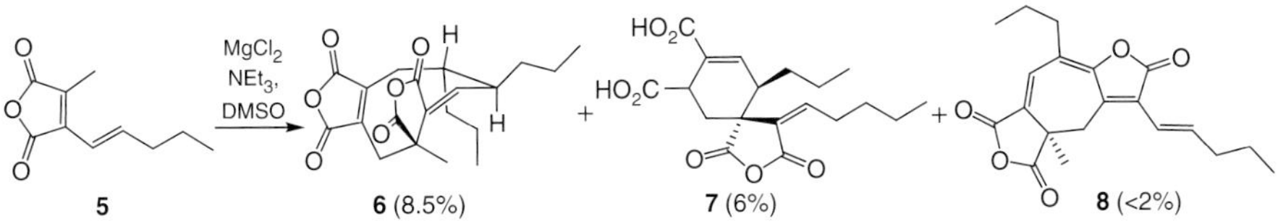

Scheme 14.2 Dimerization route toward the nonadride **6** and side-products **7** and **8**.

explained by postulation, in the latter case, of an isomerization of the anion **1a** to the anion **1b**, which would then attack the neutral form **1**. This isomerization would not happen *in vivo*.

More recently the mechanism of this reaction was reinvestigated by Baldwin, who suggested a stepwise dimerization of the anhydride unit (Scheme 14.2) rather than a [6π+4π] cycloaddition [11]. After testing different reaction conditions (solvent, temperature, base, addition of a metal salt) to optimize the dimerization of the *homo*-derivative **5**, it was found that a mixture dimethyl sulfoxide (DMSO)/NEt$_3$ (0.66 equiv.)/MgCl$_2$ (0.5 equiv.) gave the best yield (8.5%) of iso-glaucanic acid analog **6**. It was accompanied with two regioisomers of cyclization (**7** and **8**), the presence of which suggested a stepwise mechanism via Michael addition for the formation of **6**. Tethering both anhydride units through their lateral alkenyl chain, giving dimeric bis(anhydrides) with various tether lengths, allowed for 1,8-diazabicyclo[5.4.0]undec-7-ene (DBU)-catalyzed intramolecular cyclizations that occurred in better yields (14–17%) and with variable stereoselectivities at the cyclononene substitutions [12].

14.2.2
The Unresolved Case of CP-225917 and CP-263114

The CP-225917 and CP-263114 molecules, or phomoidrides A and B (**10**, **11**), have been isolated from an unidentified fungus (Scheme 14.3) [13]. They are oxygenated and bridged nonadrides with inhibitory activities against squalene synthase and *ras*-farnesyltransferase, thus providing lead structures for the development of anticancer and cholesterol-lowering agents. Compared to glaucanic acid (**2**), these compounds share longer alkyl chains and a trans double bond embedded within the common cyclononane ring, which is also bridged by a new C–C bond between C10 and C26. New structural challenges thus arise with regard to a biomimetic strategy. Biogenetically, **10** would be formed by the controlled dimerization and functionalization of anhydride **9**.

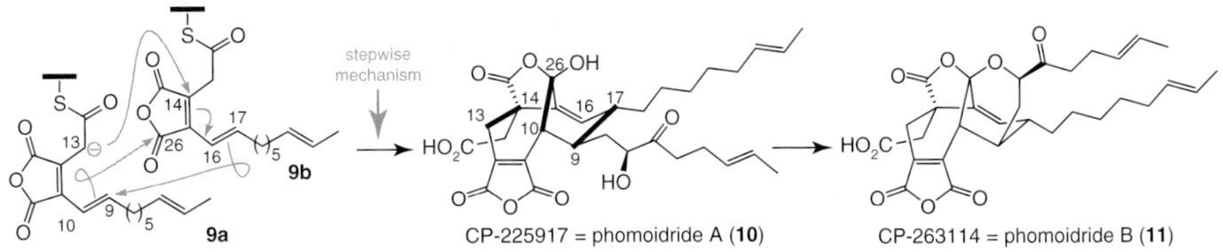

Scheme 14.3 Biosynthetic origin of the CP-225917 (**10**) and CP-263114 (**11**).

With regard to polyketide biosyntheses, Sulikowski and coworkers speculated that both monomers (**9**) were covalently attached to an enzyme active site (thioester linkages) during the dimerization process [14]. This would serve to govern the topology of the reaction toward the cyclononene ring through a stepwise mechanism involving two Michael additions between the anhydride units **9** (C13 → C14, C17 → C9). Conformational restrictions (*exo* topology) for the cyclization would be imposed by the thioester linkages. Then, a transannular Dieckmann cyclization (C10 → C26) and a decarboxylation at C13 [15] would lead to the CP–225917 skeleton **10**.

Sulikowski attempted to use this biosynthetic hypothesis in a biomimetic strategy for the synthesis of **10** and **11** [16]. The thioester templating effect was mimicked by the use of 1,n-diols with variable chain lengths ($n = 1-6$) to connect both anhydrides units through their acetyl part. Unfortunately, when basic cyclization conditions were applied to these dimeric anhydrides (DBU, MeCN, 80 °C, 0.001 M), the desired carbon connection C13–C14 was not observed [16a] – instead the C13–C17 bond was formed. Condensation derivatives were described for longer tethers ($n = 3-6$). The biomimetic synthesis of phomoidrides thus remains unresolved.

14.3
Biomimetic Syntheses Involving the Diels–Alder Reaction

First reported in 1928, the Diels–Alder reaction [17] is, to date, one of the mostly used methodologies in organic synthesis and often constitutes a key step in the total synthesis of natural products [18], among which polyketides are far from being an exception. Additionally, even though few "Diels–Alderases" have yet been identified, the enzymatic Diels–Alder reaction stands as the biogenetic hypothesis of numerous polyketides (see also Chapter 21) [19]. Chemists have attempted to demonstrate the relevance of biosynthetic Diels–Alder reactions in numerous biomimetic syntheses of natural products, some of which are described in the following section.

14.3.1
Biomimetic Diels–Alder Reactions Affording Decalin Systems

Lovastatin (**14**) (= mevinolin, Scheme 14.4) was isolated from a culture of *Aspergillus terreus* [20] and from *Monascus ruber* [21]. It is a potent inhibitor of cholesterol biosynthesis in humans, after hydrolysis of the lactone into the active β-hydroxy-acid form. The biosynthesis of **14** originates from the polyketide metabolism, through the triene (**12**), which undergoes an *endo*-selective Diels–Alder addition to form the decalin core **13** [22]. It was then suggested that this cycloaddition would be driven by a biological process that involves a Diels–Alderase [23].[2]

---

2) For a discussion on Diels–Alderases see Chapter 21.

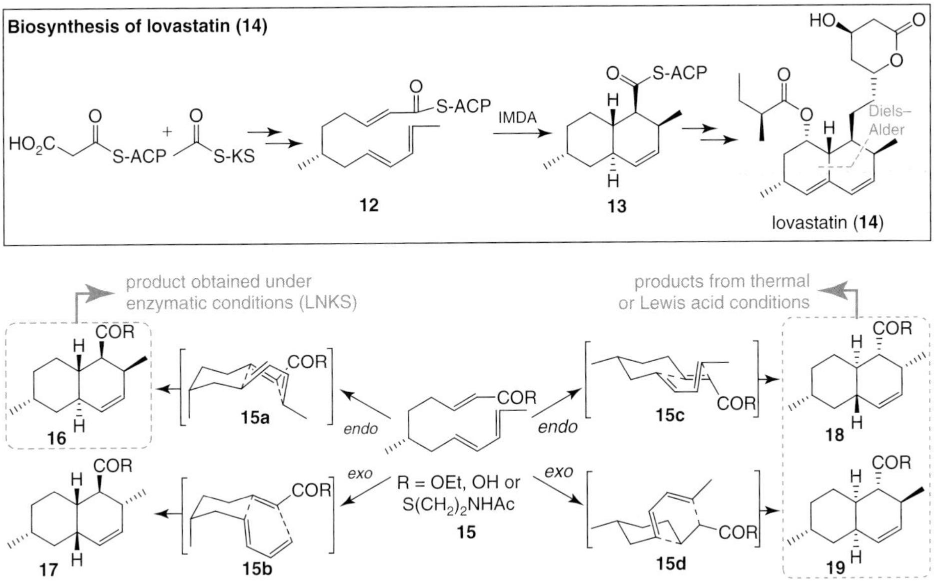

14.3 Biomimetic Syntheses Involving the Diels–Alder Reaction

Biosynthesis of lovastatin (14)

Scheme 14.4 Diels–Alder cyclization of trienes **15**.

Vederas and coworkers studied the relevance of this biosynthetic Diels–Alder cyclization by testing the reactivity of various synthetic trienes (**15**) (Scheme 14.4) under thermal or Lewis acid conditions [24]. It only resulted in the formation of decalins **18** and **19**, respectively, by *endo* and *exo* addition from chair like transition states **15c** and **15d** with the methyl substituent in an equatorial position. The *endo* compound **16**, which contains the required stereochemistry, was not formed under these conditions. This suggested implication of an enzymatic Diels–Alder reaction to promote the *endo* transition state **15a** *in vivo* towards the desired cycloadduct **16**, with the methyl substituent axially positioned. In 2000, Vederas isolated and characterized the lovastatin nonaketide synthase (LNKS), a type I iterative PKS and the first naturally occurring Diels–Alderase [25]. The desired decalin (**16**), with the stereochemistry of lovastatin **14**, was formed exclusively when the linear substrate **15** was submitted to LNKS. This Diels–Alderase would indeed stabilize the transition state **15a**.

Solanapyrones (**23**, **27**, Scheme 14.5) are phytotoxic polyketides and inhibitors of DNA polymerase β and λ, produced by the fungi *Alternaria solani* and *Ascochyta rabiei* [26]. Their biosynthesis would involve an IMDA (intramolecular Diels–Alder reaction) reaction leading to the decalin system [27, 28]. Studies indicated that the biosynthesis of most of these compounds proceeds via an *exo* selective Diels–Alder cycloaddition that would be catalyzed by an oxidase involved in the biosynthetic route [29]. The *exo* cycloadduct corresponds to the *cis*-decalin ring of solanapyrones A and B, while the *endo* cycloadduct holds the *trans*-fusion of solanapyrones D and E. The first biomimetic synthesis of (±)-solanapyrone A (**23**) was achieved in 1987 by Ichihara and coworkers (Scheme 14.5a) [30]. Later it was shown by Oikawa

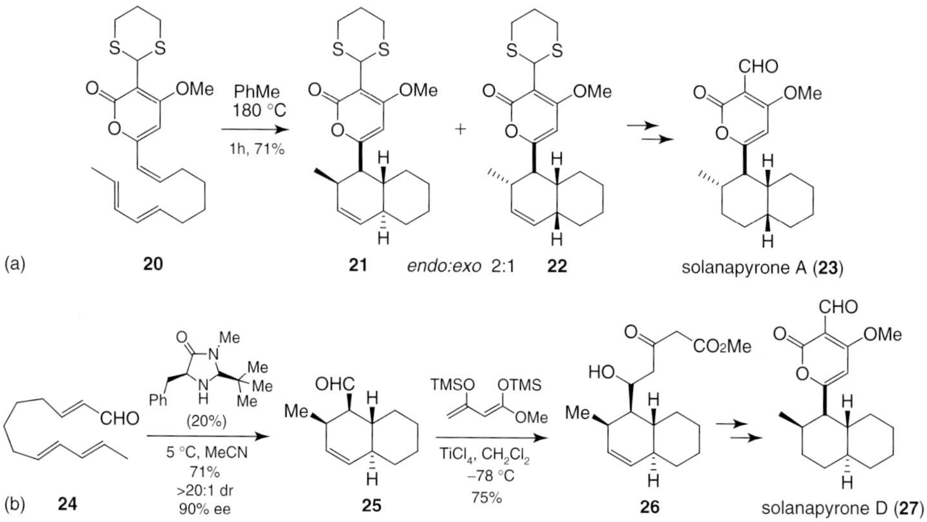

508 | *14 Biomimetic Synthesis of Non-Aromatic Polycyclic Polyketides*

Scheme 14.5 Biomimetic syntheses of solanapyrones: (a) Ichihara's synthesis of solanapyrone A (**23**); (b) MacMillan's organocatalytic synthesis of solanapyrone D (**27**).

and coworkers that thermal conditions and *A. solani* cell-free extracts promoted different stereoselectivity. The same team succeeded in the enantioselective and *exo*-selective synthesis of (−)-**23** utilizing a crude enzyme preparation [31]. Recently, the purification and identification of the Diels–Alderase solanapyrone synthase has been achieved [32]. The facile cyclization of the linear prosolanapyrone intermediate in enzyme-free aqueous buffer suggested that this Diels–Alderase attends only to control the stereochemistry of the cyclization rather than serving as a true enzymatic catalyst. In 2005, the enantioselective synthesis of solanapyrone D (**27**) involving an organocatalytic biomimetic IMDA was achieved by MacMillan and coworkers (Scheme 14.5b) [33]. The cycloaddition proceeded with 71% yields (>20 : 1 dr, 90% ee).

Nargenicin A_1 (**28**, Scheme 14.6) is an antibiotic isolated from *Nocardia argentinensis* [34]. Cane and coworkers established that nargenicin is derived from four propionate and five acetate units [35]. Later, intact incorporation of a series of postulated chain elongation intermediates into nargenicin demonstrated that the oxygen atom at C13 is not derived from propionate. This implied that the C4–C13 bond is not formed by an aldol-type condensation [36]. Cane suggested that the *cis*-decalin system of **28** may arise from an IMDA cycloaddition of a linear precursor. In 1989 Roush and coworkers investigated the Diels–Alder reaction towards **28** from decatrienone substrates such as **29** [37]. The stereoselectivity of the IMDA reaction was improved in the transannular version of this reaction (Scheme 14.6) [38]. The authors succeeded in the stereoselective synthesis of the tricyclic lactone core (**31**) of nargenicin A_1 thanks to the transannular Diels–Alder (TADA) reaction of the transient 18-membered macrolide **30**.

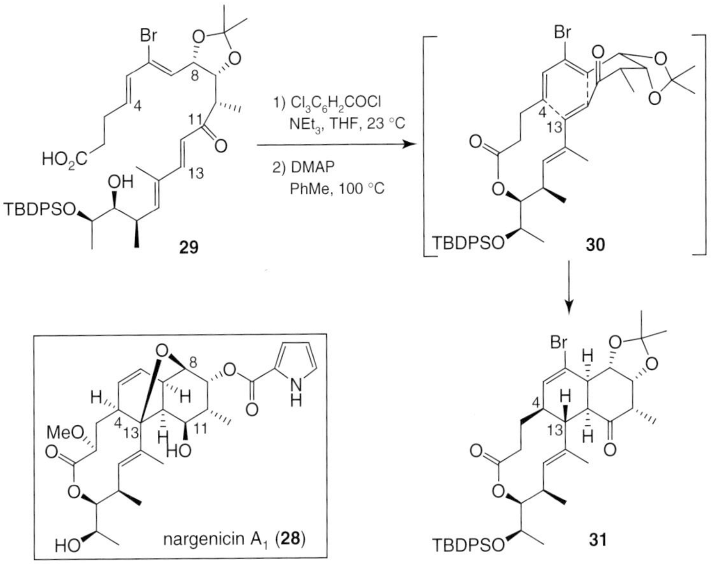

Scheme 14.6 Roush's studies toward the biomimetic synthesis of nargenicin A₁ (**28**).

Superstolide A (**34**) is a cytotoxic macrolide isolated from the New Caledonian marine sponge *Neosiphonia superstes* [39]. In 2008, Roush and coworkers reported the total synthesis of (+)-superstolide A by stereocontrolled TADA reaction of macrolactone **32** (Scheme 14.7) [40]. On this occasion, the comparison between IMDA and TADA strategies showed that higher diastereoselectivity was obtained in the transannular version. Indeed, while the IMDA reaction gave a mixture of three cycloadducts, the TADA cycloaddition of **32** provided the sole cycloadduct **33**. This highly diastereoselective cycloaddition was attributed to conformational effects within the macrocycle.

Himbacine (**38**), himbeline (**39**), and himandravine (**40**) are members of *Galbulimina* alkaloids, isolated from the bark of *Galbulimina baccata* (Scheme 14.8) [41]. Their biosynthesis has polyketide origins and would involve a Diels–Alder reaction of the linear precursor **36**, where an iminium ion would activate the reaction as postulated by Baldwin and coworkers [42]. In 2005, this hypothesis was supported by Baldwin's biomimetic synthesis of **38**–**40** (Scheme 14.8) [43]. The one-pot *N*-Boc deprotection of intermediate **35** and subsequent iminium formation (**36**) induced IMDA cycloaddition to give the *endo* cycloadduct **37**. This hypothetical intermediate in the biosynthesis was then converted into *Galbulimina* alkaloids **38**–**40**.

14.3.2
Biomimetic Diels–Alder Reactions Affording Tetrahydroindane Systems

Spiculoic acid A (**42**) is a cytotoxic metabolite isolated from the Caribbean marine sponge *Plakortis angulospiculatus* [44]. The polyketide biogenetic origin of **42** would involve four butyrate and one propionate units incorporated into the linear

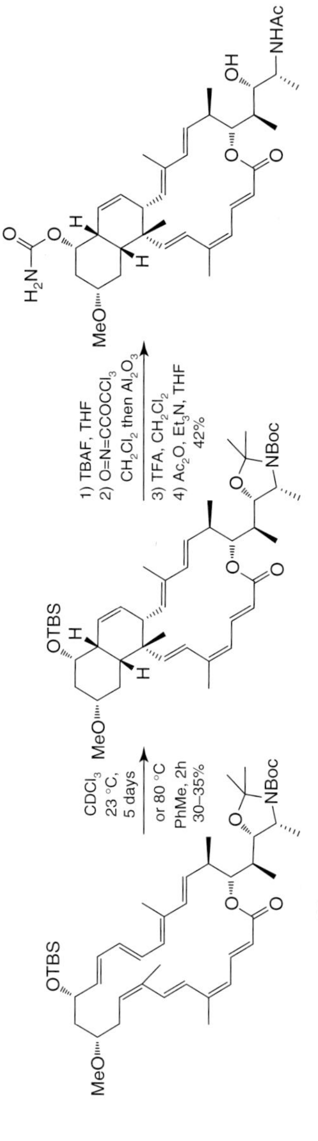

510 *14 Biomimetic Synthesis of Non-Aromatic Polycyclic Polyketides*

Scheme 14.7 Roush's total synthesis of superstolide A (**34**).

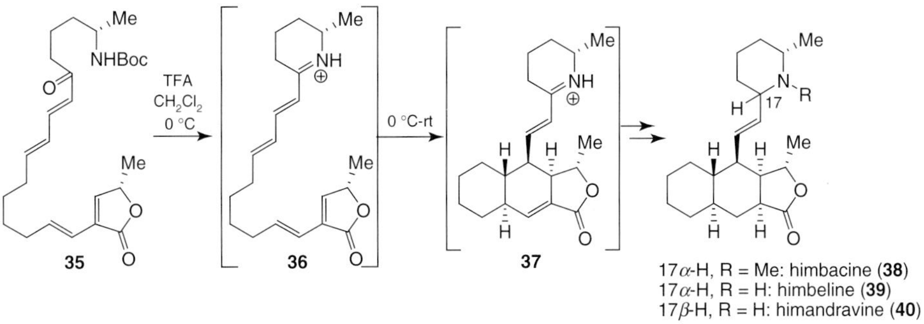

Scheme 14.8 Baldwin's biomimetic synthesis of *Galbulimina* alkaloids **38–40**.

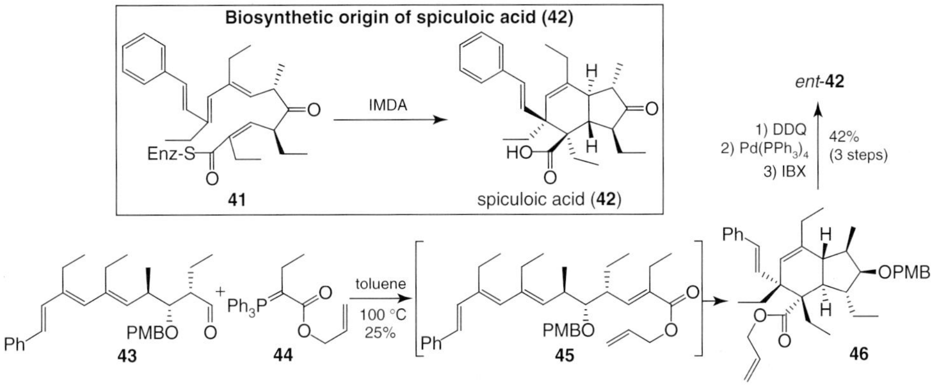

Scheme 14.9 Baldwin's biomimetic synthesis of *ent*-spiculoic acid (**42**).

precursor **41**, the IMDA reaction of which would afford the spiculane skeleton (Scheme 14.9). In 2006 Baldwin and coworkers reported the biomimetic total synthesis of *ent*-spiculoic acid **42** [45]. This work revealed the absolute configuration of the natural product. In fact, the Wittig reaction between the aldehyde **43** and the phosphoranylidene **44** furnished directly the cycloadduct **46**, through IMDA reaction of the presumed linear intermediate **45**. *ent*-Spiculoic acid (*ent*-**42**) was finally obtained upon three steps of deprotection and oxidation. This achievement, as well as Perkin's conclusions regarding the stereoselectivity of the thermal IMDA cyclization towards spiculoic acids [46], argues in favor of a biosynthetic Diels–Alder step.

(−)-Galiellalactone (**49**, Scheme 14.10) is a hexaketide isolated from the ascomycete *Galiella rufa* [47], with selective and potent inhibition of IL-6 [48] and potential antitumor activities [49]. The biosynthesis would involve an IMDA reaction from the biosynthetic intermediate (−)-pregaliellalactone (**47**) into (+)-deoxygaliellalactone (**48**), followed by enzymatic hydroxylation into **49** [50]. This unusual cycloaddition with inverse electron demand was supported by feeding experiments of the substrate **47** to the fungus [51]. On this occasion, Sterner and coworkers showed that the reaction was selective, providing the natural (+)-deoxygaliellalactone (**48**) as a sole isomer. Since spontaneous or thermally induced cyclizations also provided stereoselectively the natural product, it was suggested that hypothetical galiellalactone Diels–Alderase would only serve to enhance the cyclization rate [51]. In 2007, a successful biomimetic synthesis of **48** from **47** was reported by Lebel and Parmentier (Scheme 14.10) [52]. The precursor **47** was involved in IMDA cyclization in the presence of AlCl$_3$ under microwave irradiation, providing (+)-deoxygaliellalactone (**48**). A one-pot sequence towards **48** from the aldehyde **51** was also reported by the authors.

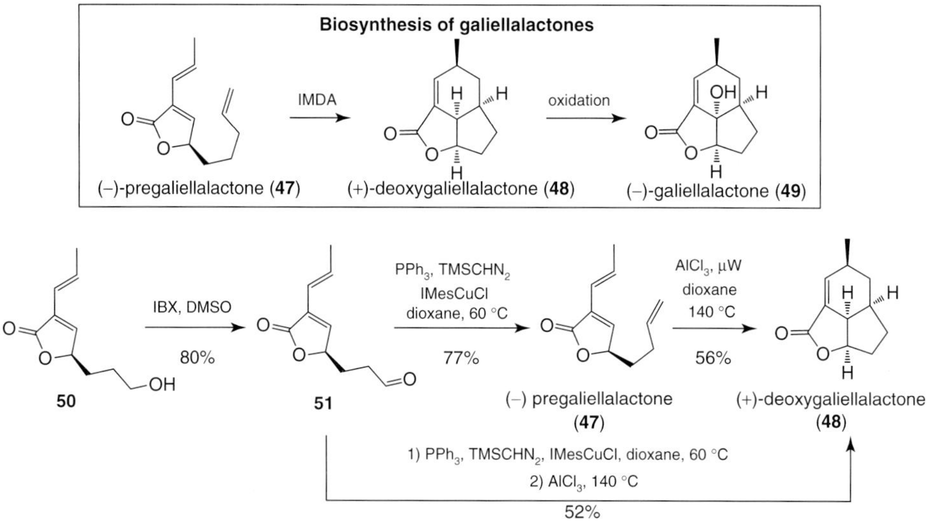

Scheme 14.10 Biosynthesis of galiellalactones and Lebel's biomimetic synthesis.

14.3.3
Biomimetic Diels–Alder Reactions Affording Spiro Systems

Abyssomicin C (**54**) is an antibiotic isolated in 2004 from the marine actinomycete *Verrucosispora* AB 18-032 [53]. Sorensen and coworkers suggested in 2005 that the biosynthesis of **54** involves an IMDA reaction [54]. Their biomimetic enantioselective synthesis of (−)-abyssomicin C via a tandem β-elimination/diastereoselective Diels–Alder macrocyclization from the linear intermediate **52** provided an argument in favor of the postulated biosynthesis (Scheme 14.11a) [54]. The reaction was accomplished in the presence of a catalytic amount of La(OTf)$_3$. Abyssomicin C (**54**) was obtained from the cycloadduct **53** in three steps: stereoselective epoxidation, demethylation, and intramolecular epoxide opening. In 2007, Nicolaou reported another total synthesis of **54** [55]. On this occasion, he demonstrated that it could be equilibrated with the atropisomer *atrop*-abyssomicin C (**55**) in the presence of ethereal HCl (Scheme 14.11b). After L-Selectride reduction of **55** and acidic workup, abyssomicin D (**57**) was obtained. These experiments gave strong support to the biosynthetic conversion of **54** into **57** through **55** in the microorganism.

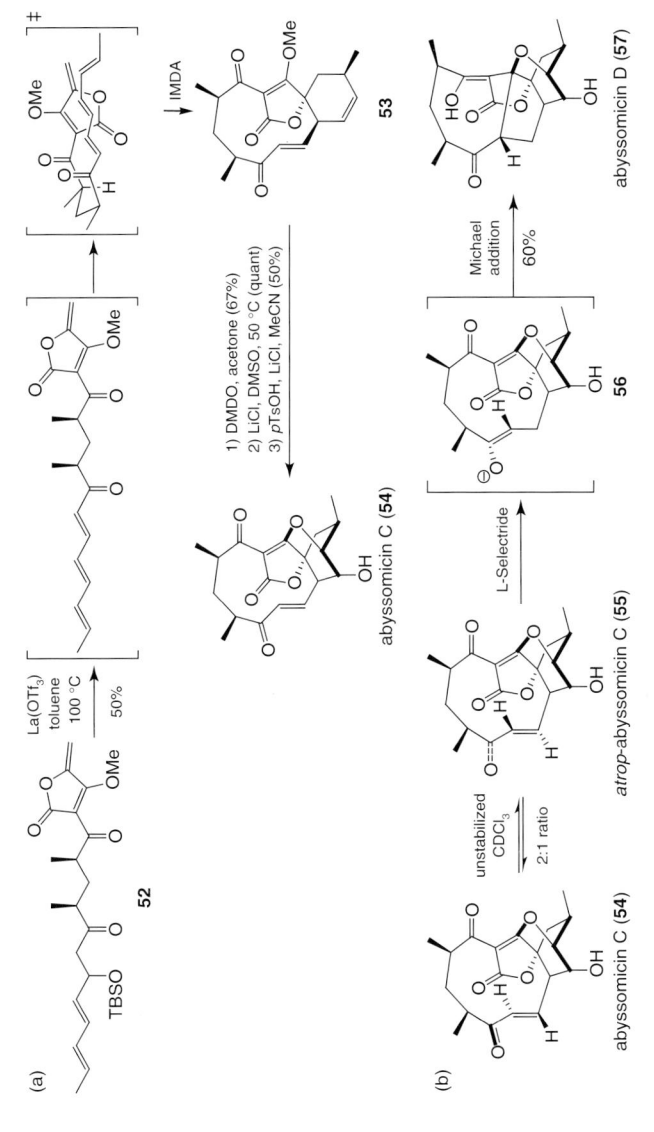

Scheme 14.11 Biomimetic synthesis of abyssomicins C and D according to Sorensen (a) and Nicolaou (b).

Last but not least, *atrop*-abyssomicin C (**55**) was finally isolated in 2007 as the major atropisomer in the strain *Verrucosispora* AB 18-032 [56], giving a successful conclusion to this synthetic work.

Gymnodimine (**58**) is a member of a large class of natural products bearing a cyclic imine fused to a cyclohexene ring and a macrocarbocycle, isolated from oysters contaminated by a Dinoflagellate [57]. Kishi and coworkers suggested that a biosynthesis apparented to pinnatoxin A [58], that is, an intramolecular reaction involving an iminium dienophile, would lead to **58** [59]. They conducted a synthetic study using the biomimetic precursor **59**, which was converted into the gymnodimine core **61** in a two-step procedure, consisting of the formation of a cyclic imine (**60**) and then IMDA reaction (Scheme 14.12).

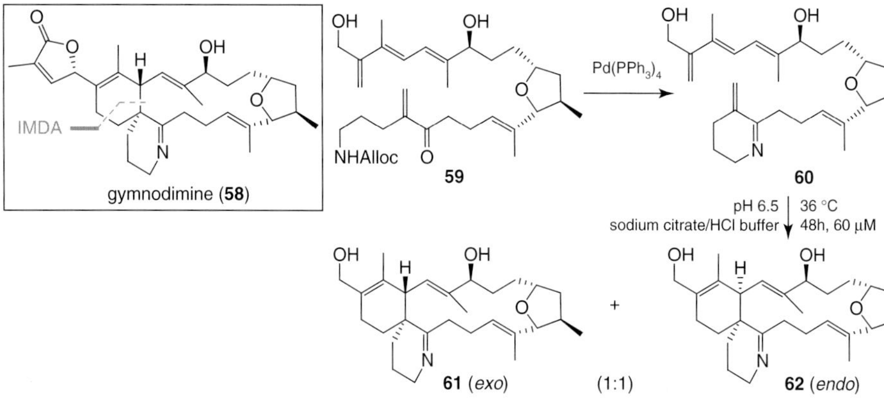

Scheme 14.12 Kishi's synthetic studies toward the biomimetic synthesis of gymnodimine (**58**).

This reaction proceeded under aqueous conditions at pH 6.5 at 36 °C, giving the *exo* product **61** with the desired diastereoselectivity and the *endo* product **62** (dr 1 : 1). This result is in contrast with the cycloaddition made on enone dienophiles similar to **59**, which only gave undesired *endo* products. These results suggested that the hypothetical biosynthesis of gymnodimine (**58**), wherein the Diels–Alder precursor reacts to form the natural product, could occur spontaneously without the aid of an enzyme.

14.3.4
Biomimetic TADA Reactions toward FR182877, Hexacyclinic Acid and the Parent Cochleamycin A and Macquarimicin A

The compound (−)-FR182877 (**64**, also named *cyclostreptin*) was isolated by Sato and coworkers from a *Streptomyces* strain (Scheme 14.13) [60, 61]. It showed potent cytotoxicity with irreversible microtubule stabilizing properties. Soon after this, the very closely related hexacyclinic acid (**65**) was reported by Zeeck and coworkers, who investigated its biosynthetic origin [62]. Using ^{13}C labeling, it was demonstrated that **65** originated from the type I polyketide biogenetic route, incorporating six acetate

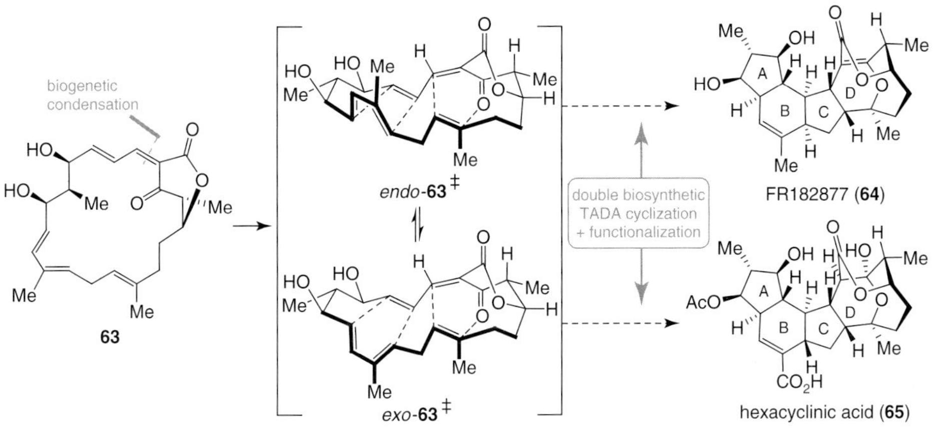

Scheme 14.13 Common biosynthetic origin for FR182877 (**64**) and hexacyclinic acid (**65**).

and four propionate units. It was supposed that an IMDA reaction and further aldol reaction, lactonization, and ketalization would provide all the functionalities of **65**. The parent FR182877 (**64**) would have the same biosynthetic origin. In fact both products would be formed from the double TADA reaction of a simpler macrocyclic polyketide precursor. Evans and Starr reasoned that they could even share the common precursor **63**, the cyclization of which would lead to **64** or **65**, depending on whether the transition state of the TADA reaction is *endo* or *exo*, respectively [63]. Macrocyclic intermediates incorporating a β-keto-δ-lactone like that in **63** have been invoked in the biosynthesis of other complex natural products such as cochleamycin A and macquarimicin A, the biomimetic synthesis of which is discussed below.

The synthesis of FR182877 (**64**) has been investigated by several authors, giving support to the biosynthetic considerations discussed above. Sorensen and coworkers first performed the biomimetic synthesis of (+)-**64** (Scheme 14.14) after

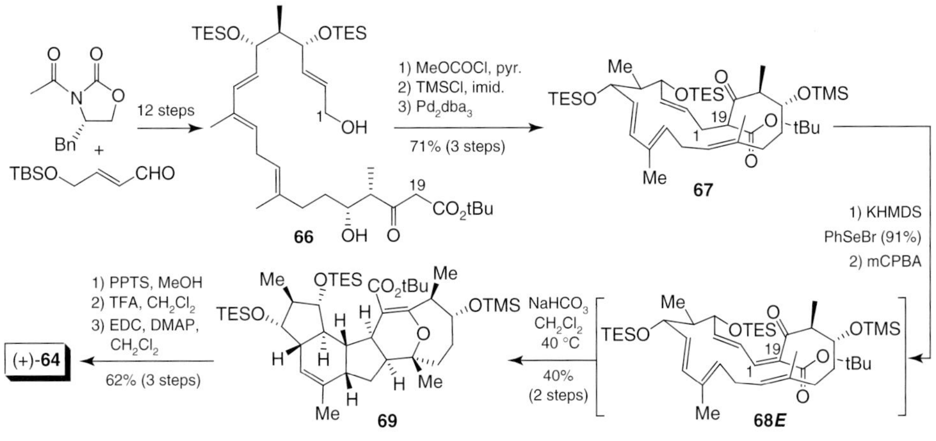

Scheme 14.14 Sorensen's biomimetic synthesis of (+)-FR182877 [(+)-**64**].

considerable work that highlighted the particular reactivity of some intermediates [64]. The intermediate **66** was obtained after 12 steps, beginning with an asymmetric aldol reaction with an Evans chiral auxiliary. This linear compound bears all the needed functionalities, that is, an allylic alcohol and a β-ketoester, to envisage the macrocyclic ring closure. After methoxycarbonylation of the primary alcohol at C1 and protection of the secondary one, a Tsuji–Trost reaction in the presence of Pd_2dba_3 allowed for the C1–C19 ring closure of **66** smoothly and in very good yield and diastereoselectivity. The macrocyclic biomimetic precursor [(E)-**68**] of FR182877 was finally revealed by selenylation and oxidative deselenylation at C19 from **67**, giving an equimolar ratio of (E) and (Z) isomers [(E)-**68** and (Z)-**68**]. The high reactivity of these macrocyclic pentaenes precluded all isolation attempts. The transannular cycloaddition was also observed at ambient temperature. Complete conversion of the isomeric pentaenes into three major products was observed on warming a chloroform solution buffered with sodium bicarbonate for 4 h at 40 °C. The major product of the sequence was the pentacycle **69**, bearing the same relative stereochemistry as FR182877. To complete the synthesis, deprotection of silyl ethers and final lactone formation were performed, giving (+)-**64** in 62% yield over three steps.

However, it had been reported that the correct structure of the natural product FR182877 was the opposite enantiomer, (−)-**64** [61]. Therefore in the same final report [64d], Sorensen and coworkers described the total synthesis of (−)-**64**, a work made on a large scale for biological purposes. It was possible to synthesize 24 g of the cyclic intermediate *ent*-**67**, which was transformed into 5.4 g of the lactone **70** (Scheme 14.15), a stable compound on storage. The conversion of **70** into the less stable and non-storable natural product (−)-**64** was undertaken on a 100-mg scale.

Evans and Starr undertook a total synthesis of (−)-**64** using a similar double TADA process [63]. This work confirmed the absolute configuration of FR182877. According to the authors,

> "the macrocyclic conformation of the reactive intermediate appears to predispose the reacting olefin/carbonyl faces to a single orientation, and the resulting high stereoselectivity of the double TADA sequence, exclusively leading to FR182877, stands in stark contrast to the results obtained in similar acyclic systems (**71**)."

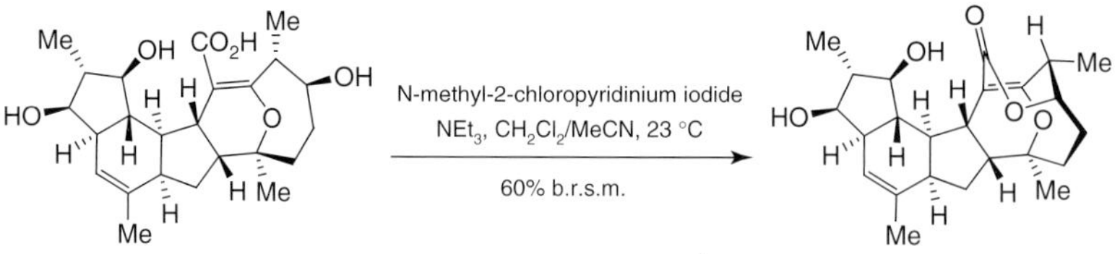

Scheme 14.15 Final step in the large-scale synthesis of the natural enantiomer of FR182877 (b.r.s.m.: based on recovered starting material).

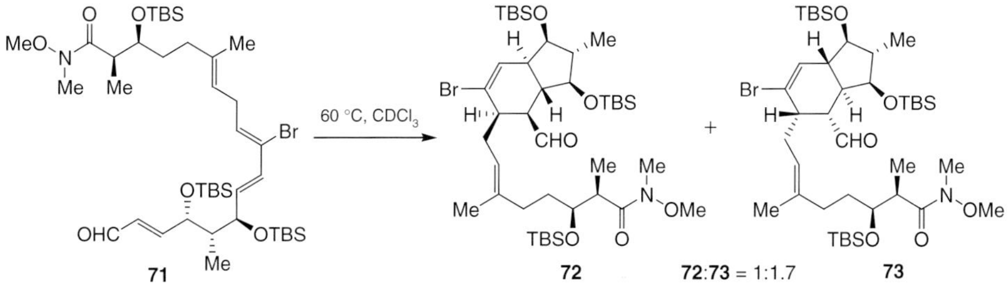

Scheme 14.16 Stereoselectivity in the IMDA reaction of an acyclic system related to FR182877.

In the acyclic IMDA reaction of **71**, only 1 : 1.7 diastereofacial selectivity (**72** : **73**) was indeed realized (Scheme 14.16). Similar results were reported by Sorensen and coworkers [64b]. During these works, no product from an *exo* transition state resembling *exo*-**63**‡ (Scheme 14.13) was observed, leaving the synthesis of hexacyclinic acid **65** as still elusive at this stage. Internal control elements such as stereocenters within the molecular architecture might be responsible for the orientation toward FR182877 (**64**).

However, **65** has, as an extra particularity, a carboxylic acid function on ring B. This may add some electronic effects in the stereocontrol of the first TADA reaction, disfavoring the *endo* transition state. This carboxylic acid could thus be present at early stages during the biosynthesis of **65**, especially on the biogenetic macrocyclic precursor.

Cochleamycin A (**77**) was isolated in 1992 from *Streptomyces* DT136 by Shindo [65]. It showed antimicrobial activity against Gram-positive bacteria and was cytotoxic against various tumor cell lines [66]. Shindo suggested a biosynthetic route to cochleamycin A (**77**) and B (**80**) from a polyketide linear product (**74**) by an IMDA reaction (Scheme 14.17) [67]. This IMDA would give the 5,6-fused ring system (**75**) with the desired stereochemistry, through an *endo* transition state of a (9*E*,11*Z*)-diene, or an *exo* transition state of a (9*Z*,11*E*)-diene (not shown), giving in both alternative routes the *cis* junction of the natural products. An intramolecular condensation and further functionalization would achieve biosynthesis of the natural product.

It is remarkable that macquarimicins (**78**, **79**, **81**) share the same carbocyclic skeleton and stereochemistry as cochleamycins, except the methyl substituent at C14 (Scheme 14.17). A high degree of stereocontrol may be necessary to achieve such an invariable selectivity. This could be controlled thanks to a biosynthetic TADA process on intermediate **76**, which arises from intramolecular condensation of **74**.

Paquette and coworkers drew inspiration from Shindo's biosynthetic hypothesis and succeeded in the highly enantioselective biomimetic IMDA reaction of the 5,6-fused bicyclic system of cochleamycin A (**77**) [68]. They used the (*E,Z,E*)-1,6,8-nonatriene (**82**), which reacted in toluene at 195 °C via an *endo* transition state to furnish the bicyclic intermediate **83** (Scheme 14.18a). One

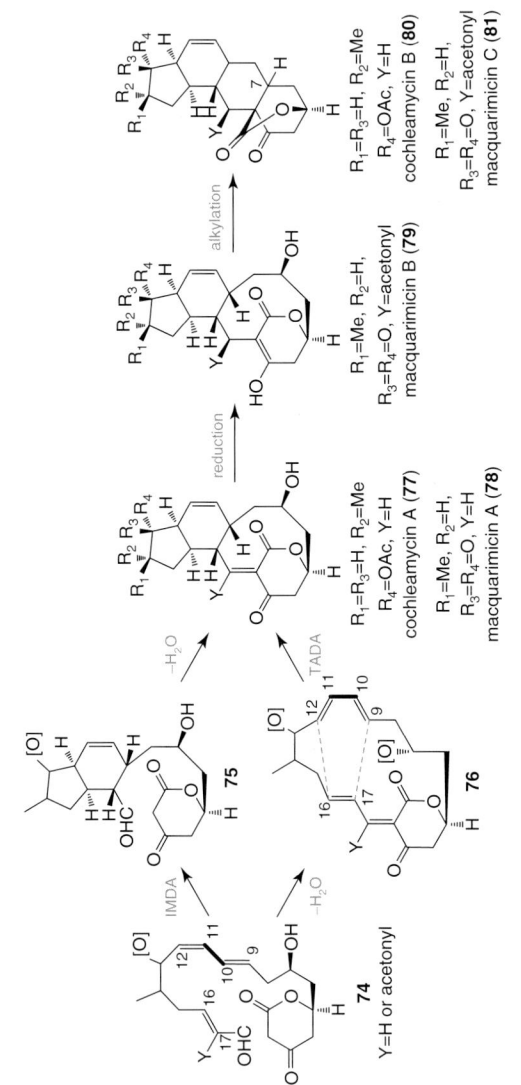

Scheme 14.17 Biosynthetic hypothesis for cochleamycins and macquarimicins.

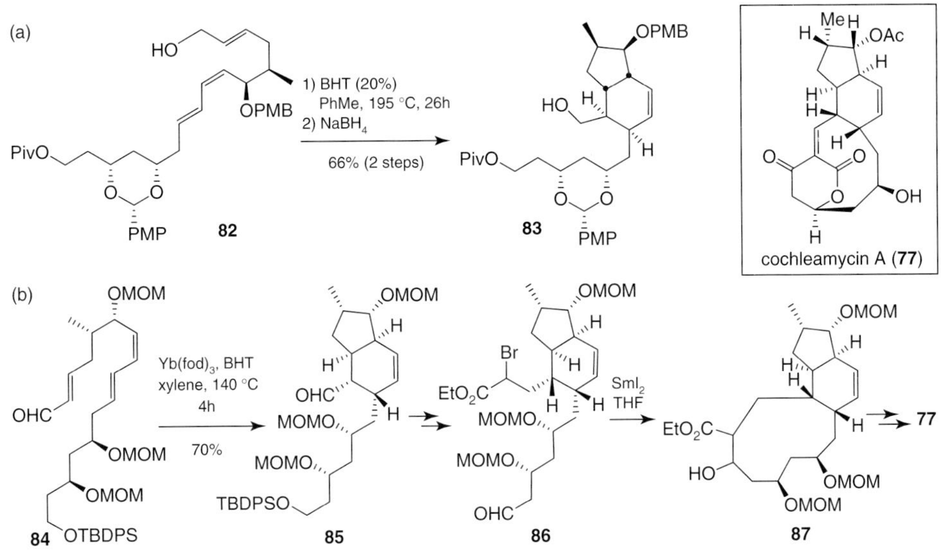

Scheme 14.18 Paquette's partial synthesis of cochleamycin A (**77**) (a) and Tatsuta's total synthesis (b).

year later, Tatsuta and coworkers achieved the total synthesis of **77** also using an IMDA reaction to form the bicyclic framework [69]. The triene (**84**) was heated at 140 °C in the presence of Yb(fod)$_3$ and BHT (butylated hydroxytoluene) in xylene, providing a single cycloadduct (**85**) towards the synthesis of **77** (Scheme 14.18b).

Roush and coworkers reported in 2004 the biomimetic total synthesis of cochleamycin A (**77**) in 2.4% yield and 23 steps from 3-butene-1-ol, accomplished by TADA reaction of the macrocyclic intermediate **88** (Scheme 14.19) [70]. The substrate was heated at 125 °C in toluene, providing the carbon skeleton **90** of the natural product as the sole cycloadduct in 69% yield, through the *endo* transition state **89**.

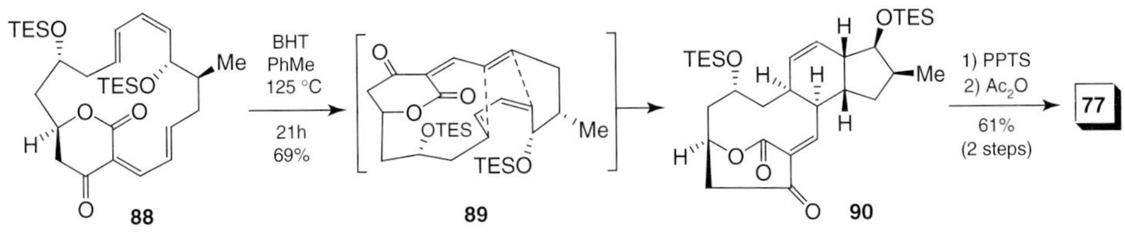

Scheme 14.19 Roush synthesis of (+)-cochleamycin A (**77**).

Macquarimicins A–C (**78**, **79**, **81**, respectively, Scheme 14.17) were isolated in 1995 from *Micromonospora chalcea* [71]. Interesting biological properties have been associated with these compounds, such as selective inhibition of membrane-bound sphingomyelinase, anti-inflammatory, or cytotoxic activities [72]. Tadano and coworkers have reported extensive investigations on biomimetic IMDA and TADA reactions applied to macquarimicins [73]. To construct the tetrahydroindane ring with *cis-anti-cis* ring fusion by IMDA cycloaddition, the reaction of a *(E,Z,E)*-1,6,8-nonatriene proceeding through an *endo* mode or the reaction of a *(E,E,Z)*-1,6,8-nonatriene through an *exo* mode are two possible routes. In this context, Tadano and coworkers embarked on the synthesis of various IMDA and TADA substrates. A model study comparing the reactivity of *(E,Z,E)*- and *(E,E,Z)*-1,6,8-nonatrienes (**91**, **92**) was performed (Scheme 14.20a). It implied the potential advantage of *(E,Z,E)*-triene **91** in the construction of the required *cis-anti-cis* ring junction (**93**). The *(E,E,Z)*-triene **92** gave indeed very poor selectivity in the IMDA reaction compared to **91** under thermal conditions. Moreover, four types of *(E,Z,E)*-trienes (**94–97**) were evaluated in the synthesis of macquarimicins, being linear or macrocyclic and incorporating a preformed lactone moiety or not (Scheme 14.20b).

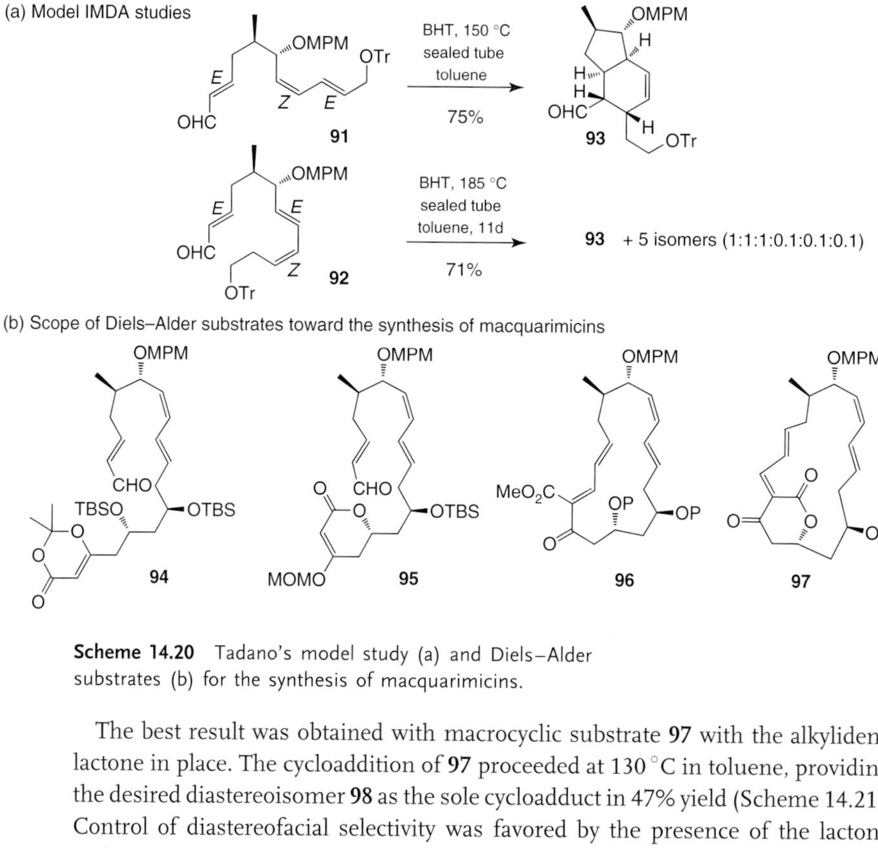

Scheme 14.20 Tadano's model study (a) and Diels–Alder substrates (b) for the synthesis of macquarimicins.

The best result was obtained with macrocyclic substrate **97** with the alkylidene lactone in place. The cycloaddition of **97** proceeded at 130 °C in toluene, providing the desired diastereoisomer **98** as the sole cycloadduct in 47% yield (Scheme 14.21). Control of diastereofacial selectivity was favored by the presence of the lactone cycle. Subsequently, macquarimicin A (**78**) was synthesized after deprotections and oxidation of the alcohol at C13. Then one-step installment of the acetonyl moiety at C17 furnished macquarimicin B (**79**). Macquarimicin C (**81**) was, finally, formed in quantitative yield by acid-catalyzed intramolecular dehydrative alkylation of macquarimicin B (**79**).

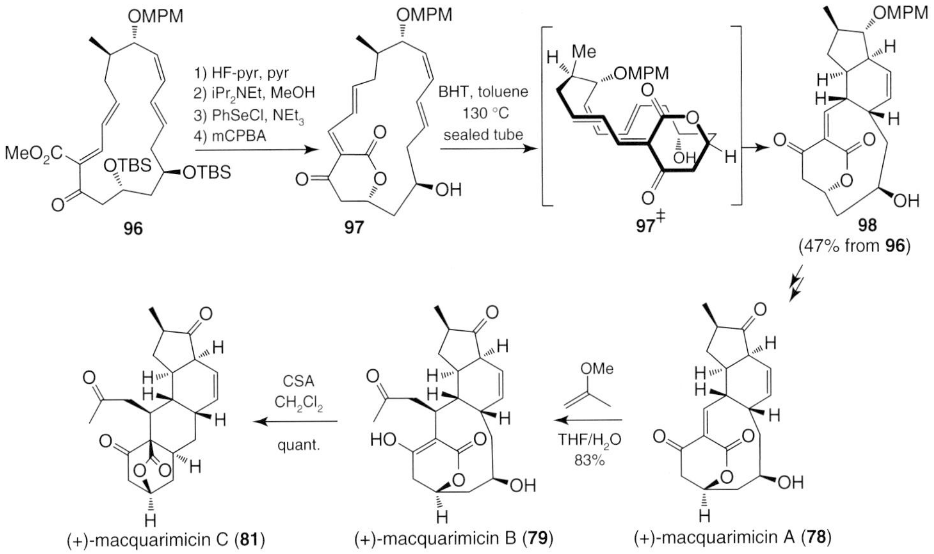

Scheme 14.21 Tadano's biomimetic synthesis of macquarimicins.

14.3.5
Biomimetic TADA Reactions toward Spinosyns

Spinosyns are produced by *Saccharopolyspora spinosa* originally collected in the Caribbean islands [74]. They are tetracyclic and contain a 12-membered macrocycle fused to a 5,6,5-tricyclic ring system. More than 25 spinosyns have been isolated to date, which constitute an important class of insecticidal polyketides. Their biosynthesis has been well studied and numerous analogs have been synthesized [75]. More recently, the entire spinosyn biosynthetic gene cluster was identified through gene sequencing and functional analysis of the gene products in *S. spinosa* [76]. The biosynthesis of spinosyns is expected to be initiated by the oxidation of the 15-OH group of the mature macrocyclic polyketide precursor **99** followed by conjugated dehydration (Scheme 14.22). The activated intermediate can then be involved in a TADA cycloaddition followed by a vinylogous Morita–Baylis–Hillman reaction, which would be mediated by an enzyme nucleophile, to deliver the tetracyclic spinosyn aglycone **100**. Subsequent glycosylation steps install the sugars of spinosyns.

The biomimetic synthesis of spinosyn A (**101**) has been studied by Roush and coworkers. It was known from Evans and Black that the diastereofacial selectivity of the IMDA reaction of linear substrates would favor an incorrect C7–C11 *trans*-fused diastereoisomer [77]. Thus, Roush first envisaged performing the TADA cyclization on a glycosylated macrocyclic intermediate **103a** bearing an additional bromine substituent at C6 (Scheme 14.23) and arising from the Wittig–Horner–Emmons reaction of phosphonate **102a** [78]. The glycosyl at C9 would allow control of the stereoselectivity at C7–C11 during cycloaddition, thanks

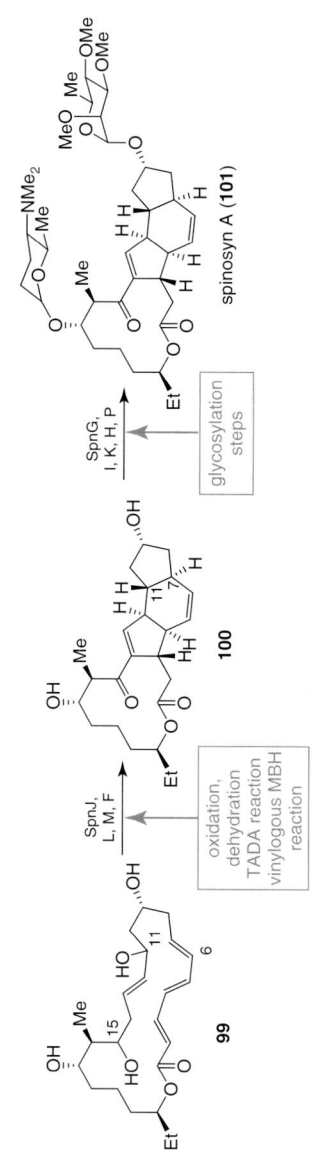

522 | *14 Biomimetic Synthesis of Non-Aromatic Polycyclic Polyketides*

Scheme 14.22 Biosynthesis of spinosyns.

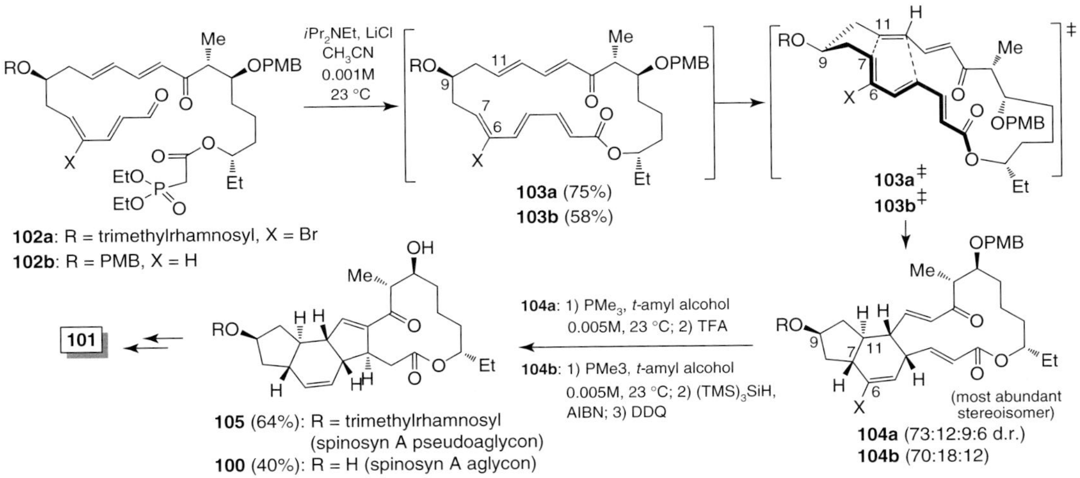

14.3 Biomimetic Syntheses Involving the Diels–Alder Reaction

102a: R = trimethylrhamnosyl, X = Br
102b: R = PMB, X = H

103a (75%)
103b (58%)

104a: 1) PMe$_3$, *t*-amyl alcohol 0.005M, 23 °C; 2) TFA
104b: 1) PMe3, *t*-amyl alcohol 0.005M, 23 °C; 2) (TMS)$_3$SiH, AIBN; 3) DDQ

104a (73:12:9:6 d.r.)
104b (70:18:12)
(most abundant stereoisomer)

105 (64%): R = trimethylrhamnosyl (spinosyn A pseudoaglycon)
100 (40%): R = H (spinosyn A aglycon)

101

Scheme 14.23 Roush's total synthesis of spinosyn A (**101**).

to steric interactions with bromine at C6 during the transition state **103a**‡. The TADA cyclization thus gave the cycloadduct **104a** in 75% yield and good selectivity (dr 73 : 12 : 9 : 6), as a consequence of strain interactions involving the C6 and C9 substituents. Eventually, the presence of a silyloxy substituent on C8 proved to be effective in increasing the steric interactions and the selectivity (dr > 95 : 5) [79]. Although the total synthesis of **101** was accomplished from **102a**, the same authors later demonstrated that a simpler substrate such as **103b** can give a very similar distribution of TADA cycloadducts. Consequently, the synthesis of the spinosyn aglycon **100** was achieved from macrolactone **103b** by TADA cycloaddition and PMe$_3$-catalyzed vinylogous Morita–Baylis–Hillman reaction on intermediate **104b**, followed by PMB (*p*-methoxybenzyl) deprotection.

14.4
Biomimetic Cascade Reactions

14.4.1
A Metalated Ionophore Template for the Biomimetic Synthesis of Tetronasin

Tetronasin (**108**) is an ionophoric antibiotic isolated from *Streptomyces longisporoflavus* (Scheme 14.24) [80]. It shows strong affinity for sodium ions, which may be linked to its antibiotic activity. This affinity would be involved in the biosynthesis of the natural product **108** from a linear polyoxygenated polyketide **106** bearing the tetronic acid and tetrahydrofuran residues at the opposite termini. The biosynthesis and the total synthesis of tetronasin have been studied by Ley and Staunton [81, 82]. Their work culminated with Ley's realization of the biomimetic step, that is, the one-step cyclization of a linear precursor analogous to **106** into the complete polycyclic core of tetronasin, using a metal-promoted folding of the chain as the directing process.

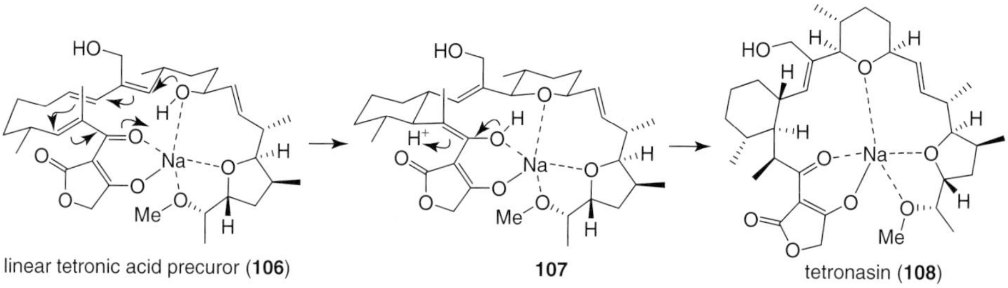

linear tetronic acid precuror (**106**)   **107**   tetronasin (**108**)

Scheme 14.24 Biosynthetic origin of tetronasin (**108**).

From the late biosynthetic precursor **106**, both tetrahydropyran and cyclohexane rings would indeed be formed simultaneously during a unique biosynthetic step [81d], which is mechanistically plausible according to stereoelectronic analysis of the cyclization process. The linear intermediate **106** would fold around the sodium salt of the tetronic acid residue with assistance of the various chelating oxygens of the polyketide. This would render the centers linked within bonding distance and the relevant orbitals correctly aligned for simultaneous cyclization of both six-membered rings. This pre-cyclization complex would adopt a low energy conformation with the metal in a central cavity, resembling that found in the natural product **108**.

This biosynthetic postulate has inspired Ley's biomimetic synthesis of **108** [82]. The polyene substrate for the cyclization was synthesized from **109** and **110** (Scheme 14.25). It was designed with an electron-deficient diene system (C10–C13) to facilitate the pyran ring formation via Michael addition. The methoxymethyl-furan ring was left intact to enhance chelation. The tetronic acid part was not installed at this stage but simplified as a methyl carboxylate. Treatment of the polyene **111** with potassium hexamethyldisilazide (KHMDS) in toluene for 30 min at 0 °C generated the potassium salt **111-K**, the conformation of which was guided by intramolecular chelation, inducing the cascade cyclization.

This challenging reaction gave one diastereoisomeric product (**112**) in very good yields (67%) by two subsequent conjugate additions, O17 → C13 then C10 → C5. The reaction generated two new rings and created four stereogenic centers simultaneously and with complete control. However, the configuration of the C4-methyl substituent in **112** was opposite to that found in tetronasin. Other conditions (metal bases, solvents, and temperature) were less satisfactory for the cyclization. Functional group interconversions finally gave the aldehyde **113**, the epimerization of which was achieved in 85% yield in the presence of morpholine and catalytic *p*-TSA (*p*-toluenesulfonic acid). The product **114** was identical to that previously prepared by Yoshii [83] as a late-stage precursor in his total synthesis of tetronasin.

In Yoshii's total synthesis of tetronasin, the reaction of **114** with the diazoacetoacetate **115** in the presence of $ZrCl_4$ (Scheme 14.26) gave the β-ketoester **116** [83]. Deprotection and Dieckmann cyclization of **116** finally installed the tetronic acid system as the sodium salt of tetronasin (**108**). This last cyclization step is also truly biomimetic.

14.4.2
The 6,5,6-Fused System and Macrocycle of Hirsutellones: Work Yet to Be Done?

The hirsutellones (e.g., **117** and **118**) constitute a group of fungal metabolites with interesting medicinal properties (Figure 14.1). They were isolated in 2005 from *Hirsutella nivea* BCC 2594, an entomopathogenic fungus, and showed significant antimycobacterial activities compared to the reference drug isoniazid [84]. The group is related to other methylated analogs sharing the same carbocyclic core (**119–121**), yet with stereochemical variations: pyrrocidines [85], pyrrospirones [86], and GKK1032 compounds [87]. Structurally, hirsutellones feature a 6,5,6-fused tricyclic system and a 12- or 13-membered *p*-cyclophane macrocycle, the construction of which may be particularly difficult. Furthermore, a γ-lactam or a succinimide motif is embedded in the macrocycle.

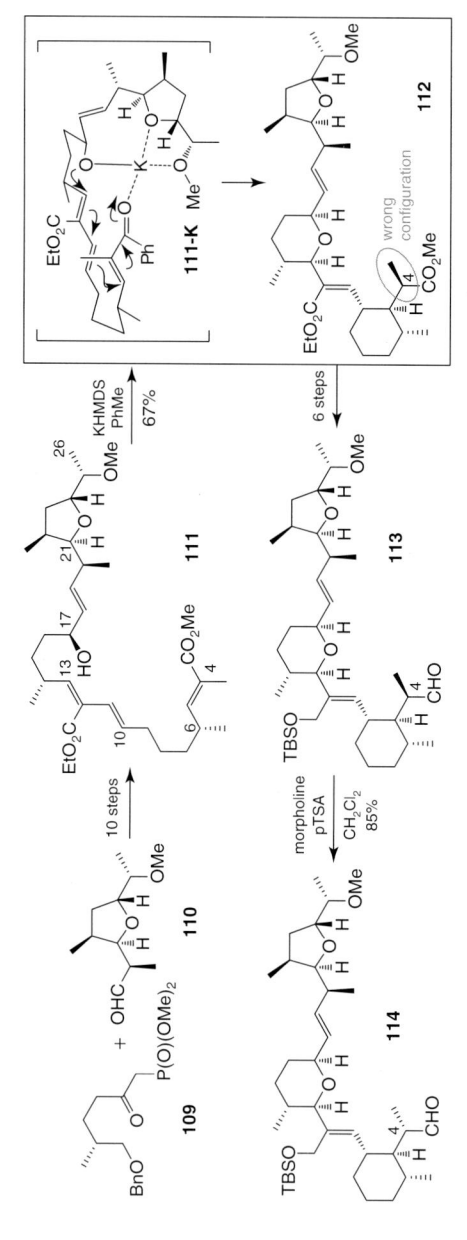

Scheme 14.25 Ley's formal synthesis of tetronasin (**108**) (carbon numbering of tetronasin).

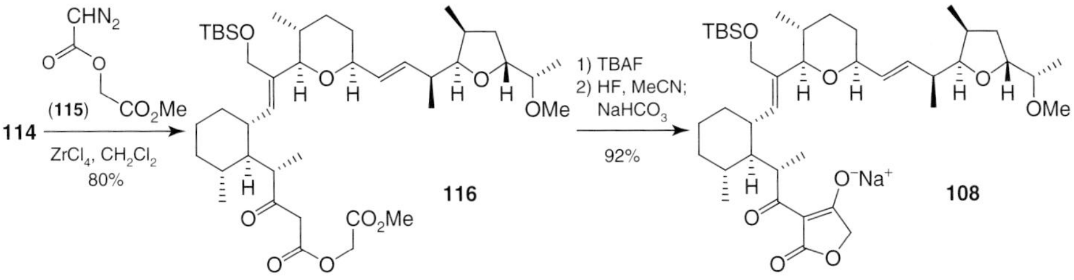

Scheme 14.26 Final steps in Yoshii's total synthesis of tetronasin (**108**).

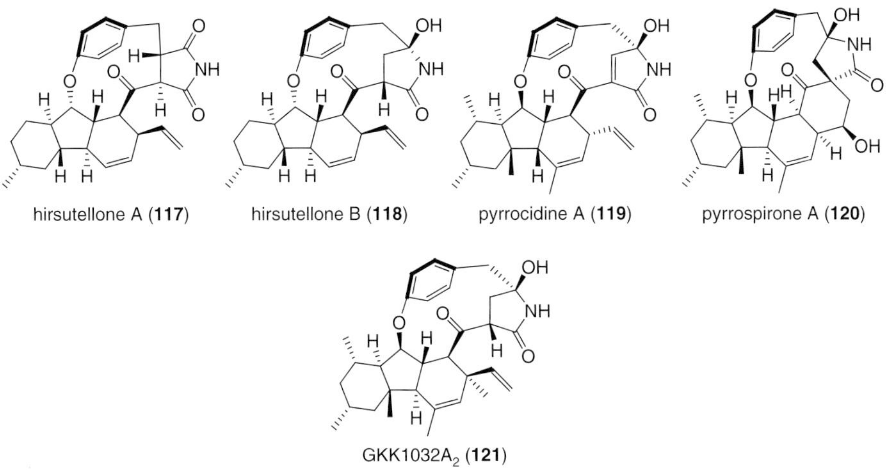

hirsutellone A (**117**)   hirsutellone B (**118**)   pyrrocidine A (**119**)   pyrrospirone A (**120**)

GKK1032A$_2$ (**121**)

Figure 14.1 Structure of hirsutellones (**117**, **118**) and related compounds (**119–121**).

The biosynthesis of this family of fungal products was described by Oikawa in 2003 [88]. The work was undertaken on the GKK1032 series (**121**) produced by *Penicillium* sp., using ^{13}C and ^{2}H labeling (Scheme 14.27). Oikawa demonstrated the nonaketide origin of the tricyclic moiety. Moreover, he showed that L-tyrosine was a precursor of the γ-lactam part while L-methionine stood for the methyl donor. A mixed PKS/NRPS (non-ribosomal peptide synthetase) enzyme would be involved. Of particular interest was the biogenetic cyclization mechanism suggested by Oikawa to make the polycyclic system from the linear precursor **122**. After oxidation of **122**, a cationic polycyclization process on intermediate **123** would occur. This may eventually be concerted. However, a stepwise mechanism including an intramolecular Diels–Alder reaction to close the cycle C is not excluded.

Several synthetic works toward hirsutellones have been reported recently. On the occasion of the isolation of the dimeric hirsutellone F (**125**), Isaka and coworkers obtained the putative biosynthetic intermediate **126** by cleavage of **125** in a 1-M NaOH solution (Scheme 14.28) [89]. This monomeric compound was poorly stable, tending to rearrange into hirsutellone A (**117**) during basic treatment, by

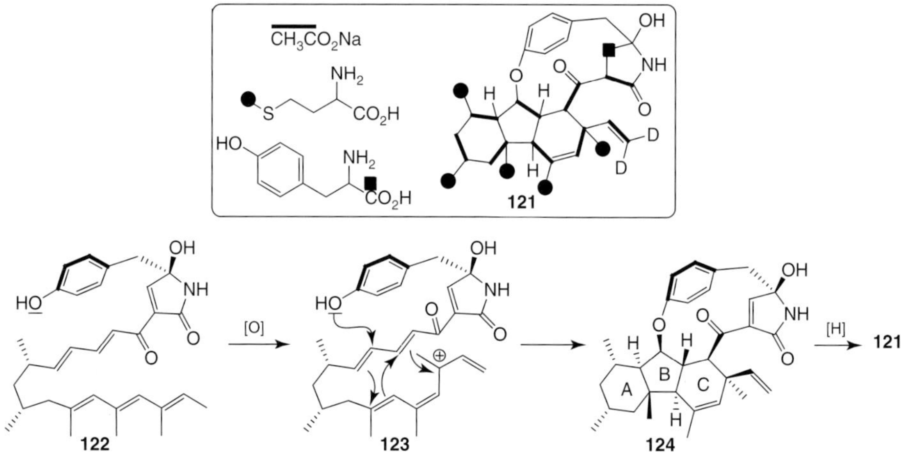

Scheme 14.27 Biosynthetic origin of the GKK1032 compounds and cyclization mechanism.

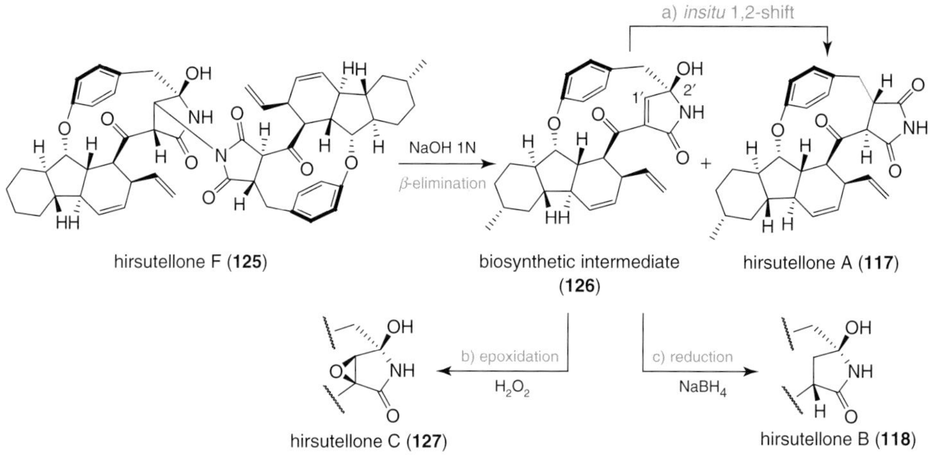

Scheme 14.28 Biomimetic conversion of intermediate **126** into hirsutellones A–C.

1,2-migration of the benzyl group to form the succinimide ring. In the presence of a reducing agent (NaBH$_4$), **126** afforded good yields of hirsutellone B (**118**), while epoxidation with H$_2$O$_2$ gave hirsutellone C (**127**). This work confirmed Oikawa's biogenetic hypotheses on the formation of the various lactam derivatives [88].

There is still no biomimetic total synthesis of hirsutellones with regard of the construction of the 6,5,6-fused tricyclic moiety. However, Nicolaou's total synthesis of hirsutellone B (**118**) [90] is remarkably reminiscent of Oikawa's cascade cyclization leading to the GKK1032A$_2$ carbocyclic core (Scheme 14.29) [88]. Starting from the epoxide intermediate **131**, electrophilic activation in the presence of Et$_2$AlCl provided diastereoselectively the tricyclic core **134** of hirsutellones in 50% yield. The reaction proceeded by intramolecular epoxide opening and

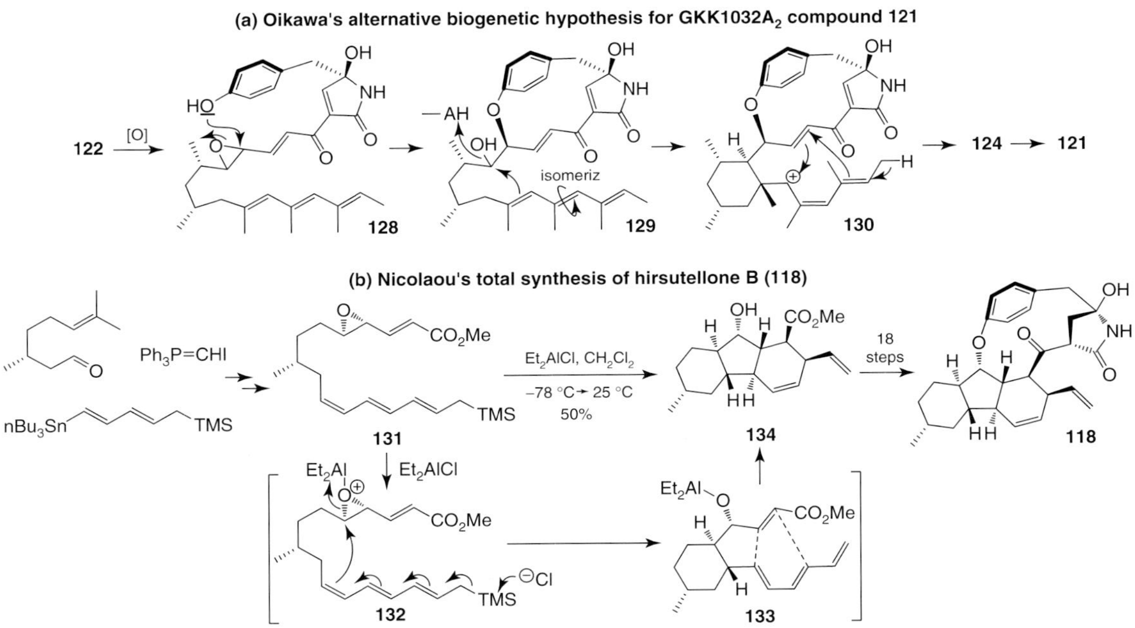

Scheme 14.29 (a) Oikawa's alternative hypothesis for the biosynthesis of GKK1032A$_2$ (**121**) in relation to (b) Nicolaou's total synthesis of hirsutellone B (**118**).

electrophilic cyclization into **133**, followed by Diels–Alder reaction through an exclusive *endo* transition state leading to **134**. Interestingly, in this mechanistic scheme, the electrophilic activation is inversed in comparison to the oxidative activation performed on the biogenetic precursor **122** (Scheme 14.27).

14.5
Conclusion

Theoretical analysis of polyketide biosynthesis has suggested that over a billion possible structures could be synthesized from PKS enzymatic models [91]. As yet, however, only 10 000 polyketide structures have been discovered. The present chapter deals only with biomimetic rearrangements in the carbocyclic non-aromatic series. Many important studies around polyketides are described in other chapters of this book. The structural complexity is an important character of polyketides and we know that it is related to chemical diversity. Considering the important biological properties of polyketides and the potential number of polyketide "privileged structures" [92], synthetic analogs will be valuable drug candidates for medicinal research of the twenty-first century. The application of diversity oriented methods to polyketide synthesis is an important issue in meeting this objective [93]. The power of this method has already been demonstrated, especially when using the biomimetic strategy [94]. Yet, an important alternative methodology emerged with the advent of the combinatorial biosynthesis of polyketides. The manipulation of gene clusters coding for each PKS modules and their combination or hybridization has provided unprecedented polyketide libraries [95]. Joining both chemical and biochemical technologies promises important discoveries in the next few decades.

References

1. (a) Chan, Y.A., Podevels, A.M., Kevany, B.M., and Thomas, M.G. (2009) *Nat. Prod. Rep.*, **26**, 90–114; (b) Smith, S. and Tsai, S.-C. (2007) *Nat. Prod. Rep.*, **24**, 1041–1072; (c) Hertweck, C., Luzhetskyy, A., Rebets, Y., and Bechthold, A. (2007) *Nat. Prod. Rep.*, **24**, 162–190; (d) Fischbach, M.A. and Walsh, C.T. (2006) *Chem. Rev.*, **106**, 3468–3496; (e) Staunton, J. and Weissman, K.J. (2001) *Nat. Prod. Rep.*, **18**, 380–416; (f) Hertweck, C. (2009) *Angew. Chem. Int. Ed.*, **48**, 4688–4716.
2. Keatinge-Clay, A.T., Maltby, D.A., Medzihradszky, K.F., Khosla, C., and Stroud, R.M. (2004) *Nat. Struct. Biol.*, **11**, 888–893.
3. (a) Collie, J.N. and Myers, W.S. (1893) *J. Chem. Soc.*, **63**, 122–128; (b) Collie, J.N. (1907) *Proc. Chem. Soc.*, **23**, 230–231.
4. Harris, T.M. and Harris, C.M. (1986) *Pure Appl. Chem.*, **58**, 283–294.
5. (a) Kobuke, Y. and Yoshida, J. (1978) *Tetrahedron Lett.*, **19**, 367–370; (b) Sakai, N., Sordé, N., and Matile, S. (2001) *Molecules*, **6**, 845–851; (c) Chen, H. and Harrison, P.H.M. (2002) *Can. J. Chem.*, **80**, 601–607; (d) Ji, Q., Williams, H.J., Roessner, C.H., and Scott, A.I. (2007) *Tetrahedron Lett.*, **48**, 8026–8028.
6. Wijkman, N. (1931) *Liebigs Ann. Chem.*, **485**, 61–73.

7. Baldwin, J.E., Barton, D.H.R., Bloomer, J.L., Jackman, L.M., Rodriguez-Hahn, L., and Sutherland, J.K. (1962) *Experientia*, **18**, 345–388.
8. For Sutherland's work on structure elucidation, see: (a) Sutherland, J.K. and Barton, D.H.R. (1965) *J. Chem. Soc.*, 1769–1771; (b) Barton, D.H.R., Jackman, L.M., Rodriguez-Hahn, L., and Sutherland, J.K. (1965) *J. Chem. Soc.*, 1772–1778; (c) Barton, D.H.R., Godinho, L.D.S., and Sutherland, J.K. (1965) *J. Chem. Soc.*, 1779–1786; (d) Baldwin, J.E., Barton, D.H.R., and Sutherland, J.K. (1965) *J. Chem. Soc.*, 1787–1798.
9. For Sutherland's work on structure elucidation, see: (a) Bloomer, J.L., Moppett, C.E., and Sutherland, J.K. (1968) *J. Chem. Soc. C*, 588–591; (b) Huff, R.K., Moppett, C.E., and Sutherland, J.K. (1972) *J. Chem. Soc., Perkin Trans. 1*, 2584–2590; (c) Moppett, C.E. and Sutherland, J.K. (1966) *J. Chem. Soc., Chem. Commun.*, 772–773.
10. Huff, R.K., Moppett, C.E., and Sutherland, J.K. (1968) *J. Chem. Soc., Chem. Commun.*, 1192–1193.
11. Baldwin, J.E., Beyeler, A., Cox, R.J., Keats, C., Pritchard, G.J., Adlington, R.M., and Watkin, D.J. (1999) *Tetrahedron*, **55**, 7363–7374.
12. Baldwin, J.E., Adlington, R.M., Roussi, F., Bulger, P.G., Marquez, R., and Mayweg, V.W. (2001) *Tetrahedron*, **57**, 7409–7416.
13. (a) Dabrah, T.T., Harwood, H.J., Huang, L.H., Jankovich, N.D., Kaneko, T., Li, J.C., Lindsey, S., Moshier, P.M., Subashi, T.A., Therrien, M., and Watts, P.C. (1997) *J. Antibiot.*, **50**, 1–50; (b) Dabrah, T.T., Kaneko, T., Massefski, W., and Whipple, E.B. (1997) *J. Am. Chem. Soc.*, **119**, 1594–1598.
14. Spencer, P., Agnelli, F., Williams, H.J., Keller, N.P., and Sulikowski, G.A. (2000) *J. Am. Chem. Soc.*, **122**, 420–421.
15. Sulikowski, G.A., Agnelli, F., Spencer, P., Koomen, J.M., and Russell, D.H. (2002) *Org. Lett.*, **4**, 1447–1450.
16. (a) Sulikowski, G.A., Agnelli, F., and Corbett, R.M. (2000) *J. Org. Chem.*, **65**, 337–342; (b) Sulikowski, G.A., Liu, W., Agnelli, F., Corbett, R.M., Luo, Z., and Hershberger, S.J. (2002) *Org. Lett.*, **4**, 1451–1454.
17. Diels, O. and Alder, K. (1928) *Justus Liebigs Ann. Chem.*, **460**, 98–122.
18. Takao, K.-I., Munakata, R., and Tadano, K.-I. (2005) *Chem. Rev.*, **105**, 4779–4807.
19. (a) Oikawa, H. and Tokiwano, T. (2004) *Nat. Prod. Rep.*, **21**, 321–352; (b) Stocking, E.M. and Williams, R.M. (2003) *Angew. Chem. Int. Ed.*, **42**, 3078–3115.
20. Alberts, A.W., Chen, J., Kuron, G., Hunt, V., Huff, J., Hoffman, C., Rothrock, J., Lopez, M., Joshua, H., Harris, E., Patchett, A., Monaghan, R., Currie, S., Stapley, E., Albers-Schönberg, G., Hensens, O., Hirschfield, J., Hoogsteen, K., Liesch, J., and Springer, J. (1980) *Proc. Natl. Acad. Sci. U.S.A.*, **77**, 3957–3961.
21. Endo, A. (1979) *J. Antibiot.*, **32**, 852–854.
22. (a) Chan, J.K., Moore, R.N., Nakashima, T.T., and Vederas, J.C. (1983) *J. Am. Chem. Soc.*, **105**, 3334–3336; (b) Moore, R.N., Bigam, G., Chan, J.K., Hogg, A.M., Takashima, T.T., and Vederas, J.C. (1985) *J. Am. Chem. Soc.*, **107**, 3694–3701; (c) Yoshizawa, Y., Witter, D.J., Liu, Y., and Vederas, J.C. (1994) *J. Am. Chem. Soc.*, **116**, 2693–2694.
23. Endo, A. and Hasumi, K. (1993) *Nat. Prod. Rep.*, **10**, 541–550.
24. Witter, D.J. and Vederas, J.C. (1996) *J. Org. Chem.*, **61**, 2613–2623.
25. Auclair, K., Sutherland, A., Kennedy, J., Witter, D.J., van den Heever, J.P., Hutchinson, C.R., and Vederas, J.C. (2000) *J. Am. Chem. Soc.*, **122**, 11519–11520.
26. (a) Ichihara, A., Tazaki, H., and Sakamura, S. (1983) *Tetrahedron Lett.*, **24**, 5373–5376; (b) Alam, S.S., Bilton, J.N., Slawin, A.M.Z., Williams, D.J., Sheppard, R.N., and Strange, R.N. (1989) *Phytochemistry*, **28**, 2627–2630; (c) Oikawa, H., Yokota, T., Sakano, C., Suzuki, Y., Naya, A., and Ichihara, A.

(1998) *Biosci. Biotechnol. Biochem.*, **62**, 2016–2022; (d) Mizushina, Y., Kamisuki, S., Kasai, N., Shimazaki, N., Takemura, M., Asahara, H., Linn, S., Yoshida, S., Matsukage, A., Koiwai, O., Sugawara, F., Yoshida, H., and Sakaguchi, K. (2002) *J. Biol. Chem.*, **277**, 630–638.

27. (a) Oikawa, H., Yokota, T., Abe, T., Ichihara, A., Sakamura, S., Yoshizawa, Y., and Vederas, J.C. (1989) *J. Chem. Soc., Chem. Commun.*, 1282–1284; (b) Oikawa, H., Yokota, T., Ichihara, A., and Sakamura, S. (1989) *J. Chem. Soc., Chem. Commun.*, 1284–1285.

28. (a) Oikawa, H., Suzuki, Y., Naya, A., Katayama, K., and Ichihara, A. (1994) *J. Am. Chem. Soc.*, **116**, 3605–3606; (b) Oikawa, H., Suzuki, Y., Katayama, K., Naya, A., Sakano, C., and Ichahara, A. (1999) *J. Chem. Soc., Perkin Trans. 1*, 1225–1232.

29. (a) Oikawa, H., Katayama, K., Suzuki, Y., and Ichahara, A. (1995) *J. Chem. Soc., Chem. Commun.*, 1321–1322; (b) Katayama, K., Kobayashi, T., Oikawa, H., Honma, M., and Ichihara, A. (1998) *Biochim. Biophys. Acta*, **1383**, 387–395.

30. Ichihara, A., Miki, M., Tazaki, H., and Sakamura, S. (1987) *Tetrahedron Lett.*, **28**, 1175–1178.

31. Oikawa, H., Kobayashi, T., Katayama, K., Suzuki, Y., and Ichihara, I. (1998) *J. Org. Chem.*, **63**, 8748–8756.

32. Katayama, K., Kobayashi, T., Chijimatsu, M., Ichihara, A., and Oikawa, H. (2008) *Biosci. Biotechnol. Biochem.*, **72**, 604–607.

33. Wilson, R.M., Jen, W.S., and MacMillan, D.W.C. (2005) *J. Am. Chem. Soc.*, **127**, 11616–11617.

34. (a) Celmer, W.D., Chmurny, G.N., Moppett, C.E., Ware, R.S., Watts, P.C., and Whipple, E.B. (1980) *J. Am. Chem. Soc.*, **102**, 4148883; (b) Celmer, W.D., Cullen, W.P., Moppett, C.E., Jefferson, M.T., Huang, L.H., Shibakawa, R., and Tone, J. (1979) US Patent 4, 148,883

35. (a) Cane, D.E. and Yang, C.-C. (1984) *J. Am. Chem. Soc.*, **106**, 784–787;

(b) Cane, D.E. and Yang, C.-C. (1985) *J. Antibiot.*, **38**, 423–426.

36. (a) Cane, D.E., Tan, W., and Ott, W.R. (1993) *J. Am. Chem. Soc.*, **115**, 527–535; (b) Cane, D.E. and Ott, W.R. (1988) *J. Am. Chem. Soc.*, **110**, 4840–4841; (c) Cane, D.E. and Luo, G. (1995) *J. Am. Chem. Soc.*, **117**, 6633–6634.

37. Coe, J.W. and Roush, W.R. (1989) *J. Org. Chem.*, **54**, 915–930.

38. Roush, W.R., Koyama, K., Curtin, M.L., and Moriarty, K.J. (1996) *J. Am. Chem. Soc.*, **118**, 7502–7512.

39. (a) D'Auria, M.V., Debitus, C., Paloma, L.G., Minale, L., and Zampella, A. (1994) *J. Am. Chem. Soc.*, **116**, 6658–6663; (b) D'Auria, M.V., Paloma, L.G., Minale, L., Zampella, A., and Debitus, C. (1994) *J. Nat. Prod.*, **57**, 1595–1597.

40. (a) Tortosa, M., Yakelis, N.A., and Roush, W.R. (2008) *J. Am. Chem. Soc.*, **130**, 2722–2723; (b) Tortosa, M., Yakelis, N.A., and Roush, W.R. (2008) *J. Org. Chem.*, **73**, 9657–9667.

41. Brown, R.F.C., Drummond, R., Fogerty, A.C., Hughes, G.K., Pinhey, J.T., Ritchie, E., and Taylor, W.C. (1956) *Aust. J. Chem.*, **9**, 283–287.

42. Baldwin, J.E., Chesworth, R., Parker, J.S., and Russell, A.T. (1995) *Tetrahedron Lett.*, **36**, 9551–9554.

43. (a) Tchabanenko, K., Chesworth, R., Parker, J.S., Anand, N.K., Russell, A.T., Adlington, R.M., and Baldwin, J.E. (2005) *Tetrahedron*, **61**, 11649–11656; (b) Tchabanenko, K., Adlington, R.M., Cowley, A.R., and Baldwin, J.E. (2005) *Org. Lett.*, **7**, 585–588.

44. Huang, X.-H., van Soest, R., Roberge, M., and Andersen, R.J. (2004) *Org. Lett.*, **6**, 75–78.

45. Kirkham, J.E.D., Lee, V., and Baldwin, J.E. (2006) *Chem. Commun.*, 2863–2865.

46. Crossman, J.S. and Perkins, M.V. (2008) *Tetrahedron*, **64**, 4852–4867.

47. (a) Hautzel, R. and Anke, H. (1990) *Z. Naturforsch., Teil C*, **45**, 68–73; (b) Köpcke, B., Johansson, M., Sterner, O., and Anke, H. (2002) *J. Antibiot.*, **55**, 36–40; (c) Johansson, M., Köpcke, B., Anke, H., and Sterner, O. (2002) *J. Antibiot.*, **55**, 104–106.

48. Weidler, M., Rether, J., Anke, T., and Erkel, G. (2000) *FEBS Lett.*, **484**, 1–6.
49. Hellsten, R., Johansson, M., Dahlman, A., Dizeyi, N., Sterner, O., and Bjartell, A. (2008) *Prostate*, **68**, 269–280.
50. Steglich, W., Eizenhöfer, T., Casser, I., Steffan, B., Rabe, U., Boeker, R., Knerr, H.J., Anke, H., and Anke, T. (1993) *DECHEMA Monograph*, vol. 129, (ed. T. Anke and U. Onken), VCH, Weinheim, pp. 3–13.
51. (a) Johansson, M., Köpcke, B., Anke, H., and Sterner, O. (2002) *Angew. Chem. Int. Ed.*, **41**, 2158–2160; (b) Johansson, M., Köpcke, B., Anke, H., and Sterner, O. (2002) *Tetrahedron Lett.*, **58**, 2523–2528.
52. Lebel, H. and Parmentier, M. (2007) *Org. Lett.*, **9**, 3563–3566.
53. (a) Bister, B., Bischoff, D., Ströbele, M., Riedlinger, J., Reicke, A., Wolter, F., Bull, A.T., Zähner, H., Fielder, H.-P., and Süssmuth, R.D. (2004) *Angew. Chem. Int. Ed.*, **43**, 2574–2576; (b) Riedlinger, J., Reicke, A., Zähner, H., Krismer, B., Bull, A.T., Maldonado, L.A., Ward, A.C., Goodfellow, M., Bister, B., Bischoff, D., Süssmuth, R.D., and Fiedler, H.-P. (2004) *J. Antibiot.*, **57**, 271–279.
54. Zapf, C.W., Harrison, B.A., Drahl, C., and Sorensen, E.J. (2005) *Angew. Chem. Int. Ed.*, **44**, 6533–6537.
55. (a) Nicolaou, K.C. and Harrison, S.T. (2007) *J. Am. Chem. Soc.*, **129**, 429–440; (b) Nicolaou, K.C., Harrison, S.T., and Chen, J.S. (2009) *Synthesis*, **129**, 33–42.
56. Keler, S., Nicholson, G., Drahl, C., Sorensen, E.J., Fiedler, H.-P., and Süssmuth, R.D. (2007) *J. Antibiot.*, **60**, 391–394.
57. Seki, T., Satake, M., Mackenzie, L., Kaspar, H.F., and Yasumoto, T. (1995) *Tetrahedron Lett.*, **36**, 7093–7096.
58. Uemura, D., Chou, T., Haino, T., Nagatsu, A., Fukuzawa, S., Zheng, S.Z., and Chen, H.S. (1995) *J. Am. Chem. Soc.*, **117**, 1155–1156.
59. Johannes, J.W., Wenglowsky, S., and Kishi, Y. (2005) *Org. Lett.*, **7**, 3997–4000.
60. (a) Sato, B., Muramatsu, H., Miyauchi, M., Hori, Y., Takase, S., Hino, M., Hashimoto, S., and Terano, H. (2000) *J. Antibiot.*, **53**, 123–130; (b) Sato, B., Nakajima, H., Hori, Y., Hino, M., Hashimoto, S., and Terano, H. (2000) *J. Antibiot.*, **53**, 204–206; (c) Yoshimura, S., Sato, B., Kinoshita, T., Takase, S., and Terano, H. (2000) *J. Antibiot.*, **53**, 615–622.
61. Yoshimura, S., Sato, B., Kinoshita, T., Takase, S., and Terano, H. (2002) *J. Antibiot.*, **55**, C1.
62. Höfs, R., Walker, M., and Zeeck, A. (2000) *Angew. Chem. Int. Ed.*, **39**, 3258–3261.
63. (a) Evans, D.A. and Starr, J.T. (2002) *Angew. Chem. Int. Ed.*, **41**, 1787–1790; (b) Evans, D.A. and Starr, J.T. (2003) *J. Am. Chem. Soc.*, **125**, 13531–13540.
64. (a) Vanderwal, C.D., Vosburg, D.A., Weiler, S., and Sorensen, E.J. (1999) *Org. Lett.*, **1**, 645–648; (b) Vanderwal, C.D., Vosburg, D.A., and Sorensen, E.J. (2001) *Org. Lett.*, **3**, 4307–4310; (c) Vosburg, D.A., Vanderwal, C.D., and Sorensen, E.J. (2002) *J. Am. Chem. Soc.*, **124**, 4552–4553; (d) Vanderwal, C.D., Vosburg, D.A., Weiler, S., and Sorensen, E.J. (2003) *J. Am. Chem. Soc.*, **125**, 5393–5407.
65. Shindo, K. and Kawai, H. (1992) *J. Antibiot.*, **45**, 292–295.
66. (a) Shindo, K., Matsuoka, M., and Kawai, H. (1996) *J. Antibiot.*, **49**, 241–243; (b) Shindo, K., Iijima, H., and Kawai, H. (1996) *J. Antibiot.*, **49**, 244–248.
67. Shindo, K., Sakakibara, M., and Kawai, H. (1996) *J. Antibiot.*, **49**, 249–252.
68. Chang, J.Y. and Paquette, L.A. (2002) *Org. Lett.*, **4**, 253–256.
69. Tatsuta, K., Narazaki, F., Kashiki, N., Yamamoto, J., and Nakano, S. (2003) *J. Antibiot.*, **56**, 584–590.
70. Dineen, T.A. and Roush, W.R. (2004) *Org. Lett.*, **6**, 2043–2046.
71. (a) Jackson, M., Karwowski, J.P., Theriault, R.J., Rasmussen, R.R., Hensey, D.M., Humphrey, P.E., Swanson, S.J., Barlow, G.J., Premachandran, U., and McAlpine, J.B.

(1995) *J. Antibiot.*, **48**, 462–466; (b) Hochlowski, J.E., Mullally, M.M., Henry, R., Whittern, D.M., and McAlpine, J.B. (1995) *J. Antibiot.*, **48**, 467–470.

72. Tanaka, M., Nara, F., Yamasato, Y., Masuda-Inoue, S., Doi-Yoshioka, H., Kumakura, S., Enokita, R., and Ogita, T. (1999) *J. Antibiot.*, **52**, 670–673.

73. (a) Munakata, R., Katakai, H., Ueki, T., Kurosaka, J., Takao, K., and Tadano, K. (2003) *J. Am. Chem. Soc.*, **125**, 14722–14723; (b) Munakata, R., Katakai, H., Ueki, T., Kurosaka, J., Takao, K., and Tadano, K. (2004) *J. Am. Chem. Soc.*, **126**, 11254–11267.

74. Mertz, F.P. and Yao, R.C. (1990) *Int. J. Syst. Bacteriol.*, **37**, 19–22.

75. For a review on the biochemistry of spinosyns, see: Huang, K.-X., Xia, L., Zhang, Y., Ding, X., and Zahn, J.A. (2009) *Appl. Microbiol. Biotechnol.*, **82**, 13–23.

76. Waldron, C., Matsushima, P., Rostek, P.R. Jr., Broughton, M.C., Turner, J., Madduri, K., Crawford, K.P., Merlo, D.J., and Baltz, R.H. (2001) *Chem. Biol.*, **8**, 487–499.

77. Evans, D.A. and Black, W.C. (1993) *J. Am. Chem. Soc.*, **115**, 4497–4513.

78. Mergott, D.J., Franck, S.A., and Roush, W.R. (2004) *Proc. Natl. Acad. Sci. U.S.A.*, **101**, 11955–11959.

79. Winbush, S.M., Mergott, D.J., and Roush, W.R. (2008) *J. Org. Chem.*, **73**, 1818–1829.

80. Davies, D.H., Snape, E.W., Suter, P.J., King, T.J., and Falshaw, C.P. (1981) *J. Chem. Soc., Chem. Commun.*, 1073–1074.

81. For biosynthetic studies of tetronasin, see: (a) Bulsing, J.M., Laue, E.D., Leeper, F.J., Staunton, J., Davies, D.H., Ritchie, G.A.F., Davies, A., Davies, A.B., and Mabelis, R.P. (1984) *J. Chem. Soc., Chem. Commun.*, 1301–1302; (b) Doddrell, D.M., Laue, E.D., Leeper, F.J., Staunton, J., Davies, A., Davies, A.B., and Ritchie, G.A.F. (1984) *J. Chem. Soc., Chem. Commun.*, 1302–1304; (c) Demetriadou, A.K., Laue, E.D., Staunton, J., Ritchie, G.A.F., Davies, A., and Davies, A.B. (1985) *J. Chem. Soc., Chem. Commun.*, 408–410; (d) Hailes, H.C., Jackson, C.M., Leadlay, P.F., Ley, S.V., and Staunton, J. (1994) *Tetrahedron Lett.*, **35**, 307–310; (e) Hailes, H.C., Handa, S., Leadlay, P.F., Lennon, I.C., Ley, S.V., and Staunton, J. (1994) *Tetrahedron Lett.*, **35**, 311–314; (f) Hailes, H.C., Handa, S., Leadlay, P.F., Lennon, I.C., Ley, S.V., and Staunton, J. (1994) *Tetrahedron Lett.*, **35**, 315–318; (g) Boons, G.-J., Clase, J.A., Lennon, I.C., Ley, S.V., and Staunton, J. (1995) *Tetrahedron*, **51**, 5417–5446; (h) Less, S.L., Leadlay, P.F., Dutton, C.J., and Staunton, J. (1996) *Tetrahedron Lett.*, **37**, 3519–3520.

82. For biomimetic synthesis studies of tetronasin, see: (a) Boons, G.-J., Lennon, I.C., Ley, S.V., Owen, E.S.E., Staunton, J., and Wadsworth, D.J. (1994) *Tetrahedron Lett.*, **35**, 323–326; (b) Ley, S.V., Brown, D.S., Clase, J.A., Fairbanks, A.J., Lennon, I.C., Osborn, H.M.I., Stokes, E.S.E., and Wadsworth, D.J. (1998) *J. Chem. Soc., Perkin Trans. 1*, 2259–2276.

83. Hori, K., Kazuno, H., Nomura, K., and Yoshii, E. (1993) *Tetrahedron Lett.*, **34**, 2183–2186.

84. Isaka, M., Rugseree, N., Maithip, P., Kongsaeree, P., Prabpai, S., and Thebtaranonth, Y. (2005) *Tetrahedron*, **61**, 5577–5583.

85. He, H.Y., Yang, H.-Y., Bigelis, R., Solum, E.H., Greenstein, M., and Carter, G.T. (2002) *Tetrahedron Lett.*, **43**, 1633–1636.

86. Shiono, Y., Shimanuki, K., Hiramatsu, F., Koseki, T., Tetsuya, M., Fujisawa, N., and Fimura, K.-I. (2008) *Bioorg. Med. Chem. Lett.*, **18**, 6050–6053.

87. Koizumi, F., Hasegawa, A., Ando, K., Ogawa, T., Hara, M., and Yoshida, M. (2001) *Jpn. Kokai Tokkyo Koho* JP 2, 001,247,574; *Chem. Abstr.*, **135** (2001) 209979.

88. Oikawa, H. (2003) *J. Org. Chem.*, **68**, 3552–3557.

89. Isaka, M., Prathumpai, W., Wongsa, P., and Tanticharoen, M. (2006) *Org. Lett.*, **8**, 2815–2817.

90. Nicolaou, K.C., Sarlah, D., Wu, T.R., and Zhan, W. (2009) *Angew. Chem. Int. Ed.*, **48**, 6870–6874.
91. Gonzáles-Lergier, J., Broadbelt, L.J., and Hatzimanikatis, V. (2005) *J. Am. Chem. Soc.*, **127**, 9930–9938.
92. Breibauer, R., Vetter, I.R., and Waldmann, H. (2002) *Angew. Chem. Int. Ed.*, **41**, 2878–2890.
93. Burke, M.D. and Figreiber, S.L. (2004) *Angew. Chem. Int. Ed.*, **43**, 46–58.
94. (a) Pelish, H.E., Westwood, N.J., Feng, Y., Kirchhausen, T., and Shair, M.D. (2001) *J. Am. Chem. Soc.*, **123**, 6740–6741; (b) Lindsley, C.W., Chan, L.K., Goess, B.C., Joseph, R., and Shair, M.D. (2000) *J. Am. Chem. Soc.*, **122**, 422–423.
95. Xue, Q., Ashley, G., Hutchinson, C.R., and Santi, D.V. (1999) *Proc. Natl. Acad. Sci. U.S.A.*, **96**, 11740–11745.

15
Biomimetic Synthesis of Polyether Natural Products via Polyepoxide Opening
Ivan Vilotijevic and Timothy F. Jamison

15.1
Introduction

The simple ether, a C–O–C motif, appears in nearly all families of oxygen-containing natural products, which are synthesized by organisms in all kingdoms of life. A subgroup of natural products characterized by the regular occurrence of multiple C–O–C motifs is designated the polyether family, and these can be broadly divided into linear and polycyclic polyethers. The latter group is of special interest due to its structural diversity and the biological activity of its members, which ranges from antibiotic, antifungal, and anticancer properties to extreme toxicity.

Distinctive structural elements of polycyclic polyethers can be used as a foundation for their classification. Depending on these particular structural features, which can generally be traced back to the biosynthetic pathways for the synthesis of these molecules, polycyclic polyethers are divided into two main groups. The first group includes molecules with multiple fused cyclic ethers that are postulated to be formed in nature via all-*endo* cascades of epoxide openings. The other group consists of molecules that are produced via all-*exo* biosynthetic cascades of epoxide-opening reactions and normally feature multiple rings that are interconnected by a carbon–carbon bond. Some polyethers are produced via cascades that feature both *endo* and *exo* epoxide openings. Further distinction between different classes of polycyclic polyethers is made based on their biosynthetic origin; these natural products are either polyketide- or terpene-derived.

For the purposes of this chapter, polycyclic polyethers will be classified in three major groups: polyether ionophores [1], squalene-derived polyethers [2], and the ladder polyethers [3]. Each of these groups will be discussed in the context of epoxide-opening cascades. Such reactions are postulated to be involved in their biosynthesis [4] and have been utilized as a method to rapidly construct polyether frameworks in the total synthesis of polycyclic polyethers [5].

Biomimetic Organic Synthesis, First Edition. Edited by Erwan Poupon and Bastien Nay.
© 2011 Wiley-VCH Verlag GmbH & Co. KGaA. Published 2011 by Wiley-VCH Verlag GmbH & Co. KGaA.

15.2
Synthetic Considerations: Baldwin's Rules

Regioselectivity and stereospecificity in epoxide-opening reactions are the main determinants of the product composition in epoxide-opening cascades and intramolecular epoxide-opening reactions in general. The vast majority of epoxide ring-opening reactions proceed with inversion of configuration. Regioselectivity, however, is far more variable and case-dependant.

Baldwin's rules of ring closure are a three-criteria classification of ring-closing reactions based on both empirical results and theoretical considerations, and are used to predict the outcome of intramolecular ring-forming reactions, including epoxide openings. The criteria are: the size of the formed ring, the position of the bond that is broken relative to the smallest formed ring, and the geometry of the electrophile [6]. If the position of the bond broken during the ring closing reaction is exocyclic, outside of the formed ring, then the reaction is classified as *exo*. If the broken bond is within the smallest formed ring, the reaction is classified as *endo* (Figure 15.1a). In Baldwin's classification, reactions involving sp^3 hybridized electrophiles are described as *tet* due to the tetragonal geometry of the electrophile, sp^2 hybridized electrophiles are *trig*, and sp electrophiles are diagonal or *dig*. With such classification in mind, Baldwin formulated a simple set of guidelines to predict the relative feasibility of different ring closing reactions [6]. Although empirical, Baldwin's rules are conceptually based on stereoelectronic considerations [7]. Favored ring-closing reactions are those in which the length and nature of the linking chain enable the terminal atoms to achieve the proper geometries for the reaction. Disfavored ring closings, on the other hand, generally require severe distortions of bond angles and bond distances. For instance, 4-*exo-trig* reactions are predicted to be favored over 5-*endo-trig* ring closing reactions (Figure 15.1a).

With few exceptions, intramolecular epoxide-opening reactions favor the smaller heterocycle (e.g., tetrahydrofuran **5**, likely arising from a spiro transition state, Figure 15.1b), not the larger one (tetrahydropyran **6**, from fused transition state, Figure 15.1b) [8]. Baldwin designates products that arise via fused and spiro

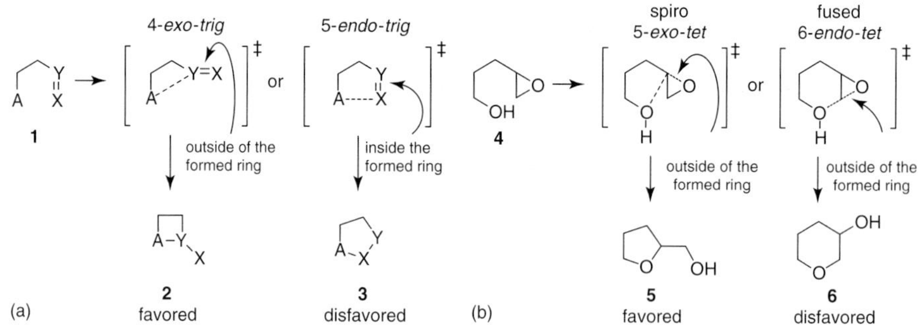

Figure 15.1 (a) Baldwin's rules: general classification; (b) Baldwin's rules in intramolecular epoxide-opening reactions [6].

transition states as *endo* and *exo*, respectively. However, because the epoxide C–O bond that breaks is outside of the newly formed ring in both cases, each can be considered an *exo* process under the same construct (Figure 15.1b). To avoid potential confusion, the distinct terms "fused" and "spiro" can be used to describe the transition states in epoxide-opening reactions [9]. Intramolecular epoxide-opening reactions tend to follow the rules that lie between those for tetrahedral and trigonal systems, generally favoring what are usually termed the *exo processes*, that is, those that proceed via a spiro transition state [6].

15.2.1
Control of Regioselectivity in Intramolecular Epoxide-Opening Reactions

The development of efficient methods for enantioselective epoxidation such as the Sharpless asymmetric epoxidation [10, 11], Jacobsen epoxidation [12, 13], and the Shi epoxidation [14–16] make epoxides attractive intermediates in asymmetric synthesis [17]. These methods have enabled syntheses of many of the polyepoxides that will be discussed herein, and have thus accelerated investigations of epoxide-opening cascades. For epoxides to be versatile synthetic intermediates, effective ways to control the regioselectivity in epoxide-opening reactions are necessary. The *exo* mode of cyclization is typically preferred; therefore, methods to facilitate *endo* cyclization have constituted a particularly active area of research.

Most of the approaches to promote the desired *endo* outcome of intramolecular epoxide openings use directing groups covalently attached to the epoxides. These directing groups either stabilize (relative to an H atom) the desired transition states, enabling regioselective nucleophilic attack, or make the undesired cyclization route less energetically favorable by changing the electronic properties of the epoxide. Currently available methods for *endo* cyclization of epoxides rely on alkenyl [20, 23–26], alkynyl [27–29], alkyl [19, 30, 31], and silyl [18, 32, 33] substituents that stabilize partial positive charge within the desired, fused transition state in the Lewis or Brønsted acid-catalyzed reactions (Scheme 15.1a). The directing groups that promote *endo* cyclization via destabilization of the undesired spiro transition state include sulfones [21, 34, 35], as well as methoxymethyl substituents in combination with a lanthanide Lewis acid [22, 36–38] (Scheme 15.1b). Catalytic antibodies [39–42] and transition-metal complexes [43, 44] can also be particularly effective in promoting *endo* cyclization by lowering the energies of fused transition states in certain cases.

15.3
Polycyclic Polyethers: Structure and Biosynthesis

15.3.1
Polyether Ionophores

Polyether ionophores are lipophilic carboxylic acids that contain multiple five- and six-membered cyclic ethers organized either as spiroketals or as linked cyclic

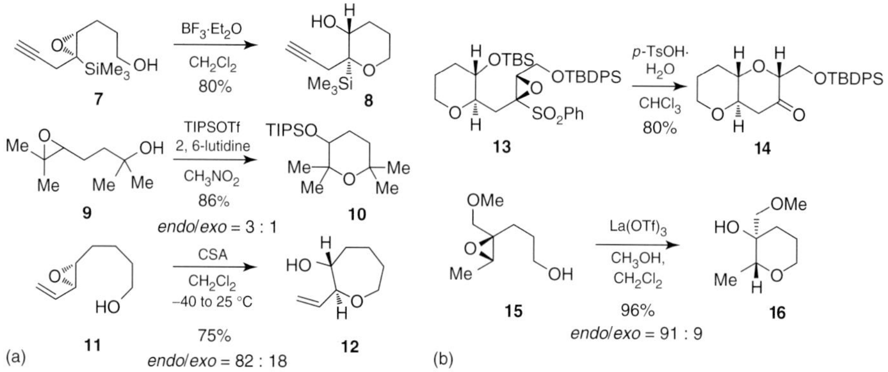

Scheme 15.1 *Endo* cyclizations via fused transition states: (a) stabilized by directing groups [18–20]; (b) enabled by deactivation of *exo* pathway [21, 22]. TIPS = triisopropylsilyl, Tf = trifluorosulfonyl, CSA = camphorsulfonic acid, Ts = toluenesulfonyl, TBS = tert-butyldimethylsilyl, and TBDPS = tert-butyldiphenylsilyl.

ethers (Figure 15.2). The first members of this family, X-206, nigericin, and lasalocid A, were isolated in 1951, but due to their toxicity did not initially draw much attention [45, 46]. It was not until 1967 – when a crystal structure of monensin A (**17**, Figure 15.2) was disclosed [47] and the cation binding abilities of these molecules were first examined [48] – that this family of natural products was thrust back into the spotlight. Subsequent discoveries of their ability to control coccidiosis [49], a devastating poultry disease, and their action as growth promoters in ruminant animals [50], both of which capitalize upon the antibiotic activity of these structures, inspired several research groups to pursue the isolation of novel members of this family, study their biosynthesis, and put effort into their total synthesis.

The biological function of polyether ionophores is directly related to their ability to selectively bind metal cations via coordination with multiple oxygen atoms and, due to their lipophilic nature, transport them through biological membranes [48]. By doing so, polyether ionophores disturb the delicate dynamic equilibria of cations across the cell membrane and thus disrupt regular cell function [51], resulting in diverse effects, including antibiotic, antimalarial, anti-obesity, and insecticide activity.

Since the isolation of the first polyether ionophores in early 1950s, well over 100 members of this family have been isolated and characterized [52–54]. Although most members of this family are produced by the *Streptomyces* genus, polyether ionophores have also been isolated from other actinomycetes [4]. A large body of experimental data on their biosynthesis [55–60] and earlier speculation by Westley [55] led the groups of Cane, Celmer, and Westley to propose a unified stereochemical model of polyether antibiotic structure and biogenesis in 1983 [61].

According to the Cane–Celmer–Westley hypothesis, in the biosynthesis of monensin A an all-*(E)* polyene precursor **25**, produced in classic type I polyketide

15.3 Polycyclic Polyethers: Structure and Biosynthesis

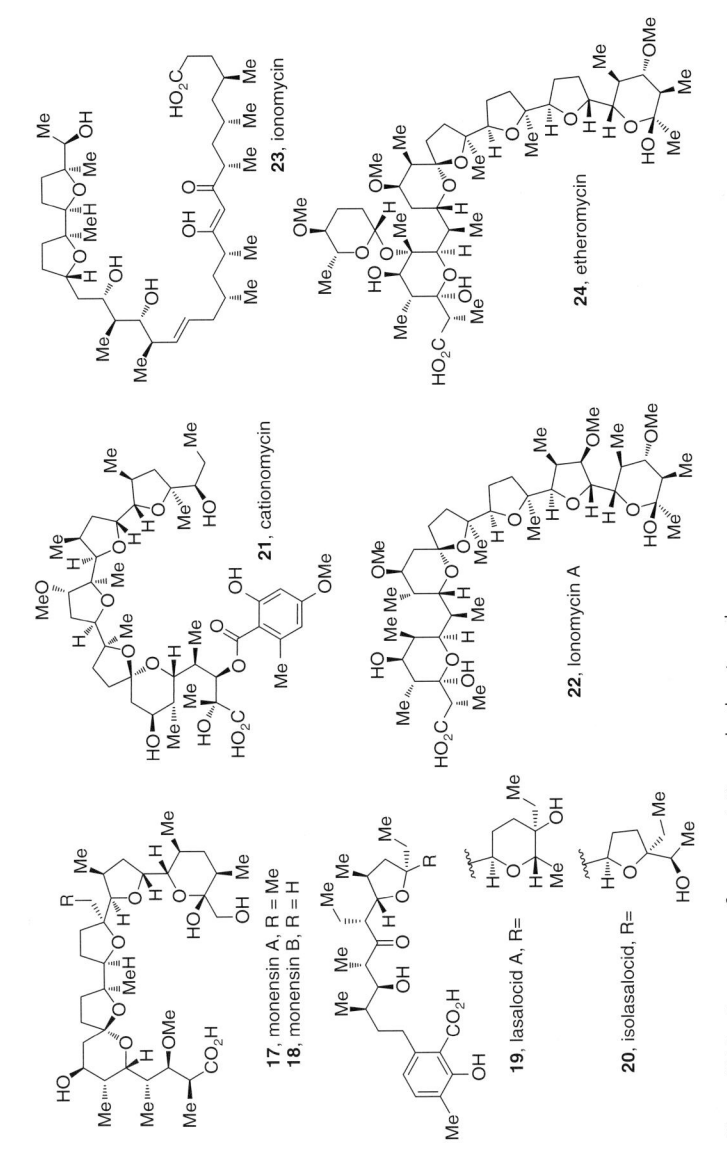

Figure 15.2 Structures of representative polyether ionophores.

synthase fashion from five acetates, seven propionates, and one butyrate unit, is oxidized to the corresponding polyepoxide **26** (Figure 15.3a). Nucleophilic addition of the C5 hydroxyl in **26** to the C9 ketone forms a hemiketal that triggers a cascade of all-*exo* epoxide-opening events leading to the formation of monensin A. Cane, Celmer, and Westley further extended this proposal to the biosynthesis of all polyether ionophores known at the time and described the requisite polyene precursors and pathways to each of these natural products [61].

Failure of the producing organism to incorporate synthetic all-*(E)* premonensin triene **25** into the biosynthetic pathway and convert it into monensin A encouraged alternative biosynthetic proposals [62, 63]. Townsend suggested that monensin A may be produced from an all-*(Z)* isomer of **25** through a series of oxidative cyclizations proceeding via a [2+2] mechanism involving an Fe-containing monooxygenase (Figure 15.3b). Related proposals were considered by McDonald [64, 65] and Leadlay [66]. Upon sequencing the monensin biosynthetic gene cluster [66, 67], Leadlay found that deletion of *monCI* from the producing organism caused accumulation of all-*(E)* premonensin triene **25**. This suggested that a single oxidase enzyme, the product of *monCI*, is involved in the production of triepoxide **26** [68]. Disruption of *monBI* and *monBII* genes led to the production of partially cyclized, chemically competent biosynthetic intermediates [69].

The Oikawa group provided direct evidence for the involvement of an enzyme-catalyzed cascade of epoxide-opening reactions in the biosynthesis of polyether ionophore lasalocid A (**19**, Figure 15.2) [70]. Significant homology of *monBI* and *monBII* with *lsd19* from the lasalocid A biosynthetic genes identified the Lsd19 as the putative epoxide hydrolase. Lsd19 was obtained in nearly pure form by cloning *las19* and expressing it in *Escherichia coli*. This enzyme was then utilized in the efficient transformation of synthetic prelasalocid diepoxide into lasalocid A *in vitro* [70, 71]. Further *in vivo* studies by Leadlay and coworkers demonstrated that the presence of Lsd19 changes the stereochemical course of polyether ring formation, channeling the polyepoxide intermediate to lasalocid A as the major product [72]. When Lsd19 is not present, as in Δ*lsd19* mutant, the formation of the second ring proceeds exclusively by the kinetically favored pathway to form isolasalocid (**20**, Figure 15.2), thus demonstrating that Lsd19 is responsible for the final biosynthetic cyclization to form lasalocid A.

15.3.2
Polyethers Derived from Squalene

Polycyclic polyethers derived from squalene have been isolated from diverse sources, including marine sponges, red algae, and tropical plants [2]. Some of these natural products, often referred to as *oxasqualenoids*, resemble polyether ionophores due to the regular occurrence of 2,5-linked oligotetrahydrofuran segments, as in glabrescol (**33**, Figure 15.4). Other oxasqualenoids, typically isolated from marine sources, such as red algae of the *Laurencia* and *Chrondria* genera, resemble ladder polyether structures and feature multiple fused rings such as those found in armatol A (**38**). The third group of oxasqualenoids, exemplified by enshuol (**36**),

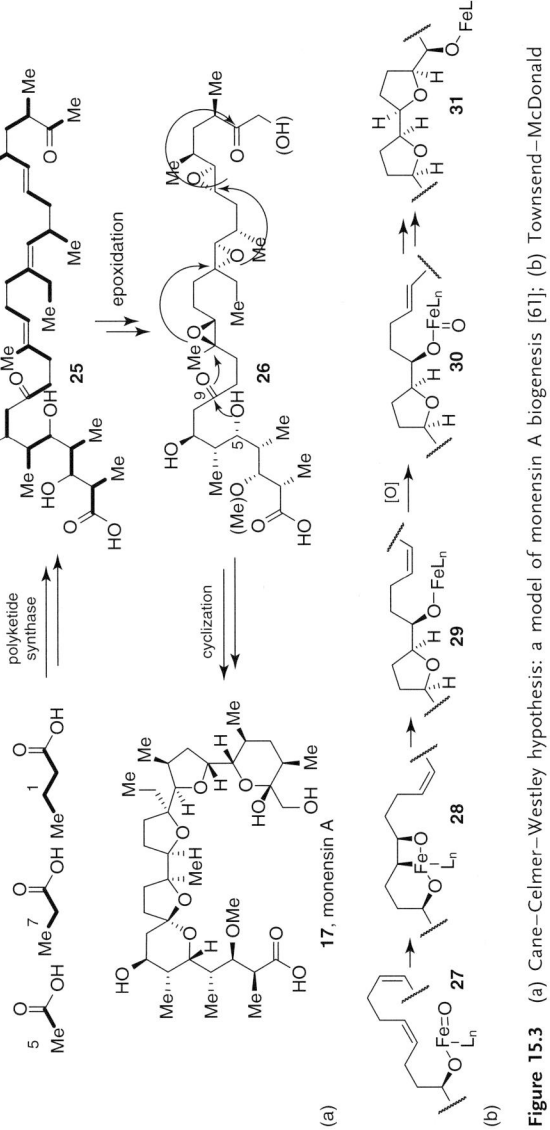

Figure 15.3 (a) Cane–Celmer–Westley hypothesis: a model of monensin A biogenesis [61]; (b) Townsend–McDonald hypothesis: biosynthesis via metal-mediated oxidative cyclization [62, 64].

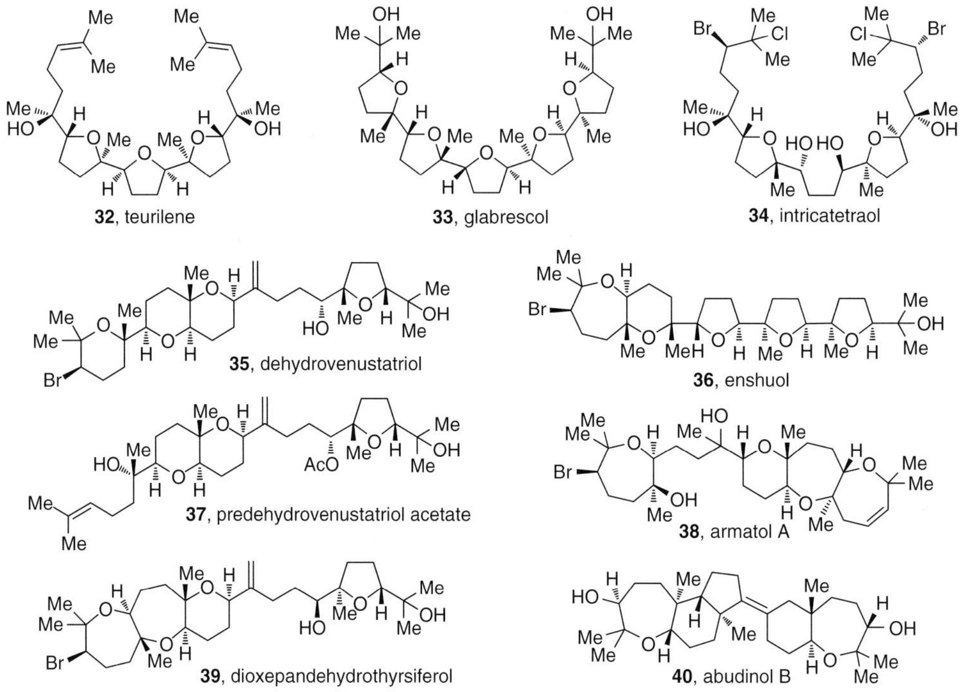

Figure 15.4 Structures of representative oxasqualenoids.

incorporates both types of polycyclic structures. In addition to these highly oxygenated triterpenes, related squalene-derived polycycles such as abudinol B (**40**) have been isolated from marine sponges. These molecules typically feature two separate cyclic systems and contain up to two cyclic ethers that are fused to other carbocycles and not to each other [2].

That these oxasqualenoids could be efficiently derived from squalene polyepoxide precursors was recognized soon after similar proposals were put forward for the polyether ionophores and ladder polyethers [2, 73, 74]. Reports of isolation of new oxasqualenoids are often accompanied by a biosynthetic proposal for the specific carbon skeleton [75–78]. The common ground for all proposals is the involvement of epoxide-opening cascades [2, 74, 79]. Biosynthetic studies on these molecules are, unfortunately, scarce. Their triterpene origin is evident due to the absence of skeletal rearrangements, but the proposed oxidation and cyclization steps remain speculative.

The important distinction between biosynthetic epoxide-opening cascades leading to polyether ionophores and those that are proposed for oxasqualenoids is the regioselectivity of epoxide-opening events. While polyether ionophores, with few exceptions, arise via all-*exo* epoxide-opening cascades, cascades that are proposed in the biosynthesis of oxasqualenoids have to incorporate both *endo* and *exo* epoxide-opening reactions depending on the structure of the oxasqualenoid natural product.

The presence of halogens in oxasqualenoid molecules (e.g., bromine at position 3 of the oxasqualenoids isolated from red algae, such as dehydrovenustatriol or armatol A) is an interesting structural feature that has itself drawn biosynthetic speculation. Formation of a bromonium species by the action of an electrophile could initiate the epoxide-opening cascade reaction, producing the complete tricycle of **39** in a single step (electrophilic initiation, Figure 15.5) [80]. However, isolation of dehydrovenustatriol and predehydrovenustatriol acetate (**35** and **37**, Figure 15.4) from the same producing organism suggests that at least two discrete steps are involved in the cyclization of the polyepoxy precursor. Predehydrovenustatriol acetate lacks the halogenated cyclic ether of dehydrovenustatriol in its otherwise fully cyclized structure, suggesting that it may be its direct precursor. If this is the case, dioxepandehydrothyrsiferol could be derived from the intermediate **42** produced in the preceding cascade (acid/base initiation, Figure 15.5).

15.3.3
Ladder Polyethers

The group of ladder polyether natural products consists of molecules featuring anywhere from 4 to 32 five- to nine-membered cyclic ethers, fused to each other in a *trans-syn-trans* arrangement. This creates a repeating C–C–O sequence that stretches throughout the polycyclic core of these molecules (Figure 15.6). The first isolated member of this family, brevetoxin B (**44**), was reported by Nakanishi and Clardy in 1981 [81] and was followed by numerous others, including maitotoxin [82–84], the largest nonpolymeric molecule isolated from natural sources to date. The minimal availability combined with the unprecedented size of ladder polyethers have inspired herculean endeavors in the isolation and structural characterization of these compounds and have pushed the limits of analytical methods, including chromatography, mass spectrometry, NMR, and X-ray diffraction [85]. The structural challenges associated with synthesizing these molecules have also stimulated development of many novel synthetic methodologies [86–89].

Ladder polyethers are notorious for their association with harmful algal blooms commonly referred to as *red tides* [90, 91]. A rapid increase in the concentration of dinoflagellate algae, for example, the brevetoxin-producing *Karenia brevis*, leads to the increased production of red tide toxins, some of which are members of ladder polyether family. The effects of red tide are devastating killings of fish and marine mammals. However, some marine species not affected by red tides accumulate and, occasionally, further elaborate the toxins [92, 93], transferring them up the food chain, resulting in human poisoning by ingestion of shellfish exposed to a red tide [94].

Despite their uniform structure, ladder polyethers exhibit diverse biological activities ranging from extreme toxicity [95–97] to anticancer [98–100] and antifungal [101, 102] properties. More recently, a member of this family, brevenal (**51**, Figure 15.6), has been shown to protect fish from the neurotoxic effects of brevetoxins [103, 104] and has been identified as a potential therapeutic for cystic fibrosis

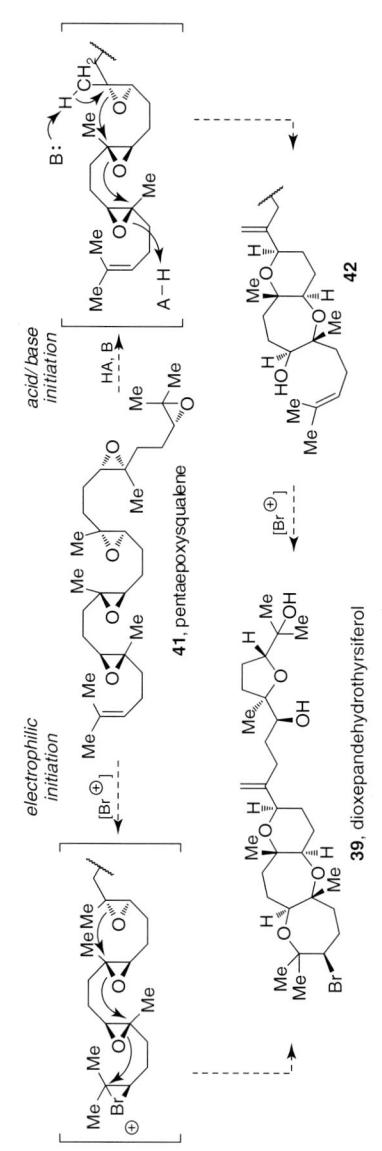

Figure 15.5 Proposed epoxide-opening cascades in the biosynthesis of dioxepandehydrothyrsiferol.

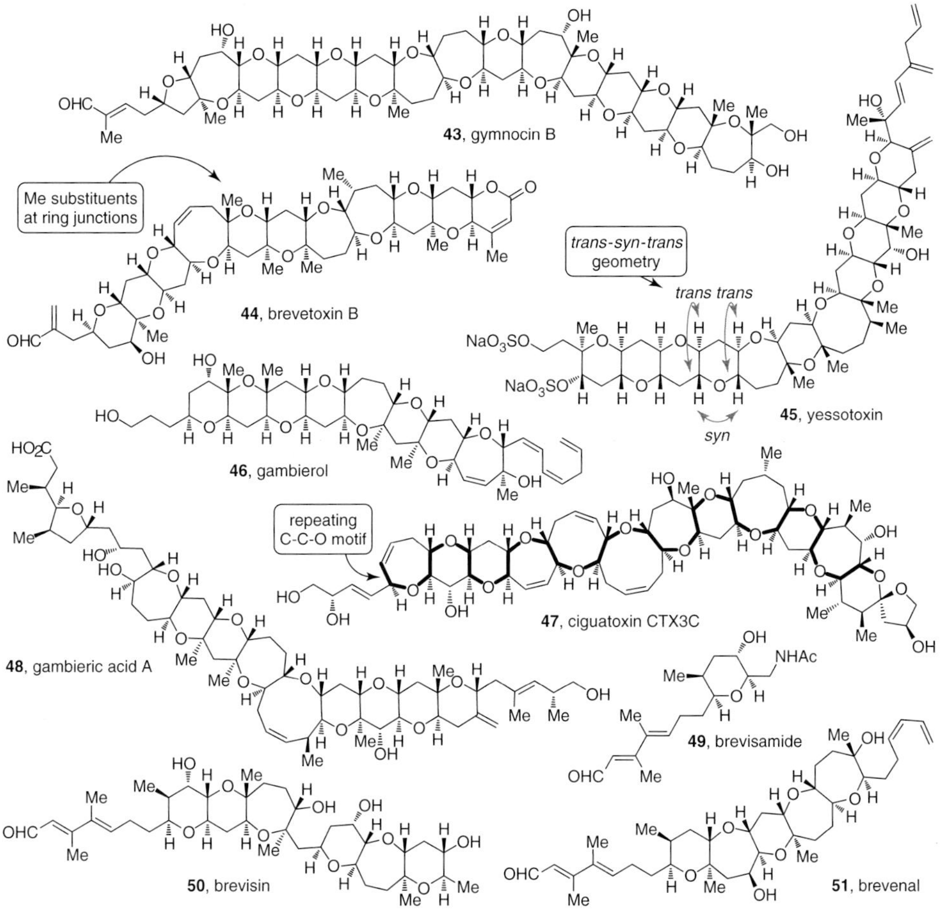

Figure 15.6 Structures of representative members of the ladder polyether family.

[105, 106]. While their mode of action is not well understood on the molecular level, it is known that brevetoxins and ciguatoxins bind and disrupt voltage-sensitive sodium channels [107–111], gambierol blocks voltage-gated potassium channels [112], and maitotoxin causes an influx of calcium ions into cells that in turn causes uncontrolled secretion of neurotransmitters and severe muscle contractions [113–116]. Binding of yessotoxin (**45**) to the transmembrane α-helix of glycophorin A causes the dissociation of oligomeric protein [117].

Soon after the structure of brevetoxin B was reported, Nakanishi [118] and Shimizu [119] hypothesized that the structural and stereochemical similarities among ladder polyethers are a direct consequence of their biosynthetic origin. Such similarity was proposed to arise through the transformation of a polyepoxide into a ladder polyether via a series or cascade of epoxide-opening events (Figure 15.7a). The oxygen and two carbon atoms of each epoxide constitute the C–C–O backbone, and, with the proviso that all of the ring openings proceed with inversion of configuration at each epoxide derived from an *(E)*-alkene, the *trans-syn* topography is explained by this mechanism. All alkenes in a hypothetical polyene precursor would require identical stereoselectivity of epoxidation to produce either an all-*(S,S)* or all-*(R,R)* polyepoxide, suggesting that a single promiscuous oxidase could be sufficient [120]. Despite its intellectual appeal, the hypothesis relies upon a ring-opening process generally regarded to be disfavored. As discussed earlier, according to Baldwin's rules [6], epoxide-opening reactions of this type typically favor the smaller heterocycle, for example, THF (tetrahydrofuran) over THP (tetrahydropyran). In the case of the proposed precursor to brevetoxin B, the biosynthetic cascade would have to overcome ten consecutive disfavored epoxide openings.

In an effort to shed some light on the validity of Nakanishi's hypothesis, labeling studies have been reported for brevetoxin A, brevetoxin B, and yessotoxin [119, 122–124]. These studies corroborated the polyketide origin of ladder polyethers, which is also supported by genetic studies [125–128], but did not illuminate any subsequent epoxidation or cyclization steps. Some remote evidence in support of this hypothesis can be taken from biosynthetic studies on a related natural product, okadaic acid [129, 130], and isolation of 27,28-epoxybrevetoxin B from *Karenia brevis* [131]. The intriguing new structures of brevisamide [132] and brevisin [133] (**49** and **50**, respectively, Figure 15.6) have recently been isolated from *K. brevis*. These structures differ from most known polyethers and may help further the understanding of biosynthetic pathways involved in production of ladder polyethers.

Wright and coworkers have demonstrated that the proposed epoxide intermediates in biosynthesis of all polyethers isolated from *Karenia brevis*, including the aberrant polyethers brevisamide (**49**) and brevisin (**50**), feature identical *(S,S)* configuration [134]. This may suggest that a single promiscuous epoxidase can be involved in the biosynthesis of not only a single ladder polyether natural product but in the biosynthesis of several related natural products within the same organism. These authors also propose that the ring-forming cascade from a polyepoxide precursor flows in the opposite direction of nascent polyketide chain biosynthesis [132, 134].

Nakanishi's hypothesis remains speculative due to the lack of strong experimental support in its favor. However, the stereochemical uniformity inferred from the polyene to polyepoxide to ladder polyether pathway has served as the basis for structural reassignment of brevenal [135, 136] and the speculative structural reassignment of the largest known natural product, maitotoxin [120, 137, 138].

According to a modification of Nakanishi's hypothesis by Spencer, biosynthesis of ladder polyethers might also proceed in an iterative fashion through the repeated action of a monooxygenase and an epoxide hydrolase with broad specificities [120]. In this scenario, rings of a ladder polyether molecule would be formed sequentially, each being formed immediately after the epoxidation of the appropriate *(E)*-alkene in the biosynthetic precursor, thus avoiding the polyepoxide intermediate proposed by Nakanishi. If Wright's proposal for the direction of the polyketide chain extension is correct, such oxidation and cyclization steps would occur after the

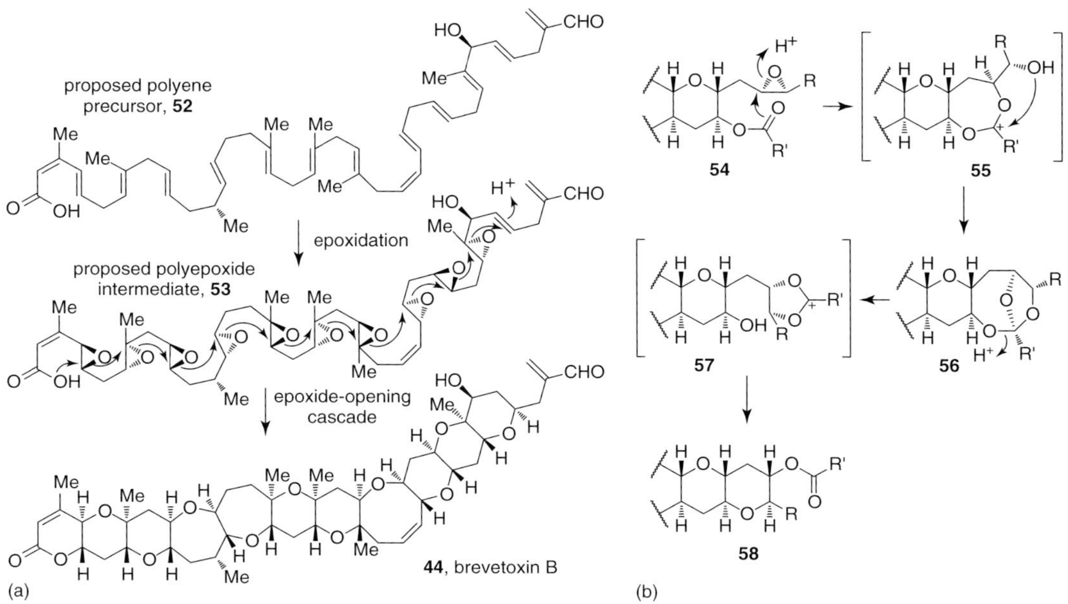

Figure 15.7 (a) Nakanishi's hypothesis: a model of brevetoxin B biosynthesis [118]; (b) Giner's proposal for biosynthesis of ladder polyethers via an epoxy ester pathway [121].

synthesis of polyene precursor, forming the polyether molecule in iterative fashion, one cyclic-ether at a time.

Giner has suggested that ladder polyethers may be derived from an all-(Z) polyene precursor [121, 139]. Giner hypothesized that an epoxy ester intermediate may undergo cyclization with the carbonyl group of the ester as nucleophile, leading to the formation of an orthoester intermediate **56** (Figure 15.7b). Upon collapse of the orthoester, attack of the alcohol nucleophile on what used to be the second electrophilic site of the starting *cis* epoxide then produces the ring of a ladder polyether and regenerates an ester for the next ring-closing reaction. The Townsend–McDonald hypothesis (Figure 15.3b) can also be extended to a proposal for biogenesis of ladder polyether via a similar all-(Z) polyene precursor. As yet, these hypotheses have not been tested experimentally.

15.4
Epoxide-Opening Cascades in the Synthesis of Polycyclic Polyethers

The first epoxide-opening cascades were disclosed in the early 1950s [140]. These early reports typically involved the rearrangement of 1,5-diepoxides that, under appropriate conditions, react with an external nucleophile to undergo a cascade of epoxide openings. Depending on the reaction conditions, these cascades may involve direct epoxide opening or formation of epoxonium ion intermediates, in either case producing tetrahydrofuran products in agreement with Baldwin's rules [6]. Other early reports on epoxide-opening cascades focused on rearrangements of topologically interesting polyepoxides [141–146], as well as the transannular epoxide-opening cascades of conformationally flexible polyepoxide substrates [147–150].

15.4.1
Epoxide-Opening Cascades in the Synthesis of Polyether Ionophores

The Cane–Celmer–Westley proposal [61] for the biosynthesis of polyether ionophores via a sequential epoxide-opening cascade quickly sparked great interest within the synthetic community, as emulation of such a biosynthetic pathway could in theory provide a rapid, straightforward approach to several natural products in this family. Especially encouraging was the agreement of the proposed cascade reactions with empirical guidelines [6] for regioselectivity in epoxide-opening reactions. For example, in their efforts toward aurodox, the Nicolaou group utilized a common approach to construct the THF backbone via epoxide-opening cascade of 1,5-diepoxide **59** and extended this methodology to triepoxide **61**. Cyclization of **61** to **62** constitutes the first epoxide-opening cascade that affords the 2,5-linked bis-tetrahydrofuran motif common for polyether ionophores (Scheme 15.2a) [151].

While working on the synthesis of uvaricin, the Hoye group reported studies on triepoxides **63–65** [152, 153]. Their strategy involved release of the alcohol nucleophile via ester hydrolysis followed by a base-promoted Payne rearrangement

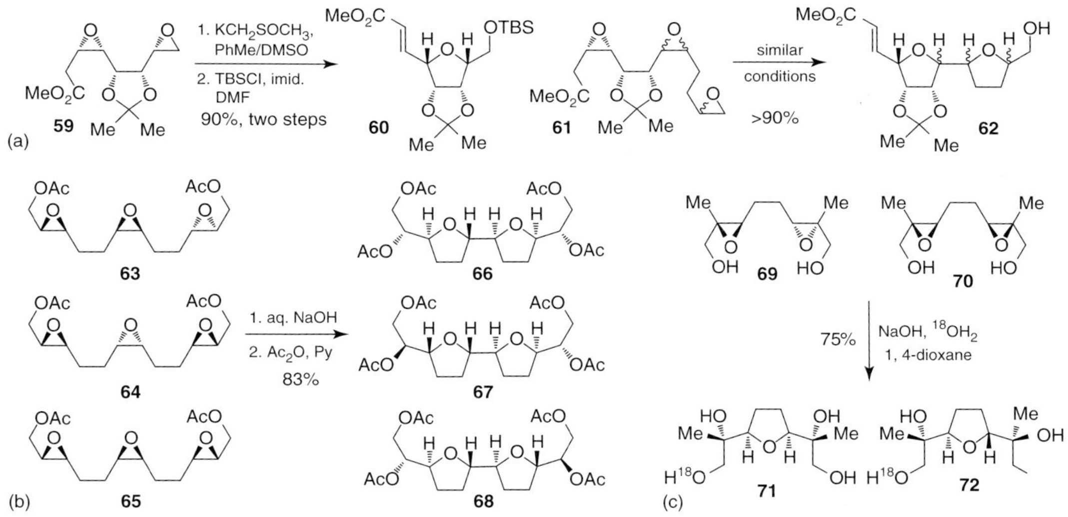

Scheme 15.2 (a) Construction of the central backbone of aurodox and related bis-tetrahydrofuran **62** [151]; (b) cascades cyclization of disubstituted 1,5,9-triepoxides reported by Hoye [152, 153]; (c) support for the Payne rearrangement mechanism [154]. imid. = imidazole; Py = pyridine.

to unveil a new secondary alcohol nucleophile that triggers a cascade of two consecutive epoxide-opening reactions to afford 2,5-linked bis-tetrahydrofuran products (**66–68**, respectively, Scheme 15.2b). To distinguish between the pathway that operates via Payne rearrangement and other possible pathways involving epoxide opening by water, Hoye and Jenkins conducted experiments on a diastereomeric mixture of diepoxides **69** and **70** in ^{18}O labeled water (Scheme 15.2c) [154]. In these experiments ^{18}O was incorporated only in the primary alcohols of **71** and **72**, which is consistent with the proposed Payne rearrangement pathway. Although the desired bis-tetrahydrofurans were formed via what appeared to be stereospecific epoxide opening, enantioenriched starting materials were transformed into racemic products due to initiation at both ends of the triepoxide substrates. This problem was addressed with application of appropriate ester protecting groups, which allowed desymmetrization of related diepoxides via unidirectional cascades enabled by the two competing pathways of different rate: hydrolysis of the ester protecting group and epoxide-opening cascade [155].

The first emulations of the alkene epoxidation/epoxide-opening cascade sequence from the proposed biosynthetic pathway to polyether ionophores came from the Schreiber [156] and Still [157] laboratories. Taking advantage of the powerful stereocontrol effects of allylic chiral centers on epoxidations of macrocyclic alkenes by peroxyacids described by Vedejs [158], Schreiber synthesized two diastereomeric diepoxides derived from a cyclic diene **73** (Scheme 15.3a). Diepoxides **74** and **75** were then subjected to a one-pot sequence of base-promoted ester hydrolysis followed by an acid-induced epoxide-opening cascade to form, respectively, 2,5-linked bis-tetrahydrofuran **76**, corresponding to C9–18 fragment of monensin B, and its diastereomer **77** after acetonide formation. [156]. In a concurrent report, the Still group described the preparation of a tricycle closely related to the C9–C23 fragment of monensin B [157]. Still used the resident chirality at C18 and C22 of **78** (carbon numbering as in premonensin B) to achieve good stereocontrol in the epoxidation reaction (Scheme 15.3b). Triepoxide **79**, isolated in 59% yield (74% corrected for purity of **78**), was then taken through a one-pot ester hydrolysis and acid-catalyzed epoxide-opening cascade sequence to afford **80**, a diastereomer of the C9–C23 fragment of monensin B. Remarkable yields in these cascades, observed irrespective of the stereochemistry of the epoxides, are likely ensured by directing effects of the appropriately positioned Me groups at each epoxide in polyepoxide precursors.

Shortly after the pioneering work of Still and Schreiber, Paterson reported initial studies on epoxide-opening cascades that afford fragments of polyether ionophores [159]. To enable more flexibility in choice of the cascade substrates, the Paterson group relied on Sharpless asymmetric epoxidation to control the stereoselectivity of epoxidation in an acyclic substrate. Under acidic conditions, diepoxyesters **81** and **83** were converted into the corresponding bis-tetrahydrofurans **82** and **84** in good yields (Scheme 15.4a). Similar to Schreiber and Still, Paterson studied trisubstituted epoxides with methyl substituents in positions where their electronic effects promote the desired outcome of the reaction. However, the stereochemical outcome of these cascades may be compromised when an electronic preference for 6-*endo* cyclization exists. Jaud and coworkers observed inversion of stereochemistry

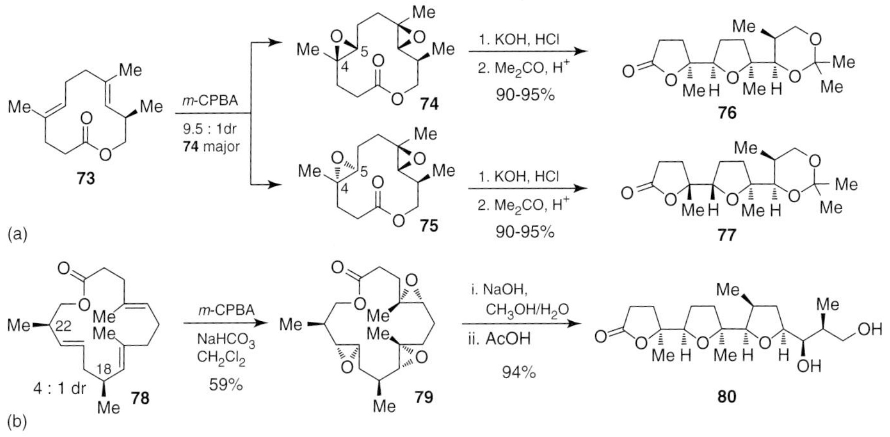

Scheme 15.3 Epoxide-opening cascades for preparation of 2,5-linked tetrahydrofurans reported by (a) Schreiber [156] and (b) Still [157]. m-CPBA = meta-chloroperbenzoic acid.

at C5 in acid-catalyzed cyclizations of **85** [160]. It was proposed that initial cyclization via a 6-*endo* pathway is operative, to produce an orthoester intermediate with inverted configuration at C5 (Scheme 15.4b). Elimination of 2-methylpropene results in the formation of a final five-membered lactone and liberates a tertiary alcohol, which is the nucleophile in the following epoxide opening.

Epoxide-opening cascades described so far have typically relied on acid or base catalysis. In these regimes, a cascade can be initiated either via activation of an alcohol nucleophile by deprotonation or, alternatively, via epoxide activation with Brønsted or Lewis acids. While nucleophile activation allows for good control over the direction of the cascade, it is limited to polyepoxide substrates with protic nucleophiles such as alcohols. Activation of the epoxide with acid, on the other hand, is typically unselective. Under these conditions any or all of the epoxides in

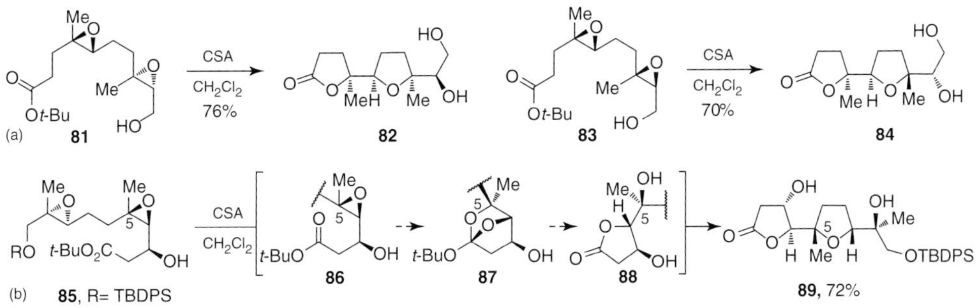

Scheme 15.4 (a) Epoxide-opening cascades for preparation of bis-tetrahydrofurans by Paterson [159]; (b) stereochemical outcome of cascades with preference for 6-*endo* cyclization [160].

the polyepoxide substrate can be activated and the cascade may proceed in both directions, with varying points of initiation. This problem is increasingly pronounced as more epoxides are added to the polyepoxide chain, limiting acid-catalyzed cascades to di- or triepoxide substrates.

Several research groups have offered alternative means of initiation in epoxide-opening cascades that address some of these issues in activation using acid and/or base. Perhaps inspired by the cleavage of the thioester linkage between the polyepoxide substrate and the acyl carrying protein in the proposed biosynthesis of polyether ionophores, several cascades have been initiated by the release of a carboxylic acid nucleophile from an ester through the action of esterase enzymes under mild conditions. Robinson and coworkers subjected esters **90** and **91** to pig liver esterase in slightly basic aqueous phosphate buffers (Scheme 15.5a) [161]. Efficient cascades afforded **82** and **92**, respectively, in good yields after prolonged exposure. These cascades appeared to proceed in a stepwise fashion, allowing detection of partially cyclized intermediates consistent with the hypothesis that the anionic carboxylate of hydrolyzed **90** initiates the cascade and is a stronger nucleophile than is a secondary alcohol in the second step.

The concept of the selective generation of a reactive epoxonium intermediate on one side of a polyepoxide substrate was introduced by the Murai group in their work towards the synthesis of ladder-type polyethers (see the discussion in Section 15.4.4.2.) [164]. Introducing the same concept for control of cascade direction to the arena of polyether ionophores, Floreancig and coworkers have demonstrated that mesolytic benzylic carbon–carbon bond cleavage in the benzylic position of the radical cations of homobenzylic ethers, such as **93**, **95**, **97**, and **99**, forms oxonium ions. These react with pendent epoxides to form epoxonium ions, which can undergo further cyclization [162, 163]. The Floreancig group has shown that both cis and trans substituted epoxides, **93** and **95**, are suitable for these reactions and that stereochemical information is preserved during the cascade (Scheme 15.5b). Several examples, in which mono- and diepoxides with a pendent acetal nucleophile are efficiently transformed into bis-tetrahydrofuran products **94**, **96**, **98**, and **100**, have established single-electron oxidation as another effective method for the initiation epoxide-opening cascade cyclizations.

Starting with bromomethylpolyepoxides (**101**, Scheme 15.6), the Marshall group has successfully initiated epoxide-opening cascades through transient formation of allylic alkoxyzinc species (Scheme 15.6) [165–167]. These reactions are reminiscent of Nicolaou's initial reports but avoid the basic conditions used in early cascades, replacing them with the generally mild metallic zinc in alcoholic solvents. Diastereomeric triepoxyfarnesyl bromides **104–112** all undergo cyclization, demonstrating the utility of this approach (Scheme 15.6).

15.4.2
Applications of Epoxide-Opening Cascades in the Synthesis of Ionophores

Epoxide-opening cascade reactions described so far have been inspired by the Cane–Celmer–Westley biosynthetic proposal for polyether natural products.

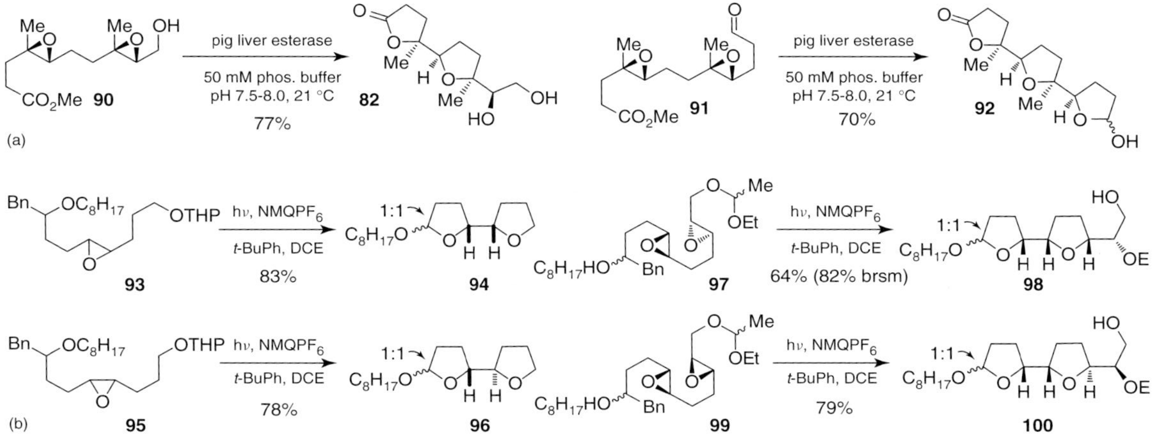

Scheme 15.5 Epoxide-opening cascades initiated by (a) enzymatic ester hydrolysis [161] and (b) single-electron oxidation of homobenzylic ethers [162, 163]. NMQ = N-methylquinolinium, DCE = 1,2-dichloroethane; brsm = based on recovered starting material.

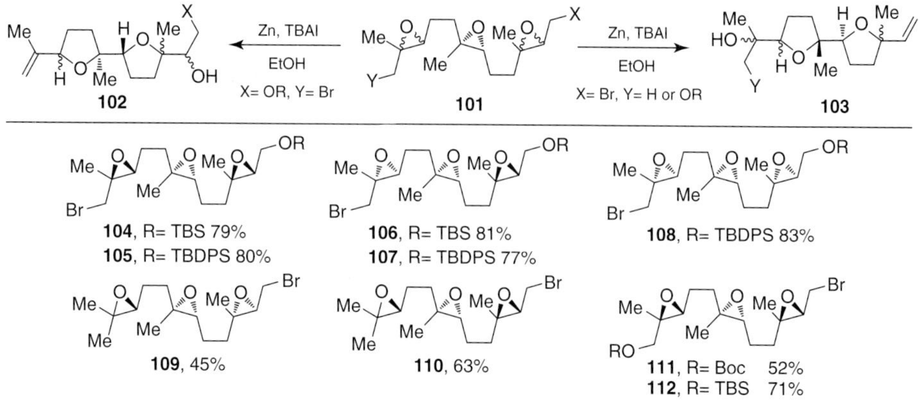

Scheme 15.6 Synthesis of bis-tetrahydrofurans via cascades of the triepoxyfarnesyl bromides reported by Marshall [165, 166]. Boc = *tert*-butoxycarbonyl, TBA = tetrabutylammonium.

Their development was typically driven by the search for chemical evidence in favor of this biosynthetic pathway. Equally important for the synthetic community, these reactions were developed with specific synthetic targets in mind, and they represent a classic example of the rapid generation of complexity from relatively simple starting materials. Although inspired by nature and developed for the synthesis of natural products, epoxide-opening cascades were used in the preparation of artificial ionophores even before the biosynthetic proposal for polyether ionophores was put forth [168, 169].

Several research groups were successful in extending these reactions to the total synthesis of polyether ionophores or fragments thereof. Notable contributions came from Paterson's group in their investigations toward etheromycin (**24**, Figure 15.2) [159]. In their first-generation approach, Paterson and coworkers extended studies on model diepoxides **81** and **83** (Scheme 15.4a) to mixtures of more complex diastereomeric triepoxides **114A** and **114B**. Triepoxides **114A** and **114B**, both featuring *tert*-butyl ester as the trapping nucleophile, underwent acid-promoted cascade reaction to afford the CDE ring system of etheromycin, **115A**, and its diastereomer **115B** (Scheme 15.7a). A more elaborate polyepoxide cyclization in the second-generation approach to etheromycin also allowed the formation of the BC spiroacetal during the cascade [170]. Exposure of diepoxide **116** to acidic conditions triggered deprotection of the secondary alcohol, which formed the hemiketal. The hemiketal nucleophile initiated the cascade of epoxide openings, resulting in formation of **117** (Scheme 15.7b). Paterson and coworkers also attempted to incorporate the appropriate substitution pattern in the starting diepoxide to install the hydroxyl at C4 of the tetrahydropyran ring in the BC spiroketal. However, when diepoxide **118** was exposed to the acidic promoter, elimination of the secondary alcohol occurred to form a trisubstituted alkene in **119** (Scheme 15.7c). In contrast, when diketone **120** was used, the desired BCD fragment of etheromycin (**122**, Scheme 15.7d) was produced in a single step [171].

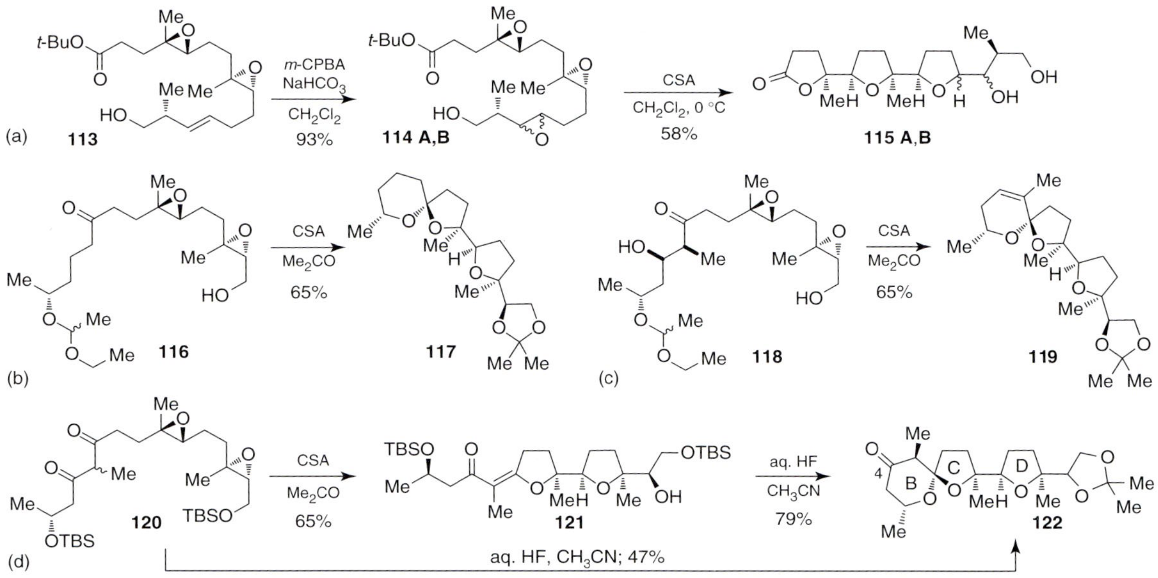

15.4 *Epoxide-Opening Cascades in the Synthesis of Polycyclic Polyethers* | 557

Scheme 15.7 (a) First-generation, (b) second-generation, and (c) and (d) third-generation cascade approaches to etheromycin reported by Paterson [159, 170, 171].

The Evans group elegantly incorporated an epoxide-opening cascade to construct the CD ring system of Ionomycin A (**22**, Figure 15.2), by way of macrocyclic diepoxide **123** (Scheme 15.8) [172]. Upon lactone hydrolysis, a cascade of epoxide openings afforded **125** in a straightforward manner. The E ring of Ionomycin A was then constructed in a fashion that mimics the following step of the proposed biosynthetic cascade to produce tricycle **126**. A cascade starting from triepoxide **127** would, in principle, be a more direct route to the Ionomycin A backbone, but stereocontrolled synthesis of such a substrate would have been considerably more complex.

Expanding on their zinc-initiated epoxide-opening cascades of terminal iodomethylepoxides [165, 166], Marshall and colleagues constructed the bis-tetrahydrofuran motif of ionomycin and transformed it into a fully elaborated C17–C32 fragment of ionomycin [167].

15.4.3
Epoxide-Opening Cascades in the Synthesis of Squalene-Derived Polyethers

As discussed earlier, oxasqualenoids feature both fused cyclic ethers and 2,5-linked oligotetrahydrofurans. This means that epoxide-opening cascades that would produce such diverse structures have to incorporate both *endo* and *exo* selective epoxide openings. Despite the development of various methods for control of regioselectivity in intramolecular epoxide-opening reactions, cascades that successfully achieve this goal are scarce. Instead, most epoxide-opening cascades are either all-*exo*, like those reported in the synthesis of polyether ionophores, or all-*endo*, like those that are to be discussed in the context of ladder polyether synthesis. Both types have been used in various ways to produce fragments of squalene derived polyethers.

In their studies toward the total synthesis of glabrescol [174], the Corey group reported a rapid synthesis of the originally proposed structure of glabrescol. They prepared pentacycle **130** from the corresponding pentaepoxide **129** in a single step under acidic conditions (Scheme 15.9a) only to find that this material had physical and spectral properties different from natural glabrescol. The authors also prepared three other diastereomers of pentaepoxide **129**, all of which cyclized to corresponding C_S-symmetric pentacyclic polyethers under the same conditions described for **129** (e.g., **131** to **132**). However, none of the produced polyethers were identical to natural glabrescol. The correct structure of glabrescol, which is in fact a C_2-symmetric molecule, was disclosed in a subsequent report by Morimoto [173]. The Corey group also investigated the possibility that glabrescol is a C_2-symmetric molecule [175]. Their synthesis of the revised structure of glabrescol relies on a bidirectional double cyclization of a tetraol tetraepoxide **133** (Scheme 15.9b). The choice of acidic reaction conditions was crucial in this case to ensure that cyclization to form the AB and A′B′ rings of glabrescol via epoxide opening at more substituted positions is faster than the rate of cyclization to form the C ring (via *exo*-opening at a less substituted position). Upon treatment of **133** with camphorsulfonic acid (CSA), the bidirectional cascade formation of AB and A′B′ was able to outcompete the undesired unidirectional cascade that would form a diastereomer of the ABCB′

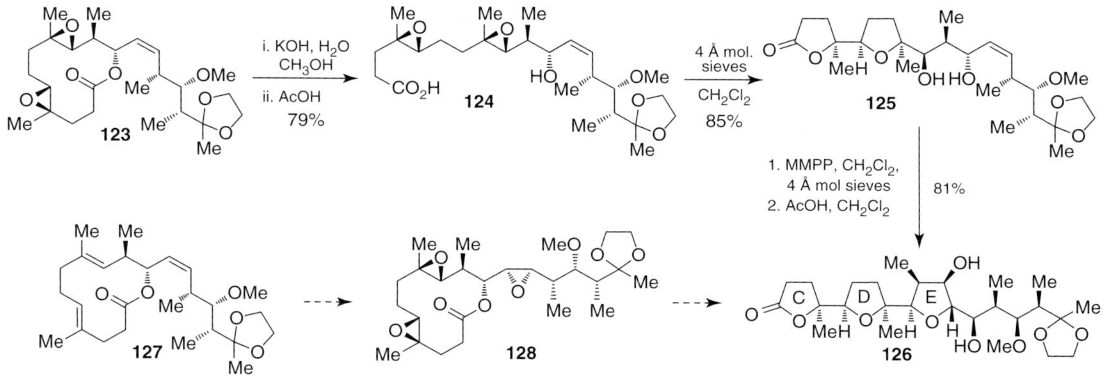

15.4 Epoxide-Opening Cascades in the Synthesis of Polycyclic Polyethers | 559

Scheme 15.8 Synthesis of the CDE ring system of Ionomycin A reported by Evans [172]. MMPP = magnesium monoperoxyphthalate.

tetracycle and afforded **134** in good yield. Tetracyclic intermediate **134** was then rapidly converted into the natural product in two steps (Scheme 15.9b).

The Morimoto group used intramolecular epoxide-openings to construct cyclic ethers of numerous oxasqualenoids. For example, a base-promoted epoxide-opening cascade on diepoxide **136** [176, 177] afforded the C ring of aurilol and enshuol (**137**, Scheme 15.10). After elaboration to the diepoxide **138**, a Brønsted acid-catalyzed epoxide-opening cascade afforded the D and E rings of enshuol. Finally, reagent-controlled, silyl triflate-promoted opening of the trisubstituted epoxide **140** with a tertiary alcohol nucleophile via 6-*endo* cyclization [19] efficiently formed **141** and **142**, containing the B ring of enshuol (Scheme 15.10) [178].

In their latest work on oxasqualenoids, the Morimoto group reported a route to omaezakianol, which includes two consecutive epoxide-opening cascades [179]. The first of the two cascades is similar to the cascade cyclization used to construct **147**, the THF-containing subunit of C_2-symmetric oxasqualenoid intricatetraol [180]. Diepoxides **143** and **144** rearranged to the functionalized tetrahydrofurans **146** and **147**, respectively, under basic conditions. Interestingly, a single bicyclic side product **145** was produced in this reaction, possibly via Payne rearrangement followed by 5-*exo* opening of the C6–C7 epoxide and 8-*endo* nucleophilic attack of the resulting tertiary alkoxide at the C7 position onto the terminal epoxide formed in the initial Payne rearrangement. Elaboration of **146** to triepoxide **148** followed by an acid-catalyzed epoxide-opening cascade afforded **149**, which was then rapidly elaborated into the natural product (Scheme 15.11a). Each of the THF rings in Morimoto's synthesis of omaezakianol is formed via an epoxide opening of a trisubstituted epoxide. Interestingly, these cascades proceed in good yield, one under basic and the other under acidic conditions, despite the potentially adverse effects of methyl substituents of the trisubstituted epoxides.

By modifying the pentaepoxide substrate **129** used in their synthesis of the originally proposed structure of glabrescol, the Corey group recently completed a short total synthesis of omaezakianol [181]. Pentaepoxidation of the racemic chlorohydrin **151** afforded the cascade substrate **152** that, upon treatment with a Brønsted acid, cyclized to the pentacyclic product **153**, which was converted into omaezakianol in a single step (Scheme 15.11b).

Elegant work on *ent*-abudinol B and the related terpenes *ent*-durgamone and *ent*-nakorone capitalizing on studies of epoxide-opening cascades directed toward synthesis of ladder polyether structures was reported by McDonald [182, 183]. In their first-generation approach to *ent*-abudinol B, McDonald and coworkers devised a convergent synthetic scheme featuring a late-stage coupling of fragments derived from *ent*-durgamone and *ent*-nakorone (**156** and **159**, Scheme 15.12a) [182]. In the synthesis of subunit **155**, a cascade of epoxide openings on diepoxide **154** was employed. Use of *tert*-butyldimethylsilyl triflate as a Lewis acid, two *endo*-selective cyclizations directed by methyl substituents, with an enol-silane as trapping nucleophile, led to formation of bicyclic compound **155** that could be further elaborated to *ent*-durgamone. An analogous strategy was utilized in the synthesis of the more complex *ent*-nakorone, but a hybrid cascade of oxacyclizations and carbocyclizations was required. Diepoxide **157**, with a terminating propargyl silane nucleophile,

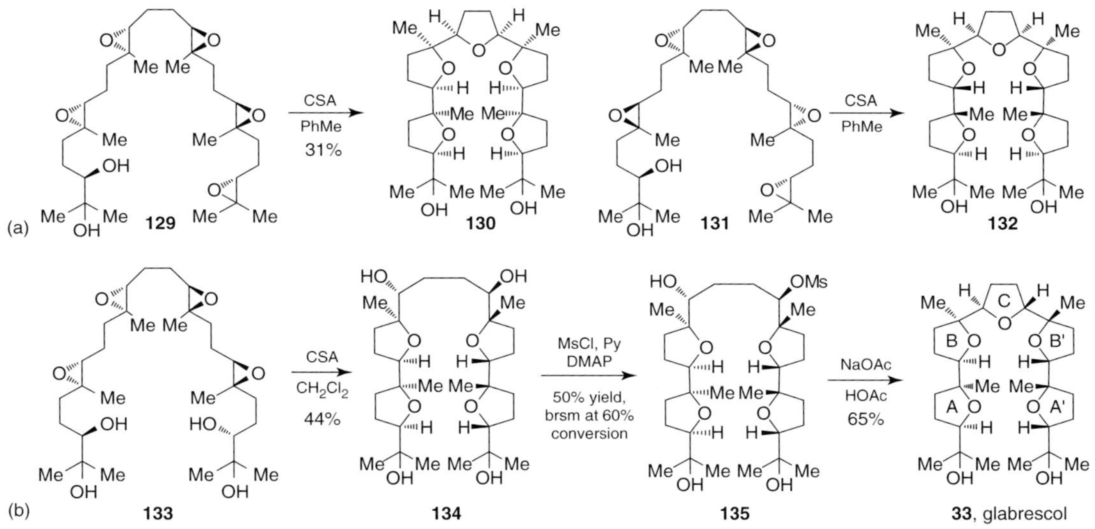

Scheme 15.9 (a) Synthesis of proposed structure of glabrescol (Corey, 2000) [174]; (b) bidirectional cyclization to form glabrescol (Corey, 2000) [175]. Ms = methanesulfonyl.

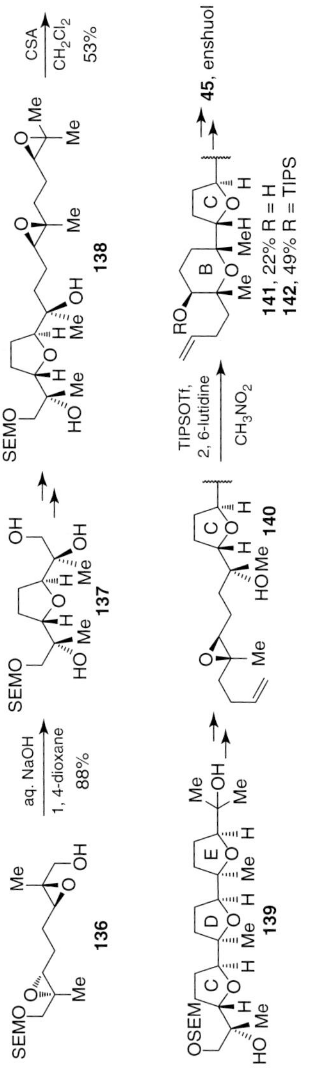

Scheme 15.10 Total synthesis of enshuol by Morimoto [178].

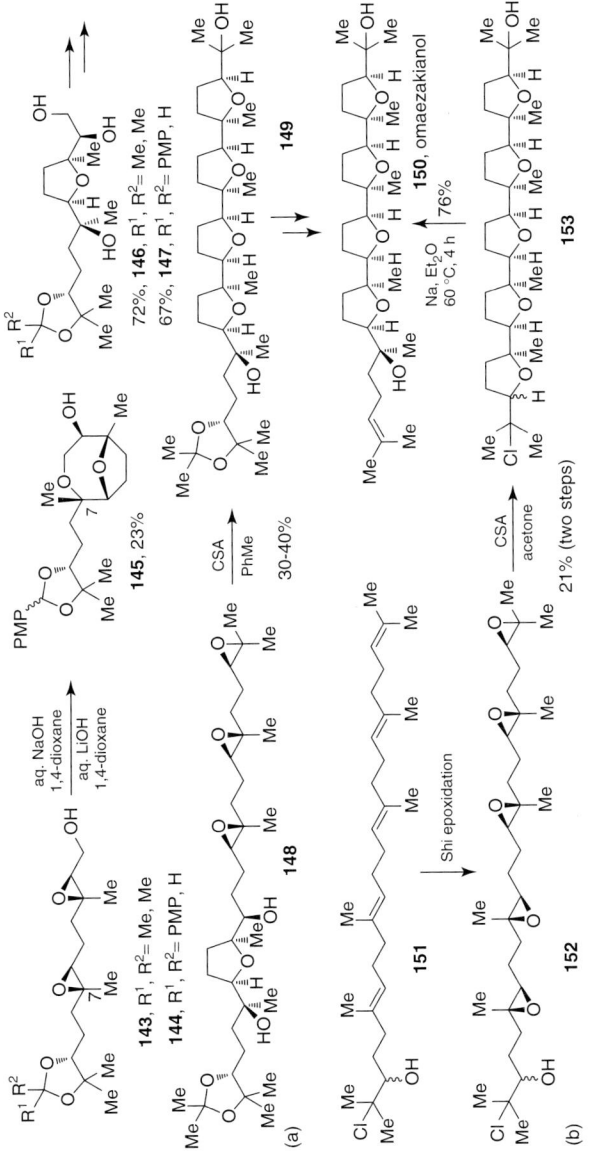

15.4 Epoxide-Opening Cascades in the Synthesis of Polycyclic Polyethers | 563

Scheme 15.11 Synthesis of omaezakianol via epoxide-opening cascades reported by (a) Morimoto's group [179] and (b) Corey's group [181]. PMP = *p*-methoxyphenyl.

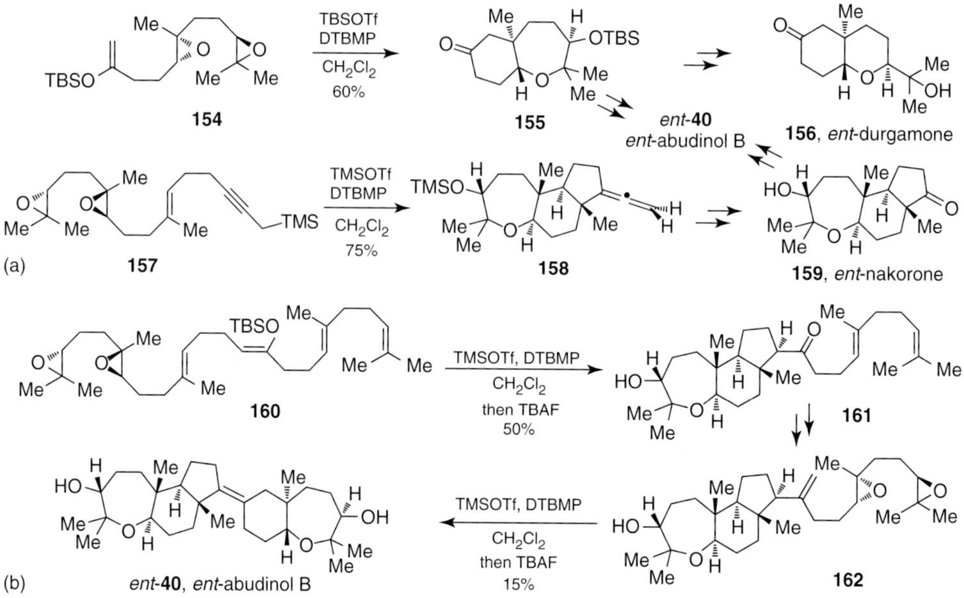

Scheme 15.12 Syntheses of *ent*-abudinol B via hybrid oxa/carbocyclizations: (a) convergent approach [182]; (b) biomimetic approach [183]. DTBMP = 2,6-di-*tert*-butyl-4-methylpyridine.

underwent efficient TMSOTf (trimethylsilyl trifluoromethanesulfonate)-promoted cyclization to tricyclic allene **158**. As in the previous cascade, the regioselectivity of cyclizations was directed by methyl substituents. Further elaboration of fragments **155** and **159** into their corresponding vinyl triflates and subsequent modified Suzuki–Miyaura coupling produced *ent*-abudinol B.

A second-generation approach was based on the proposed biosynthetic pathway to *ent*-abudinol [2, 73, 74], which involves a hybrid cascade of epoxide openings and carbocyclizations [183]. Similar to the first-generation approach, diepoxide **160** was treated with TMSOTf to produce **161**, containing the tricyclic fragment of *ent*-abudinol (Scheme 15.12b). A two-step elaboration of the cascade product **161** via Wittig methylenation and Shi epoxidation afforded diepoxide **162**. Diepoxide **162**, carrying a terminal alkene instead of an enol ether trapping nucleophile (as in **154**), was subjected to the same conditions as in the transformation of **158** into **159** to produce *ent*-abudinol B, along with several isomeric products resulting from pathways enabled by the relatively low nucleophilicity of the terminating alkene. Despite the linear nature of this route to *ent*-abudinol B, structural complexity is generated quickly, making this approach very efficient. In efforts to further improve the synthesis of abudinol B, the McDonald group has prepared (3*R*,6*R*,7*R*,18*R*,19*R*,22*R*)-squalene tetraepoxide, a putative biosynthetic precursor to various oxasqualenoids, but this material failed to cyclize to *ent*-abudinol B [184].

The Jamison group recently reported an epoxide-opening cascade-based approach to the synthesis of the enantiomer of the oxasqualenoid natural product dioxepandehydrothyrsiferol (Scheme 15.13) [80]. The presence of bromine in dioxepandehydrothyrsiferol presented the opportunity to initiate the epoxide-opening cascade with electrophilic bromine reagents. This approach promised good control over the direction of the cascade. Relying on the well-documented bromoetherification strategy to construct bromooxanes and bromooxepanes [76, 177, 178, 185], Jamison and coworkers first examined diepoxide substrates **163**, **164**, and **167** featuring various terminating nucleophiles (Scheme 15.13a). When carried out in highly polar, non-nucleophilic solvents such as hexafluoro-*iso*-propanol, cyclization reactions of **163**, **164**, and **167** afforded the desired tricyclic products **165**, **166**, and **168**, respectively, in good yields via all-*endo* epoxide-opening cascades, albeit as 1 : 1 mixtures of diastereomers resulting from non-stereoselective bromonium formation. A polar solvation environment facilitates a cationic cascade, thus maximizing the directing effects of the appropriately positioned methyl substituents, and securing good *endo* selectivity in each cyclization. The Jamison group extended these findings to the cyclization of triepoxide **169**. Treatment of **169** with *N*-bromosuccinimide (NBS) in hexafluoro-*iso*-propanol afforded the tetracycle **170**, which contains the fully elaborated tricycle of *ent*-dioxepandehydrothyrsiferol (Scheme 15.13b).

15.4.4
Epoxide-Opening Cascades in the Synthesis of Ladder Polyethers

Epoxide-opening cascades were initially, and nearly exclusively, explored in the context of the synthesis of polyether ionophores and other natural products that could arise from *exo* opening of epoxides. This is not surprising considering the breadth of data supporting Baldwin's rules. Successful *endo*-selective cascades for preparation of ladder polyether-like fragments require circumventing the inherent selectivity for smaller rings in epoxide-opening reactions.

15.4.4.1 Iterative Approaches
As discussed previously, most methods for regioselective *endo* epoxide opening rely on the effects of directing groups directly attached to the epoxide. These directing groups are typically not present in the target ladder polyethers; the fact that they are incorporated in products of such epoxide-opening reactions presents a major challenge for their successful utilization in total synthesis of ladder polyethers because of the need for their removal or extensive synthetic elaboration. If such reactions are extended to cascades of epoxide openings, multiple directing groups would be incorporated at the ring junctions of the final product, thus creating the need for selective elaboration of each of the groups into H or Me groups, the exclusive substituents found at the ring junctions of ladder polyethers. As they are good directors of regioselectivity, methyl groups would appear to be the exception; however, they are typically present at only a few ring junctions in each ladder polyether and are rarely distributed in a uniform substitution pattern.

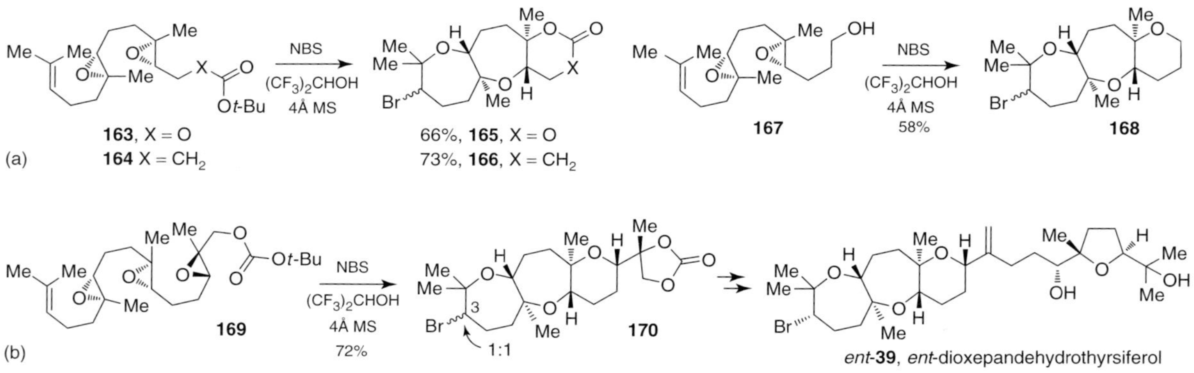

Scheme 15.13 (a) Bromonium-initiated epoxide-opening cascades; (b) synthesis of *ent*-dioxepandehydrothyrsiferol reported by Jamison [80]; NBS = N-bromosuccinimide.

Despite the problems associated with the use of directing groups in cascades of epoxide-opening reactions, they have been of tremendous value in iterative approaches to ladder polyether synthesis. Such approaches depend on the type of the directing group used in the epoxide-opening reaction and require efficient removal of this group following each iteration. If all requirements are met, a sequence of *endo* cyclization, removal of the directing group, and homologation to a new epoxide bearing the appropriate directing group for the next cyclization results in the formation of one cyclic ether per iteration.

The Nicolaou research group was the first to explore and report a successful iterative approach to ladder polyethers based on *endo*-selective epoxide opening (Scheme 15.14a) [24]. The epoxy alcohol **171** bearing an alkenyl directing group underwent Brønsted acid-catalyzed cyclization with excellent *endo*-selectivity due to the ability of alkenyl substituent to stabilize partial positive charge in the transition state for the desired cyclization. Upon elaboration of the tetrahydropyran **172** to epoxy alcohol **173**, another acid-catalyzed opening of alkenyl epoxide afforded diad **174** with excellent efficiency.

The Mori group reported a complementary approach to ladder polyethers relying on *endo*-selective opening of epoxysulfones (Scheme 15.14b) [21]. Exposure of epoxysulfone **175** to Brønsted acid led to 6-*endo* cyclization and subsequent loss of phenyl sulfonate to yield ketone **176**. A sequence involving alkylation of the sulfone-stabilized *cis*-oxiranyl anion completed the homologation process to **177**, which contains an epoxide with the appropriate directing group for the next iteration. Repeating this protocol three times furnished tetracycle **178**. Mori and coworkers also developed methods for larger oxygen heterocycles [186] and ring junction substitution patterns (Me and H) [35, 187] present in ladder polyether natural products.

Capitalizing on the effects of a silyl group attached directly to an epoxide, studied in detail by Hudrlik [188, 189] and Paquette [190], Jamison developed an iterative approach to the synthesis of *trans*-fused oligo(tetrahydropyran) fragments (Scheme 15.14c) [18]. In contrast to cyclization of epoxysulfones, in which the sulfone deactivates the undesired site of epoxide opening, silyl groups stabilize positive charge in the transition state leading to 6-*endo* epoxide opening. The directing group can easily be removed after cyclization by treatment with TBAF (tetrabutylammonium fluoride). The utility of this approach was demonstrated by synthesis of THP triad **182**.

15.4.4.2 Epoxide-Opening Cascades Leading to Fused Polyether Systems

Early work on epoxide-opening cascades by the Murai group led to the development of methods for the *endo*-selective lanthanide-promoted opening of methoxymethyl substituted epoxides [22]. Murai and coworkers prepared polyepoxides **183** and **189**, which incorporate a methoxymethyl directing group at each epoxide [191]. Under conditions described for substrates containing one epoxide, diepoxide **183** was converted into a THP diad **184** with methoxymethyl groups present at the ring junctions (Scheme 15.15). The side products isolated in this reaction are suggestive of a pathway that proceeds in a stepwise fashion from the primary alcohol, initially

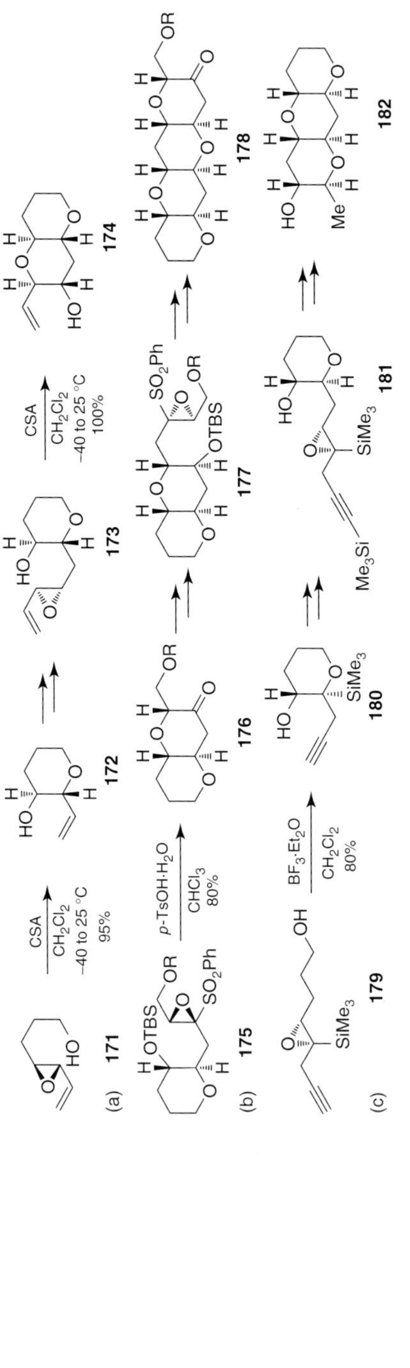

Scheme 15.14 Iterative synthesis of oligo(tetrahydropyran) fragments via 6-*endo* cyclization of (a) alkenyl epoxides [24], (b) epoxysulfones [21], and (c) epoxysilanes [18]. R = TBPDS.

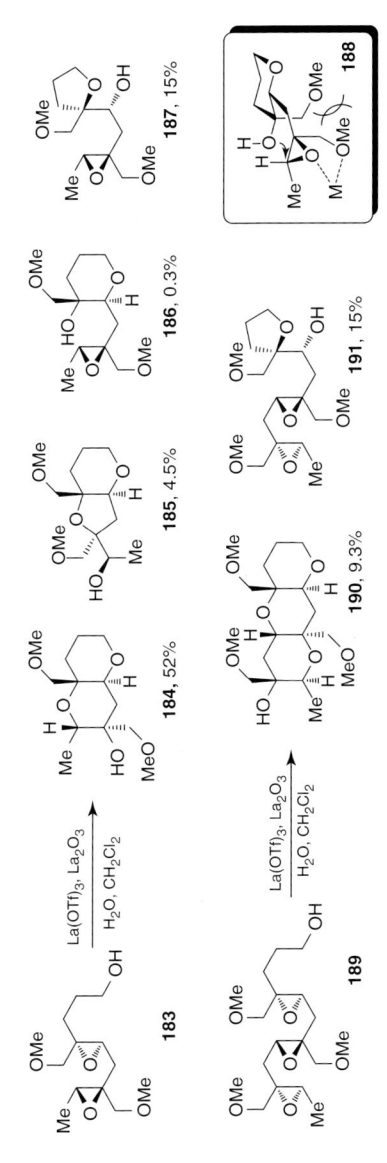

Scheme 15.15 Murai's epoxide-opening cascades directed by a methoxymethyl group [191].

forming intermediate **186**, and then **184** and **185**. The authors also reported that the other diastereomer of **183** fails to afford any of the corresponding THP diad. The postulated intermediate **188** does not react further due to strain in the requisite boat-like transition state and steric repulsions between the two methoxymethyl substituents. Murai has also demonstrated that cascades directed by methoxymethyl groups in combination with an appropriate Lewis acid can be extended to larger ladder polyether type fragments such as triad **190**, albeit in low yield.

McDonald reported the first cascade reactions that produce oxepane and *trans*-fused bisoxepane motifs via *endo*-selective epoxide opening [30, 192]. A range of terminating nucleophiles such as ketones, esters, carbonates, and acetals were examined and demonstrated to be a factor in determining regioselectivity in these Lewis acid-promoted polyepoxide cyclizations (Scheme 15.16a). McDonald and coworkers extended these reactions to polyepoxides **202–205** for the synthesis of polyoxepane systems [30, 192]. The efficiency of these reactions tends to drop as the number of epoxides in the polyepoxide precursor increases (Scheme 15.16b). A possible reason for a nonlinear decrease of yield in cascades that involve more than two epoxides may be unselective activation of any or all of the epoxides in the starting materials. If selective activation of the epoxide distal to the terminating nucleophile could be achieved, cascades would presumably proceed in one direction, and higher yields should be observed. With this in mind, the McDonald group prepared substrates **210** and **211** that feature a vinyl and a methyl substituent at the terminal epoxide of the polyepoxide chain (Scheme 15.16c) [193]. Based on Nicolaou's work on alkenyl epoxides [20, 23, 24], it was expected that the stabilization by the vinyl substituent would not only improve selectivity in epoxide-opening reactions but also lead to selective activation of the alkenyl epoxide over the interior epoxides under finely tuned conditions. Optimization revealed Gd(OTf)$_3$ and Yb(OTf)$_3$ as the best reaction promoters. Indeed, desired oxepane ring-containing products **212** and **213** were produced in higher yield than in the corresponding reactions of substrates **202** and **203**, which lack vinyl substituents.

The synthetic utility of these impressive cascades is, however, limited by the requirement for an alkyl directing group on each epoxide, resulting in the incorporation of the directing groups at every ring junction of the final cascade product. Were these cascades to be used in the synthesis of natural products, they would have to accommodate polyepoxides without directing groups and allow for various substitution patterns that would install methyl groups only at the desired positions of the final products. The McDonald group has offered two approaches to address these concerns. The first is based on the similar directing effects of silyl groups relative to alkyl substituents and the opportunity to remove them after the cascade. For example, **214** and **215** were converted into the corresponding cyclization product **216**, with efficiencies comparable to those of cascades with only methyl-substituted epoxides (Scheme 15.17) [194].

The McDonald group has also investigated cascades that would incorporate disubstituted epoxides with no directing groups present, as in **217**, **218**, **220**, and **221** (Scheme 15.17) [194]. The difference between the electronic properties of disubstituted and trisubstituted epoxides may have worked in favor of the desired

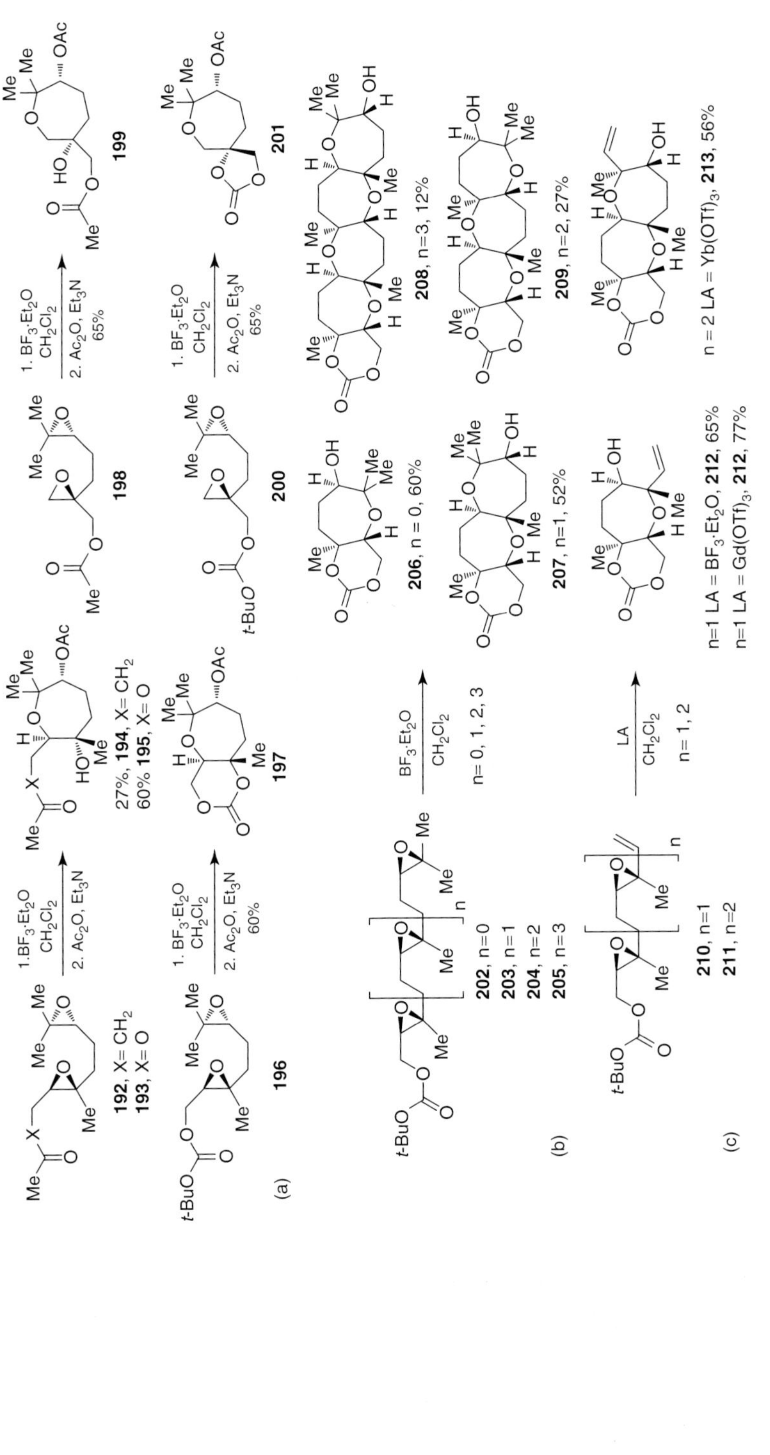

Scheme 15.16 Lewis acid catalyzed alkyl-directed cascades of (a) 1,5-diepoxides, (b) CH_2–CH_2 interrupted polyepoxides [30, 192], and (c) CH_2–CH_2 interrupted polyepoxides with terminal alkenyl epoxides reported by McDonald [193].

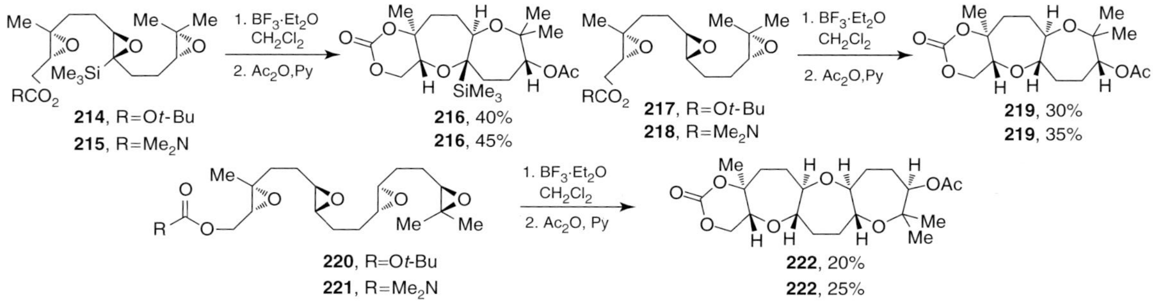

Scheme 15.17 Synthesis of polyoxepane ring systems, via cascades of polyepoxide precursors without alkyl directing groups at internal epoxides, reported by McDonald [194].

15.4 Epoxide-Opening Cascades in the Synthesis of Polycyclic Polyethers

cascade through preferential activation of the epoxide distal to the terminating nucleophile in a fashion similar to cascades on alkenyl epoxides **210** and **211**. Cascades of both triepoxides **217** and **218**, and tetraepoxides **220** and **221**, under standard Lewis acid activation proceeded to form the desired tricyclic polyether **219** and tetracyclic polyether **222** (Scheme 15.17). It was proposed that once the first epoxonium ion is formed at the distal end the transition states leading to *endo* and *exo* opening of the disubstituted epoxonium ion are different in energy, with a higher degree of ring strain associated with the bicyclo[3.1.0] intermediate than for the bicyclo[4.1.0] intermediate formed as the product of *endo* opening. A directing group was required on the epoxide proximal to the trapping nucleophile, as there is minimal strain associated with either five- or six-membered carbonates formed at the end of the cascade when carbonate or carbamate nucleophiles were used. A directing group was, therefore, necessary to ensure *endo* regioselectivity in the opening of this last epoxide.

In addition to their studies on the use of epoxide-opening cascades in the construction of oxepanes, the McDonald group has, in similar fashion, also explored cascades directed toward synthesis of oligo(tetrahydropyran)s. The effects of the terminating nucleophile on epoxide-opening cascades of 1,4-diepoxides **223–225** were examined first [31]. Depending on the nucleophile, these reactions can proceed with either retention or inversion of configuration at the ring junction (Scheme 15.18a). Stronger nucleophiles at elevated temperatures favored the inversion of stereochemistry in the opening of internal epoxide, thus setting a trans geometry at the ring junction of **226**. In contrast, less nucleophilic carbonates favored the production of diastereomeric product **227**, corresponding to retention of configuration. These were explained by the mechanism outlined in the Scheme 15.18b, in which the terminating nucleophile intercepts a cationic intermediate at different points in the continuum between the extremes of epoxonium ion **230** and tertiary alkyl carbocation **234**. McDonald proposed that *cis*-fused products arise from fast nucleophilic addition to the tertiary carbocation, whereas *trans*-fused products are favored with a stronger nucleophile, which intercepts a tight ion pair intermediate structurally related to the epoxonium ion [31]. The McDonald group extended these findings to epoxide-opening cascades of triepoxides **235** and **236**, which carry directing groups on each of the epoxides. When activated by a Lewis acid at an appropriate temperature, triepoxide **235** with the carbamate terminating nucleophile was transformed into the ladder polyether-like tricycle **238** in 31% yield. Triepoxide **236** with a carbonate nucleophile, however, failed to afford any of the desired products and, instead, at low temperatures gave **237** (Scheme 15.18c). To summarize, the choice of the terminal nucleophile not only dictates whether cyclization will proceed with retention or inversion but in the case of 1,4,7-triepoxides it also determines the regioselectivity of the cyclization onto the epoxide proximal to the carbonyl nucleophile [31].

A conceptually novel way of promoting epoxide-opening cascades to ladder polyethers was investigated by Murai *et al.* [164]. They envisioned that activation of a polyepoxy halide with a silver salt would selectively generate an epoxonium ion at one end of the polyepoxide chain. This epoxonium would then serve as an

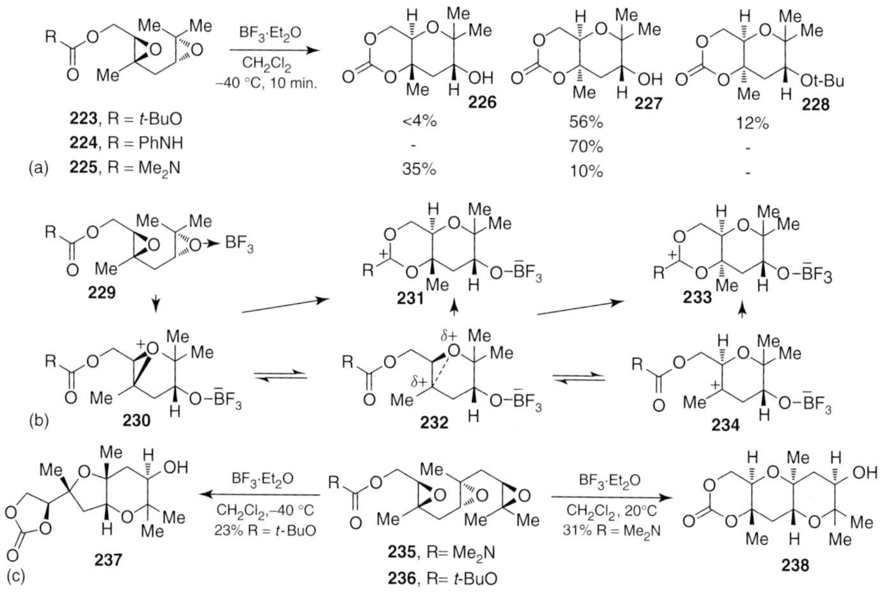

Scheme 15.18 (a) Stereochemical outcome of epoxide-opening cascades as function of the type of terminating nucleophile; (b) mechanistic rationale; (c) effects of the terminating nucleophile on the outcome of cascades leading to poly(tetrahydropyran) systems [31].

electrophile for nucleophilic attack by the neighboring epoxide, forming a new ring and a new epoxonium intermediate, thus propagating the cascade. The direction of the cascade in these reactions is therefore controlled by the position of the halide, and the need for selective activation of only one of many epoxides in polyepoxide substrates is eliminated. As Murai focused on the trans-disubstituted epoxide substrates without directing groups, these studies constituted the first efforts toward epoxide-opening cascades in the synthesis of ladder polyether fragments. While this mode of activation proved to be somewhat effective in reactions involving a single epoxide, cascade cyclizations of 1,4-diepoxides uniformly failed to produce any of the desired *trans*-fused oligo(tetrahydropyran)s [164, 195].

As described in Section 15.4.1, Floreancig and coworkers have demonstrated that mesolytic benzylic carbon–carbon bond cleavage of the radical cations of homobenzylic ethers, such as **93**, **95**, **97**, and **99** (Scheme 15.5b), forms oxonium ions that react with pendent epoxides to form epoxonium ions capable of undergoing further nucleophilic attack. This strategy is conceptually similar to Ag-promoted reactions of halo epoxides reported by Murai [164] and Jamison [195]. After their initial success in the synthesis of bis-tetrahydrofuran fragments of polyether ionophores, the Floreancig group published their experimental and computational studies, in collaboration with Houk and coworkers, on the structure–reactivity relationships for intramolecular additions to bicyclic epoxonium ions [196]. They observed that ring size has a significant impact on these processes, with *endo*-cyclizations being preferred for bicyclo[4.1.0] epoxonium ions bearing an alkyl directing group and *exo*-cyclizations being preferred for bicyclo[3.1.0] epoxonium ions, despite the presence of a directing group (Scheme 15.19a). The authors propose that these effects can be attributed to the ability of the larger ring to accommodate a looser transition state with significant S_N1 character, thereby promoting the *endo*-process regardless of solvent polarity. As they had clearly demonstrated that the epoxonium ion structure is a significant determinant of regioselectivity under these kinetic cyclization conditions, Floreancig and coworkers then designed several extended substrates that underwent cascade cyclizations to form fused tricyclic systems under the oxidative conditions described earlier (Scheme 15.19a).

Gagné and coworkers have reported another method for initiation of epoxide-opening cascades that provides good control over direction of the cascade [197]. In this work, allenyl epoxides **252** and **253** were treated with a cationic gold(I) phosphite catalyst to form epoxonium intermediates that were, in turn, opened by the pendant alcohol nucleophiles to afford bicyclic products **254** and **255** (Scheme 15.19b). Appropriately positioned methyl groups were required for the reactions to proceed with good regioselectivity.

As described earlier, the Jamison laboratory has successfully utilized the directing effects of trimethylsilyl groups to develop an iterative approach to the synthesis of oligo(tetrahydropyran) fragments (Scheme 15.14c). However, when a cascade reaction under the same Lewis acid conditions was attempted on diepoxide **256**, with suitably positioned silyl groups, the only isolable product was bis-tetrahydrofuran **257** (Scheme 15.20a) [198]. Thorough evaluation of reaction conditions revealed that the outcome of this reaction was very different when a Brønsted base in alcoholic solvents was used. Under these conditions diepoxide **256** underwent a cascade to produce THP diad **258**. Surprisingly, the trimethylsilyl directing group was absent from the ring junction in the product. Further modification to the design of polyepoxide substrates and reaction conditions resulted in the development of epoxide-opening cascades directed by "disappearing" silyl groups (Scheme 15.20b) [198]. These modifications included the pre-formation of one THP ring in substrates **259**, **262**, and **265** prior to the cascade and the addition of CsF to the reaction mixture. It was proposed that these cascades proceed as a sequence of silyl-directed epoxide opening followed by protiodesilylation via a homo-Brook rearrangement pathway. After each Brook rearrangement, removal of the silyl group by fluoride reveals the alkoxide nucleophile to propagate the cascade.

While disappearing directing groups address problems related to the removal of substituents not present in the natural targets, these reactions developed by Jamison and coworkers suffer from the inability to incorporate the methyl substituents found frequently at ring junctions. In their efforts to develop a directing group-free *endo*-selective intramolecular epoxide opening and pave the way to a directing group-free cascade capable of incorporating all types of epoxide substitution, the Jamison group reasoned that pre-organization of the substrate in an appropriate fashion could encourage the cyclization toward the *endo* pathway by altering the approach of the alcohol nucleophile to the epoxide. When designing

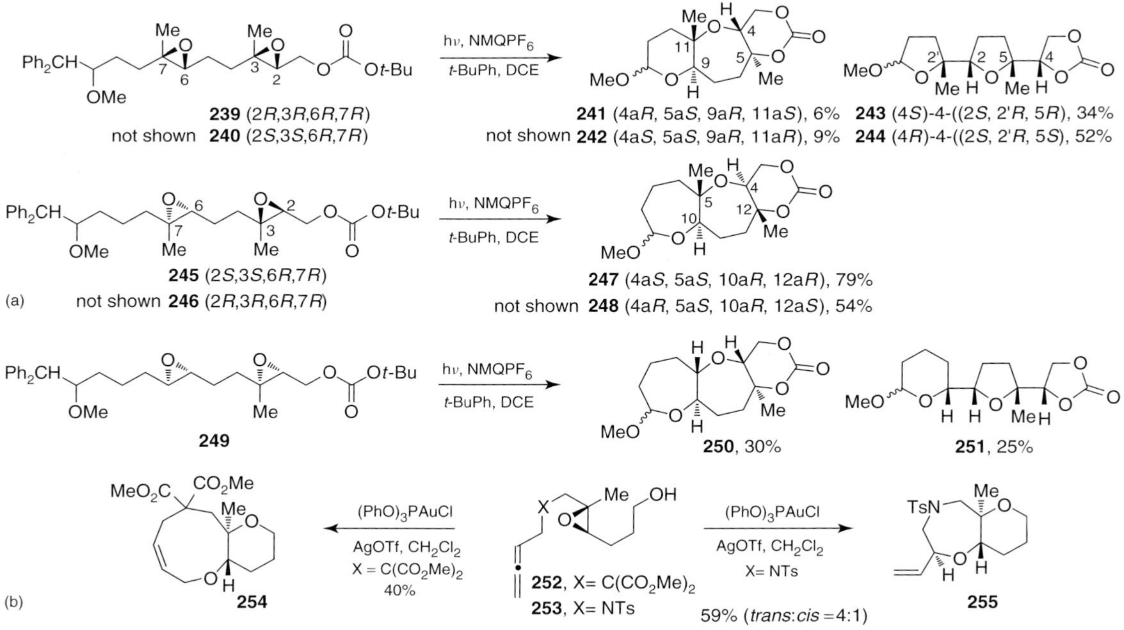

Scheme 15.19 Endo-selective epoxide-opening cascades initiated by: (a) oxidative cleavage of homobenzylic ethers [196] and (b) cationic gold(I) phosphite catalyst with allenyl epoxides [197].

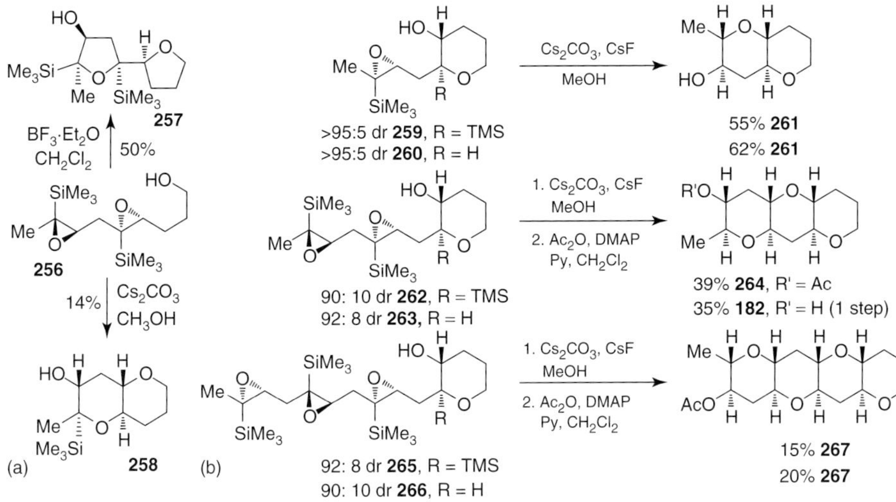

15.4 Epoxide-Opening Cascades in the Synthesis of Polycyclic Polyethers

Scheme 15.20 Epoxide-opening cascades directed by a disappearing silyl group by Jamison [198].

substrates to test this hypothesis, Jamison proposed that such a situation may exist in cyclizations of disubstituted epoxy-alcohol **268**, where one THP is formed prior to the epoxide-opening reaction (Figure 15.8). Analysis of the transition states in cyclization reaction of **268** revealed that *trans*-bicyclo[4.4.0]decane derivatives are typically less strained than the corresponding *trans*-bicyclo[4.3.0]nonanes. If the greater stability of the *endo*-product **261** compared to **269** was reflected in the energies of the transitions states leading to these products, increased *endo* selectivity would be observed under kinetic control.

While investigating this hypothesis, Jamison discovered that the regioselectivity of epoxide opening in epoxy-alcohol **268** is dependent on the pH of the aqueous medium used to promote cyclization [199]. The selectivity for the desired THP product **261** increases substantially as the pH of the reaction environment approaches neutrality (Figure 15.8a). Increased water content in the reaction mixture maps well with the increase in selectivity and rate of conversion of reactions examined in water–THF mixtures (Figure 15.8b). Aqueous cascade reactions of templated diepoxide **270** and triepoxide **271** that lack directing groups were also examined and found to proceed with good efficiency in water at elevated temperatures, affording the THP triad **182** and tetrad **272** [199].

Hydrogen-bonding interactions between the THP template, epoxide, and water molecules were proposed as the origin of *endo* selectivity in the described reactions. Kinetic studies of the cyclization reactions of epoxy alcohol **268** and its carbocyclic analog featuring a cyclohexane in place of the THP template suggest existence of at least two competing mechanisms that are first- and second-order in water, respectively [200]. It was proposed that the selective pathway is second-order in water and operable only for the THP-templated epoxy alcohol **268**. Jamison hypothesized that epoxy alcohol cyclizations in water occur for hydrated conformations that

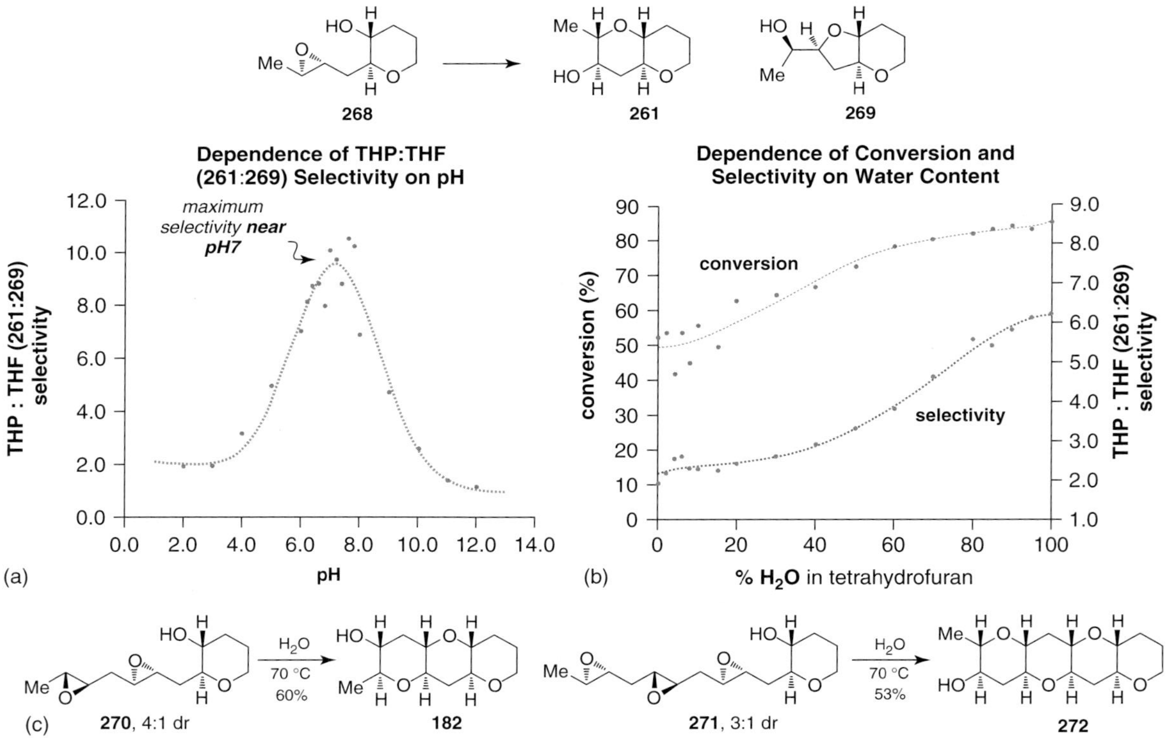

Figure 15.8 (a) THP:THF selectivity in reactions of THP-templated epoxides as a function of pH; (b) effect of water on conversion and selectivity in cyclizations of **268** in THF–water mixtures; (c) directing group-free epoxide-opening cascades promoted by water [199].

possess the appropriate geometry, which is possibly attained in the form of a twist-boat conformation that is stabilized by hydrogen-bonding interactions with water molecules. The electron-withdrawing effects of the oxygen in the THP template may also provide electronic bias for *endo* cyclization via destabilization of the positive charge at the *exo* position.

Methyl groups are commonly encountered at ladder polyether ring junctions, with approximately one in five ring junctions bearing a methyl substituent. Therefore, the putative polyepoxide precursors to ladder polyethers, in addition to disubstituted epoxides, regularly feature two types of trisubstituted epoxides. A methyl group on a trisubstituted epoxide can be situated in a position where its directing effect promotes *endo*-opening under acidic conditions (as in **273**, Scheme 15.21a) or in a position that would normally promote *exo*-selective epoxide opening under the same conditions (as in **274**, Scheme 15.21a). To incorporate methyl substituents at the ring junctions of the final products in epoxide-opening cascades, the Jamison group, in an immediate extension of their work on aqueous directing group-free *endo*-selective epoxide openings, prepared and evaluated cyclization reactions of templated epoxy alcohols **273** and **274**, which feature both types of methyl substitution on the epoxide (Scheme 15.21a) [201, 202]. Acid-catalyzed cyclizations of epoxy alcohol **273** proceeded with high *endo* selectivity due to the electronic effect of the methyl substituent at the *endo* site of attack on the epoxide. Base-promoted cyclizations of the same molecule, however, predominantly produced the *exo* product, bicycle **277**. In contrast, base-promoted cyclizations of the more challenging epoxy alcohol **274** proceeded with moderate *endo* selectivity, while acid-promoted reactions afforded *exo* product **278**. Both **273** and **274** produced *endo* products **275** and **276** with good selectivity when cyclized in deionized water. In addition to achieving high *endo* selectivity, aqueous cyclizations circumvent side reactions associated with other activation conditions, such as the rearrangement of epoxide **273** to a *iso*-propyl ketone via 1,2-hydride shifts under acidic conditions.

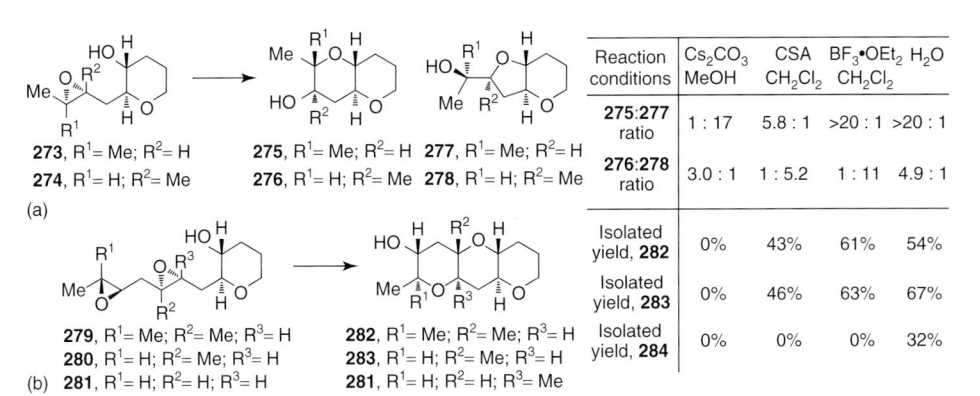

Reaction conditions	Cs_2CO_3 MeOH	CSA CH_2Cl_2	$BF_3 \cdot OEt_2$ CH_2Cl_2	H_2O
275:277 ratio	1 : 17	5.8 : 1	>20 : 1	>20 : 1
276:278 ratio	3.0 : 1	1 : 5.2	1 : 11	4.9 : 1
Isolated yield, **282**	0%	43%	61%	54%
Isolated yield, **283**	0%	46%	63%	67%
Isolated yield, **284**	0%	0%	0%	32%

273, R^1= Me; R^2= H
274, R^1= H; R^2= Me
(a)

275, R^1= Me; R^2= H **277**, R^1= Me; R^2= H
276, R^1= H; R^2= Me **278**, R^1= H; R^2= Me

279, R^1= Me; R^2= Me; R^3= H
280, R^1= H; R^2= Me; R^3= H
281, R^1= H; R^2= H; R^3= H
(b)

282, R^1= Me; R^2= Me; R^3= H
283, R^1= H; R^2= Me; R^3= H
281, R^1= H; R^2= H; R^3= Me

Scheme 15.21 (a) Cyclizations of templated, trisubstituted epoxides **273** and **274**; (b) water overcomes methyl group directing effects in epoxide-opening cascades [201].

Extending this work further, the Jamison group demonstrated that methyl substituents on epoxides are also tolerated in aqueous cascades (Scheme 15.21b) [201, 202]. Under aqueous conditions, diepoxides **279** and **280**, which incorporate methyl substituents at the positions of *endo* attack afforded the triads **282** and **283** (Scheme 15.21b). Diepoxide **281**, with a methyl substituent at the *exo* site, afforded the THP triad **284**. While diepoxides **279** and **280** also produced some of the desired products upon acidic activation, only aqueous conditions afforded any of the THP triad **284** from diepoxide **281**. Aqueous epoxide-opening cascades overcome the need for methyl substituents to be uniformly distributed on each epoxide and can accommodate both types of trisubstituted epoxides in combination with disubstituted epoxides [203].

15.4.4.3 Applications of Epoxide-Opening Cascades in the Synthesis of Ladder Polyethers

The *endo*-selective opening of alkenyl epoxides has become a standard tool for synthesis of tetrahydropyran rings that has been used by the groups of Nicolaou, Yamamoto, Nakata, Mori, and Sasaki in syntheses of hemibrevetoxin B [204–207], brevetoxin B [208–210], brevetoxin A [211], gambierol [212–216], and breveval [135, 136]. This method is generally not amenable to cascades of more than one epoxide. Nevertheless, it has been used in iterative syntheses of ladder polyethers. For example, Nicolaou's approach to the FG ring system of brevetoxin B includes an acid-catalyzed opening of alkenyl epoxides **285** and **287** to form both rings in the FG fragment (**288**, Scheme 15.22a) [208].

Other iterative approaches to oligo(tetrahydropyran)s have also been used in the synthesis of ladder polyether natural products and their fragments. Mori and coworkers have reported a total synthesis of hemibrevetoxin B (**289**, Scheme 15.22b) that relies solely on their iterative strategy for construction of *trans*-fused tetrahydropyran rings [217]. In combination with methods that allow for ring expansion of tetrahydropyran to oxepane systems [186], hemibrevetoxin B was prepared in an iterative fashion using *endo*-selective intramolecular opening of epoxysulfones (Scheme 15.22b). The Mori group was also successful in the preparation of gambierol and the ABCDEF-ring system of yessotoxins and adriatoxins using analogous strategy [35, 187, 216].

An epoxide-opening cascade was utilized in the synthesis of hemibrevetoxin B by Holton [218]. Although only one epoxide is involved in this reaction, two cyclic ethers of the natural product are produced in a single operation. Computational studies by Houk [40, 42] that suggest that alkyl group-directed 6-*endo* cyclization normally requires a loose, S_N1-like transition state prompted Holton to carry out the cascade in a strongly polar solvent. In a fashion similar to the work of the Murai [164], Jamison [80, 195], and Floreancig [196], the alkene in **294** was activated with *N*-(phenylseleno)phthalimide, and the cascade leading to formation of the 7,6-fused BC ring system of hemibrevetoxin B proceeded in high yield (Scheme 15.23).

In efforts to expand the utility of aqueous epoxide-opening cascades for synthesis of ladder polyether fragments, the Jamison group investigated oxygen-containing

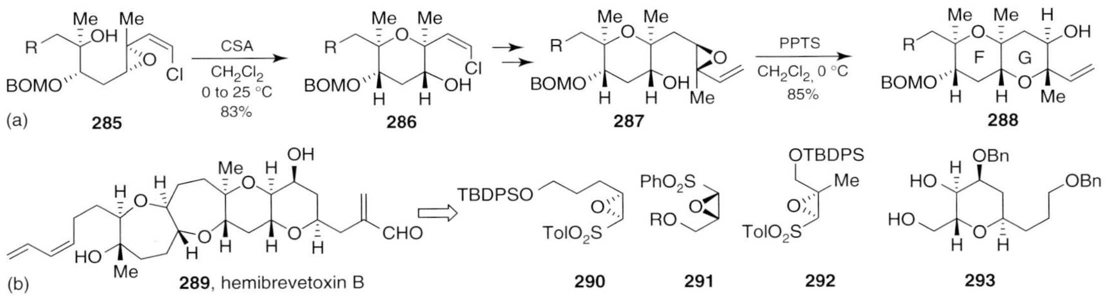

Scheme 15.22 (a) Iterative synthesis of FG fragment of brevetoxin B by Nicolaou [208]; (b) retrosynthetic analysis for the iterative synthesis of hemibrevetoxin B by Mori [217]. R = prenyl, BOM = benzyloxymethyl, PPTS = pyridinium p-toluenesulfonate.

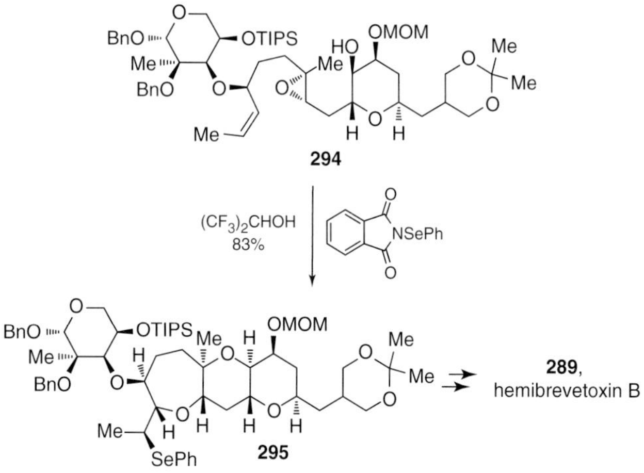

Scheme 15.23 Cascade approach to BC rings of hemibrevetoxin B reported by Holton [218].

templates that could produce fragments suitable for coupling and further elaboration to natural products [219]. Evaluation of a benzylidene acetal template revealed significant differences in the reactivity of epoxy alcohols templated by this motif (e.g., **296**, Scheme 15.24) compared to those templated by a THP. Slow cyclization rates in water and the instability of benzylidene acetals prevented these substrates from being used in aqueous reactions. However, silica gel was found to promote cyclization of benzylidene acetal templated epoxy alcohol **296** to **297** in good yield and with high *endo* selectivity. Elaboration of **297** to the triepoxy-alcohol **298**, with a functionalized THP template, set the stage for an epoxide-opening cascade. Incubation of **298** in water at 60 °C for five days afforded some of the desired THP tetrad **300** and a larger quantity of **299**, in which two THP rings had formed but the final epoxide remained intact (Scheme 15.24). More forceful conditions (80 °C, nine days) drove the cyclization reaction of **298** to completion and allowed the isolation of **300**, the HIJK fragment of gymnocin A, in 35% yield upon acetylation. THP

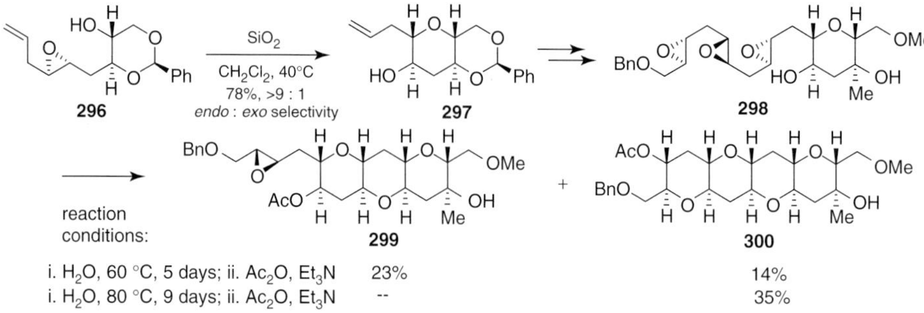

reaction conditions:

i. H$_2$O, 60 °C, 5 days; ii. Ac$_2$O, Et$_3$N   23%   14%
i. H$_2$O, 80 °C, 9 days; ii. Ac$_2$O, Et$_3$N   --   35%

Scheme 15.24 Cascade synthesis of the HIJK fragment of gymnocin A reported by Jamison [219].

tetrad **300** features four differently substituted hydroxyl groups ready for further synthetic elaboration.

15.5
Summary and Outlook

We have presented examples of the many uses of biomimetic epoxide-opening cyclizations in the synthesis of polycyclic polyethers. Epoxide-opening cascades are, however, in no way limited only to the synthesis of polycyclic polyether natural products. Cascade cyclizations that involve epoxide-opening steps have, in fact, found use in many syntheses of natural products outside the polyether families discussed herein (i.e., schweinfurthins and wortmannin) [220–224].

Epoxide-opening cascades that create 2,5-linked oligotetrahydrofurans proceed in agreement with Baldwin's rules and almost always with high selectivity for the smaller rings. While regioselectivity in these reactions is not a major challenge, further improvements are needed to accommodate more diverse substrates and better address challenges of total synthesis. Development of mild conditions to enable better functional group compatibility and methods for selective activation of specific epoxides are imperative for these reactions to proceed with higher yields. This is especially true for the cascades that involve three or more epoxides.

Epoxide-opening cascades leading to the formation of fused polyethers are harder to achieve, burdened by empirical rules of regioselectivity that generally regard *endo*-selective epoxide-opening reactions to be disfavored. Despite considerable work toward epoxide-opening cascades that would produce a ladder polyether in a single synthetic operation, this goal remains elusive. For a cascade to be successful it must be sufficiently flexible to accommodate various ring sizes and epoxide substitution patterns, among other significant challenges, thus making the design of such reactions all the more difficult. While many problems have been addressed in creative ways, further advances are necessary in several directions. For example, construction of seven-, eight-, and nine-membered rings without directing groups on the epoxide remains challenging. Furthermore, accommodating various ring sizes in a single cascade is difficult. Polyepoxide cyclization reactions for the synthesis of ladder polyethers should ideally be able to incorporate a larger number of epoxides (greater than 3–4 epoxides at a time, which is the current state of the art) to construct large polyether fragments. Avoiding side reactions and maintaining efficiency is currently a substantial hurdle for these ambitions. Finally, any such methodology needs to provide products that are amenable to rapid functionalization and elaboration into entire natural products.

Reagents and catalysts that will enable efficient control of regioselectivity in epoxide-opening cyclization hold great promise. These promoters could be either small molecules or the enzymes that are postulated to be involved in the biosynthesis of the polycyclic polyether natural products. Several reports testify to viability of this approach and suggest that research in this area may be worthwhile [39, 41, 43, 44, 70, 72]. The challenges posed by the synthesis of the fascinating family of polycyclic polyether natural products will continue to stimulate research and bring exciting new developments to the field of organic synthesis.

References

1. Dutton, C.J., Banks, B.J., and Cooper, C.B. (1995) *Nat. Prod. Rep.*, **12**, 165–181.
2. Fernandez, J.J., Souto, M.L., and Norte, M. (2000) *Nat. Prod. Rep.*, **17**, 235–246.
3. Rein, K.S. and Borrone, J. (1999) *Comp. Biochem. Physiol. Part B*, **124**, 117–131.
4. Gallimore, A.R. (2009) *Nat. Prod. Rep.*, **26**, 266–280.
5. Vilotijevic, I. and Jamison, T.F. (2009) *Angew. Chem. Int. Ed.*, **48**, 5250–5281.
6. Baldwin, J.E. (1976) *J. Chem. Soc., Chem. Commun.*, 734–736.
7. Johnson, C.D. (1993) *Acc. Chem. Res.*, **26**, 476–482.
8. Coxon, J.M., Hartshorn, M.P., and Swallow, W.H. (1973) *Aust. J. Chem.*, **26**, 2521–2526.
9. Danishefsky, S., Dynak, J., Hatch, E., and Yamamoto, M. (1974) *J. Am. Chem. Soc.*, **96**, 1256–1259.
10. Katsuki, T. and Sharpless, K.B. (1980) *J. Am. Chem. Soc.*, **102**, 5974–5976.
11. Katsuki, T. and Martin, V.S. (1996) *Org. React.*, **48**, 1–299.
12. Zhang, W., Loebach, J.L., Wilson, S.R., and Jacobsen, E.N. (1990) *J. Am. Chem. Soc.*, **112**, 2801–2803.
13. Jacobsen, E.N., Zhang, W., Muci, A.R., Ecker, J.R., and Deng, L. (1991) *J. Am. Chem. Soc.*, **113**, 7063–7064.
14. Tu, Y., Wang, Z.-X., and Shi, Y. (1996) *J. Am. Chem. Soc.*, **118**, 9806–9807.
15. Shi, Y. (2004) *Acc. Chem. Res.*, **37**, 488–496.
16. Wong, O.A. and Shi, Y. (2008) *Chem. Rev.*, **108**, 3958–3987.
17. Xia, Q.H., Ge, H.Q., Ye, C.P., Liu, Z.M., and Su, K.X. (2005) *Chem. Rev.*, **105**, 1603–1662.
18. Heffron, T.P. and Jamison, T.F. (2003) *Org. Lett.*, **5**, 2339–2342.
19. Morimoto, Y., Nishikawa, Y., Ueba, C., and Tanaka, T. (2006) *Angew. Chem. Int. Ed.*, **45**, 810–812.
20. Nicolaou, K.C., Prasad, C.V.C., Somers, P.K., and Hwang, C.K. (1989) *J. Am. Chem. Soc.*, **111**, 5335–5340.
21. Mori, Y., Yaegashi, K., and Furukawa, H. (1996) *J. Am. Chem. Soc.*, **118**, 8158–8159.
22. Fujiwara, K., Tokiwano, T., and Murai, A. (1995) *Tetrahedron Lett.*, **36**, 8063–8066.
23. Nicolaou, K.C., Duggan, M.E., Hwang, C.K., and Somers, P.K. (1985) *J. Chem. Soc., Chem. Commun.*, 1359–1362.
24. Nicolaou, K.C., Prasad, C.V.C., Somers, P.K., and Hwang, C.K. (1989) *J. Am. Chem. Soc.*, **111**, 5330–5334.
25. Suzuki, T., Sato, O., and Hirama, M. (1990) *Tetrahedron Lett.*, **31**, 4747–4750.
26. Matsukura, H., Morimoto, M., Koshino, H., and Nakata, T. (1997) *Tetrahedron Lett.*, **38**, 5545–5548.
27. Mukai, C., Ikeda, Y., Sugimoto, Y.-i., and Hanaoka, M. (1994) *Tetrahedron Lett.*, **35**, 2179–2182.
28. Mukai, C., Sugimoto, Y.-i., Ikeda, Y., and Hanaoka, M. (1994) *Tetrahedron Lett.*, **35**, 2183–2186.
29. Mukai, C., Sugimoto, Y.-i., Ikeda, Y., and Hanaoka, M. (1998) *Tetrahedron*, **54**, 823–850.
30. McDonald, F.E., Wang, X., Do, B., and Hardcastle, K.I. (2000) *Org. Lett.*, **2**, 2917–2919.
31. Bravo, F., McDonald, F.E., Neiwert, W.A., Do, B., and Hardcastle, K.I. (2003) *Org. Lett.*, **5**, 2123–2126.
32. Hudrlik, P.F., Holmes, P.E., and Hudrlik, A.M. (1988) *Tetrahedron Lett.*, **29**, 6395–6398.
33. Adiwidjaja, G., Flörke, H., Kirschning, A., and Schaumann, E. (1995) *Tetrahedron Lett.*, **36**, 8771–8774.
34. Mori, Y. (1997) *Chem. Eur. J.*, **3**, 849–852.
35. Mori, Y., Furuta, H., Takase, T., Mitsuoka, S., and Furukawa, H. (1999) *Tetrahedron Lett.*, **40**, 8019–8022.

36. Fujiwara, K., Mishima, H., Amano, A., Tokiwano, T., and Murai, A. (1998) *Tetrahedron Lett.*, **39**, 393–396.
37. Fujiwara, K., Saka, K., Takaoka, D., and Murai, A. (1999) *Synlett*, 1037–1040.
38. Tokiwano, T., Fujiwara, K., and Murai, A. (2000) *Chem. Lett.*, 272–273.
39. Janda, K.D., Shevlin, C.G., and Lerner, R.A. (1993) *Science*, **259**, 490–493.
40. Na, J., Houk, K.N., Shevlin, C.G., Janda, K.D., and Lerner, R.A. (1993) *J. Am. Chem. Soc.*, **115**, 8453–8454.
41. Janda, K.D., Shevlin, C.G., and Lerner, R.A. (1995) *J. Am. Chem. Soc.*, **117**, 2659–2660.
42. Na, J. and Houk, K.N. (1996) *J. Am. Chem. Soc.*, **118**, 9204–9205.
43. Tokunaga, M., Larrow, J.F., Kakiuchi, F., and Jacobsen, E.N. (1997) *Science*, **277**, 936–938.
44. Wu, M.H., Hansen, K.B., and Jacobsen, E.N. (1999) *Angew. Chem. Int. Ed.*, **38**, 2012–2014.
45. Berger, J., Rachlin, A.I., Scott, W.E., Sternbach, L.H., and Goldberg, M.W. (1951) *J. Am. Chem. Soc.*, **73**, 5295–5298.
46. Harned, R.L., Hidy, P.H., Corum, C.J., and Jones, K.L. (1951) *Antiobiot. Chemother.*, **1**, 594–596.
47. Agtarap, A., Chamberlin, J.W., Pinkerton, M., and Steinrauf, L.K. (1967) *J. Am. Chem. Soc.*, **89**, 5737–5739.
48. Pressman, B.C., Harris, E.J., Jagger, W.S., and Johnson, J.H. (1967) *Proc. Natl. Acad. Sci. U.S.A.*, **58**, 1949–1956.
49. Shumard, R.F. and Callender, M.E. (1967) *Antimicrob. Agents Chemother.*, **7**, 369–377.
50. Raun, A.P., Cooley, C.O., Potter, E.L., Rathmacher, R.P., and Richardson, L.F. (1976) *J. Anim. Sci.*, **43**, 670–677.
51. Russell, J.B. and Houlihan, A.J. (2003) *FEMS Microbiol. Rev.*, **27**, 65–74.
52. Westley, J.W. (1977) in *Advances in Applied Microbiology*, vol. 22 (ed. D. Perlman), Academic Press, New York, pp. 177–223.
53. Westley, J.W., Evans, R.H. Jr., Sello, L.H., Troupe, N., Liu, C.M., and Miller, P.A. (1981) *J. Antibiot.*, **34**, 1248–1252.
54. Westley, J.W., Evans, R.H. Jr., Sello, L.H., Troupe, N., Liu, C.-M., Blount, J.F., Pitcher, R.G., Williams, T.H., and Miller, P.A. (1981) *J. Antibiot.*, **34**, 139–147.
55. Westley, J.W., Evans, R.H. Jr., Harvey, G., Pitcher, R.G., Pruess, D.L., Stempel, A., and Berger, J. (1974) *J. Antibiot.*, **27**, 288–297.
56. Westley, J.W., Blount, J.F., Evans, R.H. Jr., Stempel, A., and Berger, J. (1974) *J. Antibiot.*, **27**, 597–604.
57. Cane, D.E., Liang, T.-C., and Hasler, H. (1981) *J. Am. Chem. Soc.*, **103**, 5962–5965.
58. Westley, J.W. (1981) *Antibiotics*, **4**, 41–73.
59. Cane, D.E., Liang, T.C., and Hasler, H. (1982) *J. Am. Chem. Soc.*, **104**, 7274–7281.
60. Hutchinson, C.R. (1983) *Acc. Chem. Res.*, **16**, 7–14.
61. Cane, D.E., Celmer, W.D., and Westley, J.W. (1983) *J. Am. Chem. Soc.*, **105**, 3594–3600.
62. Townsend, C.A. and Basak, A. (1991) *Tetrahedron*, **47**, 2591–2602.
63. Koert, U. (1995) *Angew. Chem., Int. Ed. Engl.*, **34**, 298–300.
64. McDonald, F.E. and Towne, T.B. (1994) *J. Am. Chem. Soc.*, **116**, 7921–7922.
65. McDonald, F.E., Towne, T.B., and Schultz, C.C. (1998) *Pure Appl. Chem.*, **70**, 355–358.
66. Leadlay, P.F., Staunton, J., Oliynyk, M., Bisang, C., Cortes, J., Frost, E., Hughes-Thomas, Z.A., Jones, M.A., Kendrew, S.G., Lester, J.B., Long, P.F., McArthur, H.A.I., McCormick, E.L., Oliynyk, Z., Stark, C.B.W., and Wilkinson, C.J. (2001) *J. Ind. Microbiol. Biotechnol.*, **27**, 360–367.
67. Oliynyk, M., Stark, C.B.W., Bhatt, A., Jones, M.A., Hughes-Thomas, Z.A., Wilkinson, C., Oliynyk, Z., Demydchuk, Y., Staunton, J., and Leadlay, P.F. (2003) *Mol. Microbiol.*, **49**, 1179–1190.
68. Bhatt, A., Stark, C.B.W., Harvey, B.M., Gallimore, A.R., Demydchuk, Y.A., Spencer, J.B., Staunton, J., and Leadlay, P.F. (2005) *Angew. Chem. Int. Ed.*, **44**, 7075–7078.
69. Gallimore, A.R., Stark, C.B.W., Bhatt, A., Harvey, B.M., Demydchuk, Y., Bolanos-Garcia, V., Fowler, D.J.,

Staunton, J., Leadlay, P.F., and Spencer, J.B. (2006) *Chem. Biol.*, **13**, 453–460.

70. Shichijo, Y., Migita, A., Oguri, H., Watanabe, M., Tokiwano, T., Watanabe, K., and Oikawa, H. (2008) *J. Am. Chem. Soc.*, **130**, 12230–12231.

71. Migita, A., Shichijo, Y., Oguri, H., Watanabe, M., Tokiwano, T., and Oikawa, H. (2008) *Tetrahedron Lett.*, **49**, 1021–1025.

72. Smith, L., Hong, H., Spencer, J.B., and Leadlay, P.F. (2008) *ChemBioChem*, **9**, 2967–2975.

73. Rudi, A., Yosief, T., Schleyer, M., and Kashman, Y. (1999) *Tetrahedron*, **55**, 5555–5566.

74. Kashman, Y. and Rudi, A. (2005) *Phytochem. Rev.*, **3**, 309–323.

75. Sakemi, S., Higa, T., Jefford, C.W., and Bernardinelli, G. (1986) *Tetrahedron Lett.*, **27**, 4287–4290.

76. Hashimoto, M., Kan, T., Nozaki, K., Yanagiya, M., Shirahama, H., and Matsumoto, T. (1990) *J. Org. Chem.*, **55**, 5088–5107.

77. Matsuo, Y., Suzuki, M., and Masuda, M. (1995) *Chem. Lett.*, 1043–1044.

78. Manriquez, C.P., Souto, M.L., Gavin, J.A., Norte, M., and Fernandez, J.J. (2001) *Tetrahedron*, **57**, 3117–3123.

79. Domingo, V., Arteaga, J.F., Moral, J.F.Qd., and Barrero, A.F. (2009) *Nat. Prod. Rep.*, **26**, 115–134.

80. Tanuwidjaja, J., Ng, S.-S., and Jamison, T.F. (2009) *J. Am. Chem. Soc.*, **131**, 12084–12085.

81. Lin, Y.-Y., Risk, M., Ray, S.M., Van Engen, D., Clardy, J., Golik, J., James, J.C., and Nakanishi, K. (1981) *J. Am. Chem. Soc.*, **103**, 6773–6775.

82. Murata, M., Naoki, H., Matsunaga, S., Satake, M., and Yasumoto, T. (1994) *J. Am. Chem. Soc.*, **116**, 7098–7107.

83. Sasaki, M., Matsumori, N., Maruyama, T., Nonomura, T., Murata, M., Tachibana, K., and Yasumoto, T. (1996) *Angew. Chem., Int. Ed. Engl.*, **35**, 1672–1675.

84. Nonomura, T., Sasaki, M., Matsumori, N., Murata, M., Tachibana, K., and Yasumoto, T. (1996) *Angew. Chem., Int. Ed. Engl.*, **35**, 1675–1678.

85. Yasumoto, T. and Murata, M. (1993) *Chem. Rev.*, **93**, 1897–1909.

86. Inoue, M. (2005) *Chem. Rev.*, **105**, 4379–4405.

87. Nakata, T. (2005) *Chem. Rev.*, **105**, 4314–4347.

88. Sasaki, M. and Fuwa, H. (2008) *Nat. Prod. Rep.*, **25**, 401–426.

89. Nicolaou, K.C., Frederick, M.O., and Aversa, R.J. (2008) *Angew. Chem. Int. Ed.*, **47**, 7182–7225.

90. Sellner, K.G., Doucette, G.J., and Kirkpatrick, G.J. (2003) *J. Ind. Microbiol. Biotechnol.*, **30**, 383–406.

91. Schrope, M. (2008) *Nature*, **452**, 24–26.

92. Murata, M., Legrand, A.M., Ishibashi, Y., Fukui, M., and Yasumoto, T. (1990) *J. Am. Chem. Soc.*, **112**, 4380–4386.

93. Flewelling, L.J., Naar, J.P., Abbott, J.P., Baden, D.G., Barros, N.B., Bossart, G.D., Bottein, M.-Y.D., Hammond, D.G., Haubold, E.M., Heil, C.A., Henry, M.S., Jacocks, H.M., Leighfield, T.A., Pierce, R.H., Pitchford, T.D., Rommel, S.A., Scott, P.S., Steidinger, K.A., Truby, E.W., Van Dolah, F.M., and Landsberg, J.H. (2005) *Nature*, **435**, 755–756.

94. Lewis, R.J. (2001) *Toxicon*, **39**, 97–106.

95. Murata, M., Legrand, A.M., and Yasumoto, T. (1989) *Tetrahedron Lett.*, **30**, 3793–3796.

96. Murata, M., Legrand, A.M., Ishibashi, Y., and Yasumoto, T. (1989) *J. Am. Chem. Soc.*, **111**, 8929–8931.

97. Murata, M., Naoki, H., Iwashita, T., Matsunaga, S., Sasaki, M., Yokoyama, A., and Yasumoto, T. (1993) *J. Am. Chem. Soc.*, **115**, 2060–2062.

98. Satake, M., Shoji, M., Oshima, Y., Naoki, H., Fujita, T., and Yasumoto, T. (2002) *Tetrahedron Lett.*, **43**, 5829–5832.

99. Ferrari, S., Ciminiello, P., Dell'Aversano, C., Forino, M., Malaguti, C., Tubaro, A., Poletti, R., Yasumoto, T., Fattorusso, E., and Rossini, G.P. (2004) *Chem. Res. Toxicol.*, **17**, 1251–1257.

100. Ronzitti, G., Callegari, F., Malaguti, C., and Rossini, G.P. (2004) *Br. J. Cancer*, **90**, 1100–1107.

101. Nagai, H., Murata, M., Torigoe, K., Satake, M., and Yasumoto, T. (1992) *J. Org. Chem.*, **57**, 5448–5453.
102. Nagai, H., Mikami, Y., Yazawa, K., Gonoi, T., and Yasumoto, T. (1993) *J. Antibiot.*, **46**, 520–522.
103. Bourdelais, A.J., Campbell, S., Jacocks, H., Naar, J., Wright, J.L.C., Carsi, J., and Baden, D.G. (2004) *Cell. Mol. Neurobiol.*, **24**, 553–563.
104. Bourdelais, A.J., Jacocks, H.M., Wright, J.L.C., Bigwarfe, P.M. Jr., and Baden, D.G. (2005) *J. Nat. Prod.*, **68**, 2–6.
105. Abraham, W.M., Bourdelais, A.J., Sabater, J.R., Ahmed, A., Lee, T.A., Serebriakov, I., and Baden, D.G. (2005) *Am. J. Respir. Crit. Care. Med.*, **171**, 26–34.
106. Baden, D.G., Abraham, W.M., Bourdelais, A.J., and Michelliza, S. (2009) Fused pentacyclic polyethers US 7638500.
107. Poli, M.A., Mende, T.J., and Baden, D.G. (1986) *Mol. Pharmacol.*, **30**, 129–135.
108. Bidard, J.N., Vijverberg, H.P.M., Frelin, C., Chungue, E., Legrand, A.M., Bagnis, R., and Lazdunski, M. (1984) *J. Biol. Chem.*, **259**, 8353–8357.
109. Gawley, R.E., Rein, K.S., Jeglitsch, G., Adams, D.J., Theodorakis, E.A., Tiebes, J., Nicolaou, K.C., and Baden, D.G. (1995) *Chem. Biol.*, **2**, 533–541.
110. Gawley, R.E., Rein, K.S., Kinoshita, M., and Baden, D.G. (1992) *Toxicon*, **30**, 780–785.
111. Trainer, V.L., Baden, D.G., and Catterall, W.A. (1994) *J. Biol. Chem.*, **269**, 19904–19909.
112. Cuypers, E., Abdel-Mottaleb, Y., Kopljar, I., Rainier, J.D., Raes, A.L., Snyders, D.J., and Tytgat, J. (2008) *Toxicon*, **51**, 974–983.
113. Takahashi, M., Ohizumi, Y., and Yasumoto, T. (1982) *J. Biol. Chem.*, **257**, 7287–7289.
114. Gusovsky, F., Yasumoto, T., and Daly, J.W. (1989) *FEBS Lett.*, **243**, 307–312.
115. Murata, M., Gusovsky, F., Yasumoto, T., and Daly, J.W. (1992) *Eur. J. Pharmacol.*, **227**, 43–49.
116. Sinkins, W.G., Estacion, M., Prasad, V., Goel, M., Shull, G.E., Kunze, D.L., and Schilling, W.P. (2009) *Am. J. Physiol. Cell Physiol.*, **297**, C1533–C1543.
117. Mori, M., Oishi, T., Matsuoka, S., Ujihara, S., Matsumori, N., Murata, M., Satake, M., Oshima, Y., Matsushita, N., and Aimoto, S. (2005) *Bioorg. Med. Chem.*, **13**, 5099–5103.
118. Nakanishi, K. (1985) *Toxicon*, **23**, 473–479.
119. Chou, H.N. and Shimizu, Y. (1987) *J. Am. Chem. Soc.*, **109**, 2184–2185.
120. Gallimore, A.R. and Spencer, J.B. (2006) *Angew. Chem. Int. Ed.*, **45**, 4406–4413.
121. Giner, J.-L., Li, X., and Mullins, J.J. (2003) *J. Org. Chem.*, **68**, 10079–10086.
122. Lee, M.S., Repeta, D.J., Nakanishi, K., and Zagorski, M.G. (1986) *J. Am. Chem. Soc.*, **108**, 7855–7856.
123. Lee, M.S., Qin, G., Nakanishi, K., and Zagorski, M.G. (1989) *J. Am. Chem. Soc.*, **111**, 6234–6241.
124. Satake, M. (2000) *Tennen Yuki Kagobutsu Toronkai Koen Yoshishu*, **42**, 259–264.
125. Snyder, R.V., Gibbs, P.D.L., Palacios, A., Abiy, L., Dickey, R., Lopez, J.V., and Rein, K.S. (2003) *Mar. Biotechnol.*, **5**, 1–12.
126. Snyder, R.V., Guerrero, M.A., Sinigalliano, C.D., Winshell, J., Perez, R., Lopez, J.V., and Rein, K.S. (2005) *Phytochemistry*, **66**, 1767–1780.
127. Rein, K.S. and Snyder, R.V. (2006) *Adv. Appl. Microbiol.*, **59**, 93–125.
128. Perez, R., Liu, L., Lopez, J., An, T., and Rein, K.S. (2008) *Mar. Drugs*, **6**, 164–179.
129. Murata, M., Izumikawa, M., Tachibana, K., Fujita, T., and Naoki, H. (1998) *J. Am. Chem. Soc.*, **120**, 147–151.
130. Izumikawa, M., Murata, M., Tachibana, K., Fujita, T., and Naoki, H. (2000) *Eur. J. Biochem.*, **267**, 5179–5183.
131. Chou, H.-N., Shimizu, Y., Van Duyne, G., and Clardy, J. (1985) *Tetrahedron Lett.*, **26**, 2865–2868.
132. Satake, M., Bourdelais, A.J., Van Wagoner, R.M., Baden, D.G., and Wright, J.L.C. (2008) *Org. Lett.*, **10**, 3465–3468.
133. Satake, M., Campbell, A., Van Wagoner, R.M., Bourdelais, A.J., Jacocks, H., Baden, D.G., and Wright,

J.L.C. (2009) *J. Org. Chem.*, **74**, 989–994.
134. Van Wagoner, R.M., Satake, M., Bourdelais, A.J., Baden, D.G., and Wright, J.L.C. (2010) *J. Nat. Prod.*, **73**, 1177–1179.
135. Fuwa, H., Ebine, M., and Sasaki, M. (2006) *J. Am. Chem. Soc.*, **128**, 9648–9650.
136. Fuwa, H., Ebine, M., Bourdelais, A.J., Baden, D.G., and Sasaki, M. (2006) *J. Am. Chem. Soc.*, **128**, 16989–16999.
137. Nicolaou, K.C. and Frederick, M.O. (2007) *Angew. Chem. Int. Ed.*, **46**, 5278–5282.
138. Nicolaou, K.C., Frederick, M.O., Burtoloso, A.C.B., Denton, R.M., Rivas, F., Cole, K.P., Aversa, R.J., Gibe, R., Umezawa, T., and Suzuki, T. (2008) *J. Am. Chem. Soc.*, **130**, 7466–7476.
139. Giner, J.-L. (2005) *J. Org. Chem.*, **70**, 721–724.
140. Wiggins, L.F. and Wood, D.J.C. (1950) *J. Chem. Soc.*, 1566–1575.
141. Benner, S.A., Maggio, J.E., and Simmons, H.E. (1981) *J. Am. Chem. Soc.*, **103**, 1581–1582.
142. Simmons, H.E. III and Maggio, J.E. (1981) *Tetrahedron Lett.*, **22**, 287–290.
143. Paquette, L.A. and Vazeux, M. (1981) *Tetrahedron Lett.*, **22**, 291–294.
144. Paquette, L.A., Williams, R.V., Vazeux, M., and Browne, A.R. (1984) *J. Org. Chem.*, **49**, 2194–2197.
145. Weitemeyer, C., Preuss, T., and de Meijere, A. (1985) *Chem. Ber.*, **118**, 3993–4005.
146. Liang, S., Lee, C.-H., Kozhushkov, S.I., Yufit, D.S., Howard, J.A.K., Meindl, K., Ruehl, S., Yamamoto, C., Okamoto, Y., Schreiner, P.R., Rinderspacher, B.C., and de Meijere, A. (2005) *Chem. Eur. J.*, **11**, 2012–2018.
147. Person, G., Keller, M., and Prinzbach, H. (1996) *Liebigs Ann.*, 507–527.
148. Guiard, S., Giorgi, M., Santelli, M., and Parrain, J.-L. (2003) *J. Org. Chem.*, **68**, 3319–3322.
149. Alvarez, E., Manta, E., Martin, J.D., Rodriguez, M.L., Ruiz-Perez, C., and Zurita, D. (1988) *Tetrahedron Lett.*, **29**, 2097–2100.
150. Alvarez, E., Diaz, M.T., Perez, R., Ravelo, J.L., Regueiro, A., Vera, J.A., Zurita, D., and Martin, J.D. (1994) *J. Org. Chem.*, **59**, 2848–2876.
151. Dolle, R.E. and Nicolaou, K.C. (1985) *J. Am. Chem. Soc.*, **107**, 1691–1694.
152. Hoye, T.R. and Suhadolnik, J.C. (1985) *J. Am. Chem. Soc.*, **107**, 5312–5313.
153. Hoye, T.R. and Suhadolnik, J.C. (1986) *Tetrahedron*, **42**, 2855–2862.
154. Hoye, T.R. and Jenkins, S.A. (1987) *J. Am. Chem. Soc.*, **109**, 6196–6198.
155. Hoye, T.R. and Witowski, N.E. (1992) *J. Am. Chem. Soc.*, **114**, 7291–7292.
156. Schreiber, S.L., Sammakia, T., Hulin, B., and Schulte, G. (1986) *J. Am. Chem. Soc.*, **108**, 2106–2108.
157. Still, W.C. and Romero, A.G. (1986) *J. Am. Chem. Soc.*, **108**, 2105–2106.
158. Vedejs, E. and Gapinski, D.M. (1983) *J. Am. Chem. Soc.*, **105**, 5058–5061.
159. Paterson, I., Boddy, I., and Mason, I. (1987) *Tetrahedron Lett.*, **28**, 5205–5208.
160. Nacro, K., Baltas, M., Zedde, C., Gorrichon, L., and Jaud, J. (1999) *Tetrahedron*, **55**, 5129–5138.
161. Russell, S.T., Robinson, J.A., and Williams, D.J. (1987) *J. Chem. Soc., Chem. Commun.*, 351–352.
162. Kumar, V.S., Aubele, D.L., and Floreancig, P.E. (2002) *Org. Lett.*, **4**, 2489–2492.
163. Kumar, V.S., Wan, S., Aubele, D.L., and Floreancig, P.E. (2005) *Tetrahedron: Asymmetry*, **16**, 3570–3578.
164. Hayashi, N., Fujiwara, K., and Murai, A. (1997) *Tetrahedron*, **53**, 12425–12468.
165. Marshall, J.A. and Chobanian, H.R. (2003) *Org. Lett.*, **5**, 1931–1933.
166. Marshall, J.A. and Hann, R.K. (2008) *J. Org. Chem.*, **73**, 6753–6757.
167. Marshall, J.A. and Mikowski, A.M. (2006) *Org. Lett.*, **8**, 4375–4378.
168. Schultz, W.J., Etter, M.C., Pocius, A.V., and Smith, S. (1980) *J. Am. Chem. Soc.*, **102**, 7981–7982.
169. Iimori, T., Still, W.C., Rheingold, A.L., and Staley, D.L. (1989) *J. Am. Chem. Soc.*, **111**, 3439–3440.
170. Paterson, I. and Boddy, I. (1988) *Tetrahedron Lett.*, **29**, 5301–5304.

171. Paterson, I. and Craw, P.A. (1989) *Tetrahedron Lett.*, **30**, 5799–5802.
172. Evans, D.A., Ratz, A.M., Huff, B.E., and Sheppard, G.S. (1995) *J. Am. Chem. Soc.*, **117**, 3448–3467.
173. Morimoto, Y., Iwai, T., and Kinoshita, T. (2000) *J. Am. Chem. Soc.*, **122**, 7124–7125.
174. Xiong, Z. and Corey, E.J. (2000) *J. Am. Chem. Soc.*, **122**, 4831–4832.
175. Xiong, Z. and Corey, E.J. (2000) *J. Am. Chem. Soc.*, **122**, 9328–9329.
176. Morimoto, Y., Iwai, T., Nishikawa, Y., and Kinoshita, T. (2002) *Tetrahedron Asymmetry*, **13**, 2641–2647.
177. Morimoto, Y., Nishikawa, Y., and Takaishi, M. (2005) *J. Am. Chem. Soc.*, **127**, 5806–5807.
178. Morimoto, Y., Yata, H., and Nishikawa, Y. (2007) *Angew. Chem. Int. Ed.*, **46**, 6481–6484.
179. Morimoto, Y., Okita, T., and Kambara, H. (2009) *Angew. Chem. Int. Ed.*, **48**, 2538–2541.
180. Morimoto, Y., Okita, T., Takaishi, M., and Tanaka, T. (2007) *Angew. Chem. Int. Ed.*, **46**, 1132–1135.
181. Xiong, Z., Busch, R., and Corey, E.J. (2010) *Org. Lett.*, **12**, 1512–1514.
182. Tong, R., Valentine, J.C., McDonald, F.E., Cao, R., Fang, X., and Hardcastle, K.I. (2007) *J. Am. Chem. Soc.*, **129**, 1050–1051.
183. Tong, R. and McDonald, F.E. (2008) *Angew. Chem. Int. Ed.*, **47**, 4377–4379.
184. Tong, R.B., Boone, M.A., and McDonald, F.E. (2009) *J. Org. Chem.*, **74**, 8407–8409.
185. Corey, E.J. and Ha, D.C. (1988) *Tetrahedron Lett.*, **29**, 3171–3174.
186. Mori, Y., Yaegashi, K., and Furukawa, H. (1997) *Tetrahedron*, **53**, 12917–12932.
187. Furuta, H., Takase, T., Hayashi, H., Noyori, R., and Mori, Y. (2003) *Tetrahedron*, **59**, 9767–9777.
188. Hudrlik, P.F., Arcoleo, J.P., Schwartz, R.H., Misra, R.N., and Rona, R.J. (1977) *Tetrahedron Lett.*, **18**, 591–594.
189. Hudrlik, P.F., Hudrlik, A.M., and Kulkarni, A.K. (1982) *J. Am. Chem. Soc.*, **104**, 6809–6811.
190. Fristad, W.E., Bailey, T.R., Paquette, L.A., Gleiter, R., and Boehm, M.C. (1979) *J. Am. Chem. Soc.*, **101**, 4420–4423.
191. Tokiwano, T., Fujiwara, K., and Murai, A. (2000) *Synlett*, 335–338.
192. McDonald, F.E., Bravo, F., Wang, X., Wei, X., Toganoh, M., Rodriguez, J.R., Do, B., Neiwert, W.A., and Hardcastle, K.I. (2002) *J. Org. Chem.*, **67**, 2515–2523.
193. Bravo, F., McDonald, F.E., Neiwert, W.A., and Hardcastle, K.I. (2004) *Org. Lett.*, **6**, 4487–4489.
194. Valentine, J.C., McDonald, F.E., Neiwert, W.A., and Hardcastle, K.I. (2005) *J. Am. Chem. Soc.*, **127**, 4586–4587.
195. Heffron, T.P. and Jamison, T.F. (2006) *Synlett*, 2329–2333.
196. Wan, S., Gunaydin, H., Houk, K.N., and Floreancig, P.E. (2007) *J. Am. Chem. Soc.*, **129**, 7915–7923.
197. Tarselli, M.A., Zuccarello, J.L., Lee, S.J., and Gagné, M.R. (2009) *Org. Lett.*, **11**, 3490–3492.
198. Simpson, G.L., Heffron, T.P., Merino, E., and Jamison, T.F. (2006) *J. Am. Chem. Soc.*, **128**, 1056–1057.
199. Vilotijevic, I. and Jamison, T.F. (2007) *Science*, **317**, 1189–1192.
200. Byers, J.A. and Jamison, T.F. (2009) *J. Am. Chem. Soc.*, **131**, 6383–6385.
201. Morten, C.J. and Jamison, T.F. (2009) *J. Am. Chem. Soc.*, **131**, 6678–6679.
202. Morten, C.J. and Jamison, T.F. (2009) *Tetrahedron*, **65**, 6648–6655.
203. Morten, C.J., Byers, J.A., Van Dyke, A.R., Vilotijevic, I., and Jamison, T.F. (2009) *Chem. Soc. Rev.*, **38**, 3175–3192.
204. Nicolaou, K.C., Reddy, K.R., Skokotas, G., Sato, F., and Xiao, X.Y. (1992) *J. Am. Chem. Soc.*, **114**, 7935–7936.
205. Morimoto, M., Matsukura, H., and Nakata, T. (1996) *Tetrahedron Lett.*, **37**, 6365–6368.
206. Mori, Y., Yaegashi, K., and Furukawa, H. (1997) *J. Am. Chem. Soc.*, **119**, 4557–4558.
207. Kadota, I. and Yamamoto, Y. (1998) *J. Org. Chem.*, **63**, 6597–6606.
208. Nicolaou, K.C., Duggan, M.E., and Hwang, C.K. (1989) *J. Am. Chem. Soc.*, **111**, 6676–6682.

209. Nicolaou, K.C. (1996) *Angew. Chem., Int. Ed. Engl.*, **35**, 589–607.
210. Matsuo, G., Kawamura, K., Hori, N., Matsukura, H., and Nakata, T. (2004) *J. Am. Chem. Soc.*, **126**, 14374–14376.
211. Nicolaou, K.C., Yang, Z., Shi, G., Gunzner, J.L., Agrios, K.A., and Gartner, P. (1998) *Nature*, **392**, 264–269.
212. Fuwa, H., Sasaki, M., and Tachibana, K. (2001) *Org. Lett.*, **3**, 3549–3552.
213. Fuwa, H., Kainuma, N., Tachibana, K., and Sasaki, M. (2002) *J. Am. Chem. Soc.*, **124**, 14983–14992.
214. Kadota, I., Takamura, H., Sato, K., Ohno, A., Matsuda, K., Satake, M., and Yamamoto, Y. (2003) *J. Am. Chem. Soc.*, **125**, 11893–11899.
215. Kadota, I., Takamura, H., Sato, K., Ohno, A., Matsuda, K., and Yamamoto, Y. (2003) *J. Am. Chem. Soc.*, **125**, 46–47.
216. Furuta, H., Hasegawa, Y., and Mori, Y. (2009) *Org. Lett.*, **11**, 4382–4385.
217. Mori, Y., Yaegashi, K., and Furukawa, H. (1998) *J. Org. Chem.*, **63**, 6200–6209.
218. Zakarian, A., Batch, A., and Holton, R.A. (2003) *J. Am. Chem. Soc.*, **125**, 7822–7824.
219. Van Dyke, A.R. and Jamison, T.F. (2009) *Angew. Chem. Int. Ed.*, **48**, 4430–4432.
220. Nicolaou, K.C., Edmonds, D.J., and Bulger, P.G. (2006) *Angew. Chem. Int. Ed.*, **45**, 7134–7186.
221. Mente, N.R., Neighbors, J.D., and Wiemer, D.F. (2008) *J. Org. Chem.*, **73**, 7963–7970.
222. Topczewski, J.J., Neighbors, J.D., and Wiemer, D.F. (2009) *J. Org. Chem.*, **74**, 6965–6972.
223. Topczewski, J.J., Callahan, M.P., Neighbors, J.D., and Wiemer, D.F. (2009) *J. Am. Chem. Soc.*, **131**, 14630–14631.
224. Sato, S., Nakada, M., and Shibasaki, M. (1996) *Tetrahedron Lett.*, **37**, 6141–6144.

16
Biomimetic Electrocyclization Reactions toward Polyketide-Derived Natural Products

James Burnley, Michael Ralph, Pallavi Sharma, and John E. Moses

16.1
Introduction

The term *biomimetic* derives from the Greek "bios," meaning life, and mimetic, the adjective for "mimesis" or mimicry, and is the application of methods observed in Nature to the design and development of synthetic systems. This methodology transfer is desirable, not only because evolutionary pressures typically force natural systems to become highly optimized and efficient, but also because such systems boast an inherent elegance [1].

When applied to chemical syntheses, biomimetic approaches often facilitate rapid access to complex structures that may otherwise require inconceivable conventional synthetic pathways. Inspired by key transformations in the known (or proposed) biosynthetic pathway, biomimetic syntheses are perhaps most applicable to those systems that do not strictly require enzymatic catalysis or control. Instead, the structure itself is predisposed to the biomimetic chemical change, following a defined reaction pathway flowing energetically "downhill." Heathcock neatly describes the biomimetic strategy:

> "The basic assumption of this approach is that Nature is the quintessential process development chemist. We think that the molecular frameworks of most natural products arise by intrinsically favorable chemical pathways – favorable enough that the skeleton could have arisen by a non-enzymatic reaction in the primitive organism. If a molecule produced in this purely chemical manner was beneficial to the organism, enzymes would have evolved to facilitate the production of this useful material" [2].

In recent times, the training of synthetic chemists has been dominated by the powerful retro-synthetic approach developed by Corey [3]. In this logical method, subskeletal functionalization generally occurs early on in the synthetic plan, and these "handles" are used as junctions for connecting and building up complex frameworks. There seems to be no limit in what can be achieved using this powerful system [4], and with the development of new reactions and catalysts, the future

Biomimetic Organic Synthesis, First Edition. Edited by Erwan Poupon and Bastien Nay.
© 2011 Wiley-VCH Verlag GmbH & Co. KGaA. Published 2011 by Wiley-VCH Verlag GmbH & Co. KGaA.

of synthesis is very exciting. However, in certain cases problems may arise in assembling together such highly functionalized sub-units into structures resembling the targeted natural products [5]. In contrast, Nature generally assembles the molecular scaffolding first with relatively few, but key, functional groups in place. Through the process of secondary metabolism, these core frameworks are then functionalized in a highly specific manner, often enzymatically [6].

Biomimetic chemistry offers alternative and complementary strategies toward total synthesis. They are often elegant and efficient processes, enlightening novel pathways to Nature's most complex architectures. From the early days of Sir Robert Robinson [7], to modern day practitioners, the biomimetic chemist is driven not only by the desire to evolve new chemistries but also at developing a deeper understanding of the processes that actually make natural products, which themselves, due to evolutionary factors, have been pre-selected for their beneficial properties and elegance of synthetic approach. As such, biomimetic syntheses are relevant, and indeed have proven to be powerful in the syntheses of a huge array of complex secondary metabolites, including, steroids [8], terpenes [9], alkaloids [2, 10], and polyketides [11].

Of the known chemical transformations available to the chemist (and indeed Nature), it is fair to say that in recent times pericyclic reactions have taken center stage in biomimetic syntheses. In particular, cycloadditions have been a cornerstone, with the Diels–Alder reaction proving to be the most powerful and well-studied transformation [12]. This is no doubt due to the remarkable efficiency and complexity-generating properties of this amazing reaction that, in one step, creates two new σ-bonds and up-to four new stereocenters. However, when considering complexity-generating processes, electrocyclization reactions have also proven valuable tools in biomimetic chemistry, not least because they often participate in reaction cascades [13]. Such cascades may involve several electrocyclizations and, in combination with cycloadditions, for instance, they can deliver spectacular molecular complexity and diversity. Although electrocyclization reactions have played a central role in the biomimetic syntheses of a wide range of natural products, this chapter is restricted to polyketide-derived natural products, where electrocyclization reactions have been a key transformation and, in some cases, have provided essential validation of biosynthetic pathways. Rather than provide a comprehensive review of all polyketide-derived natural product syntheses involving electrocyclizations, we have chosen to limit the number of case studies to provide a more detailed analysis of the thought processes behind the biomimetic strategy. Electrocyclization reactions that have featured in total syntheses but have not been proposed, or are unlikely, to be biomimetic are not discussed.

16.2
Electrocyclic Reactions

Electrocyclic reactions are pericyclic processes that involve the cyclizations of conjugated polyenes. In such reactions, the π-bonds move in a cyclic manner and a new σ-bond is formed at the terminal carbon in the polyene system, leading to a ring structure [14, 15]. Thermodynamically, electrocyclizations are driven by the formation of stronger σ-bonds at the expense of weaker π-bonds. In principle, there is no limit to the number of π-bonds that can be involved in a given electrocyclization, but systems containing 6π and 8π electrons leading to six- and eight-membered rings are most common, particularly in biomimetic syntheses. Electrocyclizations are not restricted to all-carbon systems, and both oxa- and aza-systems can also participate.

The most striking and synthetically useful feature of electrocyclization reactions is that their stereochemical outcome can be accurately predicted and controlled, depending upon the substrate choice and reaction conditions. For example, (2E,4Z,6E)-octa-2,4,6-triene (1) yields cis-5,6-dimethyl-1,3-cylohexadiene (4) selectively under thermal conditions, whereas (2E,4Z,6Z)-octa-2,4,6-triene (3) yields trans-5,6-dimethyl-1,3-cylohexadiene (2) (Scheme 16.1). This trend is reversed under photochemical conditions, and is dependent upon whether the reaction proceeds via a conrotatory or disrotatory pathway, and can be readily predicted according to the rules formulated by Woodward and Hoffman [14]. Another key feature of electrocyclization is the requirement for the reacting termini to be brought into close proximity. As such, reactive polyenes must normally contain internal double bond(s) with *(Z)*-configuration. This is an essential consideration when planning a biomimetic synthesis involving an electrocyclization reaction.

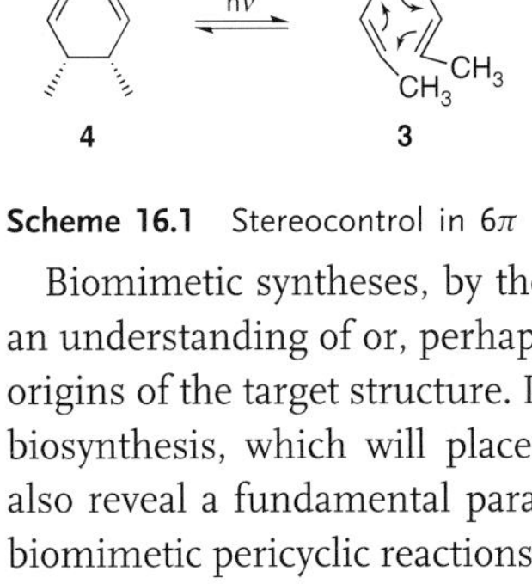

Scheme 16.1 Stereocontrol in 6π electrocyclizations.

Biomimetic syntheses, by their very nature, require the practitioner to develop an understanding of or, perhaps more correctly, a rationale about the biosynthetic origins of the target structure. It is therefore pertinent to briefly discuss polyketide biosynthesis, which will place biomimetic-inspired syntheses into context, and also reveal a fundamental paradox about the involvement of enzyme catalysis in biomimetic pericyclic reactions.

16.3
Polyketides

The polyketide-derived family of natural products has, for many years, been a rich source of biologically and chemically intriguing molecules. These diverse secondary metabolites, which are often obtained from moulds and other soil and airborne microorganisms and microbes, have provided us with many pharmaceutically significant targets. For example, one of the longest standing antibiotics erythromycin [16] has been in use for over 50 years, whereas neocarzinostatin [17], of the enediyne family of natural products, demonstrates potent anticancer properties, and epitomizing the diversity of polyketides is maitotoxin [18], one of the largest, most complicated, and toxic compounds known to man (Figure 16.1).

The struggle by scientists to manufacture and uncover the biological mechanisms for the production of these compounds has provided great insights into cell biology and synthetic methodology [19]. At first glance, structures so dissimilar seem unlikely to come from a common biosynthetic pathway. For example, comparing erythromycins' highly saturated macrolactone skeleton with doxorubicin's polyaromatic center piece exemplifies the variety of structures that originate via this pathway.

16.4
Fatty Acid Biosynthesis

By the 1950s it was realized that the polyketide biosynthetic pathway was similar to the biosynthesis of fatty acids. Through a series of radiolabeling experiments it was clear that iterative Claisen condensations followed by a series of reductive transformations were responsible for the formation of long chain fatty acids [19].

Polyketide biosynthesis takes place in the polyketide synthase (PKS), and for each turn of the cycle the fatty acid chain length is extended by two carbons. Throughout the cycle thionyl bonds are used to attach the growing fatty acid, incoming monomers and intermediates to the enzyme complexes. The fatty acid chain is attached initially to a ketosynthase (KS), and the process is initiated when an unbound KS enzyme attacks acetyl co-enzyme A (AcCoA). Once initiated, a free acyl carrier protein (ACP) acquires a malonyl group from malonyl-co-enzyme A (MCAT) obtained from the organism's primary metabolism, a process mediated by acyl transferase (AT). Claisen condensation takes place with the loss of CO_2, and by exchanging the growing fatty acid onto the ACP. The β-ketothioester is then carried through a reductive pathway until the β-ketone is fully reduced to the saturated methylene. The cycle is continued until the fatty acid chain has reached a specific length, then the thionyl esterase (TE) ejects the molecule as a long-chain fatty acid (Scheme 16.2).

After condensation, further reductive manipulations may take place as the ketide is passed around the enzyme complex or megasynthase via the ACP. First the keto-reductase (KR) reduces the β-ketone to an alcohol, which then eliminates water as a result of the dehydratase (DH) to give an olefin, and the enoyl reductase (ER) then reduces it further to the saturated thioester.

Rather than rigidly following the full reductive pathway to give saturated products polyketide synthesis can omit certain transformations to leave a more oxidized

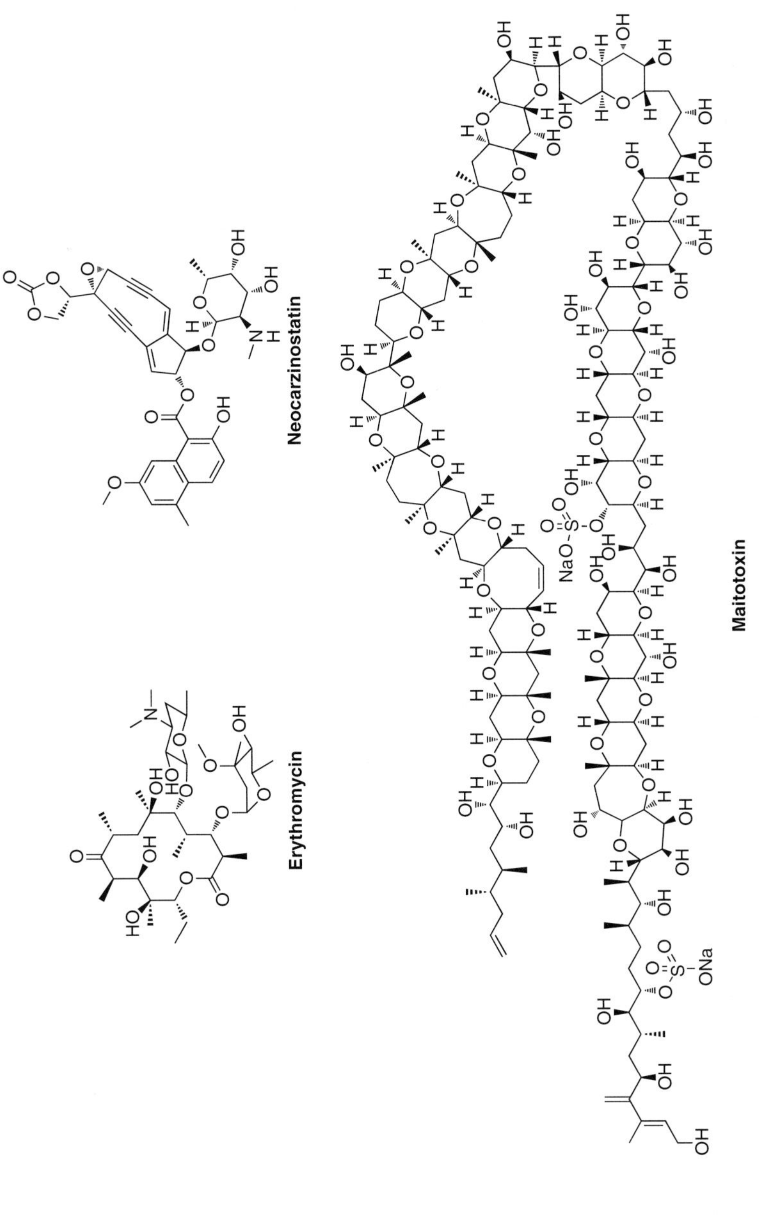

Figure 16.1 Examples of structurally diverse complex polyketides.

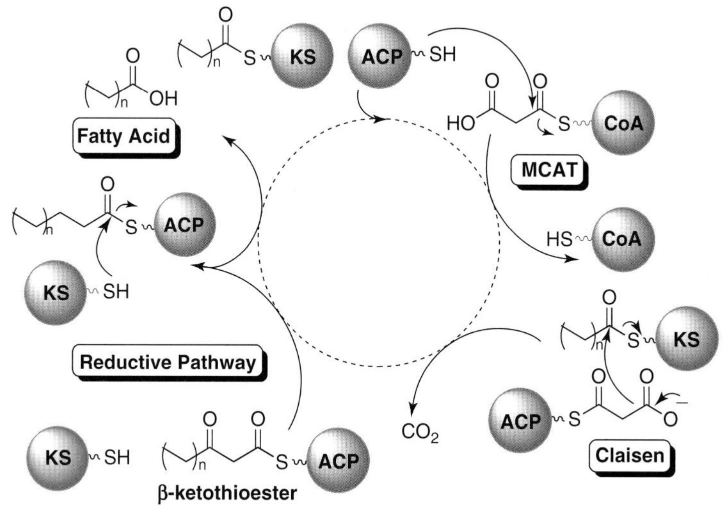

Scheme 16.2 General polyketide biosynthesis.

skeleton (Scheme 16.3). For example, the cycle can be terminated after the ketoreductase step giving a β-hydroxy ketone. The TE has a more important role, rather than just hydrolyzing the thionyl to give a carboxylate, since it can facilitate lactonization to give macrolides. Further to this, polyketide synthesis has access to a greater variety of starting materials (rather than solely MCAT) and aromatizing enzymes that produce polyaromatic structures. Often the initial product of the megasynthase is not the natural product, just the basic compound skeleton; further functionalization can be achieved through oxidation and glycosylation.

There are three classes of PKSs (the collective term for the group of enzymes), aptly named I, II, and III. Type I PKS is the most complicated of the PKSs and can be split further into two subcategories. Type I modular PKS utilizes a series of individual catalytic modules to provide each transformation. These modules will add one C2 unit of varying oxidation state and consist of individual catalytic domains responsible for condensation and reduction, often leading to reduced macrolide-type products. Type I iterative PKS exploits a single multi-enzyme to carry out the iterative steps and commonly gives reduced aromatic polyketides. This iterative complex consists of two KS, α and β, and an ACP, which associate to give the active complex. A defined number of catalytic cycles produce a poly β-ketone chain that is then transformed to give aromatic products.

Most relevant to biomimetic syntheses employing electrocyclization reactions are the products resulting from the action of the DH, which leads to the formation of polyenes (step 3). The DH that catalyzes the transformation of the β-hydroxythioester into the (E)-α,β-unsaturated species consists of histidine and asparagine residues in the active site. The histidine acts as a base removing the enolic proton and the asparagine provides a proton to release water and the unsaturated polyketide. It is these polyenes that will form the basis of discussion for

16.5 Biomimetic Analysis

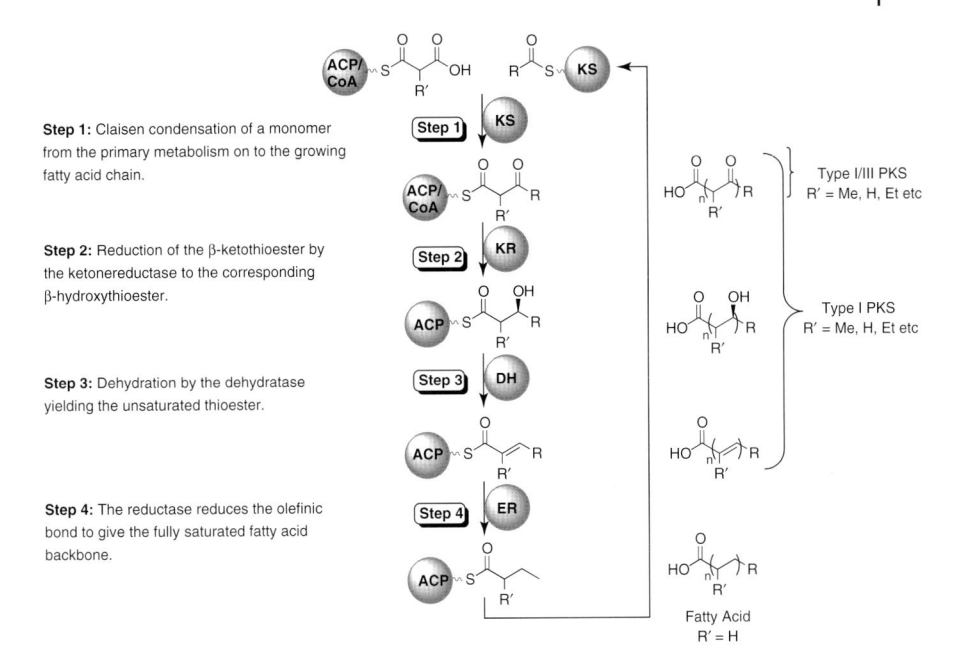

Step 1: Claisen condensation of a monomer from the primary metabolism on to the growing fatty acid chain.

Step 2: Reduction of the β-ketothioester by the ketonereductase to the corresponding β-hydroxythioester.

Step 3: Dehydration by the dehydratase yielding the unsaturated thioester.

Step 4: The reductase reduces the olefinic bond to give the fully saturated fatty acid backbone.

Scheme 16.3 Stepwise illustration of polyketide biosynthesis.

the rest of this chapter, focusing on the biomimetic syntheses on complex natural products.

One major concern of natural products chemistry is the uncertainty about whether the isolated structures were Nature's initial intention! In other words, the true natural product produced by the given organism may not always be "static," and the isolated compounds may be the product of a post-biosynthetic opportunity. As such, the complexity of the given structure has simply been endowed upon the system, as a consequence of its original function. These structures are still worthy targets for synthesis, since they are often biologically active and structurally fascinating. Since pericyclic reactions frequently require little activation, it is compelling to propose that these transformations are ideal candidates for post-biosynthetic activity and, as such, would not have pressurized the producing organism to develop enzymatic machinery. This may explain the rarity, or indeed absence, of the elusive "Diels–Alderases" and "electrocyclases," and may also explain why in many cases the isolated natural products, proposed to arise through electrocyclizations, are obtained in racemic form.

16.5
Biomimetic Analysis

Through interrogation of biosynthetic pathways and by considering each individual step, one may identify a key transformation that could form the basis of a biomimetic

Table 16.1 Polyketide biosynthesis broken down into sub-stages.

Stage in polyketide biosynthesis	Description	Comment
Stage 1	Claisen condensation/formation of the carbon chain	Under enzymatic control (ACP/KS)
Stage 2	Post-condensation transformation, for example, reduction/dehydration	Under enzymatic control (KR/DH/ER)
Stage 3 (or 4)	Complexity generation	
	Aromatization	Enzymatic control
	Cyclization	Enzymatic control
	Electrocyclization	Unknown, but most likely non-enzymatic control
Stage 4 (or 3)	Oxidation	Both enzymatic and non-enzymatic control

synthesis. Table 16.1 summarizes an approach to aid in the planning and execution of a biomimetic synthesis, which involves breaking down polyketide-derived biosynthesis mechanism into key stages. Stages 1 and 2 are classified as lower order, and are common to all polyketide derived natural products. They involve stitching together the key carbon segments, which eventually lead to fragments ripe for conversion into more complicated higher order structures through complexity generating reactions. Stages 3 or 4 are later stages and often provide the inspiration for biomimetic syntheses. Although planning a synthesis based upon one critical step may be deemed high risk, the potential rewards of developing novel transformations and validating biosynthetic hypotheses are worthwhile.

To demonstrate the power of the biomimetic approach, a selection of case studies now follows. Unfortunately, and necessarily, the work of many groups around the world who have made significant contributions to this field will be omitted. We can only apologize for this and encourage the reader to research the primary literature in detail for other examples.

16.6
6π Electrocyclizations

Thermal 6π electrocyclizations of conjugated trienes are probably the most common of all electrocyclizations employed in the biomimetic synthesis of polyketide-derived natural products. Such reactions lead to conjugated cyclohexadiene units (Scheme 16.1) and, although thermodynamically favorable, often require activation by relatively high temperatures. The cyclohexadiene products can also participate in further tandem reactions such as Diels–Alder cycloadditions, and such cascades can lead to extremely complex architectures.

16.6.1
Tridachiahydropyrones

However, we begin by examining a family of natural products whose synthesis relied upon a photochemical 6π electrocyclization. Although very powerful, photochemical electrocyclizations are rarely utilized in total synthesis compared to their thermal counterparts. In biomimetic systems, however, one may intuitively realize that Nature would in fact take full advantage of the essentially endless supply of photons at her disposal, and there is perhaps no better example of the power of photochemical electrocyclizations in polyketide synthesis than in the tridachiahydropyrone family.

Adopting a biomimetic strategy, researchers in our own laboratories were able to provide convincing evidence to corroborate the biosyntheses of oxytridachiahydropyrone (5) [formerly tridachiahydropyrone B (5′) and C (5″)], and tridachiahydropyrone (6). The successful approach to these complex metabolites also led to their structural reassignment [20].

(−)-Tridachiahydropyrone (6) (corrected structure shown in Scheme 16.4) is a marine-derived natural product isolated in 1996 by Cimino *et al.* from a sacoglossan mollusc (*Tridachia crispata*) [21], whereas tridachiahydropyrone B (5′) and C (5″) [later reassigned as oxytridachiahydropyrone (5)] were isolated a few years later from *P. ocellatus* (order Sacoglossa, family Elysioidea) along with several other propionate-derived metabolites [22].

Such opisthobranch molluscs characteristically produce polyketide (polypropionate)-derived metabolites, which commonly incorporate a γ-pyrone unit. Metabolites 5 and 6 clearly share a common core framework of differing oxidation state, providing a strong indication about their biosynthetic relationship. Scheme 16.4 illustrates the biomimetic analysis of tridachiahydropyrones: retro-oxidation of oxytridachiahydropyrone (5) (stage 4) leads to tridachiahydropyrone (6). As previously mentioned, late-stage oxidation is common in biosynthesis, and may or may not be governed by enzymes. In the present case, it was proposed that the oxidation was photochemically driven, and therefore highly unlikely to be enzyme-catalyzed. The inspiration behind this conclusion came from the studies of Ireland and Scheuer, who proposed that metabolites of this family are synthesized by the shell-less molluscs to act as natural sunscreens [23]. Understanding the larger biological system helped unravel the complexities behind the biogenesis, making it truly biomimetic. It was proposed that tridachiahydropyrone (6) itself could be derived from the (2Z,4E,6E)-triene 7, through a photochemical 6π conrotatory electrocyclization (stage 3), the key biomimetic complexity generating step. The lower order biosynthetic stages, 2 and 1, lead to polyene 8, derived from the building block propionate 9, which, in Nature, would be stitched together using the PKS machinery. In the forward synthesis, triene 7 became the primary synthetic target, with the key question being, could 7 be converted into tridachiahydropyrone (6) through a photochemical 6π electrocyclization?

The biomimetic rationale did not account for the fact that (−)-tridachiahydropyrone (6) was isolated as a single enantiomer, and it was not certain whether

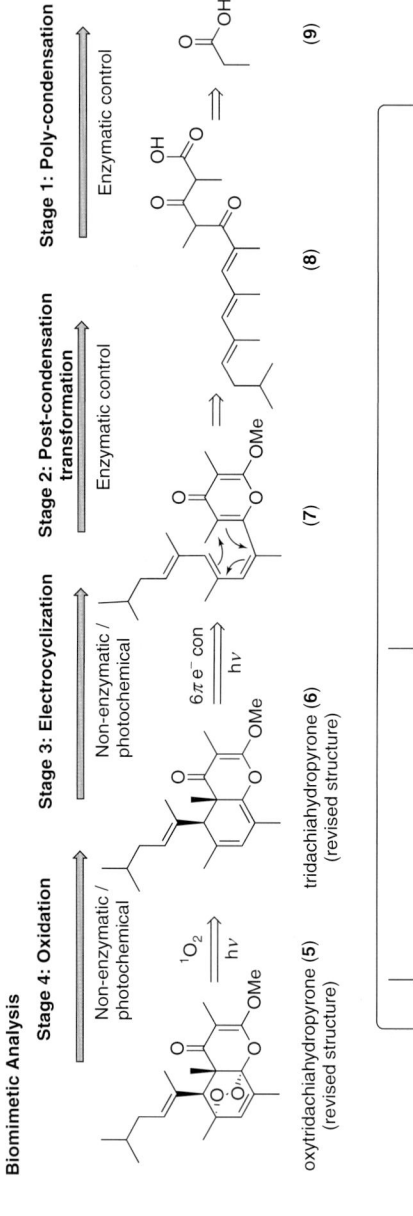

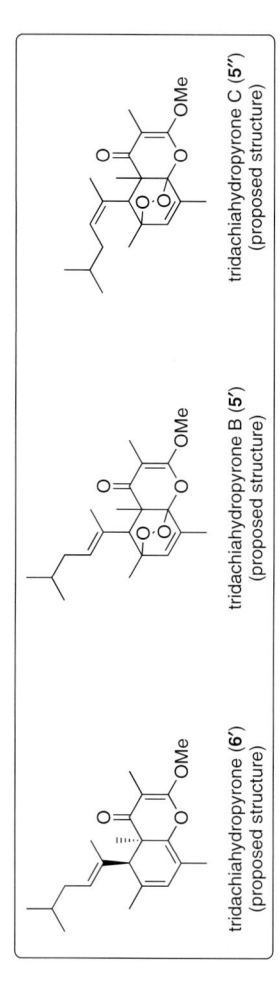

600 *16 Biomimetic Electrocyclization Reactions toward Polyketide-Derived Natural Products*

Scheme 16.4 Biomimetic analysis of the tridachiahydropyrones.

the biosynthesis was enzymatically controlled. To test the photochemical electrocyclization hypothesis, the researchers first had to synthesize the proposed (2Z,4E,6E)-polyene precursor **7**. In biosynthetic terms, **7** would be a product of stages 1 and 2, which is driven by the powerful PKS machinery. Nature is the master of such chemistry, and although a similar iterative approach could in principle be developed in the laboratory, the efficiency and overall yield would most likely be compromised. The research team opted for a convergent strategy employing a Suzuki cross coupling reaction between the boronate ester **10** and the γ-pyrone **11**, which gave the biomimetic precursor **7** in good yield (Scheme 16.5). To plan the photochemical reaction, the research team considered the marine habitat of the producing molluscs to optimize the chance of success. The practical lower limit for UV radiation in seawater is approximately 290 nm, and it was reasoned that placing the substrate in a glass vial in direct sunlight would be a good model. After exposure of a 0.018 mmol sample of **7** in methanol on a window ledge for three days, the researchers obtained target **6** in 29% yield (along with 57% starting material), and were able to corroborate the proposed reassigned structure using X-ray crystallography. When the corresponding all-*(E)*-polyene **13** was irradiated, a tandem reaction sequence also led to the target **6**. Monitoring of the reaction by ^{1}H NMR spectroscopy, specific *(E)–(Z)* isomerization about the C2=C3 double bond to give **7** was observed, which was followed by 6π electrocyclization to give **6**. It was concluded that the all-*(E)*-polyene **13** was perhaps the true biosynthetic precursor. Interestingly, an unexpected product **14**, later named phototridachiahydropyrone, was also co-isolated from the reaction mixture, whose origin was thought to arise from **6** through a photochemically allowed suprafacial [1,3]-sigmatropic shift. In fact, prolonged exposure of **6** led to the complete conversion into **14**, thus indicating that this photochemical end point of the tandem reaction sequence may also be a natural product yet to be isolated, but may also depend upon the natural habitat [24]. The laboratory synthesis of compounds that later turn out to be natural products is not an uncommon event in biomimetic chemistry, and examples will be discussed below.

The final step in the study was to determine if **6** could be converted into oxytridachiahydropyrone (**5**) [formerly tridachiahydropyrone B (**5′**) and C (**5″**)]. Both compounds were reported as an inseparable mixture of isomers differing in geometry about the C10=C11 double bond, in a ratio of 4 : 5, as determined by ^{1}H NMR spectroscopic analysis. The relative, or absolute, configuration of the chiral centers was not elucidated [22].

It was reasoned that tridachiahydropyrone B (**5′**) and C (**5″**) were derived biosynthetically from **6**, via a photochemical [4+2] cycloaddition with singlet oxygen. A mixture of diastereoisomeric compounds would be expected to arise from a facially selective addition of singlet oxygen to either the concave or convex face of tridachiahydropyrone (**6**). Tridachiahydropyrone C (**5″**) would arise either through isomerization of the C10=C11 double bond of tridachiahydropyrone B (**5′**) or its precursor. It was known that α-methoxy-γ-pyrones function as photosensitizers, and perhaps the fused pyrone ring of tridachiahydropyrone would also function analogously. When a solution of **6** was irradiated with a UV

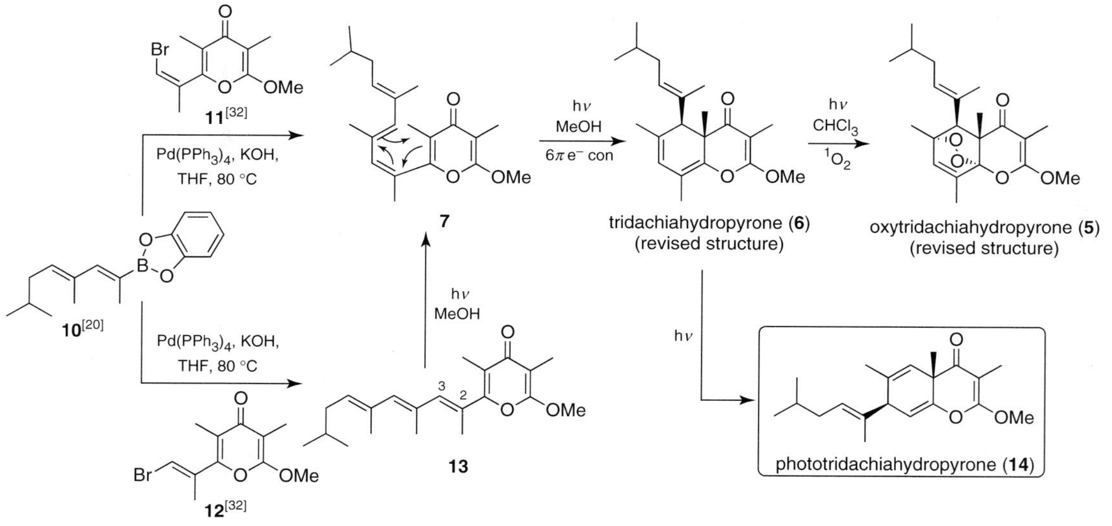

Scheme 16.5 Biomimetic synthesis of the tridachiahydropyrones by a key photochemical 6π electrocyclization.

source under a continuous flow of molecular oxygen for 4 h a new product was obtained, whose spectral data agreed with that reported for the non-separable mixture of tridachiahydropyrone B (**5′**) and C (**5″**), and in identical ratio (4 : 5). However, further analysis revealed that the mixture was in fact two rotameric forms of the related diastereoisomer, which was renamed oxytridachiahydropyrone (**5**) (Scheme 16.6) [20b]. Compound **5** was the product of exclusive *endo* addition of oxygen via the less hindered face of the diene. The biomimetic studies of the tridachiahydropyrones led not only to the successful syntheses and structural correction of two complex natural products, it also provided great insight into the occurrence of these interesting compounds. The photochemical transformations delivered unusually high levels of control. The specific *(E)–(Z)* isomerization of the C2=C3 double bond of polyene **13**, followed by a 6π conrotatory electrocyclization is a powerful and remarkable tandem sequence, yet the ease with which the transformation occurs clearly demonstrates that Natures pathways can be replicated with great success.

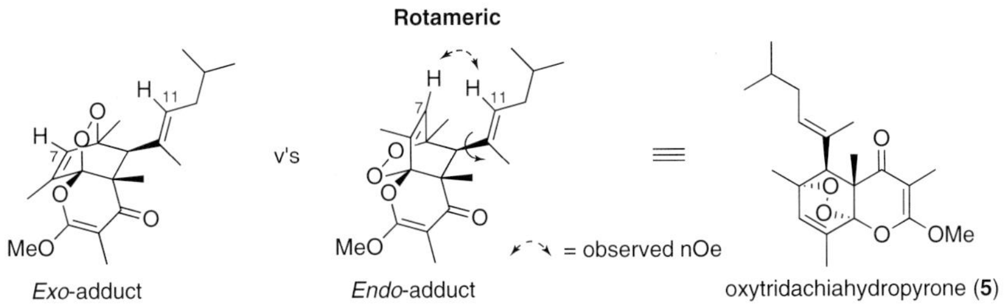

Scheme 16.6 Biomimetic synthesis of oxytridachiahydropyrone (**5**) *via* photochemical reaction with singlet oxygen.

16.6.2
Tridachione Family

Several other cyclohexadiene natural products isolated from sacoglossan molluscs can be envisaged as being derived from conjugated polyene systems through 6π electrocyclization [25, 26]. For example, 9,10-deoxytridachione (**15**), tridachiapyrone I (**16**), and tridachiapyrone A (**17**) can be readily traced back to a tetraene of type **18**, whereas deoxyisotridachione (**19**) can be thought of as arising through a photochemical 6π conrotatory electrocyclization of the tetraene **20** (Scheme 16.7) [27–29].

In some cases, the cyclohexadiene moiety of the molluscan natural products acts as staging points for further transformations. For example, pioneering photochemical studies by Ireland and Scheuer demonstrated that 9,10-deoxytridachione (**15**) could be converted *in vitro* and *in vivo* into the complex isomer photodeoxytridachione (**22**) [25]. The transformation was presumed to occur through a $[_\sigma 2_a + _\pi 2_a]$ pericyclic mechanism since the non-enzymatic reaction led exclusively to enantiomerically pure products, indicating that an open chain polyene was not involved. Furthermore, epoxidation of **15** gave tridachione (**21**) [28], which was the parent compound of the series (Scheme 16.8).

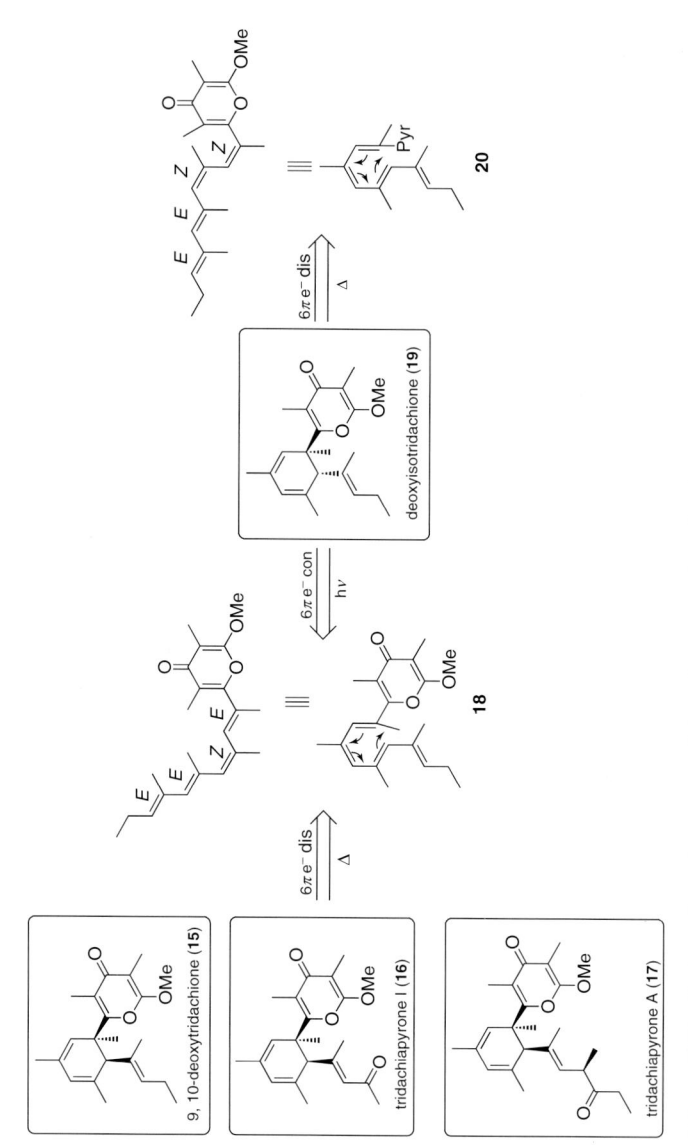

Scheme 16.7 Electrocyclic pathways to cyclohexadiene natural products.

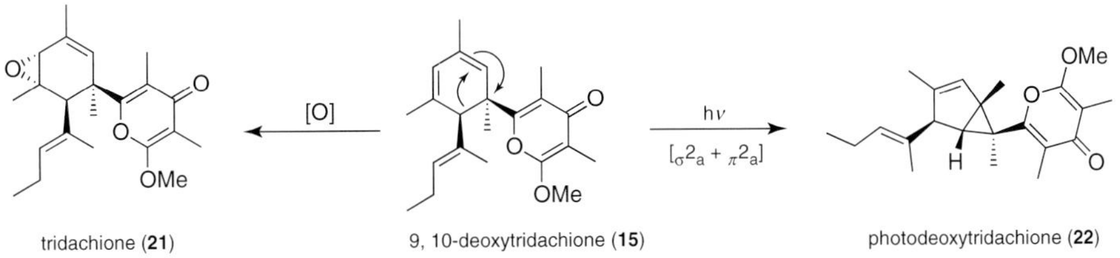

16.6 6π Electrocyclizations

Scheme 16.8 Chemical transformations of 9,10-deoxytridachione (**15**).

The biomimetic synthesis of 9,10-deoxytridachione (**15**) was independently achieved by Baldwin et al. [29] and Trauner et al. [30]. Both research groups utilized a convergent cross-coupling approach to assemble the (2E,4Z,6E,8E)-polyene-pyrone **18**, which was perfectly set up to undergo a 6π disrotatory electrocyclization to give the target cyclohexadiene **15**. Baldwin et al. opted for a Suzuki coupling to unite the boronate **23** to the vinyl iodide **24**, which gave the desired polyene **18** in 65% yield. Subsequent heating of **18** in benzene initiated thermal 6π electrocyclization to afford **15** in 31%, along with the unexpected bicyclo[4.2.0]octadiene **27** product, which was itself later isolated as a natural product and named ocellapyrone A (Scheme 16.9a). The formation of **27** occurs through an elegant cascade of electrocyclizations and will be discussed in more detail below. In Trauner's synthesis, a modified Stille coupling [31] was employed to unite the (Z)-vinyl stannane **25** to the (E)-iodo alkene **26**, to yield the polyene **18**, which then underwent electrocyclization under microwave conditions (150 °C, toluene) to give **15** and **27** in 32% and 38% yield respectively (Scheme 16.9b).

In a broader study of sacoglossan mollusc-derived polyketides, Moses et al. [32] proposed a general biosynthetic scheme to explain the formation of a range of natural products, including 9,10-deoxytridachione (**15**), from the all-(E)-tetraene pyrone **28**. The scheme could be expanded to account for all possible diastereoisomers, but for presentation purposes only a selection of compounds are illustrated. Through a sequence of selective (E)–(Z) double bond isomerizations and electrocyclizations, it was realized that several core structures could in principle be obtained from **28**. Scheme 16.10 illustrates several possible pathways, which involve chemistry at stage 3 and 4 of polyketide biosynthesis. Clearly, 9,10-deoxytridachione (**15**) plays an important role in this general biosynthetic scheme.

To test the hypothesis, the researchers again used a Suzuki coupling to synthesize the all (E)-polyene **28**. Several isomerization–electrocyclization conditions were examined, each of which lead to interesting outcomes. Studies with **28** carried out in the dark under palladium-catalyzed isomerization conditions lead to a complex mixture of products. Purification afforded a reasonable quantity of the racemic bicyclo[4.2.0]octadiene **34**, corresponding to the endo-isomer of ocellapyrone A (**27**) (Scheme 16.11). The formation of the bicyclo[4.2.0]core will be discussed later, since its origin also lies in electrocyclic chemistry. Interestingly, compound **34** was itself incorrectly proposed as the original structure of ocellapyrone A (**27**) [33], before revision and confirmation of the correct structure by synthesis. Notably,

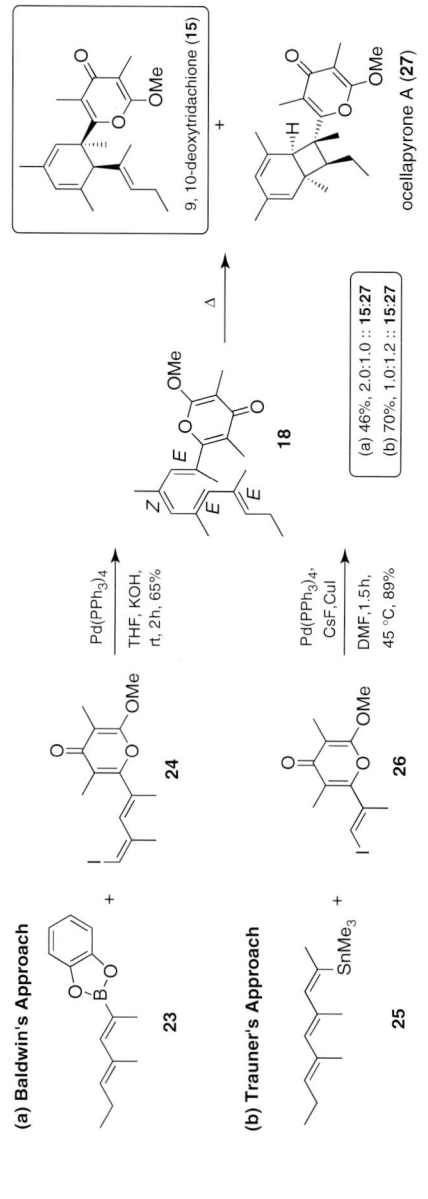

Scheme 16.9 Biomimetic syntheses of 9,10-deoxytridachione (**15**).

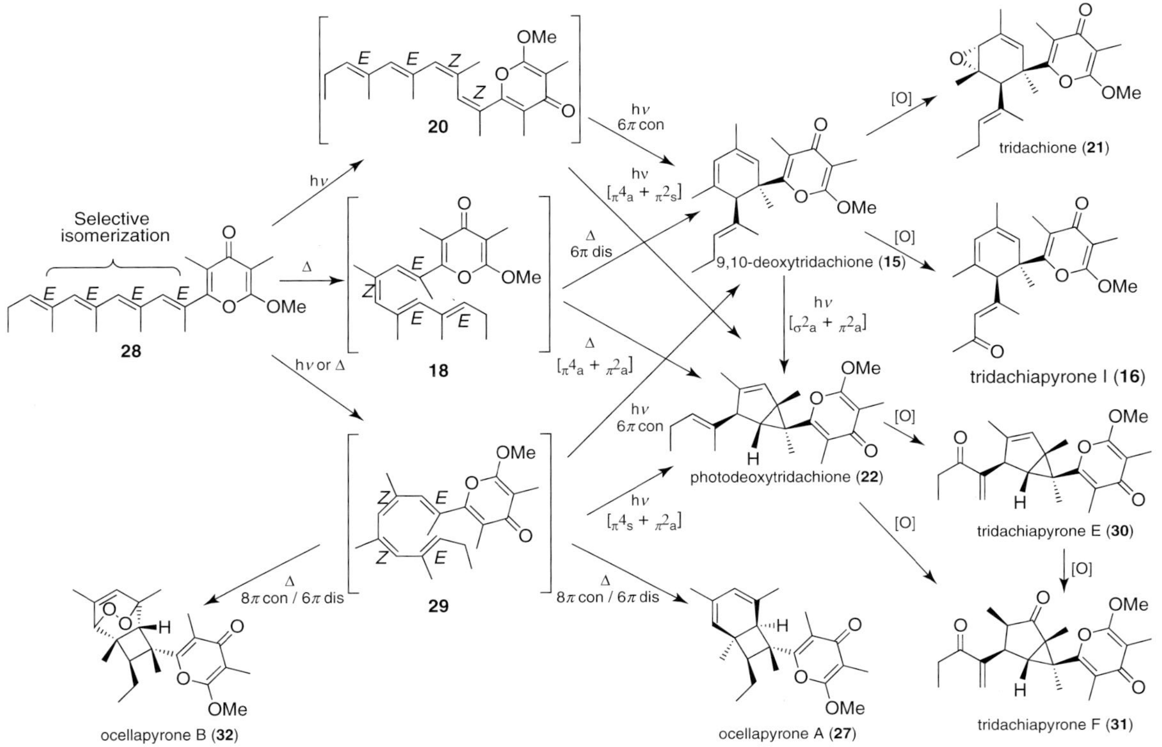

16.6 6π Electrocyclizations

Scheme 16.10 Proposed biosynthetic pathways from polyene **28**.

34 could be a precursor in the biosynthesis of ocellapyrone B (**32**). However, this intermediate has not yet been isolated from a natural source.

Thermal cyclization of **28** was carried out in xylene at 150 °C in the dark for 1.5 h. These reaction conditions afforded multiple products, including ocellapyrone A (**27**), along with the related diastereoisomer **33** and cyclohexadiene 9,10-deoxytridachione (**15**) (Scheme 16.11). Photolysis of **28** again afforded a complex mixture of crude products, which could be separated to give photodeoxytridachione (**22**) and deoxyisotridachione (**19**).

Presumably, **22** was formed through **15** in a manner consistent with the experiments of Ireland and Scheuer, although the latter was not isolated in the photochemical experiments. Alternative non-electrocyclic reaction mechanisms have been considered, including a photochemical Diels–Alder [34] and a highly specific diradical process [35]. These studies clearly demonstrate the multiple pathways that are possible from one common intermediate, but also emphasize the power of employing biomimetic strategies. Although the yields were not optimized, the work addressed the bigger question about the biosynthetic origins of these complex natural products. It is highly likely that the pathways unraveled by the researchers are similar to those that operate in Nature. However, the racemic syntheses again raise fundamental questions about the origins of the enantiopurities of the isolated compounds.

16.6.3
Pseudorubrenoic Acid A

The possible involvement of 6π electrocyclic ring closures, mediated by electrocyclase enzymes of polyunsaturated acyclic polyketide intermediates, has been raised by Rickards et al. in the biosynthesis of pseudorubrenoic acid A (**35**). This aromatic fatty acid is an antimicrobial carboxylic acid isolated from the soil bacterium *Pseudomonas fluorescens* [36]. Rickards et al. noticed that the lack of oxygen functionality may indicate that the compound is constructed through a pathway distinct from the common biosynthesis of aromatic polyketides.

The biosynthetic pathway was proposed to involve the formation of cyclohexadiene **36** through 6π electrocyclization of (7E,9Z,11E,13Z)-tetraene **37**, followed by oxidation to **35** (Scheme 16.12).

To evaluate their biomimetic hypothesis, the tetraene **43** was synthesized from phenyl sulfide **38** and (Z)-1-bromopent-2-ene (**39**), which under basic conditions afforded the enyne **40**. Oxidation of **40** to the corresponding sulfoxide and elimination next gave a mixture of (E/Z) diastereoisomers of **41**. Palladium-catalyzed cross coupling of the sp and sp^2 centers present in **41** and **42**, followed by a reduction gave a mixture of the tetraenes **43a–d**. Upon heating, this isomeric mixture of tetraenes underwent thermally induced transformation to give cyclohexadiene **44** as the major product, which was followed by subsequent aromatization to the *o*-disubstituted benzene **45**. Rickards' synthesis provided support for his electrocyclization hypothesis, and serves as a model for the possible

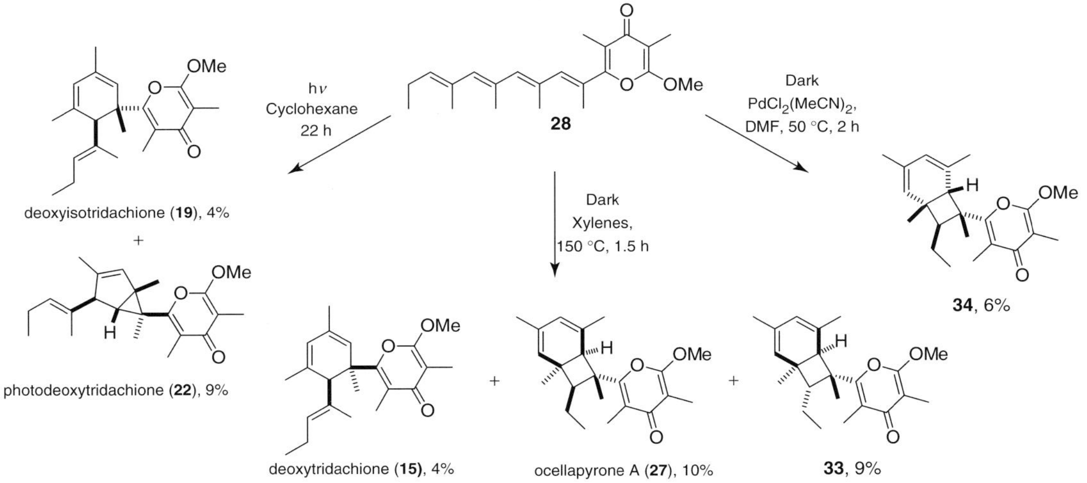

16.6 6π Electrocyclizations

Scheme 16.11 Biomimetic conversion of **28**.

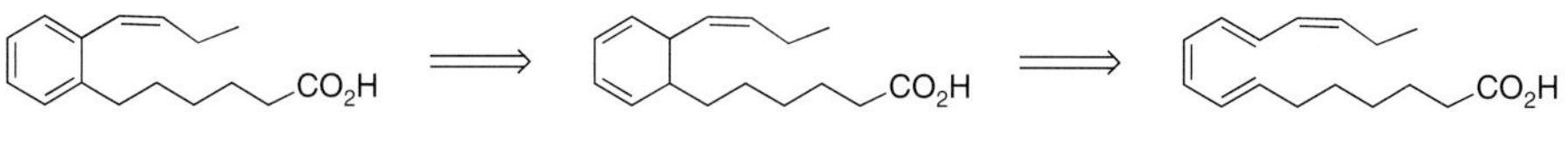

Scheme 16.12 Retrosynthetic biomimetic analysis of pseudorubrenoic acid A (**35**).

involvement of an electrocyclase in the biosynthesis of pseudorubrenoic acid A (**35**) (Scheme 16.13) [36].

16.6.4
Torreyanic Acid

As briefly mentioned earlier, electrocyclization reactions can also occur with hetero-carbon polyene systems, and it is not surprising that this reactivity has been exploited in biomimetic syntheses. In particular, oxa-6π electrocyclic ring closures are common, with such reactions being essentially thermoneutral processes with low activation barriers. The relative equilibrium concentration of the dienal/dienone **46** and the (2*H*)-pyran **47** (Scheme 16.14) depends upon the electronic properties of the system [37], and despite such reactions being common in Nature, (2*H*)-pyrans that are not trapped by further transformations are rare motifs in natural products.

Among the wide variety of polyketide-based natural products discovered to date, the synthesis of the epoxyquinone (+)-torreyanic acid (**51**) provides an excellent example of a biomimetic oxa-6π electrocyclization. Isolated from the endophytic fungus *Pestalotiopsis microspora* and characterized by Clardy and coworkers in 1996 [38], the complex acid **51** displayed potent biological activity. In the original report of its isolation and structural characterization a biosynthetic scheme toward torreyanic acid (**51**) was proposed, involving Diels–Alder dimerization of 2*H*-pyran monomers epimeric at C9 (C9′) (**50/50′**). In this *endo*-selective [4+2] cycloaddition the two pentyl side chains are oriented away from one another in the transition state. The isolation of the monomeric epoxyquinol ambuic acid (**48**) [39] from *P. microspora* further supports the proposed biosynthesis of torreyanic acid via oxidative dimerization of monomeric intermediates (Scheme 16.15).

The first total synthesis of torreyanic acid in racemic form was successfully executed by Porco *et al.* [40]. This research group later used a complementary strategy to complete the first asymmetric synthesis of **51**, following the same general biomimetic route [41]. Beginning with the 1,3-dioxane derivative **52**, the homoallylic alcohol **53** was synthesized in several steps. An interesting enantioselective epoxidation of **53**, developed by Porco, was achieved using a tartrate-mediated nucleophilic method to give **54** in excellent yield and ee (91% yield, 91% ee). Dess–Martin periodinane (DMP) oxidation of the homoallylic alcohol followed by a two-carbon homologation next afforded **55**. A cross-coupling reaction then realized **56**, which was subsequently deprotected to afford the biomimetic precursor **57**. Finally, treating **57** with DMP initiated the tandem oxidation/6π-electrocyclization/Diels–Alder dimerization to give (+)-torreyanic acid *tert*-butyl ester (**58**), which was hydrolyzed

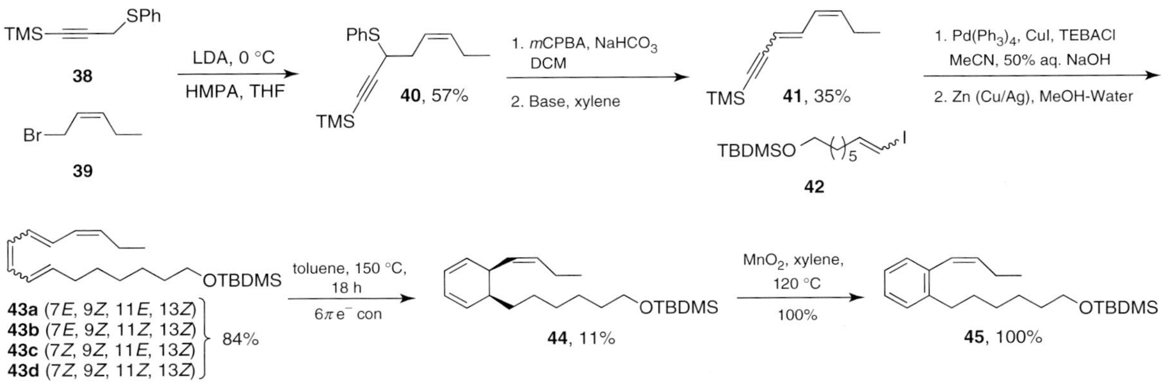

16.6 6π Electrocyclizations

Scheme 16.13 Biomimetic synthesis of a pseudorubrenoic acid A model system (**45**).

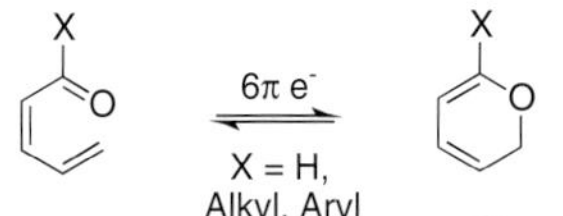

Scheme 16.14 Oxa-6π electrocyclizations.

under acidic conditions to afford free (+)-torreyanic acid (**51**). Analytical techniques confirmed the absolute configuration as that of the natural (+)-torreyanic acid (**51**) (Scheme 16.16). Interestingly, the biomimetic complexity-generating cascade demonstrated by Porco occurred without the need for catalysis. The optical activity associated with **51** was carried through from the monomeric precursor ambuic acid (**48**), which is most likely to be the case in the natural system.

Mehta *et al.* have reported a related biomimetic approach to (±)-torreyanic acid, starting from 2-allyl-*p*-benzoquinone to access the *tert*-butyl ester of ambuic acid **48**, which underwent the analogous biomimetic tandem oxidation/6π electrocyclization/Diels–Alder dimerization sequence [42].

A similar dimerization approach accounts for the formation of the polyketide-derived metabolites epoxyquinol A (**63**), B (**62**), and C (**64**), isolated from a fungi by Osada *et al.* [43]. A biosynthetic proposal for **62**–**64** from the dienal **59** was proposed by Osada, which centered upon a tandem oxa-6π electrocyclization to give the corresponding C5 epimers **60** and **61**, followed by a [4+2] cycloaddition/dimerization sequence as depicted in Scheme 16.17. Interestingly, a formal [4+4] cycloaddition of **60** produced the corresponding homodimer epoxytwinol (**65**) (Scheme 16.17) [44]. Several groups have reported asymmetric biomimetic syntheses of the epoxyquinols, including, Hayashi [45], Porco [46], Kuwahara [47], and Mehta [48], with each team developing unique routes to the key biomimetic dienal precursor **59**.

16.7
8π Systems and the Black 8π–6π Electrocyclic Cascade

16.7.1
Endiandric Acids

The 8π–6π electrocyclic cascade (sometimes referred to as *Black's cascade*) is probably the most elegant and classical display of the power of electrocyclization reactions in Nature. This tandem reaction sequence, which involves a cascade of thermally induced 8π, then 6π electrocyclizations, has proven its worth in biomimetic synthesis. The most well-known example of this amazing transformation was demonstrated by K.C. Nicolaou *et al.* in their elegant syntheses of endiandric acids A–G (**66**–**72**) (Figure 16.2) [49, 50]. Detailed accounts of this landmark study already exist [4a], and although we will refer to aspects of this pioneering work we will also focus on other compounds where this cascade has been a key transformation.

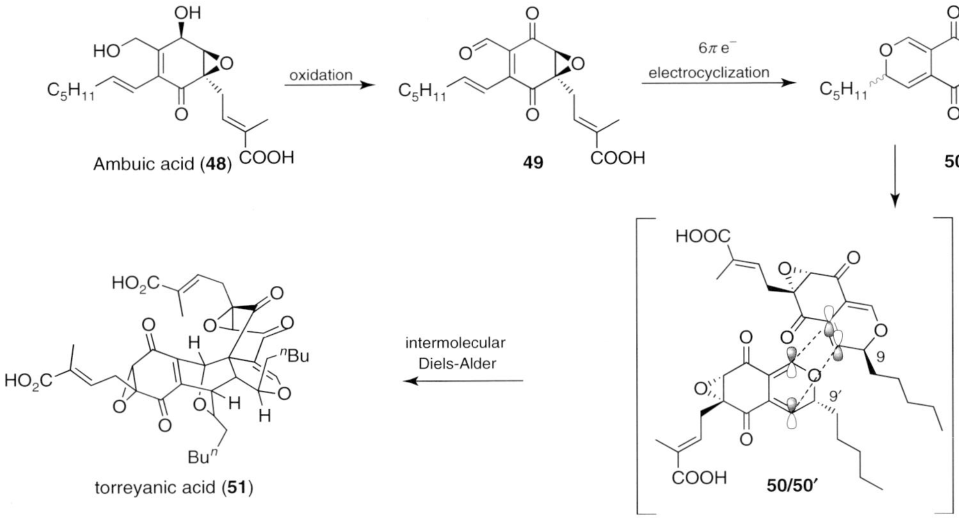

16.7 8π Systems and the Black 8π–6π Electrocyclic Cascade | 613

Scheme 16.15 Lee *et al.* biosynthetic hypothesis for the formation of (+)-torreyanic acid (**51**).

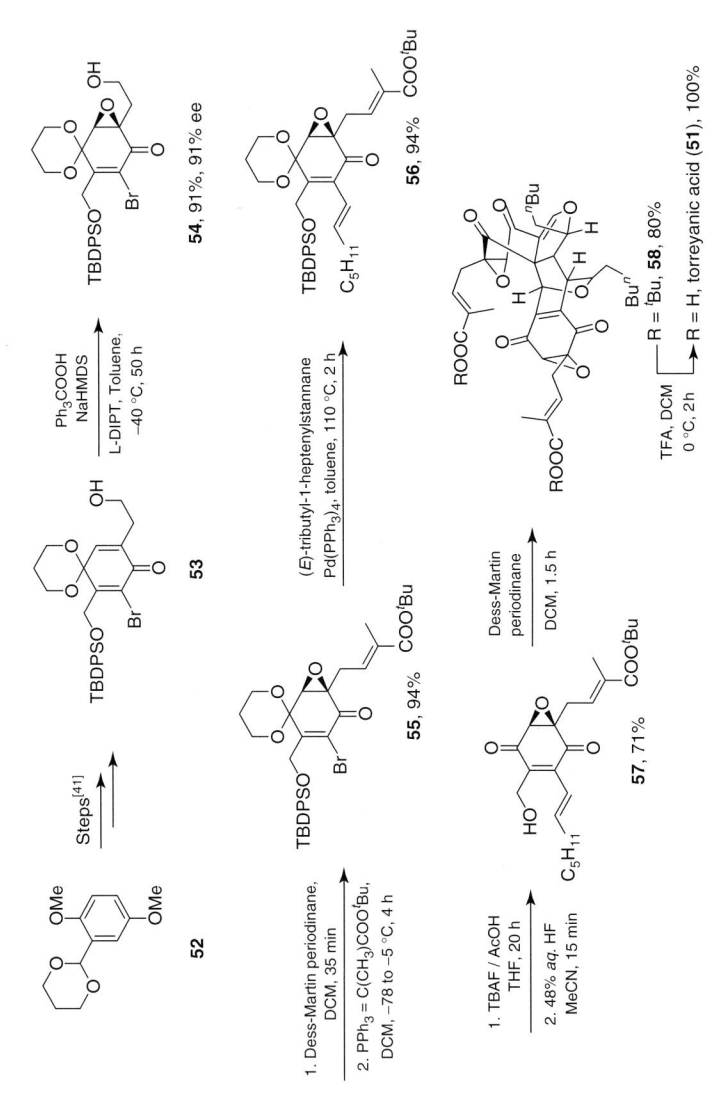

Scheme 16.16 Porco's biomimetic synthesis of (+)-torreyanic acid (**51**).

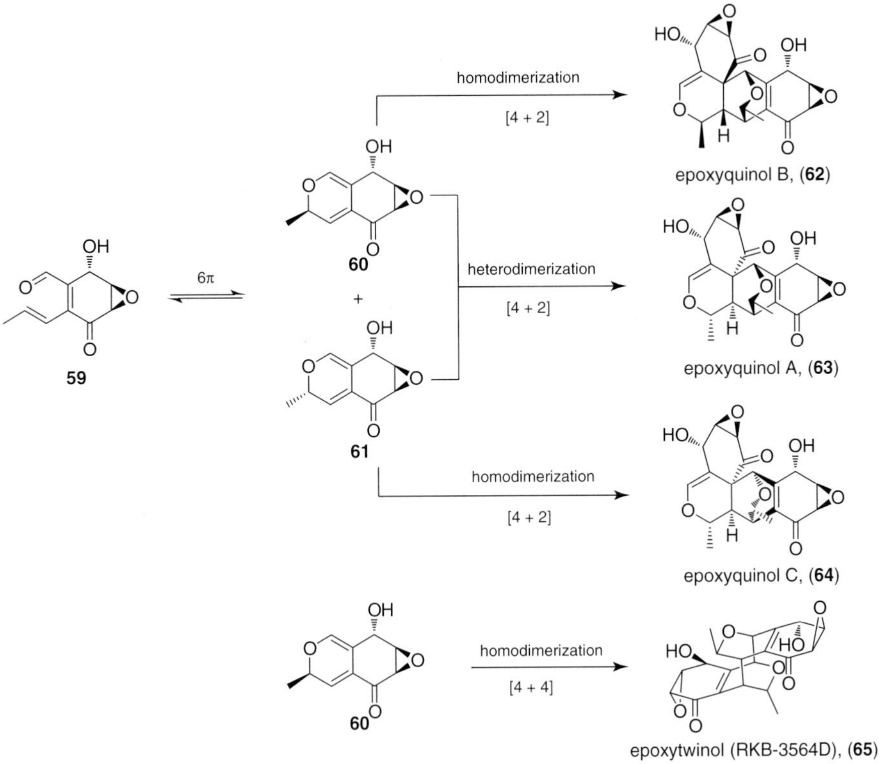

16.7 *8π Systems and the Black 8π–6π Electrocyclic Cascade* | 615

Scheme 16.17 Biosynthetic hypothesis for the epoxyquinols A–C (**62–64**) and epoxytwinol (**65**).

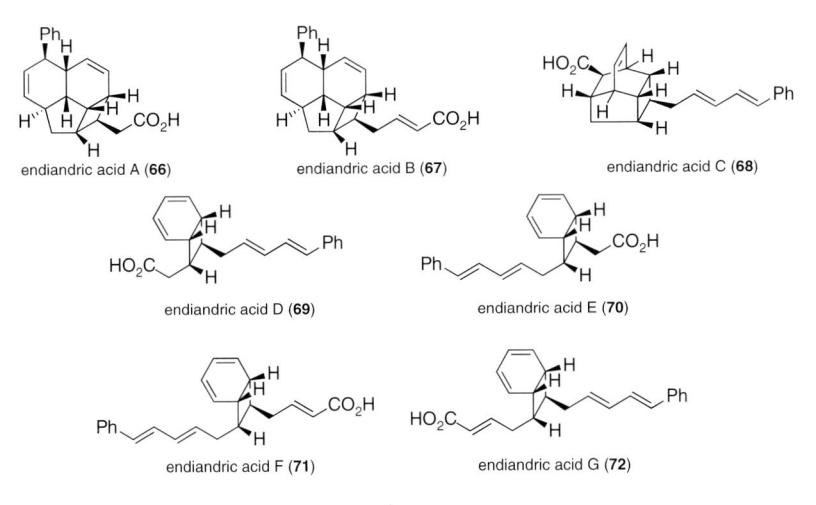

Figure 16.2 Structures of endiandric acids A–G (**66–72**).

The endiandric acids were isolated from the Australian plant *Endiandra introrsa* (Lauraceae) by Black's group in 1980 [49]. Despite containing several stereocenters, the endiandric acids are found in Nature as racemates, which is unusual for chiral center containing natural products. To explain this observation, Black proposed an intriguing hypothesis for the biosynthesis of these molecules from achiral poly-unsaturated precursors through a series of *non-enzymatic* electrocyclizations (Scheme 16.18). The Black hypothesis suggests a cascade of reactions, shown in Scheme 16.18, by which endiandric acids A–G are formed in Nature. Thus endiandric acids E (**70**), F (**71**), and G (**72**) were proposed as intermediate precursors to the tetracyclic endiandric acids A (**66**), B (**67**), and C (**68**), respectively; the conversion being facilitated by an intramolecular Diels–Alder reaction. Endiandric acid D (**69**) cannot undergo an intramolecular Diels–Alder reaction, and so it does not form a corresponding tetracycle. An additional, striking feature of the Black hypothesis is that endiandric acids D–G could arise from sequential electrocyclizations of achiral polyenes (**73–76**).

Intrigued by the structures of the endiandric acids and by Black's hypothesis for their biogenetic origin, Nicolaou *et al.* initiated a program directed towards their total synthesis. The strategy they used relied upon three sequential pericyclic reactions, two of which were electrocyclizations (8π and 6π) and one of which was an intramolecular Diels–Alder reaction. The two electrocyclic reactions are thermally allowed by the Woodward–Hoffmann rules, and proceed in a stereospecific manner as shown in Scheme 16.18b. Thus, to obtain the desired trans disubstituted [4.2.0] bicyclic product, the *(E,Z,Z,E)* or the *(Z,Z,Z,Z)* tetraenes must be used. Through a display of a series of electrocyclization reactions, Nicolaou *et al.* demonstrated the biomimetic, one-step synthesis of the endiandric acids involving the cascade of reactions proposed by Black. For example, the acetylenic precursor **81** was assembled and converted into the conjugated polyene **82** using a diastereospecific Lindlar catalyst reduction to introduce the central *(Z,Z)* geometry. Compound **82** could not be isolated but instead underwent spontaneous 8π–6π electrocyclic cascades to yield endiandric acid methyl esters D (**83**) and E (**84**). Upon heating **84** in toluene at 100 °C, an intramolecular Diels–Alder reaction then gave endiandric acid A methyl ester (**85**) (Scheme 16.18). Interestingly, although **83** cannot undergo a cycloaddition reaction it can equilibrate with **84**. In fact, a one-pot reduction of **81** followed immediately by heating the crude products gave endiandric acid A methyl ester (**85**) in 30% yield [50].

The tendency of cyclooctatrienes like **77** to undergo further pericyclic reactions means that they are not common motifs in natural products. One rare example, however, is compound **87** isolated from the brown algae *Cutleria multifida* [51]. Boland *et al.* proposed that cyclooctatriene **87** is derived *in vivo* from (3*Z*,5*Z*,7*E*)-nona-1,3,5,7-tetraene (**86**), via a thermal 8π electrocyclization, and then confirmed his proposal through synthesis. The octatriene **87** was found to be relatively stable at room temperature and did not readily undergo 6π electrocyclization, which required heating to above 50 °C. The corresponding 6π electrocyclization product **88**, and the octatriene isomer **89**, itself a product of [1,5]-hydrogen shift, were both subsequently found in Nature (Scheme 16.19) [52].

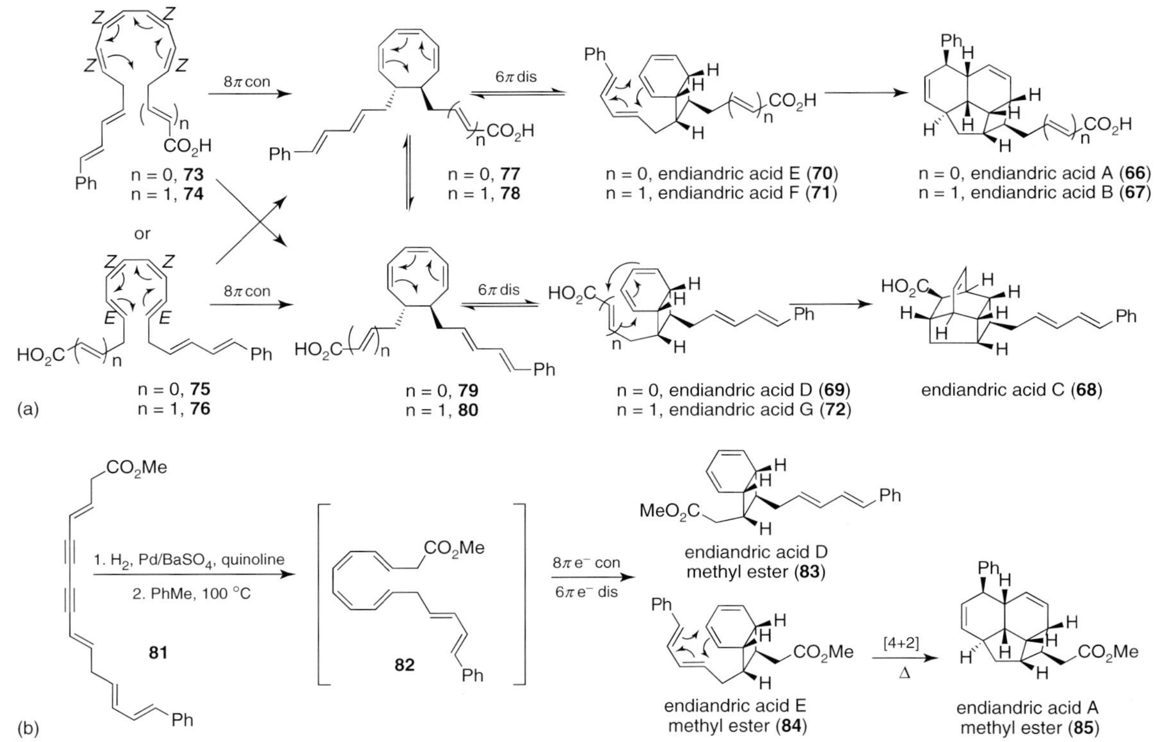

Scheme 16.18 (a) Black's electrocyclization cascade hypothesis; (b) Nicolaou's biomimetic synthesis of the endiandric acids.

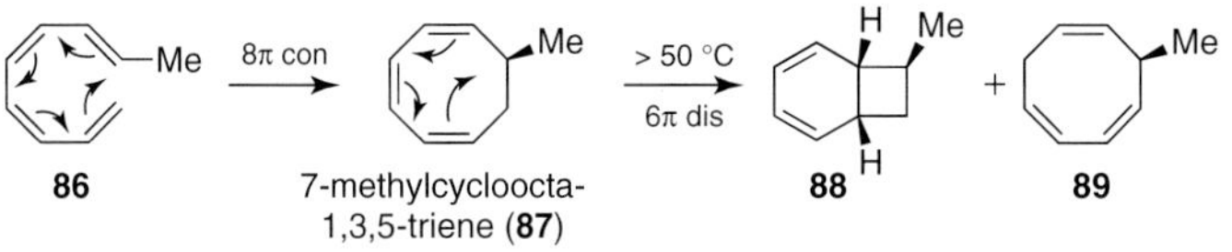

Scheme 16.19 Biomimetic synthesis of 7-methylcycloocta-1,3,5-triene (**87**).

The methodology for the construction of bicyclo[4.2.0] ring systems that evolved from studies on the endiandric acids has now become firmly established in organic synthesis. Most notably, the tandem electrocyclization sequence has been observed by Widmer et al., using synthetic analogs of (9Z,11Z)-vitamin A, (9Z,11Z)-vitamin A acetate, and palmitate [53]. The thermal instability of conjugated tetraenes with similar geometry had previously been reported [54], and the research of Nicolaou and Widmer has demonstrated the strict requirement for the internal double bonds to attain the (Z,Z) configuration, so that the terminal double bonds may be brought into close proximity to initiate the cascade reaction sequence.

16.7.2
Nitrophenyl Pyrones: SNF4435 C and D

Natural products containing, or derived from, the bicyclo[4.2.0]octadiene ring systems are, however, rare, and only in the past few years have compounds been isolated with this motif in place. Two such compounds are the diastereoisomers SNF4435 C (**90**) and SNF4435 D (**91**), which were isolated in 2001 from the culture broth of a strain of *Streptomyces spectabilis* in a [2.3 : 1.0] ratio [55]. Both compounds displayed impressive immunosuppressant activity and multidrug resistance reversal effects, selectively suppressing B-cell proliferation versus T-cell growth. This mode of action was different from known immunosuppressants such as cyclosporin A (CsA) and FK-506, which work by inhibiting T-cell activation. The SNF compounds **90** and **91** belong to the nitrophenyl pyrone family of natural products, and are identified as consisting of a bicyclo[4.2.0]octadiene core linked to a spiro-fused tetrahydrofuran ring. This tricyclic core is substituted with a γ-pyrone moiety, and contains five stereocenters, two of which are quaternary. A further distinguishing structural feature is the presence of a *para*-nitrophenyl group, which is also found in other natural products such as orinocin (**92**) [56], aureothin (**93**) [57], and spectinabilin (**94**) [58] (Figure 16.3).

The interesting stage 1 and 2 biosynthesis of the nitrophenyl pyrones **90–94** involves combining polypropionate units with *p*-aminobenzoic acid, and has been the topic of much investigation [59]. The proposed biogenesis of the endiandric acids suggested by Black, and the supporting experimental evidence provided by Nicolaou et al., inspired the groups led by Trauner [60] and Baldwin [61] to propose biomimetic routes to the SNF compounds **90** and **91** based upon stage 3 transformations. Both research groups realized that the origins of SNF4435 C (**90**) and SNF4435 D (**91**) could also be embedded in conjugated polyene chemistry,

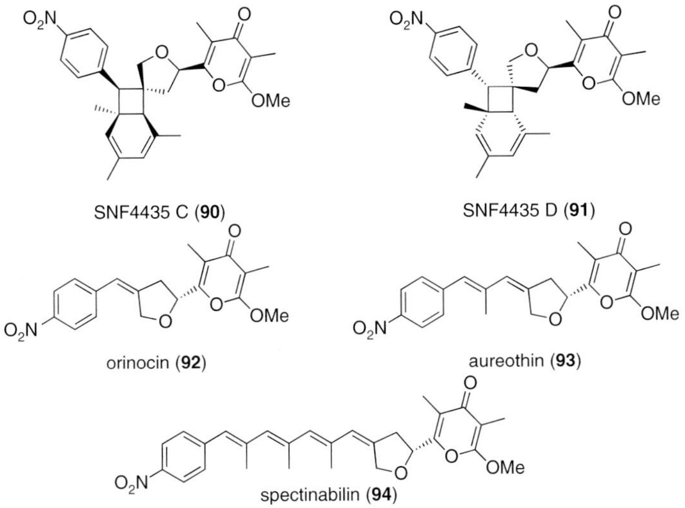

Figure 16.3 Structures of p-nitrophenyl pyrones.

with Baldwin and one of the authors suggesting that spectinabilin (**94**) could be their biosynthetic precursor in Nature.

Providing some weight to this hypothesis was an observation by Rinehart et al. in their report on the isolation of spectinabilin (**94**). It was noted that **94**, a constitutional isomer of SNF4435C (**90**) and SNF4435D (**91**), "*is not very stable, about 50% being converted to other substances during one month at room temperature*" [58]. Although the "other substances" were not identified, it was not unreasonable to propose that spectinabilin (**94**) may have undergone various thermally/photochemically-induced double bond isomerizations, which could lead to a range of products derived from pericyclic chemistry.

Thus, SNF4435 C (**90**) and SNF4435 D (**91**) were envisaged as having originated from an isomer of spectinabilin (**94**), namely, the (−Z,2Z,4Z,6E)-tetraene (**95**), which could itself arise from (**94**) by two consecutive internal double bond (E−Z)-isomerizations. The "internal" (Z,Z)-double bond geometry of tetraene **95** would allow the two "outer" double bonds of the tetraene back bone to meet and undergo the reaction cascade (stage 3), beginning with a diastereoselective 8π conrotatory electrocyclization of the tetraene **95** giving rise to both the major octatriene **96** and the minor diastereoisomer **97**. An *endo*-selective 6π disrotatory electrocyclization of the octatrienes **96** and **97** then give rise to SNF4435 C (**90**) and SNF4435 D (**91**) respectively (Scheme 16.20).

Model studies by the Trauner et al. and Baldwin et al. provided validation for this biomimetic hypothesis. The Trauner group opted to target the polyene precursor **101**, with the internal double bonds set up with (Z,Z)-geometry, utilizing a convergent Stille cross coupling of the vinyl iodide **98** with the stannane **99**. The intermediate **101** was not isolated, but instead immediately underwent a 8π–6π electrocyclic cascade leading to the cyclohexadiene **102**, with a fused bicyclo[4.2.0]

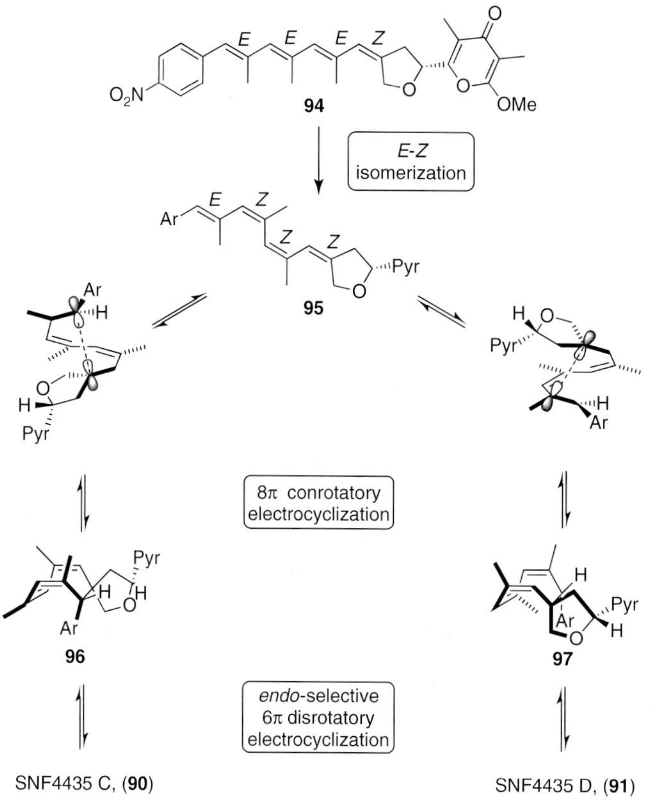

Scheme 16.20 Proposed biosynthesis of SNF compounds **90** and **91**.

core (Scheme 16.21a). Baldwin *et al.*, on the other hand, based on the spectinabilin (**94**) hypothesis, prepared the all-*(E)*-polyene **100**, which upon treatment with Pd(II) resulted in *(E–Z)*-isomerization of the two internal double bonds to give the (2*E*,4*Z*,6*Z*,8*E*)-tetraene **101**, which underwent the Black cascade in the same fashion as Trauner's example (Scheme 16.21b).

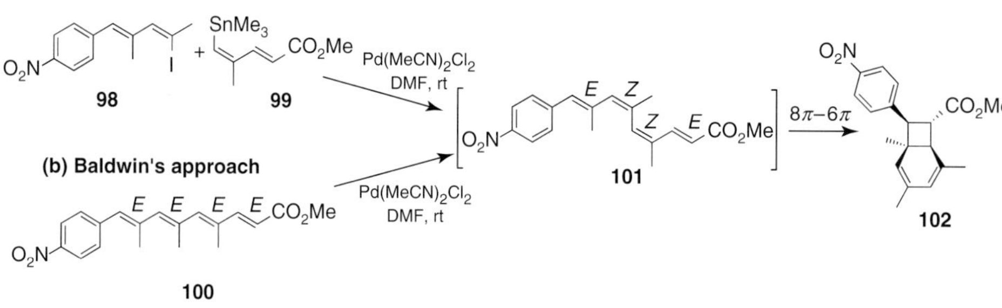

Scheme 16.21 Model studies toward SNF4435 C (**90**) and D (**91**).

Following these studies, several total syntheses of the SNF-compounds were reported. Parker *et al.* reported the first total synthesis, having used a very similar coupling strategy as Trauner to access the (−*E*,2*Z*,4*Z*,6*Z*)-tetraene **95** (Scheme 16.22a) [62]. The vinyl iodide **98** and the vinyl stannane **103** underwent Pd-catalyzed cross coupling to give the tetraene **95**, which then underwent the expected electrocyclization cascade to **90/91**. Trauner's synthesis involved a similar approach to his model studies, but in this case the stannane **104** and the iodide **105** were coupled to give the target structures in excellent yield (89%) (Scheme 16.22b) [63]. Baldwin's biosynthetic hypothesis was developed using palladium catalysis as a means of isomerizing spectinabilin (**94**) to the required (−*Z*,2*Z*,4*Z*,6*E*)-tetraene geometry [64]. The tetraene **94/94′** was constructed by sequential Suzuki and Negishi coupling reactions, beginning with the boronate ester **107** and the 1,1-dibromide **108** (Scheme 16.23). The isomerization followed by 8π–6π electrocyclic cascade was then initiated by 25 mol.% $PdCl_2(MeCN)_2$ to furnish **90** and **91** in a [2.5 : 1.0] in 22% yield. This isomer ratio was close to the [2.3 : 1.0] ratio observed in Nature, and supports the biosynthetic hypotheses.

Interestingly, diastereoisomers **110** and **111** were also isolated in 18% yield as a [2.1 : 1.0] mixture. This is consistent with the 8π–6π-electrocyclization cascade of the (−*Z*,2*Z*,4*Z*,6*Z*)- and (−*E*,2*Z*,4*Z*,6*E*)-tetraene isomers **112** and **113** (Scheme 16.24).

The double bond isomerization hypothesis to **90** and **91** has since been established as being photo-induced in Nature, by Trauner and Hertweck *et al.* [65]. The biosynthetic cascade is initiated by two *(E–Z)*-internal double bond isomerization reactions of spectinabilin (**94**) to provide the helical geometry required for thermal 8π-conrotatory and 6π-disrotatory electrocyclizations.

16.7.3
Ocellapyrones

The powerful sequence involving double bond isomerization followed by electrocyclization cascades has been further developed by the research groups of Baldwin [29] and Trauner [30] in their syntheses of the molluscan polypropionate ocellapyrone A (**27**) [33]. Although the original synthetic target of both groups was 9,10-deoxytridachione (**15**) (Scheme 16.9), the key (2*E*,4*Z*,6*E*,8*E*)-tetraene **18** arrived at by both research groups was found to undergo further double bond isomerization to give the (2*E*,4*Z*,6*Z*,8*E*)-polyene **29**. Intermediate **29** led to both ocellapyrone A (**27**) and its isomer (**34**) in varying amounts, and can be explained by examining the transition state for the 6π electrocyclization. Diastereomer **34** is the product arising from *exo*-orientation of the ethyl group, whereas compound **27** arises from *endo*-orientation (Scheme 16.25). To access the (2*E*,4*Z*,6*Z*,8*E*)-polyene **29** directly, Trauner [30] synthesized the vinyl stannane **116** and vinyl iodide **24**, which were coupled using the conditions of Mee *et al.* [31]. Polyene **29** was not isolated, and underwent the electrocyclic cascade directly to give isomers **27** and **34**. Interestingly, a stage 4 type Diels–Alder reaction of isomer **34** with singlet oxygen gave the corresponding ocellapyrone B (**32**) in excellent yield (Scheme 16.26).

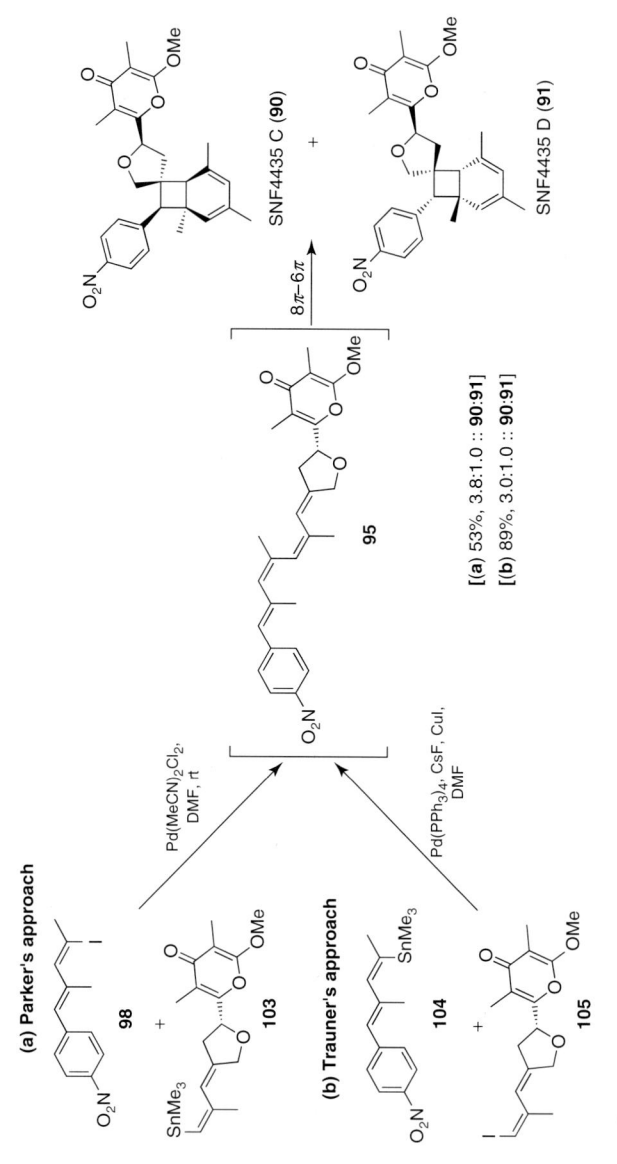

Scheme 16.22 Synthesis of SNF4435 C (**90**) and D (**91**): (a) Parker's approach; (b) Trauner's approach.

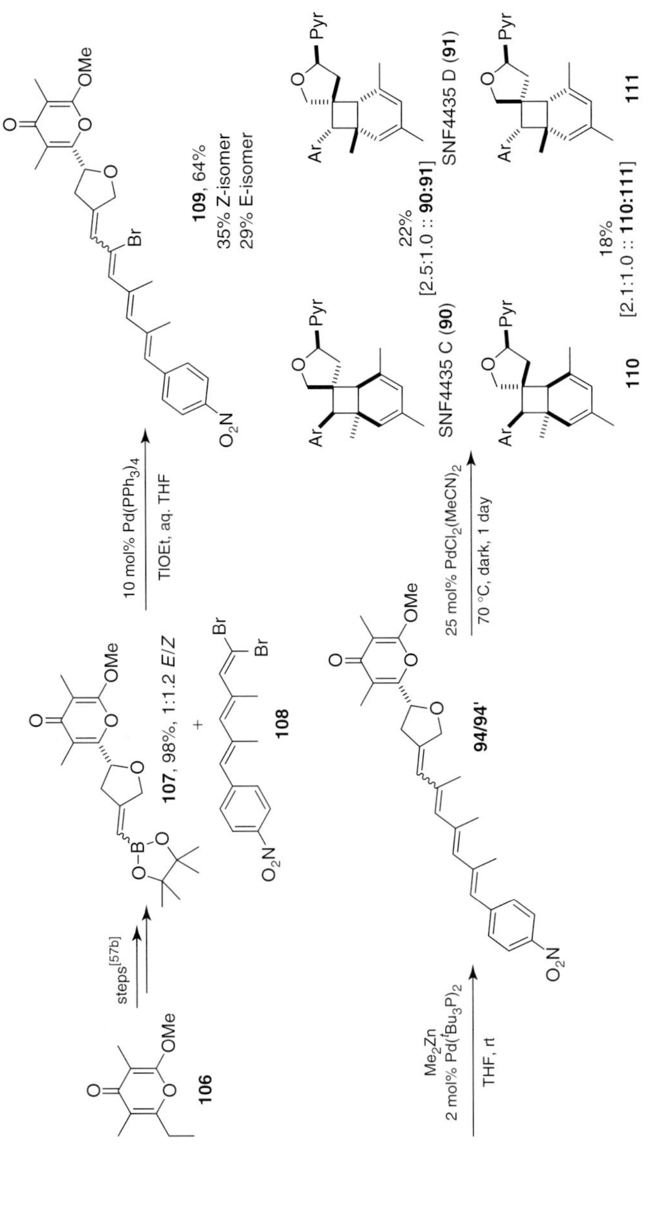

16.7 8π Systems and the Black 8π–6π Electrocyclic Cascade | 623

Scheme 16.23 Baldwin's syntheses of the SNF compounds **90** and **91**.

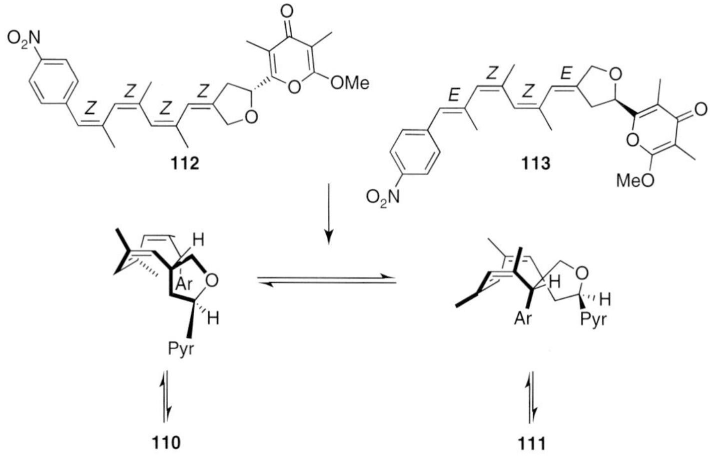

Scheme 16.24 Formation of diastereoisomers **110** and **111**.

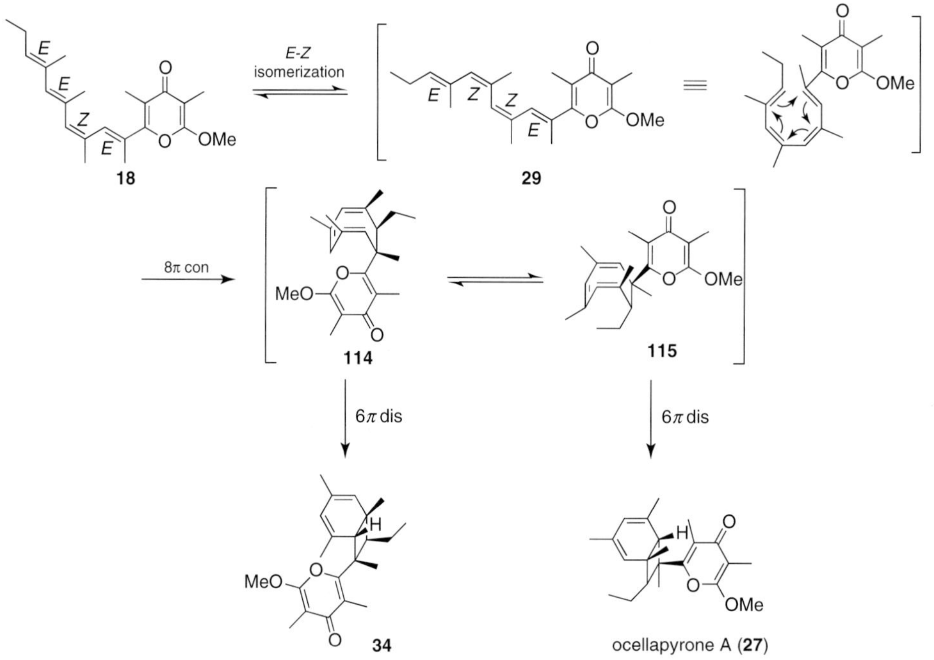

Scheme 16.25 Electrocyclic formation of ocellapyrone A (**27**).

16.7.4
Elysiapyrones

The γ-pyrone functionality is clearly a very common motif in polypropionate-derived natural products from marine molluscs [66]. It is therefore not surprising that

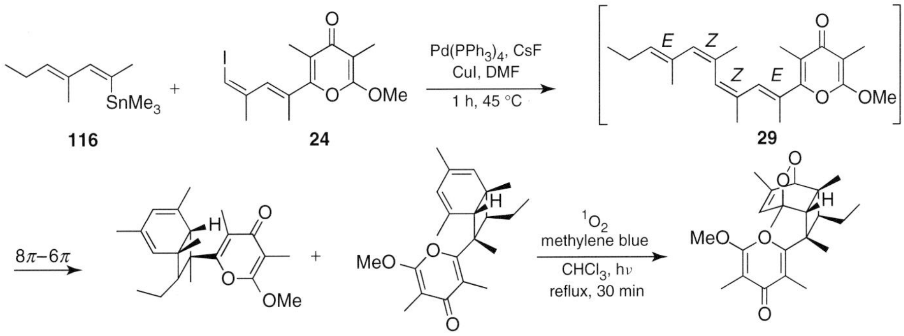

Scheme 16.26 Trauner's synthesis of ocellapyrones A (**27**) and B (**32**).

elysiapyrones A (**125**) and B (**126**), which were isolated from the mollusc *Elysia diomedea*, also have this privileged functionality. The elysiapyrones share common structural features with the ocellapyrones, consisting of a familiar bicyclo[4.2.0]octane core, but with the exception of epoxide ring decoration [67]. Darias proposed a biosynthetic hypothesis for these optically active metabolites, which involved an enzyme-mediated Black 8π–6π electrocyclization cascade (stage 3), followed by a [4+2] cycloaddition of singlet oxygen and subsequent rearrangement of the corresponding endoperoxide (stage 4) [67]. Trauner *et al.* used their experience in biomimetic electrocyclizations to investigate Darias' hypothesis, and developed a total synthesis of elysiapyrones A (**125**) and B (**126**) in racemic form [68]. They again called upon the Stille reaction to access the key (2*E*,4*Z*,6*Z*,8*E*)-polyene **118**, from fragments **117** and **24**, which was not isolated and underwent the 8π–6π electrocyclization cascade to afford the diastereoisomers **121** and **122**. Diels–Alder cycloaddition of **121** and **122** with singlet oxygen next gave *endo*-peroxides **123** and **124**, respectively, which upon treatment with a ruthenium catalyst then gave the target isomerized epoxide products **125** and **126**. This elegant synthesis followed the biosynthetic rationale closely, but then led to the products in racemic form (Scheme 16.27).

16.7.5
Shimalactones

Shimalactones A (**127**) and B (**128**) are another selection of natural products possessing a bicyclo[4.2.0]octadiene core. These interesting lactones were isolated from a marine-derived fungus, *Emericella variecolor* GF10, found at depths of 70 m off the coast of Gokasyo Gulf, Mie Prefecture, Japan [69], and display promising cytotoxic properties.

Interesting structural features include a novel oxabicyclo[2.2.1]heptane moiety linked to a bicyclo[4.2.0]octadiene core by a trisubstituted alkene. These structural features make the shimalactones a challenging molecular target for organic synthesis. In 2006, Trauner *et al.*, proposed that the biosynthesis of the shimalactones

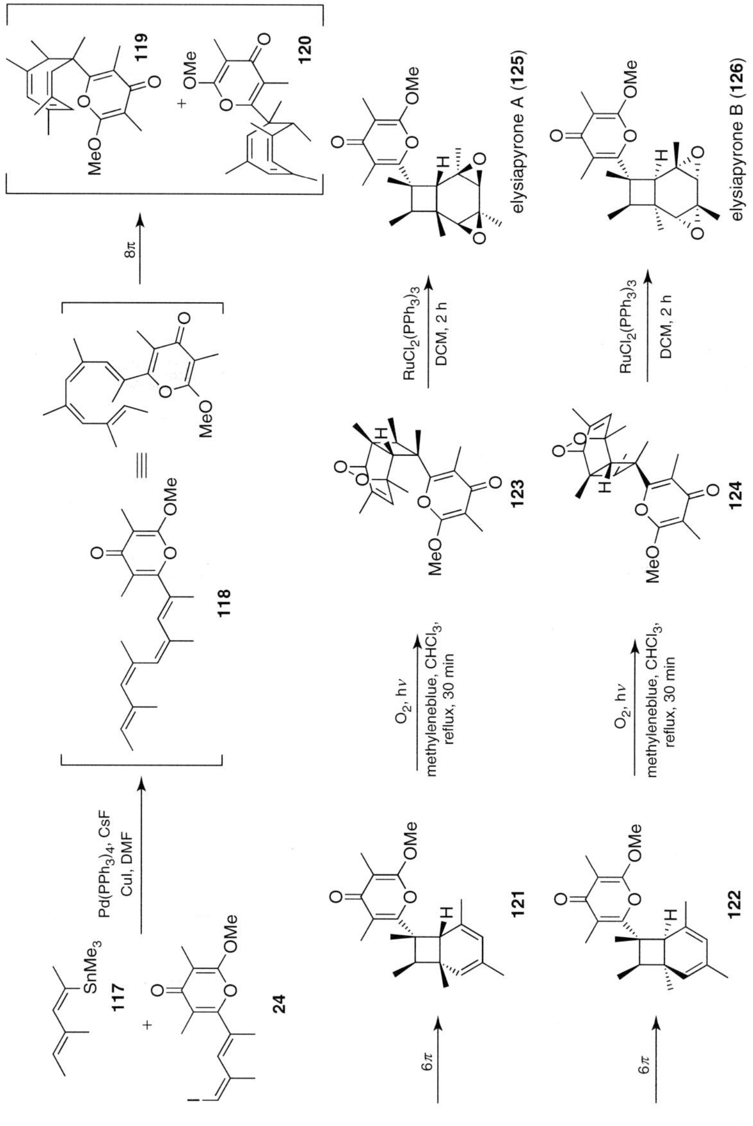

Scheme 16.27 Trauner's biomimetic synthesis of elysiapyrones (**125**) and (**126**).

could involve the heptaenyl β-ketolactone **135** undergoing enzymatic epoxidation at the penultimate double bond to render **134**, followed by acid-catalyzed epoxide opening to give undecapentaheptenyl cation **133** [70]. This cation could then isomerize to cation **132**, which is attacked by the enol form of the β-ketolactone. Notably, structures **133** and **132** are probably not true resonance structures as A1,3-strain induced by the methyl groups prevent the π-system from being planar. Polyene **132** would then need to undergo isomerization to the *(E,E,Z,Z,E)*-polyene **131** to set up the 8π–6π electrocyclizations cascade to obtain shimalactones A (**127**) and B (**128**) (Scheme 16.28).

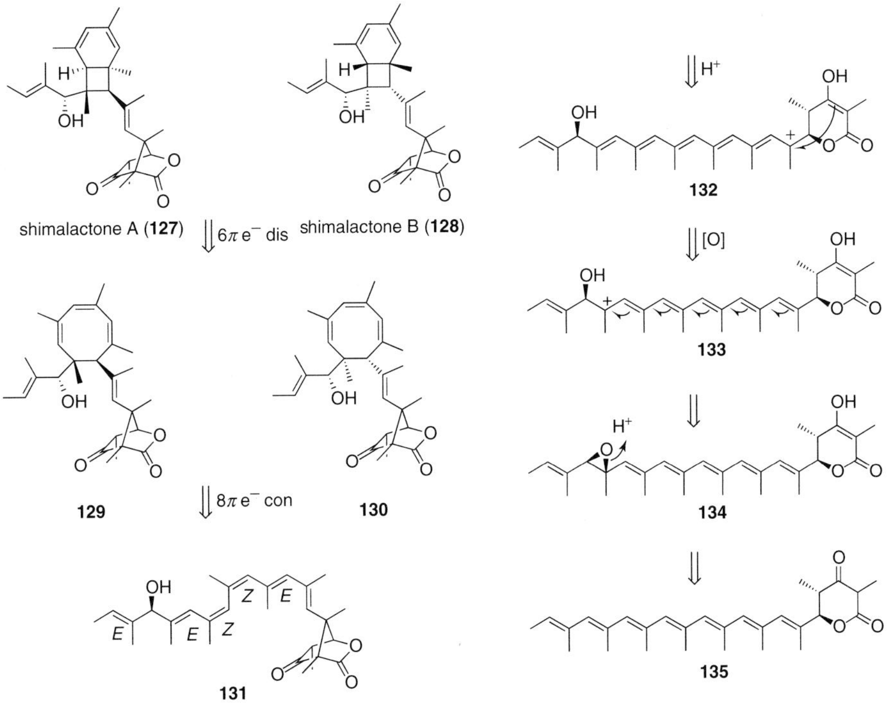

Scheme 16.28 Proposed biosynthesis of shimalactones A (**127**) and B (**128**).

Trauner *et al.* later described the biomimetic synthesis of the shimalactones using the acid-catalyzed cyclization 8π–6π electrocyclization strategy mentioned [70]. The electrocyclization precursor was assembled from the two fragments **145** and **148**. The synthesis of compound **145** began with a Heathcock anti-aldol addition of the boron enolate of Evan's reagent **136**, followed by acylation of the resulting alcohol **137** with propionic anhydride to give **138**, which underwent a Stille coupling with tributylisopropenylstannane to afford **139** (Scheme 16.29). Dieckmann cyclization using KHMDS (potassium hexamethyldisilazane) next gave the β-ketolactone **140** as a single diastereoisomer. After careful optimization it was found that camphorsulfonic acid (CSA) catalyzed the cyclization to give the oxabicyclo[2.2.1]heptane core found in the natural product. Allylic bromination of

141 under radical bromination conditions gave rise to 142 and 142′ as a 1 : 1 mixture. Compound 142 underwent smooth oxidation with IBX (2-iodoxybenzoic acid) to yield the corresponding aldehyde 143. Despite being very sensitive to basic conditions, a three-step sequence was realized to afford 144, which after a Stork–Zhao olefination gave fragment 145 (Scheme 16.29).

Fragment 148 was prepared using an asymmetric addition of 2-butenylmethyl zinc to aldehyde 146, followed by iodine–tin exchange. In a spectacular display of the power of biomimetic synthesis, when fragments 145 and 148 were exposed to Pd(0), shimalactones A (127) and B (128) were accessed directly in 66% yield as a 5.0 : 1.0 mixture, with the intermediate 131 not isolated (Scheme 16.30).

16.8
Biological Electrocyclizations and Enzyme Catalysis

The involvement of enzyme catalysis in biological electrocyclizations is uncertain. Many of the natural products that are believed to be derived through electrocyclizations have been isolated in racemic form (endiandric acids, etc.). Clearly, in such cases one can rule-out the involvement of electrocyclases. On the other hand, where optically active precursors undergo electrocyclization this would, in principle, lead to optically active products without the need for enzyme assistance. For example, the SNF compounds 90 and 91 described above are derived from enantiomerically pure spectinabilin (Scheme 16.20), and it is highly unlikely that an electrocyclase is involved. This is more important when considering photochemical electrocyclizations, which would not require enzyme assistance since they proceed through high-energy excited states.

However, there are cases that cannot be explained so easily. For instance, ocellapyrones A (27) and B (32) and elysiapyrones A (125) and B (126) have all been isolated in optically active form, despite their proposed precursors being achiral. Although it is possible that a particular enantiomer may be consumed selectively through the process of metabolism, such a "kinetic resolution" theory is unlikely to explain every case. It has been proposed that the course of the 8π–6π electrocyclization may be guided in the active site of the PKS that generates the polyketide precursors (Figure 16.4) [71]. The chiral environment could in principle influence the torquoselectivity of the electrocyclization, leading to optically enriched products. Furthermore, the synthase may also accelerate the cyclization by preorganizing the conformation of the polyene precursor, thus acting as a catalyst. This hypothesis has some weight, when one considers the biosynthesis of aromatic polyketides. In this event, polycarbonyl intermediates undergo intramolecular condensation and aromatization when covalently linked to the type II PKS [72].

A similar argument applies to several cyclohexadienes, such as 9,10-deoxytridachione (15), which have been isolated in optically active form. Such compounds are proposed to arise through 6π thermal electrocyclizations from achiral polyenes (Schemes 16.7–16.10). However, it has been shown that elevated temperatures

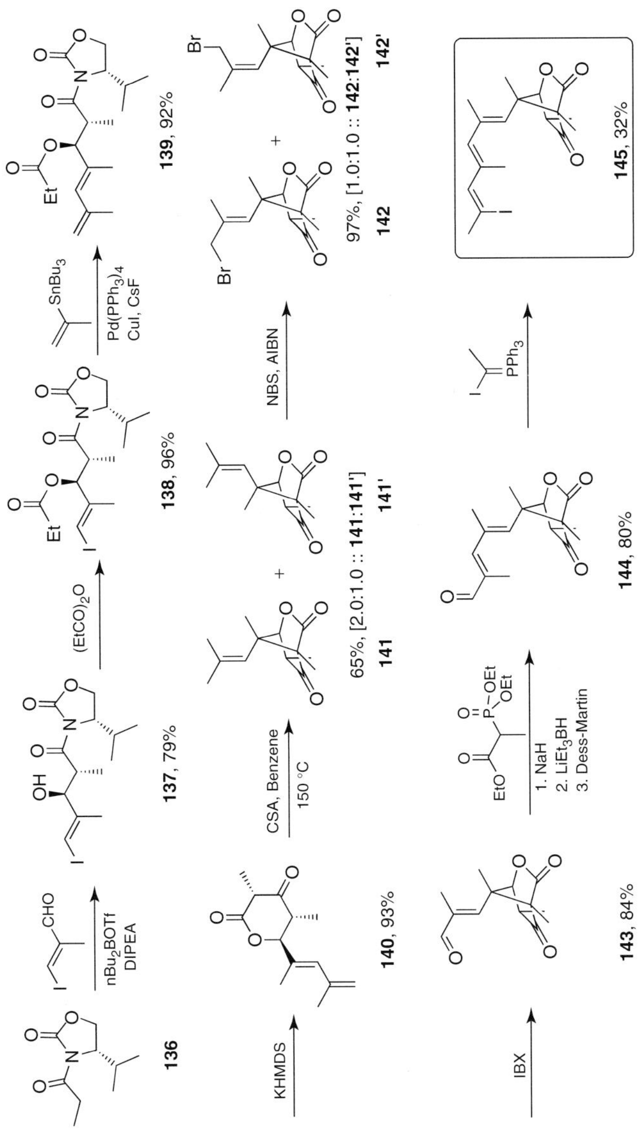

Scheme 16.29 Trauner's synthesis of key fragment **145**, *en route* to the shimalactones.

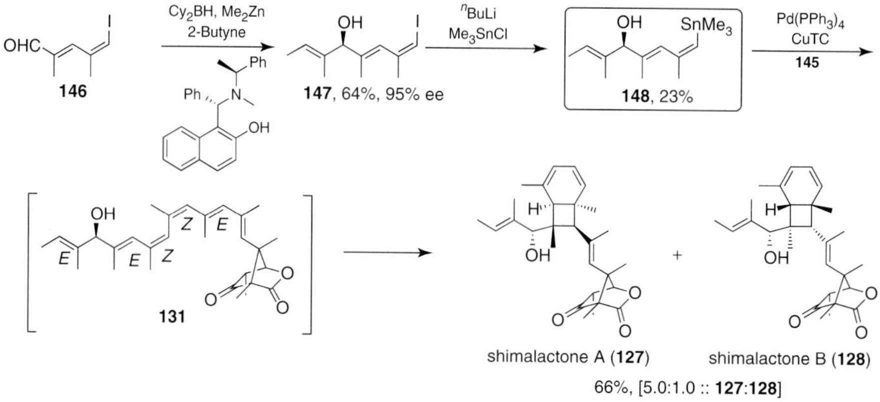

Scheme 16.30 Trauner's biomimetic synthesis of **148** and shimalactones A (**127**) and B (**128**).

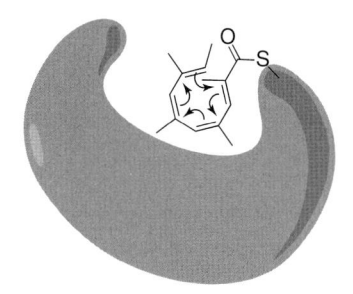

Figure 16.4 PKS-bound tetraene undergoing an asymmetric-torquoselective 8π electrocyclization.

are often required to initiate these cyclizations, and it could be the case that the reactions are actually photochemically derived, although the relative double bond configurations of the polyene precursor would need to be different. This still does not explain the origin of optical activity in these compounds. In fact, compounds that are chemically derived pose a more intriguing question, since it is highly unlikely that photoactive enzymes have evolved to effect such processes, since they are very scarce [73]. The authors would suggest an alternative explanation for the case of the mollusc metabolite tridachiahydropyrone (**6**) and other related photochemical electrocyclization products.[1]

It has been suggested that some polypropionate metabolites are biosynthesized and then translocated to the tissue of the digestive diverticula of the molluscs, where they serve as sunscreens to protect the molluscs from damaging UV-radiation

---

1) The proposal was first put forward by one of the authors, J. E. Moses, during a research lecture at The Chemistry Research Laboratory, University of Oxford on 15 October 2009.

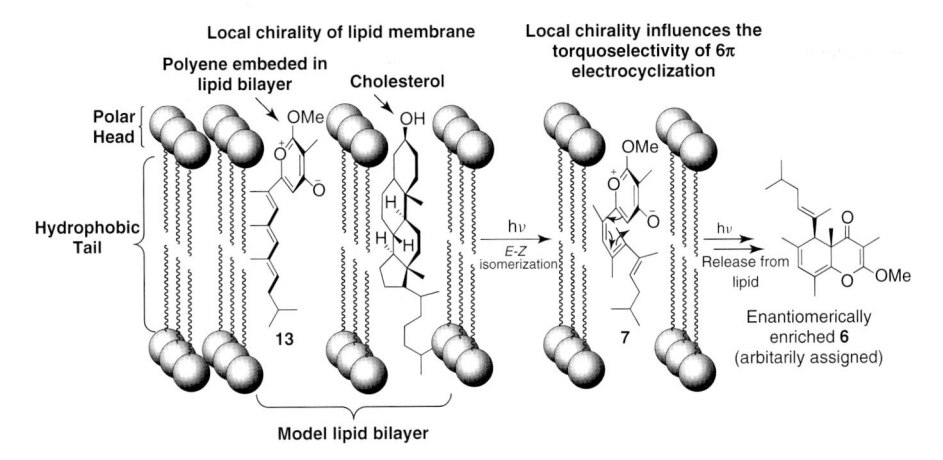

Scheme 16.31 Photochemical electrocyclizations in membranes.

[23, 25]. This hypothesis is reasonable, since these sacoglossans live in shallow lagoons where they are exposed to direct sunlight. In this respect, perhaps the complexity of the natural products becomes a consequence of the function of their predecessor's (e.g., polyene **13**, Scheme 16.5) to act as a sunscreen, and, as such, would not necessarily have evolved enzymatic assistance for their syntheses.

As sunscreens, optimal protection would result from the compounds being located in the outer region of the cell, and most likely in the membrane. Given that the key polyene **13** has similar structural properties to that of a lipid, that is, *a fatty chain and a polar head group*, then it is reasonable that **13** would sit comfortably in the lipid bilayer of the surface cell membrane (Scheme 16.31). The cell membrane consists of a thin layer of amphipathic phospholipids that forms a continuous, spherical lipid bilayer. The composition is complex and includes proteins, carbohydrates, and cholesterol among other components. The immediate chiral environment of the membrane could in principle influence the resulting enantiomeric purity of the tridachiahydropyrone family (and related compounds); that is, polyene **13** becomes embedded in the bilayer, and upon irradiation is transformed into (−)-tridachiahydropyrone (**6**). The surrounding chiral environment of the membrane may influence the reaction in a torquoselective fashion, favoring the formation of one enantiomer over the other. Studies by Hailes in the 1990s provide some support for this hypothesis. In their work, Hailes *et al.* demonstrated that chiral micellar media can be used to induce modest levels of enantiomeric enrichment in aqueous Diels–Alder reactions [74].

16.9 Conclusion

Considering the volume of evidence, it is highly likely that electrocyclic reactions play an important role in the biosynthesis of polyketide-derived natural products.

The above discussion is not exhaustive or complete, and no doubt the future will provide many more interesting examples. With very few exceptions, most notably 9,10-deoxytridachione (**15**), most of the electrocyclizations discussed occur readily without need for catalysis or excessive temperatures, which is supportive of the biomimetic hypothesis in general [1, 2]. As such, electrocyclization reactions have proven valuable and effective tools in the synthesis of highly complex and structurally interesting natural products.

The identification of enzymes that mediate electrocyclizations is still an important goal and, if they do indeed exist, such enzymes may provide inspiration for biomimetic catalysts to achieve asymmetric induction in these processes. To date, there are very few examples of asymmetric electrocyclizations [75], providing a wealth of opportunity for discovery and development [1].

Acknowledgments

We are very grateful to Professor G. Pattenden, FRS (University of Nottingham) and Dr. R. Rodriguez (University of Cambridge) for their helpful comments and suggestions during the preparation of this manuscript. We thank the EPSRC (P.S.), Pfizer (M.R.), and University of Nottingham (J.B.) for financial support.

References

1. Moses, J.E. and Adlington, R.M. (2005) *Chem. Commun.*, 5945–5952.
2. Heathcock, C.H. (1996) *Proc. Natl. Acad. Sci. U.S.A.*, **93**, 14323–14327.
3. Corey, E.J. and Cheng, X.-M. (1995) *The Logic of Chemical Synthesis*, John Wiley & Sons, Inc., New York.
4. (a) Nicolaou, K.C. and Sorensen, E.J. (1996) *Classics in Total Synthesis: Targets, Strategies, Methods*, Wiley-VCH Verlag GmbH, Weinheim; (b) Nicolaou, K.C. and Snyder, S. (2003) *Classics in Total Synthesis II: More Targets, Strategies, Methods*, Wiley-VCH Verlag GmbH, Weinheim.
5. Sierra, M.A., de la Torre, M., and Nicolaou, K.C. (2004) *Dead Ends and Detours: Direct Ways to Successful Total Synthesis*, Wiley-VCH Verlag GmbH, Weinheim.
6. The biosynthesis of Taxol provides an excellent example: Rhor, J. (1997) *Angew. Chem. Int. Ed.*, **36**, 2190–2195.
7. Sir Robert Robinson's synthesis of tropinone is generally credited as the first biomimetic synthesis: Robinson, R. (1917) *J. Chem. Soc., Trans.*, **111**, 762–768.
8. (a) Johnson, W.S., Gravestock, M.B., Parry, R.J., Myers, R.F., Bryson, T.A., and Miles, D.H. (1971) *J. Am. Chem. Soc.*, **93**, 4330–4332; (b) Johnson, W.S., Gravestock, M.B., and McCarry, B.E. (1971) *J. Am. Chem. Soc.*, **93**, 4332–4334.
9. Examples of biomimetic terpene synthesis: (a) Yoder, R.A. and Johnston, J.N. (2005) *Chem. Rev.*, **105**, 4730–4756; (b) Tang, B., Bray, C.D., and Pattenden, G. (2009) *Org. Biomol. Chem.*, **7**, 4448–4457; (c) Trauner, D., Elliott, G.I., Maimone, T.J., and Malerich, J.P. (2005) *J. Am. Chem. Soc.*, **127**, 6276–6283.
10. Examples of biomimetic Alkaloid synthesis: (a) Scholz, U. and Winterfeldt, E. (2000) *Nat. Prod. Rep.*, **17**, 349–366; (b) Amat, M., Griera, R., Fabregat, R., Molins, E., and Bosch, J. (2008) *Angew. Chem. Int. Ed.*, **47**, 3348–3351; (c) Schwartz, M.A. and Zoda, M.F. (1981) *J. Org. Chem.*, **46**, 4623–4625.

11. Examples of biomimetic polyketide synthesis that are non-electrocyclic: Gademann, K., Bethuel, Y., Locher, H.H., and Hubschwerlen, C. (2007) *J. Org. Chem.*, **72**, 8361–8370.
12. Moses, J.E., Adlington, R.M., Baldwin, J.E., Commeiras, L., Cowley, A.R., Baker, C.M., Albrecht, B., and Grant, G.H. (2006) *Tetrahedron*, **62**, 9892–9901; (b) Moses, J.E., Commeiras, L., Baldwin, J.E., and Adlington, R.M. (2003) *Org. Lett.*, **5**, 2987–2988; (c) Nicolaou, K.C., Snyder, S.A., Montagnon, T., and Vassilikogiannakis, G. (2002) *Angew. Chem. Int. Ed.*, **41**, 1668–1698.
13. For example see: Tietze, L.F. (1996) *Chem. Rev.*, **96**, 115–136.
14. Woodward, R.B. and Hoffmann, R. (1964) *J. Am. Chem. Soc.*, **87**, 395–397.
15. Fleming, I. (2010) *Molecular Orbitals and Organic Chemical Reactions: Reference Edition*, Wiley-Blackwell.
16. Staunton, J. (1997) *Chem. Rev.*, **97**, 2611–2629.
17. Shida, N., Miyazaki, K., Kumagai, K., and Rikimaru, M. (1965) *J. Antibiot.*, **18**, 68–76.
18. Murata, M., Naoki, H., Iwashita, T., Matsunaga, S., Saski, M., Yokoyama, A., and Yasumoto, T. (1993) *J. Am. Chem. Soc.*, **115**, 2060–2062.
19. Weissman, K.J. (2009) *Methods Enzymol.*, **459**, 3–16.
20. (a) Sharma, P., Griffiths, N., and Moses, J.E. (2008) *Org. Lett.*, **10**, 4025–4027; (b) Sharma, P., Lewis, W., Lygo, B., and Moses, J.E. (2009) *J. Am. Chem. Soc.*, **131**, 5966–5972.
21. Ortea, J., Cimino, G., Mollo, E., and Gavagnin, M. (1996) *Tetrahedron Lett.*, **37**, 4259–4262.
22. Fu, X., Hong, E.P., and Schimitz, F.J. (2000) *Tetrahedron*, **56**, 8989–8993.
23. Ireland, C. and Scheuer, P.J. (1979) *Science*, **205**, 922–923.
24. Sharma, P. and Moses, J.E. (2010) *Synlett*, **4**, 525–529.
25. Ireland, C. and Faulkner, J.D. (1981) *Tetrahedron*, **9**, 233–240.
26. Schmitz, F.J. and Ksebati, M.B. (1985) *J. Org. Chem.*, **50**, 5637–5642.
27. Gavagnin, M., Spinella, A., Castelluccio, F., Cimino, G., and Marin, A. (1994) *J. Nat. Prod.*, **57**, 298–304.
28. Epoxidation of deoxytridachione: Ireland, C., Faulkner, D.J., Solheim, B.A., and Clardy, J. (1978) *J. Am. Chem. Soc.*, **100**, 1002–1003.
29. (a) Rodriguez, R., Adlington, R.M., Eade, S.J., Walter, M.W., Baldwin, J.E., and Moses, J.E. (2007) *Tetrahedron*, **63**, 4500–4509; (b) Moses, J.E., Adlington, R.M., Rodriguez, R., Eade, S.J., and Baldwin, J.E. (2005) *Chem. Commun.*, 1687–1689.
30. Miller, A.K. and Trauner, D. (2005) *Angew. Chem. Int. Ed.*, **44**, 4602–4606.
31. Mee, S.P.H., Lee, V., and Baldwin, J.E. (2004) *Angew. Chem. Int. Ed.*, **43**, 1132–1136.
32. Eade, S.J., Walter, M.W., Byrne, C., Odell, B., Rodriguez, R., Baldwin, J.E., Adlington, R.M., and Moses, J.E. (2008) *J. Org. Chem.*, **73**, 4830–4839.
33. Manzo, E., Ciavatta, M.L., Gavagnin, M., Mollo, E., Wahidulla, S., and Cimino, G. (2005) *Tetrahedron Lett.*, **46**, 465–468.
34. (a) Brückner, S., Baldwin, J.E., Moses, J.E., Adlington, R.M., and Cowley, A.R. (2003) *Tetrahedron Lett.*, **44**, 7471–7473; (b) Moses, J.E., Baldwin, J.E., Marquez, R., Adlington, R.M., Claridge, T.D.W., and Odell, B. (2003) *Org. Lett.*, **5**, 661–663.
35. Zuidema, D.R., Miller, A.K., Trauner, D., and Jones, P.B. (2005) *Org. Lett.*, **7**, 4959–4962.
36. Rickards, R.W. and Skropeta, D. (2002) *Tetrahedron*, **58**, 3793–3800.
37. (a) Marvell, E.N. (1980) *Thermal Electrocyclic Reactions*, vol. 43, Academic Press, New York; (b) Ansari, F.L., Qureshi, R., and Qureshi, M.L. (1999) *Electrocyclic Reactions*, Wiley-VCH Verlag GmbH, Weinheim; (c) Woodward, R.B. and Hoffmann, R. (1970) *The Conservation of Orbital Symmetry*, Verlag Chemie, Weinheim.
38. Lee, J.C., Strobel, G.A., Lobkovsky, E., and Clardy, J. (1996) *J. Org. Chem.*, **61**, 3232–3233.
39. Li, J.Y., Harper, J.K., Grant, D.M., Tombe, B.O., Bashyal, B., Hess, W.M., and Strobel, G.A. (2001) *Phytochemistry*, **56**, 463–468.

40. Li, C., Lobkovsky, E., and Porco, J.A. Jr. (2000) *J. Am. Chem. Soc.*, **122**, 10484–10485.
41. Li, C., Johnson, R.P., and Porco, J.A. Jr. (2003) *J. Am. Chem. Soc.*, **125**, 5095–5106.
42. Mehta, G. and Pan, S.C. (2004) *Org. Lett.*, **6**, 3985–3988.
43. (a) Kakeya, H., Onose, R., Koshion, H., Yoshida, A., Kobayashi, K., Kageyama, S.I., and Osada, H. (2002) *J. Am. Chem. Soc.*, **124**, 3496–3497; (b) Kakeya, H., Onose, R., Yoshida, A., Koshino, H., and Osada, H. (2002) *J. Antibiot.*, **55**, 829–831.
44. Kakeya, H., Onose, R., Koshino, H., and Osada, H. (2005) *Chem. Commun.*, 2575–2577.
45. Shoji, M., Imai, H., Mukaida, M., Sakai, K., Kakeya, H., Osada, H., and Hayashi, Y. (2005) *J. Org. Chem.*, **70**, 79–91.
46. Li, C. and Porco, J.A. Jr. (2005) *J. Org. Chem.*, **70**, 6053–6065.
47. Kuwahara, S. and Imada, S. (2005) *Tetrahedron Lett.*, **46**, 547–549.
48. Mehta, G. and Islam, K. (2004) *Tetrahedron Lett.*, **45**, 3611–3615.
49. (a) Bandaranayake, W.M., Banfield, J.E., Black, D.St.C., Fallon, G.D., and Gatehouse, B.M. (1980) *J. Chem. Soc., Chem. Commun.*, 162–163; (b) Bandaranayake, W.M., Banfield, J.E., and Black, D.St.C. (1980) *J. Chem. Soc., Chem. Commun.*, 902–903.
50. (a) Nicolaou, K.C., Petasis, N.A., Zipkin, R.E., and Uenishi, J. (1982) *J. Am. Chem. Soc.*, **104**, 5555–5557; (b) Nicolaou, K.C., Petasis, N.A., Uenishi, J., and Zipkin, R.E. (1982) *J. Am. Chem. Soc.*, **104**, 5557–5558; (c) Nicolaou, K.C., Zipkin, R.E., and Petasis, N.A. (1982) *J. Am. Chem. Soc.*, **104**, 5558–5560; (d) Nicolaou, K.C., Petasis, N.A., and Zipkin, R.E. (1982) *J. Am. Chem. Soc.*, **104**, 5560–5562.
51. Pohnert, G. and Boland, W. (1994) *Tetrahedron*, **50**, 10235–10244.
52. Pohnert, G. and Boland, W. (2002) *Nat. Prod. Rep.*, **19**, 108–122.
53. Vogt, P., Schlageter, M., and Widmer, E. (1991) *Tetrahedron Lett.*, **32**, 4115–4116.
54. Huisgen, R., Dahmen, A., and Huber, H. (1967) *J. Am. Chem. Soc.*, **89**, 7130–7131.
55. (a) Kurosawa, K., Takahashi, K., and Tsuda, E.J. (2001) *J. Antibiot.*, **54**, 541–547; (b) Takahashi, K., Tsuda, E., and Kurosawa, K. (2001) *J. Antibiot.*, **54**, 548–553.
56. Pancharoen, O., Picker, K., Reutrakul, V., Taylor, W.C., and Tuntiwachwuttikul, P. (1987) *Aust. J. Chem.*, **40**, 455–459.
57. (a) Hirata, Y., Nakata, H., Yamada, K., Okuhara, K., and Naito, T. (1961) *Tetrahedron*, **14**, 252–274. (b) synthesis: Jacobsen, M.F., Moses, J.E., Adlington, R.M., and Baldwin, J.E. (2005) *Org. Lett.*, **7**, 641–644.
58. Kakinuma, K., Hanson, C.A., and Rinehart, K.L. Jr. (1976) *Tetrahedron*, **32**, 217–222.
59. (a) Hertweck, C. (2009) *Angew. Chem. Int. Ed.*, **48**, 4688–4716; (b) Buscha, B. and Hertweck, C. (2009) *Phytochemistry*, **70**, 1833–1840; (c) He, J. and Hertwick, C. (2004) *J. Am. Chem. Soc.*, **126**, 3694–3695; (d) Taniguchi, M., Watanabe, M., Nagai, K., Suzumura, K., Suzuki, K., and Tanaka, A. (2000) *J. Antibiot.*, **53**, 844–847; (e) Yamazaki, M., Katoh, F., Ohnishi, J., and Koyama, Y. (1972) *Tetrahedron Lett.*, **26**, 2701–2704; (f) Yamazak, M., Maebayashi, Y., Katoh, H., Ohishi, J., and Koyama, Y. (1975) *Chem. Pharm. Bull.*, **23**, 569–574; (g) Cardillo, R., Fuganti, C., Giangrasso, G.D., Grasselli, P., and Santopietro-Amisano, A. (1974) *Tetrahedron*, **30**, 459–461; (h) Cardillo, R., Fuganti, C., Ghiringhelli, D., and Giangrasso, D. (1972) *Tetrahedron Lett.*, **48**, 4875–4878.
60. Beaudry, C.M. and Trauner, D. (2002) *Org. Lett.*, **4**, 2221–2224.
61. Moses, J.E., Baldwin, J.E., Marquez, R., and Adlington, R.M. (2002) *Org. Lett.*, **4**, 3731–3734.
62. Parker, K.A. and Lim, Y.H. (2004) *J. Am. Chem. Soc.*, **126**, 15968–15969.
63. Beaudry, C.M. and Trauner, D. (2005) *Org. Lett.*, **7**, 4475–4477.
64. (a) Jacobsen, M.F., Moses, J.E., Adlington, R.M., and Baldwin, J.E.

(2006) *Tetrahedron*, **62**, 1675–1689; (b) Jacobsen, M.F., Moses, J.E., Adlington, R.M., and Baldwin, J.E. (2005) *Org. Lett.*, **7**, 2473–2476.

65. Muller, M., Kusebauch, B., Liang, G., Beaudry, C.M., Trauner, D., and Hertweck, C. (2006) *Angew. Chem. Int. Ed.*, **45**, 7835–7838.

66. (a) Wetzel, S., Wilk, W., Chammaa, S., Sperl, B., Roth, A.G., Yektaoglu, A., Renner, S., Berg, T., Arenz, C., Giannis, A., Oprea, T.I., Rauh, D., Kaiser, M., and Waldmann, H. (2010) *Angew. Chem. Int. Ed.*, **49**, 3666–3670; (b) Wilk, W., Waldmann, H., and Kaiser, M. (2009) *Bioorg. Med. Chem.*, **17**, 2304–2309.

67. Cueto, M., D'Croz, L., Mate, J.L., San-Martin, A., and Darias, J. (2005) *Org. Lett.*, **7**, 415–418.

68. Barbarow, J.E., Miller, A.K., and Trauner, D. (2005) *Org. Lett.*, **7**, 2901–2903.

69. Wei, H., Itoh, T., Kinoshita, K., Kotoku, N., Aoki, S., and Kobayashi, M. (2005) *Tetrahedron*, **61**, 8054–8058.

70. Sofiyev, V., Navarro, G., and Trauner, D. (2008) *Org. Lett.*, **10**, 149–152.

71. Beaudry, C.M., Malerich, J.P., and Trauner, D. (2005) *Chem. Rev.*, **105**, 4757–4778.

72. Korman, T.P., Hill, J.A., Vu, T.N., and Tsai, S.C. (2004) *Biochemistry*, **43**, 14529–14538.

73. Heyes, D.J., Ruban, A.V., Wilks, H.M., and Hunter, C.N. (2002) *Proc. Natl. Acad. Sci. U.S.A.*, **99**, 11145–11150.

74. Diego-Castro, M.J. and Hailes, H.C. (1998) *Chem. Commun.*, 1549–1550.

75. Maciver, E.E., Thompson, S., and Smith, M.D. (2009) *Angew. Chem. Int. Ed.*, **48**, 9979–9982.

Part IV
Biomimetic Synthesis of Polyphenols

Biomimetic Organic Synthesis, First Edition. Edited by Erwan Poupon and Bastien Nay.
© 2011 Wiley-VCH Verlag GmbH & Co. KGaA. Published 2011 by Wiley-VCH Verlag GmbH & Co. KGaA.

17
Biomimetic Synthesis and Related Reactions of Ellagitannins

Takashi Tanaka, Isao Kouno, and Gen-ichiro Nonaka

17.1
Introduction

Tannins are defined as a specific group of plant polyphenols that precipitate proteins and heavy metals from aqueous solutions. The scientific term *"tannin"* etymologically originated from leather tanning, in which collagen proteins of animal raw hides are precipitated by complexation with tannins to produce durable leather. Tannins are also responsible for the astringent taste of food and beverages caused by precipitating salivary proteins [1]. This property is related to the biological dysfunction of proteins such as inhibition of enzymes in animal digestive tracts; therefore, tannins are thought to accumulate as feeding deterrents against herbivores [2]. Ironically, it has recently been recognized that the inhibition of the digestive enzymes is linked to a decrease in the incidences of common diseases caused by diets rich in carbohydrates and fats because the uptake of sugars and fats is decreased [3]. In addition, many biological activities, such as antioxidative, anticancer, and immunomodulation activities are reported [4–6].

Tannins are basically classified into two major groups based on their biosynthetic origins: hydrolyzable tannins and condensed tannins [1]. The condensed tannins, a synonym of "proanthocyanidins," are C–C linked oligomers of flavan-3-ols (catechins). The hydrolyzable tannins are esters of gallic acid (3,4,5-trihydroxybenzoic acid) or related phenol carboxylic acids with polyalcohols, usually D-glucose. There are also some exceptional tannins possessing both structural units, which are called *complex tannins* [7]. In contrast to the wide distribution of condensed tannins in Angiospermae, Gymnospermae, and even in ferns [8], hydrolyzable tannins are found in relatively limited Dicotyledoneae plant families [9]. Nevertheless, the structural variation of hydrolyzable tannins is wider than proanthocyanidins and therefore attractive from the viewpoint of organic chemistry, because the oxidative metabolism of simple galloyl esters with glucopyranose produces a wide range of structurally different molecules, as seen in this chapter. The oxidation products are called ellagitannins, which typically possess 3,4,5,3',4',5'-hexahydroxydiphenoyl (HHDP) esters derived by oxidative C-C coupling between two galloyl esters (Scheme 17.1) [10]. The coupling reactions proceed under complete stereochemical control. The

Biomimetic Organic Synthesis, First Edition. Edited by Erwan Poupon and Bastien Nay.
© 2011 Wiley-VCH Verlag GmbH & Co. KGaA. Published 2011 by Wiley-VCH Verlag GmbH & Co. KGaA.

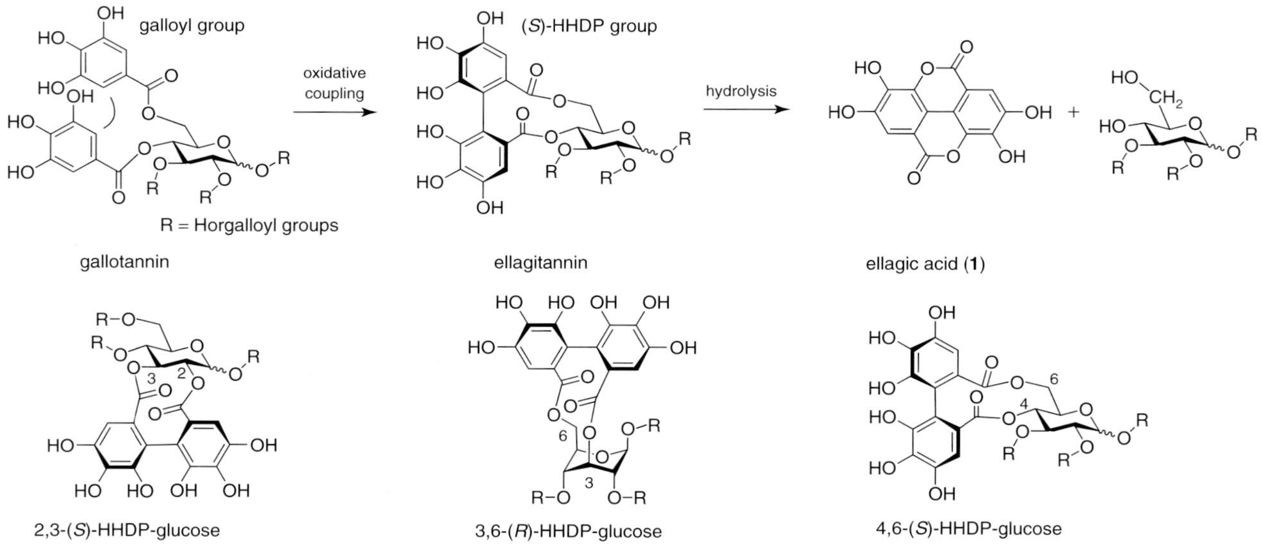

640 | *17 Biomimetic Synthesis and Related Reactions of Ellagitannins*

Scheme 17.1 Biosynthesis and decomposition of ellagitannins.

term *ellagitannins* originated from the observation that the hydrolysis of usual ellagitannins affords ellagic acid (1). Structural diversity of ellagitannins comes from the presence of positional and stereochemical isomers of the HHDP group on a glucopyranose core and further oxidative modifications of the molecules including production of quinone derivatives and oligomeric compounds. This chapter provides a concise overview of the total synthesis of ellagitannins and reactions related to semi-synthesis.

17.2 Biosynthesis of Ellagitannins

It is commonly recognized that different plant families accumulate different types of ellagitannins. This fact apparently implies that the biosynthesis of ellagitannins is controlled by enzymes specific to each plant, and this is the reason why the structural variation of ellagitannins is also important from the viewpoint of chemotaxonomy [11]. Although the details of ellagitannin biosynthesis remain unresolved, quite recent pioneering studies have been achieved by Gross and coworkers [12].

Oxidation of 1,2,3,4,6-pentagalloyl-β-D-glucose (2) with an enzyme preparation obtained from *Tellima grandiflora* (Saxifragaceae) furnished tellimagrandin II (3) [13] (syn. eugeniin [14]) as a major product (Scheme 17.2). This was the first and only successful enzymatic preparation of ellagitannins. The reaction proceeds not only regioselectively but also diastereoselectively. It was demonstrated that the enzyme belongs to the laccase subgroup of phenol oxidases (EC 1.10.3.2). In addition, another enzyme catalyzing regioselective intermolecular C–O coupling of 3 to cornusiin E (4) [15] was characterized (Scheme 17.2). This enzyme also belongs to the same subgroup of polyphenol oxidases. The reactions, however, only represent the biosynthesis of limited members of the vast ellagitannin class. Plants exemplify various intramolecular C–C couplings of galloyl groups on a glucopyranose core; that is, coupling between 1,6-; 1,3-; 2,3-; 2,4-; 3,6-; 3,4-; and 4,6-positions are known [16, 17]. Regiospecificity of intermolecular couplings in oligomerization also differs with plant species [9, 11]. Furthermore, the mechanisms of further oxidative metabolism of the HHDP groups (*vide infra*) remains a matter of speculation.

If the tannins are accumulated simply as defensive substances to disable functions of a herbivore's proteins [2] it may not be necessary to convert the pentagalloylglucose into ellagitannins because galloylglucose exhibits sufficient activity to precipitate proteins. However, the wide distribution of ellagitannins in the plant kingdom is suggestive of particular benefits of the oxidation of the gallotannins. Higher water solubility of ellagitannins compared to pentagalloylglucose may have some advantage in the accumulation of the compounds in the plant cell vacuoles [18], because protein precipitation is only effective when ellagitannins are present in high concentrations [2]. In addition, there are interesting reports from another viewpoint of plant–herbivore interactions. It is known that many herbivorous caterpillar midgut fluids have a high pH (e.g., pH 10),

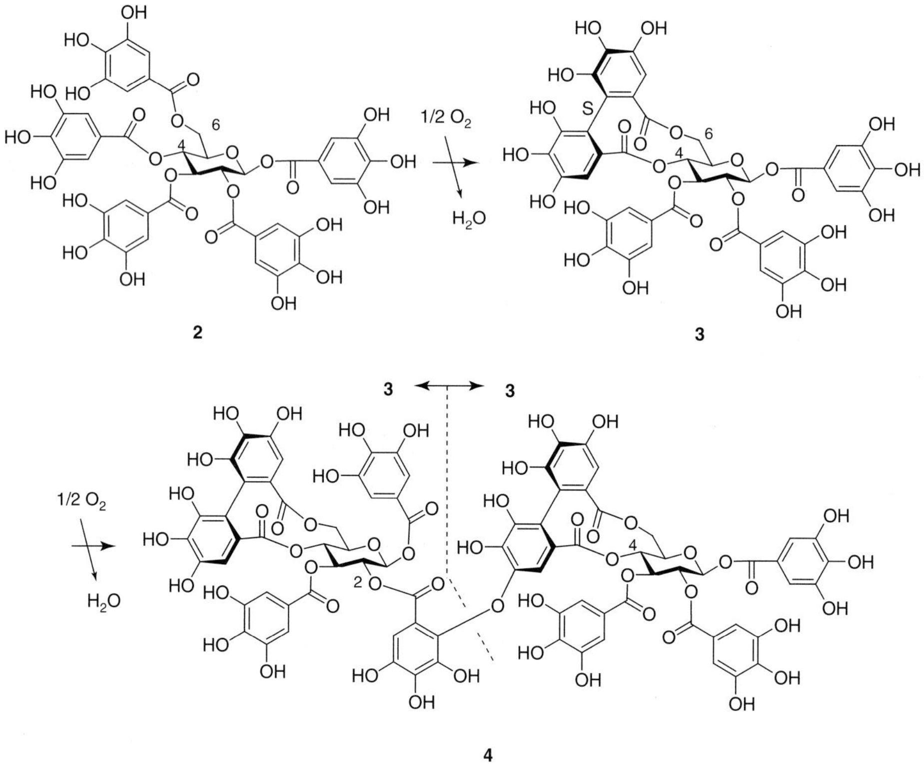

Scheme 17.2 Production of monomeric and dimeric ellagitannins from pentagalloylglucose.

and ellagitannins react to form high concentrations of harmful semiquinone radicals under alkaline conditions compared to proanthocyanidins and simple gallotannins [19].

17.3
Biomimetic Total Synthesis of Ellagitannins

17.3.1
Chemical Synthesis of Ellagitannins by Biaryl Coupling of Galloyl Esters

The first total synthesis of a natural ellagitannin was achieved by Feldman *et al.* in a manner similar to the enzymatic HHDP formation (Scheme 17.3) [20]. A synthetic precursor **5** having two partially protected galloyl groups at glucose 4,6-positions was treated with Pb(OAc)$_4$ to give a cyclized product **6** and subsequent deprotection furnished tellimagrandin I (**7**) [21]. The reaction was diastereoselective and the atropisomerism of the biphenyl bond was exclusively the *(S)*-configuration. In addition, an analogous reaction using substrate **8** having the same acyl groups

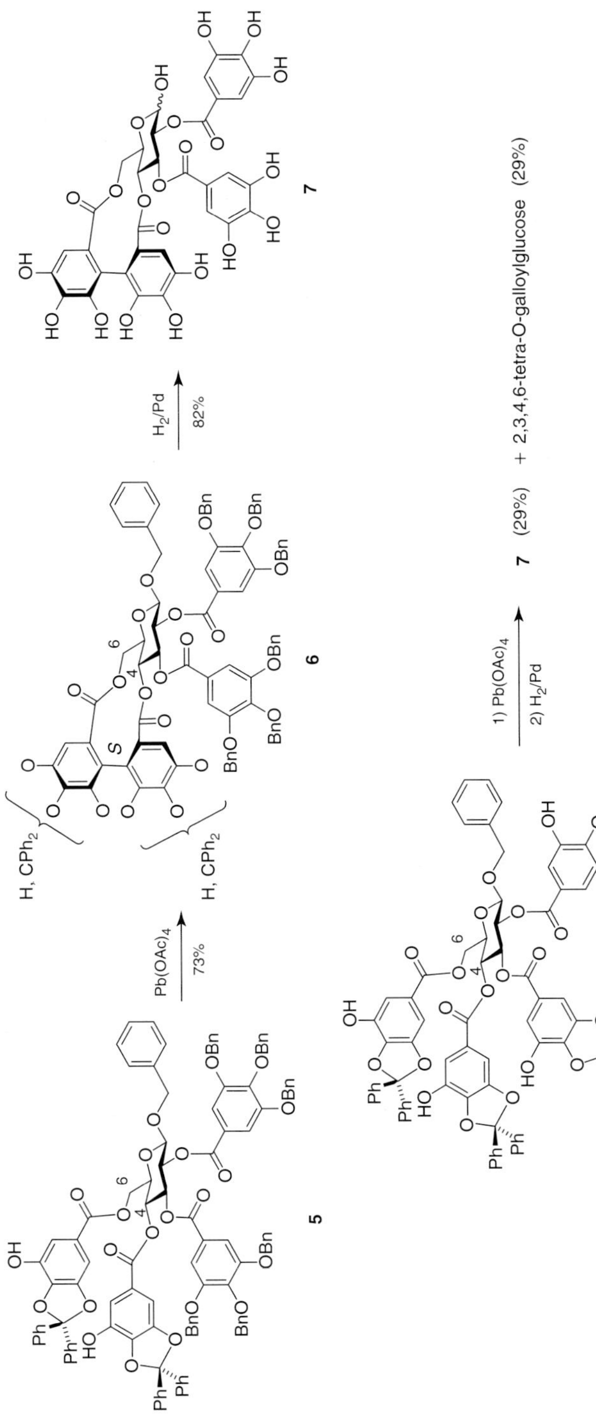

Scheme 17.3 First chemical synthesis of natural ellagitannin, tellimagrandin I (**7**).

at glucose 2,3,4,6-positions furnished **7** along with another natural gallotannin, 2,3,4,6-tetragalloyl glucose [22]. The reaction was not only diastereoselective but also regioselective, that is, coupling between the C2 and C3 galloyl groups was not observed. Tellimagrandin II (**3**), the 1-*O*-β-galloyl form of **7**, was also synthesized using a similar methodology [23]. Molecular mechanics based conformational studies suggested that the precedence of the 4,6-galloyl coupling over the 2,3- or 3,4-galloyl couplings in the reaction was due to the shorter distance between the aromatic rings [20, 24]. As for the atropisomerism of the HHDP group, the exclusive formation of the *(S)*-biphenyl bond in this synthetic study was also supported by computational chemistry calculations. The first synthesis of an ellagitannin with 2,3-HHDP glucose, sanguiin H5 (**11**), was also accomplished by Feldman and coworkers by similar Pb(OAc)$_4$ coupling of partially protected galloyl groups attached to glucose 2,3 positions. The reaction also yielded *(S)*-HHDP groups diastereoselectively [25]. Furthermore, the biaryl coupling strategy was adapted for the synthesis of 2,3-; 4,6-bis-*(S)*-HHDP glucose [26]. The results were consistent with the naturally occurring ellagitannins, that is, the HHDP groups bridged over glucose 4,6- or 2,3-positions are almost exclusively adapted to the *(S)*-configuration, with only a few exceptions [27, 28].

More recently, another synthetic study of **11** using a different atropdiastereoselective biaryl coupling strategy was reported (Scheme 17.4) [29]. Su *et al.* employed an organocuprate oxidative intramolecular biaryl-bond forming reaction to a halogenated benzoyl derivative (**9**). Treatment of **9** with isopropylmagnesium bromide, followed by transmetalation with CuBrSMe$_2$ and subsequent cuprate oxidation afforded **10** [30]. The reaction proceeded with complete diastereoselectivity without any dimeric side products.

Developments in chromatographic and spectroscopic methods during the 1970–1980s expanded the structural and chemotaxonomical knowledge of natural ellagitannins [10]. The stereochemical regularity found in the structures led to a biogenetical postulation that the regulation of the HHDP atropisomerism in natural ellagitannins reflects the conformational preferences of pentagalloylglucose (**2**), a pivotal biosynthetic precursor. In the case of 4,6- and 2,3-HHDP glucopyranoses, the postulation was supported by molecular mechanics based conformation searches [24, 28], and it was further substantiated by the chemical synthesis achieved by the Feldman's group. Dai and Martin also demonstrated experimental results that were in line with the postulation [31]. They examined Ullmann-type biaryl couplings between methylated galloyl esters of α-methyl aldopyranosides (Scheme 17.5).

The 4,6-coupling of **12** furnished a permethylated derivative of tellimagrandin I (**13**) with an *(S)*-biphenyl bond, a result in accordance with that of Feldman and coworkers. In addition, the coupling between 2,3-*cis*-located galloyl groups on D-mannopyranose (**14**) afforded an *(R)*-HHDP group (**15**), whereas the 3,4-*cis*-galloyl groups on D-galactopyranose (**16**) yielded product **17** with an opposite configuration. Although the products **15** and **17** do not represent natural ellagitannins, these diastereoselective biaryl couplings strongly suggest the importance of conformational effects in the substrate. It may also imply that the enzyme catalyzes

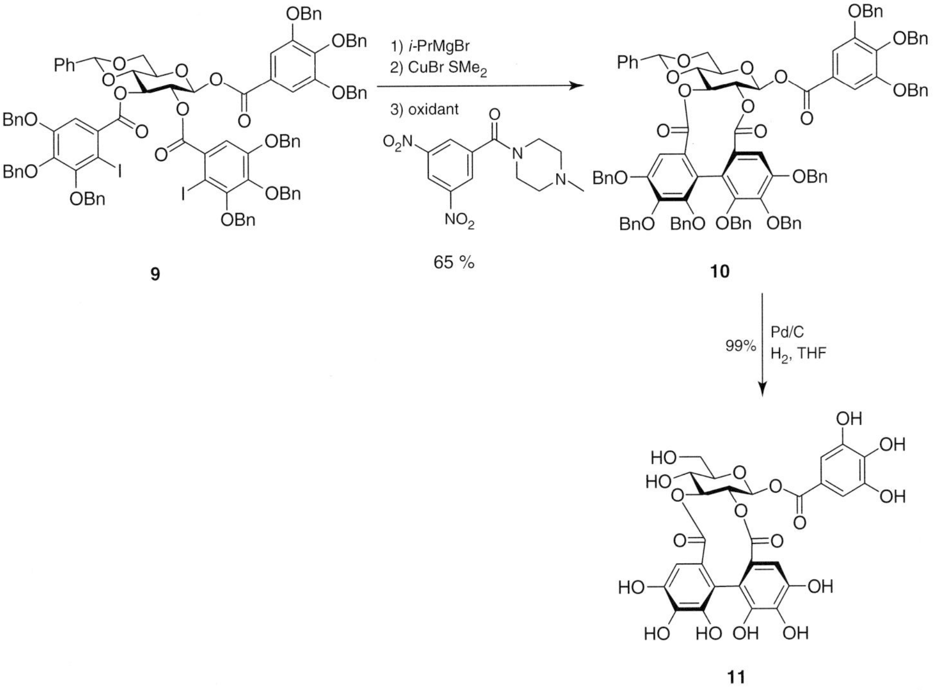

Scheme 17.4 Synthesis of sanguiin H5 (**11**).

only the oxidation of the galloyl groups in the ellagitannin biosynthesis and the stereochemical regulation of the subsequent coupling reaction occurs spontaneously in the most stable conformation. However, this hypothesis is probably not applicable to the ellagitannins with glucopyranose cores with a $^1C_4$ or related boat conformation [10, 17]. In the biosynthesis, it is required that the enzymes force the glucopyranose core to adopt an unstable conformation prior to the biaryl coupling.

17.3.2
Ellagitannins with $^1C_4$ Glucopyranose Cores

Dai and Martin [31] demonstrated lower stereoselectivity of the coupling between the 2,4-digalloyl groups in a derivative of 3,6-anhydro-D-glucose (**18**) adopting $^1C_4$ conformation (Scheme 17.5). The result indicated the greater conformational flexibility of the two galloyl groups in **18** compared to those of proximally located esters in **12, 14,** and **16**. In natural ellagitannins, only one example of 2,4-HHDP-glucose, phyllanemblinin B (**20**), was found in an extract of *Phyllanthus emblica* (Euphorbiaceae) and its atropisomerism was found to be the *(R)*-configuration [32]. The 2,4-HHDP group in ellagitannins is found as an oxidized form commonly called the dehydrohexahydroxydiphenoyl (DHHDP) ester [33, 34], such as the *(S)*-DHHDP group in granatin B (**21**) [35, 36] and the *(R)*-DHHDP group in geraniin (**22**) (Scheme 17.6) [34].

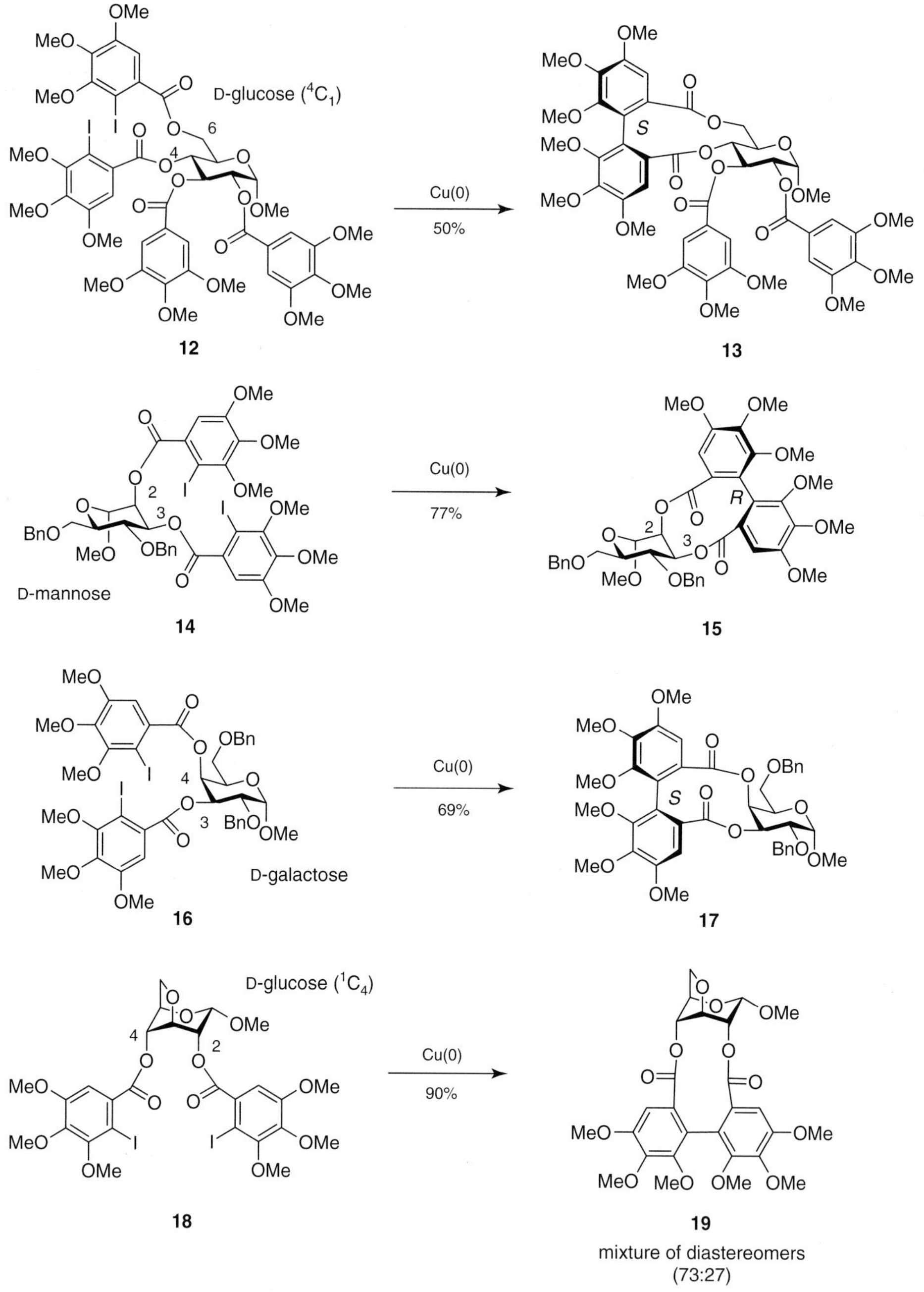

Scheme 17.5 Ullmann-type biaryl couplings between methylated galloyl groups.

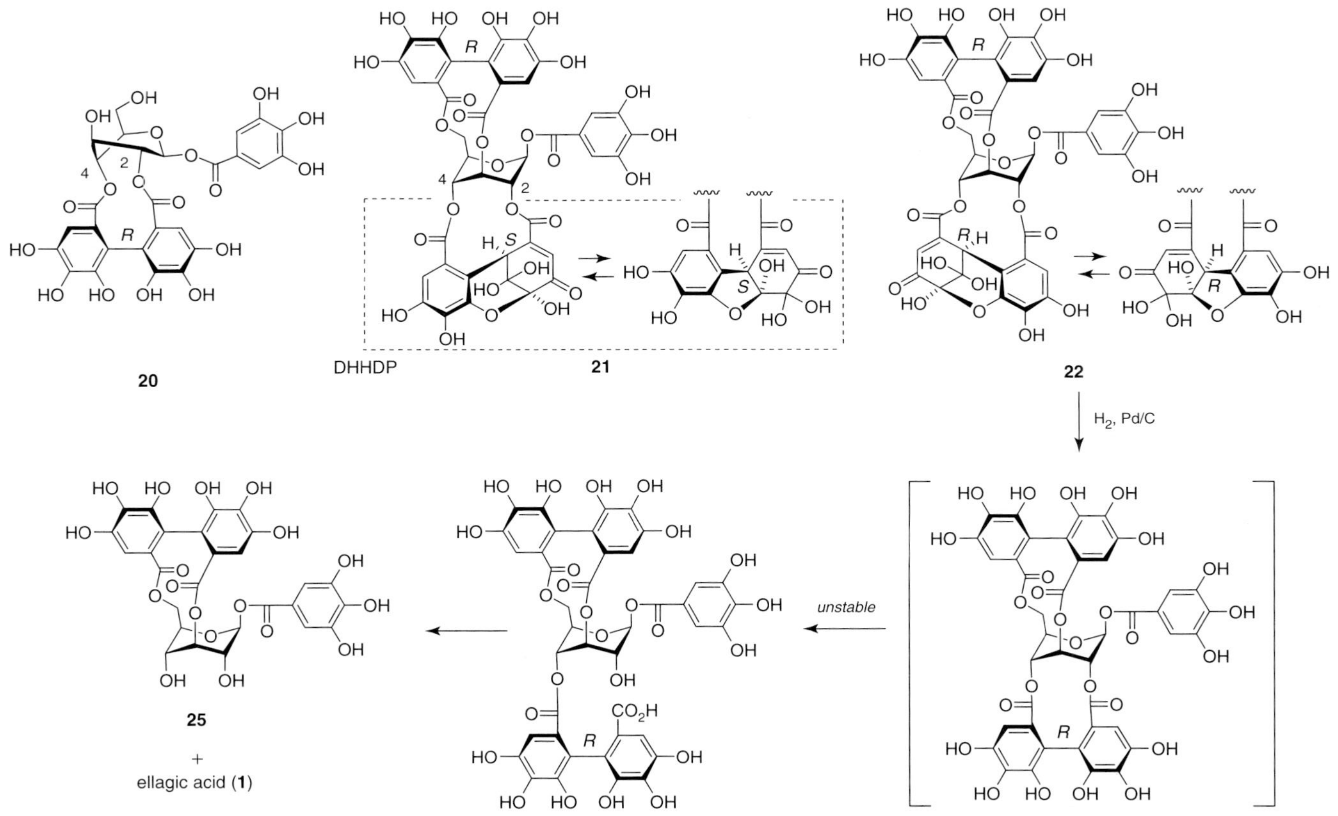

17.3 Biomimetic Total Synthesis of Ellagitannins

Scheme 17.6 Ellagitannins with 2,4-HHDP and 2,4-DHHDP esters.

22, **24**, and **25** were isolated from Geranium sp. **23** is a hypothetical biosynthetic intermediate.

These ellagitannins are called *dehydroellagitannins*. Occurrence of both *(R)*- and *(S)*-DHHDP groups may reflect the above-mentioned low diastereoselectivity of the coupling between 2,4-digalloyl derivatives. The *(R)*-DHHDP group is more common in Nature, and geraniin and related ellagitannins are widely distributed in Euphorbiaceous plants [9, 10]. Catalytic hydrogenation of the *(R)*-DHHDP group of **22** yields the corresponding *(R)*-HHDP ester **(23)** [10b]. Although **23** has not been found in Nature, its partial hydrolysis product, geraniinic acid A **(24)**, was found in *Geranium thunbergii* together with **22** [37]. The 2,4-HHDP ester attached to the $^1C_4$ glucopyranose is unstable and easily hydrolyzed in aqueous solution to give corilagin **(25)** [38] and ellagic acid **(1)**. The stability of phyllanemblinin B **(20)** is apparently contradictory; however, this can be ascribed to the conformational flexibility of the molecule due to the absence of the 3,6-HHDP ester, which allows the glucopyranose to adopt a skew boat conformation and mitigate the strain of the macrocyclic ester ring [32]. Corilagin **(25)** has been shown to adopt a $^1C_4$ conformation in acetone-d_6 and a skew-boat conformation in DMSO-d_6 [39, 40].

A persuasive explanation for the rare occurrence of 2,4-HHDP glucoses in Nature was made by Feldman and coworkers [41]. They applied the aforementioned Pb(OAc)$_4$ oxidative cyclization to a derivative **26** of 2,4-digalloyl 1,6-anhydroglucose and obtained a mixture of regioisomers of 2,4-HHDP glucose **(27)**. However, subsequent deprotection by hydrogenation gave the unexpected 2,4-digalloyl product **(30)** (Scheme 17.7). This unusual result was explained by formation of a cyclohexadienone tautomer **28**. Molecular mechanics calculations

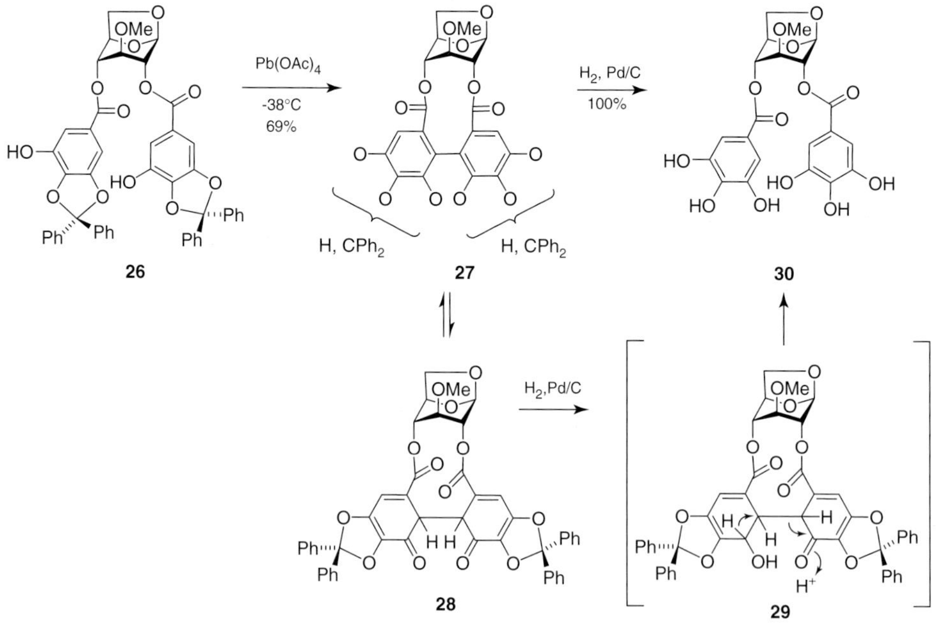

Scheme 17.7 Reductive cleavage of the biphenyl bond of a 2,4-HHDP derivative.

indicated that the non-aromatic tautomer **28** is substantially lower in energy than 2,4-HHDP-1,6-anhydroglucose **27**. Reduction of the tautomer **28** afforded an intermediate **29** and subsequent elimination of the aromatic rings yielded **30**. Although the biosynthetic mechanism of the DHHDP esters is unknown, this reaction mechanism strongly suggests that tautomerization between the aromatic ring and the cyclohexadienone structure explains the predominant disposition of the DHHDP esters at the glucose 2,4-positions in natural ellagitannins.

The bis-cyclohexadienone structure of **28** is apparently related to the acyl group found in carpinin B (**31**) isolated from *Carpinus japonica* (Beturaceae) [42]. Hydrogenolysis of **31** yielded the DHHDP ester (**32**) and finally tellimagrandin I (**7**), indicating that the 4,6-acyl group of **31** is an oxidation metabolite of the HHDP group (Scheme 17.8). Compound **32** is a desgalloyl analog of trapain [43] (1,2,3-tri-*O*-galloyl-4,6-(*S*)-DHHDP-β-D-glucose, syn. isoterchebin [44]). In addition, the coexistence of **31** and **32** with **7** in the same plant source supports their biosynthetic relationship.

The DHHDP hydrated quinone structure of **22** shows close similarity to dehydrotheasinensin A (**34**) produced by the enzymatic oxidation of (−)-epigallocatechin 3-*O*-gallate (**33**) during black tea fermentation (Scheme 17.9) [45]. The quinone dimer **34** is unstable and moderately undergoes oxidation–reduction dismutation

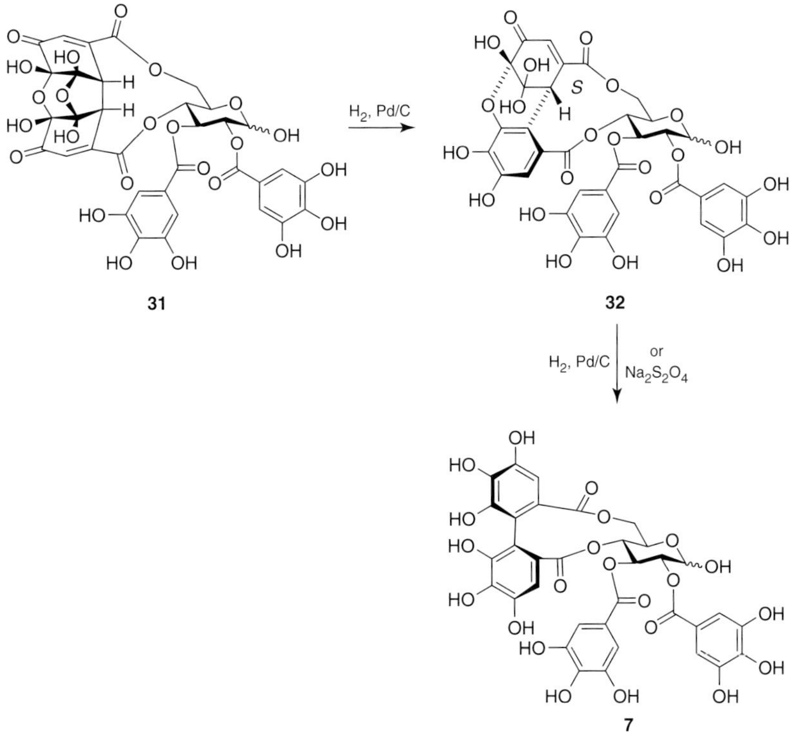

Scheme 17.8 Reduction of carpinin B (**31**) and production of DHHDP and HHDP esters.

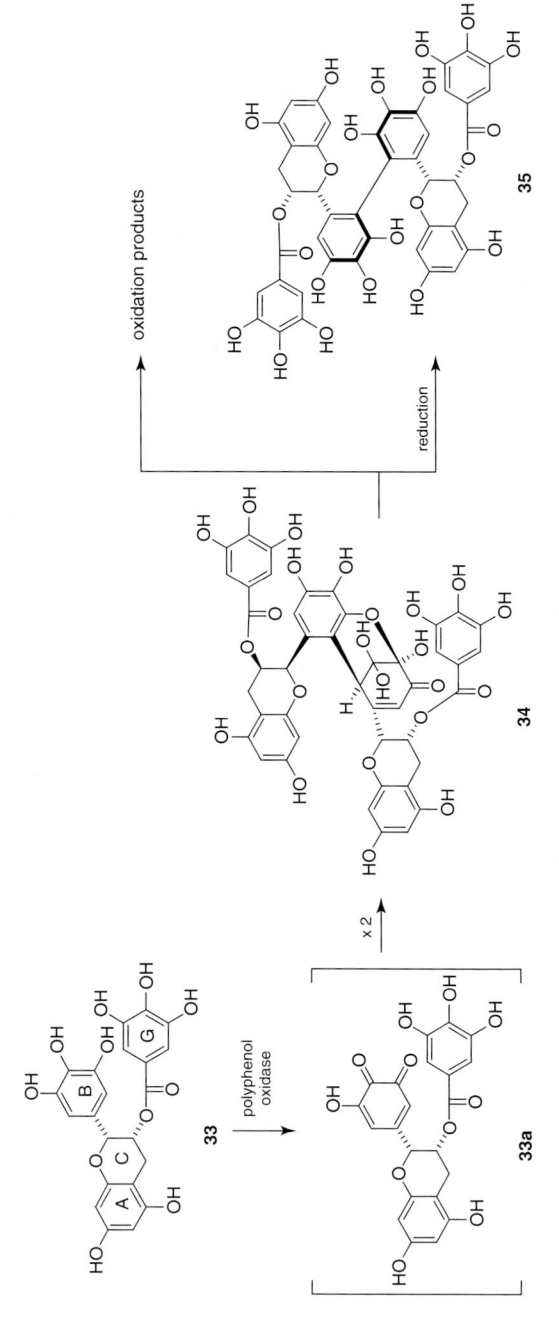

Scheme 17.9 Oxidation of epigallocatechin gallate during tea fermentation.

to furnish a stable biphenyl dimer, theasinensin A (**35**), along with a mixture of oxidation products. Theasinensin A seems to be a final product because enzymatic oxidation of **35** did not afford **34**. Despite their structural similarity, there has been no experimental evidence of similar reductive conversion of the DHHDP ester into the HHDP group (e.g., **32** to **7**) in plant metabolism. Oxidation of the galloyl group of **33** scarcely occurred due to the higher redox potential of the galloyl group than that of the B-ring. Only small amounts of the trimers were produced by the oxidative coupling between the galloyl group of **35** and the B-ring quinone **33a** [46, 47].

Although total synthesis of ellagitannins bearing the DHHDP esters has not been accomplished, it was reported that oxidation of methyl gallate (**36**) with o-chloranil produced a product **38** equivalent to an imaginary DHHDP dimethyl ester (**37**, **37a**) (Scheme 17.10) [48]. The product **38** is formed by the intramolecular addition within the quinone dimer **37**. Treatment of **38** with o-phenylenediamine and $Na_2S_2O_4$ affords phenazine (**39**) and HHDP derivatives (**40**), respectively, which are also expected to be derived from **37a** by similar treatment.

17.3.3
Synthesis of an Allagitannin with 3,6-(R)-HHDP Group

Corilagin (**25**) usually coexists with the dehydroellagitannins, such as **21** and **22**, and may be a partial hydrolysate of the biosynthetic precursor of the dehydroellagitannins (Scheme 17.6). Total synthesis of the 3,6-(R)-HHDP glucose structure was achieved by a biaryl coupling methodology; however, the strategy is different from those of preceding studies.

Yamada et al. [49] first cleaved the pyranose ring of a protected glucopyranose **41** by Wittig olefination at the anomeric position to give **42**. Partially protected galloyl groups were then introduced at the 3,6-positions (**43**) and $CuCl_2$-amine mediated oxidative coupling between the acyl groups furnished the desired 3,6-bridged macrocyclic bi-ester **44** with an (R)-biphenyl bond. Subsequent regeneration of the aldehyde at the anomeric position by oxidative cleavage of the olefinic bond formed the glucopyranose ring (**46**). Finally, acylation at the anomeric position and deprotection yielded **25** (Scheme 17.11). This is the first success in constructing a 3,6-bridged (R)-HHDP glucose by the biphenyl coupling of non-proximal galloyl groups.

17.3.4
Synthesis of Ellagitannins by Double Esterification of Hexahydroxydiphenic Acid

The aforementioned biaryl coupling methodology is "biomimetic" in the sense that the key reaction is apparently the same as the plant biosynthesis. In addition, there is another successful strategy, in which firstly a protected hexahydroxydiphenic acid is prepared; secondly, the dicarboxylic acid is docked onto a diol derivative of the glucopyranose; lastly, the protecting groups are removed. Nelson and Meyers synthesized an enantiomerically pure HHDP derivative (**50**) from bromide **48** by

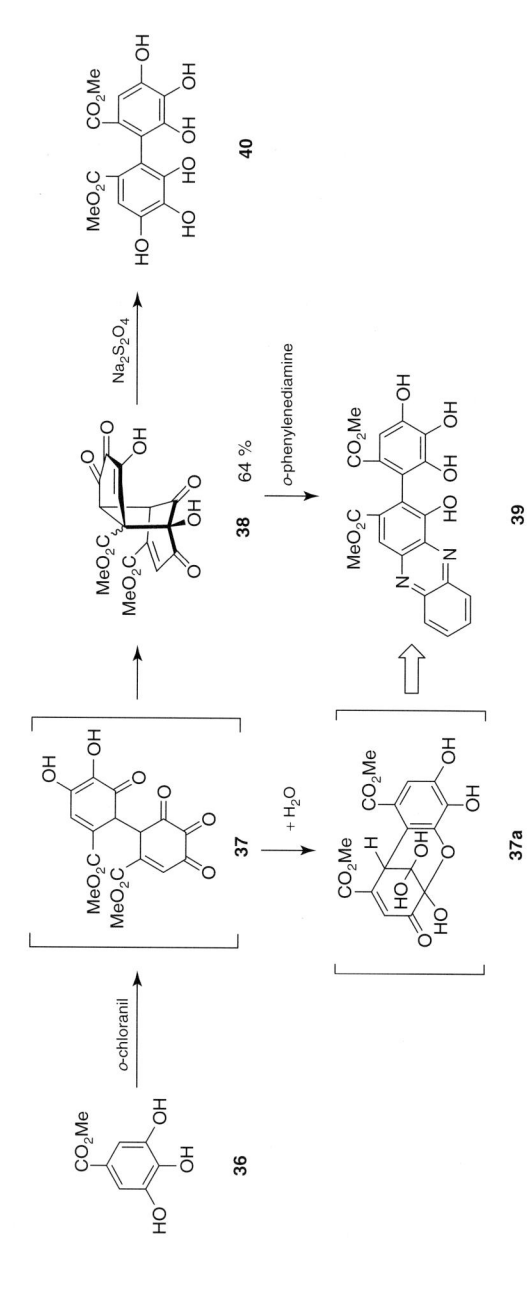

Scheme 17.10 Synthesis of dehydrohexahydroxydiphenic acid esters from methyl gallate.

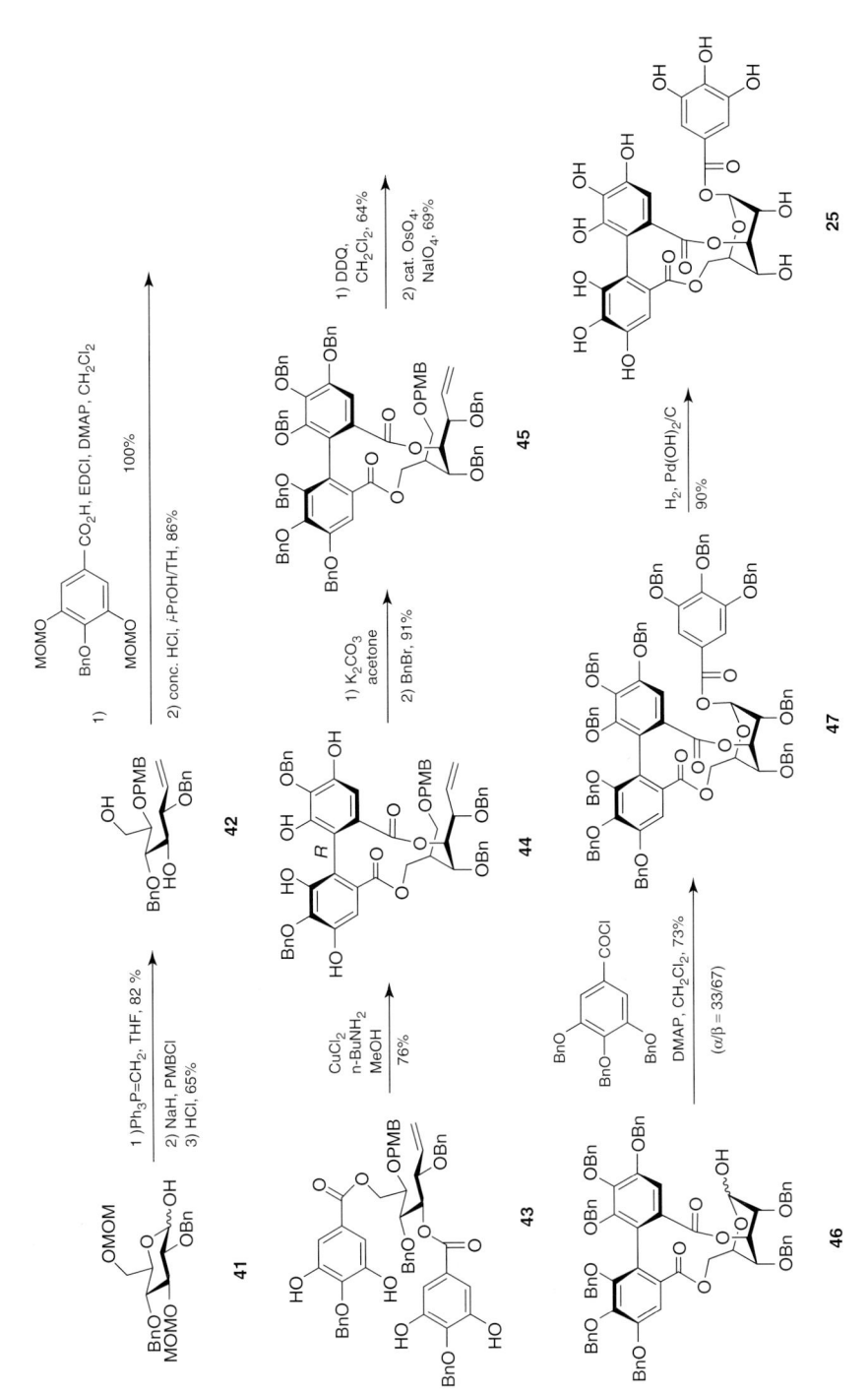

Scheme 17.11 Total synthesis of corilagin (**25**).

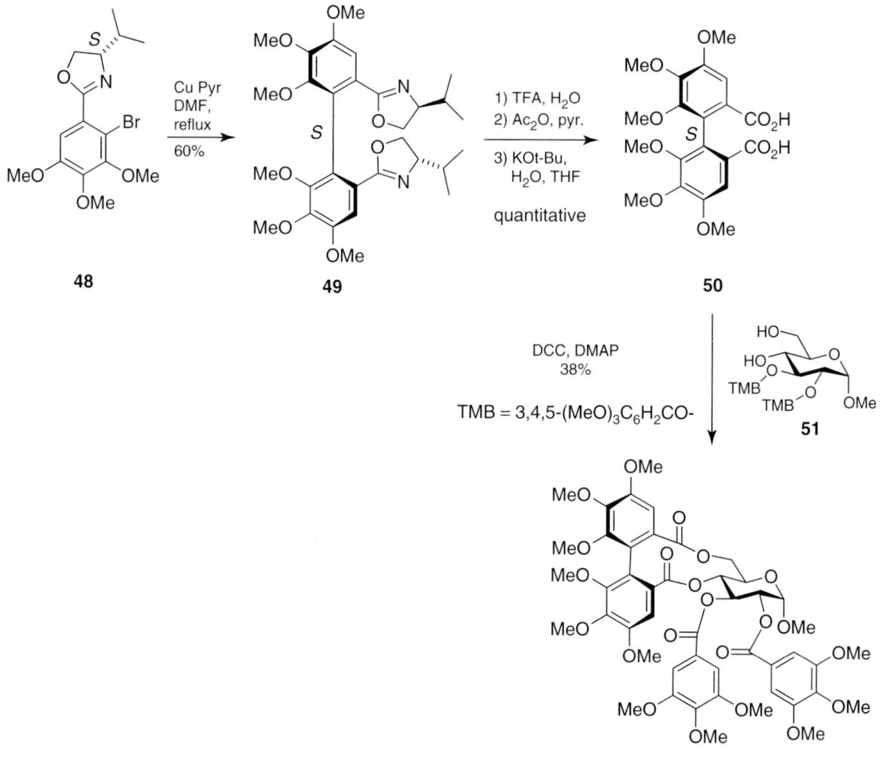

Scheme 17.12 Synthesis of a permethylated derivative of tellimagrandin I (**13**).

oxazoline-mediated asymmetric Ullmann coupling (Scheme 17.12) [50]. Further coupling of **50** with diol **51** furnished a permethylated derivative of tellimagrandin I (**13**).

Lipshutz et al. synthesized enantiopure **50** in a different manner (Scheme 17.13) [51]. The aryl bromide precursor **54** was prepared by coupling of bromide **52** with (S,S)-stilbene diol **53**. Dilithiation of **54** and subsequent treatment with CuCN forms diarylcuprate **55**. Oxidation of **55** with molecular oxygen yielded a single diastereomer (**56**). Finally, hydrogenation and oxidation produced (S)-**50**. The acid **50** was used for the synthesis of a permethyl derivative of tellimagrandin II (**3**).

The synthesis using the enantiomerically pure HHDP acid was followed by a double esterification strategy using "racemic" HHDP dicarboxylic acid. Itoh et al. demonstrated diastereoselective esterification of racemic biaryl acid chloride **57** with a 2,3-diol of the glucose derivative (**58**) [52]. The diastereoselectivity was dependent on the base and solvent. A NaH–toluene combination predominantly afforded 2,3-(R)-diastereomer **60**, which is thermodynamically less stable (Scheme 17.14). It was shown that the esterification occurred first at the C2 position (**59**), and the product **60** was produced by kinetically-controlled intramolecular ester cyclization.

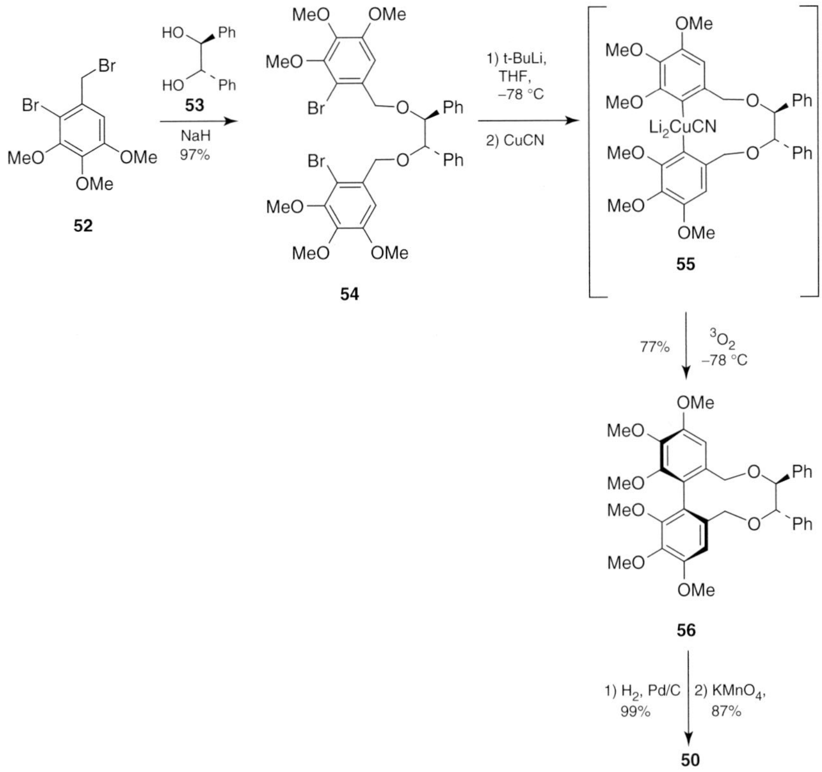

Scheme 17.13 Preparation of (S)-hexamethoxydiphenic acid by Lipshutz et al.

In contrast, when Et$_3$N was used as the base in THF, the (S)-diastereomer was mainly produced (S : R = 4.4 : 1). The results were applied to the synthesis of a permethyl derivative (**61**) of pedunculagin [2,3-; 4,6-bis-(S)-HHDP glucose] [53].

The total synthesis of the corresponding unprotected ellagitannin is based on the double esterification strategy established by Khanbabaee et al. [54]. The synthesis used racemic HHDP dicarboxylic acid and succeeded in complete diastereoselective esterification with a glucose 4,6-diol derivative; that is, esterification of the hexabenzyl HHDP acid (**62**) to the 4,6-diols of a partially protected glucose (**63**) yielded **64** with (S)-chirality. Subsequent deprotection and acylation at the anomeric position followed by hydrogenolysis of the benzyl protective groups furnished the natural ellagitannin strictinin (**65**) (Scheme 17.15).

Esterification of the same racemic dicarboxylic acid (**62**) to a 2,3-diol derivative of glucose (**66**) afforded an almost equal amount of diastereomers (S)-**67** and (R)-**67**, which were separated using silica-gel chromatography (Scheme 17.15). Total synthesis of two naturally occurring ellagitannins, praecoxin B [2,3-(S)-HHDP-4,6-digalloyl-glucose] and pterocarinin C [1,4,6-trigalloyl-2,3-(S)-HHDP-β-glucose], was accomplished from (S)-**67** [55]. In addition, unnatural ellagitannin analogs, that is, 4,6-digalloyl- and 1(β),4,6-trigalloyl-2,3-(R)-HHDP-β-glucoses, were synthesized

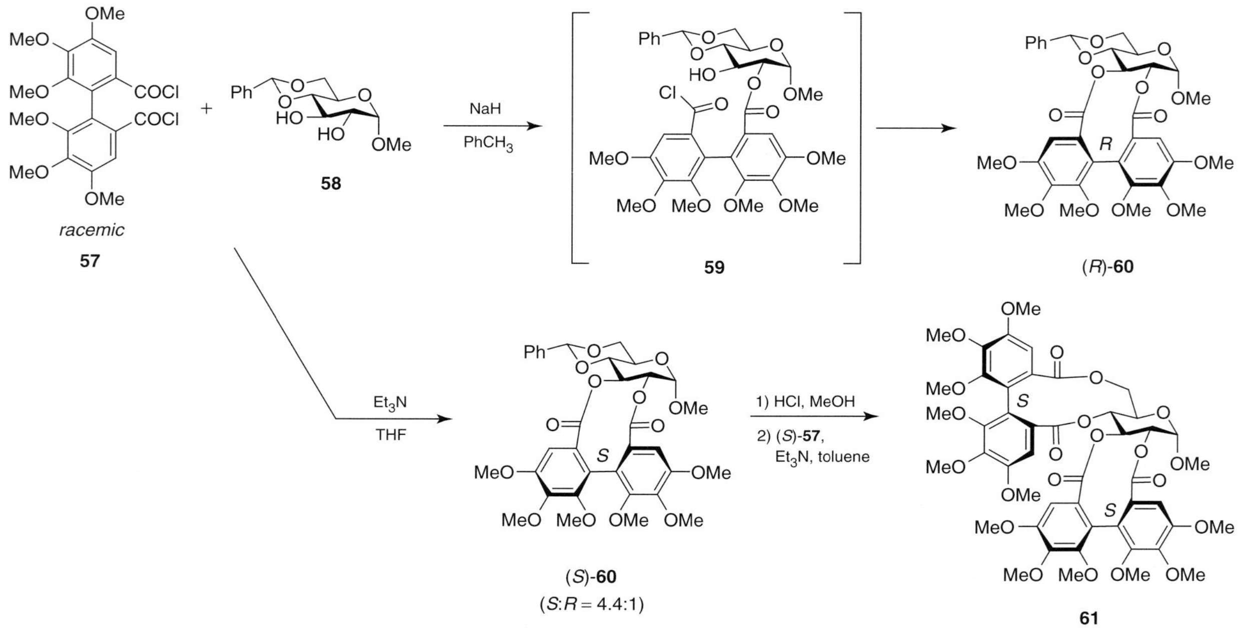

Scheme 17.14 Kinetic resolution of HHDP derivatives.

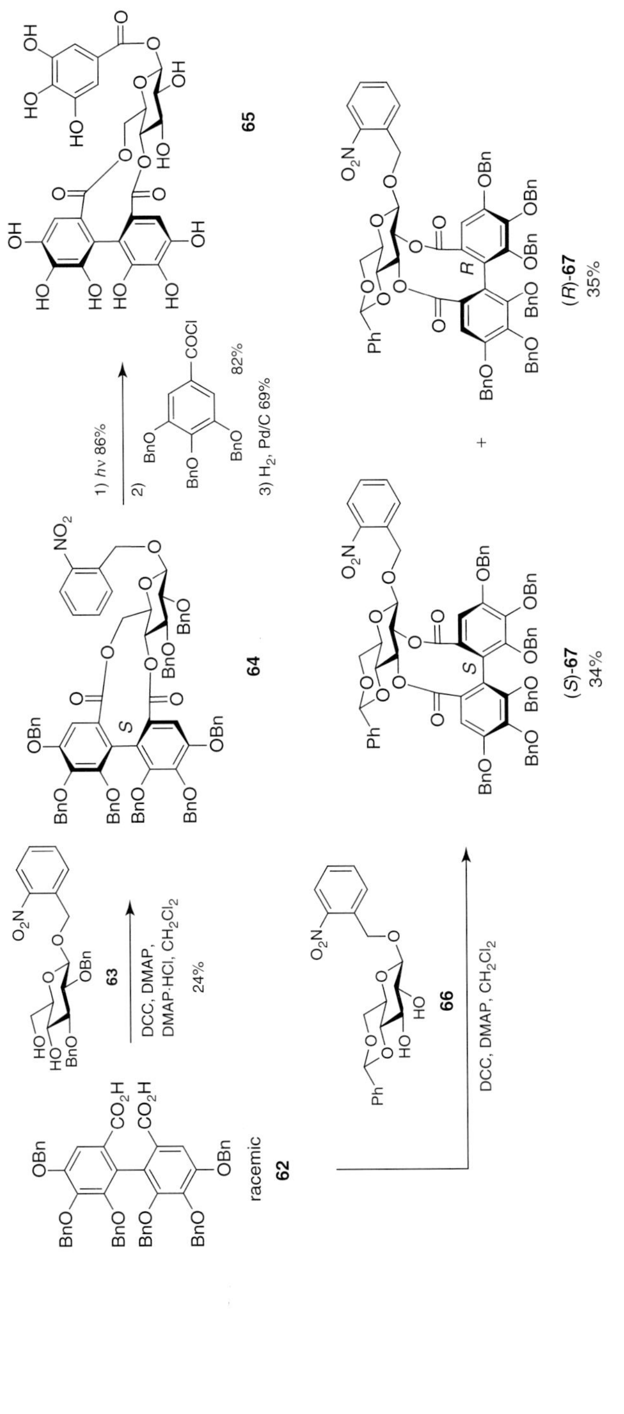

Scheme 17.15 Synthesis of strictinin (**65**) and diastereomers of 2,3-HHDP-glucoses.

from diastereomer (*R*)-**67** [56]. Furthermore, Khanbabaee *et al.* applied the methodology to the synthesis of gemin D [3-*O*-galloyl-4,6-*(S)*-HHDP-D-glucose], hippomenin A [2-*O*-galloyl-4,6-*(S)*-HHDP-D-glucose] [57], and a synthetic precursor of 5-*O*-galloyl-4,6-*(S)*-HHDP-D-gluconic acid [58].

A simple total synthesis was demonstrated by Khanbabaee *et al.* (Scheme 17.16). The 2,3,4,6-tetrahydroxy derivative of glucose (**68**) was directly esterified with enantiomerically pure *(S)*-**62** to give 2,3;4,6-bisdiphenoyl esters in good yields (60%). Undesired products (35%) were shown to be produced by an intermolecular esterification. Subsequent removal of the protecting groups furnished the natural ellagitannin pedunculagin (**69**) [59].

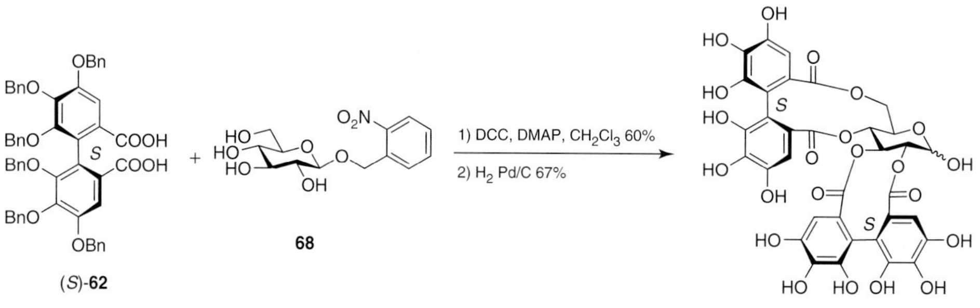

Scheme 17.16 Synthesis of pedunculagin (**69**).

17.3.5
Biomimetic Synthesis of Dimeric Ellagitannin

There is a distinctive class of ellagitannins called *oligomeric ellagitannins*, which are produced by intermolecular oxidative couplings of simple HHDP esters of glucopyranoses [9, 11]. Besides a few C–C linked dimers [60], the oligomers are produced by intermolecular C–O couplings between two pyrogallol rings of galloyl or HHDP esters, as exemplified by cornusiin E (**4**) (Scheme 17.2) [15]. As the sole success of ellagitannin total synthesis of this class, Feldman *et al.* [23, 61] reported the synthesis of coriariin A (**70**), which is biosynthesized by oxidative coupling of two tellimagrandin II molecules (**3**) (Figure 17.1) [62]. From a different viewpoint, the molecule can be deemed to be an ester composed of dehydrodigallic acid and two tellimagrandin I units. The synthetic study started from the preparation of the dehydrodigallic acid unit.

The authors applied a hetero-Diels–Alder cycloaddition/reductive rearrangement sequence to the synthesis of the dehydrodigalloyl unit (Scheme 17.17) [63]. The *o*-quinone **71** was treated with a Lewis acid to give Diels–Alder adducts (**72a** and **72b**) and subsequent elimination with NaOAc/HOAc to give a mixture of *o*-quinones (**73a** and **73b**). Reduction of the quinones with hydrosulfite yielded a mixture of *meta*- and *para*-isomers **74a** and **74b** of dehydrodigallic acid. Upon the

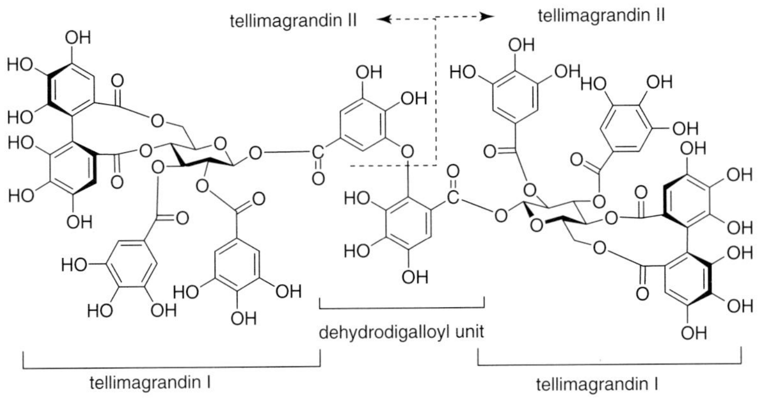

Figure 17.1 Structure of coriariin A.

usual benzylation, the *para*-isomer **74b** underwent a Smiles rearrangement and the *meta*-derivative **75** was obtained as the sole product.

Unfortunately, direct application of this methodology to the synthesis of **70** by dimerization of the protected tellimagrandin II units failed due to the instability of the glucose-linked *o*-quinone derivative corresponding to **71**. Therefore, an alternative three-component coupling strategy was adopted to synthesize **70** (Scheme 17.18) [61a,b]. Firstly, two equivalents of protected glucose **77** were connected to dicarboxylic acid **76**. Secondly, benzyl-protected galloyl residues were introduced at the C2 and C3 positions of the glucose residue, and lastly the (*S*)-HHDP esters at the 4,6-positions were constructed according to the Feldman's diastereomeric coupling method (Scheme 17.18).

17.4
Conversion of Dehydroellagitannins into Related Ellagitannins

17.4.1
Reduction of DHHDP Esters

As mentioned before, dehydroellagitannins are characterized by the DHHDP esters derived from HHDP groups (Scheme 17.6). These tannins are widely distributed in the plant families of Euphorbiaceae, Geraniaceae, Elaeocarpaceae, Aceraceae, Combretaceae, Punicaceae, Trapaceae, and Leguminosae [9], and particular tannins are important as constituents of oriental medicinal plants. The most typical and widely distributed dehydroellagitannin is geraniin (**22**); this tannin is sometimes accompanied by structurally related tannins probably produced by further metabolism of the DHHDP group. Some ellagitannins are chemically derived from **22**.

Heating of **22** with pyridine in CH_3CN [36] yielded mallotusinin (**81**) [64], 1-*O*-galloyl-2,4; 3,6-bis-(*R*)-HHDP-β-D-glucose (**23**), and acalyphidin M_1 (**82**) [65]

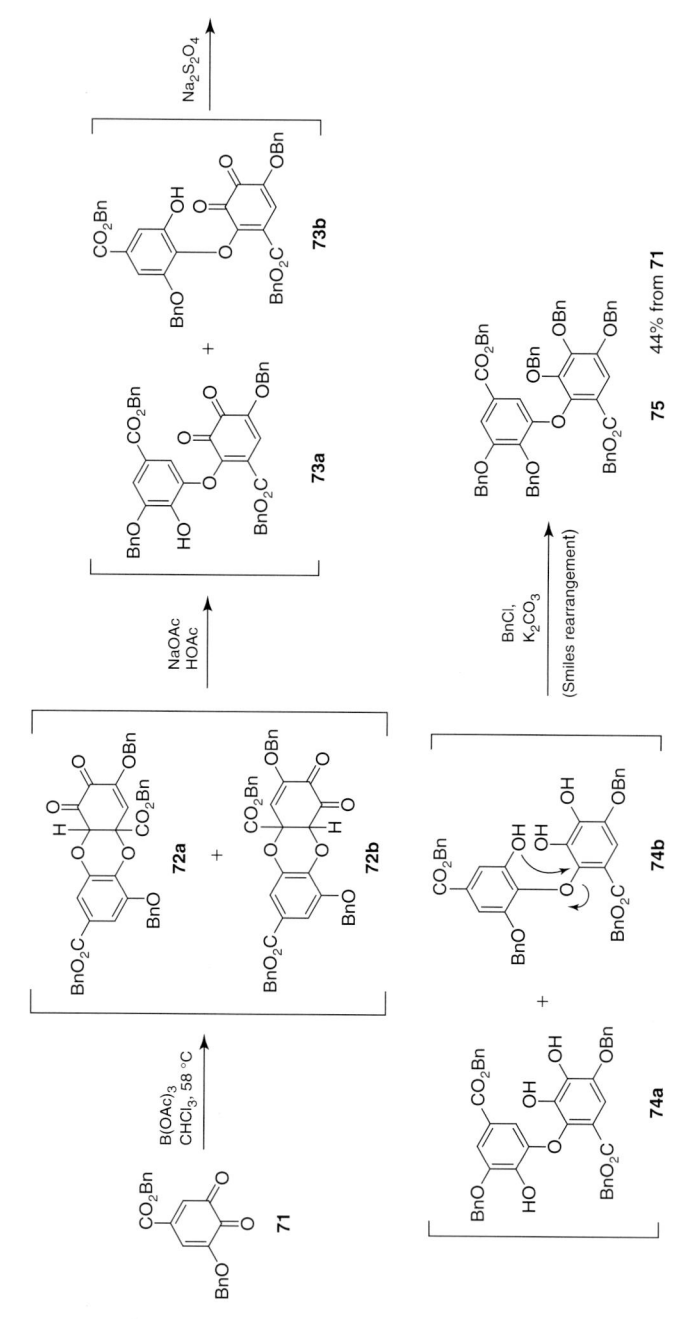

Scheme 17.17 Synthesis of benzyl-protected dehydrodigallic acid.

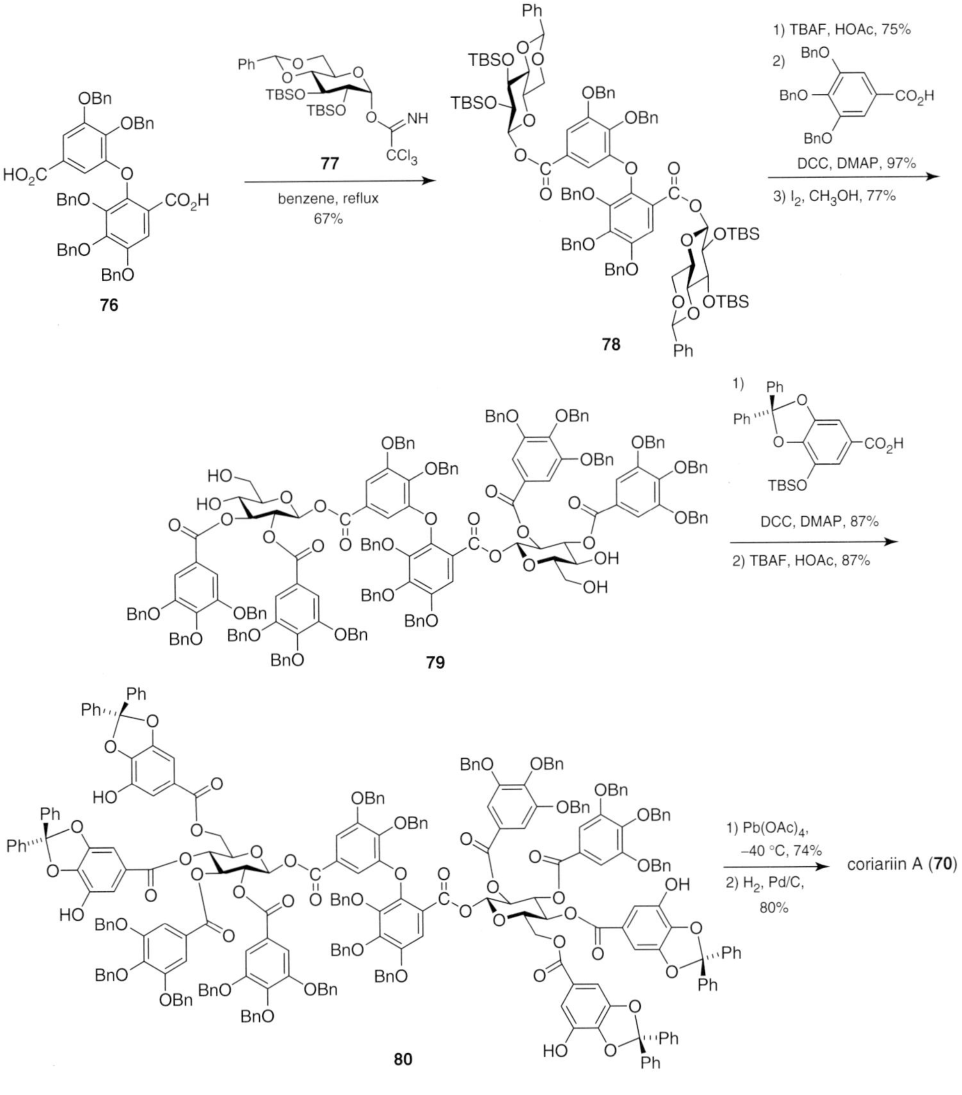

Scheme 17.18 Total synthesis of dimeric ellagitannin coriariin A.

(Scheme 17.19), of which **81** and **82** were isolated from *Euphorbia* species. In this experiment, the production of **81** and **23** indicated the occurrence of reduction/oxidation dismutation; however, the oxidation products could not be identified.

This reaction resembles decomposition of the quinone dimer **34** during production of black tea polyphenols (Scheme 17.9). A similar reductive reaction of quinone derivatives was also reported by Feldman *et al.* [63]. Treatment of a mixture of Diels–Alder adducts of gallic acid quinone **83** with DBU (1,8-diazabicyclo [5.4.0]undec-7-ene) afforded dehydrodigallic acid derivative **84** without the

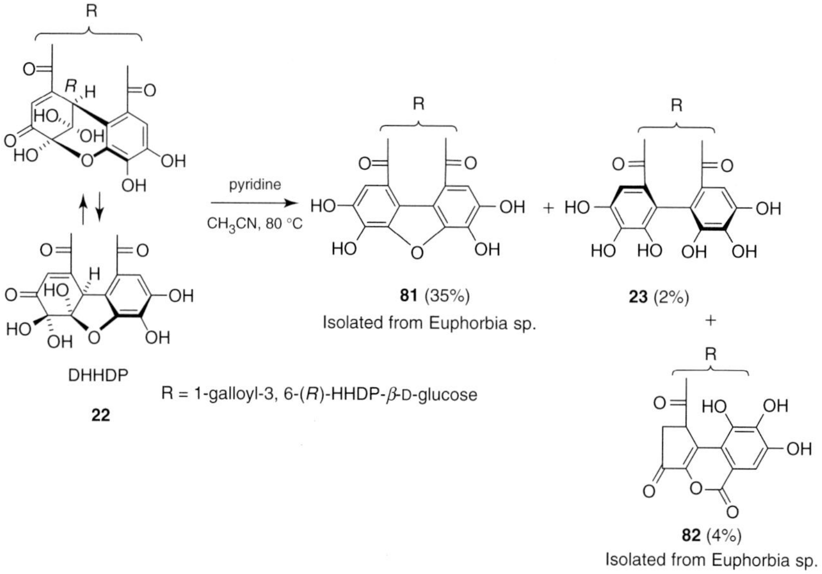

Scheme 17.19 Production of mallotusinin (**81**) and acalyphidin M1 (**82**) from geraniin (**22**).

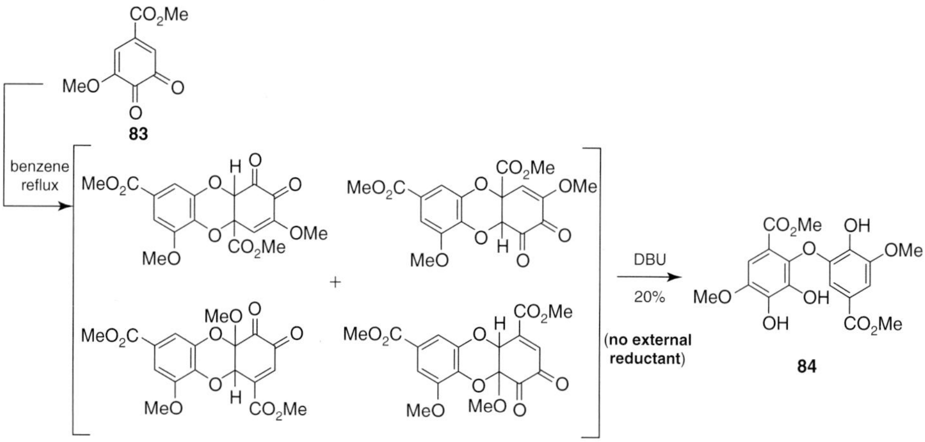

Scheme 17.20 Synthesis of the dehydrodigallic acid derivative **84**.

requirement of an external reducing agent (Scheme 17.20). These results indicate the importance of redox dismutation in the reactions of non-protected derivatives.

17.4.2
Reaction with Thiol Compounds and the Biomimetic Synthesis of Chebulagic Acid

Thiol compounds readily react with DHHDP esters in aqueous solution at room temperature. Reaction of geraniin (**22**) with L-cysteine methyl ester afforded adduct **85**, which retains the cyclohexene ring [66]. In contrast, cleavage of the cyclohexenone ring structure occurred on reaction with glutathione, a physiologically important tripeptide (Scheme 17.21) [67]. The products were determined to be sulfide compounds **86** and **87**, and an α-ketocarboxylic acid **88**. Reductive desulfurization of products **86** and **87** yielded natural ellagitannins chebulagic acid (**89**) and neochebulagic acid (**90**), respectively [68]. The reactions with thiol compounds may have some physiological significance because the thiol groups of proteins and peptides play important roles in biological systems.

The structure of **88** resembles the natural ellagitannin euphormisin M$_2$ (**91**) [69]. The production mechanism of **88** was revealed by the reaction of **22** with N-acetyl-L-cysteine, which afforded an isolable intermediate **92**. This intermediate was decomposed to give **90** through the carboxylic acid **93** (Scheme 17.22) [67]. The reaction may be related to the biosynthesis of **91**.

17.4.3
Other Reactions of DHHDP Esters

Other semi-synthetic reactions of the DHHDP esters have been reported. Brief treatment of **22** with aqueous sodium hydroxide (pH 9) yielded repandusinic acid A (**94**) [70] along with acalyphidin M$_1$ (**82**) (Scheme 17.23) [36]. More interestingly, simple 1,4-addition of ascorbic acid at the α,β-unsaturated carbonyl group occurs in aqueous solution to give elaeocarpusin (**95**) [71], which is often found in plants together with **22**. The reaction is reversible [38] and, probably, **22** and **95** exist as an equilibrium mixture in plant cells. Phylanthusiin D (**96**) [72], a simple 1,2-adduct of acetone, was shown to be prepared in aqueous acetone solution in the presence of ammonium formate or ammonium acetate [73]. This reaction serves as a method to prepare stable derivatives of dehydroellagitannins. This is because the equilibrium between six- and five-membered ring hemiacetal structures hampers the interpretation of spectroscopic data.

17.5
Reactions of C-Glycosidic Ellagitannins

C-Glycosidic ellagitannins with an open-chain glucose core are distributed in plant families of Fagaceae, Myrtaceae, Combretaceae, Casuarinaceae, and Rosaceae. Although biosynthesis remains unresolved, the close biogenetical relationship with of 2,3;4,6-bis-(S)-HHDP glucoses is apparent from the structural similarities and their co-occurrence in plants. The most important C-glycosidic ellagitannins are probably vescalagin (**99**) and castalagin (**100**) found in oak wood [74]. Because the barrels used for the aging of wine, brandy, and whisky arc made from oak wood the tannins and their degradation products contribute to the taste and flavor of the liquors [75]. A plausible biogenesis of these tannins through aldehydes **97** and **98** is shown in Scheme 17.24 [76, 77]. Precursor **98** was isolated from *Liquidambar formosana* [76] and galls of *Carpinus tschonoskii* [78].

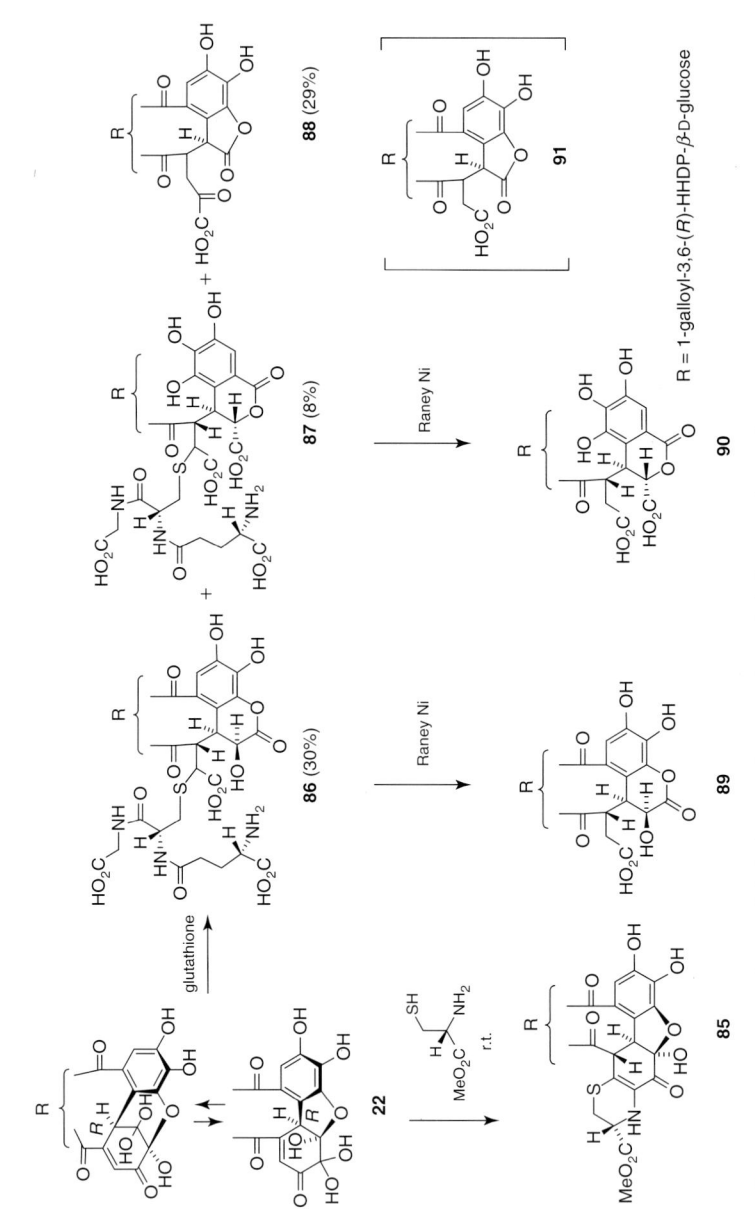

Scheme 17.21 Semi-synthesis of chebulagic acid (**89**) and neochebulagic acid (**90**) from geraniin (**22**).

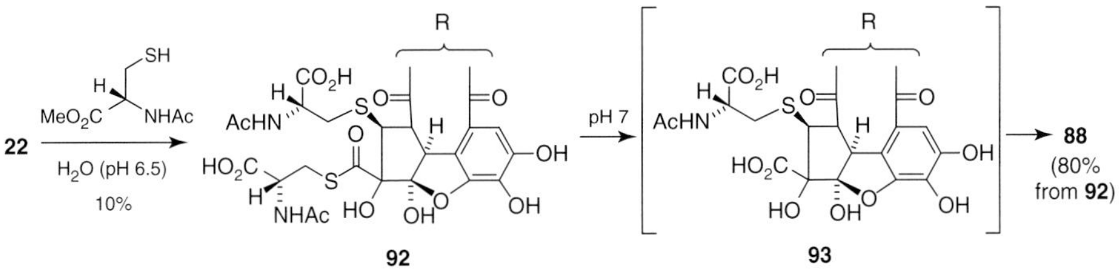

17.5 Reactions of C-Glycosidic Ellagitannins

Scheme 17.22 Production mechanism of the unnatural ellagitannin **88** from **22**.

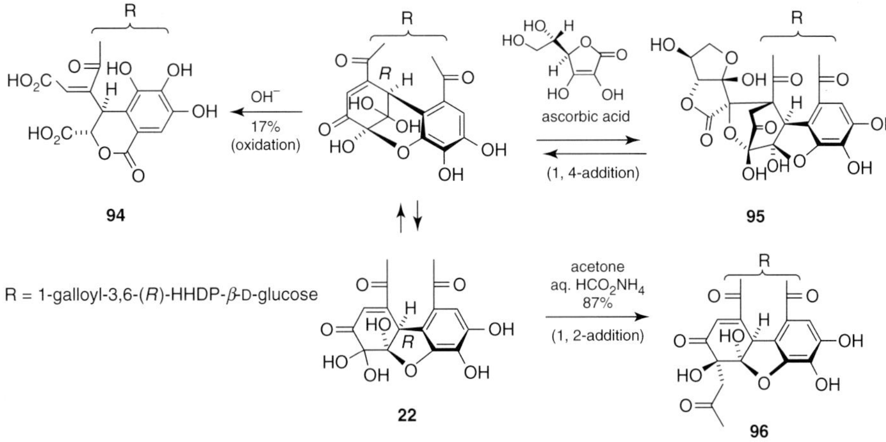

Scheme 17.23 Production of repandusinic acid A (**94**), elaeocarpusin (**95**), and phyllanthusiin D (**96**) from geraniin (**22**).

R = 1-galloyl-3,6-(R)-HHDP-β-D-glucose

17.5.1
Conversion between Pyranose-Type Ellagitannins and C-Glycosidic Ellagitannins

Conversion of pyranose-type ellagitannin pedunculagin (**69**) into the corresponding C-glycosidic ellagitannins, 5-desgalloyl stachyurin (**101**) and casuariin (**102**), occurs in aqueous solutions at pH 7.5 [26]. The reverse reaction, C-glycosidic ellagitannin to pyranose-type ellagitannin, was observed upon acid hydrolysis of a punicacortein C (**103**) [79] to produce punicalagin (**104**) [80] (Scheme 17.25).

17.5.2
Reaction at the C1 Positions of C-Glycosidic Ellagitannins

In the center of the heartwood of *Castanea crenata*, degradation products of C-glycosidic ellagitannins, castacrenins A (**106**), B (**107**), and C (**108**), were detected. These products are produced from castalin (**105**) under acidic conditions by intramolecular cyclization or dehydration at the glucose C1 position (Scheme 17.26) [81].

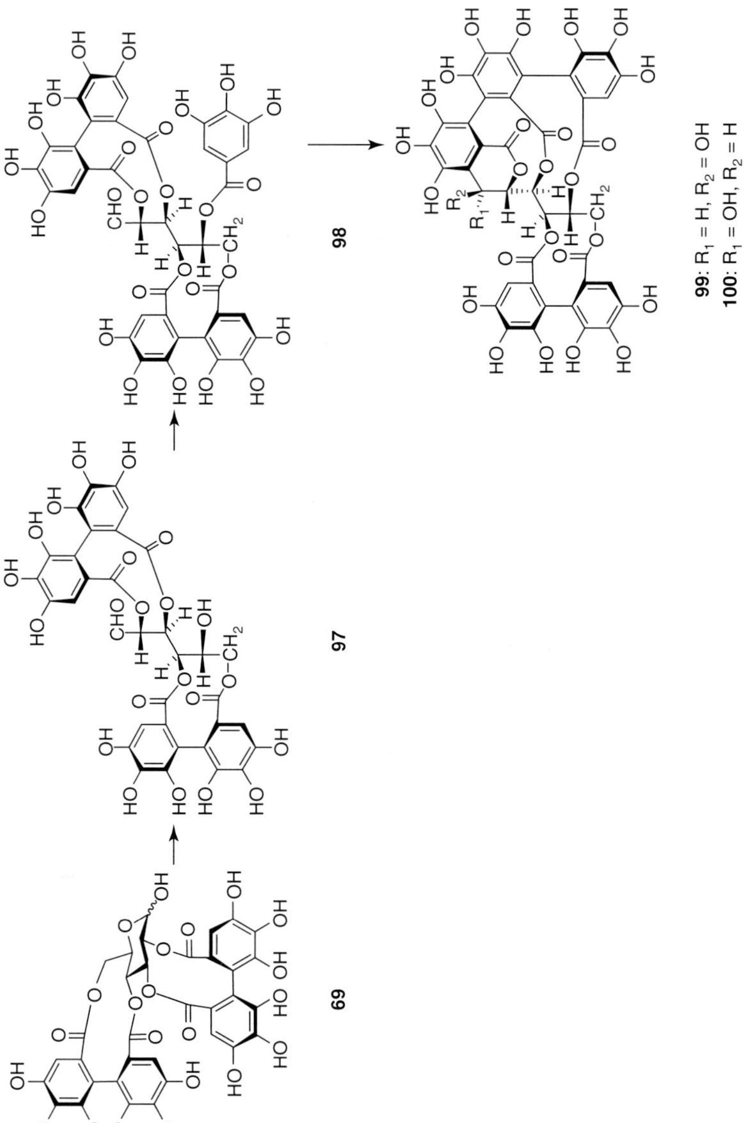

666 | *17 Biomimetic Synthesis and Related Reactions of Ellagitannins*

99: $R_1 = H, R_2 = OH$
100: $R_1 = OH, R_2 = H$

Scheme 17.24 Possible biogenesis of vescalagin (**99**) and castalagin (**100**).

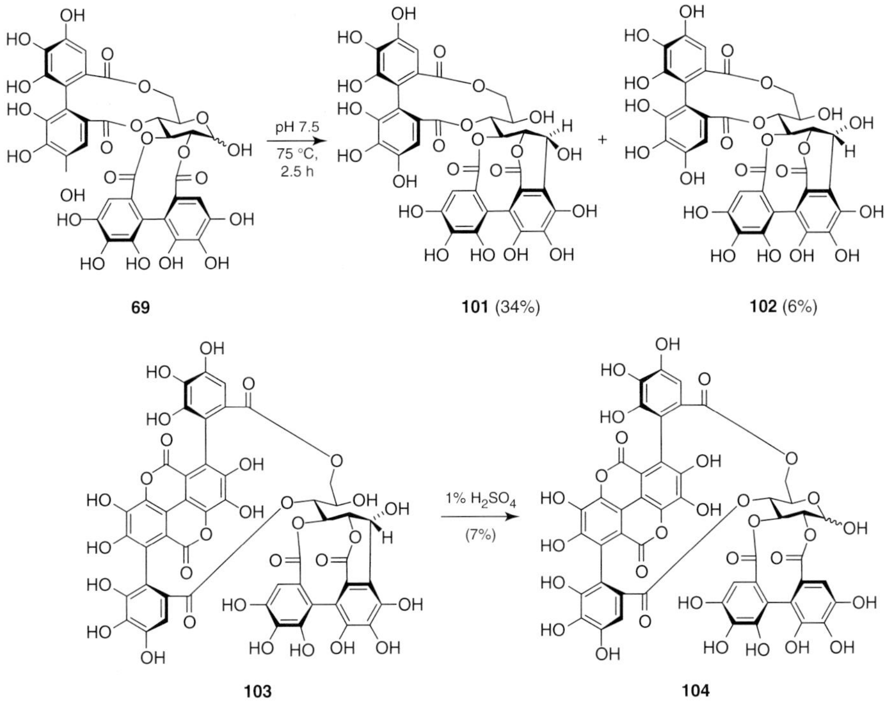

Scheme 17.25 Conversion between pyranose-type ellagitannins and C-glycosidic ellagitannins.

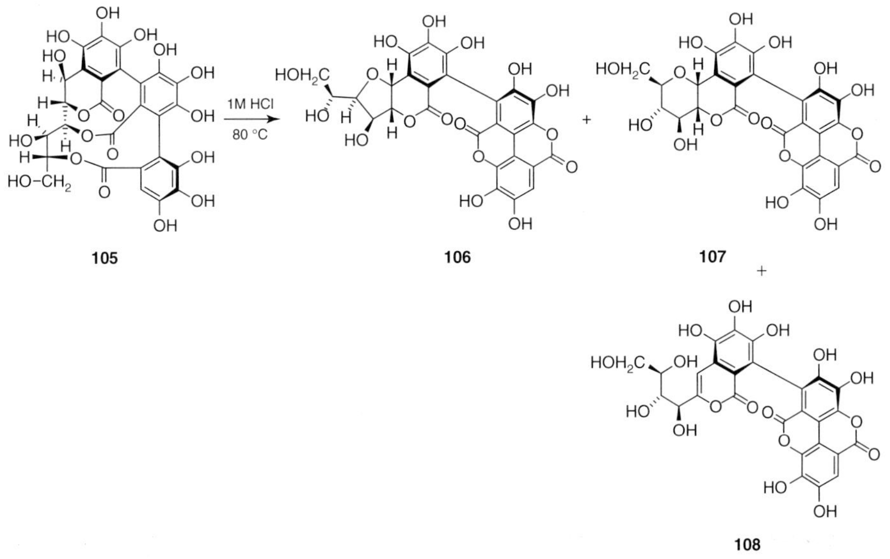

Scheme 17.26 Conversion of castalin (**105**) into castacrenins.

17 Biomimetic Synthesis and Related Reactions of Ellagitannins

A group of natural ellagitannins are produced from **99** and **100** by substitution of the C1 hydroxyl group with various nucleophilic compounds, and semi-synthetic studies have been demonstrated (Scheme 17.27). Substitution with (+)-catechin and ascorbic acid afforded acutissimin A (**109**) [82] and grandinin (**110**) [83], respectively, which are found in some Fagaceous plants together with **99** and **100**. Dimerization of **99** under acidic conditions produces roburin A (**111**) [84]. The high substrate concentration in water (100 mg ml^{-1}) was critical for the dimerization reaction, suggesting the possible involvement of hydrophobic self-association in the regioselective coupling. Roburin A and related oligomeric C-glycosidic ellagitannins are found in the wood of *Quercus robur* [85] and *Castanea crenata* [83] and the tannins accumulate at the sapwood–heartwood transition region in high concentrations. The oligomers may be produced by non-enzymatic self-dimerization. The substitution reactions of ellagitannins in liquors leached from the oak wood barrel are also important in food chemistry. More recently, S-glutathionyl vescalagin (**112**) was synthesized [82b] and the presence of the product in red wine aged in oak barrels was confirmed by LC/MS [86]. Reaction with wine anthocyanidins was also expected, and malvidin-8-C-vescalagin and oenin-8-C-vescalagin (**113**) were synthesized [86]. Experiments commonly showed the same stereochemical features: substitution at the C1 position of **99** exclusively gave products with a vescalagin-type configuration, and this stereoselectivity also seems to be the same in Nature. This selectivity was explained by the stabilization of the C1 hydroxyl group of castalagin (**100**) by the formation of hydrogen bonding with an adjacent phenolic hydroxyl group [82b, 86].

17.5.3
Oxidation of C-Glycosidic Ellagitannins

From a commercial whisky, an oxidized form of ellagitannin originating from oak wood castalagin (**100**) was isolated and named whiskytannin B (**115**) [87]. When oak wood was extracted with 60% EtOH for three months, tannin **115** was gradually generated together with **114**. The reaction was accelerated by raising the pH to 7.0, and pure **99** and **100** were converted into **114** and **115**, respectively, within 24 h (Scheme 17.28). The reaction involves autoxidation and subsequent rearrangement along with the addition of an ethanol molecule. Comparison of the spectral data suggested that these products are identical to previously reported autoxidation products of **99** and **100** [88]. The rate of the generation and subsequent degradation of **114** was much faster than the rates for **115**; this may be the reason why only **115** was detected in the commercial whisky. The difference in stability is probably due to the formation of intramolecular hydrogen bonding between the C1 hydroxyl group and the adjacent carbonyl group.

Scheme 17.27 Biomimetic synthesis of acutissimin A (**109**), grandinin (**110**), roburin A (**111**), S-glutathionyl vescalagin (**112**), and 1-deoxyvescalagin-(1β → 8)-oenin (**113**).

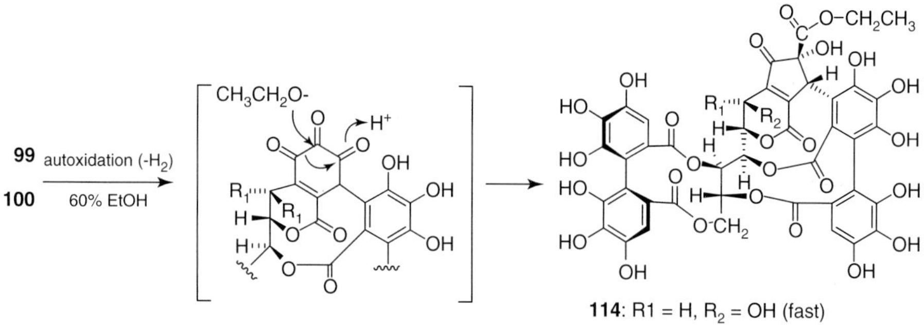

Scheme 17.28 Autoxidation of vescalagin (**99**) and castalagin (**100**) in 60% EtOH.

114: R1 = H, R₂ = OH (fast)
115: R1 = OH, R₂ = H (slow)

The ellagitannins also undergo pyrolytic degradation at the process stage of charring or toasting in barrel production, and the degradation products were shown to impact on the astringent sensations of the liquors [75, 87]. Interestingly, the pyrolysis of vescalagin (**99**) and castalagin (**100**) furnished different products, namely, deoxyvescalagin (**116**) and dehydrocastalagin (**117**), respectively (Scheme 17.29). On pyrolysis of **99**, dehydration and subsequent reduction furnished **116**. In contrast, oxidation of the C1 aromatic ring of **100** occurs predominantly to yield the ketone **117**.

17.6
Conclusions and Perspectives

Beginning with the initial coupling between two galloyl groups in a molecule of the same biosynthetic precursor, pentagalloyl-β-D-glucose (**2**), diverse structures of ellagitannins are biosynthesized. The products are classified into several subgroups, such as simple HHDP esters, dehydroellagitannins, oligomeric ellagitannins, and C-glycosidic ellagitannins, and the structural differences have sometimes had chemotaxonomical significance. The structures indicate that the biosynthesis is both regio- and stereoselectively regulated by enzymes specific to each plant species. However, most of the biosynthetic mechanisms are still unresolved and many of the current biogenetic discussions are speculative. Nonetheless, biogenetic consideration and many findings observed during biomimetic total synthesis described in this chapter suggest that ellagitannin molecules are constructed as a consequence of reasonable chemical requirements. Hopefully, biomimetic synthetic studies will lead to an understanding of the mechanism for the regulation of the regio- and stereoselective intramolecular and intermolecular oxidative couplings in ellagitannin biosynthesis. Taking the distribution of these compounds in Nature and their importance in herbal or food chemistry into account, production of dehydroellagitannins with a $^1C_4$-glucopyranose core

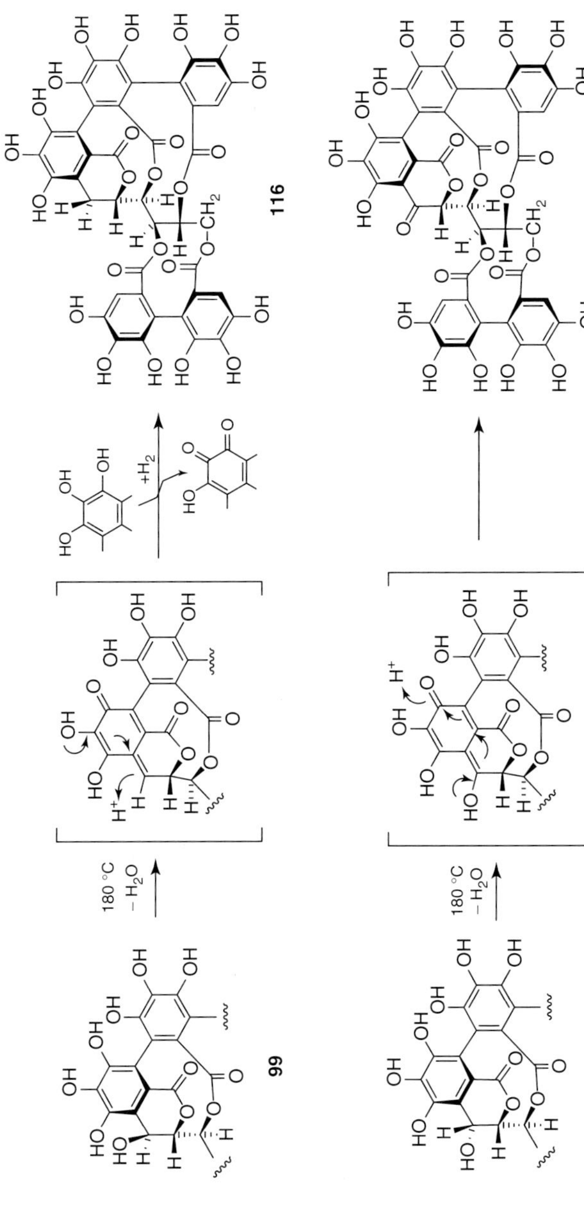

17.6 Conclusions and Perspectives | 671

Scheme 17.29 Pyrolytic degradations of vescalagin and castalagin.

and C-glycosidic ellagitannins may represent the most challenging organic synthesis and plant biochemistry targets. In addition, the biomimetic conversion of non-protected **2** into ellagitannins in aqueous solution should be considered in future studies.

References

1. Haslam, E. (1996) *J. Nat. Prod.*, **59**, 205–215.
2. Zucker, W.V. (1983) *Am. Nat.*, **121**, 335–365.
3. Guy, K., Jaekyung, K., Klaus, H., Yanyan, C., and Xiaozhuo, C. (2007) *Evid. Based Complement Altern. Med.*, **4**, 401–407.
4. Okuda, T. (2005) *Phytochemistry*, **66**, 2012–2031.
5. Heber, D. (2008) *Cancer Lett.*, **269**, 262–268.
6. (a) Feldman, K.S. (2005) *Phytochemistry*, **66**, 1984–2000; (b) Coca, A., Feldman, K.S., and Lawlor, M.D. (2009) in *Chemistry and Biology of Ellagitannins, an Underestimated Class of Bioactive Plant Polyphenols* (ed. S. Quideau), World Scientific, Hackensack NJ, pp. 203–272.
7. Nonaka, G., Nishimura, H., and Nishioka, I. (1985) *J. Chem. Soc., Perkin Trans. 1*, 163–172.
8. Porter, L.J. (1988) in *Flavonoids* (ed. J.B. Harborne), Chapman & Hall, London, pp. 21–62.
9. Okuda, T., Yoshida, T., and Hatano, T. (2000) *Phytochemistry*, **55**, 513–529.
10. (a) Gupta, R.K., Al-Shafi, S.M.K., Layden, K., and Haslam, E. (1982) *J. Chem. Soc., Perkin Trans. 1*, 2525–2534; (b) Haddock, E.A., Gupta, R.K., and Haslam, E. (1982) *J. Chem. Soc., Perkin Trans. 1*, 2535–2545; (c) Haslam, E. (1982) *Prog. Chem. Org. Nat. Prod.*, **41**, 1–46; (d) Haddock, E.A., Gupta, R.K., Al-Shafi, S.M.K., Layden, K., Haslam, E., and Magnolato, D. (1982) *Phytochemistry*, **21**, 1049–1062; (e) Haslam, E. and Cai, Y. (1994) *Nat. Prod. Rep.*, **11**, 41–66.
11. Yoshida, T., Amakura, Y., and Yoshimura, M. (2010) *Int. J. Mol. Sci.*, **11**, 79–106.
12. (a) Niemetz, R. and Gross, G.G. (2005) *Phytochemistry*, **66**, 2001–2011; (b) Niemetz, R. and Gross, G.G. (2003) *Phytochemistry*, **64**, 1197–1201; (c) Niemetz, R., Schilling, G., and Gross, G.G. (2003) *Phytochemistry*, **64**, 109–114; (d) Niemetz, R. and Gross, G.G. (2003) *Phytochemistry*, **62**, 301–306; (e) Niemetz, R., Gross, G.G., and Schilling, G. (2001) *Chem. Commun.*, 35–36.
13. Wilkins, C.K. and Bohm, B.A. (1976) *Phytochemistry*, **15**, 211–214.
14. Nonaka, G., Harada, M., and Nishioka, I. (1980) *Chem. Pharm. Bull.*, **28**, 685–687.
15. Hatano, T., Yasuhara, T., and Okuda, T. (1989) *Chem. Pharm. Bull.*, **37**, 2665–2669.
16. Tanaka, T., Nonaka, G., Ishimatsu, M., Nishioka, I., and Kouno, I. (2001) *Chem. Pharm. Bull.*, **49**, 486–487.
17. Lee, S.-H., Tanaka, T., Nonaka, G., and Nishioka, I. (1991) *Chem. Pharm. Bull.*, **39**, 630–638.
18. Tanaka, T., Zhang, H., Jiang, Z., and Kouno, I. (1997) *Chem. Pharm. Bull.*, **45**, 1891–1897.
19. (a) Barbehenn, R.V., Jones, C.P., Hagerman, A.E., Karonen, M., and Salminen, J.-P. (2006) *J. Chem. Ecol.*, **32**, 2253–2267; (b) Barbehenn, R.V., Jaros, A., Lee, G., Mozola, C., Weir, Q., and Salminen, J.-P. (2009) *J. Insect Physiol.*, **55**, 297–304.
20. Feldman, K.S., Ensel, S.M., and Minard, R.D. (1994) *J. Am. Chem. Soc.*, **116**, 1742–1745.
21. Okuda, T., Yoshida, T., and Ashida, M. (1981) *Heterocycles*, **16**, 1681–1685.
22. Gramshaw, J.W., Haslam, E., Haworth, R.D., and Searle, T. (1962) *J. Chem. Soc.*, 2944–2947.
23. Feldman, K.S. and Sahasrabudhe, K. (1999) *J. Org. Chem.*, **64**, 209–216.

24. Quideau, S. and Feldman, K.S. (1996) *Chem. Rev.*, **96**, 475–503.
25. Feldman, K.S. and Sambandam, A. (1995) *J. Org. Chem.*, **60**, 8171–8178.
26. Feldman, K.S. and Smith, R.S. (1996) *J. Org. Chem.*, **61**, 2606–2612.
27. Nonaka, G., Ishimatsu, M., Ageta, M., and Nishioka, I. (1989) *Chem. Pharm. Bull.*, **37**, 50–53. The structures of cercidinins A and B in this reference were later revised. See Reference 88.
28. Tanaka, T., Kirihara, S., Nonaka, G., and Nishiokoa, I. (1993) *Chem. Pharm. Bull.*, **41**, 1708–1716.
29. Su, X., Surry, D.S., Spandl, R.J., and Spring, D.R. (2008) *Org. Lett.*, **10**, 2593–2596.
30. Surry, D.S., Su, X., Fox, D.J., Franckevicius, V., Macdonald, S.J.F., and Spring, D.R. (2005) *Angew. Chem. Int. Ed.*, **44**, 1870–1873.
31. Dai, D. and Martin, O.R. (1998) *J. Org. Chem.*, **63**, 7628–7633.
32. Zhang, Y.-J., Abe, T., Tanaka, T., Yang, C.-R., and Kouno, I. (2001) *J. Nat. Prod.*, **64**, 1527–1532.
33. Schmidt, O.T., Schanz, R., Wurmb, R., and Groebke, W. (1967) *Liebigs Ann. Chem.*, **706**, 154–168.
34. Okuda, T., Yoshida, T., and Hatano, T. (1982) *J. Chem. Soc., Perkin Trans 1*, 9–14.
35. Okuda, T., Hatano, T., Nitta, H., and Fujii, R. (1980) *Tetrahedron Lett.*, **21**, 4361–4364.
36. Tanaka, T., Nonaka, G., and Nishioka, I. (1990) *Chem. Pharm. Bull.*, **38**, 2424–2428.
37. Okuda, T., Yoshida, T., Hatano, T., Ikeda, Y., Shingu, T., and Inoue, T. (1986) *Chem. Pharm. Bull.*, **34**, 4075–4082.
38. Schmidt, O.T., Schmidt, D.M., and Herok, J. (1954) *Liebigs Ann. Chem.*, **587**, 67–74.
39. Seikel, M.K. and Hillis, W.E. (1970) *Phytochemistry*, **9**, 1115–1128.
40. Jochims, J.C., Taigel, G., and Schmidt, O.T. (1968) *Liebigs Ann. Chem.*, **717**, 169–185.
41. Feldman, K.S., Iyer, M.R., and Liu, Y. (2003) *J. Org. Chem.*, **68**, 7433–7438.
42. Nonaka, G., Mihashi, K., and Nishioka, I. (1990) *Chem. Commun.*, 790–791.
43. Nonaka, G., Matsumoto, Y., Nishioka, I., Nishizawa, M., and Yamagishi, T. (1981) *Chem. Pharm. Bull.*, **29**, 1184–1187.
44. Okuda, T., Hatano, T., and Yasui, T. (1981) *Heterocycles*, **16**, 1321–1324.
45. (a) Tanaka, T., Watarumi, S., Matsuo, Y., Kamei, M., and Kouno, I. (2003) *Tetrahedron*, **59**, 7939–7947; (b) Tanaka, T., Mine, C., Watarumi, S., Fujioka, T., Mihashi, K., Zhang, Y.-J., and Kouno, I. (2002) *J. Nat. Prod.*, **65**, 1582–1587.
46. Tanaka, T., Matsuo, Y., and Kouno, I. (2005) *J. Agric. Food Chem.*, **53**, 7571–7578.
47. Li, Y., Tanaka, T., and Kouno, I. (2007) *Phytochemistry*, **68**, 1081–1088.
48. Quideau, S. and Feldman, K.S. (1997) *J. Org. Chem.*, **62**, 8809–8813.
49. (a) Yamada, H., Nagao, K., Dokei, K., Kasai, Y., and Michihata, N. (2008) *J. Am. Chem. Soc.*, **130**, 7566–7567; (b) Ikeda, Y., Nagao, K., Tanigakiuchi, K., Tokumaru, G., Tsuchiya, H., and Yamada, H. (2004) *Tetrahedron Lett.*, **45**, 487–489.
50. Nelson, T.D. and Meyers, A.I. (1994) *J. Org. Chem.*, **59**, 2577–2580.
51. Lipshutz, B.H., Liu, Z.-P., and Kayser, F. (1994) *Tetrahedron Lett.*, **35**, 5567–5570.
52. Itoh, T. and Chika, J. (1995) *J. Org. Chem.*, **60**, 4968–4969.
53. Itoh, T., Chika, J., Shirakami, A., Ito, H., Yoshida, T., Kubo, Y., and Uenishi, J. (1996) *J. Org. Chem.*, **61**, 3700–3705.
54. Khanbabaee, K., Schulz, C., and Lötzerich, K. (1997) *Tetrahedron Lett.*, **38**, 1367–1368.
55. Khanbabaee, K. and Lötzerich, K. (1997) *Liebigs Ann. Recl.*, 1571–1575.
56. Khanbabaee, K. and Lötzerich, K. (1998) *J. Org. Chem.*, **63**, 8723–8728.
57. Khanbabaee, K., Lötzerich, K., Borges, M., and Großer, M. (1999) *J. Prakt. Chem.*, **341**, 159–166.
58. Khanbabaee, K. and Lötzerich, K. (1999) *Eur. J. Org. Chem.*, **11**, 3079–3083.

59. Khanbabaee, K. and Großer, M. (2003) *Eur. J. Org. Chem.*, 2128–2131.
60. (a) Jiang, Z.-H., Tanaka, T., and Kouno, I. (1995) *Chem. Commun.*, 1467–1468; (b) Tanaka, T., Jiang, Z.-H., and Kouno, I. (1997) *Chem. Pharm. Bull.*, **45**, 1915–1921.
61. (a) Feldman, K.S., Lawlor, M.D., and Sahasrabudhe, K. (2000) *J. Org. Chem.*, **65**, 8011–8019; (b) Feldman, K.S. and Lawlor, M.D. (2000) *J. Am. Chem. Soc.*, **122**, 7396–7397.
62. Hatano, T., Hattori, S., and Okuda, T. (1986) *Chem. Pharm. Bull.*, **34**, 4092–4097.
63. Feldman, K.S., Quideau, S., and Appel, H.M. (1996) *J. Org. Chem.*, **61**, 6656–6665.
64. Saijo, R., Nonaka, G., and Nishioka, I. (1989) *Chem. Pharm. Bull.*, **37**, 2063–2070.
65. Amakura, Y., Miyake, M., Ito, H., Murakaku, S., Araki, S., Itoh, Y., Lu, C.-F., Yang, L.-L., Yen, K.-Y., Okuda, T., and Yoshida, T. (1998) *Phytochemistry*, **50**, 667–675.
66. Tanaka, T., Fujisaki, H., Nonaka, G., and Nishioka, I. (1992) *Heterocycles*, **33**, 375–383.
67. Tanaka, T., Kouno, I., and Nonaka, G. (1996) *Chem. Pharm. Bull.*, **44**, 34–40.
68. (a) Schmidt, O.T. and Nieswandt, W. (1950) *Liebigs Ann. Chem.*, **568**, 165–173; (b) Schmidt, O.T., Hensler, R.H., and Stephan, P. (1957) *Liebigs Ann. Chem.*, **609**, 186–191; (c) Yoshida, T., Fujii, R., and Okuda, T. (1980) *Chem. Pharm. Bull.*, **28**, 3713–3715.
69. Yoshida, T., Amakura, Y., Liu, Y.-Z., and Okuda, T. (1994) *Chem. Pharm. Bull.*, **42**, 1803–1807.
70. Saijo, R., Nonaka, G., and Nishioka, I. (1989) *Chem. Pharm. Bull.*, **37**, 2624–2630.
71. Tanaka, T., Nonaka, G., Nishioka, I., Miyahara, K., and Kawasaki, T. (1986) *J. Chem. Soc., Perkin Trans. 1*, 369–376.
72. (a) Yoshida, T., Itoh, H., Matsunaga, S., Tanaka, R., and Okuda, T. (1992) *Chem. Pharm. Bull.*, **40**, 53–60; (b) Foo, L.Y. and Wong, H. (1992) *Phytochemistry*, **31**, 711–713.
73. Tanaka, T., Fujisaki, H., Nonaka, G., and Nishioka, I. (1992) *Chem. Pharm. Bull.*, **40**, 2937–2944.
74. (a) Mayer, W., Gabler, W., Riester, A., and Korger, H. (1967) *Liebigs Ann. Chem.*, **707**, 177–181; (b) Nonaka, G., Sakai, T., Tanaka, T., Mihashi, K., and Nishioka, I. (1990) *Chem. Pharm. Bull.*, **38**, 2151–2156
75. (a) Glabasnia, A. and Hofmann, T. (2007) *J. Agric. Food Chem.*, **55**, 4109–4118; (b) Glabasnia, A. and Hofmann, T. (2006) *J. Agric. Food Chem.*, **54**, 3380–3390.
76. Okuda, T., Hatano, T., Kaneda, T., Yoshizaki, M., and Shingu, T. (1987) *Phytochemistry*, **26**, 2053–2055.
77. (a) Vivas, N., Laguerre, M., Glories, Y., Bourgeois, G., and Vitry, C. (1995) *Phytochemistry*, **39**, 1193–1199; (b) Quideau, S., Varadinova, T., Karagiozova, D., Jourdes, M., Pardon, P., Baudry, C., Genova, P., Diakov, T., and Petrova, R. (2004) *Chem. Biodivers.*, **1**, 247–258.
78. Ono, T. and Shigemori, H. (2009) *Heterocycles*, **78**, 1993–2001.
79. Tanaka, T., Nonaka, G., and Nishioka, I. (1986) *Chem. Pharm. Bull.*, **34**, 656–663.
80. (a) Tanaka, T., Nonaka, G., and Nishioka, I. (1986) *Chem. Pharm. Bull.*, **34**, 650–655; (b) Mayer, W., Goerner, A., and Andrae, K. (1977) *Liebigs Ann. Chem.*, 1976–1986; (c) Schilling, G. and Schick, H. (1985) *Liebigs Ann. Chem.*, 2240–2245.
81. Tanaka, T., Ueda, N., Shinohara, H., Nonaka, G., Fujioka, T., Mihashi, K., and Kouno, I. (1996) *Chem. Pharm. Bull.*, **44**, 2236–2242.
82. (a) Ishimaru, K., Nonaka, G., and Nishioka, I. (1987) *Chem. Pharm. Bull.*, **35**, 602–610; (b) Quideau, S., Jourdes, M., Lefeuvre, D., Montaudon, D., Saucier, C., Glories, Y., Pardon, P., and Pourquier, P. (2005) *Chem. Eur. J.*, **11**, 6503–6513.
83. Nonaka, G., Ishimaru, K., Azuma, R., Ishimatsu, M., and Nishioka, I. (1989) *Chem. Pharm. Bull.*, **37**, 2071–2077.

84. Viriot, C., Scalbert, A., Hervé du Penhoat, C., and Moutounet, M. (1994) *Phytochemistry*, **36**, 1253–1260.
85. Herve du Penhoat, C.L.M., Michon, V.M.F., Peng, S., Viriot, C., Scalbert, A., and Gage, D. (1991) *J. Chem. Soc., Perkin Trans. 1*, 1653–1660.
86. (a) Jourdes, M., Lefeuvre, D., and Quideau, S. (2009) in *Chemistry and Biology of Ellagitannins, an Underestimated Class of Bioactive Plant Polyphenols* (ed. S. Quideau), World Scientific, Hackensack, NJ, pp. 320–365; (b) Chassaing, S., Lefeuvre, D., Jacquet, R., Jourdes, M., Ducasse, L., Galland, S., Grelard, A., Cédric Saucier, C., Teissedre, P., Dangles, O., and Quideau, S. (2010) *Eur. J. Org. Chem.*, 55–63.
87. Fujieda, M., Tanaka, T., Suwa, Y., Koshimizu, S., and Kouno, I. (2008) *J. Agric. Food Chem.*, **56**, 7305–7310.
88. Puech, J.-L., Mertz, C., Michon, V., Le Guernevé, C., Doco, T., and Hervé du Penhoat, C. (1999) *J. Agric. Food Chem.*, **47**, 2060–2066.

18
Biomimetic Synthesis of Lignans
Craig W. Lindsley, Corey R. Hopkins, and Gary A. Sulikowski

18.1
Introduction to Lignans

Lignans was a term coined by Haworth in 1936 to describe a group of phenylpropanoid dimeric natural products linked through the $\beta-\beta'$ (8–8′) carbons of the propanyl side-chains [1, 2]. Biosynthetically, lignans are derived from the shikimate pathway, and these secondary metabolites from the Plant Kingdom form the basis of chemical defense in plants and have been found to possess a broad spectrum of biological activities, including anticancer, antimitotic, antiangiogenesis, and antiviral, to list a few [3–10]. In Nature, oxidases (such as laccases, peroxidases, and other oxidases) initiate a one-electron oxidation of an electron-rich cinnamyl residue **1**, wherein the number and location of ethers and phenol moieties is varied and R_1 can be CH_2OH, $COOH$, $COOR$, or $CONHR$, which ultimately generates a reactive C8 quinone methide radical **2** [3–13]. Coupling of two of these 8–8′ radicals often occurs with an *anti-* (or trans) orientation, to diminish steric interactions, delivering the $\beta-\beta'$ coupled intermediate **3** (Scheme 18.1). Based on the functionalization of the aromatic rings (Ar and Ar′) and the oxidation state of R_1, enormous structural diversity exists for the resulting lignan natural products **4–14** (stereochemistry not shown) upon addition of oxygen nucleophiles into the quinone methide core and/or cyclization (Figure 18.1) [3–14]. Notable examples of lignans include podophyllotoxin (**15**), etoposide (**16**), (+)-pinoresinol (**17**), and carpanone (**18**) [3–15].

In 1972, Gottlieb extended this family to include phenylpropanoid dimers with linkages other than 8–8′, which were named "*neolignans*" [16]. An attempt by Gottlieb to restrict the term neolignan to natural products only derived from the oxidative dimerization of allyl and/or propenylphenols met with resistance and, in 2000, IUPAC recommended defining neolignans as dimers between two phenylpropanoid (cinnamyl residues **1**) units different from 8–8′ [17, 18]. Indeed, Scheme 18.1 was an oversimplification. For example, (*E*)-coniferyl alcohol (**19**), upon one-electron oxidation by endogenous oxidases (Scheme 18.2), generates the coniferyl alcohol radical **20**, which subsequently exists in several mesomeric forms (the C5 radical **21**, the C1 radical **22**, and the C8 radical **23**) [12, 13]. All of

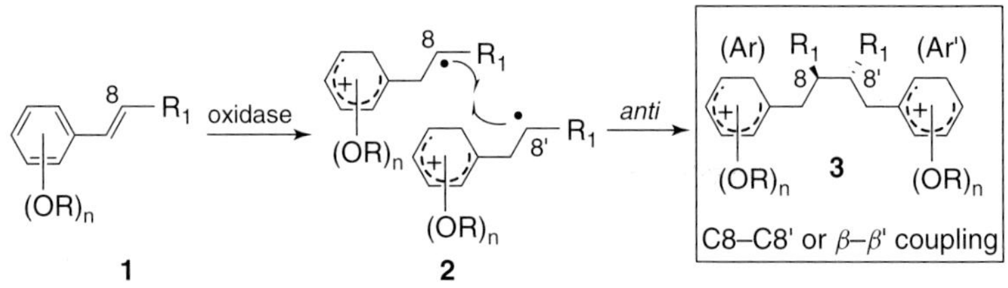

Scheme 18.1 One-electron oxidation of electron-rich cinnamyl residue **1** to generate the quinone methide C8 radical **2**, which undergoes a β–β' (8–8') coupling to deliver the trans-**3**.

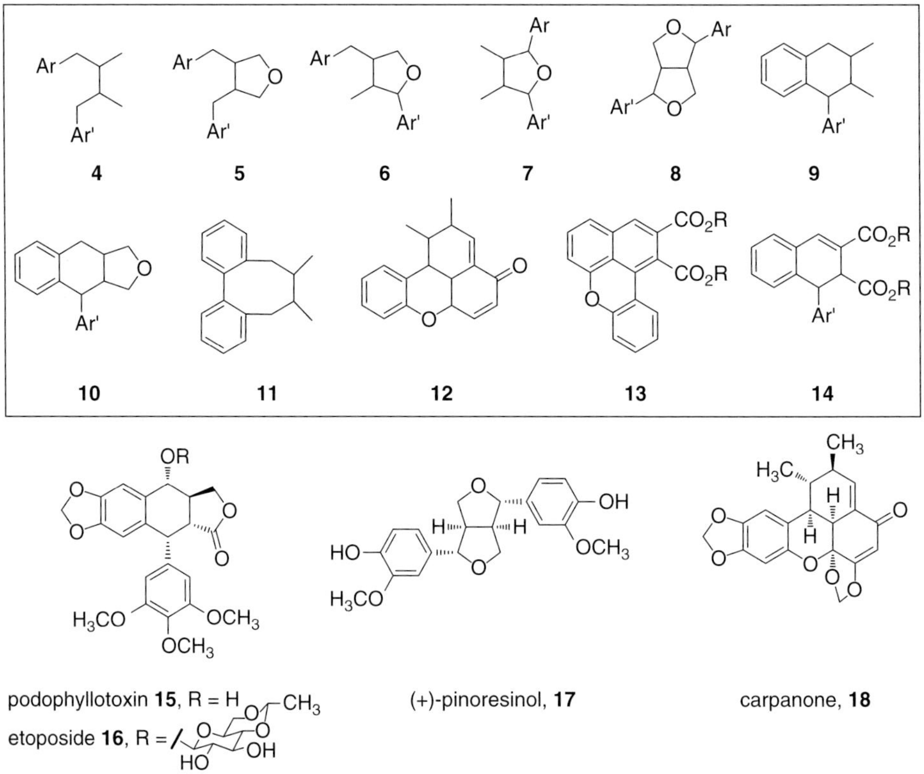

Figure 18.1 Representative chemotypes **4–14** of lignans derived from β–β' oxidative phenolic couplings (stereochemistry not shown); structures of well-known lignans **15–18**.

these mesomeric forms can combine with either themselves or with any of the other mesomeric radicals **21–23**, leading to diverse neolignan chemotypes **24–29** (Figure 18.2). This chapter will focus on the biomimetic synthesis of lignans.

In Nature, lignan chemotypes **4–14** are produced as either racemates (though usually a single diastereomer) or, in some cases, as single enantiomers [3–14]. In

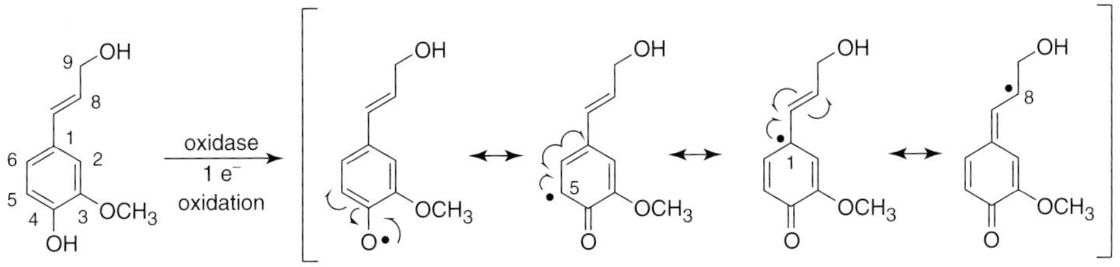

19 20 21 22 23

Scheme 18.2 One-electron oxidation of *(E)*-coniferyl alcohol (**19**) to multiple quinone methide mesomeric radicals **21–23**, which combine to give rise to neolignans (coupling products other than 8–8′).

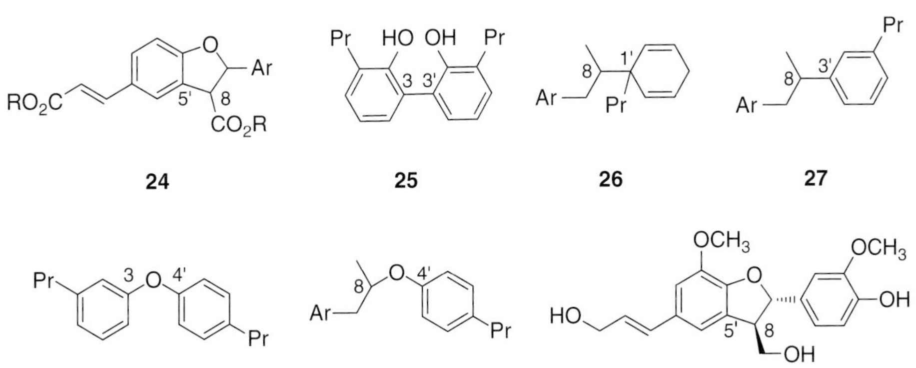

24 25 26 27

28 29 (+)-dehydrodiconiferyl alcohol, **30**

Figure 18.2 Representative chemotypes **24–29** of neolignans derived from oxidative phenolic couplings (stereochemistry not shown) other than 8–8′; structure of the well-known neolignan **30**.

the laboratory, biomimetic radical coupling approaches for lignan synthesis are characterized by a lack of stereochemical control, and provide racemic products (though usually a single diastereomer). This lack of stereocontrol is observed with both metal-catalyzed (Ag^{I}, Mn^{II}, I^{III}, Cu^{II}, Co^{II}, Fe^{II}) oxidative phenolic couplings as well as enzyme-mediated [horse radish peroxidase (HRP), laccases, oxidases] oxidative phenolic couplings [3–14]. In Nature, the stereochemical outcome of these radical couplings is somehow controlled to provide enantiopure lignans. In 1997, Davin and coworkers identified the source – a 78-kDa protein, devoid of catalytic activity – which they termed a "*dirigent protein*," a protein that dictates the stereochemistry of a compound synthesized by other enzymes [19]. This so-called auxiliary "dirigent" protein, in the presence of either a one-electron oxidant, $(NH_4)_2S_2O_8$, or an oxidase, flavin mononucleotide (FMN), effected an enantioselective β–β' phenolic coupling of *(E)*-coniferyl alcohol (**19**) to (+)-pinoresinol

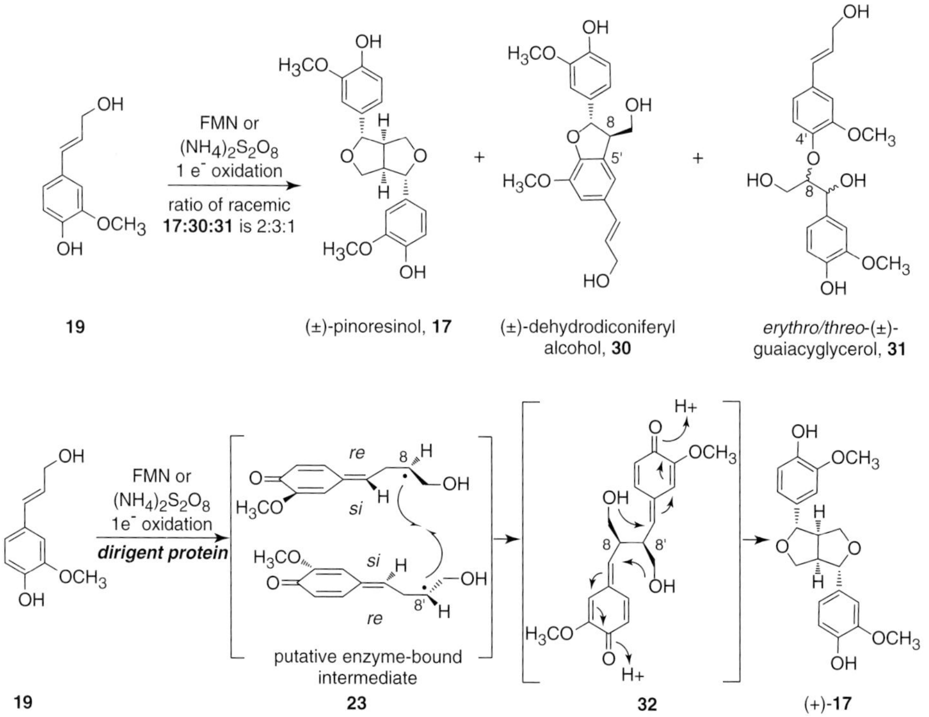

Scheme 18.3 One-electron oxidation of (E)-coniferyl alcohol (19) in the absence of a dirigent protein affords racemic (±)-pinoresinol (17) along with neolignans 30 and 31 in a 2 : 3 : 1 ratio. One-electron oxidation of (E)-coniferyl alcohol (19) in the presence of a dirigent protein provides enantiopure (+)-pinoresinol (17) via a putative enzyme-bound intermediate wherein radicals 23 are oriented "Si-face to Si-face" for the oxidative phenolic coupling. Only trace amount of neolignans 30 and 31 are detected in the presence of the dirigent protein.

(17) in high yield and with only traces of other lignan and neolignan products. In contrast, without the auxiliary "dirigent" protein, racemic (±)-pinoresinol was obtained in both cases along with neolignans 30 (C8–C5′) and 31 (8–O–4′) in 2 : 3 : 1 ratio. The mechanism of action of the "dirigent" protein is postulated as capture of the free radical 23 into enzyme binding sites and orientation of two 23 "Si-face to Si-face" to direct enantioselective coupling to provide 32 and, upon conjugate addition into the quinone methide, (+)-pinoresinol 17 (Scheme 18.3) [19].

Despite the similarity in nomenclature, lignans should not be confused with lignins. Lignins are heterogeneous cell-wall biopolymers (>10 000 amu) that consist of various electron-rich cinnamyl residues 1 and account for over 30% of all non-fossil organic carbon [20]. Thus, lignins are second only to cellulose as the most abundant biopolymer and the second most abundant natural product on earth. Lignins play critical roles in conducting water in plant stems and in taking atmospheric carbon into the living plant tissues and highly lignified wood is a valuable material for multiple applications [20].

18.1.1
Biomimetic Synthesis of Lignans

While numerous synthetic routes have been developed for the total synthesis of lignans representing all the key chemotypes 4–14, biomimetic approaches here are restricted to oxidative phenolic coupling reactions that are either metal-catalyzed (Ag^I, Mn^{II}, I^{III}, Cu^{II}, Co^{II}, Fe^{II}, Tl^{III}) or enzyme-mediated (HRP, laccases, oxidases). Of these, Ag_2O and Ag_2CO_3 are the most commonly employed metal catalysts, and $HRP–H_2O_2$ is the most common enzyme-mediated system [3–14].

18.1.1.1 Biomimetic Synthesis of Podophyllotoxin-Like Lignans

Podophyllotoxin (15) is unique as an aryltetralin lactone lignan due to its unique structure and as a precursor to clinical oncology drugs such as etoposide (16) [21]. While many synthetic strategies have been developed to circumvent the growing demand on the plant resource, few are truly biomimetic. Indeed, most synthetic routes toward 15 rely on tandem conjugate addition approaches [22–24], Diels–Alder approaches [25–27], or, more recently, enantioselective sequential conjugate addition–allylation reactions [28]. It is well established that 15 is biosynthesized by the oxidative cyclization of the dibenzylbutyrolactone yatein (33), itself a lignan derived from 8–8' heterocoupling [29]. Upon treatment of 33 with Tl_2O_3 in trifluoroacetic acid (TFA) a C1 epimeric, des-4-hydroxypodophyllotoxin (34) results (Scheme 18.4). Under hypervalent iodine or ruthenium catalysis, an unexpected eight-membered ring (35) forms exclusively [30]. Thus, with the proper metal catalysis, a biomimetic approach to podophyllotoxin is possible, but other routes have proven more direct and, importantly, enantioselective [28].

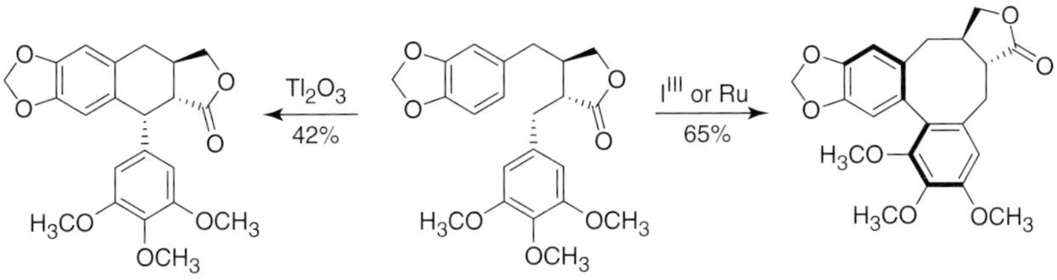

Scheme 18.4 Biomimetic approaches to podophyllotoxin (15). Oxidative cyclization of the biosynthetic precursor of 15, yatein (33), yields the basic epimeric core of 15 under Tl^{III} Lewis acid catalysis; however, hypervalent iodine or ruthenium catalysis leads to the eight-membered ring congener 35.

18.1.1.2 Biomimetic Synthesis of Furofuran Lignans

The furofuran class of lignans, exemplified by (+)-pinoresinol 17 discussed earlier, represents one of the largest groups of lignan chemotypes, and is characterized by a 3,7-dioxabicyclic core with diverse electron-rich aromatics in the 2- and

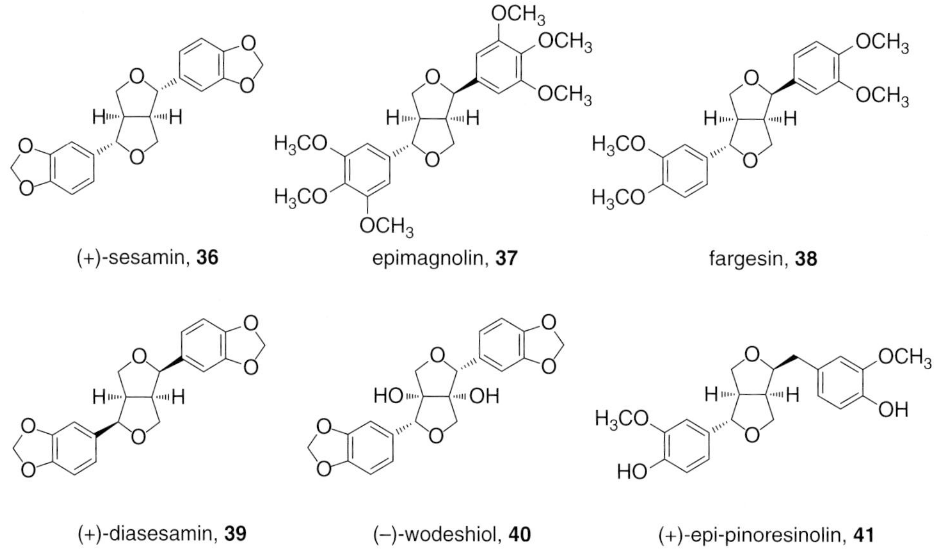

(+)-sesamin, **36**

epimagnolin, **37**

fargesin, **38**

(+)-diasesamin, **39**

(−)-wodeshiol, **40**

(+)-epi-pinoresinolin, **41**

Figure 18.3 Representative furofuran lignans **36–41**.

6-positions [12, 31]. Further diversity emanates from the degree of core oxidation and the stereochemistry of pendant aryl groups (*exo* or *endo* face). Representative furofurans **36–41** (Figure 18.3) are highlighted, and several excellent reviews have been written [32–37]. This class of lignans has demonstrated a wide range of biological activities with therapeutic relevance, and is found in traditional medicines [32, 38].

Like podophyllotoxin (**15**) and related compounds, numerous synthetic routes have been developed to access furofurans based on acyl anion additions, aldol chemistry, cycloaddition/rearrangements, and radical/photochemistry [30–36]. Here, we discuss the biomimetic, oxidative (β–β') dimerization of cinnamyl derivatives, which follows the synthetic route illustrated in Scheme 18.3 for the biomimetic synthesis of (+)-pinoresinol (**17**). Enzyme-mediated biomimetic syntheses have been reported for racemic (±)-pinoresinol (**17**) and (±)-syringaresinol (**43**, Scheme 18.5) [32, 39]. In the latter case, exposure of (*E*)-coniferol (**19**) to *Caldariomyces fumago* open to air for 16 h provided a 1 : 1 ratio of racemic **17** (8–8′) and racemic **30** (8–5′) in low yield (10%). Exposure of **42** to a crude emulsion open to air for 11 days generated racemic **43** in low yield (12%), but without the production of the neolignan 8–5′ product [32, 39].

Metal-catalyzed approaches have also been employed. For example, prior to 1945, both Haworth [40] and Erdtman [41] demonstrated that treatment of ferulic acid **44** with either $FeCl_3/O_2$ or $Fe_2(SO_4)_3$ followed by an acidic work-up delivered the bis-lactone **45** in 20–22% yields. This powerful biomimetic sequence to arrive at the bis-lactone precursor drove furofuran chemistry, and in 1955 the first demonstration of lithium aluminium hydride (LAH) reduction of a bis-lactone **47** led to the total synthesis of (±)-**43** (Scheme 18.6) [42]. Subsequently, biomimetic

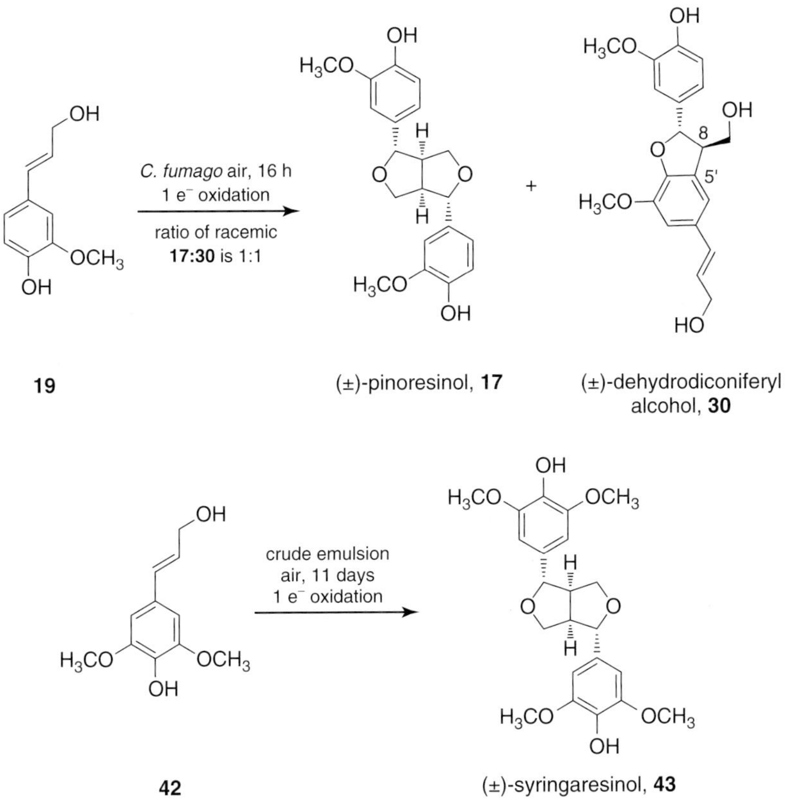

Scheme 18.5 Enzyme-mediated, biomimetic synthesis of racemic (±)-pinoresinol (**17**) and (±)-syringaresinol (**43**).

efforts moved from the use of cinnamyl acids, such as **44** and **46**, to cinnamyl alcohols, such as **42**, to avoid the reduction step. In this instance, the application of CuSO$_4$ under an O$_2$ atmosphere in aqueous acetone perfected the reaction to deliver racemic (±)-**43** in 90% yield (Scheme 18.6) [43]. In recent years, Tl^{III} and Mn^{III} metal species have proven equally effective in promoting high yielding β–β′ phenolic couplings for the biomimetic synthesis of racemic furofurans [3–14, 32].

18.1.1.3 Biomimetic Synthesis of Benzoxanthenone Lignans

The benzoxanthenone class of lignans was first discovered in 1969 by Brophy and coworkers, from the light petroleum extract of the Carpano tree, *Cinnamomum* sp., and is exemplified by a rigid tetracyclic core with five contiguous stereocenters [15]. To date, all benzoxanthenones identified have been isolated as single diastereomers, and produced in Nature as racemates. The first member of this class, carpanone (**18**) was isolated along with an *ortho*-methoxystyrene named carpacin (**48**). Brophy and coworkers then proposed a biosynthesis for capanone that involved demethylation

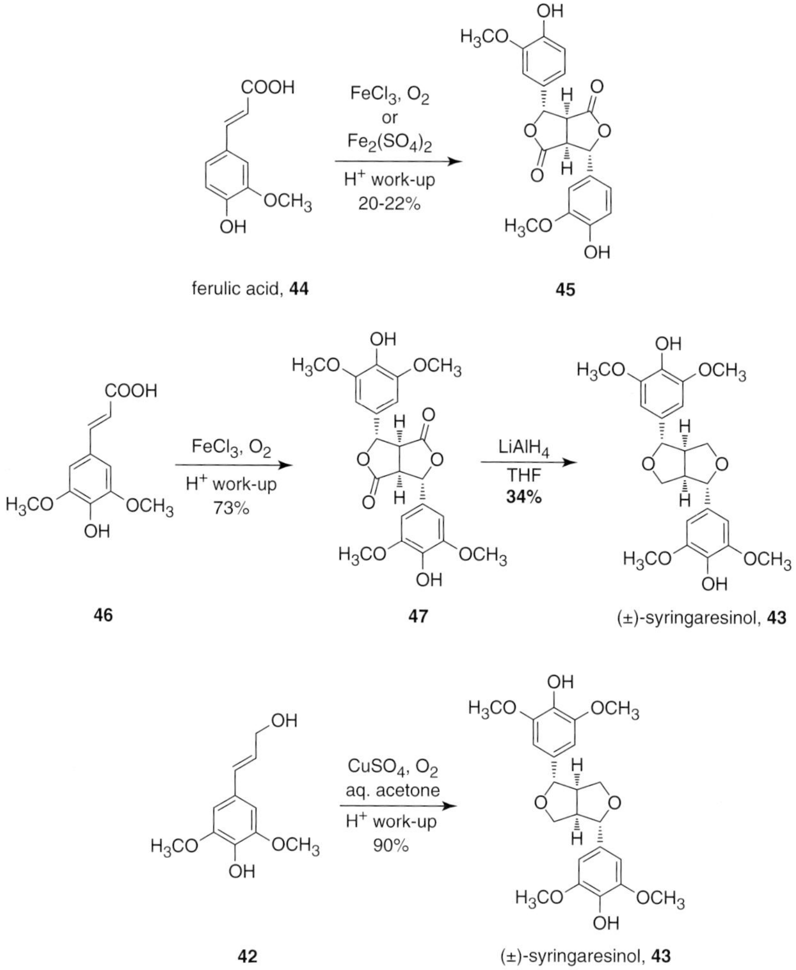

Scheme 18.6 Metal-catalyzed, biomimetic synthesis of racemic bis-lactone **45** and (±)-syringaresinol (**43**).

of carpacin to produce **49**, β–β′ phenolic coupling to deliver *trans*-ortho-quinone methide **50**, followed by an *endo*-selective inverse electron demand Diels–Alder to afford carpanone (**18**) (Scheme 18.7).

Now considered a classic in total synthesis, Chapman and coworkers validated the biosynthetic proposal with the first total synthesis of carpanone (**18**) [44]. Following the rationale in Scheme 18.4, Chapman decided that the β–β′-phenolic coupling must occur trans and that this configuration would dictate the subsequent inverse-electron demand Diels–Alder reaction. While must phenolic couplings employed one-electron oxidants, there was precedent for two-electron oxidants; therefore, Chapman utilized Pd^{II} to bring the two styrenyl phenols **49** together, delivering **51** *en route* to **50**, and then carpanone **18** (Scheme 18.8). Chapman's

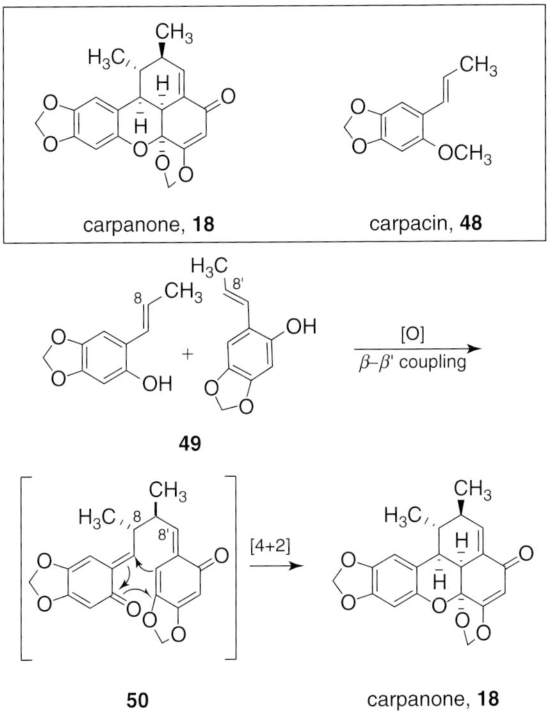

Scheme 18.7 Structures of carpanone (**18**) and carpacin (**48**); biosynthetic proposal by Brophy and coworkers to account for the synthesis of lignan carpanone.

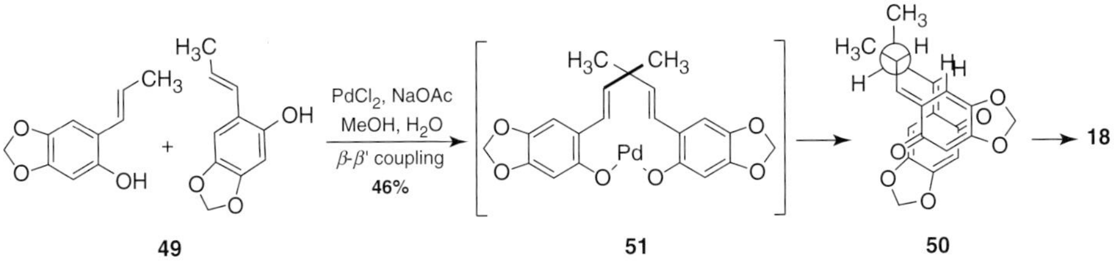

Scheme 18.8 The now classical biomimetic total synthesis of carpanone (**18**).

strategy worked, providing carpanone (**18**) in 46% yield as a single diastereomer, as confirmed by single-crystal X-ray [44]. From starting materials devoid of stereocenters, the one-pot construction of a tetracyclic scaffold with complete stereocontrol of five contiguous stereocenters highlights the power of biomimetic synthesis. After this initial report, several laboratories disclosed additional oxidative systems, both stoichiometric and catalytic, to produce carpanone, including metal(II) salen/O_2 (metal = Co, Mn, Fe), O_2 (hv, Rose Bengal), AIBN (azobisisobutyronitrile), dibenzoyl peroxide, and AgO in yields ranging from 14 to 94% [45, 46]. In 2001, Ley reported on the total synthesis of carpanone employing only solid-supported

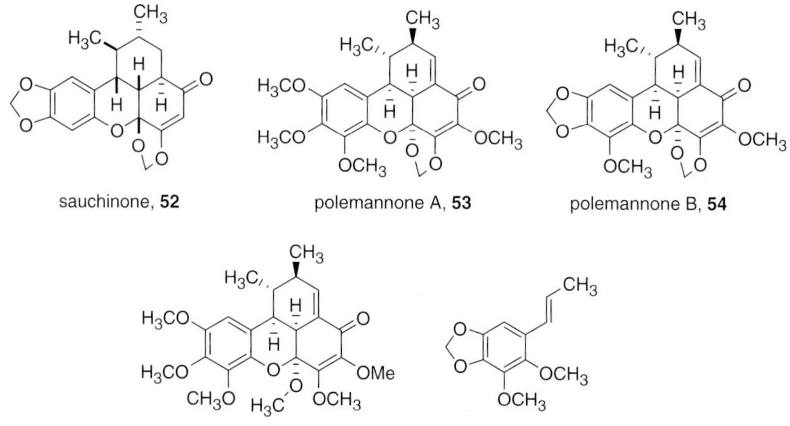

sauchinone, **52** polemannone A, **53** polemannone B, **54**

polemannone C, **55** isoapiol, **56**

Figure 18.4 Representative benzoxanthone lignans **52–55**, and a biosynthetic precursor to **53–55**, isoapiol (**56**).

reagents and scavengers [47]. Around the same time, Lindsley and Shair described a hetero-β–β'-phenolic coupling reaction, facilitated by $IPh(OAc)_2$, to deliver hetero-tetracyclic analogs of carpanone; however, this oxidant system was unable to produce carpanone itself, but it was able to produce less electron-rich homodimers [48].

Since the initial account of the discovery of carpanone (**18**) and the benzoxanthone class of lignans, several other congeners have been reported, including the therapeutically relevant sauchinone (**52**) [49], a 1′,7′-dihydro-stereoisomer of carpanone, and the highly oxygenated polemannones A–C (**53–55**, respectively) [50]. Interestingly, along with **53–55**, Jakupovic also isolated a more highly oxygenated congener of carpacin, isoapiol (**56**), suggesting a similar biosynthetic pathway (Figure 18.4).

Lindsley and coworkers recently developed a novel, catalytic $CuCl_2$/(−)-sparteine oxidative β–β'-phenolic coupling reaction of styrenyl phenols which, after a rapid inverse-electron demand Diels–Alder reaction, affords the benzoxanthanone natural product carpanone (**18**) and related unnatural congeners in yields exceeding 85% (Scheme 18.9) as single diastereomers [51]. Less than 5% ee was observed under these conditions, suggesting that the reaction does not take place in the copper coordination sphere due to rapid dissociation of the intermediate keto-radical leading to no enantioselection [49]. However, the catalytic $CuCl_2$/(−)-sparteine oxidative system enabled the first biomimetic total syntheses of polemannones B and C, **54** (79% yield) and **55** (90% yield), from presumed biosynthetic precursors **57** and **58**, respectively [52].

18.1.1.4 **Biomimetic Synthesis of Benzo[*kl*]xanthene Lignans**

Benzo[*kl*]xanthenes such as rufescidride (**59**), mongolicumin A (**60**), and yunnaneic acid H (**61**, Figure 18.5) represent a small and rare class of lignans [53–55], but

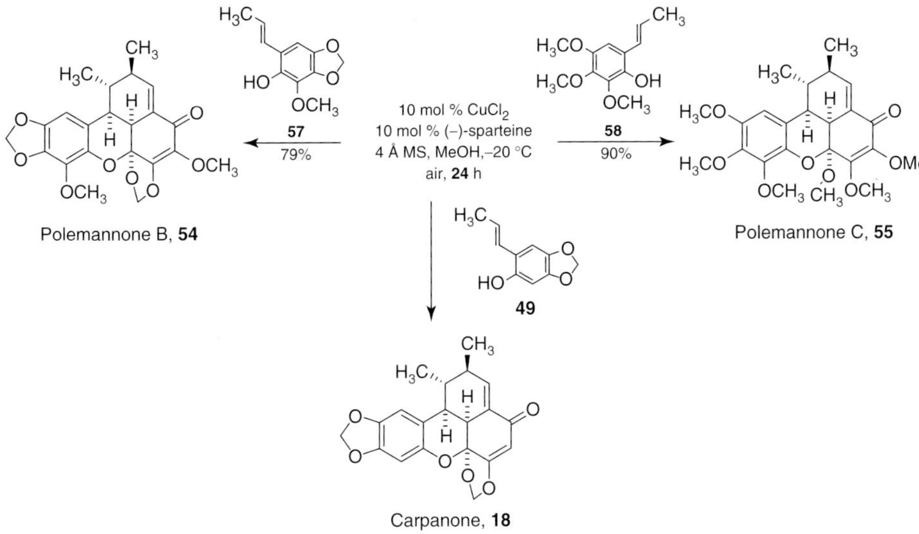

Scheme 18.9 Application of a CuCl$_2$/(−)-sparteine oxidative catalysts system for the total synthesis of carpanone (**18**), polemannone B (**54**) and polemannone C (**55**).

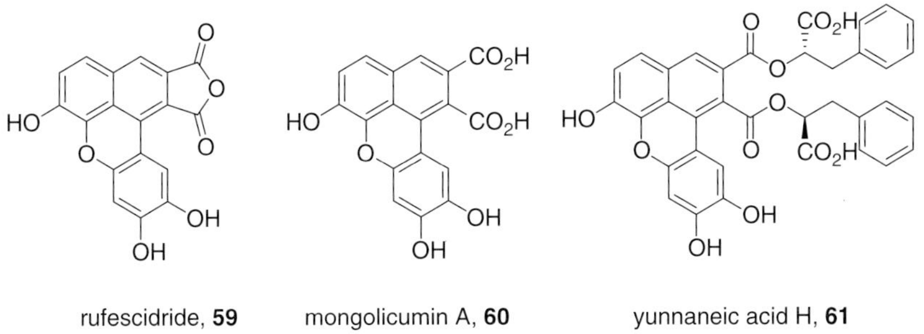

Figure 18.5 Representative benzo[*kl*]xanthene lignans **59–61**.

recent biomimetic synthesis reports have renewed interest in them [56]. In 2009, Tringali and coworkers initiated a study of the products from the biomimetic, metal-mediated oxidative phenolic coupling of caffeic acid phenethyl ester (CAPE, **62**) [56]. Treatment of CAPE (**62**) with MnO$_2$ in CH$_2$Cl$_2$ at room temperature for 4 h generated two lignan products, **63** (48%) and **64** (16.5%), with the remaining mass balance being un-reacted CAPE (**62**). This represents the first biomimetic synthesis of the benzo[*kl*]xanthene class of lignans (Scheme 18.10) [56].

The mechanisms illustrated in Scheme 18.11 to account for the formation of **63** and **64** differ only in the *exo* versus *endo* orientation of the *meta*-OCH$_3$ group in **B** (*exo*) and **B′** (*endo*). MnO$_2$ catalyzes a one-electron oxidation and, ultimately, the C8 radical quinone methides (**65**). An 8–8′ oxidative phenolic coupling provides regioisomers **A** and **A′**. Elimination provides either **B** or **B′**. A 2–7′ intramolecular

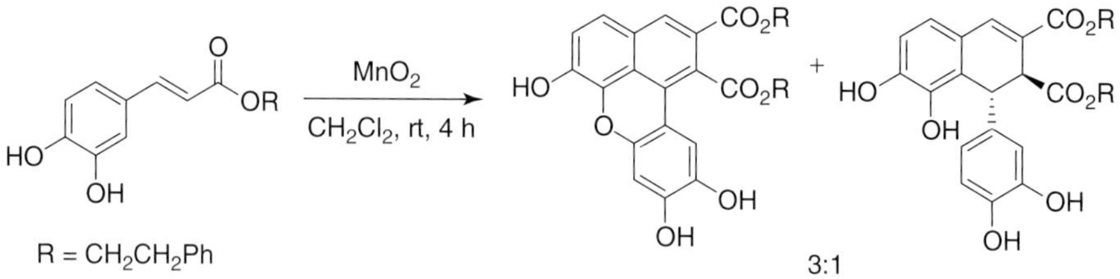

Scheme 18.10 Manganese-mediated, biomimetic synthesis of benzo[*kl*]xanthenes **63** and **64**.

cyclization of the electron-rich aromatic into the quinone methide **B** accounts for the formation of **64**; note that **B** and **B′** are conformational isomers and rapidly interconvert. A similar 2–7′ intramolecular cyclization with **B′** leads to **C′**. Two additional oxidation steps lead to **D′** and **E′**. Intramolecular nucleophilic 3–6′ attack of the *meta*-OH into the quinone methide affords **F′**, which can then tautomerize, accounting for the formation of **63** [56].

Interestingly, an earlier report by Lemiere and coworkers found that methyl caffeate (**66**) afforded predominantly **71**, a 5–8′ neolignan (Scheme 18.12), as opposed to the two lignans **63** and **64** produced under MnO_2 catalysis [56, 57]. Mechanistically, a 5–8′ phenolic coupling between **67** and **68** provides **69**, which can eliminate to generate **70**, followed by 4–7′ intramolecular cyclization to deliver racemic **71**. Tringali and coworkers then treated CAPE (**62**) with Ag_2O in CH_2Cl_2 at room temperature for 2 h and similarly formed the analogous 8–5′ product **71**. To determine if the Mn^{2+} ions are responsible for the observed divergent product pathways, CAPE (**62**) was again treated with Ag_2O, but this time in the presence of Mn(acac)$_2$ in CH_2Cl_2 at room temperature for 5 h, affording a 54% conversion of CAPE (**62**) into the benzo[*kl*]xanthene lignan **63** (51% yield), without any detectable 8–5′ neolignan **71** (Scheme 18.13). These results highlight the influence of the metal in directing the formation of either lignans or neolignans; moreover, these data suggest Mn^{2+} ions can somehow stabilize quinone methide radicals in a manner that favors 8–8′ oxidative phenolic couplings. It will be interesting to see if this ion selectivity (Ag^+ versus Mn^{2+}) is general for other cinnamyl substrates, providing control for the formation of various lignan and neolignan chemotypes [56].

18.2
Conclusion

While this chapter could not cover the full spectrum of lignan biomimetic syntheses reported, we focused on key syntheses and mechanistic understandings from the major classes of lignans. In the 75 years since "lignans" was coined by Haworth

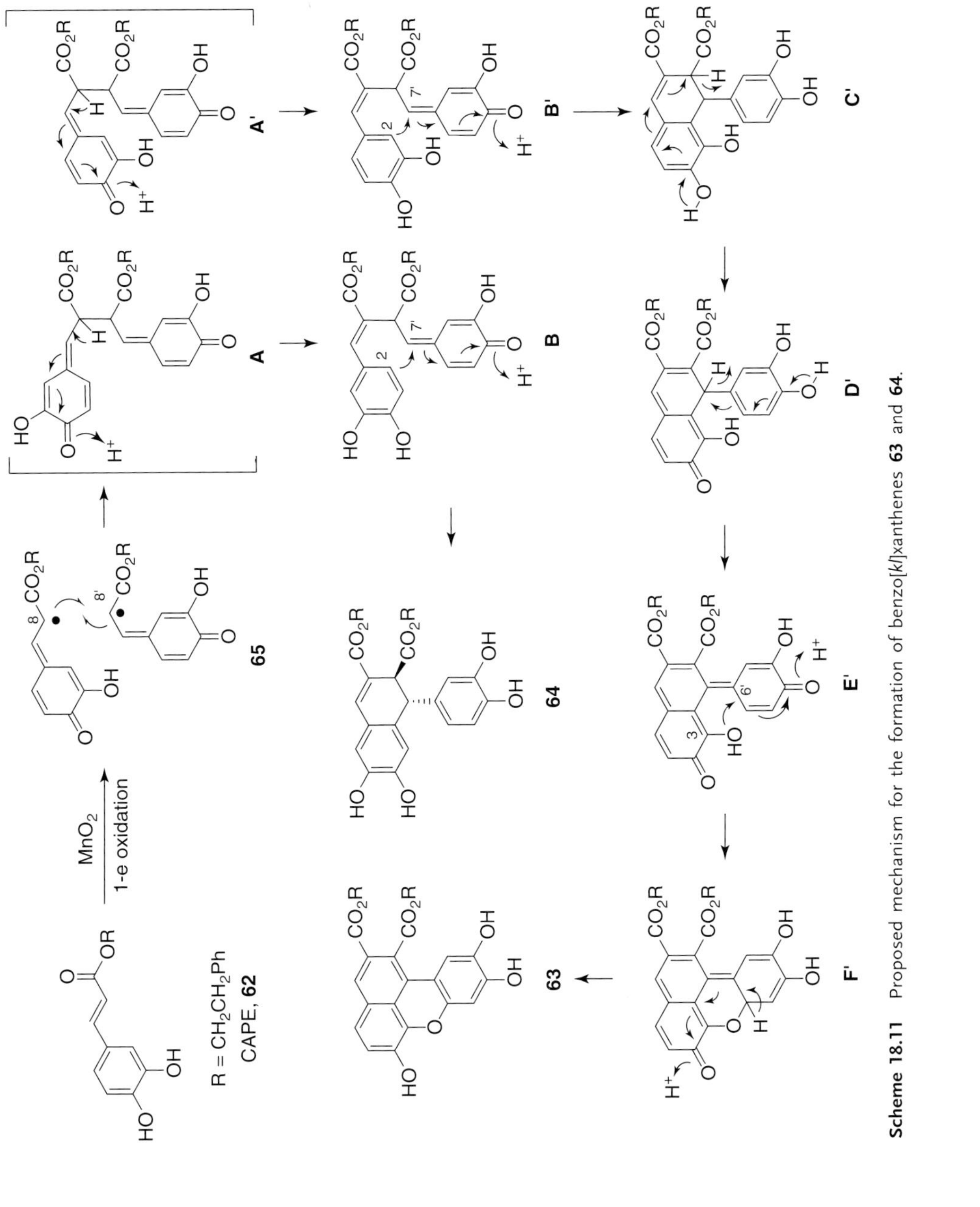

Scheme 18.11 Proposed mechanism for the formation of benzo[k/l]xanthenes **63** and **64**.

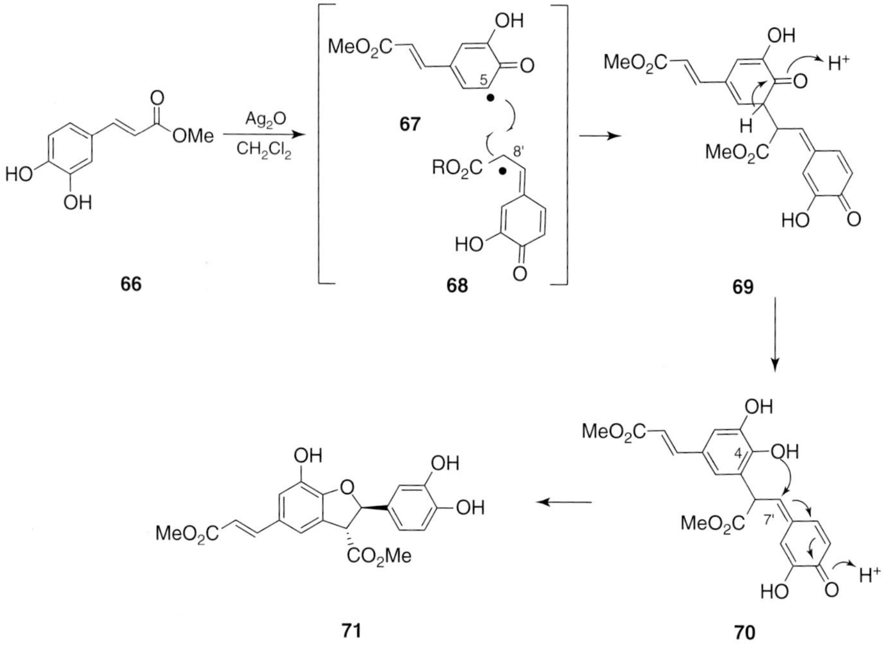

Scheme 18.12 Ag$_2$O-mediated oxidative phenolic coupling of methyl caffeate (**66**) forms racemic neolignan **71** through the proposed mechanism.

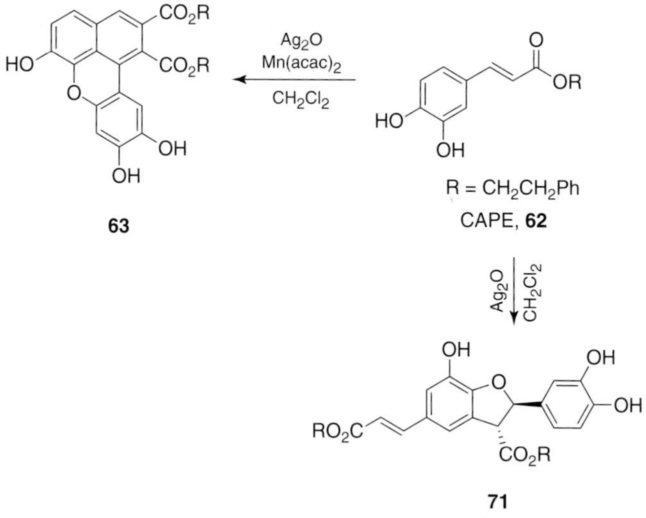

Scheme 18.13 Differential product formation in the presence of Mn^{2+} ions in the Ag$_2$O-mediated oxidative phenolic coupling of CAPE (**62**) to form either racemic neolignan **71** or benzo[*kl*]xanthene **63**.

to describe a group of phenylpropanoid dimeric natural products linked through the $\beta-\beta'(8-8')$ carbons of the propanyl side-chains, our understanding has grown tremendously in terms of the origin of enantioselectivity, of biomimetic approaches for $\beta-\beta'$ oxidative phenolic couplings (both metal- and enzyme-mediated), and "metal-tuning" to access distinct chemotypes via stabilization of certain radical pathways. Lignans truly highlight the power of biomimetic synthesis, accessing diverse stereochemically rich scaffolds from starting materials devoid of stereocenters. With a more complete arsenal of metal and enzyme catalysts available, the next decade holds even greater promise for the biomimetic synthesis of lignans.

References

1. Haworth, R.D. (1936) Heterocyclic compounds. *Annu. Rep. Prog. Chem.*, **33**, 310–334.
2. Haworth, R.D. (1942) Chemistry of the lignan group of natural products. *J. Chem. Soc.*, 448–456.
3. Ayres, D.C. and Loike, J.D. (1990) *Lignans – Chemistry, Biology and Clinical Properties*, Cambridge University Press, Cambridge.
4. Ward, R.S. (1995) Lignans, neolignans, and related compounds. *Nat. Prod. Rep.*, **12**(2), 183–205.
5. Wallis, A.F.A. (1998) *Lignin and Lignan Biosynthesis* (eds N.G. Lewis and S. Sarkanen) ACS Symposium Series, vol. 697, American Chemical Society, Washington DC, pp. 323–333.
6. Bose, J.S., Gangan, V., Prakash, R., Jain, S.K., and Manna, S.K. (2009) A dihydrobenzofuran lignin induces cell death by modulating mitochondrial pathway and G2/M cell cycle arrest. *J. Med. Chem.*, **52**(10), 3184–3190.
7. Van Miert, S., Van Dyck, S., Schmidt, T.J., Brun, R., Vlietinck, A., Lemiere, G., and Pieters, L. (2005) Antileishmanial activity, cytotoxicity and QSAR analysis of synthetic dihydrobenzofuran lignans and related benzofurans. *Bioorg. Med. Chem.*, **13**(3), 661–669.
8. Apers, S., Paper, D., Buergermeister, J., Baronikova, S., Van Dyck, S., Lemiere, G., Vlietinck, A., and Pieters, L. (2002) Antiangiogenic activity of synthetic dihydrobenzofuran lignans. *J. Nat. Prod.*, **65**(5), 718–720.
9. Charlton, J.L. (1998) Antiviral activity of lignans. *J. Nat. Prod.*, **61**(11), 1447–1451.
10. Syah, Y.M. and Ghisalberti, E.L. (1996) Biologically active cyanogenetic, iridoid and lignan glycosides from Eremophila maculata. *Fitoterapia*, **67**(5), 447–451.
11. Ward, R.S. (2000) in *Bioactive Natural Products (Part E)*, (ed. A.U. Raman, Studies in Natural Products Chemistry, vol. 24, Elsevier, Amsterdam, pp. 739–798.
12. Spatafora, C. and Tringali, C. (2007) in *Targets in Heterocyclic Synthesis*, vol. 11 (eds O.A. Attanasi and D. Spinelli), Societa Chimica Italiana, Italy, pp. 284–312.
13. Dewick, P.M. (2009) *Medicinal Natural Products: A Biosynthetic Approach*, 3rd edn, John Wiley & Sons, Ltd, Chichester.
14. Pan, J.-P., Chen, S.-L., Yang, M.-H., Wu, J., Sinkkonen, J., and Zou, K. (2009) An update on lignans: natural products and synthesis. *Nat. Prod. Rep.*, **26**, 1251–1292.
15. Brophy, G.C., Mohandas, J., Slaytor, M., Sternhell, M., Watson, T.R., and Wilson, L.A. (1969) Novel lignans from a *Cinnamomum* sp. from Bougainville. *Tetrahedron Lett.*, **10**, 5159–5162.
16. Gottlieb, O.R. (1972) Plant chemosystematics and phylogeny. III. Chemosystematics of the Lauraceae. *Phytochemistry*, **11**(5), 1537–1570.
17. Moss, G.P. (2000) Nomenclature of Lignan and Neolignan. IUPAC-IUB Joint Commission on Biochemical

Nomenclature, LG-1.2. Available online at http://www.chem.qmul.ac.uk/iupac/lignan/LG0n1.html#p12 (accessed 22 November 2010).

18. Ward, R.S. (1993) Lignans, neolignans, and related compounds. Nat. Prod. Rep., **10**(1), 1–28.
19. Davin, L.B., Wang, H.-B., Crowell, A.L., Debgar, D.L., Martin, D.M., Sarkanen, S., and Lewis, N.G. (1997) Stereoselective bimolecular phenoxy radical coupling by an auxiliary (dirigent) protein without an active center. Science, **275**, 362–366.
20. Davin, L.B., Jourdes, M., Pattern, A.M., Kim, K.-W., Vassao, D.G., and Lewis, N.G. (2008) Dissection of lignin macromolecular configuration and assembly: comparison to related biochemical processes in allyl/propenyl phenol and lignin biosynthesis. Nat. Prod. Rep., **25**, 1015–1090.
21. Ward, R.S. (1992) Synthesis of podophyllotoxin and related compounds. Synthesis, **8**, 719–730.
22. Ward, R.S. (2003) Different strategies for the chemical synthesis of lignans. Phytochem. Rev., **2**, 391–400.
23. Ziegler, F.E. and Schwartz, J.A. (1978) Synthetic studies on lignan lactones: aryl dithiane route to (±)-podorhizol and (±)-isopodophyllotoxin and approaches to the stagnane skeleton. J. Org. Chem., **43**, 985–991.
24. Pelter, A., Ward, R.S., Pritchard, M.C., and Kay, I.T. (1988) A short versatile synthesis of aryltetralin lignans including deoxypodophyllotoxin and epiisopodophyllotoxin. J. Chem. Soc., Perkin Trans. 1, 1603–1614.
25. Rodrigo, R. (1980) A stereo- and regiocontrolled synthesis of Podophyllum lignans. J. Org. Chem., **45**, 4538–4540.
26. Rajapaksa, D. and Rodrigo, R. (1981) A stereocontrolled synthesis of antineoplastic Podophyllum lignans. J. Am. Chem. Soc., **103**, 6208–6209.
27. Forsey, S.P., Rajapaksa, D., Taylor, N.J., and Rodrigo, R. (1989) Comprehensive synthetic route to eight diastereomeric Podophyllum lignans. J. Org. Chem., **54**, 4280–4290.
28. Wu, Y., Zhao, J., Chen, J., Pan, C., and Zhang, H. (2009) Enantioselective sequential conjugate addition-allylation reactions: a concise total synthesis of (+)-podophyllotoxin. Org. Lett., **11**, 597–600.
29. Broomhead, A.J., Rahman, M.M.A., Dewick, P.M., Jackson, D.E., and Lucas, J.A. (1991) Matairesinol as a precursor of Podophyllum lignans. Phytochemistry, **30**, 1489–1492.
30. Cambie, R.C., Craw, P.A., Rutlegde, P.S., and Woodgate, P.D. (1988) Oxidative coupling of lignans III. Non-phenolic oxidative coupling of deoxypodorhizon and related compounds. Aust. J. Chem., **41**, 897–918.
31. Ward, R.S. and Hughes, D.D. (2001) Oxidative cyclisation of cis- and trans-2,3-dibenzyl-butyrolactone using phenyl iodonium bis(trifluoroacetate) and 2,3-dichloro-5,6-dicyano-1,4-benzoquinone. Tetrahedron, **57**, 5633–5639.
32. Brown, R.C.D. and Swain, N.A. (2004) Synthesis of furofuran lignans. Synthesis, 811–827.
33. MacRae, W.D., Hudson, J.B., and Towers, G.H. (1989) The antiviral action of lignans. Planta Med., **55**(6), 531–535.
34. Anjaneyulu, A.S.R., Ramaiah, P.A., Row, L.R., Venkateswarlu, R., Pelter, A., Ward, R.S. (1981) New lignans from the heartwood of Cleistanthus collinus. Tetrahedron, **37**(21), 3641–3652.
35. Whiting, D.A. (1985) Lignans, neolignans, and related compounds. Nat. Prod. Rep., **2**, 191–212.
36. Whiting, D.A. (1987) Lignans, neolignans, and related compounds. Nat. Prod. Rep., **4**, 499–525.
37. Whiting, D.A. (1990) Lignans, neolignans, and related compounds. Nat. Prod. Rep., **7**, 349–360.
38. Whiting, D.A. (2001) Natural phenolic compounds 1900–2000: a bird's eye view of a century's chemistry. Nat. Prod. Rep., **18**(6), 583–606.
39. Sih, C.J., Ravikumar, P.R., Huang, F.C., Buckner, C. and Whitlock, H. Jr. (1976) Isolation and synthesis of pinoresinol diglucoside, a major antihypertensive principle of Tu-Chung

(*Eucommia ulmoides* Oliver). *J. Am. Chem. Soc.*, **98**(17), 5412–5413.
40. Cartwright, N.J. and Haworth, R.D. (1944) Constituents of natural phenolic resins. XIX. Oxidation of ferulic acid. *J. Chem. Soc.*, 535–537.
41. Erdtman, H. (1935) Dehydrogenation of phenols. *Svensk Kemisk Tidskrift*, **47**, 223–230.
42. Freudenberg, K. and Schraube, H. (1955) Sinapyl alcohol and the synthesis of syringaresinol. *Chem. Ber.*, **88**, 16–23.
43. Vermes, B., Seligmann, O., and Wagner, H. (1991) Synthesis of biologically active tetrahydrofurofuranlignan (syringin, pinoresinol) mono- and bisglucosides. *Phytochemistry*, **30**(9), 3087–3089.
44. Chapman, O.L., Engel, M.R., Springer, J.P., and Clardy, J.C. (1971) The total synthesis of carpanone. *J. Am. Chem. Soc.*, **93**, 6696–6698.
45. Matsumoto, M. and Kuroda, K. (1981) Transition metal (II) Schiff's base complexes catalyzed oxidation of trans-2-(propenyl)-4,5-methylenedioxyphenol to carpanone by molecular oxygen. *Tetrahedron Lett.*, **22**, 4437–4440.
46. Iyer, M.R. and Trivedi, G.K. (1992) Silver(I) oxide catalyzed oxidation of o-allyl- and o-(1-propenyl)phenols. *Bull. Chem. Soc. Jpn.*, **65**, 1662–1664.
47. Baxendale, I.R., Lee, A.I., and Ley, S.V. (2001) A concise synthesis of the natural product carpanone using solid-supported reagents and scavengers. *Synlett*, 1482–1484.
48. Lindsley, C.W., Chan, L.K., Goess, B.C., Joseph, R., and Shair, M.D. (2001) Solid-phase biomimetic synthesis of carpanone-like molecules. *J. Am. Chem. Soc.*, **122**, 422–423.
49. Sung, S.H. and Kim, Y.C. (2000) Hepatoprotective diastereomeric lignans from *Saururus chinensis* herbs. *J. Nat. Prod.*, **63**, 1019–1021.
50. Jakupovic, J. and Eid, F. (1987) Benzoxanthenone derivatives from *Polemannia montana*. *Phytochemistry*, **26**, 2427–2429.
51. Daniels, N.R., Fadeyi, O., and Lindsley, C.W. (2008) A new catalytic Cu(II)/sparteine oxidant system for β,β-phenolic couplings of styrenyl phenols: total synthesis of carpanone and unnatural analogues. *Org. Lett.*, **10**, 4097–4100.
52. Fadeyi, O., Daniels, R.N., DeGuire, S., and Lindsley, C.W. (2009) Total synthesis of polemannone B and C. *Tetrahedron Lett.*, **50**, 3084–3087.
53. Da Silva, A.A.S., Souto, A.L., Agra, M.F., da-Cunha, E.V.L., Barbosa-Filho, J.M., da Silva, M.S., and Braz-Filho, R. (2004) A new arylnaphthalene type lignan from *Cordia rufescens* A. DC. (Boraginaceae). *Arkivoc*, **6**, 54–58.
54. Shi, S., Zhang, Y., Huang, K., Liu, S., and Zhao, Y. (2008) Application of preparative high-speed counter-current chromatography for separation and purification of lignans from *Taraxacum mongolicum*. *Food Chem.*, **108**, 402–406.
55. Tanaka, T., Nishimura, A., Kouno, Y., Nonaka, G., and Yang, C.-R. (1997) Four new caffeic acid metabolites, yunnaneic acids A-H, from *Salvia yunnanensis*. *Chem. Pharm. Bull.*, **45**, 1596–1600.
56. Daquino, C., Rescifina, A., Spatafora, C., and Tringali, C. (2009) Biomimetic synthesis of natural and 'unnatural' lignans by oxidative coupling of caffeic esters. *Eur. J. Org. Chem.*, 6289–6300.
57. Pieters, L., Van Dyck, S., Gao, M., Bai, R., Hamel, E., Vlietinck, A., and Lemiere, G. (1999) Synthesis and biological evaluation of dihydrobenzofuran lignans and related compounds as potential antitumor agents that inhibit tubulin polymerization. *J. Med. Chem.*, **42**(26), 5475–5481.

19
Synthetic Approaches to the Resveratrol-Based Family of Oligomeric Natural Products

Scott A. Snyder

> Look deep into nature, and then you will understand everything better.
>
> *Albert Einstein*

19.1
Introduction

Although the natural product resveratrol (**1**, Figure 19.1) is quite simple from the standpoint of molecular architecture, its two aromatic rings, double bond, and three phenol residues confer an ability to accomplish a profound array of biochemistry [1]. Indeed, recent screens in various *in vivo*, mouse-based assays have shown that it can serve as a potent anti-oxidant and antitumor agent as well as improve neuronal functioning, slow the aging process, and even effect weight loss [2]. In addition, although unconfirmed, many have extended these preliminary findings into a hypothesis that resveratrol impacts human health in similar ways, with its main source in our diets being the consumption of wine, particularly red variants [3]. The goal of this chapter, however, is not to evaluate these claims nor focus on the biological properties of resveratrol, profound as they may be. Rather, it is to discuss the main reason that many plants create this intriguing natural product and explore what can then be done chemically with its atoms in terms of bond constructions.

Like most natural products, resveratrol (**1**) is produced largely to enable organismal survival. In this case, it is a phytoalexin whose protective role is potent antifungal activity that enables the over 70 species of plants throughout the world that synthesize it to combat (and hopefully survive) a number of exogenous attacks [3]. Grapevines, for instance, utilize resveratrol to manage the effects of the *Botrytis* fungus, a disease that many vintners welcome since it leads to grapes with intensified flavor but which comes with the risk of total harvest loss if not properly managed [4]. Resveratrol, however, is just the tip of the iceberg in terms of any plant's total antifungal arsenal. Indeed, once infected, plants will also use resveratrol as a powerful synthetic building block, one that can be joined in myriad ways to create new and architecturally distinct polyphenolic natural products such

Biomimetic Organic Synthesis, First Edition. Edited by Erwan Poupon and Bastien Nay.
© 2011 Wiley-VCH Verlag GmbH & Co. KGaA. Published 2011 by Wiley-VCH Verlag GmbH & Co. KGaA.

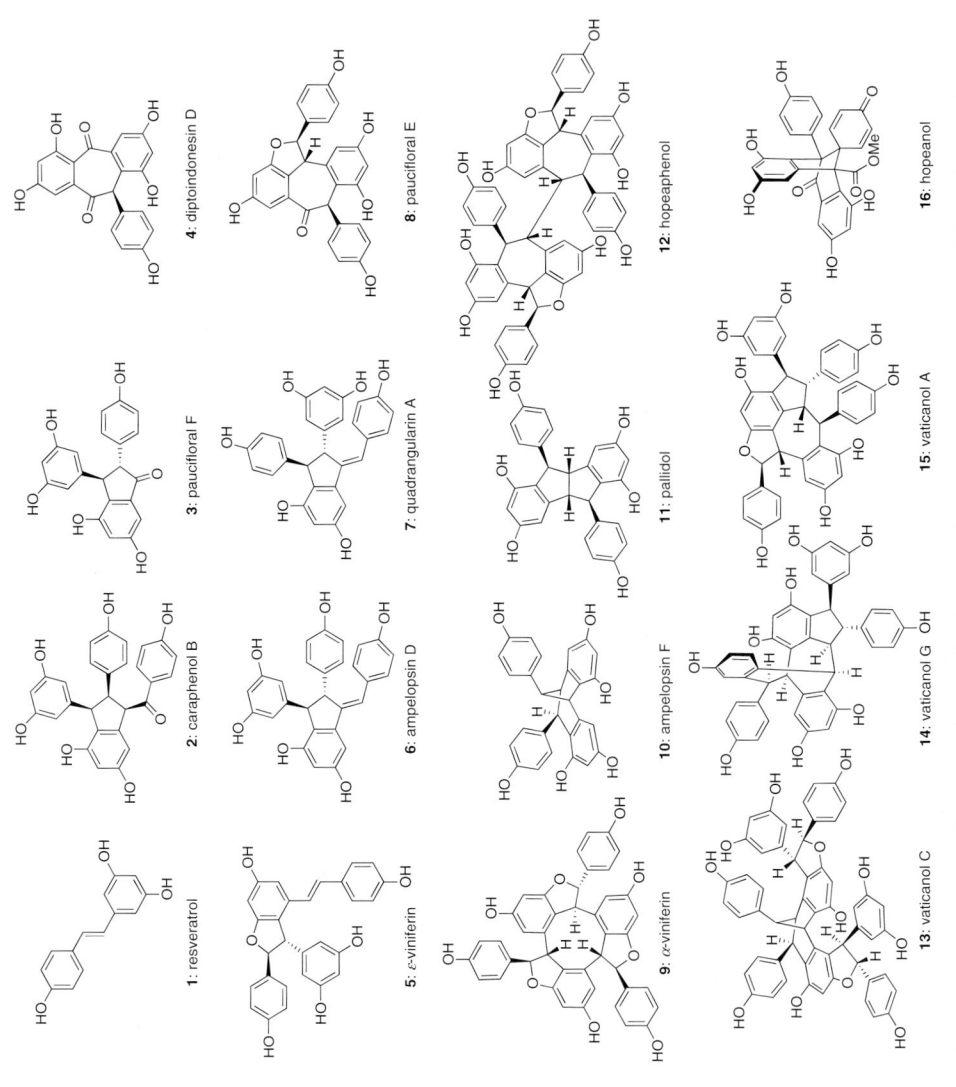

Figure 19.1 Structures of resveratrol (1) and selected oligomers (2–16) representative of the architectural complexity of the family.

as **2–16** [5]. These structures, along with the approximately 500 others [6] that have been isolated to date, encompass diverse carbocyclic and heterocyclic systems that display fascinating stereochemical complexity; they also possess broad spectrum biological activity that includes antifungal properties, of course, but also leads to materials that could, in some cases, fight cancer, HIV, and bacteria as well [5]. In this way, resveratrol is the lynchpin for a cohesive, immune-like, chemical response wherein an array of structurally unique natural products (anywhere from ten or more from one plant) are marshaled concurrently to thwart an invading pathogen in the hope that at least one of the molecules has enough power to enable organismal survival [7].

In the ensuing sections, we present the current level of understanding regarding how these oligomers are formed in Nature as well as the synthetic efforts that have been directed toward fashioning their fascinating chemical diversity. As will be seen, much work remains to be done, but, hopefully, as several studies over the past few years have demonstrated, chemists are finally beginning to gain a handle on how to tackle the chemical complexity presented by the family, complexity that the parent molecule's structure belies. In fact, in our opinion, that degree of diversity is without parallel in terms of any other oligomeric natural product family in existence.

19.2
Biosynthetic Approaches

Resveratrol is the product of an enzyme-based synthesis, one that sequentially adds three molecules of malonyl-CoA to 4-coumaroyl-CoA with a terminating enzymatic cyclization using "stilbene synthase" to fashion its 3,5-dihydroxybenzene ring [8]. Beyond the building block itself, however, far less is known with regard to how the dimeric and higher-order structures are fashioned; enzymatic participation is often invoked because the oligomers are obtained as single enantiomers, though at what point that involvement occurs, and how it occurs (i.e., whether to create proper reactive intermediates, ensure correct presentation of monomers to create a specific dimer, etc), is unknown. Heightening that mystery is the fact that certain plants will produce the antipode of a given natural product forged enantioselectively by another species [9].

Nevertheless, despite such uncertainty, the structures within Figure 19.1 hint at several possibilities for their general mode of construction based on certain homologies. For instance, if resveratrol (**1**) was dimerized in an uncontrolled way into a range of structures such as **5–8**, these molecules could then serve as launching points for the generation of many of the higher-order and/or more unique structures due to their possession of reactive functionality. For instance, oxidative cleavage of the double bond within ampelopsin D (**6**) could lead to the triaryl-containing natural product pauciforal F (**3**), while regioselective oxidation of that function followed by enol equilibration could afford caraphenol B (**2**). Alternatively, electrophilic activation of the double bond within **6**, followed by a Friedel–Crafts reaction with the neighboring 3,5-dihydroxybenzene system, could lead to a structure such as ampelopsin F (**10**). These cores, along with all of the others, could then be further elaborated with additional resveratrol monomers; for instance, vaticanol C (**13**) formally has two resveratrol units attached stereospecifically onto ampelopsin F (**10**). Thus, if we were to encapsulate these ideas within the context of the opening discussion of this section, the higher-order structures could potentially result as single enantiomers through relative stereocontrol if Nature had prepared its given dimeric precursor with optical purity.

The pathway by which most of the bond-forming chemistry occurs in Nature appears to be through radical bond constructions. More specifically, only partially controlled, or even non-controlled, radical-based constructions [3]. Scheme 19.1 shows a typical biosynthetic proposal for these higher-order structures, using ε-viniferin (**5**) for purposes of illustration. As indicated, initial oxidation of resveratrol (**1**) gives rise to two different radicals, one resonance stabilized (**17**) and one on the 3,5-dihydroxybenzene system (**18**). These species then unite to form **19**, which, following rearomatization and phenol attack onto the remaining quinone methide, leads to the target structure (**5**). The *trans*-orientation of the two rings, a hallmark of nearly all the dihydrofuran rings within the resveratrol family, is likely the result of equilibration through ring-opening to produce the most thermodynamically stable isomer [10]. Of course, though the mechanism as shown suggests high control in radical positioning (as many biosynthetic hypotheses for this molecule class similarly implicate), such control is quite difficult to achieve in practice. Indeed, **18** is more difficult to form than **17** and the mechanistic pathway assumes that only the specific radicals drawn, and not their varied resonance alternatives, react. Thus, many pathways in addition to that shown for **5** are likely to occur concomitantly, thereby leading to many products. This statement is reflected further in several interesting biosynthetic explorations undertaken during the past three decades, studies that have revealed that (i) synthesizing resveratrol-based natural products in a controlled way is nearly impossible to achieve through biosynthetic approaches (in line with Nature's outcome) and (ii) when a single structure is generated in the laboratory with some measure of control (i.e., in yields greater than 50%) it is often an analog/structural isomer of a natural product rather than a natural product.

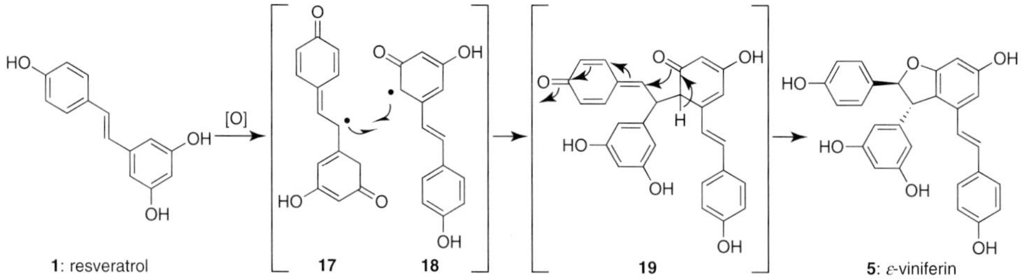

Scheme 19.1 Representative biosynthetic scheme showing the putative generation of ε-viniferin (**5**) from resveratrol (**1**).

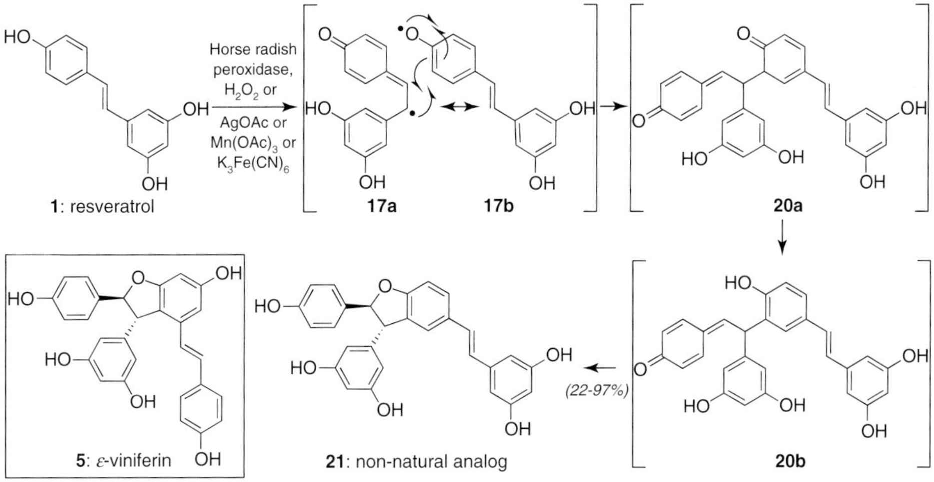

Scheme 19.2 Efforts to effect direct resveratrol (**1**) dimerization typically result in non-natural analogs of resveratrol dimers (such as **21**).

The first reported exploration into the reactivity of resveratrol was disclosed by Langcake and Pryce in 1977 [11], investigations that evaluated the products generated from exposure of resveratrol (**1**, Scheme 19.2) to horseradish peroxidase in the presence of H_2O_2. Of the materials formed, only compound **21**, the likely product of O- and C-centered radicals uniting through the most readily oxidized phenol within resveratrol, was characterized. As can be seen with a cursory inspection, though its dihydrofuran ring is reminiscent of the family, the architectural connectivity does not match that of ε-viniferin (**5**), the form most commonly found in Nature. A more recent study of the same process by Sako and coworkers using single-electron transfer reagents as initiators instead of an enzymatic system obtained similar results [12]. In fact, certain reagents led to **21** quantitatively, but no ε-viniferin (**5**) was ever characterized from these experiments.

Of course, in most investigations of this type, the yield is not quantitative, and as such it is certainly possible that within the collection of additional materials produced several natural products can be found. For instance, in a study of resveratrol oligomerization using *Anthromyces ramosus* in the presence of H_2O_2, the investigators were able to obtain pure **22** (Scheme 19.3) as well as the natural product pallidol (**11**), albeit in low yield relative to the common, major product (i.e., **21**) presented above [13]. To date, only one report describes a chemical synthesis of ε-viniferin (**5**) with any measure of efficacy; that work, shown in Scheme 19.4, found that if resveratrol was exposed to $FeCl_3 \cdot 6H_2O$ in methanol as solvent, **5** could be obtained in a 30% yield [14]. Thus, these examples (Schemes 19.2–19.4) collectively reveal that with enough trial and error it is possible to forge a given scaffold with some preference over others through radical-based chemistry starting with resveratrol, but, nevertheless, not with any real degree of control or *de novo* predictability based on first principles of chemical reactivity.

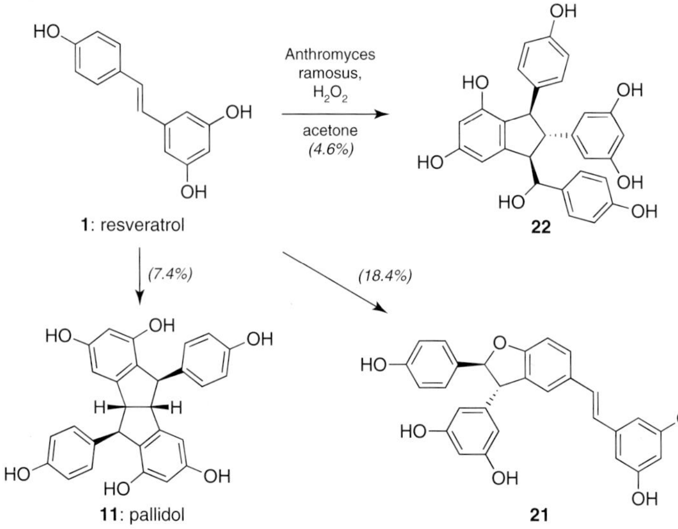

Scheme 19.3 A second example of enzyme-induced oxidative dimerizations of resveratrol (1).

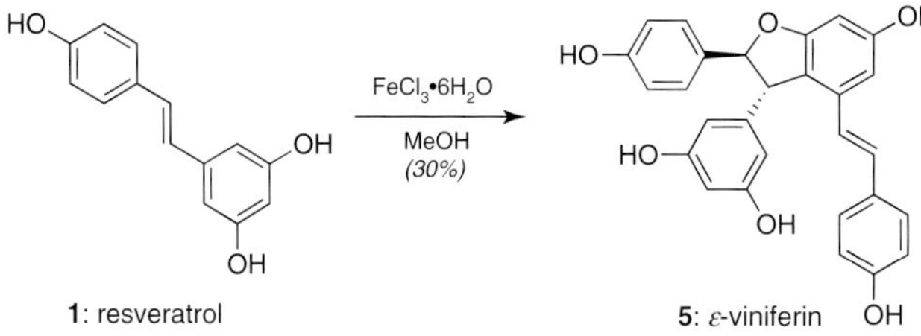

Scheme 19.4 Conditions that are reported to provide ε-viniferin (5) from resveratrol (1).

More recent efforts, however, have taken a different approach to addressing the challenge of controlled radical generation on the resveratrol framework. Rather than utilizing resveratrol itself as the starting material, these investigations have asked what products would arise from derivatives of resveratrol and/or selectively protected forms through similar oxidation chemistry. From the standpoint of biomimetic chemistry, these investigations might make more inherent sense in that Nature could certainly temporarily engage a given phenol in resveratrol with an acetate, phosphate, sulfonate, or carbohydrate group, thereby altering the radicals that could be generated and thus leading to different product distributions [15]. Indeed, since each plant seems to synthesize different arrays of resveratrol-based oligomers (in terms of specific structures and their respective amounts), such tuning might reflect the mechanism whereby plants have evolved their compound collections in certain directions based on selective evolutionary pressure. Natural product isolation procedures may be overly masking the actual presence of these groups by effecting their excision from the phenol subunits prior to characterization.

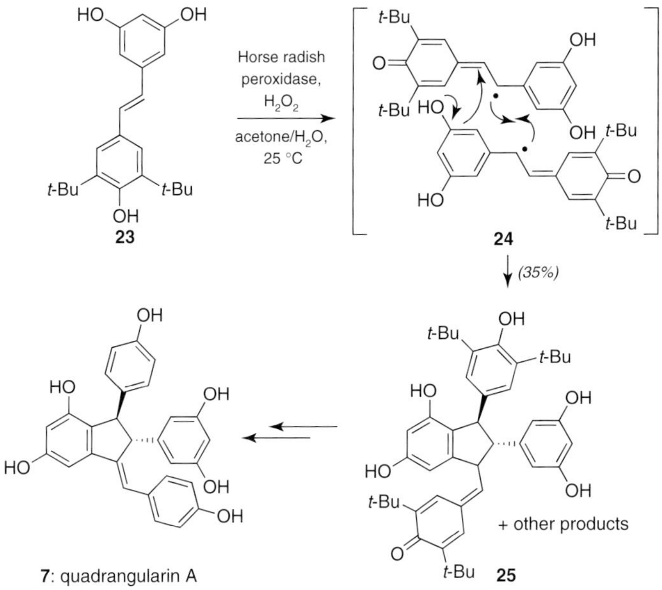

Scheme 19.5 Use of resveratrol derivative **23** to achieve a total synthesis of quadrangularin A (**7**).

The first landmark study along these lines was reported by the Hou group in 2006, in which the *t*-butylated resveratrol analog **23** (Scheme 19.5) was exposed to horseradish peroxidase and H_2O_2 in aqueous acetone [16]. Such a design took advantage of the more facile generation of radicals at the *para*-disposed phenol that the earlier reports had consistently mentioned, but used the bulky *t*-butyl groups to ensure that it would be the carbon-centered resonance-forms of the original radical that would be the primary reactive species. As a result, they were able to obtain indane **25** in 35% yield. Subsequently, these investigators were able to remove the *t*-butyl "protective" groups through exposure to strong acid and ultimately access the natural product quadrangularin A (**7**). Whether this design will enable access to other resveratrol-based cores remains the subject of additional investigations, but it certainly illustrates the power of a different approach as relates to the controlled resveratrol oligomerization problem, which did not appear possible starting from resveratrol itself.

Similarly, work by Velu and coworkers sought to examine the reactivity of dimethylated resveratrol analog **26** (Scheme 19.6), a material that can only form radicals from one phenol [17]. Their study showed a remarkable ability to fashion unique chemical diversity as a result of this structural alteration, principally just by changing the solvent in which they used a given oxidant. For instance, if **26** was exposed to $FeCl_3 \cdot 6H_2O$ in a mixture of CH_2Cl_2–MeOH (7 : 3), the unique tetrahydrofuran **28** was produced as the predominant product, while exposure of the same material to the same oxidant in CH_2Cl_2 alone gave rise to a mixture that contained both **29** and **30**, protected forms of the natural products ampelopsin

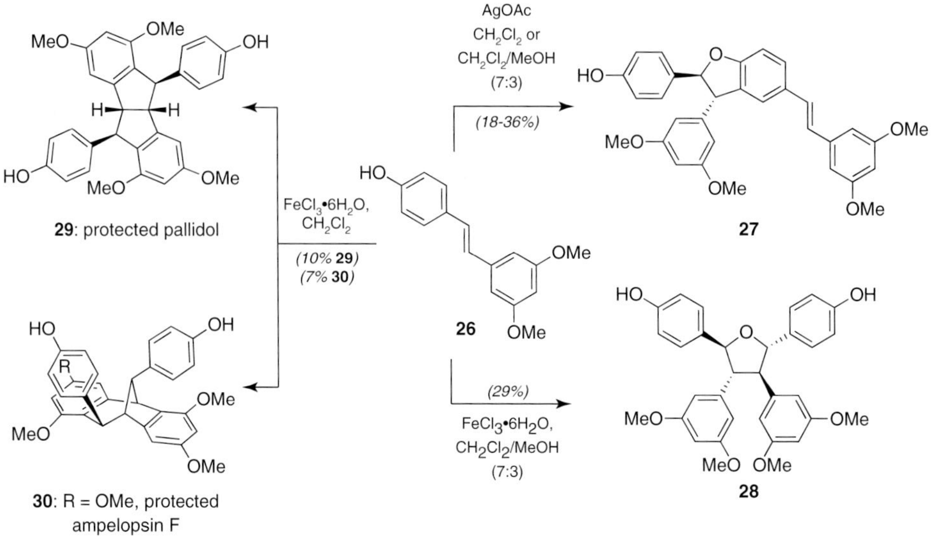

Scheme 19.6 Explorations into the effect of partial phenol protection of resveratrol (i.e., **26**) on dimeric product formation.

F (**10**) and pallidol (**11**, cf. Figure 19.1). In general, however, most compounds formed are not natural products, though perhaps the most important conclusion from such an outcome to date is that there are many frameworks in addition to those isolated from Nature that can be formed from resveratrol's atoms and functional groups.

In addition to altered resveratrol fragments, an intriguing biomimetic approach has been to examine what can be produced from a given higher-order structure. In other words, the question is not what materials can be made from resveratrol itself but rather what frameworks could result from a given dimer through exposure to different conditions. The Niwa group has been the main contributor to such endeavors, and Scheme 19.7 illustrates some of their seminal experiments. Here, ε-viniferin (**5**) was viewed as a critical starting point, and, in fact, upon exposure to various acid sources, these investigators obtained isoampelopsin D (**33**), ampelopsin D (**6**), and ampelopsin F (**10**) along with several other natural products in differing amounts [18]. Mixtures were always obtained, largely as a result of the different stereochemistry that can result from the ring-opening and reclosing process proposed (i.e., access to either **32a** or **32b**), but their study highlights the provocative notion that ε-viniferin (**5**), not resveratrol itself, could be the real lynchpin to many of the higher-order structures of the family since much of the rest of the family, as noted above, could derive from this array of new dimeric structures.

In addition to exploring the reactions of ε-viniferin (**5**) under rearrangement conditions, the Niwa group has also evaluated what it can do when oxidized in the presence of resveratrol [9b]. As one might expect, complicated mixtures are

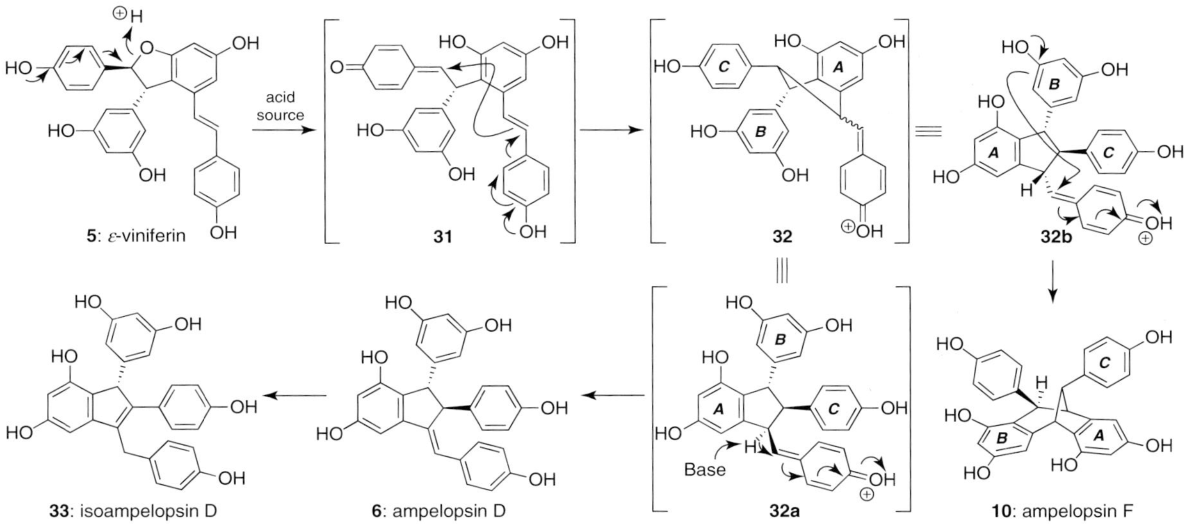

Scheme 19.7 Biogenetic explorations using ε-viniferin (**5**) instead of resveratrol as the starting material.

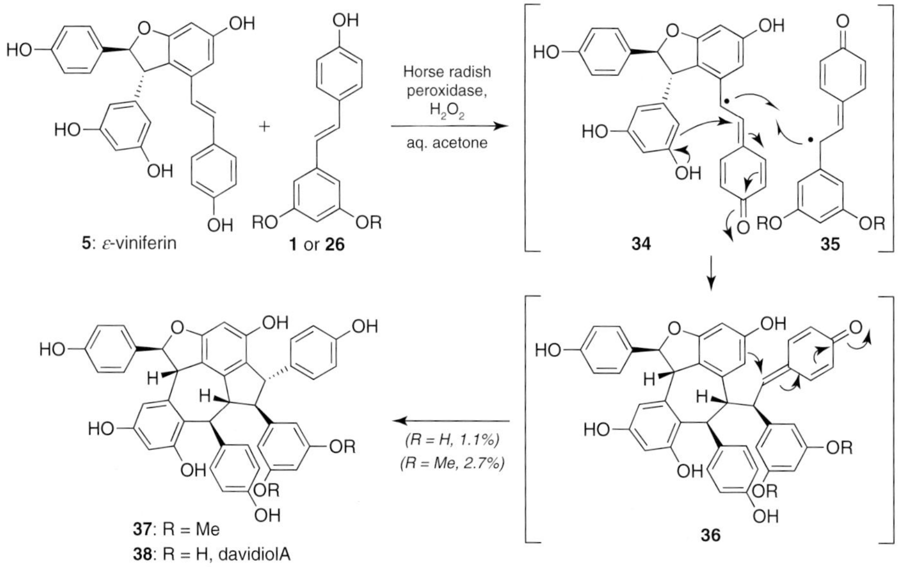

Scheme 19.8 Biomimetic total synthesis of davidiol A (**38**) starting from ε-viniferin (**5**) and resveratrol (**1**).

formed. Nevertheless, out of this milieu they were recently able to isolate the natural product davidiol A (**38**, Scheme 19.8), albeit in only 2.7% overall yield if a protected form of resveratrol (**26**) was used in the process and in 1.1% yield if it was resveratrol itself. This natural product is one of many diastereomers possessing a fused 7,5-bicyclic system, of which vaticanol A (**15**, cf. Figure 19.1) is another example.

Finally, outside of radical chemistry, there are other ways to envision the biogenetic synthesis of resveratrol-like frameworks. Carbocation-induced reactions, not so unlike the work discussed above in the context of Scheme 19.7, are one such option. The key results of this approach are shown in Scheme 19.9, wherein select examples of stilbene derivatives of general structure **39** have been shown to lead to various indane (**43**) and tetralin (**42**) ring systems reminiscent of the resveratrol family following regioselective cation generation formed through a retro-Ritter reaction [19]. To date, however, the extension of such studies to the exact resveratrol phenol patterning has yet to afford the same carbon skeletons. Thus, it remains to be seen whether such acid-induced rearrangements can afford pertinent resveratrol-based oligomers [20]. There is at least, in our opinion, one other mechanistic possibility for resveratrol oligomer formation: photochemical resveratrol dimerization followed by subsequent rearrangements. To our knowledge this particular approach has never been pursued, though it is one additional alternative that, at least based on simple mechanistic pictures, could afford many of the desired frameworks as well. Recent synthesis work targeting a marine-based family of varied dimeric structures provides the impetus behind this idea [21].

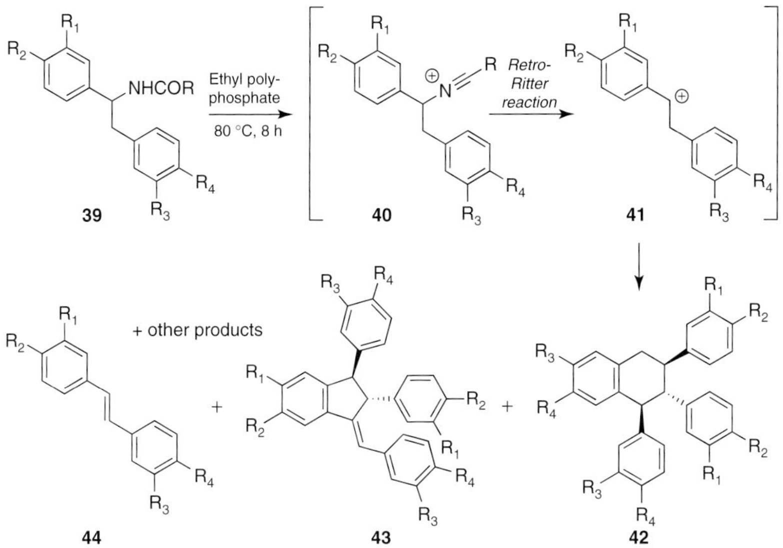

Scheme 19.9 Example of non-radical-based dimerizations; the use of this chemistry with appropriately functionalized resveratrol derivatives has not yet proven effective.

19.3
Stepwise Synthetic Approaches

19.3.1
Work toward Single Targets within the Resveratrol Family

While much has clearly been learned from the biogenetic approaches above, the resveratrol family has also served as a testing ground for the power of various synthetic methods, as well as more stepwise, non-biomimetic approaches. In this section, we will present three recent studies along such lines [22].

The first was performed by Jeffrey and Sarpong, wherein various palladium-catalyzed processes were evaluated for their ability to produce resveratrol-like dimeric structures [23]. As indicated in Scheme 19.10, materials possessing a brominated 3,5-dimethoxybenzene ring system provided the critical handle by which to add additional carbogenic complexity, with a Heck-type reaction cascade between **45** and **46** affording indene **47** in 53% yield and a Larock annulation between **49** and the same alkyne (**46**) leading to a 1 : 1 mixture of **50** and **51**. Although none of these frameworks themselves are natural products, **47** is an oxidized form of quadrangularin A (**7**, cf. Figure 19.1) and could be converted into a similarly overoxidized form of pallidol (**11**, cf. Figure 19.1), while **51** could be transformed into paucifloral F (**3**) in two additional steps.

The second approach was executed by Kim and Choi in an effort to prepare the seven-membered carbocyclic ring system possessed by various members of the resveratrol family [24]. These investigators targeted the natural product

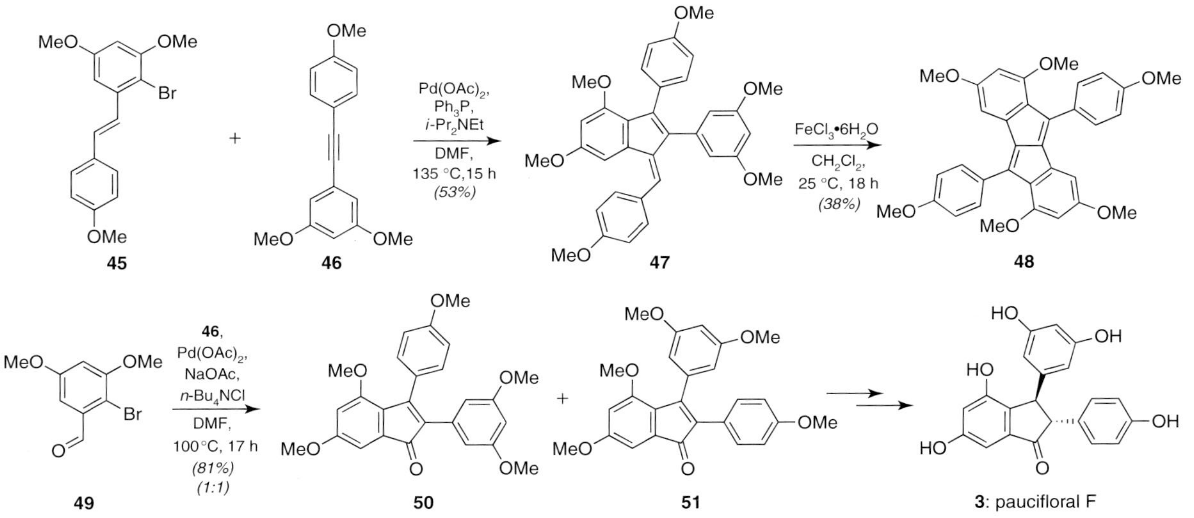

Scheme 19.10 Use of palladium-based reactions to forge resveratrol-based cores as well as to synthesize pauciforal F (**3**).

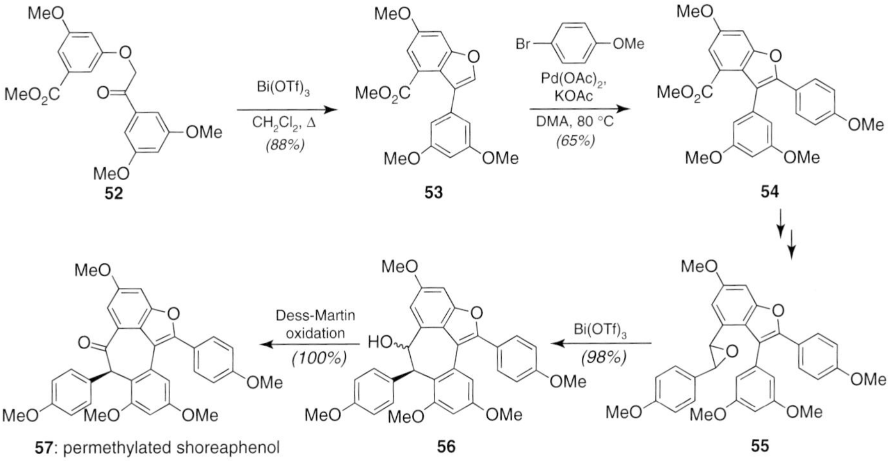

Scheme 19.11 Synthesis of a permethylated form of shoreaphenol (**57**) based on two key Bi(OTf)$_3$-catalyzed reactions.

shoreaphenol specifically [25], which they were able to access in protected form (**57**, Scheme 19.11) through a series of Bi(OTf)$_3$-catalyzed reactions [26]. The first of these Lewis acid-induced operations proved critical in forging the furan ring of the target through a Friedel–Crafts cyclization onto the ketone group within **52** followed by dehydration. Then, following the attachment of an aryl ring onto the open C2 position of the resultant benzofuran (i.e., **53**) and the execution of several additional steps to generate **55**, Bi(OTf)$_3$ was utilized again to promote ring-opening of the lone epoxide and final C–C bond construction. This final step is certainly biomimetically inspired as it reflects a possible mode for Nature's formation of seven-membered rings within the family.

Our final example targeted a very unique set of natural products within the resveratrol family, molecules whose biogenetic origin is unclear as their architectural core is unlike any other structures in the family. These compounds are hopeanol (**16**) and hopeahainol A (**64**, Scheme 19.12), first isolated by Tan and coworkers in 2008 and shown to possess modest antitumor activity profiles [27]. Structurally, these polycycles differ primarily in terms of their aryl oxidation patterning, a difference that then determines either the presence or absence of a lactone function as well as an additional C–C bond. In early 2009, Nicolaou, Chen, and coworkers accomplished the total synthesis of these two targets, illustrating in the process that hopeahainol A (**64**) is likely Nature's precursor to hopeanol (**16**), as a final exposure of the first natural product to NaOMe in MeOH afforded the other smoothly in 80% yield [28]. Critical to the sequence were two sequential cascade-based reactions, the first starting from **59**, which used *p*-TsOH to effect a multistep conversion affording the rearranged polycycle **61**, and the second being a ring-opening/lactonization sequence which then converted that new material into **63**. Epoxidation of the lone double bond within **63**, followed by Friedel–Crafts cyclization on the resultant

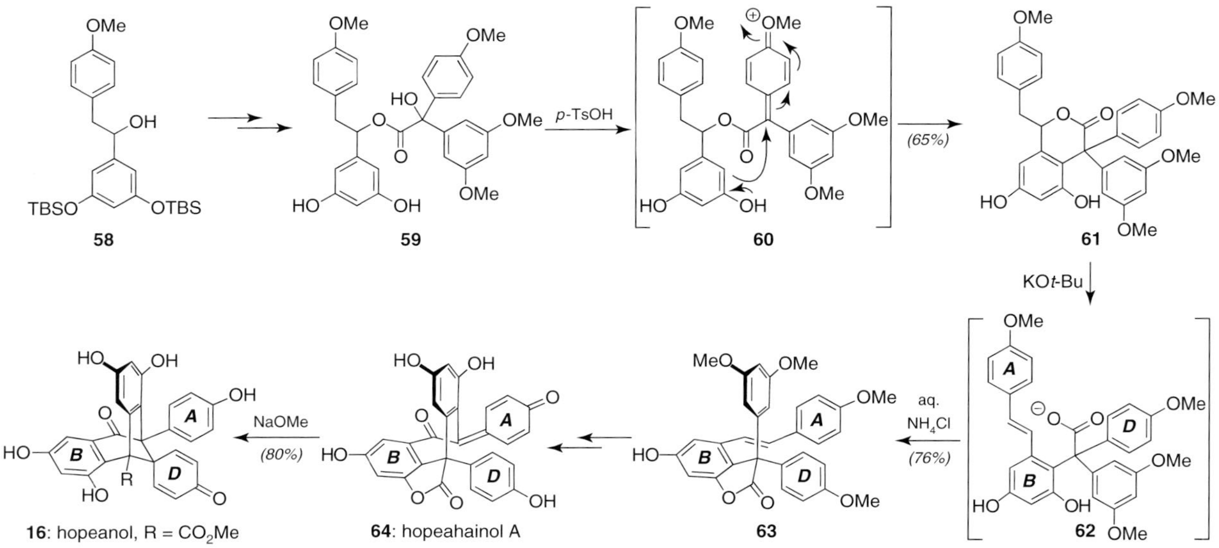

Scheme 19.12 Nicolaou and Chen's synthesis of hopeahainol A (**64**) and hopeanol (**16**) via iterative cascade-based rearrangements.

product, forged the final bonds needed to reach hopeahainol A (**64**), much in the manner described above for the Kim and Choi synthesis of protected shoreaphenol.

19.3.2
Towards a Universal, Controlled Synthesis Approach

Despite the elegance of the biosynthetic studies described earlier and the utility of the stepwise approaches just presented in the preceding section, neither of these avenues proved capable of selectively delivering all of the main resveratrol-derived frameworks in a controlled manner. Over the course of the past three years, our research group has sought to develop a solution to this challenge. Specifically, we wondered if an additional synthetic approach might exist that, even if stepwise in design, could enable access to every carbocyclic framework within the family. Given the failure of biosynthetic studies to take resveratrol itself, or variously protected forms thereof, into single natural product structures in high yields, we felt that a different starting material was required, one which would possess some of the reactivity of the parent molecule for the family, but which could be tempered appropriately to allow for controlled reactions upon exposure to different reagents in ways that the other approaches could not achieve.

Our search for that alternate starting material began by examining all the known structures of the family, looking for any anomalies that might serve as a potential clue for its design. Those anomalies were found in the form of molecules such as paucifloral F (**3**, cf. Figure 19.1) in that it, like several others in the family, has an odd number of aromatic rings. Although these structures likely result from normal resveratrol dimers, as mentioned earlier, it was these structures that prompted the idea that perhaps the key building block needed to be one that possessed three aromatic rings; specifically, something like compound **65** (Scheme 19.13),

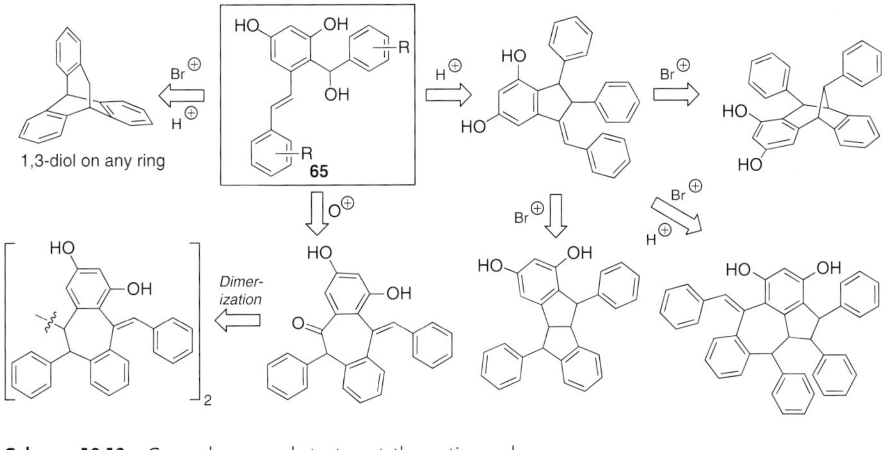

Scheme 19.13 General approach to target the entire carbocyclic diversity of the resveratrol class based on the use of key intermediate **65**.

with alteration of the bottom, resveratrol-like, ring coupled with an additional functional group and modifiable third aryl system. Our hope was that this piece and its derivatives, following exposure to various different electrophiles (bromine, oxygen, and proton), could enable the controlled generation of all of the frameworks indicated, carbon cores that reflect much of the diversity of the entire family. As the remainder of this section will reveal, those efforts have borne fruit as nearly all of the pathways defined within Scheme 19.13 have been reduced to practice. In fact, over 15 natural products and dozens of analogs encompassing eight distinct architectures have been prepared from this critical building block to date [29].

We begin our discussion of these efforts by describing our approach to indane-containing resveratrol oligomers. As shown in Scheme 19.14, compound **66** was exposed to an acid source under controlled temperature conditions, which led to the cyclized carbocation **68** via initial ionization followed by nucleophilic attack of the proximal olefin. What happened next, however, depended on the nature of the acid counterion. When that ion was trifluoroacetate, that species itself served as a nucleophile, attacking **68** to forge, following basic work-up, the secondary alcohol **69** in 75% yield. This material could then be converted into paucifloral F (**3**) in just two additional steps. However, if an acid with a less nucleophilic counterion, such as *p*-TsOH, was used instead in the cyclization leading to **68**, then it proved possible to add other nucleophiles and access different complexity. For instance, with chloride present, molecules such as **73** were accessed smoothly while the addition of excess *p*-methoxybenzyl sulfide afforded compound **70**. Use of a Ramberg–Bäcklund olefination [30] with the latter of these materials (i.e., **70**), followed by global methyl ether cleavage, completed a synthesis of ampelopsin D (**6**) in 18% overall yield for these final steps. From here, a final treatment with HCl in MeOH at elevated temperature (80 °C) provided the means to access isoampelopsin D (**39**) as well; intriguingly, only the specific tetra-substituted olefin was formed through this isomerization process. Thus, simple alteration of reaction conditions allowed for the controlled synthesis of three different indane-containing resveratrol oligomers. Pleasingly, application of the same chemistry to alternative forms of the key building block such as **71** led to the smooth formation of related natural products with their pendant aryl rings switched (i.e., **72** and **7**) as well despite the fact that the electron density of the phenols is directing differently.[1]

With these syntheses in hand, we next turned our attention to accessing the complexity posed by molecules such as ampelopsin F (**10**) and pallidol (**11**), compounds that are, in principle, one intramolecular Friedel–Crafts cyclization away from ampelopsin D (**6**) and quadrangularin A (**7**), respectively. Indeed, the idea for executing such a transformation was revealed by the earlier work from the Niwa group in the context of Scheme 19.7 [18]. As noted, however, the reaction

---

1) Notably, use of a protecting group other than methyl for the phenols prevented the initial addition of the third aryl ring, likely due to the added steric bulk. Also, the ketone within molecules such as paucifloral F (**3**) could not be olefinated under any set of conditions attempted; isomerization of the adjacent stereocenter was observed instead. Thus, only the sulfide nucleophile incorporated through the cascade provided a means to incorporate a fourth aryl ring to access the desired targets.

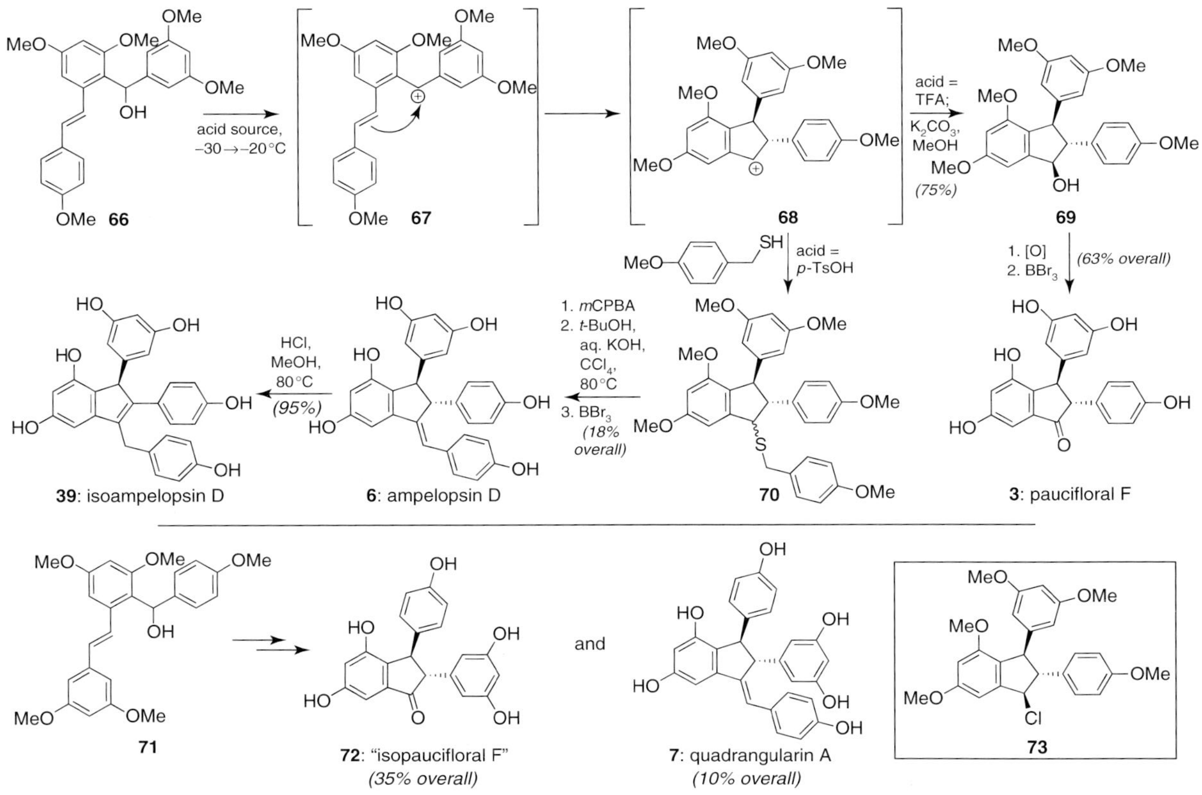

Scheme 19.14 Cascade-based approach to fashion indane-containing members of the resveratrol family of natural products from key triaryl intermediates.

was not selective in their case. Their electrophile added to both faces of the alkene and afforded access to many products. Our hope was that if we used an alternative electrophile that could add reversibly to the double bond (unlike simple proton), then perhaps higher control could be achieved if the desired C–C bond construction terminated the sequence. Our choice for this purpose was bromine, and, as shown in Scheme 19.15, that approach proved successful.

In the event, we separately treated both **74** and **77** with an electrophilic bromine source [either Br_2 or *N*-bromosuccinimide (NBS)] and found that the reaction commenced in both cases with smooth and regioselective bromination of the A-ring, followed by a second halogenation of the pendant 3,5-dimethoxybenzene ring system.[2] Both of the respective mono-brominated and di-brominated species could be isolated in near quantitative yield. With the addition of a little more halogen, however, the double bond was finally engaged by the electrophile, and cyclization to either **76** or **79** proceeded smoothly to deliver the desired carbocyclic cores with three halogens attached. Importantly, the reversibility of bromonium formation is critical to the end of these sequences, as in the lower reaction (i.e., **78** → **79**) product generation requires the addition of bromine onto the more hindered olefin face based on the stereochemistry of its direct aryl C-ring neighbor; thus, though this mode of addition is less favored on steric grounds, the reversibility of bromonium formation [31] in advance of the terminating cyclization was likely the reason why this process proceeded as smoothly as was observed. Pleasingly, we were able to perform similar chemistry with more highly oxidized forms of the same starting materials, accessing two members of the cararosinol family of natural products (i.e., **80** and **81**) [32] as well. In addition, although one can argue from the standpoint of atom economy [33] that adding extra bromine atoms into a structure only to remove them later is not ideal [34], it is worth noting that the aryl halides with both **76** and **79** are positioned perfectly to incorporate the extra carbon frameworks needed to access natural products such as nepalensinol B (**82**) [35] or vaticanol C (**13**, Figure 19.1). Work to address these more complex tetramers through such strategies is ongoing, and hopefully will be reported soon.

The final major element of carbogenic complexity within the resveratrol family is seven-membered rings as typified by such structures as diptoindonesin D (**4**), paucifloral E (**8**), hopeaphenol (**12**), and both vaticanols G (**14**) and A (**15**). We have recently found that we can access such systems as well from our common precursors; the key, however, was to adjust that intermediate slightly through an oxidation; subsequent exposure to electrophilic oxygen in the form of 1,1,1-trifluorodimethyldioxirane [36] afforded the means to access the seven-membered carbocycle of fully protected hemsleyanol E (**83**, Scheme 19.16).

2) As a general trend based on a broad reading of the literature, electrophilic bromine will add para to protected phenols; the observed chemoselectivity here follows that trend exclusively, with the order of halogenation likely reflecting steric biases.

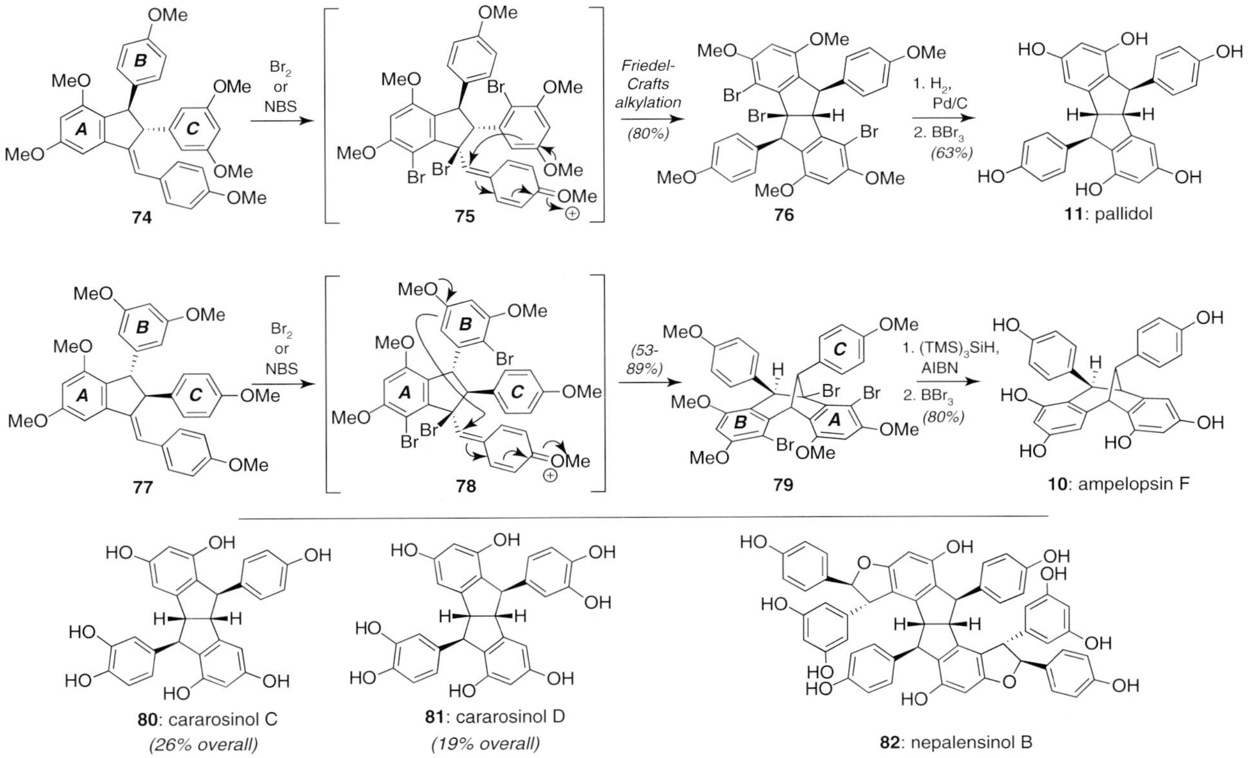

Scheme 19.15 Bromine-induced cyclization cascades to fashion pallidol (**11**) and ampelopsin F (**10**) from appropriate precursors.

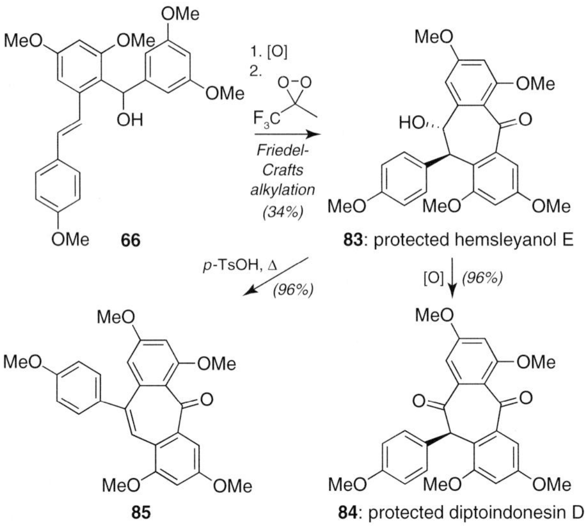

Scheme 19.16 Preparation of seven-membered rings for protected natural products such as diptoindonesin D (**84**) from key intermediate **66**.

Although this reaction appears simple, it proved exceedingly difficult to achieve. Indeed, no other epoxidation reagent afforded the desired product, and only through the use of *in situ* generated 1,1,1-trifluorodimethyldioxirane was the reaction successful. We surmise that methyl ether protection of the phenols may be responsible, as efforts with biologically-derived materials have proven easier to epoxidize [18]. Nevertheless, through further simple oxidation chemistry we have been able to produce a protected form of diptoindonesin D (**84**) and are well on the way to many of the other seven-membered ring natural products mentioned above. Intriguingly, if one exposes molecules like **83** to strong acid in non-polar solvents, the alcohol can be ionized and the neighboring aryl group will rearrange through a phenonium shift to afford natural product analog structures such as **85**; to date, no such structure has been isolated from Nature even though the free-phenol form of **83** is a known natural product.[3] Thus, the sequence affords access to additional structural diversity that, at least at present, Nature does not seem to sample.

Current work within our group is directed towards evaluating what additional frameworks can be built starting from the same intermediates, and, interestingly, some other research teams have already begun to explore such pathways with our key materials. For instance, the Kraus group [37] has recently used the oxidized form of **66** (i.e., **86**, Scheme 19.17) to fashion the dihydrofuran ring of amurensin H (**90**) [38] through a phosphine-base catalyzed condensation reaction.

---

3) This phenonium shift only occurs at elevated temperatures in non-polar solvents, which may explain why Nature does not appear to have such rearranged materials though the presence of the alcohol itself implies the potential for facile ionization.

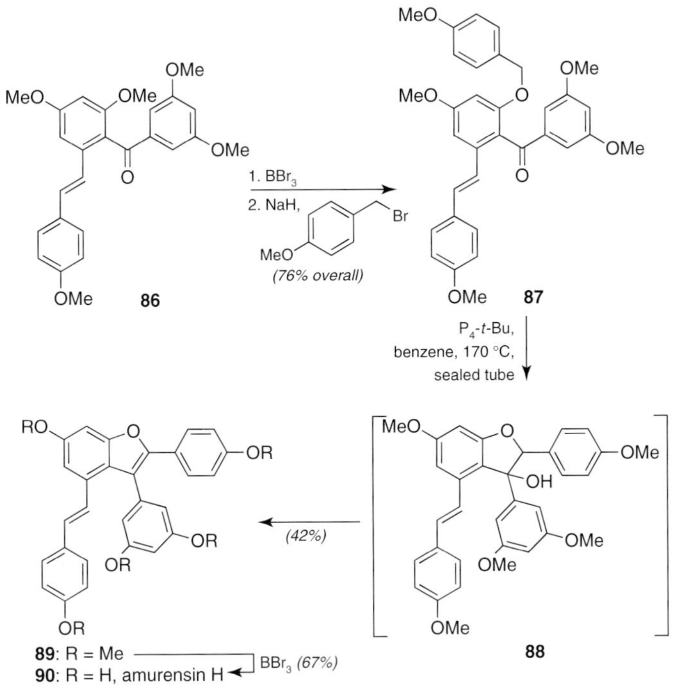

Scheme 19.17 Use of the same key intermediate by the Kraus group to accomplish a total synthesis of the furan-containing natural product amurensin H (**90**).

Beyond resveratrol-based oligomers, however, we have also initiated several programs seeking to determine what other polyphenolic natural products can be prepared from the same key starting materials (i.e., structures of general form **66**, cf. Scheme 19.13). Schemes 19.18 and 19.19 present some of our more advanced work, efforts that have been directed toward three non-resveratrol-based natural products. In the first of these schemes, a series of Friedel–Crafts cyclizations provided the means to convert **91** into the natural product cassigarol B (**97**) [39] in two to four steps, while other chemoselective sequences afforded the means to access two other [3.2.2]-bicycles (**98** and **99**) [29b]. In the second, the dalesconols (**104** and **105**) [40], two potent immunosuppressive natural products, were synthesized in several steps from key building block **100**; this work featured a cyclization cascade that forged the entire polycyclic core from that critical starting material using the same carbon atom on the B-ring as nucleophile and electrophile, respectively [41]. Alteration in reaction conditions with the same types of starting material afforded the means to access other unique structures, such as **106**, in a controlled manner as well. Thus, based on this work it could be argued that intermediates of general form **66** are "privileged" materials for the creation of much structural diversity, both of the resveratrol class and of many other scaffolds of potential biochemical significance [42].

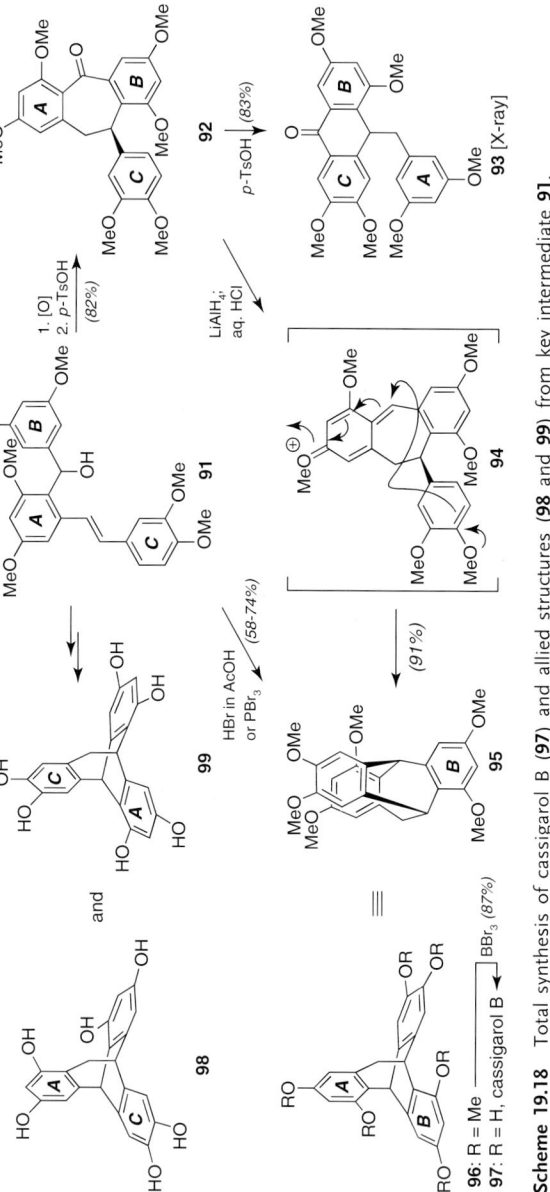

Scheme 19.18 Total synthesis of cassigarol B (**97**) and allied structures (**98** and **99**) from key intermediate **91**.

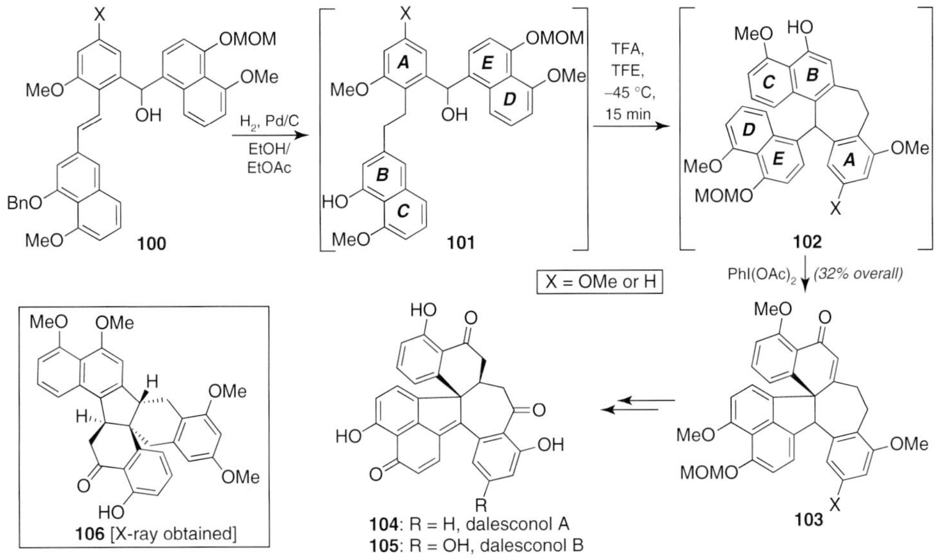

Scheme 19.19 Use of key precursor **100** to access dale-sconols A and B via a cascade-based cyclization sequence, as well as other unique non-natural frameworks (such as **106**).

19.4 Conclusions

As this chapter has, hopefully, demonstrated, there is a wealth of chemistry that can be performed with the reactive functionality of resveratrol, chemistry that leads to an impressive array of stereochemical and architectural complexity. Biomimetic studies have offered insights into how that complexity is made in Nature, and rational synthetic approaches have begun to afford various perspectives into how that complexity can be selectively achieved, largely based on the selection of appropriate building blocks. As such, the future for the synthesis of resveratrol-based oligomers certainly seems bright. In addition, what may even be brighter is the wealth of biochemical knowledge garnered from the thorough evaluation of resveratrol's larger and arguably more complex oligomeric forms which those synthetic approaches can now fuel.

Acknowledgments

The author would like to thank all of the undergraduate, graduate, and postdoctoral students within his group (past, present, and future) for their work on this exciting family of natural products and for their many contributions, both experimental and intellectual, which has made this project a true joy to pursue. Special thanks are noted to Mr. Adel ElSohly, Ms. Yunqing Lin, and Mr. Daniel Wespe for their comments on the manuscript as well as assistance in its preparation. Financial support for our endeavors in the resveratrol field from the National Institutes of Health (R01-GM84994), the Research Corporation for Science Advancement (Cottrell Scholar award to S.A.S.), the Alfred P. Sloan Foundation (Research Fellowship to S.A.S.), Eli Lilly (Grantee award to S.A.S.), and Bristol-Myers Squibb are gratefully acknowledged.

References

1. Resveratrol was originally isolated from the roots of white hellebore, and later, from the roots of the Japanese knotweed: (a) Takaoka, M. (1940) *Proc. Imp. Acad. Tokyo*, **16**, 405–407; (b) Hillis, W.E. and Inoue, T. (1967) *Phytochemistry*, **6**, 59–67.
2. For seminal and representative work, see the following, and references cited therein: (a) Jang, M., Cai, L., Udeani, G.O., Slowing, K.V., Thomas, C.F., Beecher, C.W.W., Fong, H.H.S., Farnsworth, N.R., Kinghorn, A.D., Mehta, R.G., Moon, R.C., and Pezzuto, J.M. (1997) *Science*, **275**, 218–220; (b) Howitz, K.T., Bitterman, K.J., Cohen, H.Y., Lamming, D.W., Lavu, S., Wood, J.G., Zipkin, R.E., Chung, P., Kisielewski, A., Zhang, L.-L., Scherer, B., and Sinclair, D.A. (2003) *Nature*, **425**, 191–196; (c) Walle, T., Hsieh, F., DeLegge, M.H., Oatis, J.E., and Walle, U.K. (2004) *Drug Metab. Dispos.*, **32**, 1377–1382; (d) Wood, J.G., Rogina, B., Lavu, S., Howitz, K., Helfand, S.L., Tatar, M., and Sinclair, D. (2004) *Nature*, **430**, 686–689; (e) Szewczuk, L.M., Forti, L., Stivala, L.A., and Penning, T.M. (2004) *J. Biol. Chem.*, **279**, 22727–22737; (f) Szewczuk, L.M., Lee, S.H., Blair, I.A., and Penning, T.M. (2005) *J. Nat. Prod.*, **68**, 36–42; (g) Chen, G., Shan, W., Wu, Y., Ren, L., Dong, J., and Ji, Z. (2005) *Chem. Pharm. Bull.*, **53**, 1587–1590; (h) Baur, J.A., Pearson, K.J., Price, N.L., Jamieson, H.A., Lerin, C., Kalra, A., Prabhu, V.V., Allard, J.S., Lopez-Lluch, G., Lewis, K., Pistell, P.J., Poosala, S., Becker, K.G., Boss, O., Gwinn, D., Wang, M., Ramaswamy, S., Fishbein, K.W., Spencer, R.G., Lakatta, E.G., Le Couteur, D., Shaw, R.J., Navas, P., Puigserver, P., Ingram, D.K., de Cabo, R., and Sinclair, D.A. (2006) *Nature*, **444**, 337–342; (i) Milne, J.C., Lambert, P.D., Schenk, S., Carney, D.P., Smith, J.J., Gagne, D.J., Jin, L., Boss, O., Perni, R.B., Vu, C.B., Bemis, J.E., Xie, R., Disch, J.S., Ng, P.Y., Nunes, J.J., Lynch, A.V., Yang, H., Galonek, H., Israelian, K., Choy, W., Iffland, A., Lavu, S., Medvedik, O., Sinclair, D.A., Olefsky, J.M., Jirousek, M.R., Elliott, P.J., and Westphal, C.H. (2007) *Nature*, **450**, 712–717; (j) Boocock, D.J., Faust, G.E.S., Patel, K.R., Schinas, A.M., Brown, V.A., Ducharme, M.P., Booth, T.D., Crowell, J.A., Perloff, M., Gescher, A.J., Steward, W.P., and Brenner, D.E. (2007) *Cancer Epidemiol. Biomarkers Prev.*, **16**, 1246–1252; (k) Kim, D., Nguyen, M.D., Dobbin, M.M., Fischer, A., Sananbenesi, F., Rodgers, J.T., Delalle, I., Baur, J.A., Sui, G., Armour, S.M., Puigserver, P., Sinclair, D.A., and Tsai, L.-H. (2007) *EMBO J.*, **26**, 3169–3179; (l) Heiss, E.H., Schilder, Y.D.C., and Dirsch, V.M. (2007) *J. Biol. Chem.*, **282**, 26759–26766.
3. For selected reviews, see: (a) Saiko, P., Szakmary, A., Jaeger, W., and Szekeres, T. (2007) *Rev. Mutation Res.*, **658**, 68–94; (b) Athar, M., Back, J.H., Tang, X., Kim, K.H., Kopelovich, L., Bickers, D.R., and Kim, A.L. (2007) *Toxicol. Appl. Pharmacol.*, **224**, 274–278; (c) Kopp, P. (1998) *Eur. J. Endocrinol.*, **138**, 619–620; (d) Soleas, G.J., Diamandis, E.P., and Goldberg, D.M. (1997) *Clin. Biochem.*, **30**, 91–113.

4. For recent work, see: (a) Favaron, F., Lucchetta, M., Odorizzi, S., Pais da Cunha, A.T., and Sella, L. (2009) *J. Plant Pathol.*, **91**, 579–588; (b) van Baarlen, P., Legendre, L., and van Kan, J.A.L. (2004) *Botrytis: Biology, Pathology and Control* (eds Y. Elad, B. Williamson, P. Tudzynski, and N. Delen), Kluwer Academic Publishers, Dordrecht, pp. 143–161

5. For selected references regarding the isolation and biological activity of the molecules in Figure 19.1, see: (a) Langcake, P. and Pryce, R.J. (1977) *Experientia*, **33**, 151–152; (b) Diyasena, M.N.C., Sotheeswaran, S., Surendrakumar, S., Balasubramanian, S., Bokel, M., and Kraus, W. (1985) *J. Chem. Soc., Perkin Trans. 1*, 1807–1809; (c) Oshima, Y., Ueno, Y., Hikino, H., Yang, L.L., and Yen, K.Y. (1990) *Tetrahedron*, **46**, 5121–5126; (d) Kurihara, H., Kawabata, J., Ichikawa, S., and Mizutani, J. (1990) *Agric. Biol. Chem.*, **54**, 1097–1099; (e) Kitanaka, S., Ikezawa, T., Yasukawa, K., Yamanouchi, S., Takido, M., Sung, H.K., and Kim, I.H. (1990) *Chem. Pharm. Bull.*, **38**, 432–435; (f) Dai, J.-R., Hallock, Y.F., Cardellina, J.H., and Boyd, M.R. (1998) *J. Nat. Prod.*, **61**, 351–353; (g) Ohyama, M., Tanaka, T., Ito, T., Iinuma, M., Bastow, K.F., and Lee, K.-H. (1999) *Bioorg. Med. Chem. Lett.*, **9**, 3057–3060; (h) Tanaka, T., Ito, T., Nakaya, K., Iinuma, M., and Riswan, S. (2000) *Phytochemistry*, **54**, 63–69; (i) Tanaka, T., Iliya, I., Ito, T., Furusawa, M., Nakaya, K., Iinuma, M., Shirataki, Y., Matsuura, N., Ubukata, M., Murata, J., Simozono, F., and Hirai, K. (2001) *Chem. Pharm. Bull.*, **49**, 858–862; (j) Luo, H.-F., Zhang, L.-P., and Hu, C.-Q. (2001) *Tetrahedron*, **57**, 4849–4854; (k) Tanaka, T., Ito, T., Nakaya, K., Iinuma, M., Takahashi, Y., Naganawa, H., and Riswan, S. (2001) *Heterocycles*, **55**, 729–740; (l) Huang, K.-S., Lin, M., and Cheng, G.-F. (2001) *Phytochemistry*, **58**, 357–362; (m) Takaya, Y., Yan, K.-X., Terashima, K., Ito, J., and Niwa, M. (2002) *Tetrahedron*, **58**, 7259–7265; (n) Ito, T., Tanaka, T., Iinuma, M., Iliya, I., Nakaya, K., Ali, Z., Takahashi, Y., Sawa, R., Shirataki, Y., Murata, J., and Darnaedi, D. (2003) *Tetrahedron*, **59**, 5347–5363; (o) Ito, T., Tanaka, T., Iinuma, M., Nakaya, K., Takahashi, Y., Sawa, R., Naganawa, H., and Chelladurai, V. (2003) *Tetrahedron*, **59**, 1255–1264; (p) Ito, T., Akao, Y., Yi, H., Ohguchi, K., Matsumoto, K., Tanaka, T., Iinuma, M., and Nozawa, Y. (2003) *Carcinogenesis*, **24**, 1489–1497; (q) Ito, T., Tanaka, T., Iinuma, M., Nakaya, K., Takahashi, Y., Sawa, R., Murata, J., and Darnaedi, D. (2004) *J. Nat. Prod.*, **67**, 932–937; (r) Supudompol, B., Likhitwitayawuid, K., and Houghton, P.J. (2004) *Phytochemistry*, **65**, 2589–2594; (s) Sahadin Hakim, H.H., Juliawaty, L.D., Syah, Y.M., bin Din, L., Ghisalberti, E.L., Latip, J., Said, I.M., and Achmad, S.A. (2005) *Z. Naturforsch., Teil C: J. Biosci.*, **60**, 723–727; (t) Ohguchi, K., Akao, Y., Matsumoto, K., Tanaka, T., Ito, T., Iinuma, M., and Nozawa, Y. (2005) *Biosci. Biotechnol. Biochem.*, **69**, 353–356; (u) Wang, S., Ma, D., and Hu, C. (2005) *Helv. Chem. Acta*, **88**, 2315–2321; (v) Guebailia, H.A., Chira, K., Richard, T., Mabrouk, T., Furiga, A., Vitrac, X., Monti, J.-P., Delaunay, J.-C., and Merillon, J.-M. (2006) *J. Agric. Food Chem.*, **54**, 9559–9564; (w) Yamada, M., Hayashi, K., Ikeda, S., Tsutsui, K., Tsutsui, K., Ito, T., Iinuma, M., and Nozaki, H. (2006) *Biol. Pharm. Bull.*, **29**, 1504–1507.

6. For selected examples of recent isolations, see: (a) He, S., Jiang, L., Wu, B., Li, C., and Pan, Y. (2009) *J. Org. Chem.*, **74**, 7966–7969; (b) Ito, T., Abe, N., Oyama, M., Tanaka, T., Murata, J., Darnaedi, D., and Iinuma, M. (2009) *Helv. Chim. Acta*, **92**, 1203–1216.

7. For a recent review discussing the generation of varied structures in the terpene collection of structures, see: Fischbach, M.A. and Clardy, J. (2007) *Nat. Chem. Biol.*, **3**, 353–355.

8. For leading references, see: (a) Zhang, Y., Li, S.-Z., Li, J., Pan, X., Cahoon, R.E., Jaworski, J.G., Wang, X., Jez, J.M.,

Chen, F., and Yu, O. (2006) *J. Am. Chem. Soc.*, **128**, 13030–13031; (b) Lee, M.S. and Pyee, J.H. (2004) *Nat. Prod. Sci.*, **10**, 248–251.

9. For example, investigators have been able to separately isolate both antipodes of the natural product davidiol A from different plant species: (a) Tanaka, T., Ito, T., Ohyama, M., Ichise, M., and Tateishi, Y. (2000) *Phytochemistry*, **53**, 1009–1014; (b) He, Y.-H., Takaya, Y., Terashima, K., and Niwa, M. (2006) *Heterocycles*, **68**, 93–100.

10. For selected examples of the proposed equilibration step, see: (a) Kurosawa, W., Kobayashi, K., Kan, T., and Fukuyama, T. (2004) *Tetrahedron*, **60**, 9615–9628; (b) Lian, Y. and Hinkle, R.J. (2006) *J. Org. Chem.*, **71**, 7071–7074.

11. Langcake, P. and Pryce, R.J. (1977) *J. Chem. Soc., Chem. Commun.*, 208–210.

12. Sako, M., Hosokawa, H., Ito, T., and Iinuma, M. (2004) *J. Org. Chem.*, **69**, 2598–2600.

13. Takaya, Y., Terashima, K., Ito, J., He, Y.-H., Tateoka, M., Yamaguchi, N., and Niwa, M. (2005) *Tetrahedron*, **61**, 10285–10290.

14. Yao, C.-S., Lin, M., and Wang, Y.-H. (2004) *Chin. J. Chem.*, **22**, 1350–1355.

15. For one example of a resveratrol-derived oligomer with carbohydrates attached, see: Ito, T., Abe, N., Oyama, M., and Iinuma, M. (2009) *Tetrahedron Lett.*, **50**, 2516–2520.

16. (a) Li, W., Li, H., and Hou, Z. (2006) *Angew. Chem. Int. Ed.*, **45**, 7609–7611; for a recent attempt to further explore such starting materials in the synthesis of resveratrol oligomers, see: (b) Li, W., Li, H., Luo, Y., Yang, Y., and Wang, N. (2010) *Synthesis*, 1247–1250; (c) Li, W., Li, H., Wang, A.X., Luo, Y., and Zang, P. (2010) *J. Chem. Res. (S)*, 118–120.

17. Velu, S.S., Thomas, N.F., Buniyamin, I., Ching, L.K., Feroz, F., Noorbatcha, I., Gee, L.C., Awang, K., Abd. Wahab, I., and Weber, J.F.F. (2008) *Chem. Eur. J.*, **14**, 11376–11384.

18. (a) Takaya, Y., Yan, K.-X., Terashima, K., He, Y.-H., Niwa, M. (2009) *Tetrahedron*, **58**, 9265–9271; (b) Niwa, M., Ito, J., Terashima, K., Koizumi, T., Takaya, Y., and Yan, K.-X. (2000) *Heterocycles*, **53**, 1475–1478.

19. (a) Aguirre, J.M., Alesso, E.N., and Moltrasio Iglesias, G.T. (1999) *J. Chem. Soc., Perkin Trans. 1*, 1353–1358; (b) Aguirre, J.M., Aleso, E.N., Ibañez, A.F., Tombari, D.G., and Moltrasio Iglesias, G.Y. (1989) *J. Heterocycl. Chem.*, **26**, 25–27.

20. For selected examples of other, resveratrol-like, biomimetic dimerization attempts, see: (a) Li, X.-C. and Ferreira, D. (2003) *Tetrahedron*, **59**, 1501–1507; (b) Thomas, N.F., Lee, K.C., Paraidathathu, T., Weber, J.F.F., Awang, K., Rondeau, D., and Richomme, P. (2002) *Tetrahedron*, **58**, 7201–7206; (c) Thomas, N.F., Lee, K.C., Paraidathathu, T., Weber, J.F.F., and Awang, K. (2002) *Tetrahedron Lett.*, **43**, 3151–3155; (d) Engler, T.A., Gfesser, G.A., and Draney, B.W. (1995) *J. Org. Chem.*, **60**, 3700–3706; (e) Engler, T.A., Draney, B.W., and Gfesser, G.A. (1994) *Tetrahedron Lett.*, **35**, 1661–1664.

21. (a) Seiple, I.B., Su, S., Young, I.S., Lewis, C.A., Yamaguchi, J., and Baran, P.S. (2010) *Angew. Chem. Int. Ed.*, **49**, 1095–1098; (b) Su, S., Seiple, I.B., Young, I.S., and Baran, P.S. (2008) *J. Am. Chem. Soc.*, **130**, 16490–16491; (c) O'Malley, D.P., Yamaguchi, J., Young, I.S., Seiple, I.B., and Baran, P.S. (2008) *Angew. Chem. Int. Ed.*, **47**, 3581–3583.

22. For selected syntheses of resveratrol itself, see: (a) Botella, L. and Nájera, C. (2004) *Tetrahedron*, **60**, 5563–5570; (b) Takaya, Y., Terashima, K., Ito, J., He, Y.-H., Tateoka, M., Yamaguchi, N., and Niwa, M. (2005) *Tetrahedron*, **61**, 10285–10290; (c) Chen, G., Shan, W., Ren, L., Dong, J., and Ji, Z. (2005) *Chem. Pharm. Bull.*, **53**, 1587–1590

23. (a) Jeffrey, J.L. and Sarpong, R. (2009) *Org. Lett.*, **11**, 5450–5453; (b) Jeffrey, J.L. and Sarpong, R. (2009) *Tetrahedron Lett.*, **50**, 1969–1972.

24. Kim, I. and Choi, J. (2009) *Org. Biomol. Chem.*, **7**, 2788–2795.

25. (a) Saraswathym, A., Puroshothaman, K.K., Patra, A., Dey, A.K., and Kundu,

A.B. (1992) *Phytochemistry*, **31**, 2561–2562; (b) Ge, H.M., Huang, B., Tan, S.H., Shi, D.H., Song, Y.C., and Tan, R.X. (2006) *J. Nat. Prod.*, **69**, 1800–1802.

26. For a recent review on bismuth(III) chemistry, see: Salvador, J.A.R., Pinto, R.M.A., and Silvestre, S.M. (2009) *Curr. Org. Synth.*, **6**, 426–470.

27. (a) Ge, H.M., Zhu, C.H., Shi, D.H., Zhong, L.D., Xie, D.Q., Yang, J., Ng, S.W., and Tan, R.X. (2008) *Chem. Eur. J.*, **14**, 376–381; (b) Ge, H.M., Xu, C., Wang, X.T., Huang, B., and Tan, R.X. (2006) *Eur. J. Org. Chem.*, 5551–5554.

28. Nicolaou, K.C., Wu, T.R., Kang, Q., and Chen, D.Y.-K. (2009) *Angew. Chem. Int. Ed.*, **48**, 3440–3443; (b) for the full account of this work, see: Nicolaou, K.C., Kang, Q., Wu, T.R., Lim, C.S., and Chen, D.Y.-K. (2010) *J. Am. Chem. Soc.*, **132**, 7540–7548.

29. (a) Snyder, S.A., Zografos, A.L., and Lin, Y. (2007) *Angew. Chem. Int. Ed.*, **46**, 8186–8191; (b) Snyder, S.A., Breazzano, S.P., Ross, A.G., Lin, Y., and Zografos, A.L. (2009) *J. Am. Chem. Soc.*, **131**, 1753–1765.

30. For a recent review on this reaction, see: Taylor, R.J.K. (1999) *Chem. Commun.*, 217–227.

31. For a leading introduction to the reversibility of bromonium formation, see: Brown, R.S. (1997) *Acc. Chem. Res.*, **30**, 131–137.

32. Yang, G., Zhou, J., Li, Y., and Hu, C. (2005) *Planta Med.*, **71**, 569–571.

33. For a review on the concept of atom economy, see: Trost, B.M. (1991) *Science*, **254**, 1471.

34. For an interesting commentary on utilizing bromine as a protective device, see: Effenberger, F. (2002) *Angew. Chem. Int. Ed.*, **41**, 1699–1700.

35. Yamada, M., Hayashi, K., Hayashi, H., Ikeda, S., Hoshino, T., Tsutsui, K., Tsutsui, M., Iinuma, M., and Nozaki, H. (2006) *Phytochemistry*, **67**, 307–313.

36. These conditions were taken from past efforts to synthesize the epothilones: Nicolaou, K.C., He, Y., Vourloumis, D., Vallberg, H., Roschanger, F., Sarabia, F., Ninkovic, A., Yang, Z., and Trujillo, J.I. (1997) *J. Am. Chem. Soc.*, **119**, 7960–7973.

37. Kraus, G.A. and Gupta, V. (2009) *Tetrahedron Lett.*, **50**, 7180–7183.

38. Li, Y., Yao, C., Bai, J., Lin, M., and Cheng, G. (2006) *Acta Pharm. Sinica*, **27**, 735–740.

39. Baba, K., Maeda, K., Tabata, Y., Doi, M., and Kozawa, M. (1988) *Chem. Pharm. Bull.*, **36**, 2977–2983.

40. (a) Zhang, Y.L., Ge, H.M., Zhao, W., Dong, H., Xu, Q., Li, S.H., Li, J., Zhang, J., Song, Y.C., and Tan, R.X. (2008) *Angew. Chem. Int. Ed.*, **47**, 5823–5826; (b) Wen, L., Cai, X., Xu, F., She, Z., Chan, W.L., Vrijmoed, L.L.P., Jones, E.B.G., and Lin, Y. (2009) *J. Org. Chem.*, **74**, 1093–1098.

41. Snyder, S.A., Sherwood, T.C., and Ross, A.G. (2010) *Angew. Chem. Int. Ed.*, **49**, 5146–5150.

42. For a recent review on this concept, see: Welsch, M., Snyder, S.A., and Stockwell, B.R. (2010) *Curr. Opin. Chem. Biol.*, **14**, 347–361.

20
Sequential Reactions Initiated by Oxidative Dearomatization. Biomimicry or Artifact?

Stephen K. Jackson, Kun-Liang Wu, and Thomas R.R. Pettus

20.1
Overview

Robert Robinson first introduced the concept of biomimetic synthesis in 1917 with a one-pot preparation of tropinone [1]. This remarkable transformation joins three reactants together, capitalizing upon the reactive proclivities of polyketides. Because aromatic compounds themselves are the end point of polyketide biosynthetic sequences, which are followed by the action of a few specific tailoring decarboxylative and oxidative enzymes, it is hard to conclude that sequences commencing with dearomatization are indeed "biomimetic" [2]. However, sequential reactions initiated by oxidative phenol dearomatization can often appear as "biomimetic," reaching very complex molecular architectures in short order because the functionality and the stereo- and regio-chemical tendencies that are programmed into the polyketide intermediates are still efficacious in the corresponding phenol derivatives and in the reactions stemming from these compounds.

In this chapter, we examine some of the sequential one-pot reactions initiated by oxidative dearomatization. This discussion is by no means a comprehensive treaty on this topic. However, we attempt to untangle this subject for the benefit of interested chemists by focusing on one-pot oxidative dearomatization sequences commencing from phenols that result in the formation or cleavage of three or more C–C, C–O, or C–N bonds. In all of the following schemes, the intermediate formed by the initial oxidative dearomatization undergoes further chemical reactions involving the formation and breakage of two or more bonds. Thus, the examples presented are different from most rudimentary oxidative dearomatization reactions that most often result in the usual quinol and ketal adducts [3].

20.2
Oxidative Dearomatization Sequences and the Initial Intermediate

The processes that we will describe involve some type of phenoxonium cation [4], a *p*-quinone methide [5], an *o*-quinone methide [6], or their corresponding *p*- and

Biomimetic Organic Synthesis, First Edition. Edited by Erwan Poupon and Bastien Nay.
© 2011 Wiley-VCH Verlag GmbH & Co. KGaA. Published 2011 by Wiley-VCH Verlag GmbH & Co. KGaA.

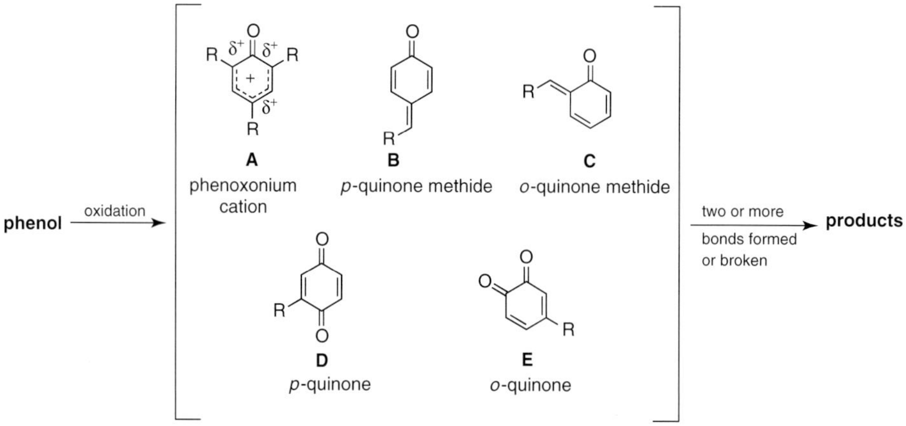

Scheme 20.1 The initial species formed by phenol oxidation that can lead to further events.

o-quinones (Scheme 20.1, reactive intermediates A–E). These species can lead to a lengthy sequence of reactions, just so long as the subsequent intermediates are themselves compatible with the initial oxidative conditions. The reactivity of each species can be further expanded by an attached π-framework of conjugated sp^2 R-substituents so as to augment the availability of reaction manifolds.

In the following sections, we discuss some sequences leading from the above five species that follow oxidative dearomatization. These subsequent transformations include intermolecular dimerizations, as well as successive intermolecular [4+2], [3+2], and [5+2] cycloadditions (Sections 20.3 and 20.4). In addition, there are numerous examples of intramolecular cascade sequences including 1,4-additions, [5+2], [4+2], [3+2] cycloadditions, and subsequent aldol-like processes (Sections 20.5 and 20.6). There are also several examples where oxidative dearomatization to one of the above species is then followed by a series of tautomerizations, sigmatropic shifts (Section 20.7), and further reaction. We will also present examples where oxidation is followed by ring rupture, resulting in either smaller rings or larger rings (Sections 20.8 and 20.9), which can even incorporate additional intramolecular and intermolecular events (Section 20.10). Finally, we present some natural products presumed to form through phenol oxidative pathways (Section 20.11). Our concluding opinions on the biomimetic authenticity of presented schemes and future of this area of research are provided in the last section.

20.3
Intermolecular Dimerizations

The most memorable example of a phenol oxidative cascade is likely Chapman's 1971 synthesis of carpanone (4) [7]. The cascade sequence can be viewed as commencing with a palladium-promoted oxidation of the starting vinyl phenol **2**

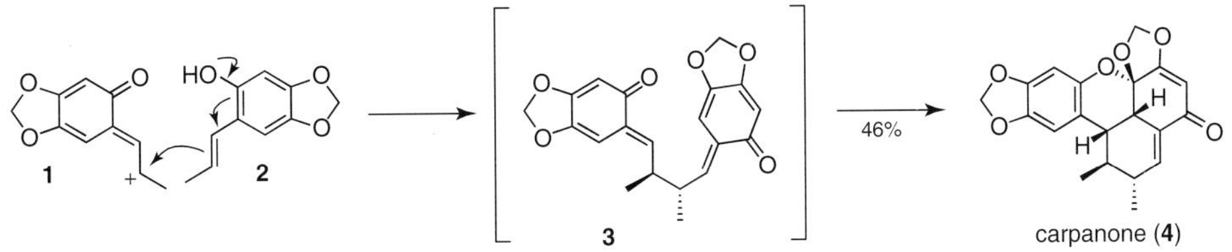

Scheme 20.2 Chapman's synthesis of carpanone.

to its corresponding resonance stabilized phenoxonium cation **1**, which succumbs to a vinylogous diastereoselective aldol-like process with the original phenol **2** (Scheme 20.2). This reaction affords the dimeric o-quinone methide intermediate **3**, which undergoes a subsequent diastereoselective intramolecular Diels–Alder reaction to produce carpanone (**4**). The cascade is somewhat unusual in that it involves both inter- *and* intra-molecular components: a phenoxonium undergoing remote addition and, subsequently, the dimerization of two o-quinone methides resulting in the formation of two carbon–carbon bonds and one carbon–oxygen bond.

In a more recent related example, Antus has shown that oxidative dimerization of isoeugenol (**5**) leads to dehydrodiisoeugenol (**9**) in a single operation (Scheme 20.3) [8]. Treatment of **5** with phenyl iodine diacetate (PIDA) results in the formation of p-phenoxonium **6**, the resonance structure (**7**) of which is intercepted by a second equivalent of isoeugenol (**5**). Nucleophilic ring closure then gives rise to the formal [3+2] addition product, dehydrodiisoeugenol (**9**).

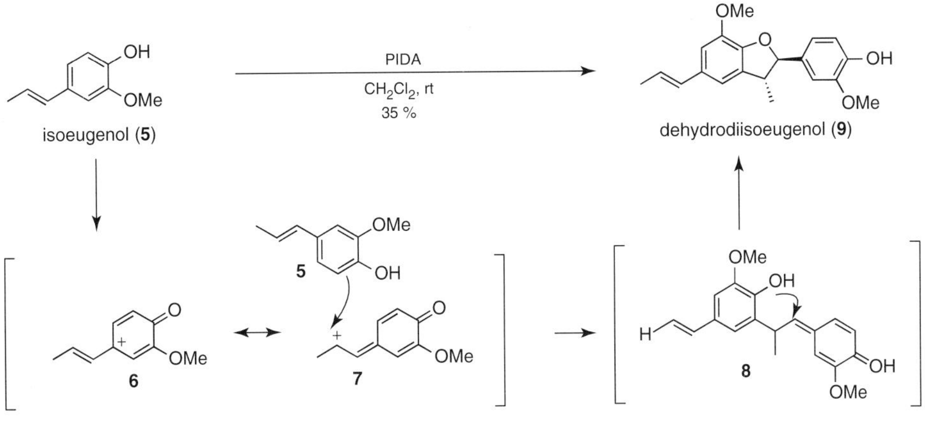

Scheme 20.3 Antus' carpanone related net [3+2] sequence for dehydrodiisoeugenol.

The more common oxidative dimerization cascade processes involve subsequent Diels–Alder dimerization of an intermediate formed by addition of a nucleophile to the phenoxonium species. In particular, these products capitalize upon the inherent preference for the p-quinol and o-quinol intermediate to undergo subsequent dimerization of the same side as the polar C–O bond that is formed from dearomatization. This trait is likely imbued through asymmetric orbital development. Within

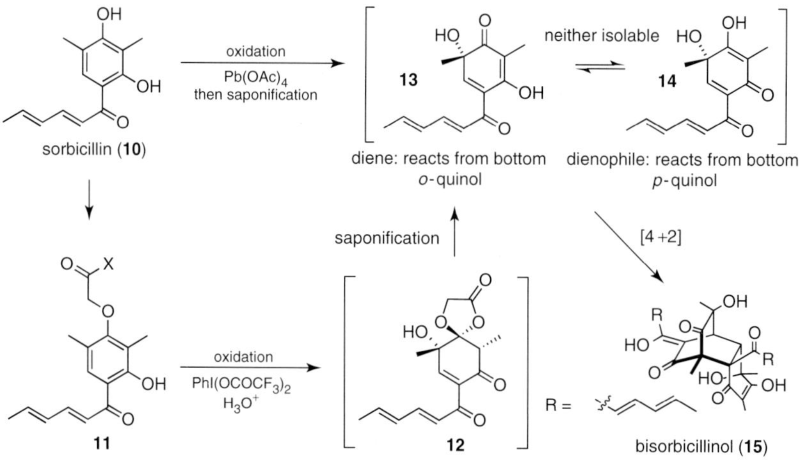

Scheme 20.4 Nicolaou and Pettus' syntheses of bisorbicillinol from sorbicillin.

this class, bisorbicillinol (**15**), a dimer of the oxidative dearomatization of sorbicillin (**10**), is perhaps the most recognized putative natural product involving this hypothetical oxidative dearomatization biomimicry (Scheme 20.4). In the Nicolaou synthesis, sorbicillin undergoes oxidation and saponification to form a tautomeric mixture of non-isolable *o*- and *p*-quinols (**13** and **14**), which then succumb to a regioselective Diels–Alder reaction on the same side of the quinol as the newly installed tertiary alcohol to afford bisorbicillinol (**15**) in 16% yield from sobicillin (**10**) [9]. Pettus, on the other hand, installs a glycolic acid derived disposable linker to produce **11** [10]. Oxidation and sequential treatment with aqueous base and acid affords the manageable spiroketal intermediate **12**. Further saponification proceeds to the same tautomeric quinol mixture of **13** and **14**, producing bisorbicillinol in a somewhat improved 33% yield.

Other examples of this kind of dearomatization and [4+2] dimerization can be found throughout the literature [11]. For example, Pettus has shown that *ortho* oxidation of xylenol **16** with the I$^{(V)}$ reagent **17**, also known as *o*-iodoxybenzoic acid (IBX), affords the I$^{(III)}$ *o*-quinol derivative **18** (Scheme 20.5) [12]. This *o*-quinol intermediate undergoes [4+2] dimerization to afford the corresponding dione **19**, which succumbs to further reduction to produce **20** in 51% overall yield.

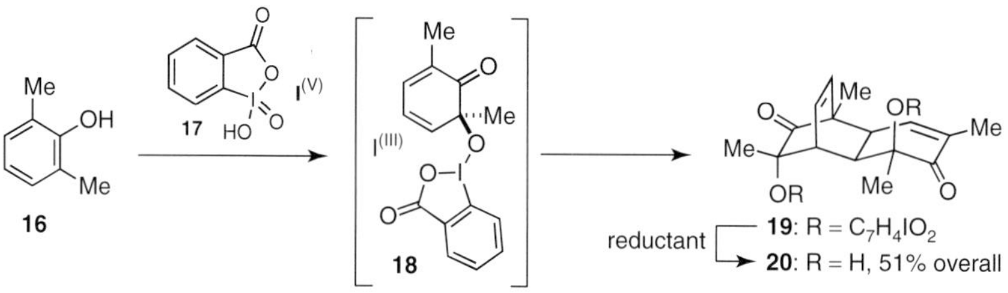

Scheme 20.5 Pettus' oxidative dearomatization and dimerization of 2,6-xylenol with IBX.

Although it remains debatable as to whether products of oxidative dimerization actually constitute legitimate natural products or are themselves merely artifacts of isolation, many examples exist in the literature. For instance, Takeya has shown that the o-quinone **22** generated from oxidation of demethylsalvicanol (**21**) affords grandione (**23**) when heated in the solid state, through a [4+2] cycloaddition pathway (Scheme 20.6) [13].

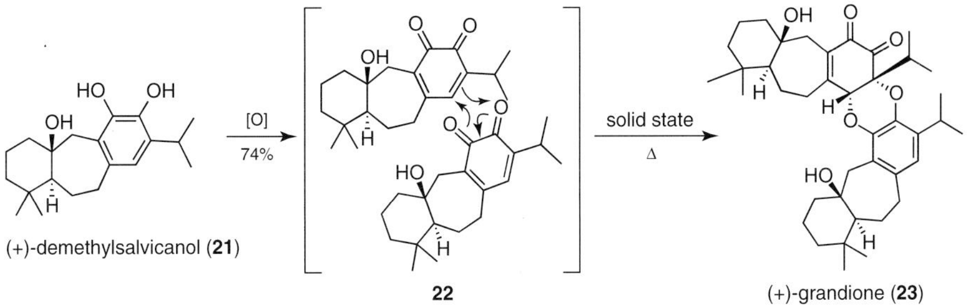

Scheme 20.6 Takeya's o-quinone dimerization.

20.4
Successive Intermolecular Reactions

Intermediates arising from oxidative dearomatization can also serve as reacting partners in intermolecular cycloadditions. For example, Singh has demonstrated that the Adler–Becker oxidation product **25** of **24** undergoes smooth cycloaddition with methyl acrylate to yield the Diels–Alder adduct **26** in 67% yield in a regio- and diastereoselective fashion (Scheme 20.7) [14]. Further manipulation of adduct **26** resulted in assembly of the diquinane skeleton of ptychanolide.

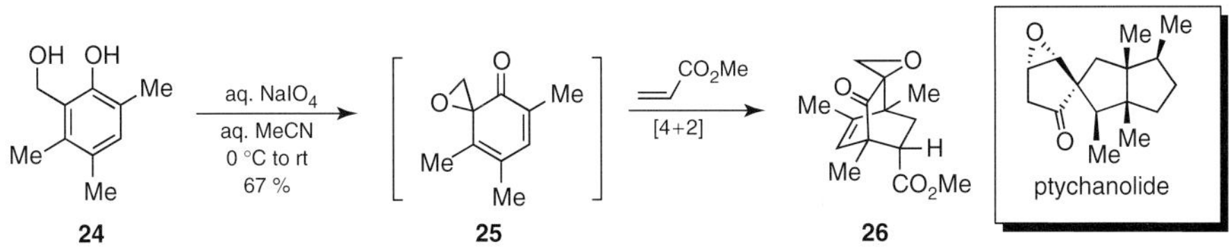

Scheme 20.7 Singh's application of Adler–Becker oxidation and successive cycloaddition.

A novel oxidative cascade involving an o-quinone methide intermediate was reported recently by Sigman [15]. Exposure of **27** to catalytic oxidation with Pd and the i-PrQuinox ligand affords o-quinone methide **29** through allylic oxidation (Scheme 20.8). The o-quinone methide generated under these very mild conditions then undergoes a [4+2] cycloaddition with enol-ether **28**, furnishing adduct **30** with excellent dr and er.

Phenoxonium species generated from oxidative dearomatization can often enter multiple reaction pathways depending on the given conditions. For example, Yamamura has reported that anodic oxidation of phenol **31** in the presence of

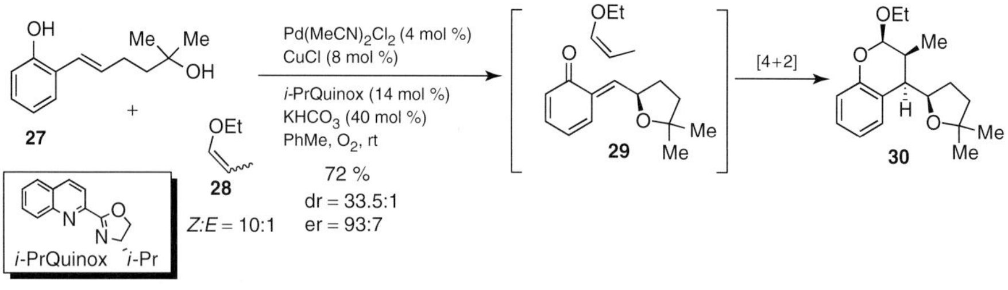

Scheme 20.8 Sigman's enantioselective oxidative dearomatization and cycloaddition.

cis-isosafrole (**32**) affords the [5+2] exo-product **33** in 25% yield (Scheme 20.9) [16]. Interestingly, the isolation of a second product, futoenone (**34**), from the reaction mixture implies a tandem cationic cascade sequence as a competing event. The use of the trans form of olefin **32**, however, affords the corresponding [5+2] product exclusively in 81% yield.

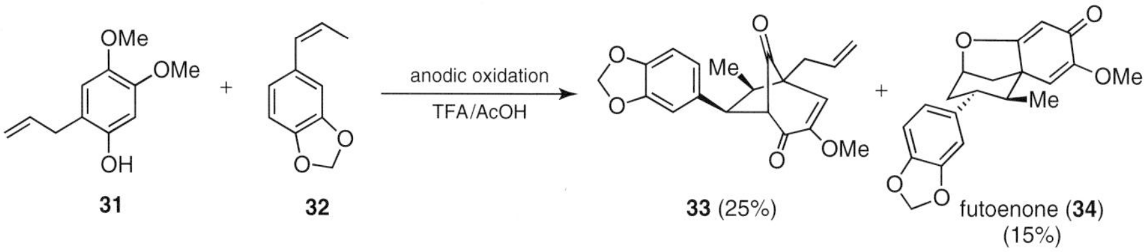

Scheme 20.9 Yamamura's [5+2] cycloaddition.

Several examples of intermolecular [3+2] reactions initiated by oxidation are known. Recently, Morrow reported that phenyl iodine(bis)trifluoroacetate (PIFA) could promote a [3+2] reaction between the chromene **35** and phenol **36** (Scheme 20.10) [17]. Only para-substituted phenols can be used by this method to generate pterocarpans, such as **37**, in appreciable yield. This is a rather unfortunate limitation of this chemistry in light of the fact that naturally occurring pterocarpans typically posses both C8 and C9 oxidation.

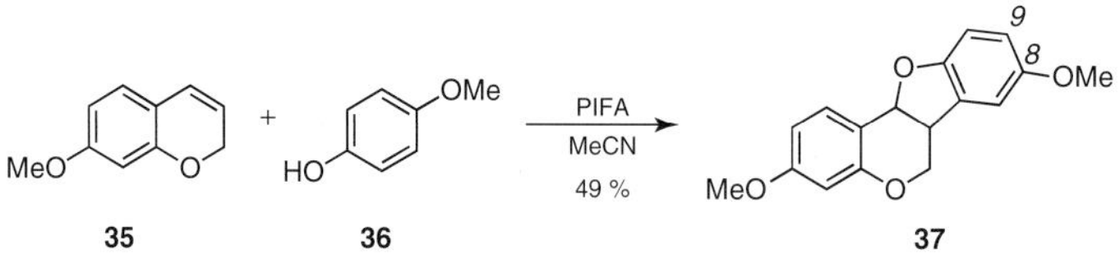

Scheme 20.10 Morrow's oxidative dearomatization and successive [3+2] reaction.

Successive oxidation of a heavily substituted phenol was reported by Stoltz in the total synthesis of the norditerpenoid, dichronanone (**40**) [18]. In the final step of the synthesis, a sequence of oxidations takes place, initiated by IBX. During these events, an o-quinone intermediate stemming from **38** is trapped by pentafluorothiophenol to generate catechol **39**, which undergoes further oxidation and addition of hydroxide to afford the p-quinone natural product (Scheme 20.11).

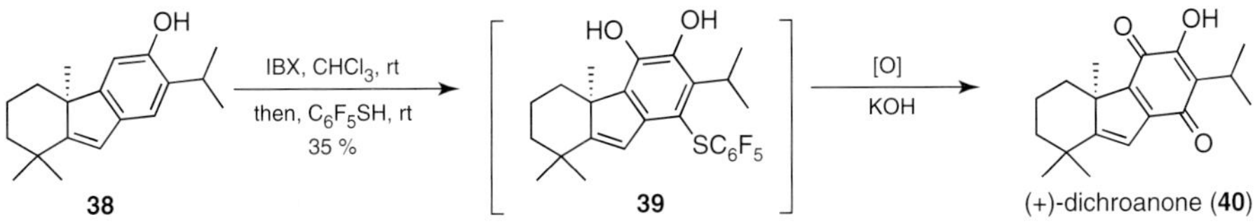

Scheme 20.11 Stoltz' successive phenol oxidation.

A related successive oxidation of a catechol substrate has been reported by Sarpong in the synthesis of abrotanone (**43**) from brussonol (**41**) [19]. This process involves a four-electron oxidation with the addition of two methanol equivalents in a regioselective fashion. Thus, treatment of brussonol with $Cu(NO_3)_2$ in the presence of morpholine affords the mixed quinone amino acetal **42**, which upon subsequent exposure to methoxide gives rise to the natural product **43** (Scheme 20.12).

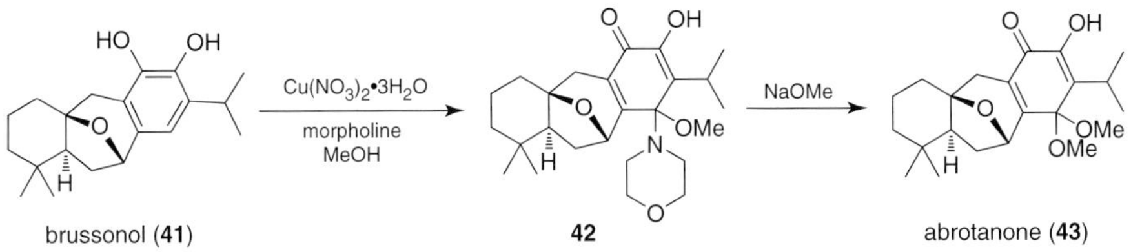

Scheme 20.12 Sarpong's successive catechol oxidation.

20.5
Intramolecular Cycloadditions

Liao was among the first to popularize successive oxidation and intramolecular cycloaddition. For instance, the penicillones were accessed through this strategy, beginning with phenol **44** [20]. Treatment of the latter with PIDA in the presence of *trans*-crotyl alcohol affords *o*-quinone acetal **45**, which slowly undergoes an intramolecular Diels–Alder reaction, giving rise to adduct **46** (Scheme 20.13). The latter was then elaborated to the penicillone B natural product (**47**).

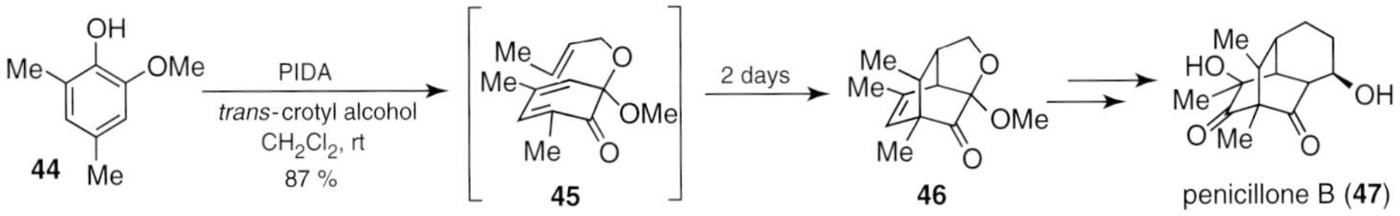

Scheme 20.13 Liao's oxidative dearomatization and intramolecular cycloaddition.

Wood has also reported an intramolecular cycloaddition prompted by oxidative dearomatization in the context of studies toward CP-263,115 [21]. Phenol **48**, upon oxidation with PIDA, forms a phenoxonium that is intercepted by propargyl alcohol to form the intermediate diene **49**, which undergoes a subsequent intramolecular Diels–Alder reaction to afford **50** (Scheme 20.14).

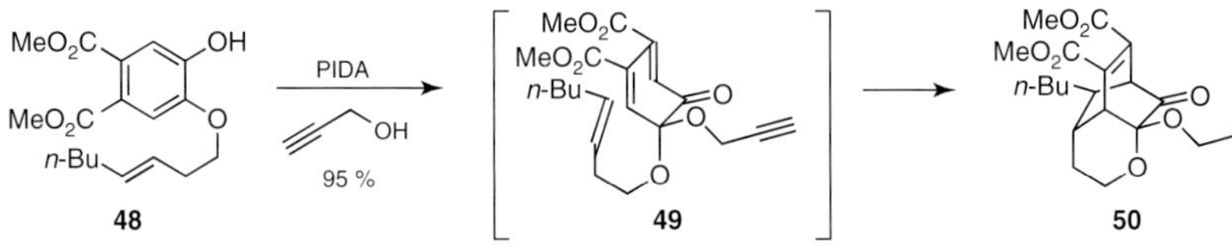

Scheme 20.14 Wood's oxidative dearomatization and intramolecular cycloaddition.

The oxidative dearomatization step may also proceed intramolecularly, as demonstrated by Danishefsky in studies toward the synthesis of the sesquiterpenoid tashironin [22]. Treatment of phenol **51** with PIDA in toluene presumably affords the intermediate *o*-quinone acetal **52**, which undergoes a transannular Diels–Alder reaction to give **53** as a single diastereomer (Scheme 20.15). Interestingly, the benzylic stereocenter in **51** controls the diastereofacial selectivity of the oxidative dearomatization step and, thereby, secures the stereoselectivity of the cycloaddition.

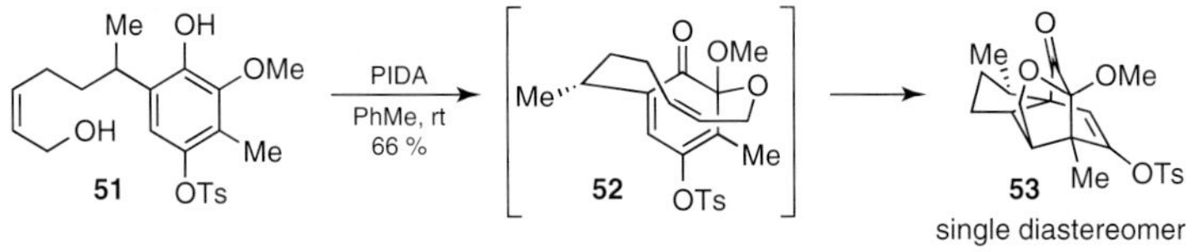

Scheme 20.15 Danishefsky's oxidative dearomatization and transannular cycloaddition.

A remarkable intramolecular cascade was reported recently by Njardarson during studies toward the synthesis of vinigrol [23]. Exposure of phenol **54** to PIDA in hexafluoro-isopropanol (HFIP) led to the isolation of the unexpected compound **55** and the desired Wessely product **56** (Scheme 20.16). Formation of the bicyclic compound **55** is postulated to arise from a polar cationic [5+2] reaction, facilitated by the electron-donating methoxy moiety of the phenol.

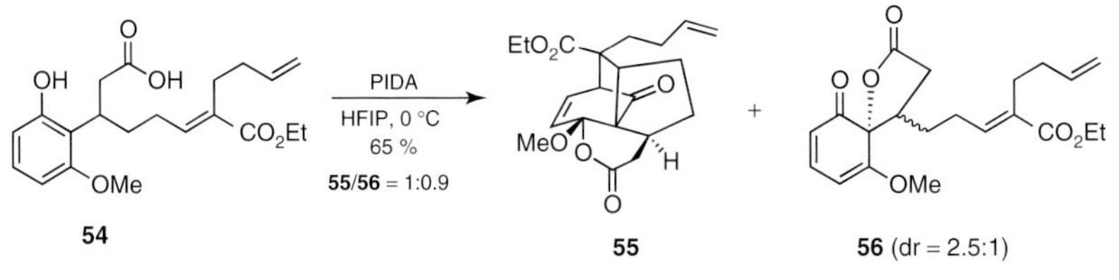

Scheme 20.16 Njardarson's [5+2] intramolecular cascade.

Other intramolecular oxidative dearomatization cascades involving a [3+2] manifold have also been reported. For example, Ciufolini has shown that aldoximes lead to [3+2] adducts via phenol oxidative dearomatization due to concurrent nitrile oxide formation [24]. Employing an elaborate substrate, Sorensen utilized this method for the construction of the pentacyclic core of the cortistatins [25]. Treatment of compound **57** with PIDA results in the formation of presumed nitrile oxide **58**, which undergoes an intramolecular [3+2] cycloaddition, affording adduct **59** in good yield (Scheme 20.17).

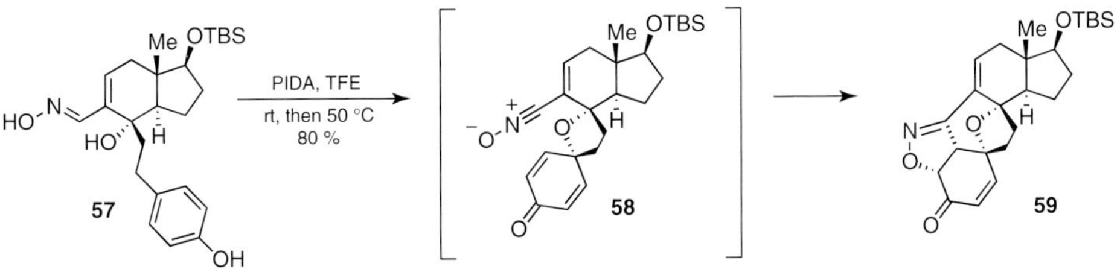

Scheme 20.17 Sorensen's oxidative dearomatization and intramolecular cycloaddition.

20.6
Other Successive Intramolecular Cascade Sequences

Pettus has reported a cascade involving the oxidation of various resorcinol amide derivatives, such as **60** with PIFA [26]. The sequence proceeds to the lactone intermediate **61**, presumably by addition of the amide carbonyl to the phenoxonium followed by hydrolysis of the intermediate iminium species by addition of water at the conclusion of the reaction (Scheme 20.18). In this instance, however, an oxygen atom belonging to the neighboring nitro substituent adds in 1,4-fashion to the vinylogous ester to produce the propeller structure **62** [27].

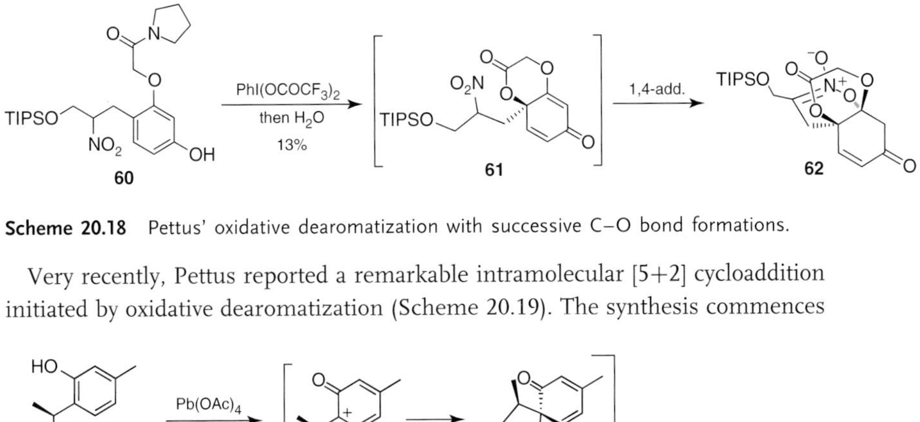

Scheme 20.18 Pettus' oxidative dearomatization with successive C–O bond formations.

Very recently, Pettus reported a remarkable intramolecular [5+2] cycloaddition initiated by oxidative dearomatization (Scheme 20.19). The synthesis commences

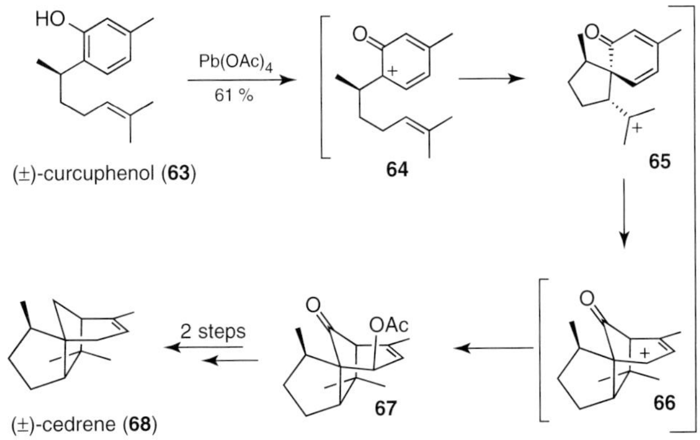

Scheme 20.19 Pettus' dearomatization with successive [5+2] and C–O bond formation.

with a one-pot synthesis of curcuphenol (**63**) from 4-methylsalicaldehyde via an *o*-quinone methide intermediate [28]. Subsequent exposure of phenol **63** to the oxidant lead tetraacetate [Pb(OAc)$_4$] proceeds through the intermediacy of phenoxonium **64**, the tertiary cation **65** and the allyl cation **66** to afford the tricyclic compound **67** by eventual steric controlled interception by a residual acetate ion. This material has been successfully converted into cedrene (**68**) in two further operations.

Heathcock has reported a facile synthesis of styelsamine B (**71**) from aniline **69** and catechol **70** (Scheme 20.20) [29]. The sequence, which undoubtedly resembles the biosynthesis, involves an *o*-quinone intermediate that succumbs to successive imine formation and intramolecular 1,4-addition of the aniline nitrogen atom. Further oxidation, aldol reaction, and dehydration produce **71** (Scheme 20.20). This compound undergoes a further series of oxidations to produce cystodytin J (**72**).

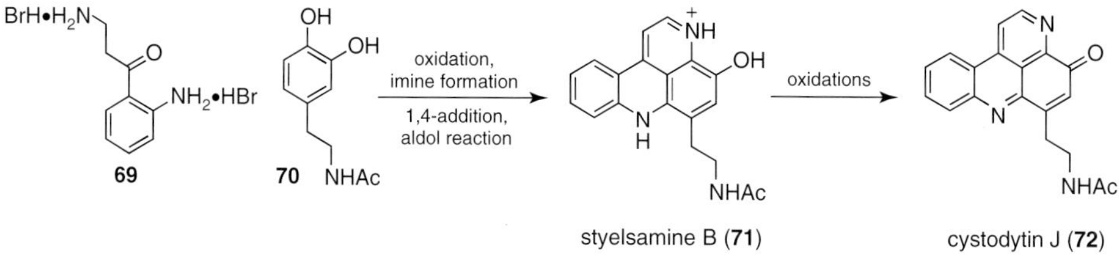

Scheme 20.20 Heathcock's synthesis of styelsamine B.

The oxidative dearomatization of symmetrical *p*-substituted phenols produces *meso*-cyclohexadienones. Various two-step methods have been reported to desymmetrize these compounds into enantioenriched products [30]. An impressive one-pot, catalytic protocol was reported by Gaunt that makes use of a pyrrolidine-catalyzed intramolecular desymmetrizing 1,4-addition to the *meso*-intermediate [31]. Phenolic substrates of type **73** are treated with PIDA in the presence of pyrrolidine catalyst **74** to afford the dearomatized dienone **75** (Scheme 20.21). Subsequent enamine formation and Michael addition affords annulated products of type **76**, many of which in high yield and stereoselectivity.

Rodríguez and coworkers have recently shown that coatline B (**77**) undergoes a fast and irreversible reaction in slightly alkaline water to afford matlaline (**82**) in near quantitative yield [32]. This fascinating process – the historic backdrop of

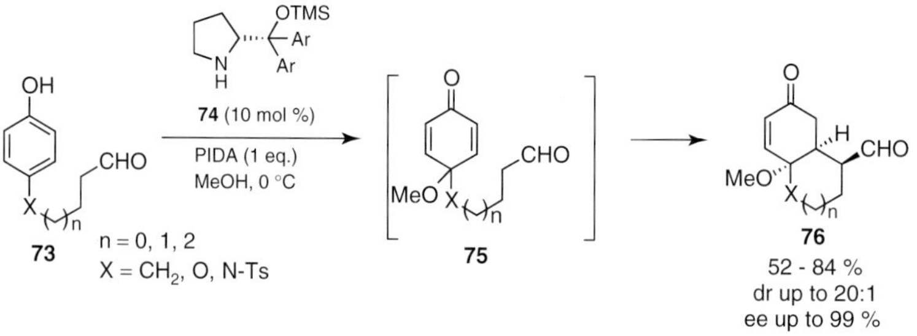

Scheme 20.21 Gaunt's oxidative dearomatization and *in situ* desymmetrization.

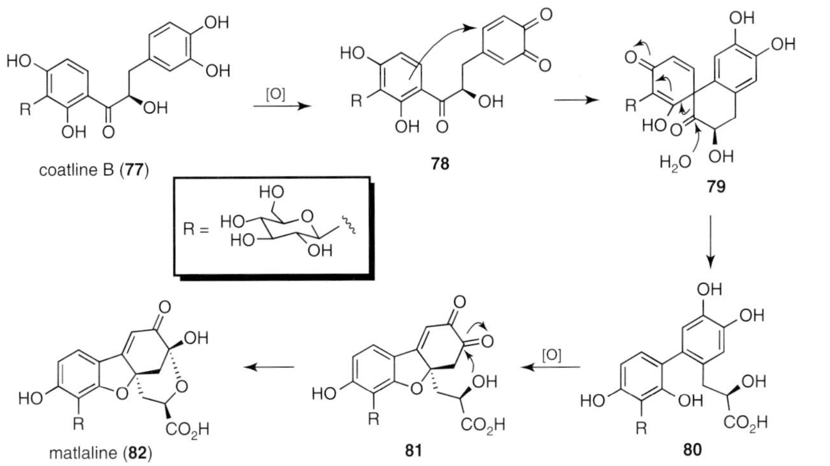

20.7 Successive Tautomerizations and Rearrangements

Scheme 20.22 Rodríguez's oxidative biaryl formation followed by furan and pyran formation.

which is equally intriguing – is thought to occur with a molecular oxygen induced oxidation of the catechol moiety in **77** to the o-quinone **78** (Scheme 20.22). This oxidation event was shown to be extremely facile, an unusual observation given that catechols are generally oxidized either enzymatically or with strong oxidant. Nucleophilic attack on the reactive quinone of **78** affords putative spiro-annulated intermediate **79**, which undergoes a retro-Friedel–Crafts reaction to give **80**. A second oxidation would generate o-quinone **81**, leading to the blue fluorescent "natural" product matlaline (**82**). This process is also entirely stereoselective, apparently driven by the α-hydroxy stereogenic center of coatline B (**77**) and not by the glucopyranosyl (R) residue.

20.7
Successive Tautomerizations and Rearrangements

There are several examples in the literature where the first intermediate derived from phenolic dearomatization undergoes further rearrangement prior to succumbing to a second reaction event. For example, Porco has shown that treatment of phenoxide **83** with a sparteine/Cu reagent produces o-quinol intermediate **84**, which undergoes a stereospecific 1,2-shift to isomeric o-quinol **85** (Scheme 20.23) [33]. The latter then spontaneously dimerizes in a [4+2] sense affording bicyclo[2.2.2]octenone **86** with superb enantioselectivity. In this case, the monomer **84** resists dimerization until rearrangement to the more reactive unsubstituted olefin in **85** takes place.

During studies on a biomimetic ring closure towards the synthesis of the morphinan alkaloids, Feldman observed an interesting series of tautomeric shifts of the oxidized products of stilbene derivatives, such as **87** [34]. Rather than succumbing

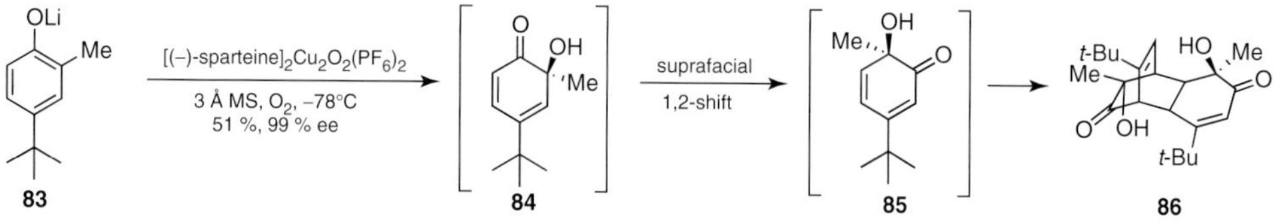

Scheme 20.23 Porco's oxidative dearomatization, suprafacial shift, and [4+2] dimerization.

to a desired 6π electrocyclization leading to the morphinan architecture, it was proposed that the bis-o-quinone ketal **88** suffers first a reversible [1,7] sigmatropic H-shift leading to intermediate **89**, followed by a reversible [1,5] H-shift affording intermediate **90**. Tautomer **91** then cyclizes irreversibly to afford compounds **92** and **93** (Scheme 20.24).

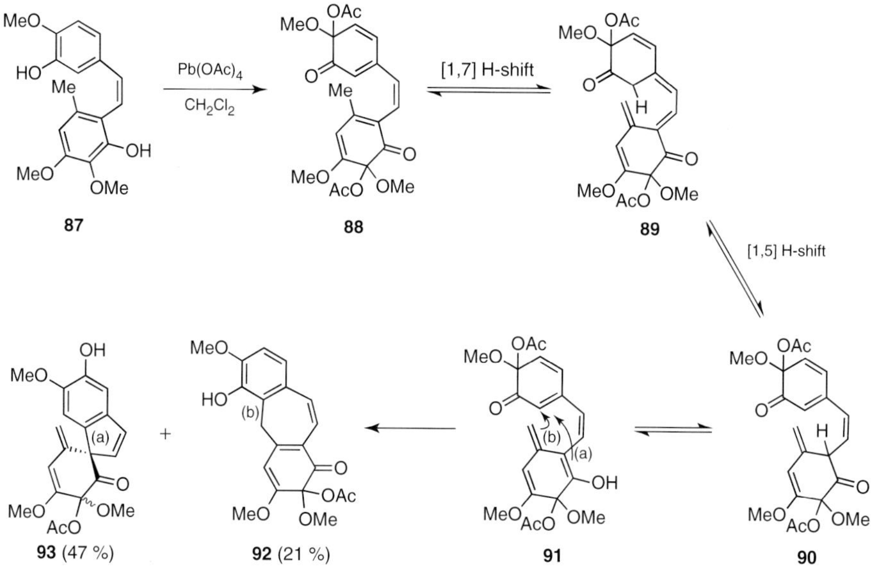

Scheme 20.24 Feldman's successive [1,7] and [1,5] H-shifts and electrocyclization.

Canesi has recently shown that the phenoxonium generated from phenol **94** can be intercepted by a tethered olefin leading to further rearrangement of functionality about the carbon chain to give **95** (Scheme 20.25) [35]. The term "homo-Wagner–Meerwein" transposition has been applied to this type of rearrangement, in what is in essence a heretofore unreported 1,3-allylic transposition, likely occurring via a concerted process.

The use of IBX to convert phenols into their corresponding o-quinones was first reported by Pettus [36]. It was further shown that o-quinone **97**, which is afforded by oxidation of **96**, could be tautomerized into the corresponding p-quinone methide **98** (Scheme 20.26) [37]. This intermediate undergoes intramolecular arylation to produce the benzylated brazilin derivative **99**, deprotection of which affords brazilin

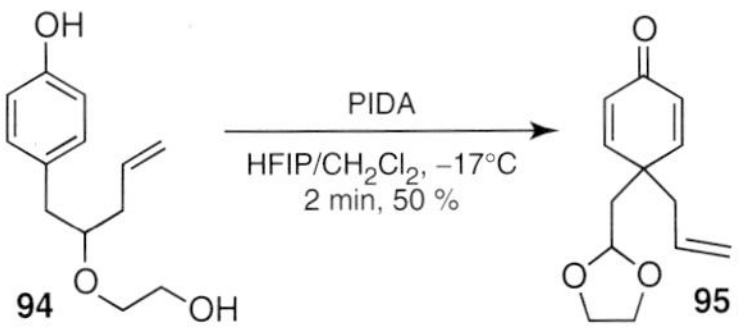

Scheme 20.25 Canesi's dearomatization, allylic transposition, and ketal formation.

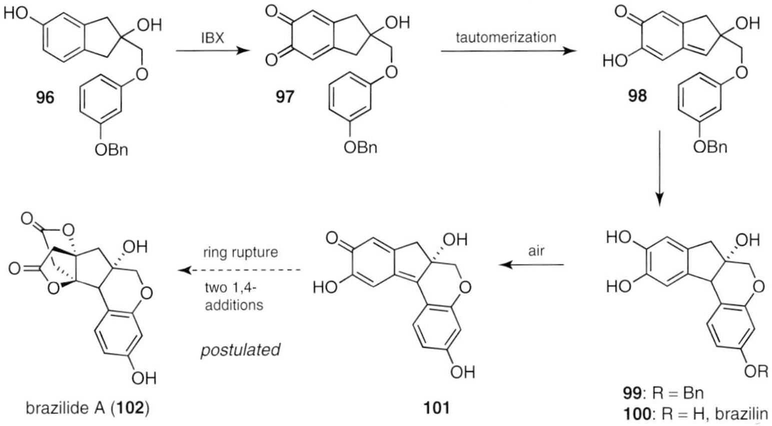

Scheme 20.26 Pettus' oxidative dearomatization followed by tautomerization, cyclization, and presumed oxidative cleavage to brazilide A.

(**100**). The catechol in this compound undergoes further oxidation upon standing to form *p*-quinone methide **101**, the oxidized form of brazilin that has often been used as a stain. It has been speculated that further oxidative rupture within this *p*-quinone methide gives rise to brazilide A (**102**) by a successive series of lactonizations [38].

Trauner has exploited the equilibrium between a *p*-quinone and its *o*-quinone methide tautomer in the synthesis of rubioncolin B (**106**) [39]. Exposure of advanced intermediate **103** to PIDA, in the presence of the desilylating reagent tris(dimethylamino)sulfonium difluorotrimethylsilicate (TASF), is presumed to initially gives rise to a *p*-quinone that is in equilibrium with *o*-quinone methide **104** (Scheme 20.27). Density functional theory calculations have shown the latter to be the energetically favored tautomer, facilitating the ensuing [4+2] cycloaddition that affords rubioncolin B derivative **105**.

In a biomimetic synthesis of icetexane-based diterpenes, Majetich reinvestigated Takeya's unusual solid state [4+2] dimerization of the *o*-quinone derived from (+)-demethylsalvicanol (**21**) (Scheme 20.6) [40]. It is thought that the high negative activation entropy associated with the [4+2] dimerization of *o*-quinone **22** is overcome by pre-organization of the components in the solid state, thereby facilitating the regioselective cycloaddition. Majetich reported that the use of water

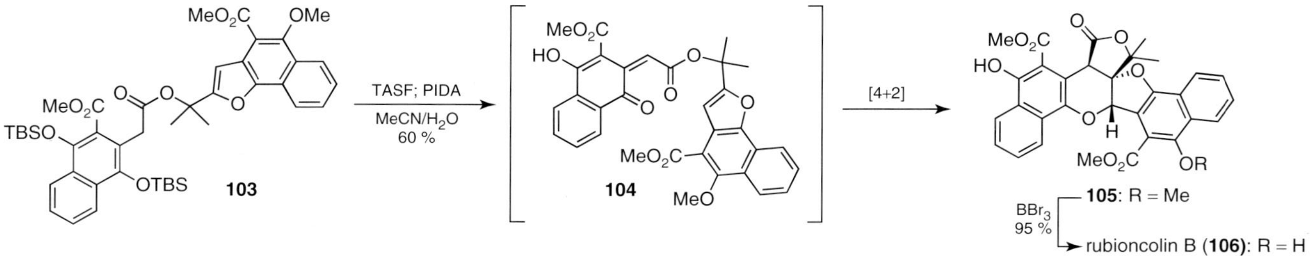

Scheme 20.27 Trauner's oxidative dearomatization, tautomerization, and [4+2] cycloaddition.

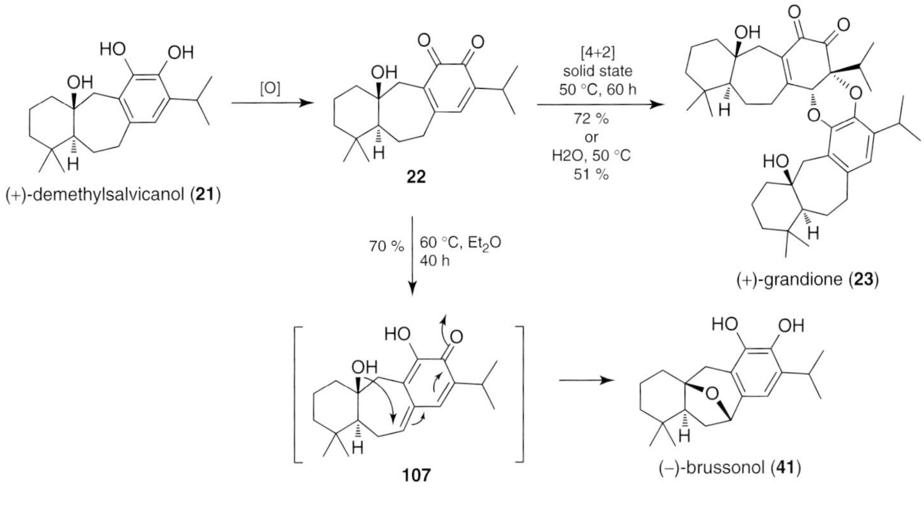

Scheme 20.28 Majetich's oxidation, tautomerization, and 1,6-addition synthesis of brussonol.

as a solvent was found to enable the dimerization, in what is speculated to be due to a high effective-molarity between reacting partners, a consequence of hydrophobic interactions that mimic the solid state. The use of ether as solvent, however, favors a tautomeric shift to enol form **107** – due to the lower dielectric constant of the solvent – which upon nucleophilic ring closure leads to (−)-brussonol (**41**) as the favored product (Scheme 20.28).

In a beautiful example of a biomimetic synthesis, Trauner has shown that hydantoin **108** can be converted into two natural products through a postulated common intermediate [41]. Exposure of **108** to AgO produces exiguamine A (**111**) or its oxygenated derivative, exiguamine B (**116**), depending on the number of equivalents of oxidant employed. The proposed mechanism for this transformation begins with the o-quinone oxidation product **109** (Scheme 20.29). Tautomerization would afford quinone methide **110**, which is poised for an oxa-6π electrocyclization, affording exiguamine A. o-Quinone **109** can also tautomerize to the isomeric quinone methide **112**, which can then undergo an oxa-6π electrocyclization to intermediate **113**. The latter is then intercepted by an irreversible oxidation to quinone **114**. Further tautomerization to **115** and electrocyclization gives rise to exiguamine B. This proposed mechanism is consistent with the observation that exposure of exiguamine A to excess AgO fails to garner its oxygenated derivative.

20.8
Sequential Ring Rupture and Contraction

The general instability of some phenol oxidation products can lead to further rearrangement through ring rupture followed by either contraction or expansion. For

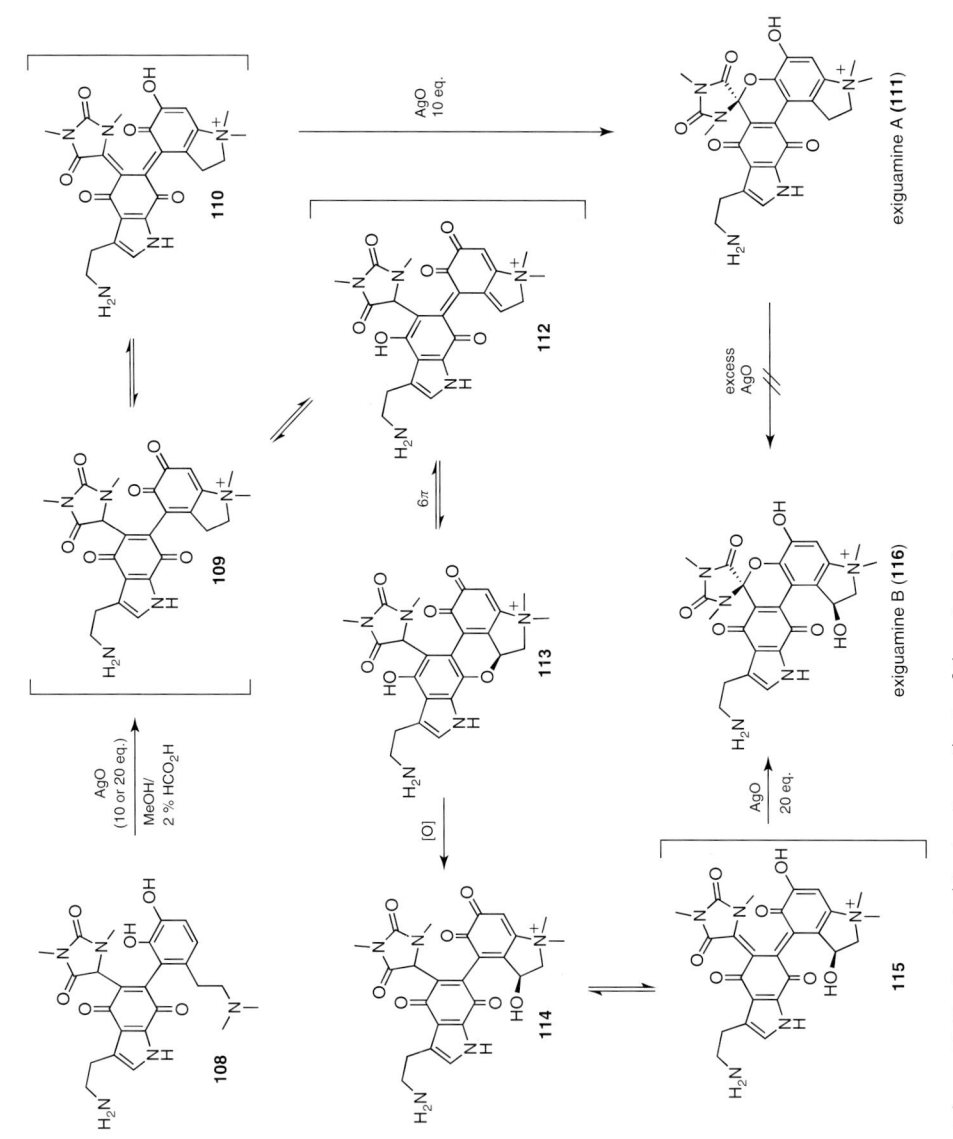

Scheme 20.29 Trauner's biomimetic synthesis of the exiguamines.

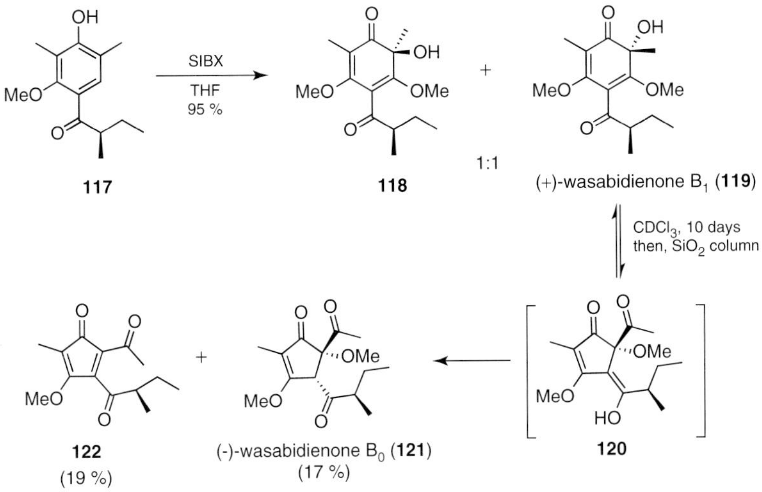

Scheme 20.30 Quideau's oxidative dearomatization and successive ring contraction.

example, Quideau observed the instability of wasabidienone B_1 (**119**) in solution and found it slowly underwent ring contraction to wasabidienone B_0 (**121**) [42]. Subjection of phenol **117** to a "stabilized" iodoxybenzoic acid (SIBX) derivative afforded an equimolar mixture of o-hydroxylated products **118** and (+)-wasabidienone B_1 (Scheme 20.30). Upon standing in deuteriochloroform for ten days and then purification on silica gel, wasabidienone B_0, and cyclopentadienone **122** were isolated, presumably through the intermediacy of enol **120**. Wasabidienone B_0 could also be obtained in higher yield by heating a solution of **119** in benzene for five days, since by-product **122** likely arises due to the acidity of the deuteriochloroform.

20.9
Sequential Ring Rupture and Expansion

The coupling of catechins of type **123** and **124** has been known for some time to yield the benzotropolone nucleus common to the theaflavins. However, it was not until Nakatsuka isolated the putative bicyclo[3.2.1] intermediate **126** that the exact nature of the mechanism was elucidated [43]. Although the details of the oxidative coupling of **123** and **124** is a matter of debate, the rearrangement and ring expansion of the adduct **125** would give rise to compound **126** (Scheme 20.31). Hydrolysis affords ene-diol **127** and further oxidation leads to diketone **128**, which undergoes a decarboxylative tautomerization affording benzotropolone **129**.

During the course of studies on the copper-mediated oxidation of catechols, Rogić reported the fascinating oxidation of catechol **130** by methoxy(pyridine)copper(II) chloride under strict anaerobic and anhydrous conditions [44]. The immediate oxidation product is an o-quinone of type **131**, which undergoes carbon–carbon bond cleavage in an extradiol sense and is reminiscent of the reaction manifold

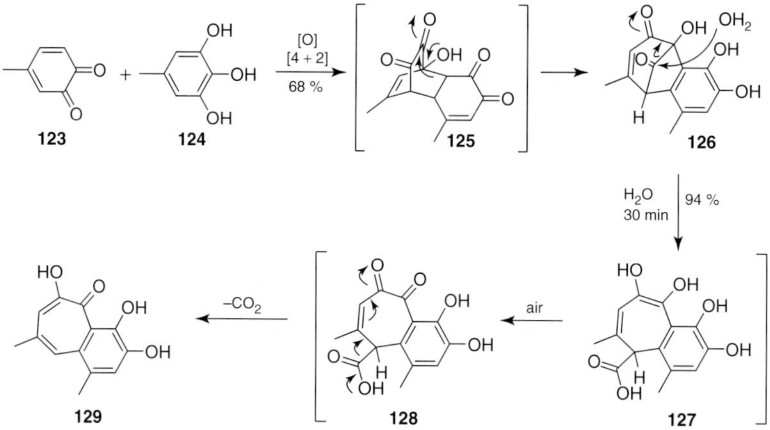

Scheme 20.31 Nakatsuka's dearomatization and successive ring expansion.

many enzyme oxygenases follow (Scheme 20.32). Rogić further suggests that the fact that this oxidation takes place in the absence of water may lead one to speculate that it may be no coincidence that the redox active sites in metalloenzymes are typically found in hydrophobic regions of the enzyme.

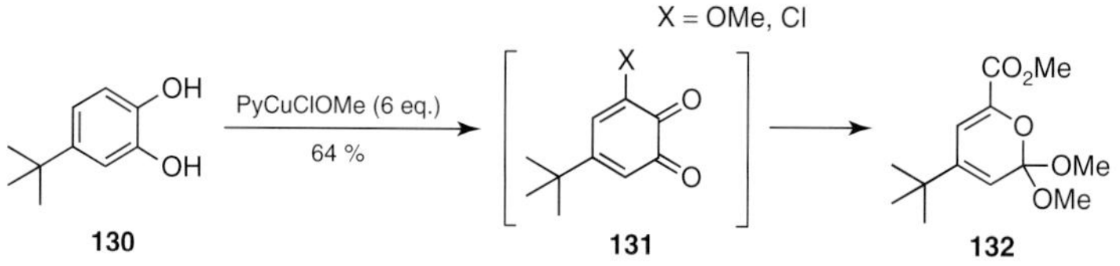

Scheme 20.32 Rogić's Cu-mediated extradiol catechol cleavage.

In an attempt to mimic the proposed biosynthetic pathway of rosmanol and isorosmanol from carnosol (**133**), Tejera observed an intriguing ring expansion of the catechol moiety [45]. Treatment of carnosol with *meta*-chloroperbenzoic acid (*m*-CPBA) and base led to the isolation of anhydride **135** (Scheme 20.33). The elucidated mechanism involves an initial oxidation to an *o*-quinone followed by peracid addition to give intermediate **134**. A subsequent Baeyer–Villiger rearrangement then affords anhydride **135**.

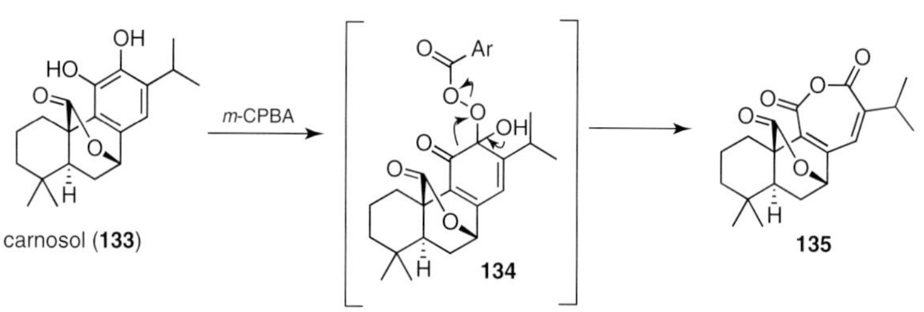

Scheme 20.33 Tejera's oxidative ring expansion.

20.10
Successive Intramolecular and Intermolecular Reactions

Some interesting combinations of intra- *and* inter-molecular reactions have also been reported. Pettus observed that oxidation of resorcinol derivative **136** with 2 equiv. of PIFA produces the epoxide **138** in a single pot (Scheme 20.34). This compound was further elaborated in two more operations so as to complete the synthesis of epoxysorbicillinol (**139**) [10]. The cascade sequence is thought to first produce intermediate **137**, which displays an extremely electron-deficient double bond. Furthermore, PIFA is thought to be in equilibrium with iodosylbenzene (PhIO) and the corresponding trifluoroacetic anhydride. The PhIO serves as a nucleophilic source of oxygen and subsequently produces the epoxide and iodobenzene in the same pot. Pettus has shown this reactivity of PhIO in various electron-deficient systems and the reagent was subsequently employed by MacMillan for the catalytic epoxidation of unsaturated iminium intermediates [46, 47].

20.11
Natural Products Hypothesized to Conclude Phenol Oxidative Cascades

Besides some of the cascades previously described for the natural products carpanone (Scheme 20.2) – dehydrodiisoeugenol (Scheme 20.3), bisorbicillinol (Scheme 20.4), grandione (Scheme 20.6), futoenone (Scheme 20.9), cedrene (Scheme 20.19), matlaline (Scheme 20.22), brazilide A (Scheme 20.26), rubioncolin B (Scheme 20.27), brussonol (Scheme 20.28), exiguamine B (Scheme 20.29), wasabidienones (Scheme 20.30), theaflavins (Scheme 20.31), carnosol (Scheme 20.33), and epoxysorbicillinol (Scheme 20.34) – there are undoubtedly other natural products that are the conclusion of cascades and sequences initiated by phenol oxidation. For example, in the biosynthesis of sclerocitrin (**144**), Steglich proposes a catechol dimerization to **140**, followed by oxidative ring cleavage to the anhydride **141** and ring contraction to the δ-lactone **142** (Scheme 20.35). Further retro-electrocyclization produces **143**, whereupon a vinylogous aldol ring closure is proposed to complete the biosynthesis of sclerocitrin (**144**) [48].

Zhao has proposed that przewalskin A (**147**) emerges from catechol **145** by oxidation to the corresponding *o*-quinone, followed by addition of acetoacetyl-CoA to both carbonyl residues to afford cyclopropane **146** (Scheme 20.36) [49]. This material is speculated to undergo ring expansion to produce the corresponding seven-membered carbocycle, which upon further oxidation affords przewalskin A (**147**).

Hertweck has proposed that oxidative tailoring enzymes are encoded in gene clusters along with specific polyketide synthetases. For example, the hexacyclic aromatic **148** is thought to emerge from 13 acetyl Co-A fragments (Scheme 20.37). Further phenol oxidation is speculated to result in dehydrocollinone (**149**), which undergoes successive oxidative rupture of the center ring so as to provide the aryl-oxy spiroketal motif found in griseorhodin (**150**) and related natural products [50].

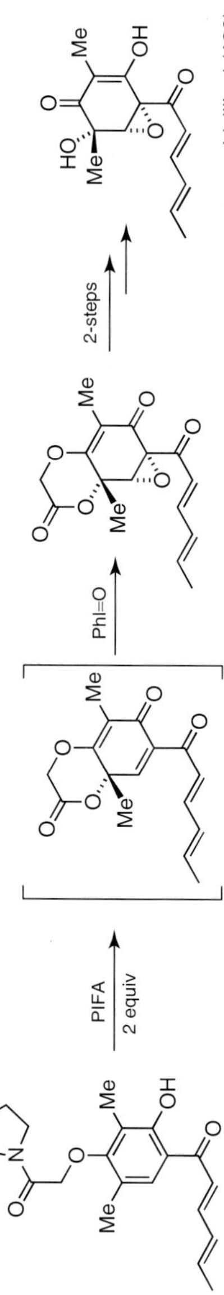

136

137

138

epoxysorbicillinol (139)
28 %

Scheme 20.34 Pettus' oxidative dearomatization and successive epoxidation.

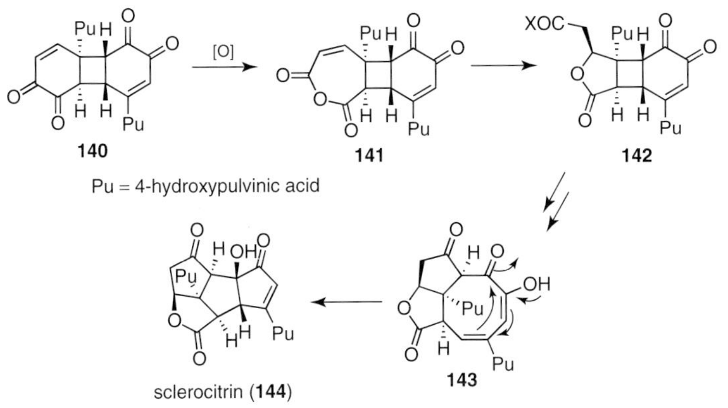

Scheme 20.35 Steglich's proposed biosynthetic pathway for sclerocitrin.

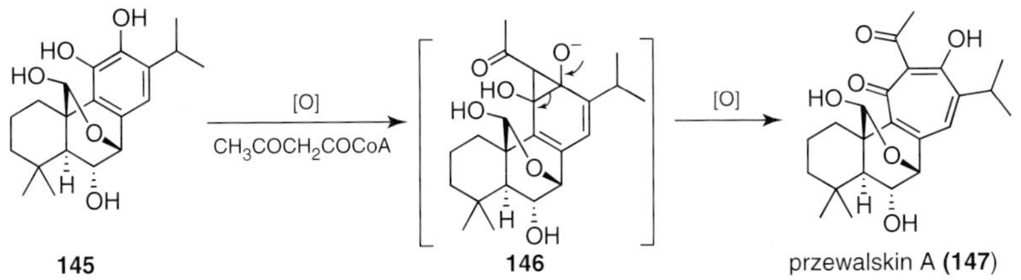

Scheme 20.36 Zhao's proposed biosynthetic pathway for przewalskin A.

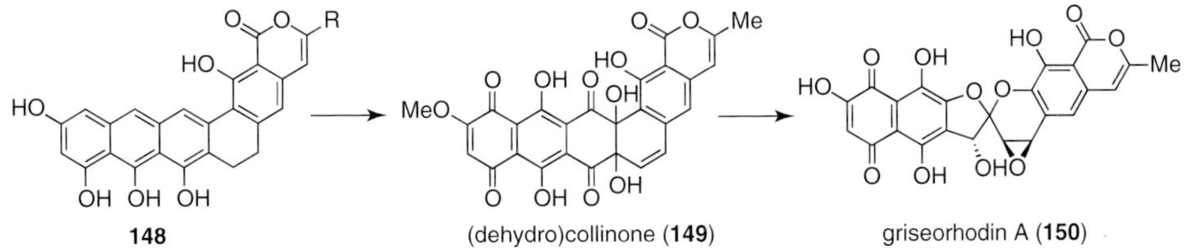

Scheme 20.37 Hertweck's proposed biosynthetic pathway for griseorhodin A.

The fredericamycins are thought to emerge from the identical hexacyclic compound **148** (Scheme 20.38) [51]. In this instance, oxidation affords fredericamycin C_1 (**151**), which is intercepted with ammonia to afford fredericamycin B (**152**). Further oxidation and rupture of the central ring in this instance proceeds through the intermediacy of the diones **153**–**155**, whereupon ring contraction affords **156** and loss of CO_2 provides the 5,5-spirocyclic array found in fredericamycin A (**157**).

Although to the best of our knowledge it has never before been postulated in the chemical literature, we suspect that many of the compounds in the frondosin family of natural products are themselves the consequence of various phenol oxidative dearomatization cascades (Scheme 20.39) [52]. For example, we believe that frondosin A (**158**) likely oxidizes to *o*-quinone methide **159**, which succumbs to a vinylogous aldol reaction to produce frondosin C (**160**). Alternatively, the

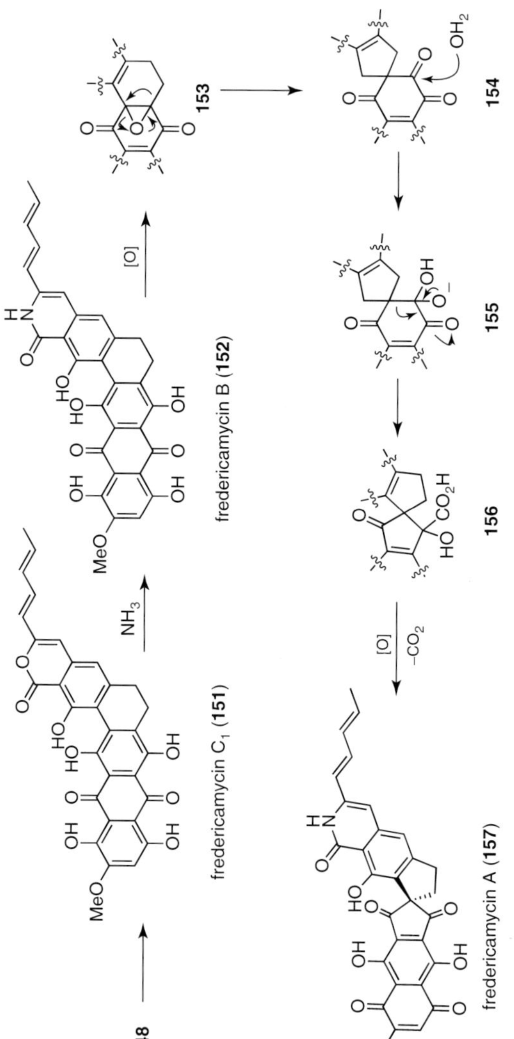

Scheme 20.38 Shen's proposed biosynthetic pathway for fredericamycin A.

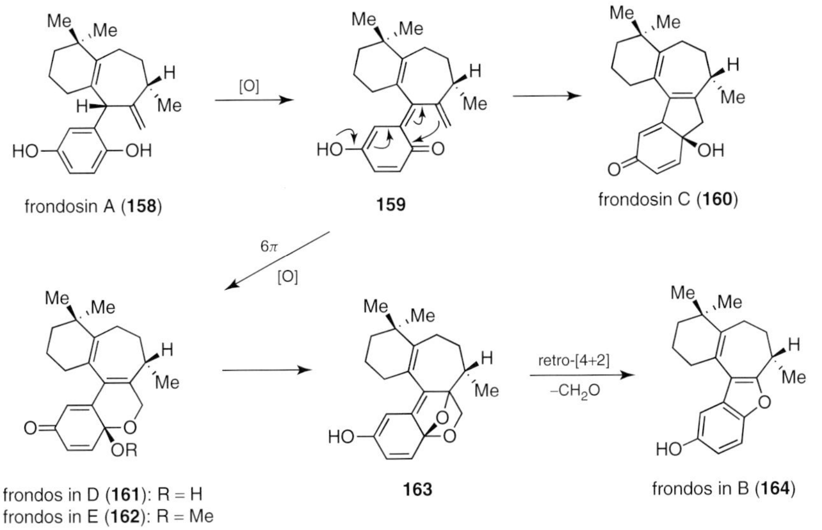

20.11 Natural Products Hypothesized to Conclude Phenol Oxidative Cascades

Scheme 20.39 Pettus' proposed biosynthetic pathway for frondosins.

o-quinone methide intermediate **159** can undergo a 6π electrocyclization and further oxidation in the presence of water or methanol to produce frondosin D (**161**) and E (**162**), respectively. A remote 1,6-conjugate addition in **161** to give **163**, followed by a retro [4+2] cycloaddition to liberate formaldehyde, would then produce frondosin B (**164**).

Porco theorized that the tetracyclic tetrapetalone skeleton would emerge from a cascade initiated by phenol oxidation of an *ansa* phenolic system **165** (Scheme 20.40) [53]. It was speculated that a phenoxonium would be intercepted by a neighboring olefin to generate the allyl cation **166**. The nitrogen atom was anticipated to intercept its conformer **167** to afford the tetracycle **168**, which would undergo further oxidation to afford tetrapetalone C (**169**). However, this proposal failed in the laboratory; the hydroquinone was simply converted into the corresponding quinone.

In addition to the previously described natural products, there are many others that appear to be the result of some sort of phenolic oxidation and subsequent series of reactions. The isariotins (**170** and **171**) [54], TK-57-164A (**172**) [54, 55], scyphostatin (not shown) [56], and aranorosin (**173**) [57] all seem to arise from various sorts of tyrosine oxidations (Figure 20.1). Hugonone A (**174**) appears to stem from a dimerization of an *o*-quinol that would be accessible from the corresponding α-methylated phenol [58], whereas malettinin A (**175**) would seem to arise from oxidation of the properly outfitted naphthol [59]. On the other hand, the heliespirones, such as **176**, emerge from oxidation of a properly attenuated hydroquinone followed by a 1,4-conjugate addition reaction [60]. Crews has proposed that curcuphenol (**63**) (Scheme 20.19) is the progenitor of many of the heliananes, such as **180**, through a sequence of reactions that presumably involves phenolic oxidation [61]. We speculate that kushecarpin (**177**) is the product of phenol

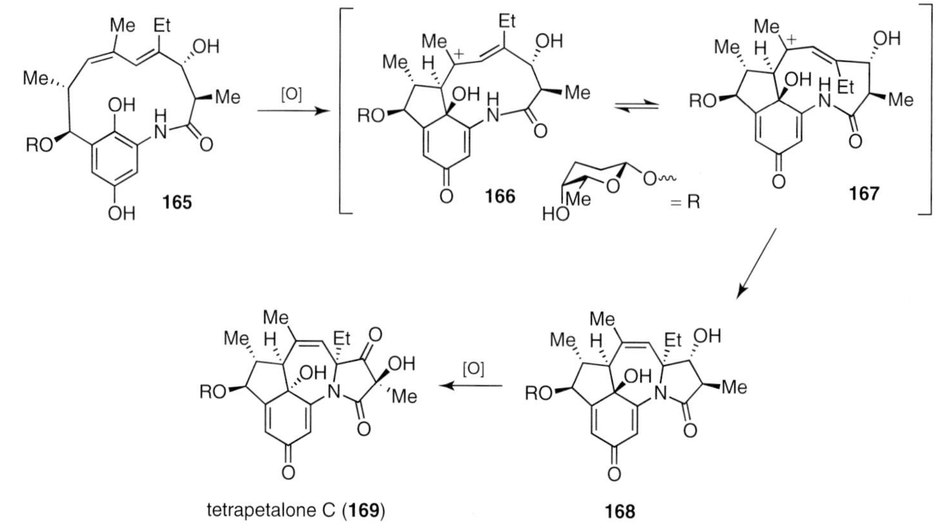

Scheme 20.40 Porco's proposed biosynthetic pathway for tetrapetalone C.

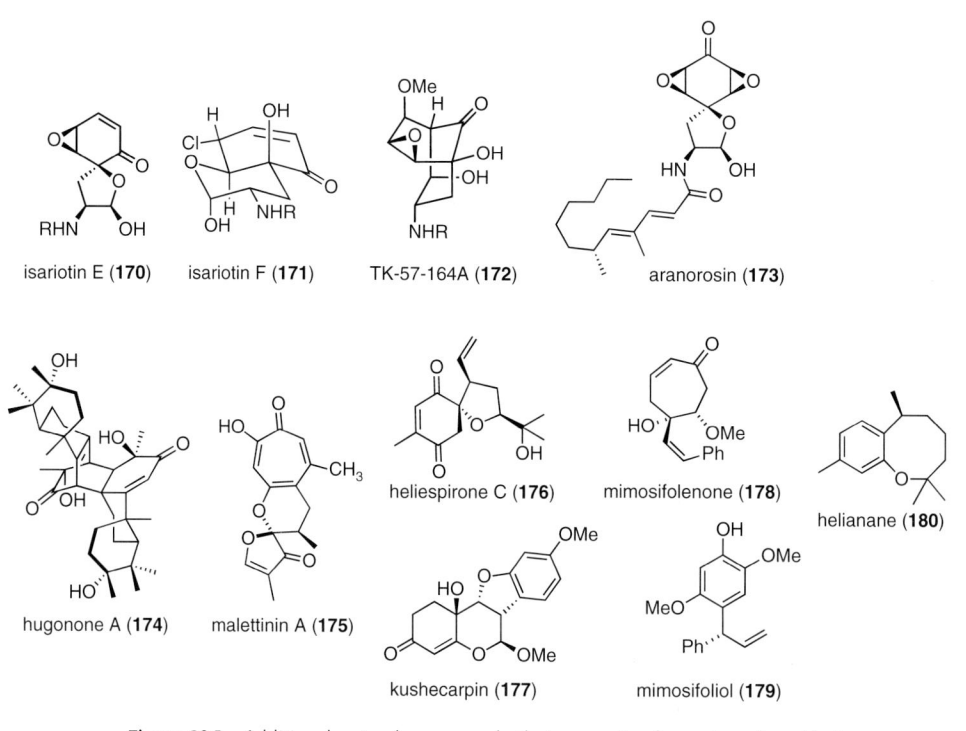

Figure 20.1 Additional natural compounds that may arise from phenolic oxidation.

isariotin E (**170**)
isariotin F (**171**)
TK-57-164A (**172**)
aranorosin (**173**)
hugonone A (**174**)
malettinin A (**175**)
heliespirone C (**176**)
kushecarpin (**177**)
mimosifolenone (**178**)
mimosifoliol (**179**)
helianane (**180**)

oxidation of some appropriately substituted pterocarpan. Finally, the coincident isolation of mimosifolenone (**178**) along with mimosifoliol (**179**) from the root wood of *Aeschynomene mimosifolia* would lead us to suppose that mimosifolenone is the result of some as yet undetermined oxidative dearomatization cascade that begins with mimosifoliol [62].

20.12
Conclusion

We hope that the foregoing discussions have given the reader pause so as to cautiously consider biomimetic claims involving oxidative phenol dearomatization as well as any future "biomimetic" suppositions on this topic. In our opinion, in the absence of actual isotopic feeding studies, these kinds of speculations are merely fanciful postulations, even if based upon enzymes that are conjectured to exist because of specific genomic sequences. Since time immemorial, man has ascribed any event leading from a simple beginning to a complex end as the result of some grand design. We believe that in the case of many of the foregoing molecules, Nature's role may be overstated. Indeed, many of these compounds may in fact be the result of specific post transformations outside of Nature's omniscient tutelage. However, no one can deny that these reaction sequences, irrespective of the origins of their creation, are beautiful, even if they are indeed artifacts invented by man's mind and brought into being by his hands. In this regard, we have been particularly inspired by the work of Rodríguez and Trauner, among others, in the area of cascades emanating for oxidative dearomatization and anticipate that these researchers will continue to inspire us in the future with their creative synthetic designs.

References

1. Robinson, R. (1917) *J. Chem. Soc., Trans.*, **111**, 762–768.
2. Leeper, F.J. and Vederas, J.C. (eds) (2000) *Biosynthesis: Aromatic Polyketide Isoprenoids, Alkaloids*, Topics in Current Chemistry, vol. 209, Springer-Verlag, Berlin, Heidelberg, New York.
3. Magdziak, D., Meek, S.J., and Pettus, T.R.R. (2004) *Chem. Rev.*, **104**, 383–1429.
4. Novak, M., Brinster, A.M., Dickhoff, J.N., Erb, J.M., Jones, M.P., Leopold, S.H., Vollman, A.T., Wang, Y.-T., and Glover, S.A. (2007) *J. Org. Chem.*, **72**, 9954–9962.
5. Angle, S.R., Rainier, J.D., and Woytowicz, C. (1997) *J. Org. Chem.*, **62**, 5884–5892.
6. (a) Van de Water, R.W. and Pettus, T.R.R. (2002) *Tetrahedron*, **58**, 5367–5405; (b) Ferreira, S.B., da Silva, F.deC., Pinto, A.C., Gonzaga, D.T.G., and Ferreira, V.F. (2009) *J. Heterocycl. Chem.*, **46**, 1080–1097.
7. (a) Chapman, O.L., Engel, M.R., Springer, J.P., and Clardy, J.C. (1971) *J. Am. Chem. Soc.*, **93**, 6696–6698; (b) Lindsley, C.W., Chan, L.K., Goess, B.C., Joseph, R., and Shair, M.D. (2000) *J. Am. Chem. Soc.*, **122**, 422–423.

8. Juhász, L., Kürti, L., and Antus, S. (2000) *J. Nat. Prod.*, **63**, 866–870.
9. Nicolaou, K.C., Vassilikogiannakis, G., Simonsen, K.B., Baran, P.S., Zhong, Y.-L., Vidali, V.P., Pitsinos, E.N., and Couladouros, E.A. (2000) *J. Am. Chem. Soc.*, **122**, 3071–3079.
10. Pettus, L.H., Van de Water, R.W., and Pettus, T.R.R. (2001) *Org. Lett.*, **3**, 905–908.
11. (a) Mehta, G. and Maity, P. (2007) *Tetrahedron Lett.*, **48**, 8865–8868; (b) Ganepain, J., Castet, F., and Quideau, S. (2007) *Angew. Chem. Int. Ed.*, **46**, 1533–1535.
12. Magdziak, D., Rodriguez, A.A., Van De Water, R.W., and Pettus, T.R.R. (2002) *Org. Lett.*, **4**, 285–288.
13. Aoyagi, Y., Takahashi, Y., Satake, Y., Fukaya, H., Takeya, K., Aiyama, R., Matsuzaki, T., Hashimoto, S., Shiina, T., and Kurihara, T. (2005) *Tetrahedron Lett.*, **46**, 7885–7887.
14. Singh, V., Chandra, G., and Mobin, S.M. (2008) *Synlett*, 3111–3114.
15. Jensen, K.H., Pathak, T.P., Zhang, Y., and Sigman, M.S. (2009) *J. Am. Chem. Soc.*, **131**, 17074–17075.
16. (a) Yamamura, S., Shizuri, Y., Shigemori, H., Okuno, Y., and Ohkubo, M. (1991) *Tetrahedron*, **47**, 635–644; (b) Yamamura, S. and Nishiyama, S. (2002) *Synlett*, 533–543.
17. Mohr, A.L., Lombardo, V.M., Arisco, T.M., and Morrow, G.W. (2009) *Synth. Commun.*, **39**, 3845–3855.
18. McFadden, R.M. and Stoltz, B.M. (2006) *J. Am. Chem. Soc.*, **128**, 7738–7739.
19. Simmons, E.M., Yen, J.R., and Sarpong, R. (2007) *Org. Lett.*, **9**, 2705–2708.
20. Hsu, D.-S. and Liao, C.-C. (2007) *Org. Lett.*, **9**, 4563–4565.
21. (a) Njardarson, J.T., McDonald, I.M., Spiegel, D.A., Inoue, M., and Wood, J.L. (2001) *Org. Lett.*, **3**, 2435–2438; (b) Bérubé, A., Drutu, I., and Wood, J.L. (2006) *Org. Lett.*, **8**, 5421–5424.
22. (a) Cook, S.P., Gaul, C., and Danishefsky, S.J. (2005) *Tetrahedron Lett.*, **46**, 843–847; (b) Polara, A., Cook, S.P., and Danishefsky, S.J. (2008) *Tetrahedron Lett.*, **49**, 5906–5908;
(c) Cook, S.P. and Danishefsky, S.J. (2006) *Org. Lett.*, **8**, 5693–5695.
23. Morton, J.G.M., Kwon, L.D., Freeman, J.D., and Njardarson, J.T. (2009) *Synlett*, 23–27.
24. Mendelsohn, B.A., Lee, S., Kim, S., Teyssier, F., Aulakh, V.S., and Ciufolini, M.A. (2009) *Org. Lett.*, **11**, 1539–1542.
25. Frie, J.L., Jeffrey, C.S., and Sorensen, E.J. (2009) *Org. Lett.*, **11**, 5394–5397.
26. (a) Wenderski, T.A., Huang, S., and Pettus, T.R.R. (2009) *J. Org. Chem.*, **74**, 4104–4109; (b) Mejorado, L.H. and Pettus, T.R.R. (2006) *J. Am. Chem. Soc.*, **128**, 15625–15631; (c) Mejorado, L.H. and Pettus, T.R.R. (2006) *Synlett*, 3209–3214; (d) Wang, J. and Pettus, T.R.R. (2004) *Tetrahedron Lett.*, **45**, 5895–5899; (e) Wang, J., Pettus, L.H., and Pettus, T.R.R. (2004) *Tetrahedron Lett.*, **45**, 1793–1796; (f) Mejorado, L.H., Hoarau, C., and Pettus, T.R.R. (2004) *Org. Lett.*, **6**, 1535–1538; (g) Van de Water, R.W., Hoarau, C., and Pettus, T.R.R. (2003) *Tetrahedron Lett.*, **44**, 5109–5113.
27. Marsini, M.A., Huang, Y., Van de Water, R.W., and Pettus, T.R.R. (2007) *Org. Lett.*, **9**, 3229–3232.
28. Green, J.C. and Pettus, T.R.R. (2010) *J. Am. Chem. Soc.*, asap, doi: 10.1021/ja109925.
29. Skyler, D. and Heathcock, C.H. (2001) *Org. Lett.*, **3**, 4323–4324.
30. (a) Imbos, R., Minnaard, A.J., and Feringa, B.L. (2002) *J. Am. Chem. Soc.*, **124**, 184–185; (b) Hayashi, Y., Gotoh, H., Tamura, T., Yamaguchi, H., Masui, R., and Shoji, M. (2005) *J. Am. Chem. Soc.*, **127**, 16028–16029; (c) Liu, Q. and Rovis, T. (2006) *J. Am. Chem. Soc.*, **128**, 2552–2553.
31. Vo, N.T., Pace, R.D.M., O'Hara, F., and Gaunt, M.J. (2008) *J. Am. Chem. Soc.*, **130**, 404–405.
32. Acuña, A.U., Amat-Guerri, F., Morcillo, P., Liras, M., and Rodríguez, B. (2009) *Org. Lett.*, **11**, 3020–3023.
33. Dong, S., Zhu, J., and Porco, J.A. Jr. (2008) *J. Am. Chem. Soc.*, **130**, 2738–2739.
34. Feldman, K.S. (1997) *J. Org. Chem.*, **62**, 4983–4990.

35. Guérard, K.C., Chapelle, C., Giroux, M.-A., Sabot, C., Beaulieu, M.-A., Achache, N., and Canesi, S. (2009) *Org. Lett.*, **11**, 4756–4759.
36. Magdziak, D., Rodriguez, A.A., Van de Water, R.W., and Pettus, T.R.R. (2002) *Org. Lett.*, **4**, 285–288.
37. Huang, Y., Zhang, J., and Pettus, T.R.R. (2005) *Org. Lett.*, **7**, 5841–5844.
38. Yang, B.O., Ke, C.-Q., He, Z.-S., Yang, Y.-P., and Ye, Y. (2002) *Tetrahedron Lett.*, **43**, 1731–1733.
39. Lumb, J.-P., Choong, K.C., and Trauner, D. (2008) *J. Am. Chem. Soc.*, **130**, 9230–9231.
40. Majetich, G. and Zou, G. (2008) *Org. Lett.*, **10**, 81–83.
41. Volgraf, M., Lumb, J.-P., Brastianos, H.C., Carr, G., Chung, M.K.W., Münzel, M., Mauk, A.G., Andersen, R.J., and Trauner, D. (2008) *Nat. Chem. Biol.*, **4**, 535–537.
42. Pouységu, L., Marguerit, M., Gagnepain, J., Lyvinec, G., Eatherton, A.J., and Quideau, S. (2008) *Org. Lett.*, **10**, 5211–5214.
43. (a) Yanase, E., Sawaki, K., and Nakatsuka, S.-I. (2005) *Synlett*, 2661–2663; (b) Takino, Y., Imagawa, H., Horikawa, H., and Tanaka, A. (1964) *Agric. Biol. Chem.*, **28**, 64–71.
44. Demmin, T.R. and Rogić, M.M. (1980) *J. Org. Chem.*, **45**, 4210–4214.
45. Marrero, J.G., Tejera, L.S.A., Luis, J.G., and Rodríguez, M.L. (2002) *Synlett*, 1517–1519.
46. McQuaid, K.M. and Pettus, T.R.R. (2004) *Synlett*, 2403–2405.
47. Lee, S. and MacMillan, D.W.C. (2006) *Tetrahedron*, **62**, 11413–11424.
48. Winner, M., Giménez, A., Schmidt, H., Sontag, B., Steffan, B., and Steglich, W. (2004) *Angew. Chem. Int. Ed.*, **43**, 1883–1886.
49. Xu, G., Hou, A.-J., Wang, R.-R., Liang, G.-Y., Zheng, Y.-T., Liu, Z.-Y., Li, X.-L., Zhao, Y., Huang, S.-X., Peng, L.-Y., and Zhao, Q.-S. (2006) *Org. Lett.*, **8**, 4453–4456.
50. Xu, Z., Schenk, A., and Hertweck, C. (2007) *J. Am. Chem. Soc.*, **129**, 6022–6030.
51. Wendt-Pienkowski, E., Huang, Y., Zhang, J., Li, B., Jiang, H., Kwon, H., Hutchinson, C.R., and Shen, B. (2005) *J. Am. Chem. Soc.*, **127**, 16442–16452.
52. Pettus, T.R.R. and coworkers (2008) NSF Grant Application, CHE-0806357.
53. (a) Wang, X. and Porco, J.A. Jr. (2005) *Angew. Chem. Int. Ed.*, **44**, 3067–3071; (b) Wang, X. and Porco, J.A. Jr., *corrigenda* (2006) *Angew. Chem. Int. Ed.*, **45**, 6607.
54. Bunyapaiboonsri, T., Yoiprommarat, S., Intereya, K., Rachtawee, P., Hywel-Jones, N.L., and Isaka, M. (2009) *J. Nat. Prod.*, **72**, 756–759.
55. (a) Nakagawa, A., Nishikawa, N., Takahashi, S., and Yamamoto, K. (2004) Patent WO2004074269; (b) Cha, J.Y., Huang, Y., and Pettus, T.R.R. (2009) *Angew. Chem. Int. Ed.*, **48**, 9519–9521.
56. Tanaka, M., Nara, F., Suzuki-Konagai, K., Hosoya, T., and Ogita, T. (1997) *J. Am. Chem. Soc.*, **119**, 7871–7872.
57. Roy, K., Mukhopadhyay, T., Reddy, G.C., Desikan, K.R., Rupp, R.H., and Ganguli, B.N. (1988) *J. Antibiot.*, **41**, 1780–1784.
58. Mdee, L.K., Waibel, R., Nkunya, M.H.H., Jonker, S.A., and Achenbach, H. (1998) *Phytochemistry*, **49**, 1107–1113.
59. (a) Angawi, R.F., Swenson, D.C., Gloer, J.B., and Wicklow, D.T. (2005) *J. Nat. Prod.*, **68**, 212–216; (b) Angawi, R.F., Swenson, D.C., Gloer, J.B., and Wicklow, D.T. (2003) *Tetrahedron Lett.*, **44**, 7593–7596.
60. Macías, F.A., Galindo, J.L.G., Varela, R.M., Torres, A., Molinillo, J.M.G., and Fronczek, F.R. (2006) *Org. Lett.*, **8**, 4513–4516.
61. Harrison, B. and Crews, P. (1997) *J. Org. Chem.*, **62**, 2646–2648.
62. Fullas, F., Kornberg, L.J., Wani, M.C., Wall, M.E., Farnsworth, N.R., Chagwedera, T.E., and Kinghorn, A.D. (1996) *J. Nat. Prod.*, **59**, 190–192.

Part V
Frontiers in Biomimetic Chemistry: From Biological to Bio-inspired Processes

21
The Diels–Alderase Never Ending Story
Atsushi Minami and Hideaki Oikawa

21.1
Introduction

The Diels–Alder reaction is one of the most widely used reactions in organic synthesis, forming a six-membered ring from a 1,3-diene and a dienophile with high regio- and stereoselectivity under relatively mild conditions [1]. In addition, by creating four chiral centers or quaternary stereogenic centers in organic synthesis, the Diels–Alder reaction is a powerful tool that has been applied to the synthesis of complex pharmaceutical and biologically active compounds [2].

Natural products presumably biosynthesized via a [4+2] cycloaddition are frequently encountered in the literature [3]. Several reviews on natural Diels–Alder type cycloadducts have covered more than 300 cycloadducts, including polyketides, terpenoids, phenylpropanoids, alkaloids, and natural products, formed via mixed biosynthetic pathways. Several observations indicate that natural products may be biosynthesized by biological Diels–Alder reactions:

1) co-isolation of an adduct and of the corresponding precursor;
2) co-occurrence of adducts and their regio- and diastereoisomers;
3) dimeric structure of a single component;
4) unusual combination of biosynthetic components such as monoterpene and triterpene.

Careful examination of the structures of natural [4+2] adducts sometimes provides useful information on their biosynthesis. However, retrosynthetic analysis of plausible adducts incorporating a cyclohexene moiety or its equivalent sometimes misled their biosynthetic pathway. For example, the flavonolignan silydianin [4] is a plausible [4+2] adduct that could be biosynthesized from coniferyl alcohol and an ortho-quinone derived from the flavanone taxifolin (Scheme 21.1, [4+2] route). Usually, radical coupling of phenylpropanoids affords various lignan products (radical coupling route) [5]. Along this line, the skeleton of silydianin is therefore most likely constructed by radical coupling followed by nucleophilic addition of the resultant enolate (path A). Co-occurrence of the lignan silychristin in the same plant supported this biosynthetic pathway (path B). In previous reviews,

Biomimetic Organic Synthesis, First Edition. Edited by Erwan Poupon and Bastien Nay.
© 2011 Wiley-VCH Verlag GmbH & Co. KGaA. Published 2011 by Wiley-VCH Verlag GmbH & Co. KGaA.

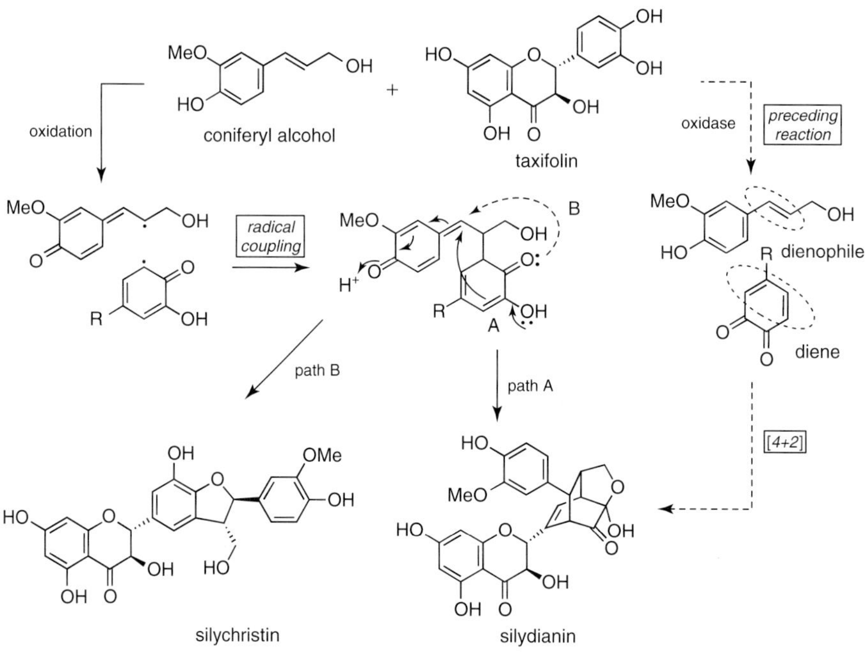

Scheme 21.1 Biosynthetic pathway of silydianin and its related metabolite.

we can find several representative examples of plausible [4+2] adducts that could be biosynthesized via alternative routes [3]. Because it is difficult to distinguish between the involvement of a Diels–Alderase or of alternative enzymes, we have collected candidates with intriguing natural [4+2] adducts and will discuss their enzymatic formation in this chapter.

21.2
Diels–Alderases Found in Nature

We reported the enzymatic activity of solanapyrone synthase in 1995 as the first Diels–Alderase [6]. To date, two additional Diels–Alderases, lovastatin nonaketide synthase [7] and macrophomate synthase (MPS) [8], have been purified and characterized. The first two catalyze intramolecular Diels–Alder (IMDA) reactions while the last one catalyses an intermolecular Diels–Alder reaction. We have reported the detailed reaction pathway [9] of MPS and its catalytic mechanism based on the crystal structure [9]. However, only three Diels–Alderases have been characterized up to now. The reason why the number of characterized Diels–Alderase is limited may be attributed to the difficulty in identifying the actual substrate and the enzyme responsible for the biosynthesis of the target molecule. In this section, we describe these three natural Diels–Alderases and discuss the mechanism of their catalysis.

21.2.1
Lovastatin Nonaketide Synthase

The biosynthesis of the cholesterol-lowering drug lovastatin isolated from *Aspergillus terreus* has been investigated extensively by Vederas and coworkers [10]. Incorporation experiments with multiple labeled acetate and ^{18}O-oxygen suggested that it is biosynthesized via the polyketide pathway. Based on feeding studies and co-occurrence of 4a,5-dihydromonacolin L (**3**), this compound was speculated as an intermediate. This was confirmed by the successful conversion of **3** into lovastatin, using a blocked mutant of *A. terreus* (Scheme 21.2) [11]. Because lovastatin does not have an electron-withdrawing group in the dienophile moiety, it was proposed that the requisite Diels–Alder reaction occurred at the hexaketide stage. However, all efforts to convert ^{13}C-labeled hexaketide precursor **1b** into lovastatin were unsuccessful due to non-enzymatic cycloadditions affording a 1 : 1 mixture of undesired diastereomers **4b** (*endo*) and **4c** (*exo*) in aqueous media (half life of **1b**: two days) (Scheme 21.2). This clearly showed significant rate acceleration for the cycloaddition of **1b** in aqueous media [12].

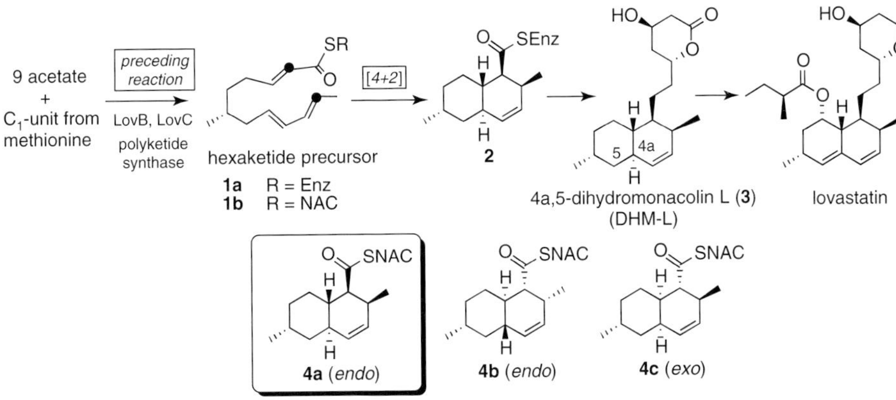

Scheme 21.2 Biosynthetic pathway of lovastatin.

In 1999, the biosynthetic gene cluster of lovastatin was cloned by Hutchinson's group [13]. They succeeded in achieving heterologous expression of polyketide synthase (PKS) genes (*lovB* and *lovC*) in *Aspergillus nidulans* to produce **3**. Collaborating with the Hutchinson group, the Vederas group started enzymatic studies using lovastatin nonaketide synthase (LovB), which is responsible for the construction of the lovastatin backbone. Using the purified LovB, hexaketide triene precursor **1b** was incubated without co-factors and substrates to give the three adducts **4a**–**c** in a ratio of 1 : 15 : 15. Minor *endo*-product **4a** was confirmed to be the one with the same stereochemistry as natural **3** [7]. In lovastatin biosynthesis, hexaketide precursor **1b** should load on the corresponding ketosynthase domain of LovB; it is then processed downstream to yield **3**. Since adducts **4a**–**c** were obtained as *N*-acetyl-cysteamine (NAC) thioesters, the obligatory thioester exchange did not occur in the Diels–Alder reaction.

Recently, LovB and LovC were successfully overexpressed in yeast cell and their catalytic functions were fully reconstituted in the presence and absence of co-factors to produce **3** and various aberrant products [14]. These results clearly demonstrated that PKS LovB is responsible for the corresponding Diels–Alder reaction.

21.2.2
Macrophomate Synthase

The phytopathogenic fungus *Macrophoma commelinae* has the ability to transform 2-pyrone **5** into the corresponding benzoate analog macrophomic acid (Figure 21.1a) [15]. Later, we succeeded in the purification of MPS, which is responsible for this complex aromatic conversion with oxaloacetate as a substrate for the C3-unit precursor [16].

Throughout all studies on MPS, the most difficult task was the identification of the C3-unit precursor. As in the case of lovastatin, establishing substrate and enzymatic activity are always tough work in a Diels–Alderase study. The catalytic mechanism of the MPS reaction was investigated extensively, showing that the reaction involves three separate steps: (i) decarboxylation; (ii) two carbon–carbon bond formations; and (iii) decarboxylation with concomitant dehydration [8, 17]. Recently, it was found that the third step of the MPS reaction is decarboxylation to afford intermediate **10** and that the subsequent dehydration is not catalyzed by MPS [18]. Unstable intermediate **10** was released from active site and dehydrated to give macrophomate in a non-enzymatic manner. In the absence of **5**, MPS simply acts as a decarboxylase with high catalytic efficiency (Figure 21.1a).

The crystal structure of the MPS complexed with pyruvate and Mg^{2+} was determined with a resolution of 1.70 Å (Figure 21.1b) [9]. On the basis of this structural information, the pathway of the MPS reaction can be outlined as follows (Figure 21.1a): oxaloacetate is incorporated into the active site of MPS in a similar way to that of pyruvate. Lewis acidity of the magnesium promotes decarboxylation to form the enolate anion, which is stabilized by an electron sink provided by the divalent cation [8, 9]. As shown in Figure 21.1b, the 2-pyrone molecule is fixed in place through two hydrogen bonds between the carbonyl oxygen of 2-pyrone, Arg101, the C5-acyl oxygen, and Tyr169. These hydrogen bonds act not only in substrate recognition but also enhance reactivity in the inverse electron demand Diels–Alder reaction by reducing the lowest unoccupied molecular orbital (LUMO) energy of the diene. The binding model explains the substrate specificity and stereochemical course of whole reaction pathway [16]. Based on formation of the aberrant adduct with pyrone **8** and the observation that dehydration proceeds formally in an anti-sense [8, 16], it was proposed that the higher energy [4+2] adducts **9** and **7** are transformed into either macrophomic acid or the rearranged product **8** as shown in Figure 21.1a.

Extensive point mutation experiments on MPS identified essential amino acid residues [19] and remarkable tolerance for mutation. Mixed quantum and molecular mechanics calculations on the MPS-catalyzed reaction suggested that the transition state of an alternative Michael–aldol route is more stable than that of the concerted

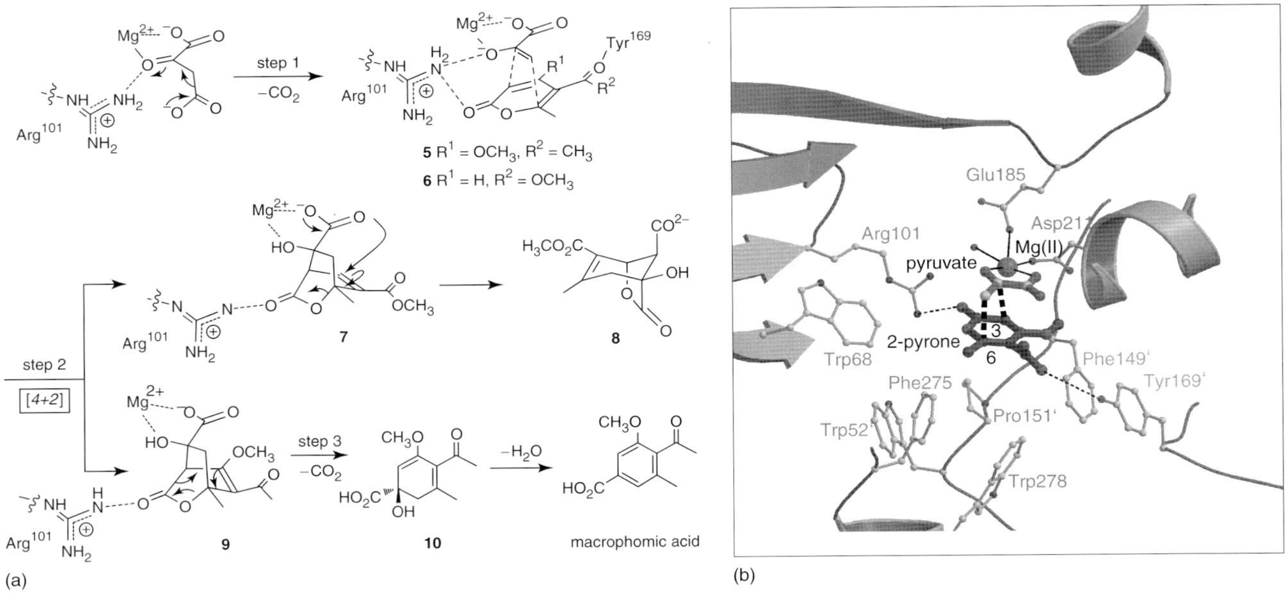

Figure 21.1 Detailed reaction mechanism of macrophomate synthase (a) and model of MPS active site (b).

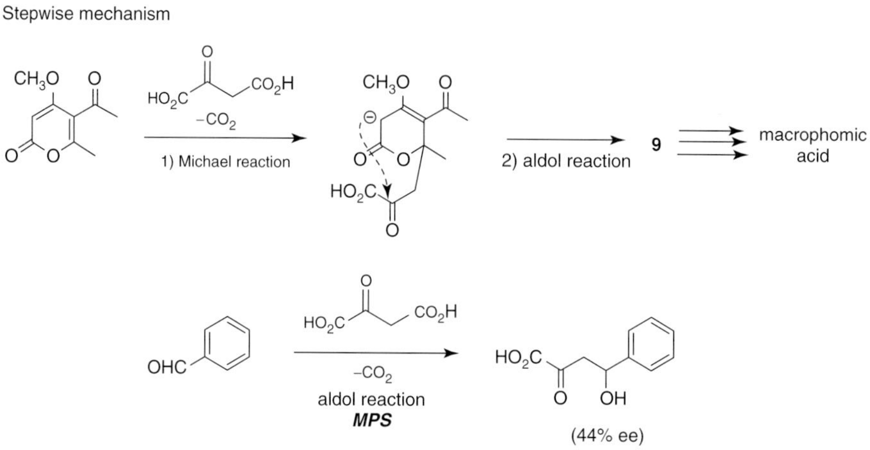

Stepwise mechanism

Scheme 21.3 Alternative reaction mechanism (Michael–aldol route) and the aldol reaction catalyzed by macrophomate synthase.

Diels–Alder route [20], indicating that the two-step route is the energetically preferred process. Although intriguing observations that MPS itself can catalyze the aldol reaction have been reported (Scheme 21.3) [21], experimental evidence must be provided to evaluate the validity of calculation results. The recent finding that the first decarboxylation, the second C–C bond formation, and the third decarboxylation into **10** are rapid processes [18] suggested that it is difficult to distinguish a concerted from a stepwise process experimentally.

21.2.3
Solanapyrone Synthase

Solanapyrones were isolated as phytotoxic substances from various fungi [22]. This family consists of diastereomers A (**15**), D (**16**), and their reduced forms B (**17**) and E (**18**) [23]. A series of feeding experiments with simple isotopically labeled precursors established that solanapyrones are biosynthesized via a polyketide pathway and that the diastereomers **15** and **16** are produced at the later stage [23, 24]. Isolation of these substances as optically active forms strongly indicates that **15** and **16** are biosynthesized from the achiral linear triene precursor prosolanapyrone III **14** via an enzyme-catalyzed Diels–Alder reaction. Although there are many candidates such as lovastatin hexaketide for the intriguing cycloaddition, incorporation of isotopically labeled biosynthetic precursors, prosolanapyrones I (**12**) and II (**13**) into **15** and **16** confirmed the biosynthetic pathway of solanapyrones as shown in Scheme 21.4 [25].

To support the involvement of a Diels–Alderase in this reaction, enzymatic conversion of **14** was examined. Using crude enzyme, we found that both **13** and **14** were converted into **15** and **16** [6]. Although purification of solanapyrone synthase (SPS) was hampered by its instability, it has been purified as a single

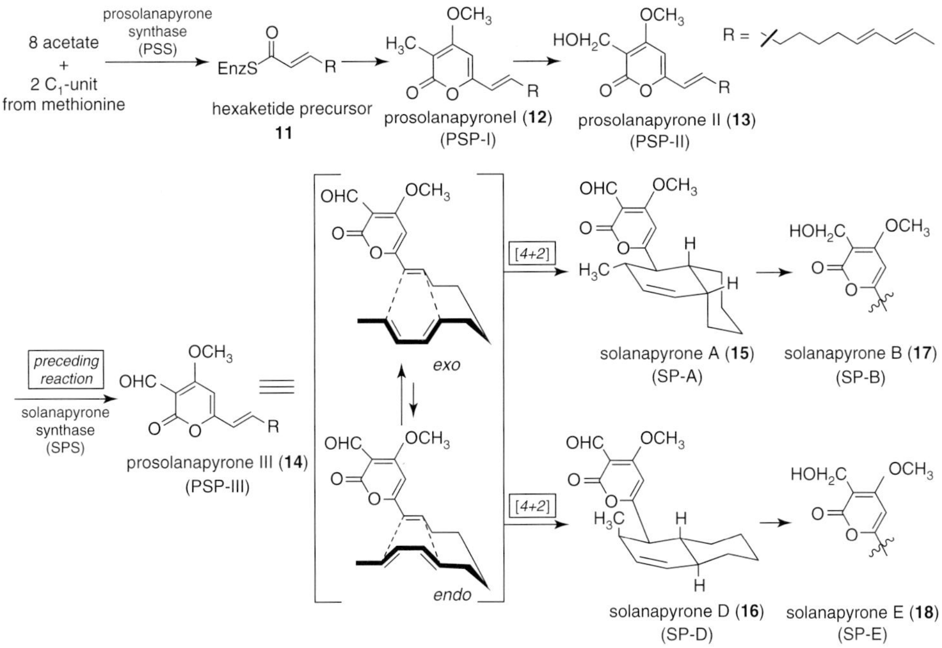

Scheme 21.4 Biosynthetic scheme of solanapyrones.

band on SDS-PAGE (sodium dodecyl sulfate polyacrylamide gel electrophoresis) [26]. Recently, the gene cluster responsible for biosynthesis of solanapyrones has been identified by homology based PCR (polymerase chain reaction) and genome walking [27]. The gene cluster contains PKS gene (*sol1*) for backbone construction and four genes for modification enzymes, including FAD-dependent monooxygenase gene (*sol5*). This monooxygenase Sol5 possessing a flavin-binding domain was overexpressed in yeast and purified as a single band on SDS-PAGE. The enzyme thus obtained showed enzymatic activity catalyzing both oxidation (**13** to **14**) and Diels–Alder reaction (**14** to **15** and **16**) with high optical activity and *exo/endo* ratio (7:1). The character of this enzyme was identical to that of crude solanapyrone synthase from *A. solani*.

To assess the diastereoselectivity and the intrinsic reactivity of prosolanapyrones, Diels–Alder reactions were examined under various conditions [28]. In less polar solvents, heating was required for the effective cycloaddition of **12** to **14**. An increase in the oxidation level of the 3-substituent in the prosolanapyrones enhances rate acceleration. This can be rationalized in terms of the LUMO energy of the dienophile moiety in the pyrone precursors. *Endo/exo*-selectivities with **12–14** were essentially the same in various organic solvents while the *endo*-selectivity was increased with increasing solvent polarity. The slight preference for *endo*-selectivity in less polar solvents suggested that there is little steric congestion in both *endo*- and *exo*-transition states (Scheme 21.4) as reported in the reactions of simple decatriene systems [29].

In aqueous medium, the non-enzymatic reactions of **12–14** were accelerated and gave *endo*-adducts with high selectivity (**15** : **16** = 3 : 97) [28]. These effects were observed significantly in the reaction of **14** but not in that of **13**, indicating that the oxidation of **13** significantly enhanced the reactivity for the Diels–Alder reaction. Similar rate accelerations and the predominant formation of *endo*-adducts have been reported as a hydrophobic effect [30]: water forces the substrate to form the more compact *endo*-transition state, reducing its molecular surface exposed to the aqueous medium. In addition, molecular orbital calculations indicated the importance of hydrogen bonding between the water and dienophile carbonyl group to reduce the LUMO energy of the dienophile [31]. Owing to the effects described above, the background reaction could not be ignored in the enzymatic reaction under standard conditions. Contrary to the non-enzymatic reaction, the enzymatic conversion of **13** provided preferentially *exo*-adduct **15**. These observations indicate that the major role of solanapyrone synthase is the oxidation of **13** to the more reactive **14** and the stabilization of the *exo*-transition state. Considering the significant rate acceleration for the cycloaddition of aldehyde **14** in aqueous medium [28], hydrogen bonding between the carbonyl group and some amino acid residues in the active site might contribute to catalyzing the cycloaddition as observed in the case of Diels–Alder catalytic antibodies [32]. SPS is an example of a Diels–Alderase that has a bifunctional role. This finding is important if we are to discuss the nature of the Diels–Alderases described in this chapter.

21.3
Intramolecular Diels–Alder Reactions Possibly Catalyzed by Dehydratase or DH-Red-Domain of PKS or Hybrid PKS-NRPS

Type-I PKS can catalyze the carbon-chain elongation process using several extender units such as malonyl-CoA and methylmalonyl-CoA to afford a β-ketoacyl intermediate [33]. The resultant β-ketoacyl product is subjected to additional modifications toward β-hydroxy, enoyl, or saturated compounds prior to the next elongation step. Because these modification reactions stop in the arbitrary stage, both *(E,E)*-conjugated diene and dienophile such as α,β-unsaturated carbonyl system can easily be biosynthesized from this elongation process. This indicates that natural products produced by type-I PKS and PKS-non-ribosomal peptide synthethases (NRPS) hybrid are better candidates as potential substrates for the IMDA reaction than other types of natural products (Scheme 21.5). Actually, polyketides can display many [4+2] cycloadducts such as decalin, spirotetronate, *as*-indacene, indane, and isoindolone skeletons [3]. Among these, the most common carbocycle is decalin in the metabolites produced by fungi and actinomycetes, while the feasibility of such biosynthetic cycloadditions has been supported by several total syntheses of decalin polyketides. Actual involvement of lovastatin PKS as Diels–Alderases has been proven, as discussed in Section 21.2.1 [7].

Recently identified biosynthetic gene clusters revealed plausible candidates to catalyze those Diels–Alder reactions. As discussed in the previous section, we

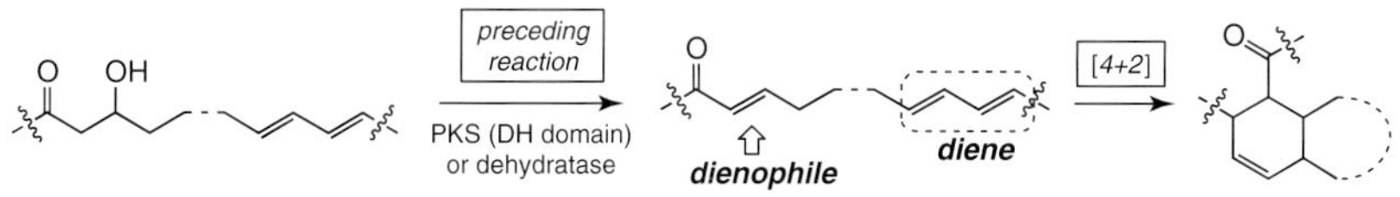

Scheme 21.5 Typical biosynthetic pathway of [4+2] adducts via type-I polyketide with PKS (dehydratase domain) or dehydratase.

assume that these putative Diels–Alderases might be bifunctional enzymes, catalyzing not only the formation of reactive species but also their [4+2] cycloaddition the same active site. In this section, we discuss recent advances on the biosynthetic machinery of polyketide natural products incorporating [4+2] cycloadducts.

21.3.1
Equisetin and Chaetoglobosin (Compactin, Lovastatin, Solanapyrone)

Equisetin, a fungal metabolite, is also a typical polyketide [4+2]-adduct with a decalin scaffold. An incorporation study with isotope labeled precursors established that the molecular skeleton of equisetin is constructed by an octaketide and L-serine (Scheme 21.6) [34]. Recently, the biosynthetic gene cluster of equisetin has been identified, which is similar to the highly reducing type-I modular PKS of various fungal polyketides [35]. An intriguing cycloaddition with PKS EqiS is proposed to occur at the heptaketide intermediate stage. At the final stage of equisetin biosynthesis, the reduction (Red) domain of EqiS catalyzes an unusual intramolecular Claisen condensation to give a tetramate moiety.

The cytochalasins constitute one of the largest family of intramolecular biological Diels–Alder adducts, which possess a characteristic isoindolone moiety fused to a 11-, 13-, and 14-membered macrocycle. Chaetoglobosin A, a specific inhibitor of myosin microtubule formation, is a representative example of a [4+2]-macrocyclic polyketide with an isoindolone moiety [36]. Based on feeding experiments of labeled precursors and accumulation of intermediates with treatment of cytochrome

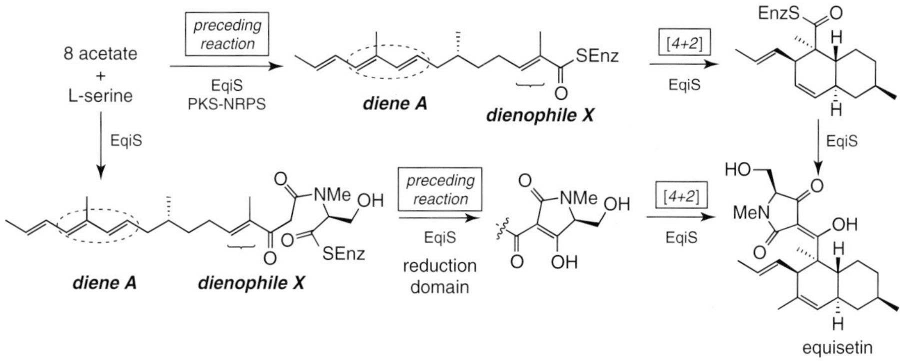

Scheme 21.6 Proposed biosynthetic pathway of equisetin.

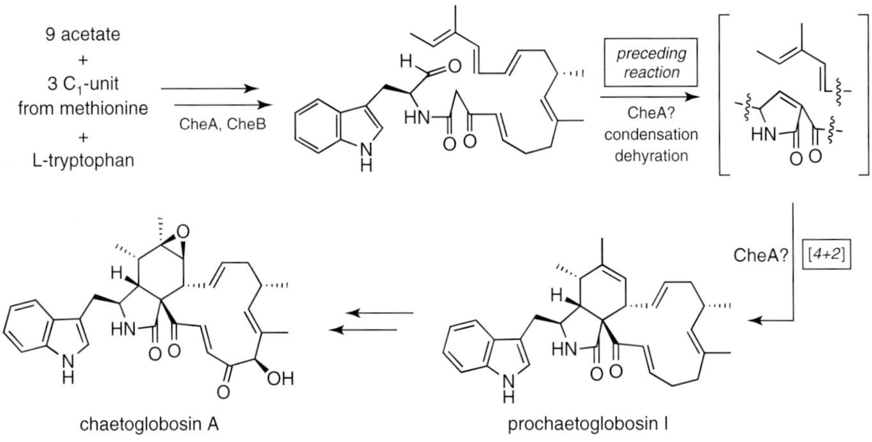

Scheme 21.7 Proposed biosynthetic pathway of chaetoglobosin.

P-450 inhibitor [37], it was proposed that the biosynthetic pathway goes via the plausible [4+2] adduct prochaetoglobosin I (Scheme 21.7). Recently, the biosynthetic gene cluster of chaetoglobosin has been identified [38]. It consists of the PKS-NRPS hybrid gene (*cheA*), stand-alone enoyl reductase gene (*cheB*), and a series of oxidation enzyme genes (*cheD*, *cheG*, and *cheE*), which correlate with the transformation proposed in the biosynthetic pathway. This suggested that the backbone of chaetoglobosin is constructed by CheA in cooperation with CheB, probably through hypothetical intramolecular *endo*-selective cycloaddition, which was supported by the retro-Diels–Alder reaction of prochaetoglobosin I [19]. Biomimetic synthesis of cytochalasin D [39] using a similar IMDA via a putative deoxytetramate intermediate provides further support for this hypothesis (Scheme 21.8).

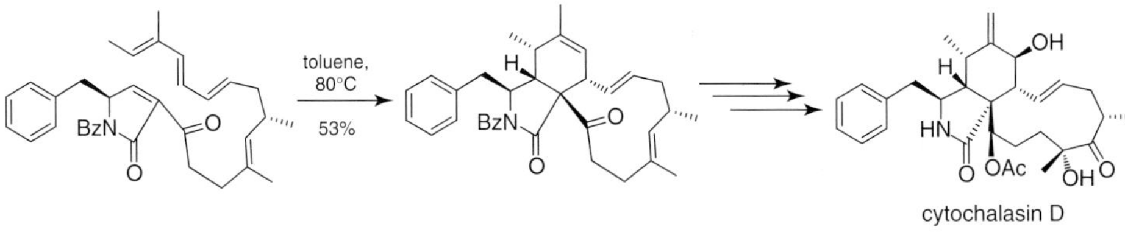

Scheme 21.8 Biomimetic construction of cytochalasan skeleton.

Accumulation of genetic and biochemical data has now allowed us to discuss Diels–Alderase in the biosynthesis of the fungal polyketides lovastatin [13], compactin (a demethylated analog of lovastatin) [40], solanapyrone [27], equisetin [9], and chaetoglobosin [38]. Figure 21.2 shows the domain organization of the PKSs. Heterologous expression of PKSs, LovB, and the solanapyrone PKS (PSS) led to the production of dihydromonacolin L [13] and desmethylprosolanapyrone I [27], respectively. These observations clearly indicated that the cycloaddition occurred during chain extension in LovB (Scheme 21.2). However, PSS did not catalyze the

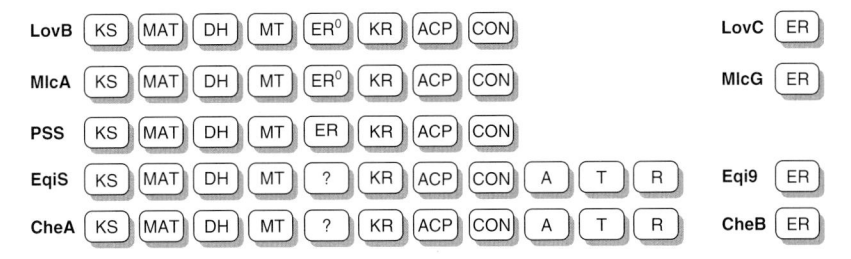

LovB	KS	MAT	DH	MT	ER⁰	KR	ACP	CON				LovC	ER
MlcA	KS	MAT	DH	MT	ER⁰	KR	ACP	CON				MlcG	ER
PSS	KS	MAT	DH	MT	ER	KR	ACP	CON					
EqiS	KS	MAT	DH	MT	?	KR	ACP	CON	A	T	R	Eqi9	ER
CheA	KS	MAT	DH	MT	?	KR	ACP	CON	A	T	R	CheB	ER

Figure 21.2 Domain organization of several [4+2] adducts.

Diels–Alder reaction (Scheme 21.4), although its domain architecture is almost identical to the corresponding PKSs, LovB, and compactin PKS MlcA, and common hexaketide precursors are involved in their chain-elongation steps. The findings on PSS raised a question on the timing of the enzymatic IMDA reaction. In the biosynthesis of lovastatin and compactin, the carbonyl group of the hexaketide intermediate is activating the dienophile and is then lost by further chain extension steps after cycloaddition. Thus, cycloaddition timed at just after introduction of the dienophile moiety is considered to be required in these PKS reactions.

Domain architectures of both PKS Che and Eqi are nearly identical and showed significant similarity to those of decalin PKS shown above. Thus, it is reasonable to speculate that decalin PKS and cytochalasin PKS are closely related to each other. The structures of other [4+2]-decalin and cytochalasan polyketides indicate that these adducts usually retain carbonyl groups adjacent to the dienophile moieties. In these compounds, [4+2] cycloaddition could proceed either during the extension steps or after the extension steps. Interesting examples are found in the biosynthesis of the octaketides equisetin [35] and aspochalasin Z [41], the cycloaddition precursors of which are closely related (Scheme 21.9). While the timing of cycloaddition in equisetin may be optional, the cycloaddition of aspochalasin Z obviously must proceed in a step following chain extension. Since these metabolites have carbonyl groups adjacent to putative dienophiles, it is likely that the intriguing cycloaddition occurs after the extension steps. This proposal can explain the formation of both aspochalasin-type and equisetin-type cycloadducts, depending on the dienophile being involved (A or B in Scheme 21.9) and selected by the corresponding Diels–Alderase. No experimental evidence is currently available on the timing and players involved in the cycloaddition in these biosyntheses.

21.3.2
Kijanimicin, Chlorothricin, and Tetrocarcin A

The spirotetronate antibiotics such as kijanimicin [42], chlorothricin [43], and tetrocarcin A [44] possess not only decalin ring scaffold but also the spirotetronate moiety, which could be assembled by IMDA. The construction of decalins is proposed to proceed in a similar way to those of lovastatin biosynthesis. In this section, we thus focus on the unique spirotetronate scaffold that is proposed to be biosynthesized from the IMDA between a diene moiety and a γ-methylene-acyltetronate

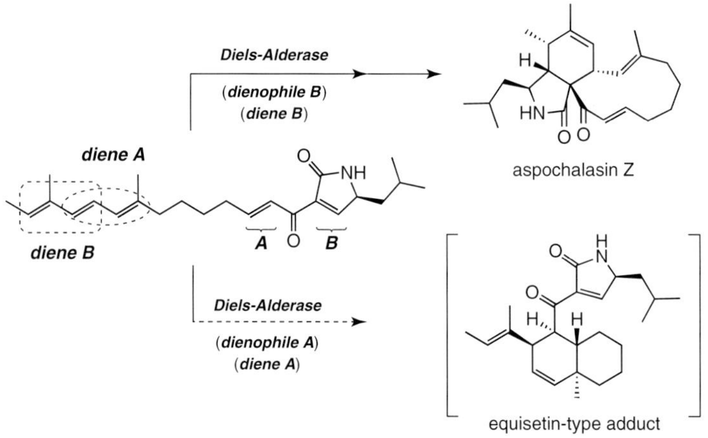

Scheme 21.9 Proposed biosynthetic pathway of aspochalasin Z.

dienophile on the putative intermediate **21** (Scheme 21.10). An efficient synthesis of this spirotetronate scaffold found in the chlorothricin-type skeleton, using IMDA, supports the proposed biosynthetic pathway.

In tetronate moiety formation (**19** to **20**), a three-carbon unit derived from glycerol is necessary to attach the linear polyketide chain as shown in the Scheme 21.10. Comparative gene analysis between kijanimicin (*kij*) [45], chlorothricin (*chl*) [46], and tetrocarcin A (*tca*) [47] biosynthetic gene clusters indicates that the tetronate moiety might be biosynthesized by specifically conserved enzymes. A recent study on the tetronate biosynthetic pathway [48] allowed speculation that the tetronate formation is catalyzed by the FkbH homolog KijC, the stand-alone acyl carrier protein (ACP) KijD, the acyltransferase KijE, and the ketoacyl ACP synthase III homolog KijB. Currently, it is proposed that the dehydrative dienophile formation (**20** to **21**) would possibly proceed with a flavin-dependent monooxygenase/oxidoreductase homolog KijA, followed by the [4+2] cycloaddition (**21** to **22**) occurring spontaneously to form the spirotetronate moiety.

21.3.3
Indanomycin

Indanomycin, produced by *Streptomyces antibioticus* NRRL 8167, is a hybrid compound of polyketide and nonribosomal peptide with an indane scaffold (Scheme 21.11) [49]. Feeding experiments with ^{13}C-labeled precursors such as acetate, propionate, and butyrate have supported the involvement of a [4+2] cycloaddition for the indane ring formation [50]. Because the diene part is installed by normal PKS functions, the formation of a reactive dienophile is a key step for cycloaddition. Bioinformatic data on the identified biosynthetic gene cluster suggested that PKS lacks a key dehydratase (DH) domain to construct the dienophile moiety [51]. Therefore, two routes can be considered (Scheme 21.11):

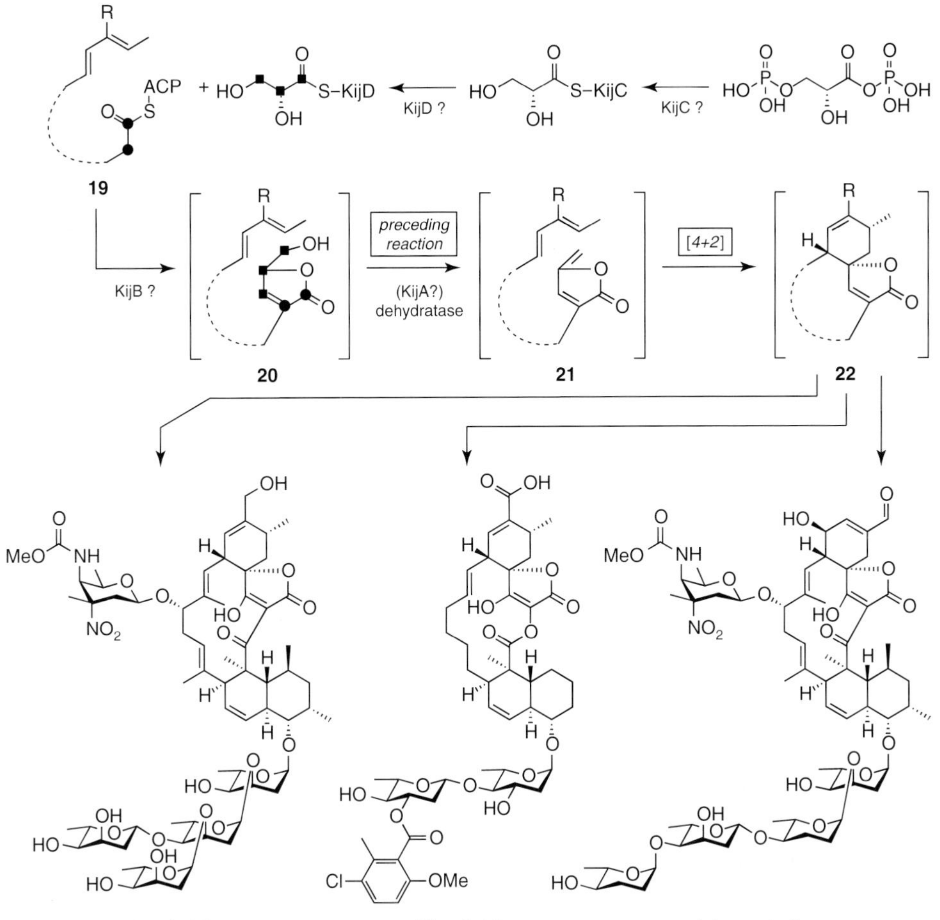

Scheme 21.10 Proposed biosynthetic pathway of kijanimicin, chlorothricin, and tetrocarcin A.

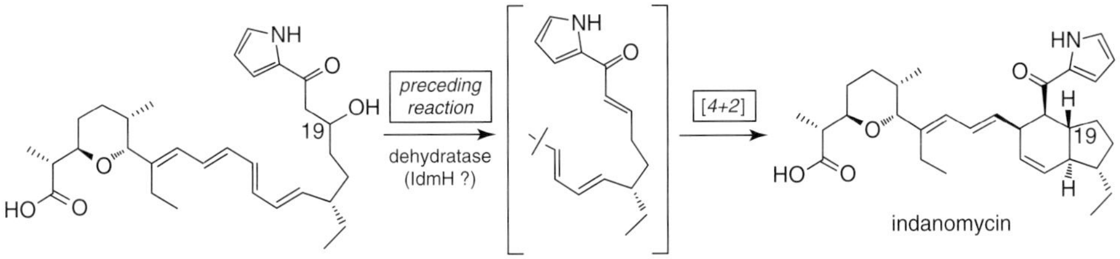

Scheme 21.11 Proposed biosynthetic pathway of indanomycin.

1) Dehydration at C19 occurs during the polyketide chain elongation, catalyzed by an adjacent DH domain such as module 3, and the cycloaddition takes place to give the adduct.
2) Alternatively, after completion of chain assembly, a putative DH (IdmH) would give the dehydration product followed by the cycloaddition at the same active site to give indanomycin, as proposed in the kijanimicin biosynthesis.

Since the former route requires an unusual participation of neighboring DH domain on the downstream module, the latter route is more likely. Again, this fits our proposal that enzymes producing a reactive intermediate catalyze the Diels–Alder reaction. Currently, the biosynthetic gene clusters for other indane-containing antibiotics such as stawamycin, plakotenin, amaminol, cafamycin, and cochleamycin have not been identified yet [50]. Thus, additional biochemical or genetic characterization of IdmH is required for detailed analysis of indane ring formation.

21.3.4
Spinosyn

A different type of Diels–Alder adduct exists in the core skeleton of spinosyn. Spinosyn possesses a 22-membered macrolide with unusual *as*-indacene scaffold, D-forosamine, and methylated L-rhamnose [52]. Bioinformatic data on the biosynthetic gene cluster suggested the biosynthetic pathway of spinosyn (Scheme 21.12) [53]. Successful bioconversion of aglycone using the strains disrupting four genes (*spnF, J, L, M*) firmly established biosynthetic transformations at the later stage, suggesting that these genes are involved in *as*-indacene formation. The detailed function of SpnJ, a flavin-dependent dehydrogenase, was identified by *in vitro* analysis to catalyze the oxidation of the hydroxy group at C15 in macrocyclic intermediate **23** [54]. Because the possible linear polyketide precursor was not accepted by SpnJ, this oxidation (**23** to **24**) occurs after the macrolactone ring formation. The remaining transformations into spinosyn may require

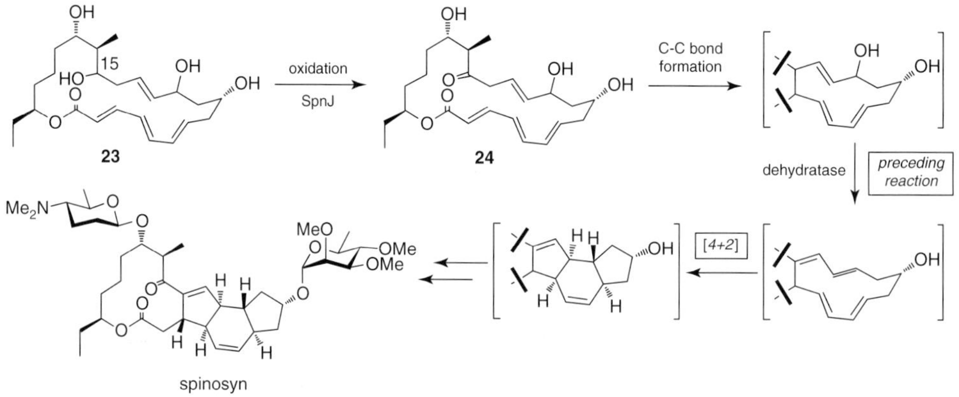

Scheme 21.12 Proposed biosynthetic pathway of spinosyn.

C–C bond formation, then dehydrative isomerization forming reactive diene, followed by the cycloaddition. The exact sequence for *as*-indacene formation must await further experimental characterizations of the remaining three genes (*spnF, L, M*).

21.4
Diels–Alder Reactions after Formation of Reactive Substrates by Oxidation Enzymes

Considering that an oxidase can convert an unactivated precursor into a reactive Diels–Alder substrate in the case of the Diels–Alderase solanapyrone synthase (Section 21.2.3), we speculated that a similar mechanism can operate in the biosynthesis of other Diels–Alder adducts. Among highly diverse structures of natural [4+2] adducts, there are significant numbers of adducts that are derived by the oxidative transformation preceding the Diels–Alder reaction [3]. For example, oxidation of catechol provides highly reactive ortho-quinones that have two dienes and two dienophiles in the same molecule (Scheme 21.13). Either non-enzymatic or enzymatic cycloaddition proceeds in various modes to give structurally unique adducts as either racemates or optically active forms. Similarly, chiral ortho-quinols derived from phenols are transformed into structurally complex molecules. Besides phenol oxidation, oxidative transformation of oligoprenyl side chain and terpene gives reactive conjugated dienes that react with suitable dienophiles to furnish [4+2] adducts. In this section, we introduce several putative examples of Diels–Alder reactions found in Nature, along with biomimetic syntheses and biological conversions of putative Diels–Alder precursors.

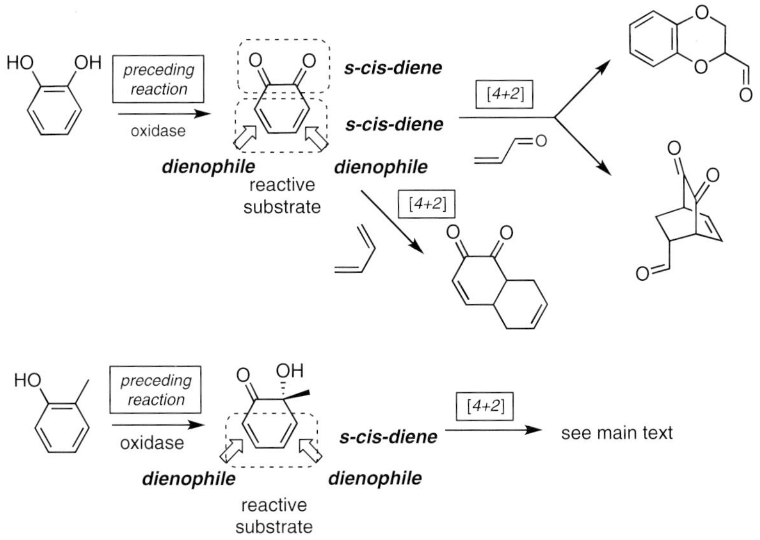

Scheme 21.13 Model of ortho-quinone and ortho-quinol formation as a Diels–Alderase.

21.4.1
Oxidation of Phenol and Catechol to Reactive Dienone and Orthoquinone

Quinones, probably derived from catechol or hydroquinone, are common precursors of natural [4+2] adducts. Bazan *et al.* proposed the involvement of a Diels–Alder reaction in the biosynthesis of phenylphenalenone lachanthocarpone based on the conversion of **25** into lachanthocarpone, with NaIO$_4$, via the ortho-quinone **27** [55] (Scheme 21.14). To examine the intermediacy of the diarylheptanoid in the biosynthesis of anigorufone, Hölscher and Schneider fed the ^{13}C-labeled **26** to the cultured root of *Anigozanthos preissii* [56]. The anigorufone isolated showed significant incorporation, establishing involvement of oxidation of **26** to **27**, followed by a Diels–Alder reaction. In this case, an oxidase would provide the reactive precursor **27** for a cycloaddition. Then, the cycloaddition proceeded after releasing it from the active site of the oxidase to give achiral products. Involvement of a Diels–Alderase is not essential in this case.

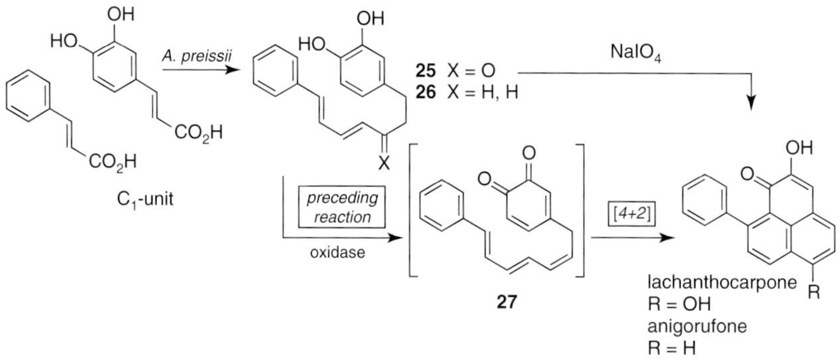

Scheme 21.14 Biosynthetic pathway of lachanthocarpone and anigorufone.

Grandione from *Torreya grandis* is a dimer formed by an *endo*-selective hetero-Diels–Alder reaction between two ortho-quinones derived from the diterpene monomer demethylsalvicanol. Biomimetic oxidation of demethylsalvicanol in the solid state at room temperature afforded a single [4+2] adduct that was identical to grandione (Scheme 21.15) [57]. In grandione biosynthesis, enzymatic oxidation would convert the catechol moiety of the monomer

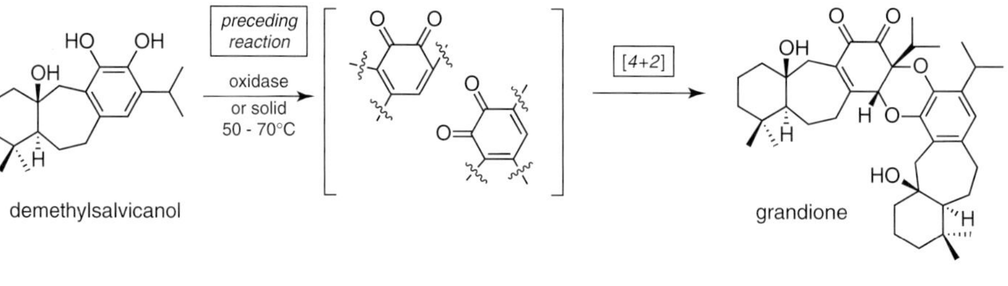

Scheme 21.15 Proposed biosynthetic pathway of grandione.

demethylsalvicanol into a reactive ortho-quinone that could afford a [4+2] adduct non-enzymatically.

A Diels–Alder reaction between different types of terpene units is plausible in several cases. Perovskone from the heartwood of *Chamaecyparis obtusa* [58] is regarded as an adduct between a chiral para-quinone (**28**) derived from the modified diterpene barbatusol and the linear monoterpene *trans*-β-ocimene from geranyl diphosphate [59] (Scheme 21.16). In the total synthesis of perovskone [60], the Diels–Alder reaction of **28** and *trans*-α-ocimene with Lewis acid Eu(fod)$_3$ proceeded in good regio- and diastereoselectivity to afford the desired diastereomer **29** as a major product. Since this biomimetic [4+2] cycloaddition is accompanied by the formation of the unnatural *endo*-adduct, a Diels–Alderase that could catalyze phenol oxidation into quinone **28** might be responsible for the construction of the perovskone skeleton during biosynthesis. A similar example involving phenol oxidation was reported in the biosynthesis of heliocides H$_1$ and H$_4$ [61]. Hemigossypol could be oxidized to a para-quinone that underwent cycloaddition to afford regioisomers H$_1$ and H$_4$ in an *endo*-selective manner.

The involvement of ortho-quinones in the biosynthesis of natural products is recognized in many cases (Figure 21.3). The ortho-quinone dimers of the plant diterpenes hongencaotone from *Salvia prionitis* [62] and actephilol A from *Actephila excelsa* are regarded as hetero-Diels–Alder adducts [63]. In addition, *endo*- and *exo*-triterpene dimers xuxuarine Eα and Eβ [64] and also trimer triscutin A [65] are found in the South American medicinal plant *Maytenus blepharodes* and *M. scutioides*, with several related metabolites. DDQ (2,3-dichloro-5,6-dicyanobenzoquinone) oxidation of the monomer pristimerin gave the diastereomeric dimers xuxuarines Eα and Eβ. This biomimetic synthesis provides circumstantial evidence for the involvement of a Diels–Alder reaction in the biosynthesis though an alternative mechanism involving ortho-quinone **30**.

Ortho-quinol, another reactive species, is oxidatively derived from phenols. The dimer of the neolignan asatone was isolated from the plant *Asarum teitonense* (Scheme 21.17) [66]. Later, two closely related trimers, **33a** and **33b**, were also isolated [67]. Based on the oligomeric structure of neolignans, biosynthetic pathways of these metabolites were proposed as shown in Scheme 21.17. Oxidation of phenol **31** produced a modified ortho-quinol (**32**) that dimerized to give asatone through a Diels–Alder reaction. Further cycloadditions of asatone with **32** yield **33a** and **b**. This proposal was supported by the result that the anodic oxidation of phenol **31** produced a reactive quinone methide that underwent cycloaddition to give asatone quantitatively [68]. Having no optical activity, all these lignans would be formed by spontaneous cycloadditions of the ortho-quinol **32**. Observations on asatone biosynthesis strongly suggest that an oxidase produces the reactive ortho-quinol, which is released from the enzyme active site before dimerizing to afford racemic [4+2] adducts.

Similar oxidase-catalyzed phenolic oxidation could produce a [4+2] adduct in the biosynthesis of monoterpene aquaticol (Scheme 21.18) [69]. A recent total synthesis of aquaticol [70] showed that a mixture of ortho-quinols **35a** and **b** prepared by oxidation of chiral monomer **34** with SIBX (stabilized 2-iodoxybenzoic

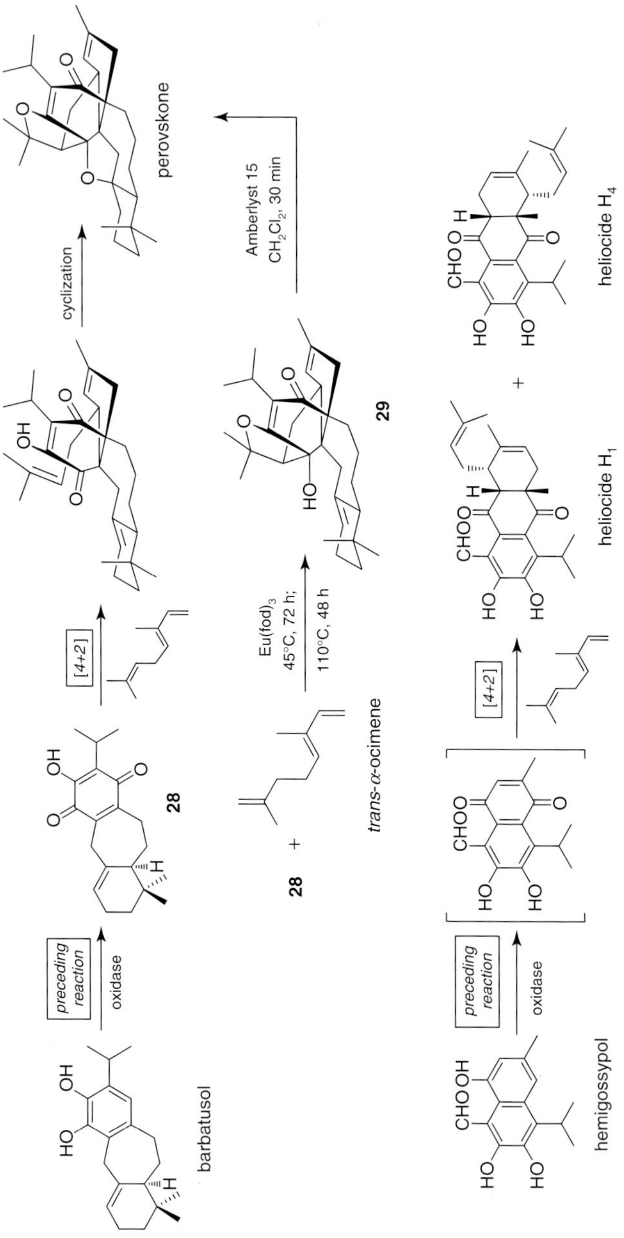

Scheme 21.16 Proposed biosynthetic pathways of perovskone and heliocides.

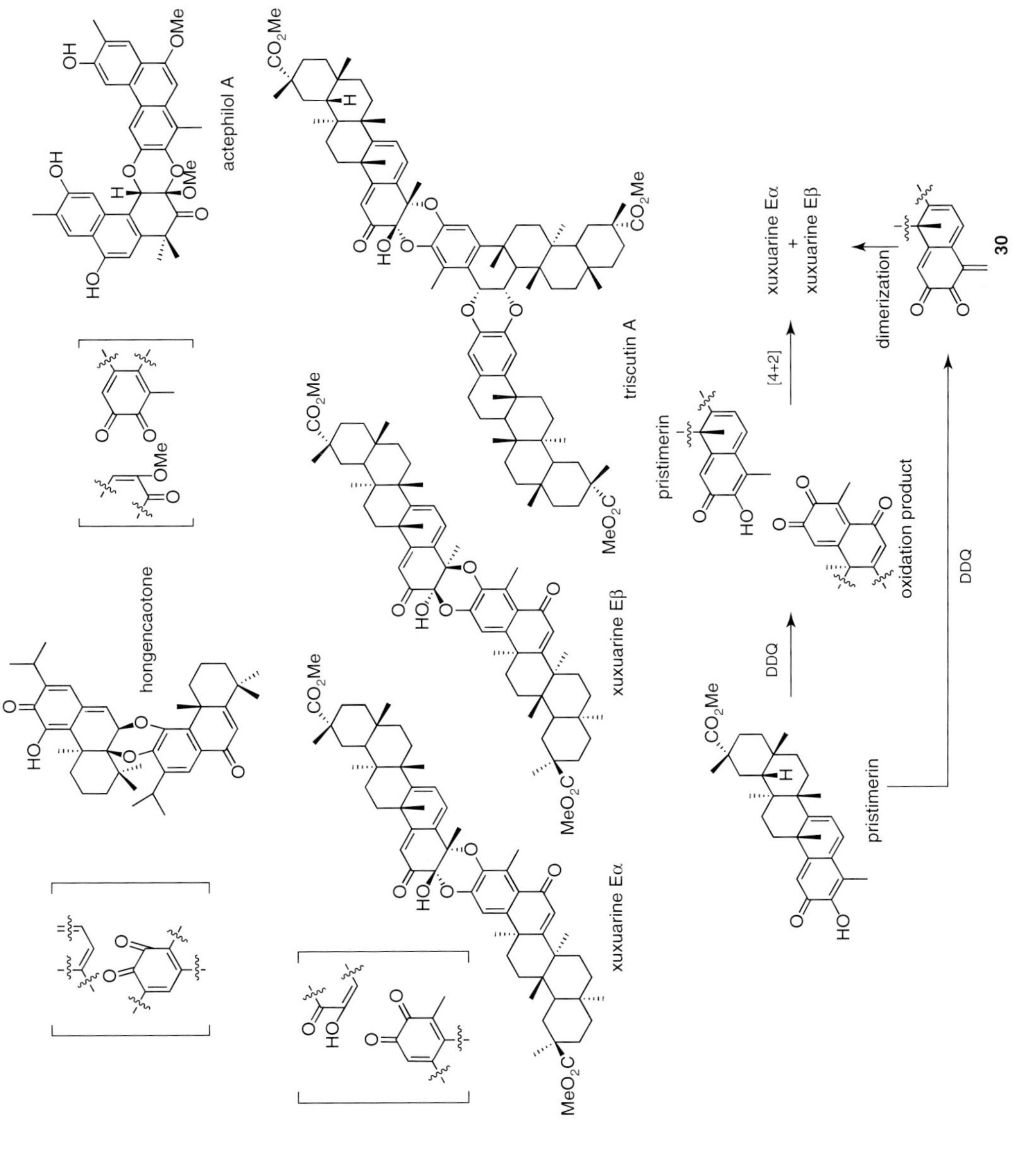

21.4 Diels–Alder Reactions after Formation of Reactive Substrates by Oxidation Enzymes

Figure 21.3 Representative ortho-quinone oligomers of natural [4+2] adducts.

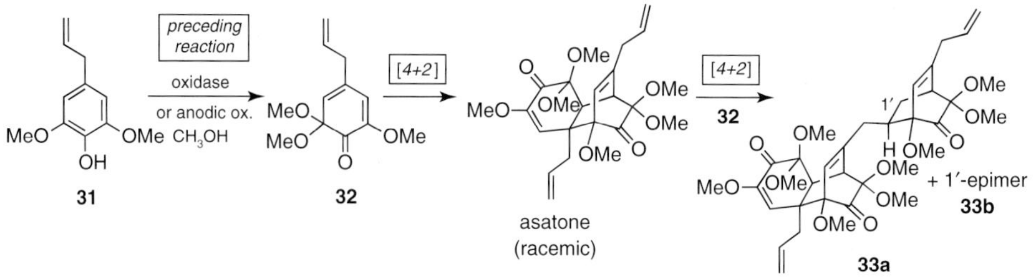

Scheme 21.17 Proposed biosynthetic pathway of asatone and its analogs.

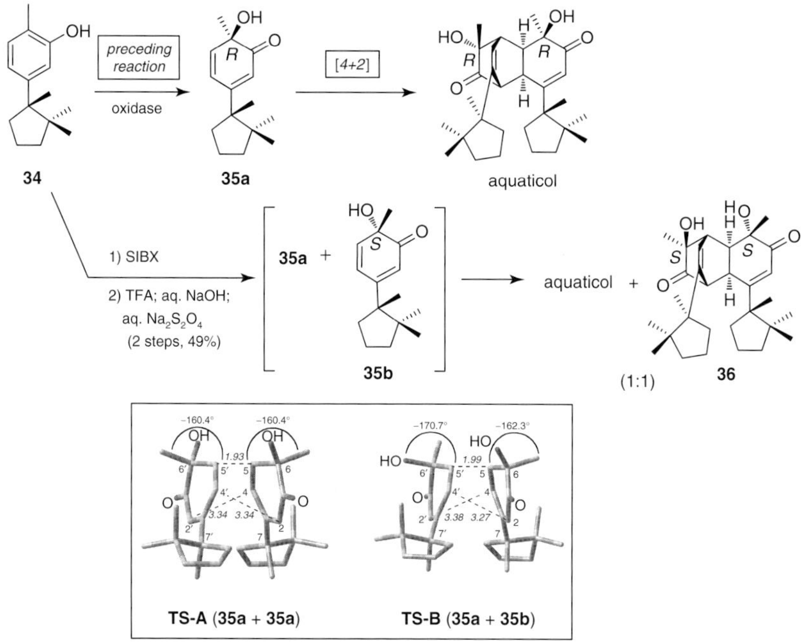

Scheme 21.18 Proposed biosynthetic pathway of aquaticol.

acid) provided only two diastereomers, aquaticol and **36**, out of four expected products (Scheme 21.18). This indicated that chiral ortho-quinol **35a** gave aquaticol as a single diastereoisomer. On the basis of computational results, the energy difference between transition states A (TS-A, aquaticol-like) and B (TS-B, **36**-like) is more than 9.9 kcal mol^{-1}, which was explained by hyperconjugative effects. In this case, reactivity and remarkable diastereoselectivity originates from the substrate, suggesting that involvement of a Diels–Alderase is not essential in this case.

The bisorbicillinoids are a family of structurally diverse fungal metabolites represented by bisorbicillinol, which shows DPPH (1,1-diphenyl-2-picrylhydrazyl) radical scavenging activity [71]. Based on extensive studies of the structure elucidation and the biosynthesis of the bisorbicillinoids, Abe et al. proposed that the

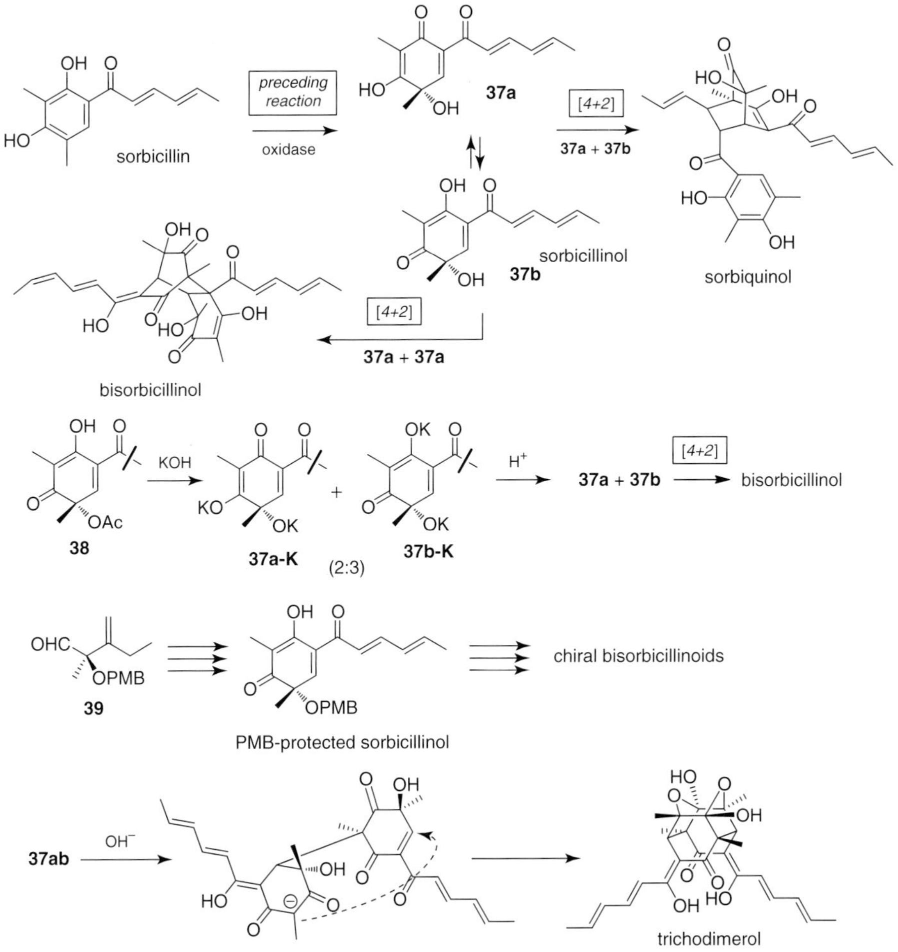

21.4 Diels–Alder Reactions after Formation of Reactive Substrates by Oxidation Enzymes | 773

Scheme 21.19 Proposed biosynthetic pathway and biomimetic synthesis of bisorbicillinoids.

stable aromatic monomer sorbicillin could be enantioselectively oxidized into the reactive ortho-quinol sorbicillinol (**37a** and **b**), which dimerizes via two modes of [4+2] cycloadditions to provide bisorbicillinol and sorbiquinol (Scheme 21.19) [72]. During the purification of **37ab**, it was found that the concentration of a solution of **37ab** caused a [4+2] cycloaddition to give sorbiquinol, indicating that **37ab** is highly reactive and that the conversion is non-enzymatic and occurs under mild conditions with complete regio- and stereoselectivity [73]. In synthetic studies of the bisorbicillinoids [74], basic hydrolysis of acetate **38** gave two discrete quinolates (bis-deprotonated forms of **37a-K** and **b-K**) that underwent cycloaddition after subsequent acidification. More recently, enantioselective total synthesis of the bisorbicillinoids has been achieved via protected sorbicillinol, which was prepared by a versatile route using readily available intermediate **39** [75]. In addition, a sequential

Michael addition–ketalization sequence of reactive ortho-quinols **37ab** gave the highly complex cyclization product trichodimerol by either simple evaporation [72] or base-catalyzed condensation [76].

Involvement of a Diels–Alderase is not necessary in this case because a non-enzymatic reaction provided a single product and because enantioselective oxidation of sorbicillin introduced the chirality in **37ab**, thus determining the stereochemistry in bisorbicillinol. Detection of a significant amount of the monomer sorbicillin in the bisorbicillinol-producing fungus indicated that an oxidase would provide the chiral reactive substrate **37ab**, and that the Diels–Alder reaction of **37ab** would be promoted in aqueous medium without a Diels–Alderase. This example suggests that conversion of an achiral substrate into a chiral cycloadduct is not conclusive evidence for distinguishing non-enzymatic and enzymatic Diels–Alder reactions.

A literature search allowed us to find many natural [4+2] adducts derived from the oxidation of phenol to reactive ortho-quinols described above. Dimeric abietane diterpenoid maytenone from *Crossopetalum rhacoma*, a dimeric phenanthrenoid from *Juncus acutus* [77], obtunone (aromatic monoterpene–linear monoterpene) from the heartwood of *Chamaecyparis obtusa* [78], and vulbilide [79] (triterpene–abietane diterpene) are representative examples (Figure 21.4). To date, there is no report dealing with the corresponding oxidase. However, in the case of fungal metabolites bisorbicillinoids, the biosynthetic gene could be identified by genomic analysis, a bioinformatic search and its functional analysis because the substrate and products have been unambiguously determined. Thus, the detailed mechanism of the enzymatic Diels–Alder reaction could be elucidated in the near future.

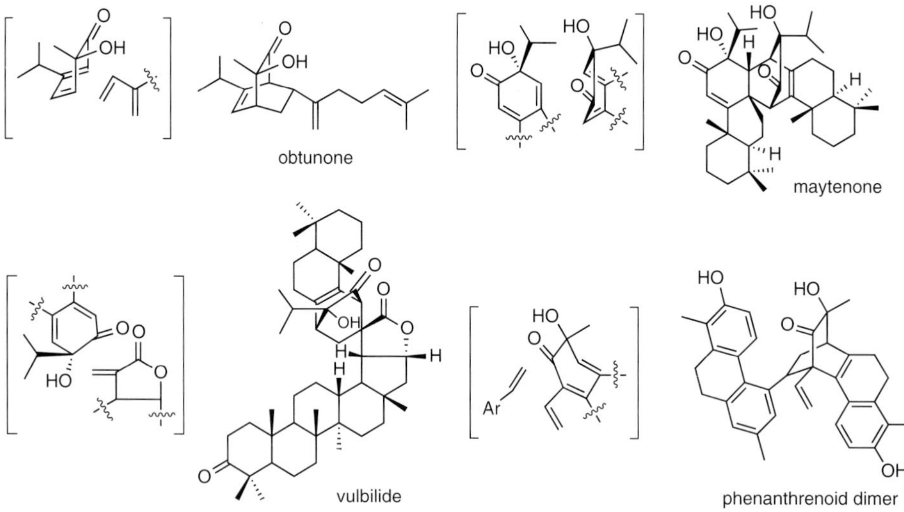

Figure 21.4 Representative ortho-quinol [4+2] adducts.

21.4.2
Conjugated Diene Derived from Dehydrogenation of Prenyl Side Chain

Prenyl side chains are frequently converted into conjugated dienes that serve as a glue to combine reactive dienophiles such as chalcone, stilbene, coumarin, and substituted prenyl diene (Scheme 21.20). Thus, in this case, a dehydrogenase providing prenyl diene is a candidate of Diels–Alderase.

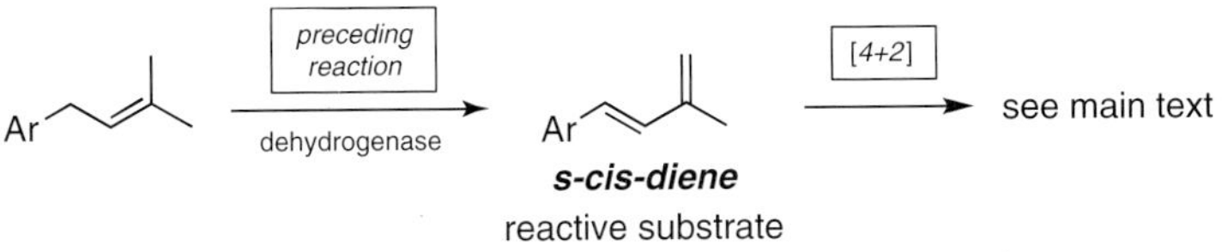

Scheme 21.20 Model of conjugated diene formation for a Diels–Alderase.

Kuwanons X, Y, I, and J, constituents of moraceous plants, are phytoalexins consisting of pairs of diastereomers derived from chalcones and stilbenes [80]. Based on their structures, kuwanons I and J are regarded as dimeric adducts of a chalcone precursor with the diene derived from the prenyl side chain (Scheme 21.21). Isolation of these diastereomers as optically active forms strongly indicated that the achiral precursors afford *endo*- and *exo*-adducts via an enzymatic Diels–Alder reaction [79]. The chemical feasibility of the corresponding [4+2] cycloaddition between chalcone and flavonoid with prenyl diene was confirmed [81]. In a feeding experiment with the non-natural methoxychalcone **40** into callus tissue of *Morus alba*, the dimeric adducts **42a** and **b** were obtained, indicating that dehydrogenation of the prenyl group followed by [4+2] cycloaddition between **40** and **41** yielded the non-natural adduct **42b** (Scheme 21.21) [82]. A plausible dehydrogenase involving kuwanon biosynthesis may convert **40** into the reactive diene **41** but in this case the same enzyme may provide an active site for the intermolecular Diels–Alder reaction between **40** and **41** to afford chiral **42b**. Although a cell-free system from *Morus bombycis* showed the catalytic activity of a similar transformation observed in cell culture [83], unfortunately no characterization of the corresponding enzyme has been reported.

Plausible dehydrogenases could produce various elaborated structures such as multicaulisin [84], sanggenon O [85], palodesagren A [86], and sorocein L [87]. Various dimeric coumarins and quinonolones derived from the corresponding monomers with prenyl dienes are found (Figure 21.5). Representative coumarin dimers are thamnosin [88], isothamnosin A [89], and toddasin [90]. In addition, dimeric quinolones and mixed dimers of coumarin–quinolone, toddacoumalone [91], mexolide [92], microcybin [93], and vepridimerine A [94] have been found from various plants (Figure 21.6). While kuwanon-type adducts were obtained as optically active forms, dimeric coumarines and quinonolones were isolated as racemic forms, indicating that dehydrogenases would release the products before the cycloaddition occurs. [4+2]-Adducts from plants are indeed frequently obtained in racemic forms [3].

776 *21 The Diels–Alderase Never Ending Story*

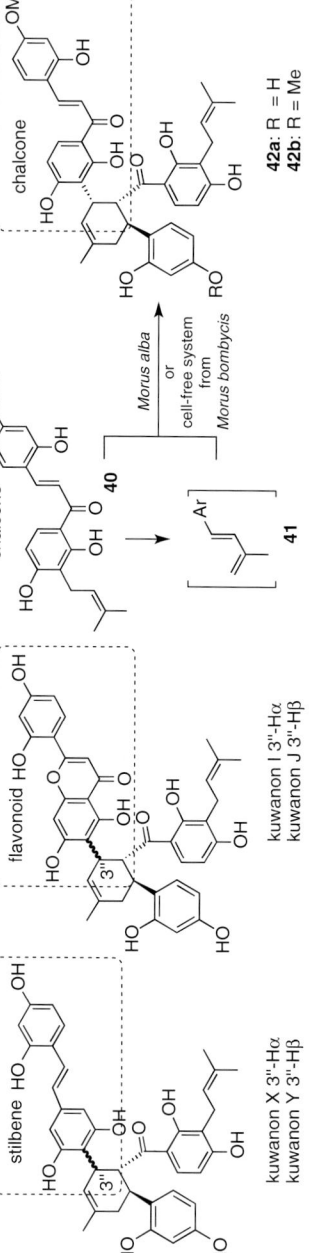

Scheme 21.21 Biosynthetic formation of kuwanons.

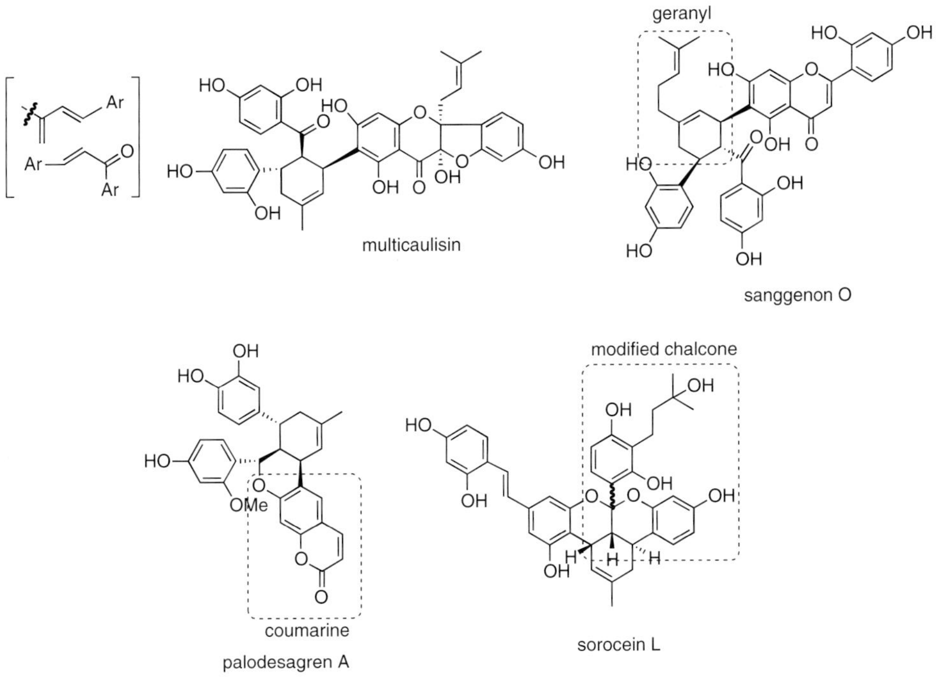

21.4 Diels–Alder Reactions after Formation of Reactive Substrates by Oxidation Enzymes

Figure 21.5 Representative examples of [4+2] adducts derived from chalcone and various metabolites with prenyl dienes.

Similar plausible Diels–Alder type cycloadditions following dehydrogenation are found in the biosynthesis of natural products with polyprenyl side chains. The unique prenylated [12]-paracyclophane quinone dimer longithorone A was isolated from the tunicate *Aplidium longithorax* as a cytotoxic agent, and is proposed to originate from sequential cycloaddition reactions between precursors **44** and **45** (Scheme 21.22) [95]. Later, a series of novel farnesylated benzoquinone monomers possessing meta-bridged structures were isolated from the same tunicate, indicating the possible involvement of a Diels–Alderase. Again, it is possible that either dehydrogenase forming the diene system on the bridged farnesylated core **44** or oxidase forming the α,β-unsaturated aldehyde **45** may proceed prior to the intriguing inter- and intramolecular cycloadditions. A biomimetic total synthesis of longithorone A [96] indicated that the intermolecular reaction shows inherent low reactivity and facial selectivity, suggesting the involvement of a Diels–Alderase.

The enantiomer of ircinianin, a plausible intramolecular *endo*-adduct [97], was isolated from the marine sponge *Ircinia* sp. Based on isolation of linear furanosesterterpene-tetronic acids such as variabilin and the successful biomimetic synthesis of ircinianin via **46** [98], it was suggested that the linear oligo-isoprenyl precursor would be dehydrogenated to give a conjugated triene, serving as a

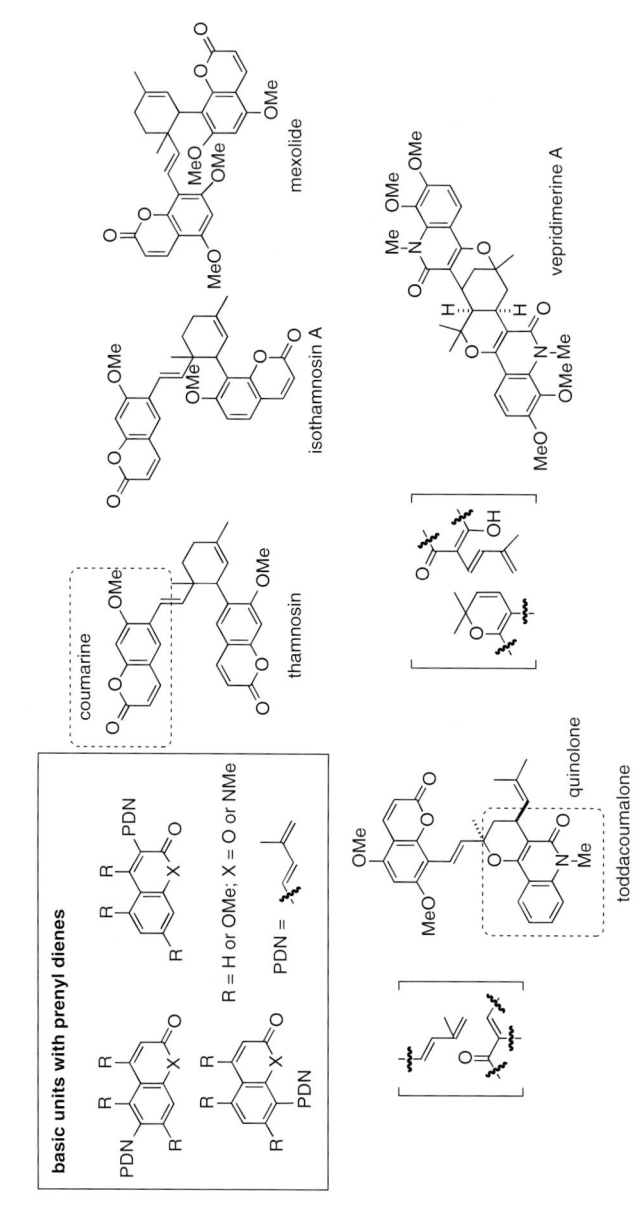

Figure 21.6 Representative examples of dimeric [4+2] adducts derived from coumarines and quinolones with prenyl dienes.

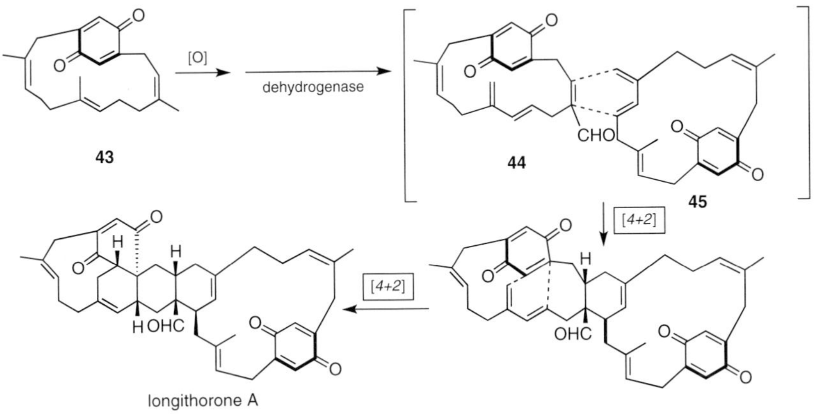

Scheme 21.22 Proposed biosynthetic pathway of longithorone.

substrate for a biosynthetic Diels–Alder reaction (Scheme 21.23). Isolation of metabolites without a diene system suggested that the dienophile part is likely introduced prior to diene formation [98]. Similar transformation could be involved in the biosynthesis of methyl isosarcotortuoate [99] and methyl sarcophytoate [100].

21.4.3
Cyclopentadiene Formation Derived from Dehydrogenation

Sesquiterpenes with a cyclopentadiene system are plausible key intermediates in the biosynthesis of related plant metabolites. Although putative intermolecular [4+2] adducts are frequently found in plant terpenoids (Scheme 21.24) [3], the corresponding monomer with cyclopentadiene is relatively rare, suggesting that dehydrogenation or hydroxylation–dehydration that installs a reactive diene precedes the Diels–Alder reaction (Figure 21.7). Along this line, the biogenetic pathway of plagiospirolide A was proposed [101]; this pathway was supported by the synthesis featuring a Diels–Alder reaction of diplophyllolide A and diene 47 [102]. Biogenetically related adducts between an α-methylene butenolide and cyclopentadiene are plagiospirolide E [103], biennin C [104], arteminolide B [105], ornativolide A [106], artemisolide [107], and stolonilactone [108]. Cyclopentadiene dimers absinthin [109] and vielanin A [110] and guaiane sesquiterpene dimer [111] are also found.

21.5
Summary

In this chapter, we introduced the concept of natural Diels–Alderases and provided reasonable speculation on potential Diels–Alderases such as PKS, DH, oxidase, and dehydrogenase. Diels–Alderases and their candidates described in this chapter

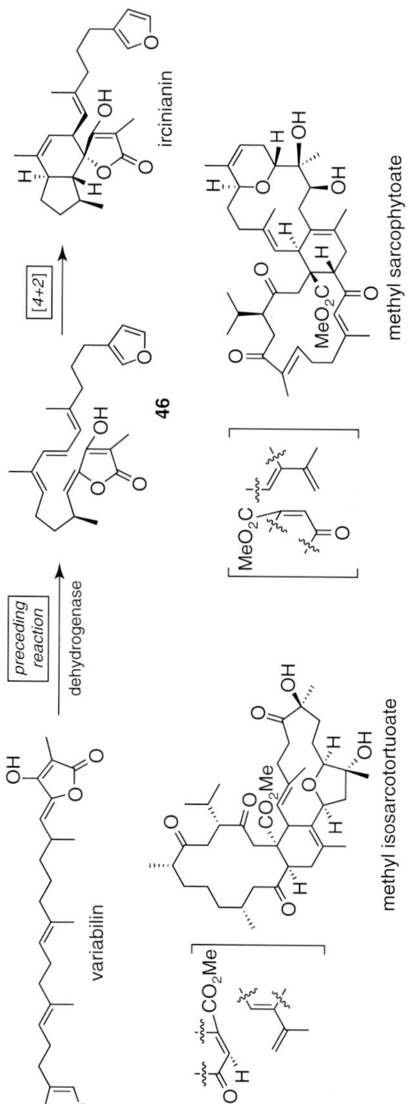

Scheme 21.23 Representative examples of [4+2] adducts derived from terpenoid dienes.

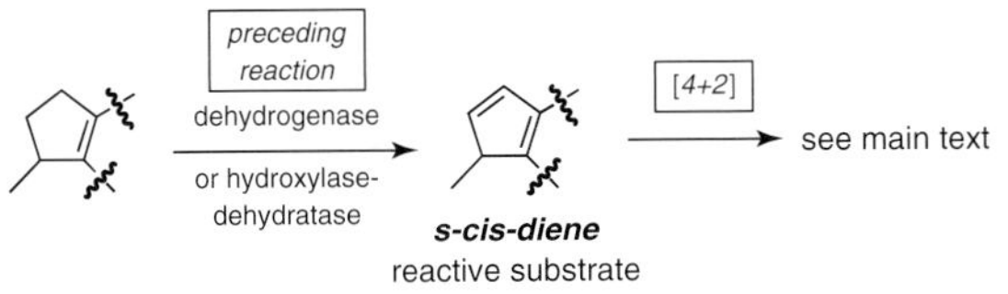

Scheme 21.24 Model of cyclic diene formation as a Diels–Alderase.

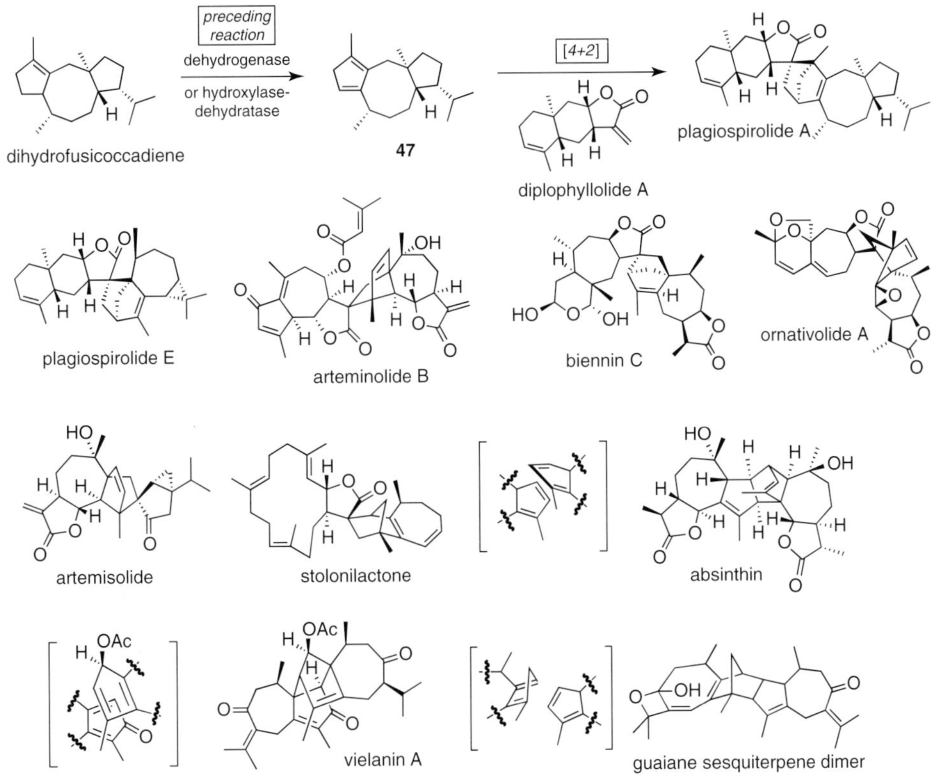

Figure 21.7 Proposed biosynthetic pathway of plagiospirolide A, and various [4+2] adducts derived from terpenes with reactive cyclopentadiene.

are not categorized as a single enzyme group because they do not show any characteristic catalytic mechanism. Therefore, we propose that any enzyme providing reactive dienes and dienophiles could be a Diels–Alderase. In our extensive survey of natural [4+2] adducts we have frequently encountered reactions related to oxidation (phenol oxidation, dehydrogenation) and dehydration (PKS or other) forming reactive substrates (Scheme 21.25). Currently, only limited information is available on the Diels–Alderases. However, innovative methodologies recently developed will provide information on gene clusters of many natural [4+2] adducts.

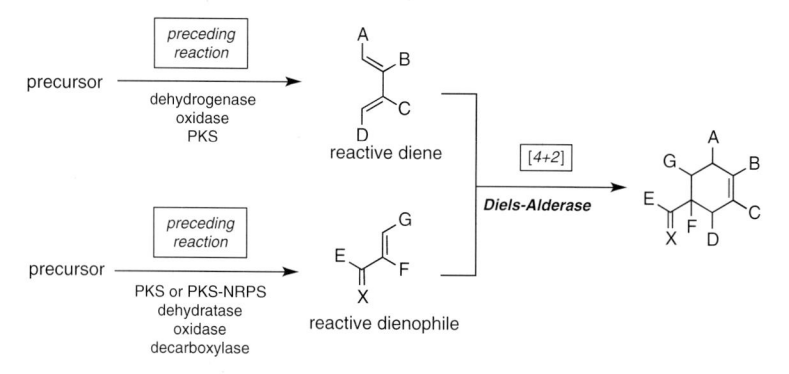

Scheme 21.25 General biosynthetic pathway involving skeletal construction with putative Diels–Alderase.

This allows biochemical investigation of Diels–Alderases, and we anticipate that the outline of this intriguing enzyme should be elucidated within the next ten years.

References

1. Carruthers, W. (1990) *Cycloaddition Reactions in Organic Synthesis*, Pergamon, Oxford.
2. (a) Nicolaou, K.C., Edmonds, D.J., and Bulger, P.G. (2006) *Angew. Chem. Int. Ed.*, **45**, 7134–7186; (b) Takao, K., Munakata, R., and Tadano, K. (2005) *Chem. Rev.*, **105**, 4779–4807.
3. (a) Ichihara, A. and Oikawa, H. (1998) *Curr. Org. Chem.*, **2**, 365–394; (b) Oikawa, H. and Tokiwano, T. (2004) *Nat. Prod. Rep.*, **21**, 321–352; (c) Oikawa, H. (2005) *Bull. Chem. Soc. Jpn.*, **78**, 537–554; (d) Oikawa, H. (2010) in *Comprehensive Natural Products II: Chemistry and Biology*, vol. 8 (eds L. Mander and H.-W. Liu), Elsevier, Oxford, pp. 277–314; (e) Stocking, E.M. and Williams, R.M. (2003) *Angew. Chem. Int. Ed.*, **42**, 3078–3115.
4. (a) Alikaridis, F., Papadakis, D., Pantelia, K., and Kephalas, T. (2000) *Fitoterapia*, **71**, 379–384; (b) Lee, J.I., Hsu, B.H., Wu, D., and Barrett, J.S. (2006) *J. Chromatogr. A*, **1116**, 57–68.
5. Dewick, P.M. (2008) *Medicinal Natural Products: A Biosynthetic Approach*, 3rd edn, John Wiley & Sons, Ltd., Chichester.
6. Oikawa, H., Katayama, K., Suzuki, Y., and Ichihara, A. (1995) *J. Chem. Soc., Chem. Commun.*, 1321–1322.
7. Auclair, K., Sutherland, A., Kennedy, J., Witter, D.J., Van den Heever, J.P., Hutchinson, C.R., and Vederas, J.C. (2000) *J. Am. Chem. Soc.*, **122**, 11519–11520.
8. Watanabe, K., Mie, T., Ichihara, A., Oikawa, H., and Honma, M. (2000) *J. Biol. Chem.*, **275**, 38393–38401.
9. Ose, T., Watanabe, K., Mie, T., Honma, M., Watanabe, H., Yao, M., Oikawa, H., and Tanaka, I. (2003) *Nature*, **422**, 185–189.
10. (a) Moore, R.N., Bigam, G., Chan, J.K., Hogg, A.M., Nakashima, T.T., and Vederas, J.C. (1985) *J. Am. Chem. Soc.*, **107**, 3694–3701; (b) Yoshizawa, Y., Witter, D.J., Liu, Y., and Vederas, J.C. (1994) *J. Am. Chem. Soc.*, **116**, 2693–2694.
11. Auclair, K., Kennedy, J., Hutchinson, C.R., and Vederas, J.C. (2001) *Bioorg. Med. Chem. Lett.*, **11**, 1527–1531.

12. Sorensen, J.L., Auclair, K., Kennedy, J., Hutchinson, C.R., and Vederas, J.C. (2003) *Org. Biol. Chem.*, **1**, 50–59.
13. Kennedy, J., Auclair, K., Kendrew, S.G., Park, C., Vederas, J.C., and Hutchinson, C.R. (1999) *Science*, **284**, 1368–1372.
14. Ma, S.M., Li, J.W.H., Choi, J.W., Zhou, H., Lee, K.K.M., Moorthie, V.A., Xie, X.K., Kealey, J.T., Da Silva, N.A., Vederas, J.C., and Tang, Y. (2009) *Science*, **326**, 589–592.
15. Sakurai, I., Miyajima, H., Akiyama, K., Shimizu, S., and Yamamoto, Y. (1988) *Chem. Pharm. Bull.*, **36**, 2003–2011.
16. (a) Watanabe, K., Oikawa, H., Yagi, K., Ohashi, S., Mie, T., Ichihara, A., and Honma, M. (2000) *J. Biochem.*, **127**, 467–473; (b) Oikawa, H., Watanabe, K., Yagi, K., Ohashi, S., Mie, T., Ichihara, A., and Honma, M. (1999) *Tetrahedron Lett.*, **40**, 6983–6986.
17. (a) Watanabe, K., Mie, T., Ichihara, A., Oikawa, H., and Honma, M. (2000) *Biosci. Biotechnol. Biochem.*, **64**, 530–538; (b) Watanabe, K., Mie, T., Ichihara, A., Oikawa, H., and Honma, M. (2000) *Tetrahedron Lett.*, **41**, 1443–1446.
18. Serafimov, J.M., Westfeld, T., Meier, B.H., and Hilvert, D. (2007) *J. Am. Chem. Soc.*, **129**, 9580–9581.
19. Serafimov, J.M., Lehmann, H.C., Oikawa, H., and Hilvert, D. (2007) *Chem. Commun.*, 1701–1703.
20. Guimaraes, C.R.W., Udier-Blagovic, M., and Jorgensen, W.L. (2005) *J. Am. Chem. Soc.*, **127**, 3577–3588.
21. Serafimov, J.M., Gillingham, D., Kuster, S., and Hilvert, D. (2008) *J. Am. Chem. Soc.*, **130**, 7798–7799.
22. (a) Ichihara, A., Tazaki, H., and Sakamura, S. (1983) *Tetrahedron Lett.*, **24**, 5373–5376; (b) Alam, S.S., Bilton, J.M., Slawin, M.Z., Williams, D.J., Sheppard, R.N., and Strange, R.M. (1989) *Phytochemistry*, **28**, 2627–2630.
23. (a) Oikawa, H., Yokota, T., Ichihara, A., and Sakamura, S. (1989) *J. Chem. Soc., Chem. Commun.*, 1284–1285; (b) Oikawa, H., Yokota, T., Sakano, C., Suzuki, Y., Naya, A., and Ichihara, A. (1998) *Biosci. Biotechnol. Biochem.* **62**, 2016–2022
24. (a) Oikawa, H., Yokota, T., Abe, T., Ichihara, A., Sakamura, S., Yoshizawa, Y., and Vederas, J.C. (1989) *J. Chem. Soc., Chem. Commun.*, 1282–1284.
25. (a) Oikawa, H., Suzuki, Y., Naya, A., Katayama, K., and Ichihara, A. (1994) *J. Am. Chem. Soc.*, **116**, 3605–3606; (b) Oikawa, H., Suzuki, Y., Katayama, K., Naya, A., Sakano, C., and Ichihara, A. (1999) *J. Chem. Soc., Perkin Trans. 1*, 1225–1232.
26. (a) Katayama, K., Kobayashi, T., Oikawa, H., Honma, M., and Ichihara, A. (1998) *Biochim. Biophys. Acta*, **1384**, 387–395; (b) Katayama, K., Kobayashi, T., Chijimatsu, M., Ichihara, A., and Oikawa, H. (2008) *Biosci. Biotechnol. Biochem.*, **72**, 604–607.
27. Kasahara, K., Miyamoto, T., Fujimoto, T., Oguri, H., Tokiwano, T., Oikawa, H., Ebizuka, Y., and Fujii, I. (2010) *ChemBioChem*, **11**, 1245–1252.
28. Oikawa, H., Kobayashi, T., Katayama, K., Suzuki, Y., and Ichihara, A. (1998) *J. Org. Chem.*, **63**, 8748–8756.
29. Raimondi, L., Brown, F.K., Gonzalez, J., and Houk, K.N. (1992) *J. Am. Chem. Soc.*, **114**, 4796–4804.
30. Breslow, R. (1991) *Acc. Chem. Res.*, **24**, 159–164.
31. Ruiz-López, M.F., Assfeld, X., García, J.I., Mayoral, J.A., and Salvatella, L. (1993) *J. Am. Chem. Soc.*, **115**, 8780–8787.
32. Xu, J.A., Deng, Q.L., Chen, J.G., Houk, K.N., Bartek, J., Hilvert, D., and Wilson, I.A. (1999) *Science*, **286**, 2345–2348.
33. Fischbach, M.A. and Walsh, C.T. (2006) *Chem. Rev.*, **106**, 3468–3496.
34. Sims, J.W., Fillmore, J.P., Warner, D.D., and Schmidt, E.W. (2005) *Chem. Commun.*, 186–188.
35. Schumann, J. and Hertweck, C. (2006) *J. Biotechnol.*, **124**, 690–703.
36. Sekita, S., Yoshihira, K., Natori, S., and Kuwano, H. (1973) *Tetrahedron Lett.*, **14**, 2109–2112.
37. (a) Oikawa, H., Murakami, Y., and Ichihara, A. (1992) *J. Chem. Soc., Perkin Trans. 1*, 2955–2959; (b) Oikawa, H., Murakami, Y., and Ichihara, A.

(1992) *J. Chem. Soc., Perkin Trans. 1*, 2949–2953.
38. Schumann, J. and Hertweck, C. (2007) *J. Am. Chem. Soc.*, **129**, 9564–9565.
39. Thomas, E.J. and Watts, J.P. (1999) *J. Chem. Soc., Perkin Trans. 1*, 3285–3290.
40. Abe, Y., Suzuki, T., Ono, C., Iwamoto, K., Hosobuchi, M., and Yoshikawa, H. (2002) *Mol. Genet. Genom.*, **267**, 636–646.
41. Holtzel, A., Schmid, D.G., Nicholson, G.J., Krastel, P., Zeeck, A., Gebhardt, K., Fiedler, H.P., and Jung, G. (2004) *J. Antibiot.*, **57**, 715–720.
42. Mallams, A.K., Puar, M.S., Rossman, R.R., McPhail, A.T., and Macfarlane, R.D. (1981) *J. Am. Chem. Soc.*, **103**, 3940–3943.
43. Muntwyler, R. and Keller-Schierlein, W. (1972) *Helv. Chim. Acta*, **55**, 2071–2094.
44. Hirayama, N., Kasai, M., Shirahata, K., Ohashi, Y., and Sasada, Y. (1982) *Bull. Chem. Soc. Jpn.*, **55**, 2984–2987.
45. Zhang, H., White-Phillip, J.A., Melancon, C.E., Kwon, H.J., Yu, W.L., and Liu, H.W. (2007) *J. Am. Chem. Soc.*, **129**, 14670–14683.
46. Jia, X.Y., Tian, Z.H., Shao, L., Qu, X.D., Zhao, Q.F., Tang, J., Tang, G.L., and Liu, W. (2006) *Chem. Biol.*, **13**, 575–585.
47. Fang, J., Zhang, Y.P., Huang, L.J., Jia, X.Y., Zhang, Q., Zhang, X., Tang, G.L., and Liu, W. (2008) *J. Bacteriol.*, **190**, 6014–6025.
48. Sun, Y.H., Hahn, F., Demydchuk, Y., Chettle, J., Tosin, M., Osada, H., and Leadlay, P.F. (2010) *Nat. Chem. Biol.*, **6**, 99–101.
49. Westley, J.W., Evans, R.H.J., Sello, L.H., Troupe, N., Liu, C.-M., and Blount, J.F. (1979) *J. Antibiot.*, **32**, 100–107.
50. Roege, K.E. and Kelly, W.L. (2009) *Org. Lett.*, **11**, 297–300.
51. Li, C.X., Roege, K.E., and Kelly, W.L. (2009) *ChemBioChem*, **10**, 1064–1072.
52. Kirst, H.A., Michel, K.H., Martin, J.W., Creemer, L.C., Chio, E.H., Yao, R.C., Nakatsukasa, W.M., Boeck, L.D., Occolowitz, J.L., Paschal, J.W., Deeter, J.B., Jones, N.D., and Thompson, G.D. (1991) *Tetrahedron Lett.*, **32**, 4839–4842.
53. Waldron, C., Matsushima, P., Rosteck, P.R. Jr., Broughton, M.C., Turner, J., Madduri, K., Crawford, K.P., Merlo, D.J., and Baltz, R.H. (2001) *Chem. Biol.*, **8**, 487–499.
54. Kim, H.J., Pongdee, R., Wu, Q.Q., Hong, L., and Liu, H.W. (2007) *J. Am. Chem. Soc.*, **129**, 14582–14584.
55. Bazan, A.C., Edwards, J.M., and Weiss, U. (1978) *Tetrahedron*, **34**, 3005–3015.
56. Hölscher, D. and Schneider, B. (1995) *J. Chem. Soc., Chem. Commun.*, 525–526.
57. Aoyagi, Y., Takahashi, Y., Satake, Y., Fukaya, H., Takeya, K., Aiyama, R., Matsuzaki, T., Hashimoto, S., Shiina, T., and Kurihara, T. (2005) *Tetrahedron Lett.*, **46**, 7885–7887.
58. Parvez, A., Choudhary, M.I., Akhter, F., Noorwala, M., Mohammad, F.V., Hasan, N.M., Zamir, T., and Ahmad, V.U. (1992) *J. Org. Chem.*, **57**, 4339–4340.
59. Dudareva, N., Martin, D., Kish, C.M., Kolosova, N., Gorenstein, N., Faldt, J., Miller, B., and Bohlmann, J. (2003) *Plant Cell*, **15**, 1227–1241.
60. Majetich, G. and Zhang, Y. (1994) *J. Am. Chem. Soc.*, **116**, 4979–4980.
61. Bell, A.A., Stipanovic, R.D., O'Brien, D.H., and Fryxell, P.A. (1978) *Phytochemistry*, **17**, 1297–1305.
62. Li, M., Zhang, J.S., and Chen, M.Q. (2001) *J. Nat. Prod.*, **64**, 971–972.
63. Ovenden, S.P.B., Yew, A.L.S., Glover, R.P., Ng, S., Rossant, C.J., Regalado, J.C., Soejarto, D.D., Buss, A.D., and Butler, M.S. (2001) *Tetrahedron Lett.*, **42**, 7695–7697.
64. (a) Gonzalez, A.G., Rodriguez, F.M., Bazzocchi, I.L., and Ravelo, A.G. (2000) *J. Nat. Prod.*, **63**, 48–51; (b) Jacobsen, N.E., Wijeratne, E.M.K., Corsino, J., Furlan, M., Bolzani, V.D., and Gunatilaka, A.A.L. (2008) *Bioorg. Med. Chem.*, **16**, 1884–1889.
65. Gonzalez, A.G., Alvarenga, N.L., Bazzocchi, I.L., Ravelo, A.G., and Moujir, L. (1999) *J. Nat. Prod.*, **62**, 1185–1187.
66. Yamamura, S., Terada, Y., Chen, Y., Hong, M., Hsu, H., Sasaki, K., and

Hirata, Y. (1976) *Bull. Chem. Soc. Jpn.*, **49**, 1940.
67. Niwa, M., Terada, Y., Nonoyama, M., and Yamamura, S. (1979) *Tetrahedron Lett.*, **49**, 813–816.
68. Nishiyama, A., Eto, H., Terada, Y., Iguchi, M., and Yamamura, S. (1983) *Chem. Pharm. Bull.*, **31**, 2820.
69. Su, B.N., Yang, L., Gao, K., and Jia, Z.J. (2000) *Planta Med.*, **66**, 281–283.
70. Gagnepain, J., Castet, F., and Quideau, W. (2007) *Angew. Chem. Int. Ed.*, **46**, 1533–1535.
71. Abe, N., Murata, T., and Hirota, A. (1998) *Biosci. Biotechnol. Biochem.*, **62**, 661–666.
72. Abe, N. and Hirota, A. (2004) *J. Synth. Org. Chem. Jpn.*, **62**, 584–597.
73. Abe, N., Sugimoto, O., Tanji, K., and Hirota, A. (2000) *J. Am. Chem. Soc.*, **122**, 12606–12607.
74. Nicolaou, K.C., Vassilikogiannakis, G., Simonsen, K.B., Baran, P.S., Zhong, Y.L., Vidali, V.P., Pitsinos, E.N., and Couladouros, E.A. (2000) *J. Am. Chem. Soc.*, **122**, 3071–3079.
75. Hong, R., Chen, Y., and Deng, L. (2005) *Angew. Chem. Int. Ed.*, **44**, 3478–3481.
76. Barnes-Seeman, D. and Corey, E.J. (1999) *Org. Lett.*, **1**, 1503–1504.
77. Falshaw, C.P. and King, T.J. (1983) *J. Chem. Soc., Perkin Trans. I*, 1749–1752.
78. Kuo, Y.H., Chen, C.H., and Huang, S.L. (1998) *Chem. Pharm. Bull.*, **46**, 181–183.
79. Alvarenga, N.L., Ferro, E.A., Ravelo, A.G., Kennedy, M.L., Maestro, M.A., and Gonzalez, A.G. (2000) *Tetrahedron*, **56**, 3771–3774.
80. Nomura, T. and Hano, Y. (1994) *Nat. Prod. Rep.*, **11**, 205–218.
81. Nomura, T., Fukai, T., Narita, T., Terada, S., Uzawa, J., Iitaka, Y., Takasugi, M., Ishikawa, S., Nagao, S., and Masamune, T. (1981) *Tetrahedron Lett.*, **22**, 2195–2198.
82. Hano, Y., Nomura, T., and Ueda, S. (1990) *J. Chem. Soc., Chem. Commun.*, 610–613.
83. Nomura, T. (2001) *Yakugaku Zasshi*, **121**, 535–556.
84. Ferrari, F., Delle Monache, F., Suarez, A., and Compagnone, R.S. (2000) *Fitoterapia*, **71**, 213–215.
85. Shi, Y.Q., Fukai, T., and Nomura, T. (2001) *Heterocycles*, **54**, 639–646.
86. Shirota, O., Takizawa, K., Sekita, S., Satake, M., Hirayama, Y., Hakamata, Y., Hayashi, T., and Yanagawa, T. (1997) *J. Nat. Prod.*, **60**, 997–1002.
87. Ferrari, F., Filho, V.C., Cabras, T., and Messana, I. (2003) *J. Nat. Prod.*, **66**, 581–582.
88. Kutney, J.P., Inaba, T., and Dreyer, D.L. (1970) *Tetrahedron*, **26**, 3171–3184.
89. Gonzalez, A.G., Cardona, R.J., Diaz, C.E., Lopez, D.H., and Rodriguez, L.F. (1977) *An. Quim.*, **73**, 1510–1514.
90. Sharma, P.N., Shoeb, A., Kapil, R.S., and Popli, S.P. (1980) *Phytochemistry*, **19**, 1258–1260.
91. Ishii, H., Kobayashi, J., and Ishikawa, T. (1991) *Tetrahedron Lett.*, **32**, 6907–6910.
92. Chakraborty, D.P., Roy, S., Chakraborty, A., Mandal, A.K., and Chowdhury, B.K. (1980) *Tetrahedron*, **36**, 3563–3564.
93. Hasan, C.M., Kong, D.-Y., Gray, A.I., Waterman, P.G., and Armstrong, J.A. (1993) *J. Nat. Prod.*, **56**, 1839–1842.
94. Ngadjuri, T.B., Ayafor, J.F., Sandengam, B.L., Connolly, J.D., Rycroft, D.S., Khalid, S.A., Waterman, P.G., Brown, N.M.D., Grundon, M.F., and Ramachandran, V.N. (1982) *Tetrahedron Lett.*, **23**, 2041–2044.
95. (a) Fu, X., Hossain, B., Helm, D., and Schmitz, F.J. (1994) *J. Am. Chem. Soc.*, **116**, 12125–12126; (b) Fu, X., Hossain, M.B., Schmitz, F.J., and VanderHelm, D. (1997) *J. Org. Chem.*, **62**, 3810–3819.
96. Layton, M.E., Morales, C.A., and Shair, M.D. (2002) *J. Am. Chem. Soc.*, **124**, 773–775.
97. Hofheinz, W. and Schonholzer, P. (1977) *Helv. Chim. Acta*, **60**, 1367–1370.
98. Uenishi, J., Kawahama, R., and Yonemitsu, O. (1997) *J. Org. Chem.*, **62**, 1691–1701.

99. Jingyu, S., Kanghou, L., Tangsheng, P., Cun-heng, H., and Clardy, J. (1986) *J. Am. Chem. Soc.*, **108**, 177–178.
100. Kusumi, T., Igari, M., Ishitsuka, M.O., Ichikawa, A., Itezono, Y., Nakayama, N., and Kakisawa, H. (1990) *J. Org. Chem.*, **55**, 6286–6289.
101. Spörle, J., Becker, H., Gupta, M.P., Veith, M., and Huch, V. (1989) *Tetrahedron*, **45**, 5003–5014.
102. Kato, N., Wu, X., Nishikawa, H., Nakanishi, K., and Takeshita, H. (1994) *J. Chem. Soc., Perkin Trans. 1*, 1047–1053.
103. Sporle, J., Becker, H., Allen, N.S., and Gupta, M.P. (1991) *Phytochemistry*, **30**, 3043–3047.
104. Gao, F., Wang, H.P., and Mabry, T.J. (1990) *Phytochemistry*, **29**, 3875–3880.
105. Lee, S.H., Kim, H.K., Seo, J.M., Kang, H.M., Kim, J.H., Son, K.H., Lee, H., Kwon, B.M., Shin, J., and Seo, Y. (2002) *J. Org. Chem.*, **67**, 7670–7675.
106. Zdero, C. and Bohlmann, F. (1989) *Phytochemistry*, **28**, 3105–3120.
107. Kim, J.H., Kim, H.K., Jeon, S.B., Son, K.H., Kim, E.H., Kang, S.K., Sung, N.D., and Kwon, B.M. (2002) *Tetrahedron Lett.*, **43**, 6205–6208.
108. Iguchi, K., Fukaya, T., Takahashi, H., and Watanabe, K. (2004) *J. Org. Chem.*, **69**, 4351–4355.
109. Beauhaire, J., Fourrey, J.L., Vuilhorgne, M., and Lallemand, J.Y. (1980) *Tetrahedron Lett.*, **21**, 3191–3194.
110. Kamperdick, C., Phuong, N.M., Van Sung, T., and Adam, G. (2001) *Phytochemistry*, **56**, 335–340.
111. Martins, D., Osshiro, E., Roque, N.F., Marks, V., and Gottlieb, H.E. (1998) *Phytochemistry*, **48**, 677–680.

22
Bio-Inspired Transfer Hydrogenations

Magnus Rueping, Fenja R. Schoepke, Iuliana Atodiresei, and Erli Sugiono

22.1
Introduction

The demand for chiral molecules, particularly those with hydrogen as part of the stereogenic center, has increased in recent years, and intensive research has been carried out in both industry and academia to develop efficient methods for synthesizing such compounds [1]. In this context, significant attention has been devoted to the asymmetric hydrogenation of unsaturated compounds, for example, olefins, carbonyls, and imines, which is recognized as one of the most important and convenient routes to the corresponding optically active products [2]. So far, most of the enantioselective reductions rely on biological processes or transition metal catalyzed high-pressure hydrogenations, hydrosilylations, and transfer hydrogenations. Beside their high substrate specificity, the enzymatic processes sometimes suffer from undesired by-products, poor catalyst stability under the operational conditions, substrate and/or product inhibition, and problems with catalyst recovery. Likewise, despite the high reactivity and selectivity exhibited by the organometallic complexes employed in the metal-catalyzed processes, most of these protocols suffer from a limited number of substrates and difficulties with catalyst separation and recycling. Hence, an alternative approach to these chiral compounds would be of great value.

22.2
Nature's Reductions: Dehydrogenases as a Role Model

Enzymes are catalysts that evolve in Nature and one of their characteristics is high selectivity. During biochemical transformations, enzymes are assisted by non-proteinogenic molecules called *co-factors*. For instance, nicotinamide adenine dinucleotide (NADH), one of the essential co-factors in Nature, serves as a hydride source for various biological reductions. The synthesis of the amino acid glutamate by the glutamate dehydrogenase (GDH) catalyzed reductive amination of 2-ketoglutarate represents one example (Equation 22.1). The reaction is reversible

Biomimetic Organic Synthesis, First Edition. Edited by Erwan Poupon and Bastien Nay.
© 2011 Wiley-VCH Verlag GmbH & Co. KGaA. Published 2011 by Wiley-VCH Verlag GmbH & Co. KGaA.

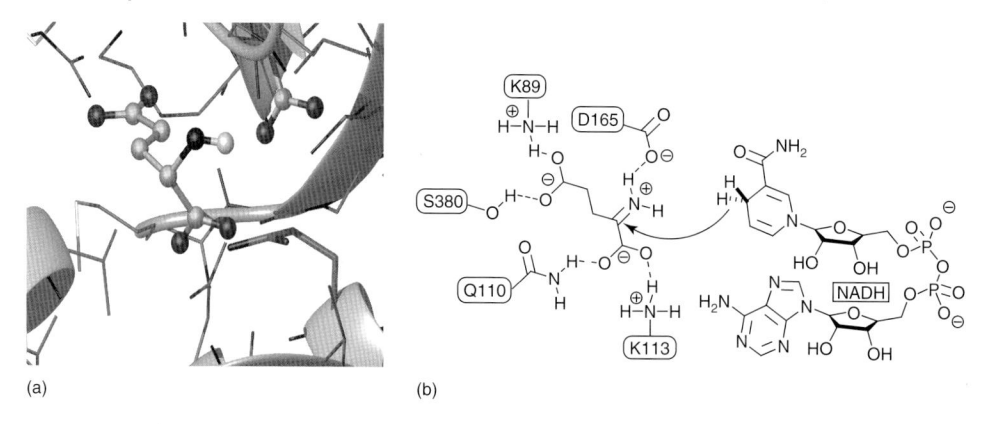

Figure 22.1 Proposed activation mechanism of GDH: (a) crystal structure of the GDH active site; (b) activation of α-imino glutarate by protonation from aspartate D165 with subsequent hydride transfer from the NADH to provide the amino acid glutamate.

although the equilibrium favors glutamate formation:

$$\text{2-ketoglutarate} + NH_3 + NAD(P)H \longleftrightarrow \text{glutamate} + NAD(P)^+ \quad (22.1)$$

On the basis of kinetic studies and X-ray crystal structure analysis, a possible mechanism for the mode of action of GDH has been proposed (Figure 22.1) [3].

It was envisaged that the aspartate D165 residue present in the catalytic active pocket of the enzyme is essential for the reductive amination of 2-ketoglutarate, being involved in the activation of α-iminoglutarate by protonation and conversion into a highly reactive iminium ion. Subsequent hydride transfer from NADH leads to the formation of the amino acid glutamate. Site-directed mutagenesis studies in which the putative catalytic aspartil residue, Asp-165, was replaced by serine support this mechanism since, despite conservation of the native folding pattern, the catalytic activity of the modified enzyme considerably decreased [4]. This clearly emphasizes the role of the aspartate in the transfer hydrogenation and indicates protonation as the key step for the reductive amination.

Based on the aforementioned activation mechanism, several groups have carried out intensive research to disclose efficient non-enzymatic systems capable of inducing asymmetry during the catalytic reduction of unsaturated substrates in a similar manner to Nature's dehydrogenases.

22.3
Brønsted Acid Catalyzed Transfer Hydrogenation of Imines, Imino Esters, and Enamines

The metal-catalyzed enantioselective reduction of olefin and carbonyl compounds represents a versatile transformation providing valuable intermediates for the

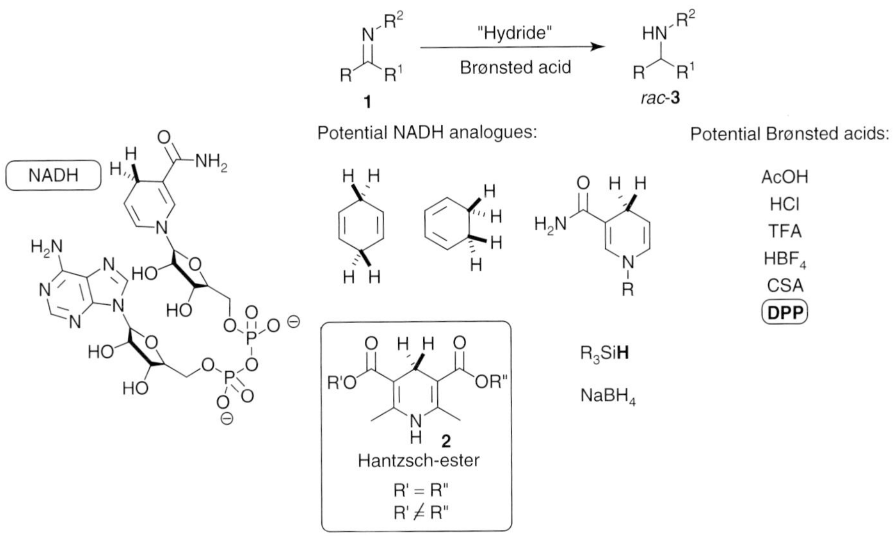

22.3 Brønsted Acid Catalyzed Transfer Hydrogenation of Imines, Imino Esters, and Enamines | 789

Figure 22.2 Towards a biomimetic transfer hydrogenation of ketimines with NADH analogs.

synthesis of natural products and biologically active substances or, when applied in the last steps, optically active target molecules. Recent comprehensive overviews on the hydrogenation of alkenes, carbonyls, and amines, respectively, are given in the chapters of books given in References [1b] and [2a,b]. Conversely, the asymmetric reduction of imines, which is a convenient route to chiral amines, still represents a challenging field. No widely applicable metal-based system has been developed to date. Therefore, as an alternative protocol, the asymmetric transfer hydrogenation of imines was one of the first reactions performed in a bio-inspired fashion, by employing a combination of NADH analog as hydride source and catalytic amounts of Brønsted acids (Figure 22.2). The strategy is based on activating the imine by catalytic protonation and formation of an iminium ion. Subsequent hydride transfer from a suitable NADH mimic yields the corresponding amine.

1,4-Dihydropyridines, also known as Hantzsch esters (HEH) [5], are synthetic analogs of NADH. Their potential as a hydrogen source was acknowledged for the first time in 1955 by Mauzerall and Westheimer [6], who showed that dihydropyridines can reduce carbonyl compounds by a direct hydrogen transfer to the substrate. Since then, a broad range of transfer hydrogenations conducted with HEH in combination with different Lewis acids and additives has been reported [7, 8]. Interestingly, whereas considerable efforts have been made to design effective HEH reagents that operate in conjunction with different metal catalysts, few metal-free asymmetric versions had been explored until recently. In the late 1980s Singh and Batra described the use of HEH as reducing agents in the metal-free acid catalyzed reduction of imines to give products in up to 62% ee [7h]. Furthermore,

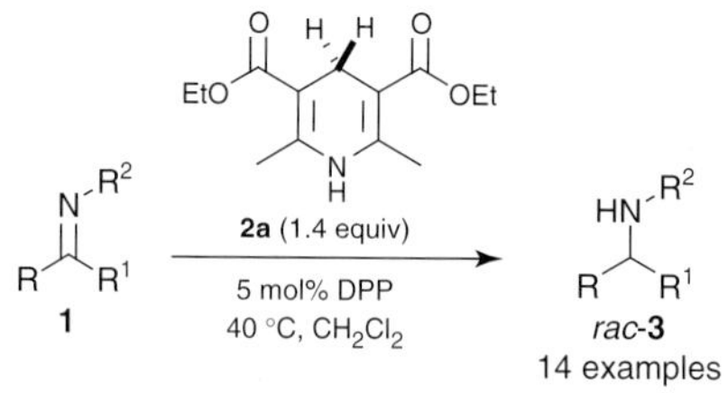

Scheme 22.1 Brønsted acid catalyzed reduction of ketimines.

List and MacMillan reported their usefulness in the organocatalytic enantioselective transfer hydrogenation of enals with chiral imidazolidinone catalysts [9, 10].

Subsequently, in 2005, Rueping described their use in the metal-free catalyzed transfer hydrogenation of imines in both an achiral as well as a chiral fashion [11]. The first attempts were devoted to the development of a viable protocol for the reduction of imines with catalytic amounts of achiral Brønsted acids (Figure 22.2) [11a]. Promising results were obtained when propiophenone ketimine **1** [R = Ph, R^1 = Et, R^2 = p-methoxyphenyl (PMP)] was treated with HEH **2a** in the presence of various Brønsted acids, to give the corresponding amine rac-**3** in moderate to good yields (Scheme 22.1). This preliminary study revealed diphenyl phosphate (DPP) as the best catalyst for this transformation. Further investigations focused on the effect of temperature and solvent on the yield and showed that the reaction could be performed well even at higher temperatures in non-polar aromatic solvents or at milder temperatures in halogenated solvents.

Accordingly, a large variety of aromatic ketimines **1**, including α-imino esters, have been reduced, under the optimized reaction conditions, to give products **3** in good to excellent yields (67–92%) (Figure 22.3).

Next, as the reduction of ketimines leads to the formation of a stereocenter, a chiral proton source [12] has been applied as catalyst. Since in the achiral version of the reaction DPP performed best in terms of yields, chiral phosphoric acids based on the BINOL core structure have been examined [13]. These had previously been used as chiral resolving agents [14], as ligands in metal catalysis [15], and more recently as organocatalysts in the enantioselective Mannich reaction [16, 17].

Notably, the steric demand and the acidity of the chiral BINOL phosphoric acids can easily be adjusted by introducing substituents with different steric

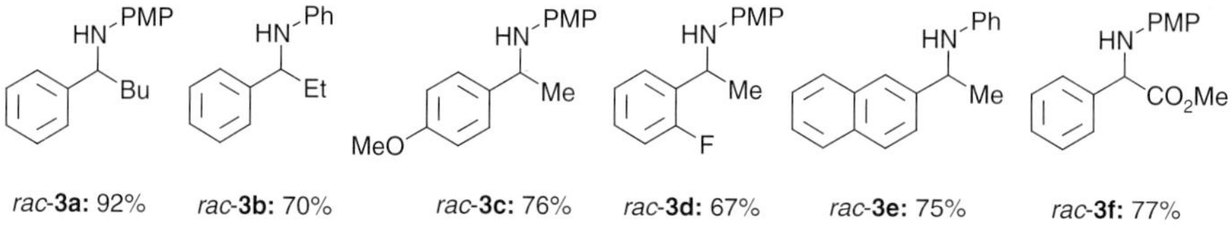

Figure 22.3 Scope of the Brønsted acid catalyzed transfer hydrogenation.

bulk and electronic properties in the 3,3′-positions of the backbone. The synthesis of the chiral BINOL phosphate derivatives is straightforward [11c–e, 16–22]. For example, Rueping reported the synthesis of **5** starting from the protected *(R)*-3,3′-dibromo-1,1′-binaphthyl-2,2′-diol and its octahydro analog derivative **4a**, which are obtained from BINOL and H_8-BINOL, respectively [11c-e] (Scheme 22.2). Suzuki coupling of **4a** with various aryl boronic acids afforded the corresponding 3,3-aryl-substituted BINOL derivatives, which were converted into the desired catalysts **5** via deprotection and subsequent phosphorylation. A similar approach towards this class of compounds was reported by Akiyama and involves Suzuki coupling of a bis(boronic acid) binaphthol derivative with different aryl halides [18], deprotection, and phosphorylation [16b]. The synthesis of the 3,3′-bis(2,4,6-triisopropylphenyl) BINOL derivative required for the preparation of **5e** can be accomplished according to Schrock [20]. The silylated Brønsted acids **6** can be synthesized following a procedure reported by Yamamoto [21] and MacMillan [22], which involves a Brook-type-rearrangement of the BINOL-derivative **4b** using *t*-BuLi and subsequent phosphorylation.

The chiral BINOL-phosphoric acid derivatives **5** were evaluated in the enantioselective transfer hydrogenation of ketimines **1** (Scheme 22.3). The best results were obtained when the reduction was carried out with 1.4 equiv. of dihydropyridine **2a** at 60 °C in benzene in the presence of 20 mol.% **5a** as catalyst [11c,d].

Under optimized conditions, a broad range of substituted aromatic ketimines could be reduced to the corresponding amines **3** in moderate to high yields (46–91%) and with good enantioselectivities (68–84% ee) (Figure 22.4). Moreover, for two selected substrates the enantiomeric excess could be increased to 94% and 98%, respectively, by simple recrystallization. This represents the first example of a highly enantioselective Brønsted acid catalyzed transfer hydrogenation.

List described the catalytic enantioselective imine reduction employing the same combination of HEH/BINOL-phosphoric acid [23]. Use of phosphoric acids with sterically demanding residues in the 3,3′-positions of the BINOL skeleton allowed shorter reaction times, milder temperatures, and lower catalyst loadings. Accordingly, a large variety of aromatic ketimines **1** were reduced in the presence of *ent*-**5e** as catalyst to give products *ent*-**3** in high yields (85–98%) and selectivities (80–92% ee) (Figure 22.5). The protocol is also suitable for the reduction of *in situ* generated imines as well as for the reduction of more challenging aliphatic substrates.

The protocol has also been extended to the reductive amination of α-branched aldehydes **7**, yielding chiral β-branched amines **9** via a dynamic kinetic resolution process (Scheme 22.4) [24]. The necessary prerequisites to this approach are (i) racemization of the *in situ* formed imine under the reaction conditions and (ii) one of the two enantiomers undergoes a much faster reaction than the other one (Scheme 22.5). Screening of reaction conditions, including different phosphoric acid derivatives, various solvents, and temperatures, in the reaction of hydratopic aldehyde **7** (R = Ph, R^1 = Me) with *p*-anisidine in the presence of HEH **2b** identified optimum conditions for a highly selective process. Use of 5 Å molecular sieves was needed to achieve a high degree of asymmetric induction. Furthermore, handling

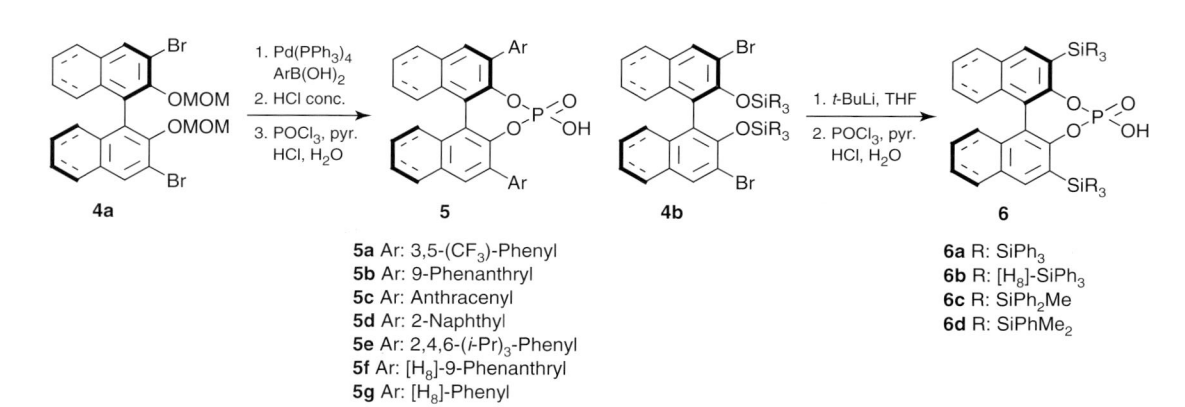

Scheme 22.2 Synthesis of BINOL-phosphoric acid derivatives **5** and **6**.

5a Ar: 3,5-(CF_3)-Phenyl
5b Ar: 9-Phenanthryl
5c Ar: Anthracenyl
5d Ar: 2-Naphthyl
5e Ar: 2,4,6-(i-Pr)$_3$-Phenyl
5f Ar: [H_8]-9-Phenanthryl
5g Ar: [H_8]-Phenyl

6a R: $SiPh_3$
6b R: [H_8]-$SiPh_3$
6c R: $SiPh_2Me$
6d R: $SiPhMe_2$

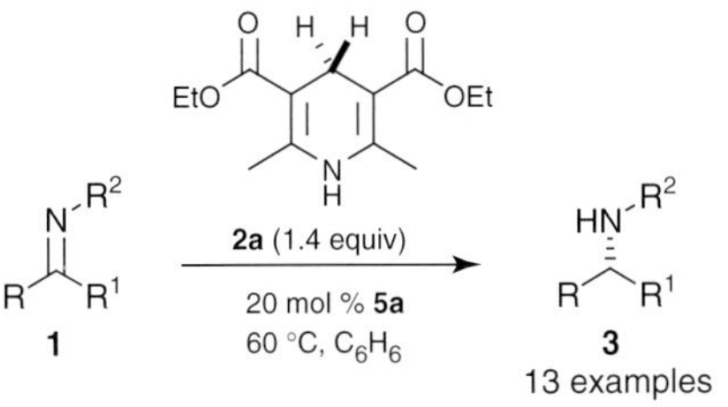

Scheme 22.3 Metal-free asymmetric transfer hydrogenation of ketimines.

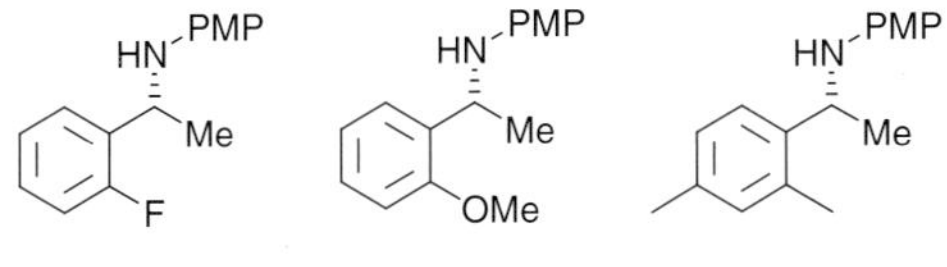

3d: 82%, 84% ee **3g:** 72%, 76% ee **3h:** 91%, 78% ee

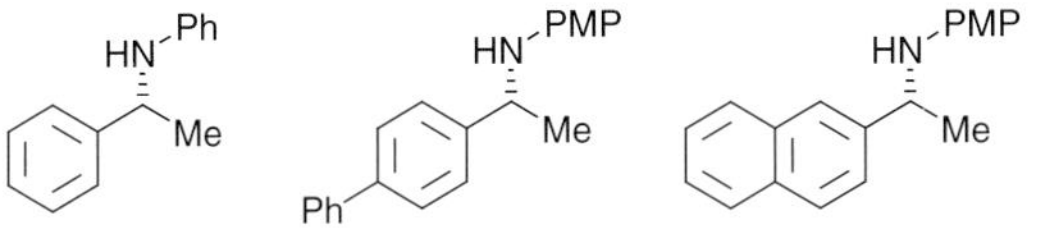

3i: 71%, 72% ee **3j:** 71%, 74% ee (98)[a] **3k:** 82%, 70% ee (94)[a]

[a] After one recrystallization from methanol

Figure 22.4 Scope of the enantioselective biomimetic transfer hydrogenation of ketimines.

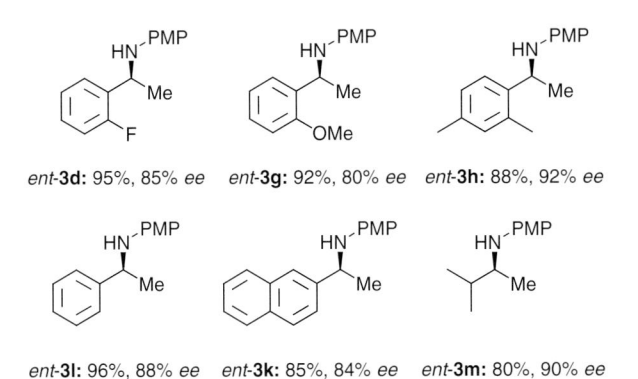

ent-3d: 95%, 85% ee **ent-3g:** 92%, 80% ee **ent-3h:** 88%, 92% ee

ent-3l: 96%, 88% ee **ent-3k:** 85%, 84% ee **ent-3m:** 80%, 90% ee

Figure 22.5 Asymmetric biomimetic transfer hydrogenation of ketimines according to List.

the reaction under oxygen-free conditions was required to circumvent formation of undesired by-products such as acetophenone and *p*-formylanisidine.

The reaction proved general with respect to aldehyde and amine structure. Substrates bearing aromatic residues with different substitution patterns and electronic properties performed best in this reaction (80–96% yield, 94–98% ee) (Figure 22.6). Aliphatic aldehydes are also tolerated although with lower selectivities

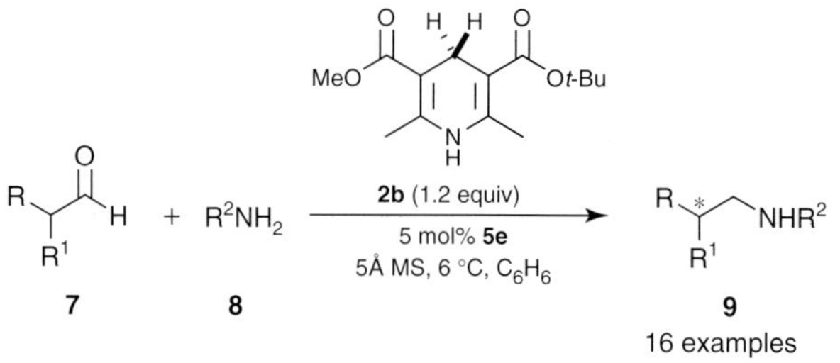

Scheme 22.4 Organocatalytic asymmetric reductive amination of α-branched aldehydes.

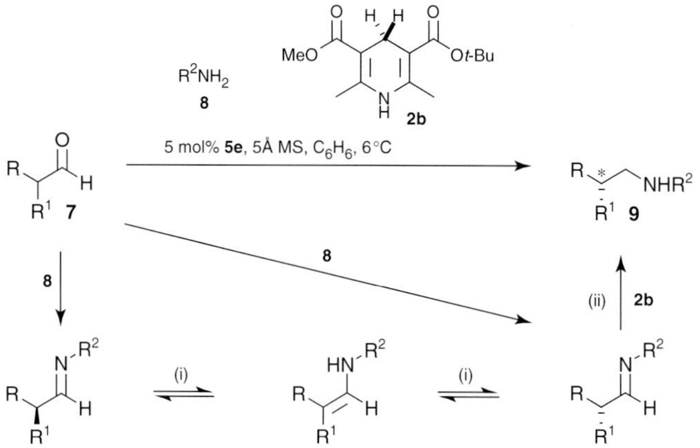

Scheme 22.5 Dynamic kinetic resolution in the reductive amination of α-branched aldehydes.

9a: 87%, 96 % ee   9b: 86%, 94% ee   9c: 89%, 94% ee

9d: 92%, 98% ee   9e: 81%, 78% ee   9f: 77%, 80% ee

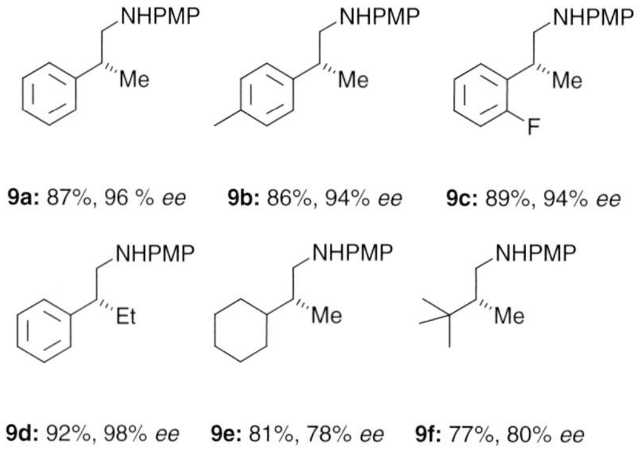

Figure 22.6 Scope of the asymmetric reductive amination of α-branched aldehydes.

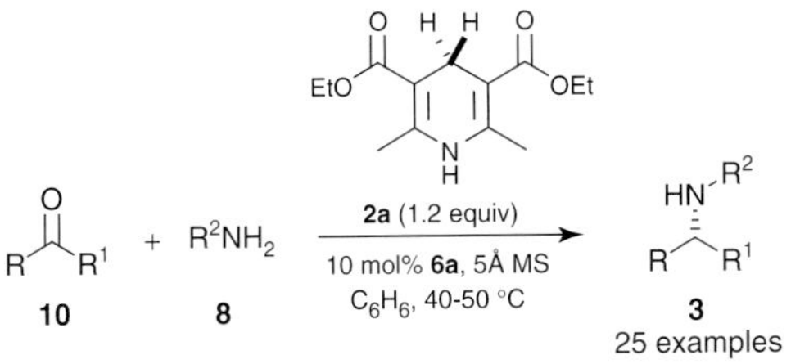

Scheme 22.6 Organocatalytic asymmetric reductive amination of ketones.

3c: 77%, 90% ee
3l: 87%, 94% ee
3n: 79%, 91% ee

3o: 71%, 95% ee
3p: 75%, 85% ee
3q: 71%, 83% ee

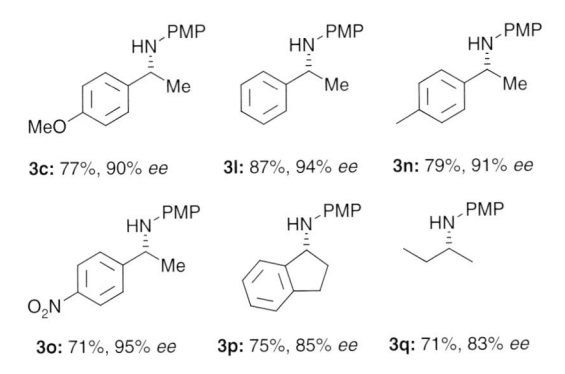

Figure 22.7 Scope of the asymmetric biomimetic reductive amination of ketones.

(40–80% ee). The absolute configuration of **9a** was determined by conversion into the known carbobenzyloxy (Cbz)-protected amine. In addition, recent mechanistic studies have rationalized the stereochemical outcome of the reaction [25].

MacMillan reported a similar catalytic system for the reductive amination of ketones (Scheme 22.6) [22]. Use of 5 Å molecular sieves during the *in situ* imine generation was crucial to attain fair levels of conversion and selectivity.

Under optimized conditions a large variety of aromatic ketones (**10**) with different electronic properties were treated with p-anisidine [**8**, R = p-MeO-C$_6$H$_4$] in the presence of HEH **2a** and catalyst **6a** to give good yields (60–87%) and excellent selectivities (83–97% ee) (Figure 22.7). The protocol has also been extended to cyclic aryl and aliphatic ketones. Furthermore, diverse aromatic as well as heteroaromatic amines are tolerated under the same reaction conditions.

Recently, Akiyama described the transfer hydrogenation of ketimines with a system based on BINOL-phosphoric acid derivatives as catalysts and benzothiazolines as alternative reducing agents [26].

You and Antilla independently reported the enantioselective reduction of α-imino esters yielding enantio-enriched α-amino acid derivatives [27, 28]. Whereas the system reported by You is based on catalysts derived from BINOL as backbone [27], Antilla reported a novel catalyst based on VAPOL as chiral backbone [28]. The best results were obtained when VAPOL derivative **13** was employed in the reduction of ethyl ester derivatives at 50 °C in non-polar solvents (Scheme 22.7).

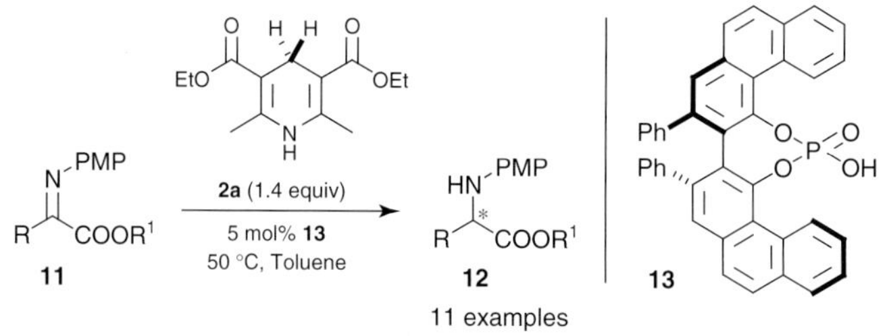

Scheme 22.7 Organocatalytic asymmetric transfer hydrogenation of α-imino esters.

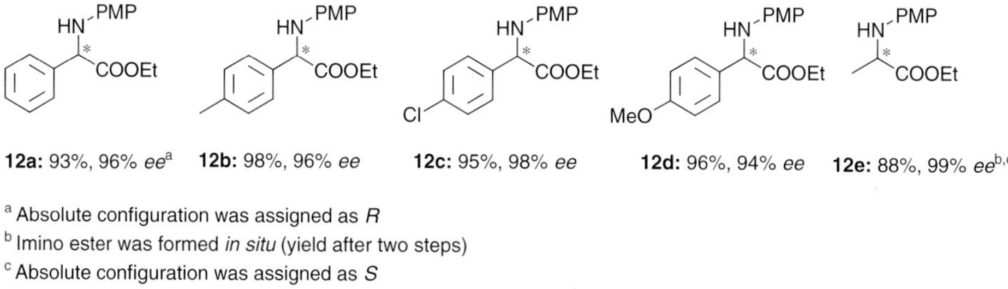

12a: 93%, 96% ee[a]   **12b:** 98%, 96% ee   **12c:** 95%, 98% ee   **12d:** 96%, 94% ee   **12e:** 88%, 99% ee[b,c]

[a] Absolute configuration was assigned as R
[b] Imino ester was formed *in situ* (yield after two steps)
[c] Absolute configuration was assigned as S

Figure 22.8 Scope of the asymmetric biomimetic transfer hydrogenation of α-imino ethyl esters.

Under the optimized conditions, various α-imino ethyl esters (**11**) bearing aromatic groups with different substitution patterns and electronic properties were reduced to give products in high yields (93–98%) and excellent selectivities (94–98% ee) (Figure 22.8). The methodology was also extended to the reductive amination of alkyl substituted α-keto esters formed *in situ* from the appropriate keto esters and *p*-anisidine.

Screening of different conditions (various catalysts, solvents, and HEH) led You and coworkers to develop an improved protocol that allows reactions with low catalyst loadings (1 mol.% *ent*-**5c**) in diethyl ether at room temperature and with the same HEH (**2a**) as hydrogen source [27]. The selectivity proved to be highly dependent on the residue of the ester group and superior enantioselectivities were obtained by increasing the steric bulk (R^1: O*t*-Bu ≈ O*i*-Pr > OBn > OEt >> OMe). A wide variety of α-imino esters were hydrogenated under these conditions to give products in high yields (78–95%) and selectivities (84–98% ee) (Figure 22.9). Only in one case was the product obtained in lower yield (R = cyclohexyl, R^1 = *i*-Pr 46%, 88% ee). To show the usefulness of the method, a representative substrate was selected and the reaction was performed on gram scale. Product, **12f**, was obtained in 85% yield and 96% ee. Surprisingly, the catalyst employed by MacMillan in the reductive amination of ketones was inactive under the conditions developed by You and Antilla.

Moreover, when β,γ-alkynyl α-imino esters were subjected to similar conditions (Scheme 22.8), both alkyne and imine moieties were reduced to give the

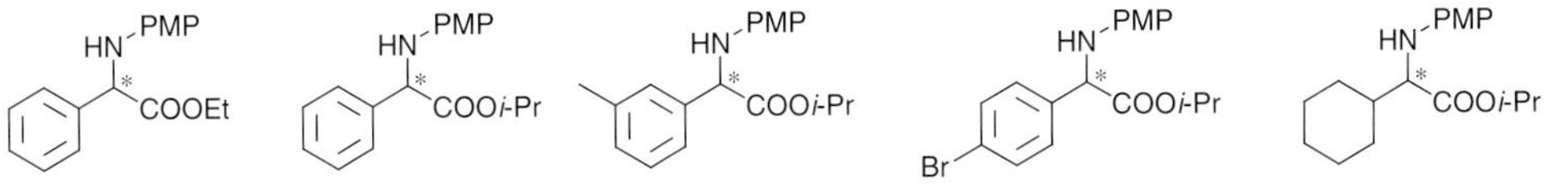

22.3 Brønsted Acid Catalyzed Transfer Hydrogenation of Imines, Imino Esters, and Enamines | 797

12a: 88%, 92% ee[a] **12f:** 87%, 97% ee[a] **12g:** 89%, 98% ee[a] **12h:** 92%, 97% (>99%) ee[a,c,d] **12i:** 46%, 88% ee[a]
  85%, 96% ee[b]

[a] Reactions were performed on a 0.2 mmol scale with 1 mol% catalyst
[b] Reaction was performed on a 4 mmol scale with 0.1 mol% catalyst
[c] After recrystallization
[d] Absolute configuration was assigned as S

Figure 22.9 Scope of the asymmetric biomimetic transfer hydrogenation of α-imino esters.

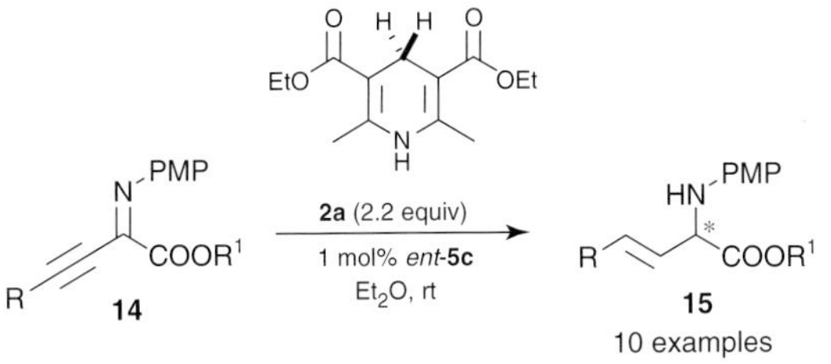

Scheme 22.8 Organocatalytic asymmetric transfer hydrogenation of β,γ-alkynyl α-imino esters.

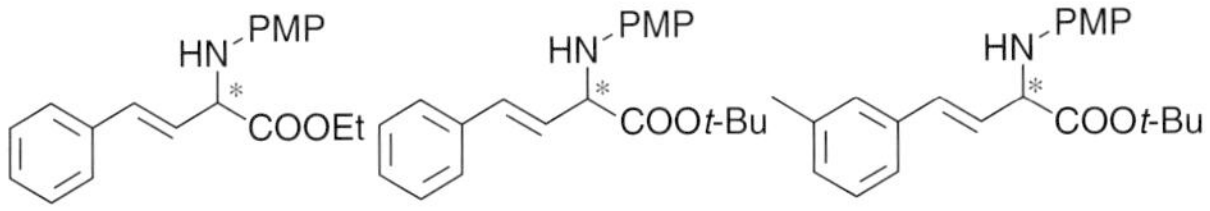

15a: 34%, 92% ee[a] **15b:** 58%, 94% ee **15c:** 60%, 93% ee

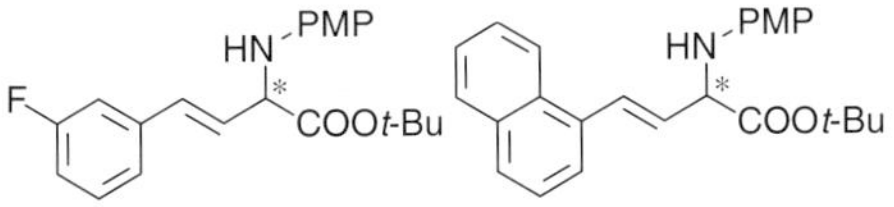

15d: 64%, 95% ee **15e:** 27%, 83% ee

[a] Absolute configuration was assigned as S

Figure 22.10 Scope of the asymmetric biomimetic transfer hydrogenation of β,γ-alkynyl α-imino esters.

corresponding β,γ-alkenyl α-amino esters **15** with very high selectivities, albeit in low yields (Figure 22.10) [29]. Nevertheless, the method is worth considering since the presence of a double bond in the products offers the possibility of further derivatization. With regard to the mechanism, control experiments have shown that reduction of the carbon–carbon triple bond is faster than that of the carbon–nitrogen double bond.

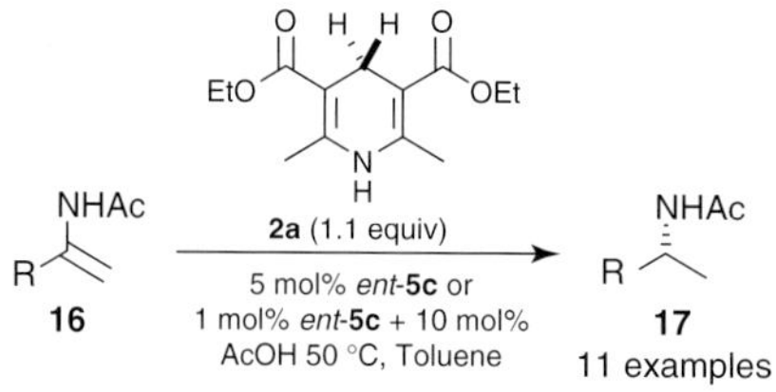

Scheme 22.9 Organocatalytic asymmetric transfer hydrogenation of enamides.

More recently, Antilla reported the enantioselective reduction of aromatic enamides with chiral phosphoric acid catalysts [30]. The product **17a** (R = Ph) was obtained in 92% yield and 91% ee when the reaction was carried out with 5 mol.% ent-**5c** as catalyst in toluene at 50 °C (Scheme 22.9). Decreasing the catalyst loading had a detrimental effect on the yield and reaction time. It was assumed that the reaction takes place through an iminium ion intermediate whose formation is affected; therefore, a dual catalytic chiral–achiral acid system was considered. The role of the achiral acid co-catalyst was to enable iminium ion formation while being inactive in the asymmetric hydrogenation step. Accordingly, the reaction times were substantially reduced and the yields considerably increased when employing an achiral acid as co-catalyst, while the selectivities remained essentially unaffected. Various enamides **16** were reduced under the optimized reaction conditions to give products **17** in moderate to excellent yields (43–99%) and moderate to high enantioselectivities (41–92% ee) (Figure 22.11). Regarding the influence of the aryl moiety, lower selectivities were obtained with increasing steric bulk. With two exceptions (α-naphthyl and o-MeO-C_6H_4 derivatives), the absolute configuration of the products has been assigned as *(R)*. Although impressive results were obtained for aromatic enamides, the protocol proved unsuitable for aliphatic analogs.

Regarding the mechanism of the imine reduction, it is assumed that the Brønsted acid catalyzed transfer hydrogenation proceeds similarly to the reaction of GDH (Scheme 22.10).

In the first step the imine substrate is activated by the chiral Brønsted acid catalyst, which generates an iminium ion with formation of the chiral-ion pair **A**.

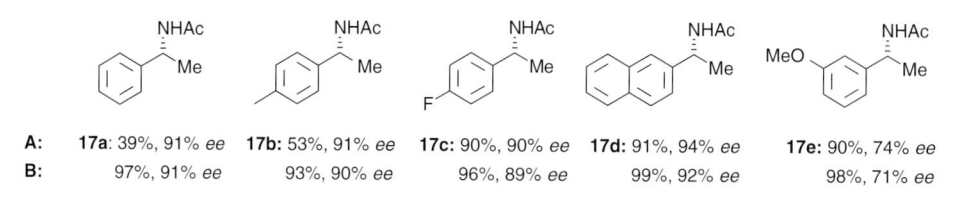

A: 17a: 39%, 91% ee   17b: 53%, 91% ee   17c: 90%, 90% ee   17d: 91%, 94% ee   17e: 90%, 74% ee
B: 97%, 91% ee   93%, 90% ee   96%, 89% ee   99%, 92% ee   98%, 71% ee

A: 5 mol% ent-**5c**
B: 1 mol% ent-**5c** + 10 mol% AcOH

Figure 22.11 Scope of the asymmetric biomimetic transfer hydrogenation of enamides.

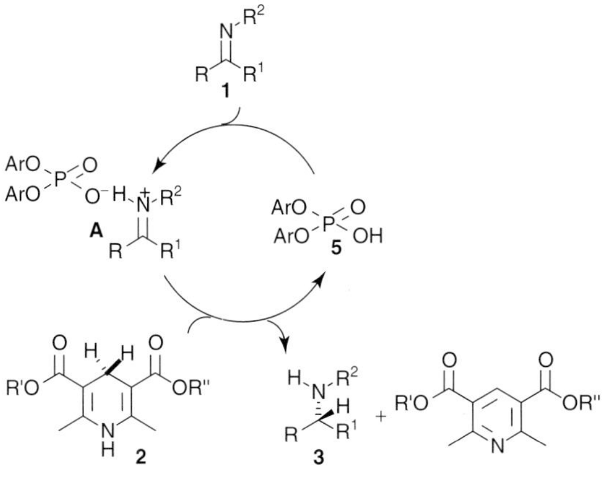

Scheme 22.10 Proposed mechanism of the organocatalyzed transfer hydrogenation of ketimines.

Subsequent hydride transfer from the HEH **2** yields the desired chiral amine **3** and regenerates the catalyst. The *(R)* absolute configuration of the amines depicted in Figure 22.4 was explained by a stereochemical model built on the basis of the X-ray crystal structure of the chiral Brønsted acid catalyst **5d**. In the transition state the Brønsted acid catalyst activates the ketimine in such a way that the *Re*-face is effectively shielded, leaving only the *Si*-face free for the nucleophilic attack (left-hand side of Figure 22.12). Based on theoretical investigations Goodman and Himo independently proposed a bifunctional activation mechanism in which the imine and the dihydropyridine are simultaneously activated by the same catalyst molecule: whereas the ketimine is protonated by the acidic proton, the dihydropyridine is activated via a hydrogen bond formed from the Lewis basic oxygen of the phosphoryl group (right-hand side of Figure 22.12) [31]. The phosphoric acid catalyst thereby acts as a Brønsted acid/Lewis base bifunctional catalyst.

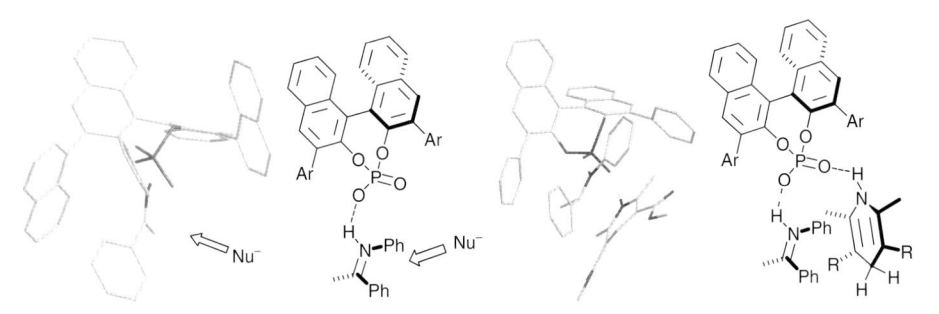

Figure 22.12 BINOL-phosphoric catalyzed activation of ketimines – a simplified stereochemical model.

22.4
Asymmetric Organocatalytic Reduction of N-Heterocycles

22.4.1
Asymmetric Organocatalytic Reduction of Quinolines

Hydrogenation of quinolines is another challenging field since it allows formation of the corresponding tetrahydroquinolines, which are common structural units in alkaloids and biologically active compounds. Given the importance of this class of molecule, considerable efforts have been made to develop improved methods for their synthesis. Although different approaches are available and well-established methods including homogeneous and heterogeneous metal-catalyzed hydrogenations, hydroborations, and transfer hydrogenations have been reported, these methods suffer from a limited substrate scope [32]. Thus, the development of efficient alternative protocols would be desirable. Given that Brønsted acid catalyzed enantioselective transfer hydrogenation proved general for the reduction of carbon–nitrogen double bonds in several acyclic systems, its application to the related cyclic analogs appeared feasible.

The first reports on metal-free quinoline hydrogenation stem from Rueping [33, 34]. Since Brønsted acids were successful in catalyzing the asymmetric reduction of imines, it was expected that the reduction of substituted quinolines would proceed in a similar manner. In the first step it was envisaged that the Brønsted acid protonates the quinoline to form a chiral ion pair **A** and **D**, respectively (Scheme 22.11a,b). A 1,4-addition was anticipated, with formation of the enamines **B** and **E**, respectively. Subsequent isomerization should yield the corresponding

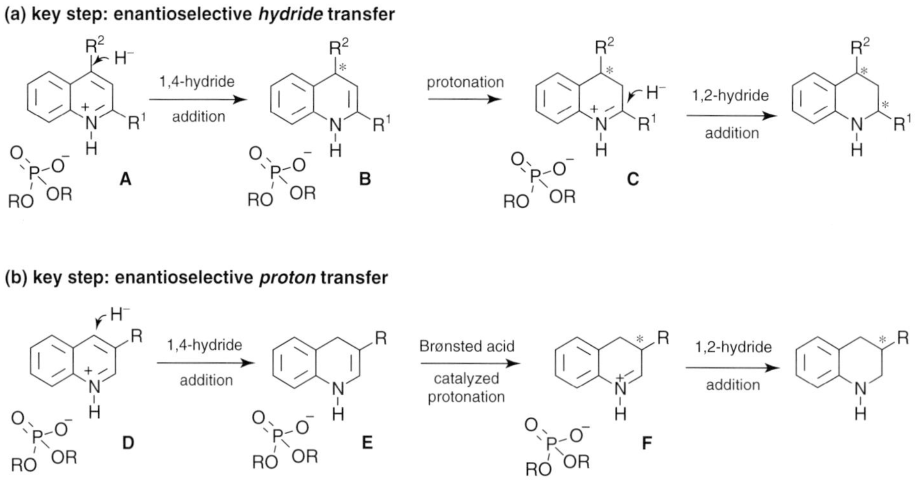

Scheme 22.11 Quinoline reduction – proposed mechanism as a function of quinoline substitution pattern: (a) enantioselective hydride transfer and (b) enantioselective protonation.

iminium ions **C** and **F**, respectively, and a second hydride transfer would give the desired tetrahydroquinolines. Obviously, the stereo-determining step of this hydrogenation cascade differs according to the quinoline substitution pattern. In the case of 2- and 4-substituted quinolines, enantioselectivity will be induced in one of the hydride transfer steps (Scheme 22.11a). In the case of 3-substituted quinolines the enantio-determining step has to be the asymmetric Brønsted acid catalyzed protonation (Scheme 22.11b).

Initial research focused on the development of an efficient achiral version with identification of suitable reaction conditions for the reduction of 2-methylquinoline as model substrate. Different Brønsted acids were capable of catalyzing the reaction; DPP gave optimum yield and reaction rate (Scheme 22.12) [33]. Overall, the best results were obtained when 2-methylquinoline (**18a**) was treated with 2.4 equiv. of dihydropyridine **2a** in the presence of 1 mol.% DPP at 60 °C in benzene for 12 h. Under these conditions a large variety of alkyl, aryl, and heteroaryl 2-substituted quinolines gave products in good to high yields (75–95%). Furthermore, 3- and 4-substituted quinolines as well as 2,3- and 2,4-disubstituted quinolines were reduced under the same conditions to give products in high yields (82–94%).

The excellent results obtained in the achiral version were the starting point for the development of an asymmetric version (Scheme 22.13) [34]. Screening of different BINOL based chiral phosphoric acids in the reduction of 2-phenylquinoline **18b** (R = Ph) revealed that large substituents in the 3,3' positions of the backbone are essential to attain high levels of enantioselectivity. Furthermore, comparable results were obtained in chlorinated as well as aromatic solvents.

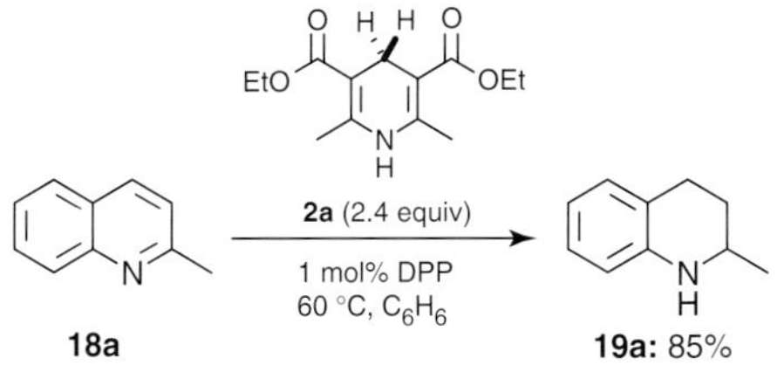

Scheme 22.12 First organocatalyzed reduction of quinolines.

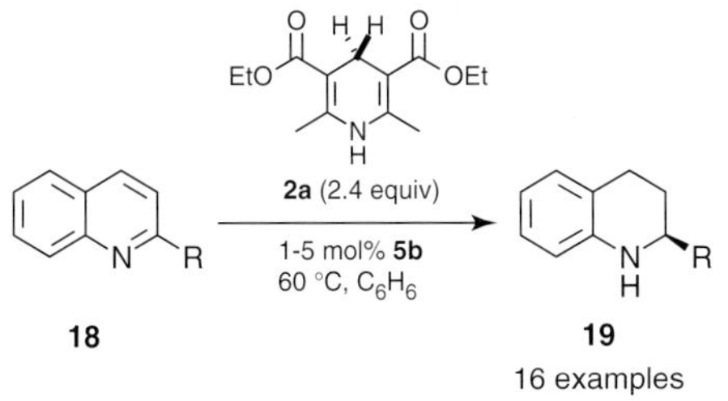

Scheme 22.13 Brønsted acid catalyzed transfer hydrogenation of 2-substituted quinolines.

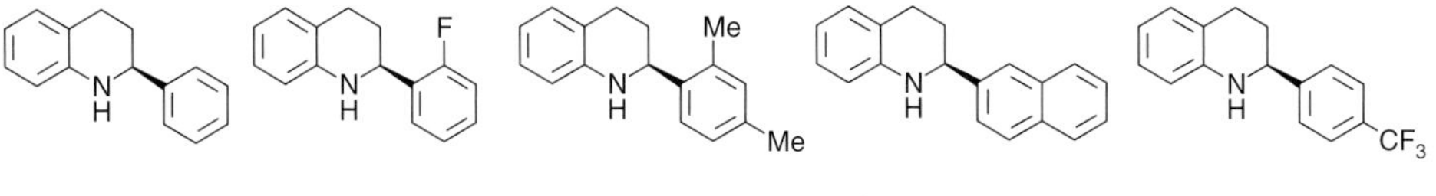

19b: 92%, 97% ee **19c**: 93%, 98% ee **19d**: 65%, 97% ee **19e**: 93%, 99% ee **19f**: 91%, 99% ee

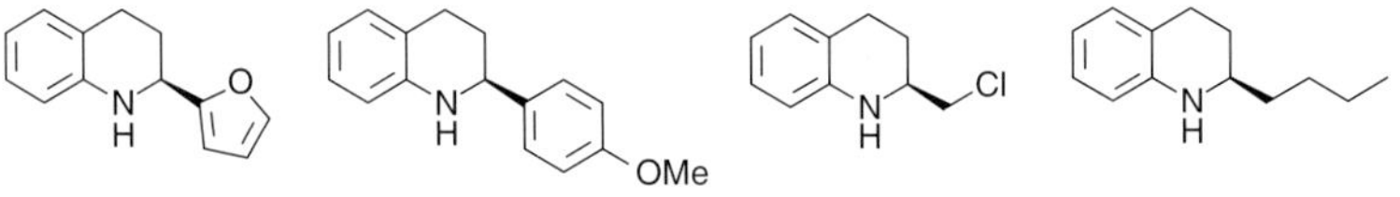

19g: 93%, 91% ee **19h**: 90%, 98% ee **19i**: 91%, 88% ee **19j**: 91%, 87% ee

Figure 22.13 Scope of the transfer hydrogenation of 2-substituted quinolines **18**.

With optimum conditions in hand, the scope of the reaction was explored and the results are summarized in Figure 22.13. Notably, a wide range of 2-substituted quinolines (**18**) bearing aliphatic, aromatic as well as heteroaromatic residues readily reacts with dihydropyridine **2a**, and the corresponding tetrahydroquinolines **19** can be isolated in moderate to good yields (54–93%) and excellent enantioselectivities (87–99% ee).

The usefulness of this highly enantioselective transfer hydrogenation process was demonstrated in a two-step synthesis of biologically active natural products with a tetrahydroquinoline core, namely, galipinine, cuspareine, and angustureine [35]. For this purpose, Brønsted acid catalyzed hydrogenation of the appropriate 2-substituted quinolines, which were available by simple alkylation of 2-methylquinoline, generated the tetrahydroquinoline derivatives with excellent enantioselectivities. Subsequent N-methylation afforded the desired natural products in good overall yields (Scheme 22.14).

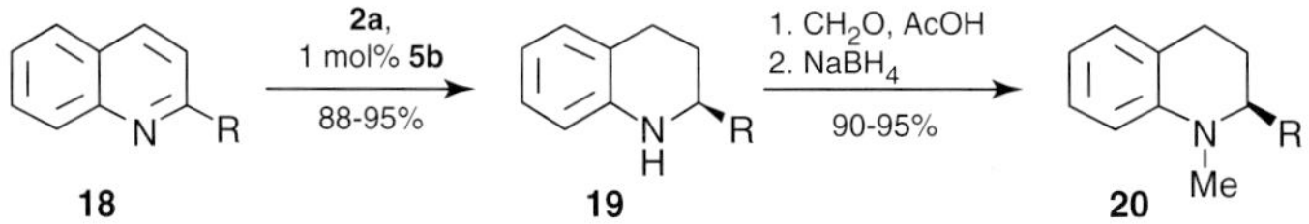

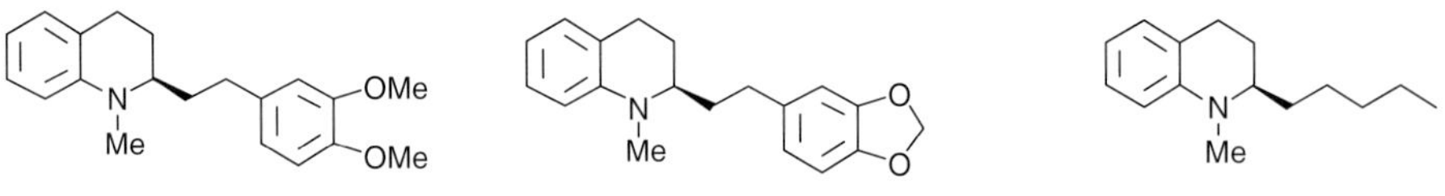

(+)-cuspareine
88%, 90% ee

(+)-galipinine
89%, 91% ee

(−)-angustureine
79%, 90% ee

Scheme 22.14 Application of the enantioselective Brønsted acid catalyzed transfer hydrogenation of 2-substituted quinolines in the synthesis of alkaloids.

Additionally, Metallinos reported a Brønsted acid catalyzed enantioselective reduction of the structurally related di-substituted 1,10-phenanthrolines [36].

Du reported a similar catalytic system, which made use of double axially chiral phosphoric acids **21** in the reduction of 2-substituted quinolines to give the corresponding tetrahydroquinolines with comparable enantioselectivities [37]. For 2-phenylquinoline, aromatic solvents and diethyl ether gave similar results, but for the *n*-butyl analog diethyl ether proved superior in terms of selectivity. Notably, lowering the catalyst loading did not affect the selectivity, and the reactions could be performed with HEH **2c** at 35 °C in diethyl ether with only 0.2 mol.% catalyst (Scheme 22.15). Compared to Rueping's system, a slight increase in the enantiomeric excess has been reported when alkyl substituted quinolines were reduced under these conditions. 2,3-Disubstituted quinolines have also been tested and products obtained with high levels of diastereo- and enantiocontrol (Figure 22.14).

Along the same lines, Rueping and coworkers also investigated the reduction of 4-substituted derivatives **23**. To date, no direct approach toward the optically active 4-substituted tetrahydroquinolines **24** has been described. Multistep sequences have been employed to synthesize these valuable chiral 4-substituted tetrahydroquinolines [38], which exhibit biological activity *in vivo* [39]. Therefore, the development of a catalytic asymmetric route would be a substantial improvement (Scheme 22.16).

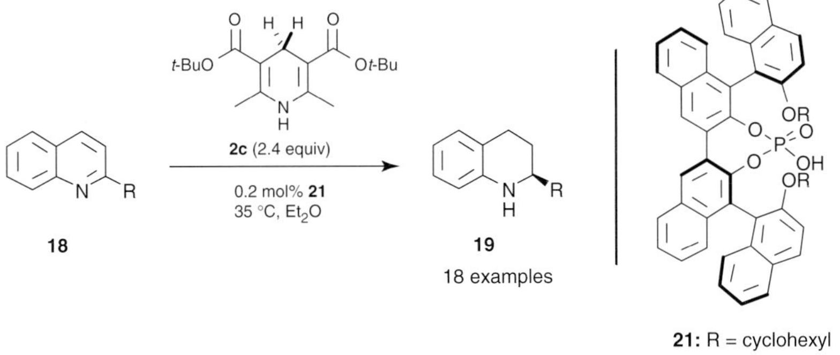

Scheme 22.15 Double axially chiral phosphoric acids in the transfer hydrogenation of 2-substituted quinolines.

22a
>99%, *cis/trans* = >20:1
cis = 82% ee

22b
93%, *trans/cis* = 94:6
trans = 91% ee

22c
>99%, *trans/cis* = >20:1
trans = 92% ee

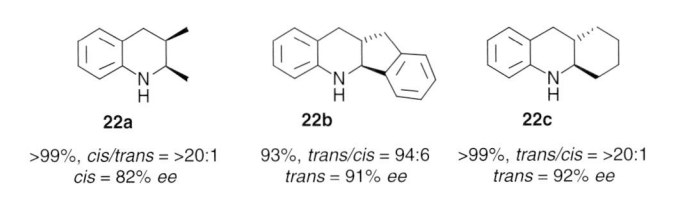

Figure 22.14 Substrate scope of the transfer hydrogenation of 2,3-substituted quinolines.

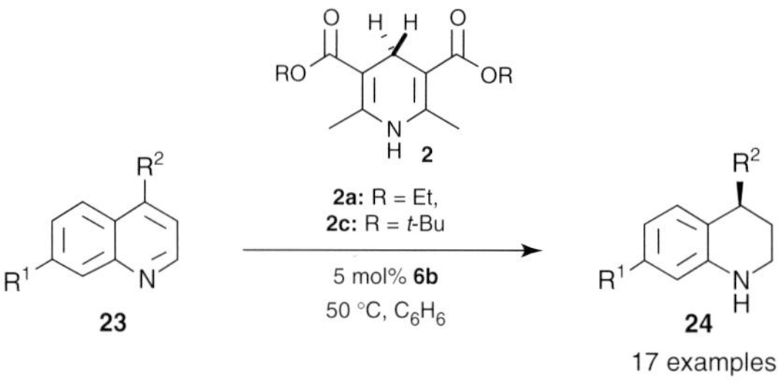

Scheme 22.16 Brønsted acid catalyzed transfer hydrogenation of 4-substituted quinolines.

Evaluation of different catalysts showed a strong correlation between the steric bulk of the substituents on the chiral backbone of the catalyst and the selectivity of the reaction, whereby larger residues lead to superior enantioselectivities [40]. This result is explained by the position of the stereocenter which is now further away from both the protonated nitrogen and the catalytic center of the phosphoric acid catalyst. After extensive optimization of the reaction parameters, the best results, with respect to both reactivity and selectivity, have been obtained with a combination of catalyst **6b** and *t*-butyl HEH **2c**. Figure 22.15 depicts selected examples, demonstrating that several 4-substituted tetrahydroquinolines can be prepared in good yields and with good to excellent enantioselectivities.

To obtain a complete picture of the transfer hydrogenation of quinolines, the reduction of 3-substituted quinolines **25** was also addressed [41]. This reaction represents the first direct access to optically active 3-substituted tetrahydroquinolines **26** (Scheme 22.17), and proceeds via an enantioselective Brønsted acid

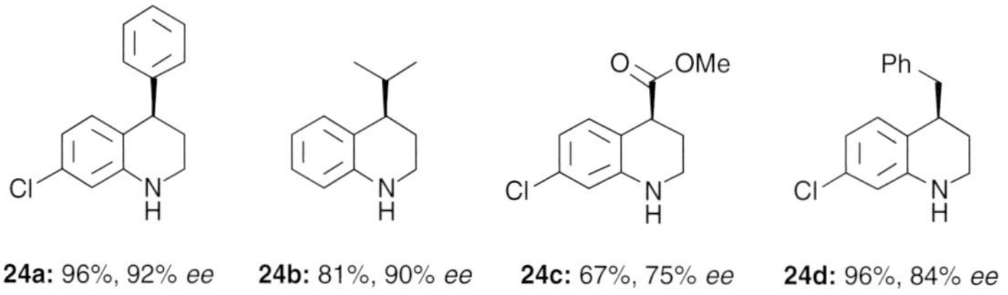

24a: 96%, 92% ee   **24b**: 81%, 90% ee   **24c**: 67%, 75% ee   **24d**: 96%, 84% ee

Figure 22.15 Scope of the organocatalyzed transfer hydrogenation of 4-substituted quinolines.

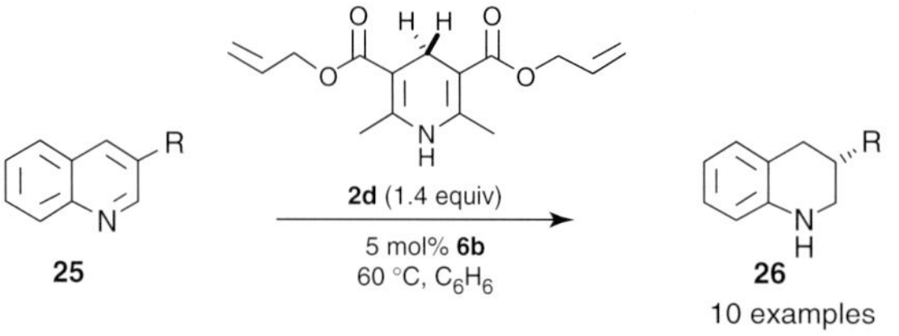

Scheme 22.17 Metal-free asymmetric reduction of 3-substituted quinolines.

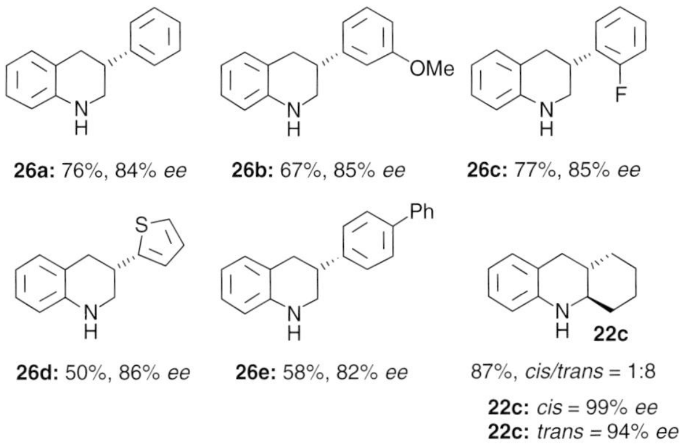

26a: 76%, 84% ee   **26b:** 67%, 85% ee   **26c:** 77%, 85% ee

26d: 50%, 86% ee   **26e:** 58%, 82% ee   87%, cis/trans = 1:8
22c: cis = 99% ee
22c: trans = 94% ee

Figure 22.16 Scope of the Brønsted acid catalyzed asymmetric protonation of 3-substituted quinolines.

catalyzed protonation (Scheme 22.11b). As in the case of related 4-substituted quinolines, application of Brønsted acid catalyst **6b** gave the best chiral induction. Among the various HEH **2** tested (R′ = R″ = Et, i-Pr, t-Bu, Bn, allyl), the allyl ester **2d** proved to be a superior hydride source, resulting in slightly better enantioselectivities.

Various 3-substituted quinolines bearing aryl and heteroaryl residues with different electronic properties could be reduced under the optimized reaction conditions to provide the corresponding tetrahydroquinolines in good yields and with high enantioselectivities (Figure 22.16). As an extension of this protocol, the octahydroacridine **22c** was obtained by reduction of the appropriate 2,3-disubstituted quinoline substrate. In contrast to the above asymmetric transfer hydrogenations of 2- and 4-substituted quinolines, which proceed through an enantioselective hydride transfer as a key step, this latest cascade involves an enantioselective Brønsted acid catalyzed proton transfer as the enantiodifferentiating step.

22.4.2
Asymmetric Brønsted Acid Catalyzed Hydrogenation of Indoles

Indolines **28** represent a common structural feature of many natural alkaloids that possess widespread biological and pharmaceutical activity. To date only the metal-catalyzed reduction of trimethyl-3H-indole has been achieved with high levels of enantiocontrol [42]. Given the structural similarities between indoles and quinolines, the application of a Brønsted acid catalyzed transfer hydrogenation to the reduction of indoles **27** seemed reasonable (Scheme 22.18).

Screening of different catalysts revealed the same Brønsted acid **5c** to be highly effective in the reduction of indoles. The reaction proved to be general and the products were obtained in high yields (54–99%) and with excellent enantioselectivities

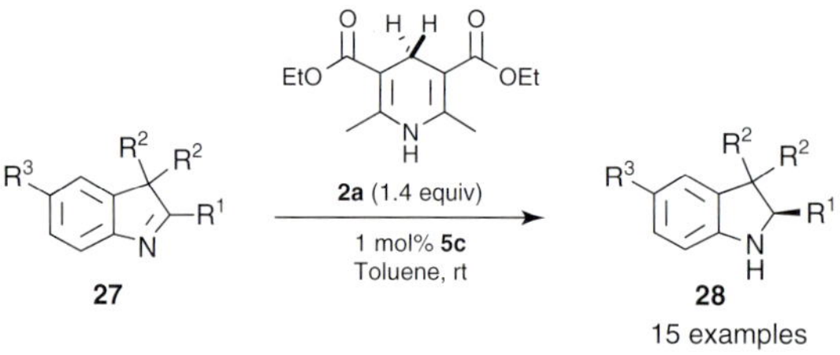

Scheme 22.18 Metal-free asymmetric transfer hydrogenation of indoles.

(70–99% ee) (Figure 22.17) [43]. In addition, lowering the catalyst loading from 1 to 0.1 mol.% resulted in a lower conversion without any significant detrimental influence on the selectivity.

22.4.3
Asymmetric Brønsted Acid Catalyzed Hydrogenation of Benzoxazines, Benzothiazines, Benzoxazinones, Quinoxalines, Quinoxalinones, Diazepines, and Benzodiazepinones

Heterocyclic compounds like dihydro-2H-benzoxazines **29**, dihydro-2H-benzothiazines **30**, dihydro-2H-quinoxalines **31** as well as dihydro-2H-benzoxazinones **32**, dihydro-2H-quinoxalinones **33**, and dihydro-2H-benzodiazepinones **34** attract considerable attention since they are common structural motifs in a large number of natural products and possess interesting biological activities (Figure 22.18). Moreover, they have often been employed as chiral building blocks in the synthesis of pharmaceuticals, for example, as promising anti-depressants, calcium antagonists, as well as anti-inflammatory, anti-nociceptive, anti-bacterial, anti-microbial agents, and as potential non-nucleoside HIV-1 reverse transcriptase inhibitors [44].

The catalytic enantioselective reduction of imine-containing heterocycles represents a direct and efficient approach to these important classes of compounds. Despite their utility, only a few catalytic systems have been reported for the reduction of these cyclic imines and they are restricted to alkyl-substituted derivates, particularly methyl- and ethyl-substituted ones [45]. Considering the success of the organocatalytic enantioselective transfer hydrogenation of imines [11a-c], quinolines [33, 34], and indoles [43], it became apparent that the bio-inspired strategy could be applicable to the transfer hydrogenation of the whole set of imine containing heterocycles (Figure 22.18). Analogous to the previously reported procedure, it was anticipated that the chiral Brønsted acid would activate the substrates through catalytic protonation, thus enabling hydride transfer from the dihydropyridine to occur (Scheme 22.19). In the case of quinoxalines **37**, a double 1,2-hydride addition has to take place to produce the desired dihydro-2H-quinoxalines **31**.

Subsequent reaction optimization showed that the highest enantioselectivities were obtained when catalyst **5b** was applied in the reduction of benzoxazines, benzothiazines, and benzoxazinones [46]. Furthermore, detailed investigation of the catalyst's activity allowed a decrease in the catalyst loading from 10 to 0.01 mol.%

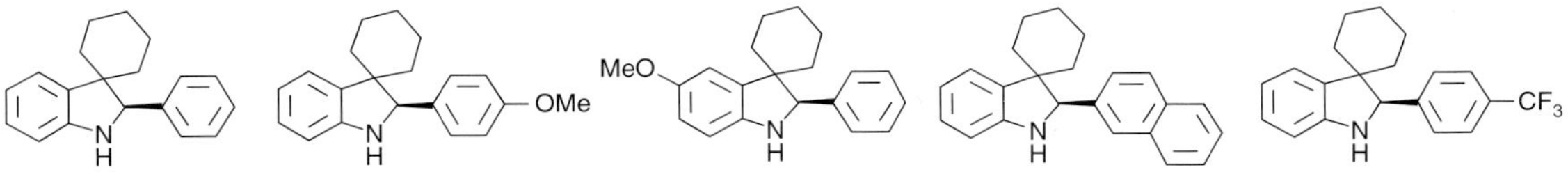

28a: 99%, 97% ee
81%, 96% ee[a]

28b: 84%, >99% ee

28c: 92%, 97% ee

28d: 86%, 99% ee

28e: 93%, >99% ee

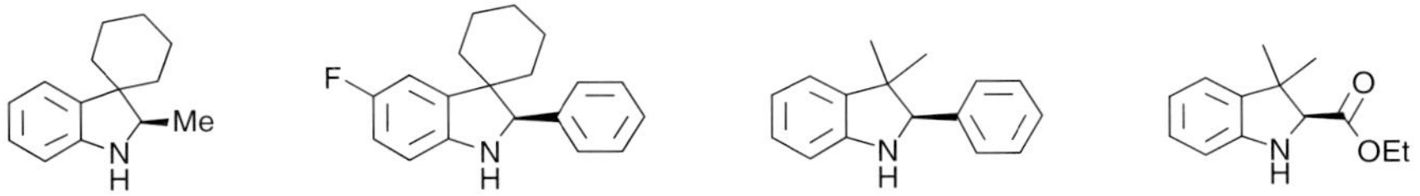

28f: 86%, 75% ee

28g: 93%, 98% ee

28h: 92%, 96% ee

28i: 87%, 90% ee

[a] Reaction performed with 0.1 mol% catalyst

Figure 22.17 Substrate scope of the Brønsted acid catalyzed transfer hydrogenation of 2,3,3-substituted indoles.

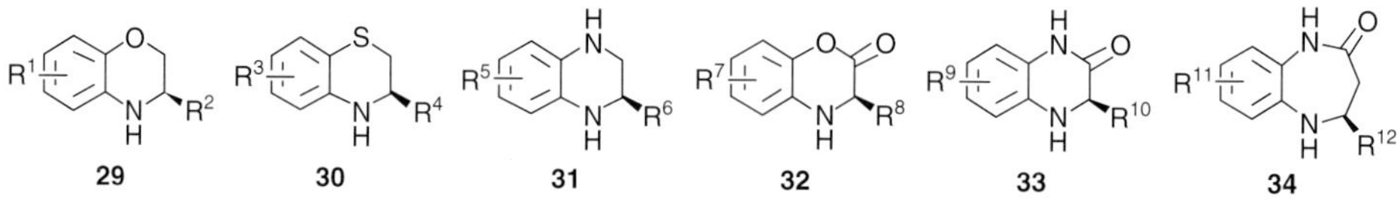

Figure 22.18 Challenging amine-containing heterocyclic classes.

(substrate : catalyst ratio = 10 000 : 1). Remarkably, the reduction of phenylbenzoxazine **35a** with only 0.01 mol.% of catalyst proceeded without much loss in reactivity and selectivity, and the corresponding 2H-dihydro-benzoxazine **29a** could be isolated in 90% yield and 93% ee. To date this is the lowest catalyst loading reported for an organocatalytic enantioselective transformation, corresponding to a turnover number (TON) of 9000 and a turnover frequency (TOF) of 500 h^{-1} (Table 22.1). Best results in terms of yield and selectivity were found in chloroform at room temperature. Accordingly, a large variety of benzoxazines were reduced at room temperature in chloroform with 0.1 mol.% catalyst to give products in high yields and excellent enantioselectivities (Figure 22.19a, **29a–f**). The excellent results obtained in this metal-free transformation along with the remarkable low catalyst loadings render this approach as a competitive alternative process to enantioselective metal-catalyzed reductions.

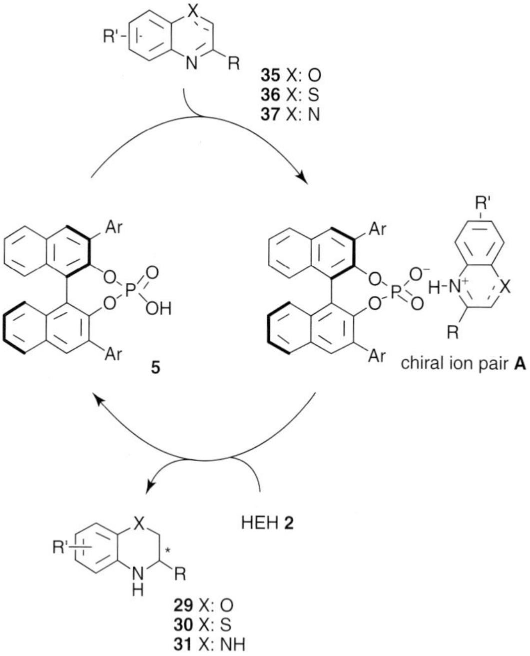

Scheme 22.19 Mechanism of the asymmetric reduction of various imine-containing heterocycles.

Table 22.1 Influence of catalyst loading on the reduction of benzoxazine **35a**.

Entry	Loading of 5b (%)	Yield (%)	ee (%)
1	10	91	96
2	5	95	96
3	2	93	96
4	1	94	96
5	0.1	95	96
6	0.01	90	93

Under similar conditions (1.25 equiv. **2a**, 1 mol.% **5b**, chloroform, room temperature) various 2-aryl-substituted benzothiazines and benzoxazinones were reduced to give products in moderate to high yields and excellent enantioselectivities (Figure 22.19b and d, **30a–f**, **32a–f**). The reduction of benzoxazinones leading to the corresponding cyclic substituted amino acid derivatives **32a–f** is of high synthetic value since the products can easily be converted into the optically active open-chain amino acids. For example, treatment of dihydrobenzoxazinone **32c** with benzylamine in the presence of pyridin-2-ol yielded the glycine derivative **38** without loss of enantiomeric excess (Scheme 22.20).

Regarding the reduction of quinoxalines, superior results were obtained when the reactions were conducted with 2.4 equiv. HEH **2a** and 10 mol.% of the Brønsted acid **5c** at 35 °C in chloroform [47]. Remarkably, a broad range of substrates, with different substitution patterns on both aryl and tetrahydroquinoxaline cores, is tolerated in the reaction (Figure 22.19c, **31a–f**). Owing to the lower solubility of quinoxalinones, a slightly modified procedure had to be applied. Products were obtained with excellent selectivities despite moderate yields (Figure 22.19e, **33a–e**).

Unfortunately, the reduction of the benzodiazepinones **39** proved to be problematic and proceeded with low conversions under standard conditions (HEH **2d**, phosphoric acid diester catalysts, toluene, 50 °C) [48]. However, this could be circumvented by employing BINOL-based N-triflylphosphoramides which, due to the strongly electron-withdrawing triflyl-group, possesses a higher reactivity (Scheme 22.21) [49a]. So far these catalysts have been used to promote asymmetric Diels–Alder reactions, dipolar cycloadditions, ene reactions, and Nazarov cyclizations in a highly efficient manner [49].

810 | 22 Bio-Inspired Transfer Hydrogenations

Figure 22.19 Substrate scope of the transfer hydrogenation of (a) benzoxazines, (b) benzothiazines, (c) quinoxalines, (d) benzoxazinones, and (e) quinoxalinones.

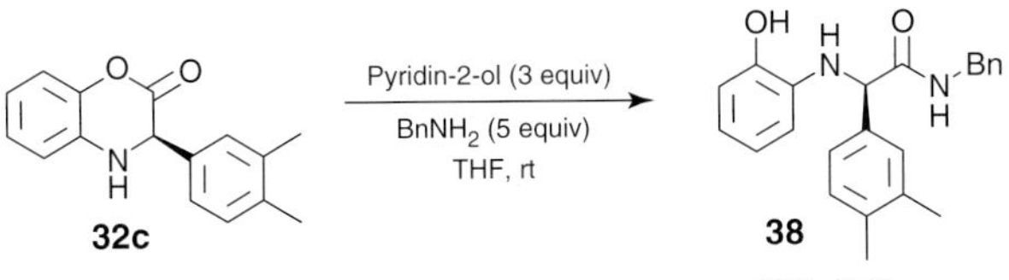

Scheme 22.20 Synthesis of the glycine derivative **38**.

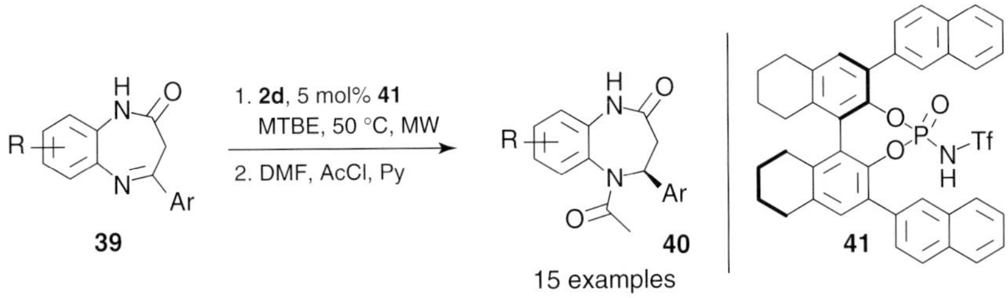

Scheme 22.21 N-Triflylphosphoramide-catalyzed asymmetric transfer hydrogenation of benzodiazepinones.

Application of 5 mol.% of the N-triflylphosphoramide **41** allowed full conversion for a limited number of substrates. A further improvement has been achieved by performing the reactions at 50 °C in methyl t-butyl ether (MTBE) under microwave irradiation. Accordingly, a broad range of benzodiazepinones **39** were reduced and subsequently acetylated to give protected dihydro-benzodiazepinones **40a–f** in moderate to excellent yields (51–95%) and with excellent enantioselectivities (83–99% ee) (Figure 22.20).

Chiral dihydro-2H-benzodiazepines have been obtained from racemic benzodiazepines via a dynamic kinetic asymmetric transfer hydrogenation process [50]. Treatment of racemic **42** with HEH **2d** in the presence of **5g** afforded products with moderate level of diastereo- and good levels of enantiocontrol (Scheme 22.22). Interestingly, the minor diastereomer has been obtained with slightly better enantioselectivity as compared to the major one. The absolute stereochemistry of the products has been assigned by X-ray crystal structure analysis.

The result has been rationalized by the pathway depicted in Scheme 22.23. It was proposed that whereas the *(S)* component of the initial racemic mixture undergoes fast reduction reaction (i) the *(R)* component participates in two distinct processes: a slow transfer hydrogenation reaction (ii) with concomitant racemization of the substrate via a sequence involving retro-Mannich and Mannich reactions (iii). The racemization was confirmed by a control experiment that showed that the starting material recovered after the reaction was racemic.

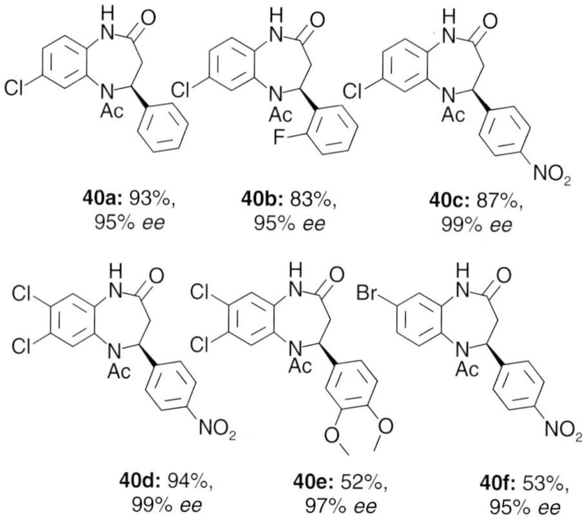

Figure 22.20 Scope of the N-triflylphosphoramide catalyzed reduction of benzodiazepinones.

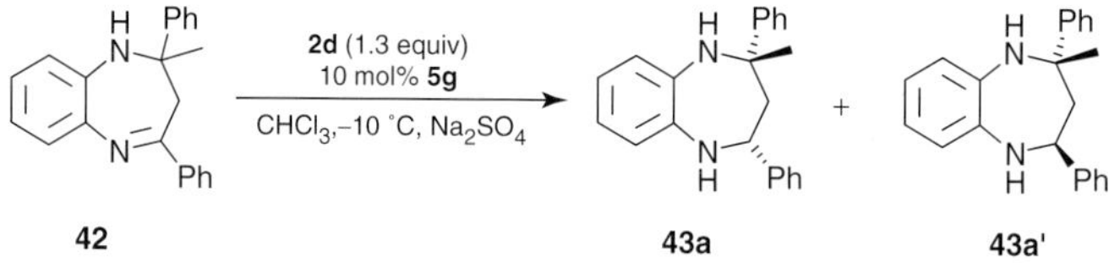

40a: 93%, 95% ee
40b: 83%, 95% ee
40c: 87%, 99% ee
40d: 94%, 99% ee
40e: 52%, 97% ee
40f: 53%, 95% ee

Scheme 22.22 Dynamic kinetic resolution in the transfer hydrogenation of benzodiazepines.

42 → (2d (1.3 equiv), 10 mol% 5g, $CHCl_3$, −10 °C, Na_2SO_4) → 43a + 43a'

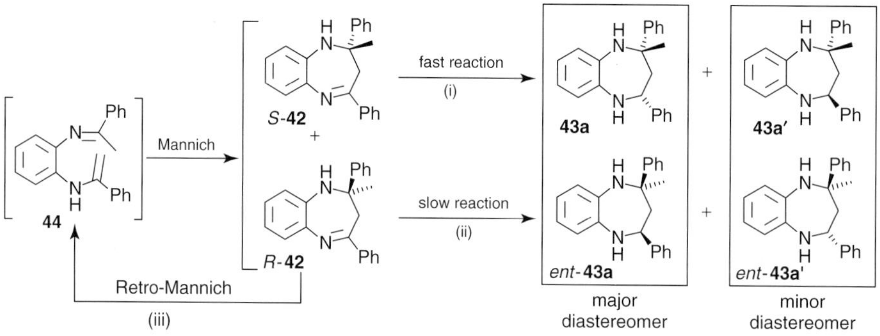

Scheme 22.23 Proposed mechanism for the dynamic kinetic transfer hydrogenation of benzodiazepines.

44 → Mannich → S-42 + R-42
S-42 → (i) fast reaction → 43a (major diastereomer) + 43a' (minor diastereomer)
R-42 → (ii) slow reaction → ent-43a (major diastereomer) + ent-43a' (minor diastereomer)
(iii) Retro-Mannich

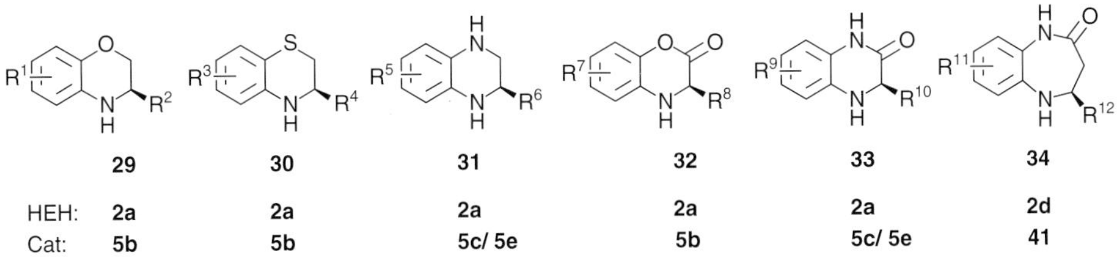

22.4 Asymmetric Organocatalytic Reduction of N-Heterocycles

[Structures 29–34 with substituents R¹–R¹² shown]

	29	30	31	32	33	34
HEH:	2a	2a	2a	2a	2a	2d
Cat:	5b	5b	5c/ 5e	5b	5c/ 5e	41

Figure 22.21 Reaction parameters for the metal-free asymmetric reduction of various imines.

Evidently, careful selection of the catalyst and reaction conditions offers access to different classes of N-heterocycles in an optically active form. Typically, enhanced reactivities and high levels of selectivities are achieved with the Hantzsch ethyl ester in combination with phenanthryl, anthracenyl, or triisopropylphenyl-substituted BINOL-phosphoric acid catalysts (Figure 22.21).

Throughout, the reductions proceed smoothly under mild conditions, providing differently alkyl-, aryl-, and heteroaryl-substituted products in good yields and with excellent enantioselectivities (Figures 22.19 and 22.20).

22.4.4
Asymmetric Organocatalytic Reduction of Pyridines

Piperidine alkaloids and derivatives belong to another important class of N-heterocycles. Piperidine is the structural core of many natural products with relevant biological and pharmaceutical properties. From a synthetic point of view, the catalytic asymmetric hydrogenation of pyridines represents the most convenient and efficient access to these compounds. However, the enantioselective reduction of substituted pyridines is a great challenge and only a few metal-catalyzed reductions are known [51].

In this context, the highly enantioselective organocatalytic protocol developed recently by Rueping constitutes a major breakthrough [52]. Screening of different Brønsted acids allowed identification of an effective catalyst for the reduction of pyridine derivatives **45** and **47** (Scheme 22.24).

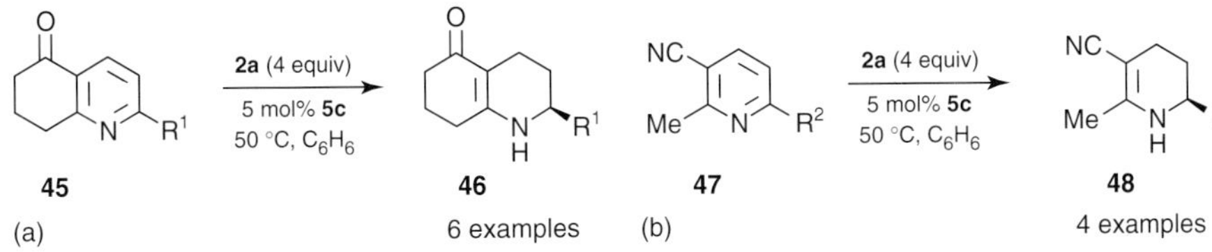

[Scheme 22.24: (a) 45 → 46 (6 examples) with 2a (4 equiv), 5 mol% 5c, 50 °C, C$_6$H$_6$; (b) 47 → 48 (4 examples) with 2a (4 equiv), 5 mol% 5c, 50 °C, C$_6$H$_6$]

Scheme 22.24 Brønsted acid transfer hydrogenation of pyridines to the corresponding piperidines.

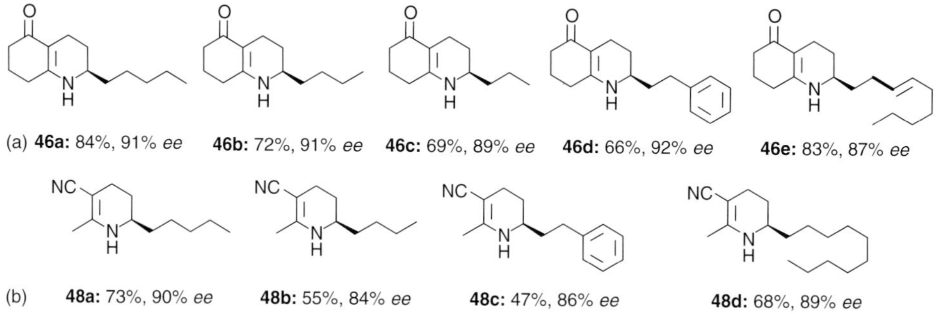

(a) 46a: 84%, 91% ee
46b: 72%, 91% ee
46c: 69%, 89% ee
46d: 66%, 92% ee
46e: 83%, 87% ee
(b) 48a: 73%, 90% ee
48b: 55%, 84% ee
48c: 47%, 86% ee
48d: 68%, 89% ee

Figure 22.22 Substrate scope of the transfer hydrogenation of pyridine-derivatives (a) **45** and (b) **47**.

A broad range of different azadecalinones **46** as well as tetrahydropyridines **48** were synthesized in good yields (47–84%) and with high enantioselectivities (84–92% ee) by employing 5 mol.% of Brønsted acid **5c** as catalyst and dihydropyridine **2a** as the hydrogen source (Figure 22.22).

Furthermore, transfer hydrogenation of pyridine **45c** with *ent*-**5c** as catalyst lead to the corresponding hexahydroquinolinone *ent*-**46c**, a valuable intermediate in the synthesis of the alkaloid diepi-pumiliotoxin C (Scheme 22.25).

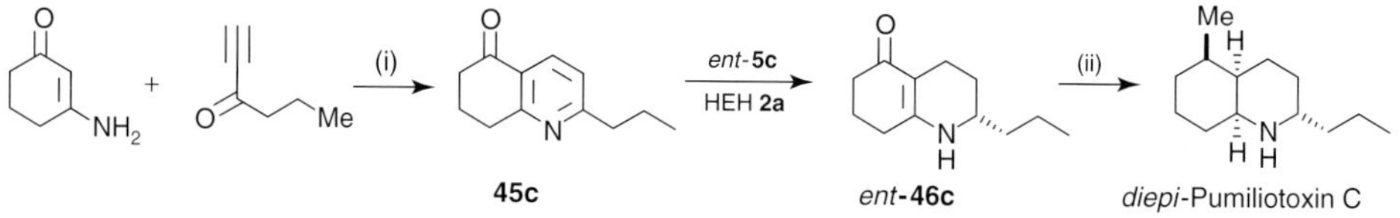

(i)
ent-5c
HEH 2a
(ii)
Me
45c
ent-46c
diepi-Pumiliotoxin C

Scheme 22.25 Transfer hydrogenation as key step in the synthesis of diepi-pumiliotoxin C: (i) EtOH, 50 °C, 12 h, then 140 °C, 2 h [53]; (ii) Reference [54e].

This new Brønsted acid catalyzed cascade reduction of pyridines gives the corresponding products in good yields and with excellent enantioselectivities. It is particularly noteworthy since it provides a simple and straightforward route to decahydroquinoline or piperidine alkaloid natural products [54, 55].

22.5 Asymmetric Organocatalytic Reductions in Cascade Sequences

As illustrated above, several efficient protocols that are able to mimic Nature's reduction by simply replacing the dehydrogenase and NADH system with a combination of a readily available Brønsted acid and a dihydropyridine were developed recently. In addition to promoting the reactions with high levels of stereocontrol, enzymes can also build up sophisticated structures starting from simple compounds. In Nature, the multistep sequences are often realized in domino- and multicomponent reactions, and are, thus, blueprints for organic synthesis [56].

22.5 Asymmetric Organocatalytic Reductions in Cascade Sequences

As part of their ongoing studies on chiral Brønsted acid catalysis, Rueping and coworkers have developed recently an asymmetric organocatalytic cascade reaction in which multiple steps are catalyzed by the same chiral Brønsted acid catalyst. This provides valuable tetrahydropyridines **48** and azadecalinones **46** with high enantioselectivities [57]. Taking advantage of their experience in the field of chiral ion pair catalysis and based on the initial biomimetic strategy, they envisioned a new organocatalytic multiple reaction cascade consisting of a one pot Michael addition–isomerization–cyclization–elimination–isomerization–transfer hydrogenation sequence in which every single step is catalyzed by the same chiral Brønsted acid (Scheme 22.26).

It was assumed that exposure of a mixture of enamine **49** and enone **50** to catalytic amounts of Brønsted acid would afford the corresponding 1,4-addition products **A** and **B** (Scheme 22.27). Subsequent Brønsted acid catalyzed cyclization

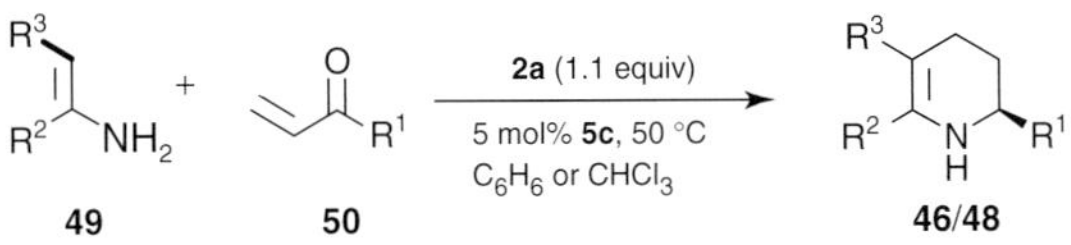

Scheme 22.26 Brønsted acid catalyzed cascade reaction.

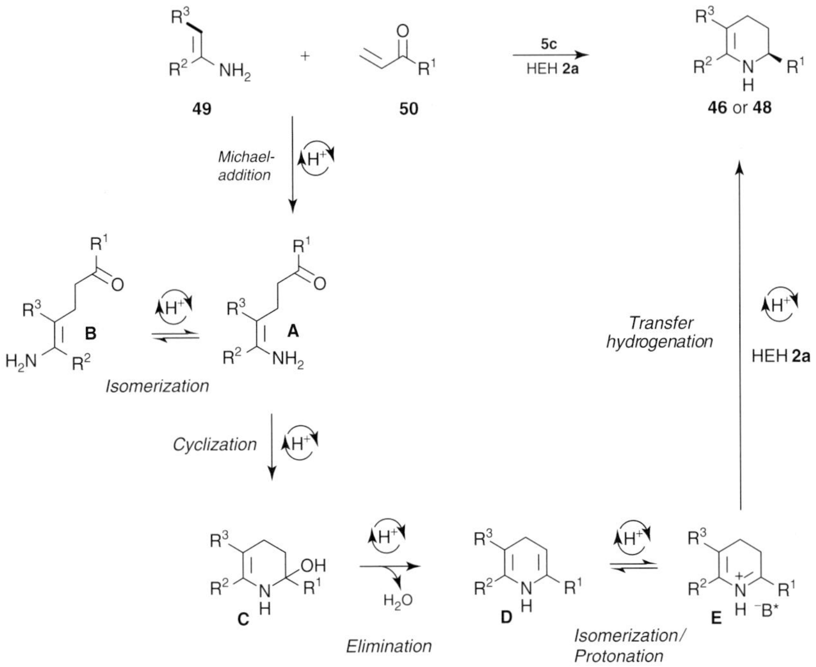

Scheme 22.27 Mechanism of the Brønsted acid catalyzed multistep reaction sequence.

of **A**, which under the reaction conditions is in equilibrium with **B**, would give the hemiaminal **C**, which is expected to readily eliminate water and form the dihydropyridine (**D**), an intermediate observed also in the asymmetric pyridine reduction. In the next step, Brønsted acid catalyzed protonation should generate an iminium ion, a chiral ion pair **E** enabling the final step, namely, enantioselective hydride transfer to give the desired product **46** or **48**, to take place. To promote this multiple reaction cascade as a powerful strategy for the asymmetric synthesis, it is necessary to obtain a high level of stereocontrol in the last step of the sequence.

Remarkably, the transformation could be accomplished by applying the ideal combination of Brønsted acid **5c** and dihydropyridine **2a** [57a]. Notably, the same chiral Brønsted acid catalyzes all six steps in this new three component reaction, allowing rapid, direct, and efficient access to valuable tetrahydropyridines and azadecalinones with excellent levels of enantiocontrol (89–99% ee) from simple readily available starting materials (Figure 22.23). This proves that the organocatalytic hydrogenation protocol is amenable to complex molecular cascading.

Another concept making use of both amine and Brønsted acid catalysis was developed by List and coworkers [58]. Starting from linear diketones **51** and achiral amines **52**, various substituted cyclohexylamines **53** were obtained with high diastereo- and enantioselectivities (2 : 1 to 99 : 1 dr, 82–96% ee) (Figure 22.24). The reaction proceeds via an amine-catalyzed intramolecular aldol condensation

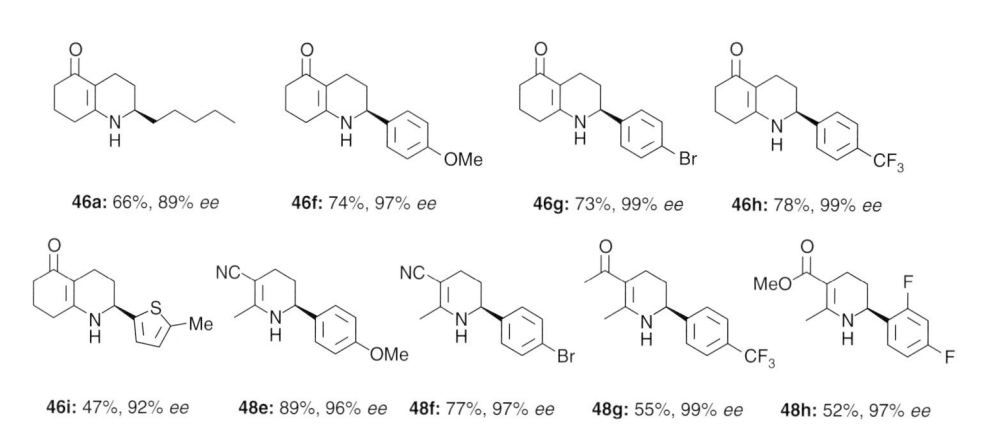

46a: 66%, 89% ee   **46f:** 74%, 97% ee   **46g:** 73%, 99% ee   **46h:** 78%, 99% ee

46i: 47%, 92% ee   **48e:** 89%, 96% ee   **48f:** 77%, 97% ee   **48g:** 55%, 99% ee   **48h:** 52%, 97% ee

Figure 22.23 Scope of the Brønsted acid catalyzed cascade reaction.

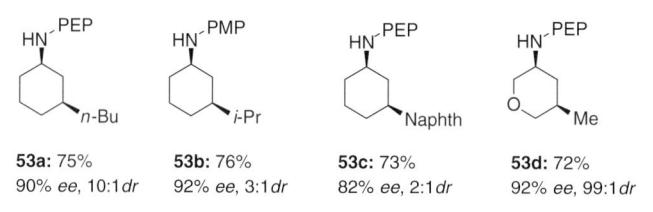

53a: 75%
90% ee, 10:1 dr

53b: 76%
92% ee, 3:1 dr

53c: 73%
82% ee, 2:1 dr

53d: 72%
92% ee, 99:1 dr

Figure 22.24 Scope of the Brønsted acid catalyzed cascade reaction to substituted amines.

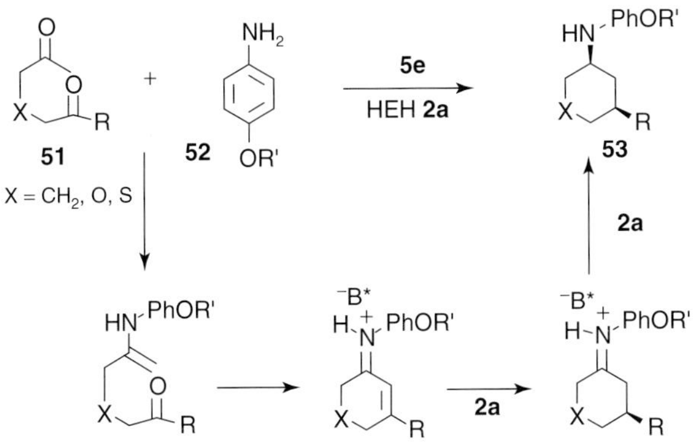

Scheme 22.28 Synthesis of amine derivatives via a Brønsted acid catalyzed cascade sequence.

and subsequent Brønsted acid catalyzed conjugate reduction–imine reduction sequence (Scheme 22.28). Recent mass spectrometry studies in which all crucial intermediates were detected strongly support the proposed catalytic cycle [59]. Additionally, this study showed that electrospray ionization mass spectroscopy (ESI-MS) is a suitable technique for the determination of reaction pathways in Brønsted acid catalysis and is likely to be used in future mechanistic investigations.

22.6 Conclusion

As illustrated in this chapter, the bio-inspired transfer hydrogenation developed by Rueping represents the key starting point for the development of many powerful reductions, and has led to the rapidly growing field of organocatalyzed transfer hydrogenation. The generality of the BINOL-phosphate/dihydropyridine combination, when compared to Nature's dehydrogenase/NADH system, renders this newly developed method as particularly noteworthy and allows a broad range of diverse cyclic and acyclic amines as well as heterocycles to be obtained with impressive levels of enantioselectivity. Notably, with these new protocols even the synthesis of challenging substrates can be successfully addressed. The mild reaction conditions and the operational simplicity and practicability render these methods as essential tools in the chemist's toolbox. Moreover, analogous to Nature's multicomponent domino reactions, the catalytic asymmetric Brønsted acid catalyzed transfer hydrogenation is effective in multistep reaction sequences, showing that this protocol can be used in complex molecular cascades. Although high to exceptional levels of selectivity and reactivity are obtained for a broad range of substrates, there is still room for improvement. It is desirable that the hydride source dihydropyridine employed in these transformations can be recycled or used in catalytic amounts in a similar way to Nature's NADH. We are confident that this metal-free transfer hydrogenation procedure will find widespread application in organic synthesis and will, at the same time, inspire the chemistry community to design more powerful organocatalytic systems.

References

1. (a) Ojima, I. (2000) *Catalytic Asymmetric Synthesis*, 2nd edn, Wiley-VCH Verlag GmbH, Weinheim; (b) Jacobsen, E.N., Pfaltz, A., and Yamamoto, H. (eds) (2000) *Comprehensive Asymmetric Catalysis*, vols 1–3, 2nd edn, Springer, Berlin; (c) Asymmetric catalysis special feature issues: (2004) *Proc. Natl. Acad. Sci. U.S.A.*, **101**(15–16); (d) Blaser, H.U. and Schmidt, E. (eds) (2004) *Asymmetric Catalysis on Industrial Scale*, Wiley-VCH Verlag GmbH, Weinheim; (e) Beller, M. and Bolm, C. (eds) (2004) *Transition Metals for Organic Synthesis Building Blocks and Fine Chemicals*, vols 1–2, 2nd edn, Wiley-VCH Verlag GmbH, Weinheim; (f) Enders, D. and Jaeger, K.-E. (eds) (2007) *Asymmetric Synthesis with Chemical and Biological Methods*, Wiley-VCH Verlag GmbH, Weinheim; (g) Mikami, K. and Lautens, M. (eds) (2007) *New Frontiers in Asymmetric Catalysis*, John Wiley & Sons, Inc., Hoboken, NJ.
2. (a) Andersson, P.G. and Munslow, I.J. (eds) (2008) *Modern Reduction Methods*, Wiley-VCH Verlag GmbH, Weinheim; (b) de Vries, J.G. and Elsevier, C.J. (eds) (2007) *The Handbook of Homogeneous Hydrogenation*, Wiley-VCH Verlag GmbH, Weinheim; (c) Tang, W. and Zhang, X. (2003) *Chem. Rev.*, **103**, 3029–3070.
3. Stillman, T.J., Baker, P.J., Britton, K.L., and Rice, D.W. (1990) *J. Mol. Biol.*, **234**, 1131–1139.
4. Dean, J.L.E., Wang, X.G., Teller, J.K., Waugh, M.L., Britton, K.L., Baker, P.J., Stillman, T.J., Martin, S.R., Rice, D.W., and Engel, P.C. (1994) *Biochem. J.*, **301**, 13–16.
5. Hantzsch, A. (1882) *Justus Liebigs Ann. Chem.*, **215**, 1–82.
6. Mauzerall, D. and Westheimer, F.H. (1955) *J. Am. Chem. Soc.*, **77**, 2261–2264.
7. (a) Steevens, J.B. and Pandit, U.K. (1983) *Tetrahedron*, **39**, 1395–1400; (b) Nakamura, K., Fujii, M., Ohno, A., and Oka, S. (1984) *Tetrahedron Lett.*, **25**, 3983–3986; (c) Vanniel, J.C.G. and Pandit, U.K. (1985) *Tetrahedron*, **41**, 6005–6011; (d) Watanabe, M., Fushimi, M., Baba, N., Oda, J., and Inouye, Y. (1985) *Agric. Biol. Chem.*, **49**, 3533–3538; (e) Vanniel, J.C.G., Kort, C.W.F., and Pandit, U.K. (1986) *Recl. Trav. Chim. Pays-Bas*, **105**, 262–265; (f) Fujii, M. (1988) *Bull. Chem. Soc. Jpn.*, **61**, 4029–4035; (g) Fujii, M.Y., Aida, T., Yoshihara, M.K., and Ohno, A.Y. (1989) *Bull. Chem. Soc. Jpn.*, **62**, 3845–3847; (h) Singh, S. and Batra, U.K. (1989) *Indian J. Chem. Sect. B: Org. Chem. Incl. Med. Chem.*, **28**, 1–2; (i) Zhu, X.Q., Liu, Y.C., and Cheng, J.P. (1999) *J. Org. Chem.*, **64**, 8980–8981; (j) Zhu, X.Q., Wang, H.Y., Wang, J.S., and Liu, Y.C. (2001) *J. Org. Chem.*, **66**, 344–347; (k) Itoh, T., Nagata, A., Kurihara, A., Miyazaki, M., and Ohsawa, A. (2002) *Tetrahedron Lett.*, **43**, 3105–3108; (l) Itoh, T., Nagata, K., Miyazaki, M., Ishikawa, H., Kurihara, A., and Ohsawa, A. (2004) *Tetrahedron*, **60**, 6649–6655; (m) Liu, Z.G., Han, B., Liu, Q., Zhang, W., Yang, L., Liu, Z.L., and Yu, W. (2005) *Synlett*, 1579–1580; (n) Liu, Z.G., Liu, Q., Zhang, W., Mu, R.Z., Yang, L., Liu, Z.L., and Yu, W. (2006) *Synthesis*, 771–774; (o) Menche, D. and Arikan, F. (2006) *Synlett*, 841–844; (p) Menche, D., Hassfeld, J., Li, J., Menche, G., Ritter, A., and Rudolph, S. (2006) *Org. Lett.*, **8**, 741–744; (q) Zhang, Z.G. and Schreiner, P.R. (2007) *Synlett*, 1455–1457; (r) Wang, D.W., Zeng, W., and Zhou, Y.G. (2007) *Tetrahedron: Asymmetry*, **18**, 1103–1107; (s) Shen, X.X., Liu, Q., Xing, R.G., and Zhou, B. (2008) **126**, 361–366; (t) Goswami, P., Ali, S., Khan, M.M.,

and Das, B. (2008) *Lett. Org. Chem.*, **5**, 659–664; (u) Liu, Q., Li, J., Shen, X.X., Xing, R.G., Yang, J., Liu, Z.G., and Zhou, B. (2009) *Tetrahedron Lett.*, **50**, 1026–1028; (v) Liu, X.Y. and Che, C.M. (2009) *Org. Lett.*, **11**, 4204–4207; (w) Richter, D. and Mayr, H. (2009) *Angew. Chem., Int. Ed.*, **48**, 1958–1961.

8. For reviews on transfer hydrogenation performed with HEH, see: (a) Ouellet, S.G., Walji, A.M., and MacMillan, D.W.C. (2007) *Acc. Chem. Res.*, **40**, 1327–1339; (b) You, S.L. (2007) *Chem. Asian J.*, **2**, 820–827; (c) Connon, S.J. (2007) *Org. Biomol. Chem.*, **5**, 3407–3417; (d) Wang, C., Wu, X.F., and Xiao, J.L. (2008) *Chem. Asian J.*, **3**, 1750–1177.

9. (a) Ouellet, S.G., Tuttle, J.B., and MacMillan, D.W.C. (2005) *J. Am. Chem. Soc.*, **127**, 32–33; (b) Huang, Y., Walji, A.M., Larsen, C.H., and MacMillan, D.W.C. (2005) *J. Am. Chem. Soc.*, **127**, 15051–15053; (c) for related enone reduction: Tuttle, J.B., Ouellet, S.G., and MacMillan, D.W.C. (2006) *J. Am. Chem. Soc.*, **128**, 12662–12663.

10. (a) Yang, J.W., Fonseca, M.T.H., and List, B. (2004) *Angew. Chem., Int. Ed.*, **43**, 6660–6662; (b) Yang, J.W., Fonseca, M.T.H., Vignola, N., and List, B. (2005) *Angew. Chem., Int. Ed.*, **44**, 108–110; for related reductions: (c) Yang, J.W., Fonseca, M.T.H., and List, B. (2005) *J. Am. Chem. Soc.*, **127**, 15036–15037; (d) Martin, N.J.A. and List, B. (2006) *J. Am. Chem. Soc.*, **128**, 13368–13369.

11. Rueping, M., Azap, C., Sugiono, E., and Theissmann, T. (2005) *Synlett*, 2367–2369; (b) Rueping, M., Sugiono, E., Azap, C., Theissmann, T., and Bolte, M. (2005) *Org. Lett.*, **7**, 3781–3783; (c) Rueping, M., Sugiono, E., Azap, C., and Theissmann, T. (2007) in *Catalysts for Fine Chemical Synthesis*, vol. 5 (eds S.M. Roberts and J. Whittall), John Wiley & Sons, Ltd, Chichester, pp. 162–170; (d) Rueping, M. and Sugiono, E. (2008) in *Ernst Schering Foundation Symposium Proceedings*, vol. 2 (eds M.T. Reetz, S. List, S. Jaroch, and H. Weinmann), Springer, Berlin, Heidelberg, pp. 207–253; (e) Rueping, M., Sugiono, E., and Schoepke, F.R. (2010) *Synlett*, 852–865.

12. (a) Schreiner, P.R. (2003) *Chem. Soc. Rev.*, **32**, 289–296; (b) Bolm, C., Rantanen, T., Schiffers, I., and Zani, L. (2005) *Angew. Chem., Int. Ed.*, **44**, 1758–1763.

13. (a) Akiyama, T., Itoh, J., and Fuchibe, K. (2006) *Adv. Synth. Catal.*, **348**, 999–1010; (b) Akiyama, T. (2007) *Chem. Rev.*, **107**, 5744–5758; (c) Connon, S.J. (2006) *Angew. Chem., Int. Ed.*, **45**, 3909–3912; (d) Terada, M. (2008) *Chem. Commun.*, 4097–4112; (e) Yamamoto, H. and Payette, N. (2009) in *Hydrogen Bonding in Organic Synthesis* (ed. P.M. Pihko), Wiley-VCH Verlag GmbH, Weinheim, pp. 73–140.

14. (a) Wilen, S.H., Qi, J.Z., and Williard, P.G. (1991) *J. Org. Chem.*, **56**, 485–487; (b) Fujii, I. and Hirayama, N. (2002) *Helv. Chim. Acta*, **85**, 2946–2960.

15. (a) Inanaga, J., Sugimoto, Y., and Hanamoto, T. (1995) *New J. Chem.*, **19**, 707–712; (b) Furono, H., Hanamoto, T., Sugimoto, Y., and Inanaga, J. (2000) *Org. Lett.*, **2**, 49–52.

16. (a) Akiyama, T., Itoh, J., Yokota, K., and Fuchibe, K. (2004) *Angew. Chem., Int. Ed.*, **43**, 1566–1568; (b) Akiyama, T. (2004) PCT Int. Appl. WO 200409675.

17. (a) Uraguchi, D. and Terada, M. (2004) *J. Am. Chem. Soc.*, **126**, 5356–5357; (b) Terada, M., Uraguchi, D., Sorimachi, K., and Shimizu, H. (2005) PCT Int. Appl. WO 2005070875.

18. Simonsen, K.B., Gothelf, K.V., and Jørgensen, K.A. (1998) *J. Org. Chem.*, **63**, 7536–7538.

19. Bartoszek, M., Beller, M., Deutsch, J., Klawonn, M., Kockritz, A., Nemati, N., and Pews-Davtyan, A. (2008) *Tetrahedron*, **64**, 1316–1322.

20. Zhu, S.S., Cefalo, D.R., La, D.S., Jamieson, J.Y., Davis, W.M., Hoveyda, A.H., and Schrock, R.R. (1999) *J. Am. Chem. Soc.*, **121**, 8251–8259.

21. Maruoka, K., Itoh, T., Araki, Y., Shirasaka, T., and Yamamoto, H. (1988) *Bull. Chem. Soc. Jpn.*, **61**, 2975–2976.

22. Storer, R.I., Carrera, D.E., Ni, Y., and MacMillan, D.W.C. (2006) *J. Am. Chem. Soc.*, **128**, 84–86.

23. Hoffmann, S., Seayad, A.M., and List, B. (2005) *Angew. Chem., Int. Ed.*, **44**, 7424–7427.
24. Hoffmann, S., Nicoletti, M., and List, B. (2005) *J. Am. Chem. Soc.*, **128**, 13074–13075.
25. Marcelli, T., Hammar, P., and Himo, F. (2008) *Adv. Synth. Catal.*, **351**, 525–529.
26. Zhu, C. and Akiyama, T. (2009) *Org. Lett.*, **11**, 4180–4183.
27. Kang, Q., Zhao, Z.A., and You, S.L. (2007) *Adv. Synth. Catal.*, **349**, 1657–1660.
28. Li, G.L., Liang, Y.X., and Antilla, J.C. (2007) *J. Am. Chem. Soc.*, **129**, 5830–5831.
29. Kang, Q., Zhao, Z.A., and You, S.L. (2007) *Org. Lett.*, **10**, 2031–2034.
30. Li, G.L. and Antilla, J.C. (2009) *Org. Lett.*, **11**, 1075–1078.
31. (a) Marcelli, T., Hammar, P., and Himo, F. (2008) *Chem. – Eur. J.*, **14**, 8562–8571; (b) Simon, L. and Goodman, J.M. (2008) *J. Am. Chem. Soc.*, **130**, 8741–8747.
32. (a) Glorius, F. (2005) *Org. Biomol. Chem.*, **3**, 4171–4175; (b) Wang, W.B., Lu, S.M., Yang, P.Y., Han, X.W., and Zhou, Y.G. (2003) *J. Am. Chem. Soc.*, **125**, 10536–10537; (c) Lu, S.M., Han, X.W., and Zhou, Y.G. (2004) *Adv. Synth. Cat.*, **346**, 909–912; (d) Yang, P.Y. and Zhou, Y.G. (2004) *Tetrahedron: Asymmetry*, **15**, 1145–1149; (e) Xu, L.K., Lam, K.H., Ji, J.X., Wu, J., Fan, Q.H., Lo, W.H., and Chan, A.S.C. (2005) *Chem. Commun.*, 1390–1392; (f) Reetz, M.T. and Li, X.G. (2006) *Chem. Commun.*, 2159–2160; (g) Han, Z.Y., Xiao, H., Chen, X.H., and Gong, L.Z. (2009) *J. Am. Chem. Soc.*, **131**, 9182–9183.
33. Rueping, M., Theissmann, T., and Antonchick, A.P. (2006) *Synlett*, 1071–1074.
34. (a) Rueping, M., Antonchick, A.P., and Theissmann, T. (2006) *Angew. Chem., Int. Ed.*, **45**, 3683–3686; (b) Rueping, M., Theissmann, T., and Antonchick, A.P. (2007) in *Catalysts for Fine Chemical Synthesis*, vol. 5 (eds S.M. Roberts and J. Whittall), John Wiley & Sons, Ltd, Chichester, pp. 170–174.
35. (a) Rakotoson, J.H., Fabre, N., Jacquemond-Collet, I., Hannedouche, S., Fouraste, I., and Moulis, C. (1998) *Planta Med.*, **64**, 762–763; (b) Jacquemond-Collet, I., Hannedouche, S., Fabre, N., Fouraste, I., and Moulis, C. (1999) *Phytochemistry*, **51**, 1167–1169; (c) Houghton, P.J., Woldemariam, T.Z., Watanabe, Y., and Yates, W. (1999) *Planta Med.*, **65**, 250–254.
36. Metallinos, C., Barrett, F.B., and Xu, S. (2008) *Synlett*, 720–724.
37. Guo, Q.S., Du, D.M., and Xu, J. (2008) *Angew. Chem., Int. Ed.*, **47**, 759–762.
38. Mani, N.S. and Wu, M. (2000) *Tetrahedron: Asymmetry*, **11**, 4687–4691.
39. (a) Higuchi, R.I., Edwards, J.P., Caferro, T.R., Ringgenberg, J.D., Kong, J.W., Hamann, L.G., Arienti, K.L., Marschke, K.B., Davis, R.L., Farmer, L.J., and Jones, T.K. (1999) *Bioorg. Med. Chem. Lett.*, **9**, 1335–1340; (b) Hamann, L.G., Mani, N.S., Davis, R.L., Wang, X.N., Marschke, K.B., and Jones, T.K. (1999) *J. Med. Chem.*, **42**, 210–212; (c) Edwards, J.P., Higuchi, R.I., Winn, D.T., Pooley, C.L.F., Caferro, T.R., Hamann, L.G., Zhi, L., Marschke, K.B., Goldman, M.E., and Jones, T.K. (1999) *Bioorg. Med. Chem. Lett.*, **9**, 1003–1008.
40. Rueping, M., Stöckel, M., and Theissmann, T. manuscript submitted for publication.
41. Rueping, M., Theissmann, T., Raja, S., and Bats, J.W. (2008) *Adv. Synth. Catal.*, **350**, 1001–1006.
42. (a) Liu, D., Li, W., and Zhang, X. (2004) *Tetrahedron: Asymmetry*, **15**, 2181–2184; (b) Qiu, L.Q., Kwong, F.Y., Wu, J., Lam, W.H., Chan, S., Yu, W.Y., Li, Y.M., Guo, R.W., Zhou, Z., and Chan, A.S.C. (2006) *J. Am. Chem. Soc.*, **128**, 5955–5965; (c) Blaser, H.U., Buser, H.P., Hausel, R., Jalett, H.P., and Spindler, F. (2001) *J. Organomet. Chem.*, **621**, 34–38.
43. Rueping, M., Brinkmann, C., Antonchick, A.P., and Atodiresei, I. (2010) *Org. Lett.*, **12**, 4604–4607.
44. (a) Belattar, A. and Saxton, J.E. (1992) *J. Chem. Soc., Perkin Trans. 1*, 679–683; (b) Krohn, H., Kirst, H.A., and Maag, H. (eds) (1993) *Antibiotics and Antiviral Compounds*, Wiley-VCH Verlag GmbH, Weinheim; (c) Kleemann, A., Engel, J., Kutscher, B., and Reichert, D. (eds) (2001) *Pharmaceutical Substances*, 4th edn, Thieme, Stuttgart, New York;

(d) Achari, B., Mandal, S.B., Dutta, P.K., and Chowdhury, C. (2004) *Synlett*, 2449–2467; (e) Fantin, M., Marti, M., Auberson, Y.P., and Morari, M. (2007) *J. Neurochem.*, **103**, 2200–2211; (f) Tenbrink, R.E., Im, W.B., Sethy, V.H., Tang, A.H., and Carter, D.B. (1994) *J. Med. Chem.*, **37**, 758–768; (g) Borrok, M.J. and Kiessling, L.L. (2007) *J. Am. Chem. Soc.*, **129**, 12780–12785; (h) Cass, L.M., Moore, K.H.P., Dallow, N.S., Jones, A.E., Sisson, J.R., and Prince, W.T. (2001) *J. Clin. Pharmacol.*, **41**, 528–535.

45. (a) Satoh, K., Inenaga, M., and Kanai, K. (1998) *Tetrahedron: Asymmetry*, **9**, 2657–2662; (b) Noyori, R. (1996) *Acta Chem. Scand.*, **50**, 380–390; (c) Zhou, Y.G., Yang, P.Y., and Han, X.W. (2005) *J. Org. Chem.*, **70**, 1679–1683; (d) Krchnak, V., Smith, J., and Vagner, J. (2001) *Tetrahedron Lett.*, **42**, 2443–2446; (e) Lee, J., Murray, W.V., and Rivero, R.A. (1997) *J. Org. Chem.*, **62**, 3874–3387; (f) Morales, G.A., Corbett, J.W., and DeGrado, W.F. (1998) *J. Org. Chem.*, **63**, 1172–1177; (g) Zaragoza, F. and Stephensen, H. (1999) *J. Org. Chem.*, **64**, 2555–2557; (h) Ilas, J., Anderluh, P.S., Dolenc, M.S., and Kikelj, D. (2005) *Tetrahedron*, **61**, 7325–7348.

46. Rueping, M., Antonchick, A.P., and Theissmann, T. (2006) *Angew. Chem., Int. Ed.*, **45**, 6751–6755.

47. Rueping, M., Tato, F., and Schoepke, F.R. (2009) *Chem. Eur. J.*, **16**, 2688–2691.

48. Rueping, M., Merino, E., and Koenigs, R.M. (2010) *Adv. Synth. Catal.*, **352**, 2629–2634.

49. (a) Nakashima, D. and Yamamoto, H. (2006) *J. Am. Chem. Soc.*, **128**, 9626–9627; (b) Rueping, M., Ieaswuwan, W., Antonchick, A.P., and Nachtsheim, B.J. (2007) *Angew. Chem., Int. Ed.*, **46**, 2097–2100; (c) Enders, D., Huttl, M.R.M., Runsink, J., Raabe, G., and Wendt, B. (2007) *Angew. Chem., Int. Ed.*, **46**, 467–469; (d) Jiao, P., Nakashima, D., and Yamamoto, H. (2008) *Angew. Chem., Int. Ed.*, **47**, 2411–2413; (e) Rueping, M., Nachtsheim, B.J., Moreth, S.A., and Bolte, M. (2008) *Angew. Chem., Int. Ed.*, **47**, 593–596; (f) Rueping, M., Theissmann, T., Kuenkel, A., and Koenigs, R.M. (2008) *Angew. Chem., Int. Ed.*, **47**, 6798–6801; (g) Rueping, M. and Ieaswuwan, W. (2009) *Adv. Synth. Catal.*, **351**, 78–84.

50. Han, Z.-Y., Xiao, H., and Gong, L.-Z. (2009) *Bioorg. Med. Chem. Lett.*, **19**, 3729–3732.

51. (a) Legault, C.Y. and Charette, A.B. (2005) *J. Am. Chem. Soc.*, **127**, 8966–8967; (b) Lei, A.W., Chen, M., He, M.S., and Zhang, X.M. (2006) *Eur. J. Org. Chem.*, 4343–4347; (c) Glorius, F., Spielkamp, N., Holle, S., Goddard, R., and Lehmann, C.W. (2004) *Angew. Chem., Int. Ed.*, **43**, 2850–2852.

52. Rueping, M. and Antonchick, A.P. (2007) *Angew. Chem., Int. Ed.*, **46**, 4562–4565.

53. (a) Bohlmann, F. and Rahtz, D. (1957) *Chem. Ber.*, **90**, 2265–2272; (b) Bagley, M.C., Brace, C., Dale, J.W., Ohnesorge, M., Phillips, N.G., Xiong, X., and Bower, J. (2002) *J. Chem. Soc., Perkin Trans. 1*, 1663–1671.

54. Synthesis of gephyrotoxin and pumiliotoxin: (a) Fujimoto, R., Kishi, Y., and Blount, J.F. (1980) *J. Am. Chem. Soc.*, **102**, 7154–7156; (b) Ito, Y., Nakajo, E., Nakatsuka, M., and Saegusa, T. (1983) *Tetrahedron Lett.*, **24**, 2881–2884; (c) Pearson, W.H. and Fang, W.-K. (2000) *J. Org. Chem.*, **65**, 7158–7174; (d) Wei, L.-L., Hsung, R.P., Sklenicka, H.M., and Gerasyuto, A.I. (2001) *Angew. Chem., Int. Ed.*, **40**, 1516–1518; (e) Sklenicka, H.M., Hsung, R.P., McLaughlin, M.J., Wie, L.-L., Gerasyuto, A.I., and Brennessel, W.B. (2002) *J. Am. Chem. Soc.*, **124**, 10435–10442.

55. (a) Daly, J.W. (1998) *J. Nat. Prod.*, **61**, 162–172; (b) O'Hagan, D. (2000) *Nat. Prod. Rep.*, **17**, 435–446; (c) Daly, J.W., Spande, T.F., and Garraffo, H.M. (2005) *J. Nat. Prod.*, **68**, 1556–1575; (d) Michael, J.P. (2005) *Nat. Prod. Rep.*, **22**, 603–626.

56. (a) Tietze, L.F. (1996) *Chem. Rev.*, **96**, 115–136; (b) Tietze, L.F., Brasche, G., and Gericke, K. (eds)

(2007) *Domino Reactions in Organic Synthesis*, Wiley-VCH Verlag GmbH, Weinheim; (c) Enders, D., Grondal, C., and Hüttl, M.R.M. (2007) *Angew. Chem., Int. Ed.*, **46**, 1570–1581; (d) Alba, A.N., Companyo, X., Viciano, M., and Rios, R. (2009) *Curr. Org. Chem.*, **13**, 1432–1474; (e) Grondal, C., Jeanty, M., and Enders, D. (2010) *Nat. Chem.*, **2**, 167–178.

57. (a) Rueping, M. and Antonchick, A.P. (2008) *Angew. Chem., Int. Ed.*, **47**, 5836–5838; For further domino reactions from the same group: (b) Rueping, M., Sugiono, E., and Merino, E. (2008) *Angew. Chem., Int. Ed.*, **47**, 3046–3049; (c) Rueping, M., Sugiono, E., and Merino, E. (2008) *Chem. – Eur. J.*, **14**, 6329–6332; (d) Rueping, M., Merino, E., and Sugiono, E. (2008) *Adv. Synth. Catal.*, **350**, 2127–2131; (e) Rueping, M., Kuenkel, A., Tato, F., and Bats, J.W. (2009) *Angew. Chem., Int. Ed.*, **48**, 3699–3702.

58. Zhou, J. and List, B. (2007) *J. Am. Chem. Soc.*, **129**, 7498–7499.

59. Schrader, W., Handayani, P.P., Zhou, J., and List, B. (2009) *Angew. Chem., Int. Ed.*, **48**, 1463–1466.

23
Life's Single Chirality: Origin of Symmetry Breaking in Biomolecules

Michael Mauksch and Svetlana B. Tsogoeva

> C'est la dissymétrie qui crée le phénomène.
> P. Curie, J. Phys. 1894, **3**, 393.

23.1
Introduction

Life on earth is based on chiral molecules: amino acids (as constituents of the proteins coded by RNA), sugars, and achiral nucleobases that form together the polymeric nucleic acids (RNA and DNA) as carriers of the genetic code [1]. Both sugars and amino acids are C-chiral molecules with one or more chiral carbon centers. The absolute configuration of all the amino acids and sugar molecules employed in the molecules of life on earth is almost exclusively uniform: "L" for amino acids and "D" for sugars – a fact called *"biological homochirality"* [2]. As an exception, D-amino acids are used as camouflage by bacteria in their cell walls, but never for functional biopolymers that are involved in proteogenesis or replication as these amino acids could be self-poisoning [3]. Many biological receptors on cell walls of various tissues, for example, those involved in the olfactory senses [4], have an explicit sensitivity for molecules with their specific handedness: for example, the cyclic monoterpene limonene has a fresh orange-like smell in the *(R)*-form, whereas the *(S)*-enantiomer has a harsh lemon-like odor (Figure 23.1) [5].

One of the biggest challenges in biomimetic chemistry is therefore to emulate Nature's selection for a single handedness of biomolecules in their neogenesis, to make single enantiomer drugs and fragrances more readily accessible.

It appears widely accepted now that life as we know it would not be conceivable without the homochirality of biomolecules [6–9]. Both left- and right-handed versions of biomolecules would, however, in principle be capable of supporting complex life. This leads us to the question whether the observation of only one form of chiral molecules is due to a deterministic process or is, alternatively, accidental, where both forms might have been present in the initial stages of life and one form had become extinct later on in evolution. While fossil amino acids with a preponderance of (not necessarily natural) L-amino acids had been found

Biomimetic Organic Synthesis, First Edition. Edited by Erwan Poupon and Bastien Nay.
© 2011 Wiley-VCH Verlag GmbH & Co. KGaA. Published 2011 by Wiley-VCH Verlag GmbH & Co. KGaA.

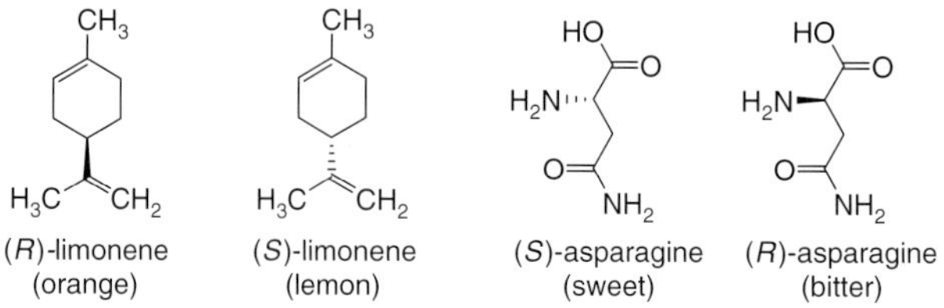

Figure 23.1 Enantiomers having different smell or taste.

in carbonaceous meteorites [10], no fossil traces of polymeric peptides or nucleic acids with an excess of the unaccustomed sense of chirality have been discovered yet. The same holds for amino acids or carbohydrates. Hence, it might appear that the origin of homochirality in the small molecules that form the building blocks of life had predated the advent of life on earth [11]. In this context, the exogenous hypothesis of organic material (enclosed in meteorites), enantioenriched by, for example, circularly polarized light in interstellar star-clouds [12], and impacting on the surface of the early earth, has some charm, because it would explain the conundrum of how life could have a both a disparate geographic origin and a deterministic single chirality. However, it is also possible that all those early monomeric chiral molecules of life, for example, those stemming possibly from outer space as well as those that resulted from hypothetical early life with opposite handedness, would mostly become fully deracemized, before the replication apparatus of life successfully sustained the now observed sense of chirality (in light of the racemization times of amino acids, which are short in geological – and probably also evolutionary – time scales [13]). This would explain today's absence of biomolecules with a greater abundance of the opposite handedness (e.g., D-amino acids) in fossils [14]. Moreover, all these precursor molecules of biopolymers and the peptides or nucleobases themselves are susceptible to oxidation [15]. The change from a probably initially reducing to an oxidizing atmosphere had presumably occurred at about the same period when life began [16]. These oxidation processes would leave hardly traces of a hypothetical extinct "antipode-life," which leaves the question of its early existence still unanswered.

Another important aspect of the homochirality of biomolecules is the distinction between the origin of the initial chiral imbalance and the mechanism of its amplification to enantiomeric purity. To explain the former, several theories have been suggested, the most popular being the proposal of forced symmetry breaking by parity violating energy differences, which are actually known to cause miniscule energetic differences between enantiomers [17]. However, these energy differences are tiny and can even be much smaller than the noise level caused by statistical (thermal) fluctuations that cause accidental chiral imbalances in the enantiomeric composition [18]. In the following we will therefore assume that an initial enantio-imbalance already exists, regardless of its cause, and concentrate on the mechanism of its amplification.

Another possibility involves enantioenrichment in biopolymers through polymerization reactions (without overall asymmetric amplification) [19]. While it appears

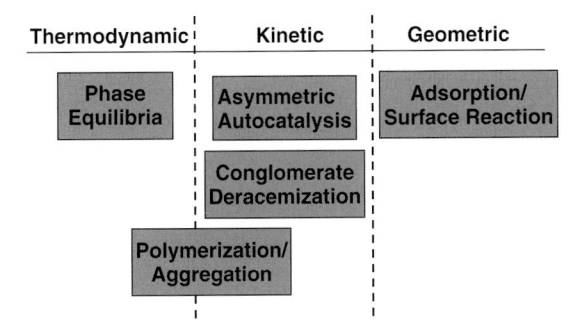

```
Thermodynamic  |  Kinetic                  |  Geometric
               |                           |
  Phase        |  Asymmetric               |  Adsorption/
  Equilibria   |  Autocatalysis            |  Surface Reaction
               |                           |
               |  Conglomerate             |
               |  Deracemization           |
               |                           |
  Polymerization/                          |
  Aggregation                              |
```

Figure 23.2 Overview of the enantiodifferentiating chemical and physicochemical processes employed in approaches to the conundrum of biological homochirality.

possible that polymerization itself led to the selection of homochiral strands of peptides or DNA – fed on a pool of racemic monomeric precursor molecules, it is unlikely that this could have occurred only at a single unique spot and only at a unique moment in time. In fact, if life itself – as an autopoietic process – could have generated its own homochiral environment we would assume that present lower organisms could feed on a racemic pool of chiral nutrient molecules. However, D-amino acids have a poisonous or at least inhibiting effect on life processes based on L-amino acids in most organisms [20]. Moreover, and because of the statistic nature of the polymerization (and depolymerization) processes it is also difficult to imagine how polymerization alone could have resulted in the deterministic formation of only left-handed or, alternatively, of only right-handed ("homochiral") DNA.

This situation has resulted in a multitude of theories that were proposed to explain either the deterministic or the accidental deracemization or, alternatively, the enantioselective formation of amino acids or carbohydrates, or, as a further alternative, the formation of homochiral biopolymers from racemic precursors. In the following, we will discuss the different approaches to the conundrum of biological homochirality, employing enantiodifferentiating chemical and physicochemical processes (see Figure 23.2 for an overview).

We focus first on the processes that could have either led to deracemization of a preformed mixture of chiral biomolecules or to the enantioselective formation of these molecules starting from an accidental miniscule chiral bias.

23.2
Autocatalytic Enantioselective Reactions

Asymmetric amplification and spontaneous mirror symmetry breaking in chemical stereoselective or stereospecific reactions appear to be the key to the solution of the homochirality problem, regardless of whether the amplification occurs during polymerization or not [21]. Spontaneous enantioenrichment (when the ee value is defined as the degree of enantiopurity of a polymer or of an ensemble of polymers) might also occur in the formation of homochiral polymers in stereospecific reactions acting on a racemic precursor pool of chiral monomers [22]. Asymmetric amplification requires less stringent mechanistic conditions than symmetry breaking in stereoselective reactions of prochiral reactants, and denotes the (usually limited) increase of enantiomeric excess from non-zero starting values, for example, in the catalyst, whereas spontaneous mirror symmetry breaking denotes the amplification of enantiomeric excess from even the tiniest initial enantio-imbalances (e.g., those caused by statistical fluctuations or parity violating energy differences) through nonlinear reaction schemes [23]. While several reaction schemes have been invented and studied that exhibit positive nonlinear effects in the ratio of product to catalyst ee (Figure 23.3) – for example, those of Noyori (with non-specific mutual inhibition due to statistical formation of homo- and heterochiral aggregates of catalyst) [24], and of Kagan (e.g., in the reservoir model of catalyst aggregation, or with catalytic ML_2 complexes built from a chiral ligand L) [25], and several examples (e.g., in the epoxidation of chalcone) have been reported – the phenomenon of true asymmetric amplification due to spontaneous mirror symmetry breaking is more elusive. In the presumably prochiral or racemic prebiotic environment, asymmetric amplification appears a prerequisite for an assumed deracemization of the biomolecule precursors.

As asymmetric amplification is implied in spontaneous symmetry breaking, and if one assumes that sizable enantio-imbalances (e.g., through enantioenriched molecular material from outer space) had not existed a priori, we concentrate in the following discussion on spontaneous mirror symmetry breaking as the crucial phenomenon and how it could have occurred endogenously on the primordial earth.

(a) $M + L_R + L_S \longrightarrow ML_RL_R + ML_SL_S \rightleftharpoons ML_RL_S$

$x \downarrow k_{RR} \quad y \downarrow k_{SS} \quad z \downarrow k_{RS}$

$ee_{ligand} \qquad ee_{max} \qquad -ee_{max} \qquad racemic$

$ee_{prod} = ee_{max} * ee_{ligand} * (1 + \beta)/(1 + g\beta)$

$\beta = z/(x+y)$
$g = k_{RS}/k_{RR}$

(b)

monomeric catalysts, active → S+S, R+S, R+R dimeric catalysts inactive

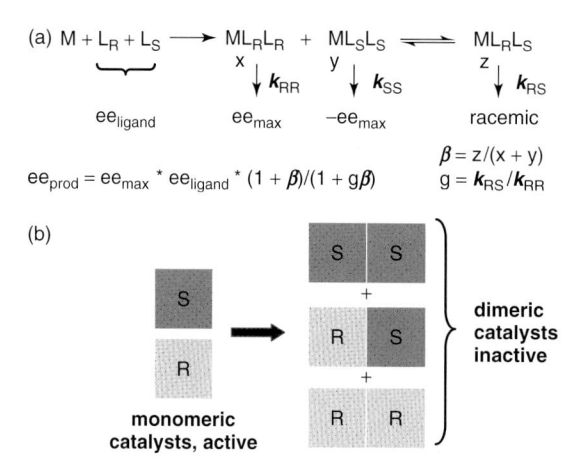

Figure 23.3 Kagan (a) and Noyori (b) models of asymmetric amplification through catalyst aggregation. Dimer formation in the Noyori model is statistic. In (a) g denotes the branching ratio for heterodimeric versus homodimeric catalyst activity, β is proportional to the equilibrium constant for the dimer equilibria, k_{RS} and k_{RR} are the rate constants for the formation of chiral product (not shown here) by heterochiral and homochiral catalysts, respectively, M is a metal center to which chiral ligands L_R and L_S could coordinate.

A possible solution to the homochirality conundrum that meets these requirements could involve complex reaction mechanisms with nonlinear kinetics. Since a landmark paper by the British crystallographer Charles Frank, published in 1953, who presented a mathematical model reaction network capable of producing homochiral solutions (with 100% ee) [26], chemists began to consider spontaneous mirror symmetry breaking reaction mechanisms in chemistry. The Frank mechanism (Figure 23.4) involves linear and irreversible autocatalytic steps followed by an irreversible nonlinear inhibition reaction in which the product enantiomers formed in the autocatalytic step recombine to give an optically and catalytically inactive heterochiral product dimer. Thereby, the enantiomeric purity of the autocatalytic product is constantly increased, once a racemic initialization phase is passed (Figure 23.4). Frank did not specify the physical boundary conditions of the system this mechanism is operated in and did not comment on thermodynamic requirements, but states confidently that a "well stirred mixture, like, for example, a reaction bulb" would be sufficient to demonstrate the process in Reference [26].

For decades, the Frank mechanism and its modifications served as a hypothetical example of spontaneous chiral symmetry breaking in homogeneous mixtures,

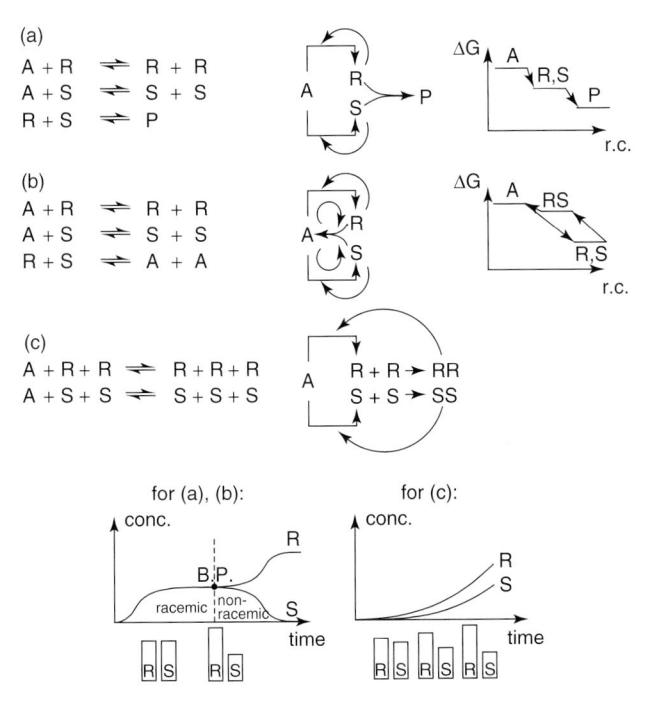

(a)
$$A + R \rightleftharpoons R + R$$
$$A + S \rightleftharpoons S + S$$
$$R + S \rightleftharpoons P$$

(b)
$$A + R \rightleftharpoons R + R$$
$$A + S \rightleftharpoons S + S$$
$$R + S \rightleftharpoons A + A$$

(c)
$$A + R + R \rightleftharpoons R + R + R$$
$$A + S + S \rightleftharpoons S + S + S$$

$$R + R \rightarrow RR$$
$$S + S \rightarrow SS$$

Figure 23.4 Basic schematic autocatalytic mechanisms of asymmetric amplification: the original Frank mechanism, depicted here with reversible reaction steps (a), and the mechanism with recycle Frank kinetics (b), which both involve mutual inhibition, and the hypercompetitive mechanism without mutual inhibition (c). A = prochiral reactant, B.P. = bifurcation point.

mostly in open flow systems. Kondepudi and Nelson investigated, for example, in 1983 a modification of the Frank mechanism with reversible autocatalytic steps and an irreversible inhibition step under flow conditions [27]. Gutman et al. studied the Frank mechanism and a modification with enantiomer racemization analytically [28]. However, with limited resources and in a closed system, complete homochirality (with 100% ee) cannot be achieved by the original Frank scheme, where a minor enantiomer recycling pathway is absent. While Avetisov first considered a hypothetical fully reversible Frank-type mechanism (Figure 23.4) [29], and Plasson proposed a complex autocatalytic recycling reaction network in 2004 [30], a simple, fully reversible and recycle modification of the Frank mechanism was first proposed in 2007 to apply in reversible organic reactions (Figure 23.4) [31]. Such a non-productive reaction cycle can, of course, not simply turn by itself – without violating the law of energy conservation, because the recycling step is necessarily non-spontaneous. Every turn of the reaction cycle is therefore accompanied by an inevitable loss in chemical energy, which has to be compensated through coupling to an irreversible process. Biochemical examples are so-called "futile" metabolic cycles, like phosphorylation/dephosphorylation or the ATP-consuming process of glycolysis combined with gluconeogenesis [32].

The proposed Frank type reaction scheme with specific minor enantiomer recycling (Figure 23.4b), and variations thereof, was further studied under the conditions of a closed system by Sugimoro et al. (using the master equation from statistical thermodynamics) [33], by Ribo and Hochberg (employing stability analysis) [34], and by Blackmond (imposing equilibrium thermodynamics) [35], who all came to the conclusion that – under assumed reversible (i.e., non-dissipative) conditions – the recycle Frank kinetics alone (with its closed reaction loops) results – in a rate equation formalism – in asymmetric depletion, rather than asymmetric amplification, because a unidirectional cyclic operation of the mechanism without a source of energy is, of course, precluded by thermodynamic restrictions [35]. This result is unsurprising because, due to the symmetry of the reaction network, the principle of microscopic reversibility demands that the concentrations of the enantiomers are equal at equilibrium: $[R]_{eq} = [S]_{eq}$, which imposes a restriction of generality at situations far from equilibrium [35, 36], letting the implication of asymmetric depletion appear as a tautology [36]. Most real physical and chemical processes are irreversible (i.e., dissipative), though. Nicolis and Prigogine suggested already decades ago that a chiral bias could be spontaneously generated and amplified in dissipative far-from-equilibrium systems [37, 38]. Indeed, important biological functions (e.g., enzyme regulation, action of molecular motor proteins) could, for example, be understood by resorting to the concept of the open system non-equilibrium steady state and implying closed reaction loops [39].

The possibility of a dissipative non-productive reaction cycle, operating at non-equilibrium concentrations under open system steady-state conditions and consisting of a sequence of reversible reaction steps, was not considered in earlier proposals for the origin of homochirality – and has apparently been overlooked; recently, the reversible and recycle modification of the original Frank mechanism

(Figure 23.4a,b) was investigated under open flow conditions and it was suggested that it might lead to mirror symmetry breaking when coupled to an energy consuming side- or follow-up reaction running inside the *experimentally* closed system [36], or, alternatively, to an external source of energy [30]. Both scenarios realize the open system conditions through exchange of chemical energy with the environment. The latter possibility was already explored in depth by Plasson and coworkers for a similar mechanism invented earlier by the same group, inspired by the hypothetical experimental prebiotic chemistry system of Wächtershäuser et al. [40]. Recently, Ribo and coworkers have demonstrated that a fully reversible Frank mechanism with a weakly exergonic inhibition step could result in spontaneous mirror symmetry breaking and (temporary) asymmetric amplification in a reversible homogeneous system, that is even closed to matter flow [41]. However, to date, there is no verified laboratory example for either the classical or recycle Frank mechanism. Nevertheless, the Frank model and its modifications have been frequently invoked to explain the observation of asymmetric autocatalysis in the Soai reaction [21], despite the absence of experimental evidence for a bifurcation mechanism. The main drawback of the Frank model though is the absence of a plausible mechanism of chirality induction in organic reactions.

As one of several discussed alternatives to the Frank scheme with its linear order autocatalytic steps [42], Decker proposed in 1975 [43] that hypercompetitive mechanisms could also lead to mirror symmetry breaking. In this class of reactions, one of the autocatalytic enantiomers is practically outrun by its antipode, the amount of which is growing faster (Figure 23.4). In these reaction schemes, the autocatalytic step has a higher than linear reaction order with respect to the monomeric product. No mutual inhibition as in the Frank mechanism is involved, and no bifurcation in the ee values occurs: the enantio-imbalances increase monotonously from a tiny starting value. It has been proposed that reactions in the coordination sphere of metal complexes could provide a plausible experimental realization of this mechanism [43]. It appeared therefore promising to look for such examples among organometallic reactions where the chiral catalyst could act as a reactant (e.g., in group transfer reactions) and which is afterwards regenerated by stereoselective oxidative addition of the transferred group from the reactant pool.

The first remarkable experimental demonstration of such spontaneous chiral symmetry breaking was achieved by Soai, who reported in 1995 the formation of chiral pyrimidyl alcohols in an irreversible organometallic reaction with high product enantiomeric excess of more than 99%, starting from very low ee values and in a sequential batch reaction protocol (Scheme 23.1) [44].

The mechanism of the Soai reaction was not elucidated until recently, when Schiaffino and Ercolani found, through density functional theory (DFT) calculations and a fit of computed and experimental data, that dimeric (at higher temperatures) or heterochiral 2 : 2 tetrameric aggregates (at lower temperatures) of the chiral Zn-alkoxide primary reaction product form the actual "autocatalytic" species (Figure 23.5) [45]. The homochiral (RR) dimer (**1**) constitutes a bidentate ligand that coordinates to two molecules of iPr_2Zn, forming a dinuclear metal complex **2** (step a, Figure 23.5). Complex **2** then rearranges to **3** through coordination of

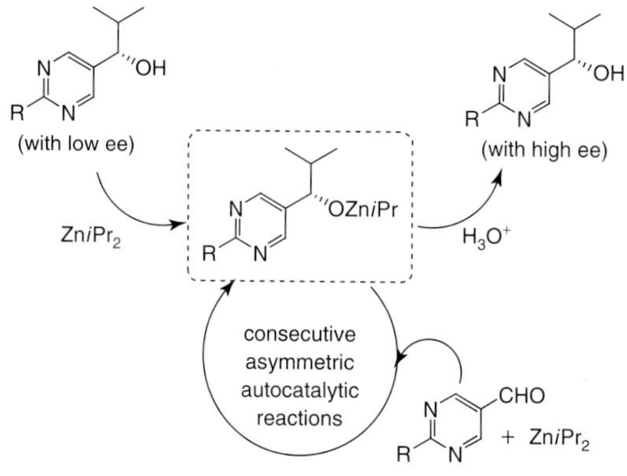

Scheme 23.1 Soai reaction: a first absolute asymmetric synthesis.

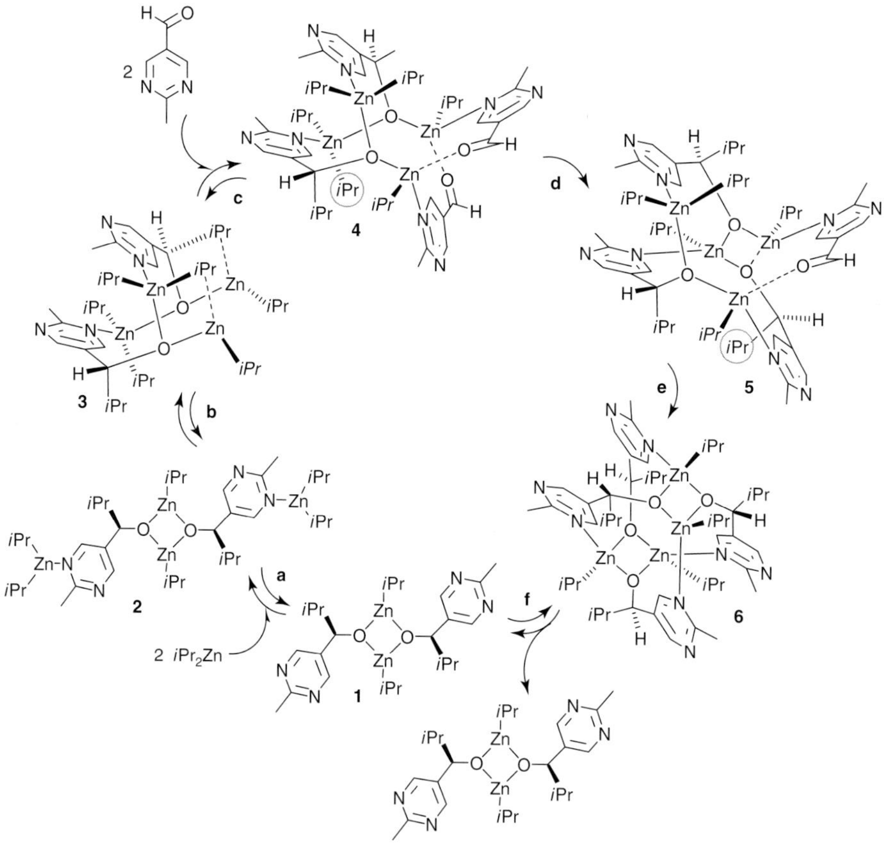

Figure 23.5 Mechanism of the Soai reaction according to DFT computations [45]. The dominant catalytic species is a homochiral dimer. Lower case letters (a–f) refer to the different reaction steps as discussed in the text. Species **1** is the homochiral dimer (RR) catalyst.

an oxygen atom of one dimer part toward the Zn metal center of the second monomer unit. Taking up two further pyrimidyl aldehyde molecules produces complex **4**, in which the aldehyde is coordinated via N and O atoms to zinc. In step d, one of the alkyl moieties (encircled) is transferred onto the carbonyl atom of one of the aldehyde molecules. Transfer of a further isopropyl unit to the second aldehyde results in complex **6** (step e), which is a homochiral tetramer (R_4). Finally, the tetramer dissociates to release the dimeric catalyst **1**. The formation of $(RS)_2$ heterodimers is further proposed to explain the observed high extent of asymmetric amplification at lower temperature.

In the year of Soai's seminal report, P. Bailey suggested the spontaneous self-amplification of enantiomeric excess in the generation of a chiral product through modification of Kagan's ML_2 model [25] to account for autocatalysis [46].

A few years later, Blackmond and Brown [47] and Buono and Blackmond [48] thoroughly investigated the kinetics of the Soai reaction with microcalorimetry, and proposed that the reaction rate depends on the enantiomeric composition of initially added product [47, 48]. They observed that the reaction proceeded twice as fast when the initially added product was enantiopure, as compared to racemic added product [48]. Hence they deduced that the catalyst should be a homochiral product dimer, rather than the monomer itself, and that the heterodimer is catalytically inactive (Figure 23.6) [47]. As a consequence, the change of the ratio of enantiomer concentrations d[R]/d[S] depends quadratically on the ratio of enantiomer concentration at a given point in time: $d[R]/d[S] = ([R]/[S])^2$ (R and S are denoted L_R and L_S, respectively, in Figure 23.6 for consistency with

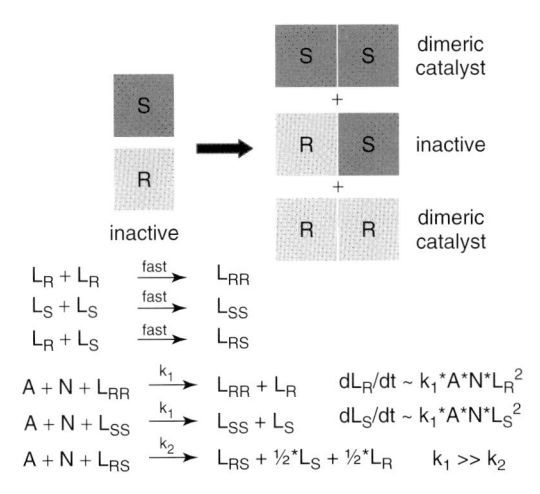

$L_R + L_R \xrightarrow{fast} L_{RR}$
$L_S + L_S \xrightarrow{fast} L_{SS}$
$L_R + L_S \xrightarrow{fast} L_{RS}$

$A + N + L_{RR} \xrightarrow{k_1} L_{RR} + L_R \quad dL_R/dt \sim k_1{}^*A{}^*N{}^*L_R{}^2$
$A + N + L_{SS} \xrightarrow{k_1} L_{SS} + L_S \quad dL_S/dt \sim k_1{}^*A{}^*N{}^*L_S{}^2$
$A + N + L_{RS} \xrightarrow{k_2} L_{RS} + ½{}^*L_S + ½{}^*L_R \quad k_1 \gg k_2$

Figure 23.6 Blackmond–Brown model of asymmetric autocatalysis in the Soai reaction. Only the homochiral dimers are catalytically active. The reaction proceeds via a second-order autocatalysis. The mechanism is a combination of Kagan's concept of dimer catalysis and Noyori's model with monomer/dimer equilibria. L_R and L_S are monomeric enantiomers from which dimers L_{RR}, L_{SS} (homochiral), and L_{RS} (heterochiral) can form. "A" is a prochiral reactant, N denotes a nucleophile.

Figure 23.3). The product enantiomeric excess is therefore continuously increasing (see also Figure 23.4c).

These mechanistic insights do not explain fully the dynamics of the Soai reaction, though. There is, for example, evidence from stochastic analysis (i.e., fitting of the likelihood in outcome for a certain enantiomeric excess to a parameterized kinetic model) that the Soai reaction proceeds via *three* cooperatively coupled autocatalytic cycles [49]. The Soai reaction serves as a paradigm example of absolute asymmetric synthesis in the absence of external asymmetric physical forces [50], and as a proof-of-principle for the possible role of autocatalytic processes in the solution of the homochirality conundrum. However, the reaction requires non-aqueous media to run in, which might not be that plausible in light of the probably aqueous conditions on the primordial earth.

Fully organic examples of asymmetric autocatalysis [51], which, similar to the Soai reaction, can also show the behavior of an absolute asymmetric synthesis, were first reported by Mauksch and Tsogoeva in 2007 [52]: an asymmetric Mannich-type product – a functionalized amino acid – was able to replicate itself and with the same absolute configuration (Figure 23.7). DFT (density functional theory) calculations showed that the formation of homochiral product dimers is kinetically preferred with respect to formation of heterochiral dimers in the autocatalytic steps [52]. Figure 23.7 shows the proposed catalytic cycle: the Mannich product is

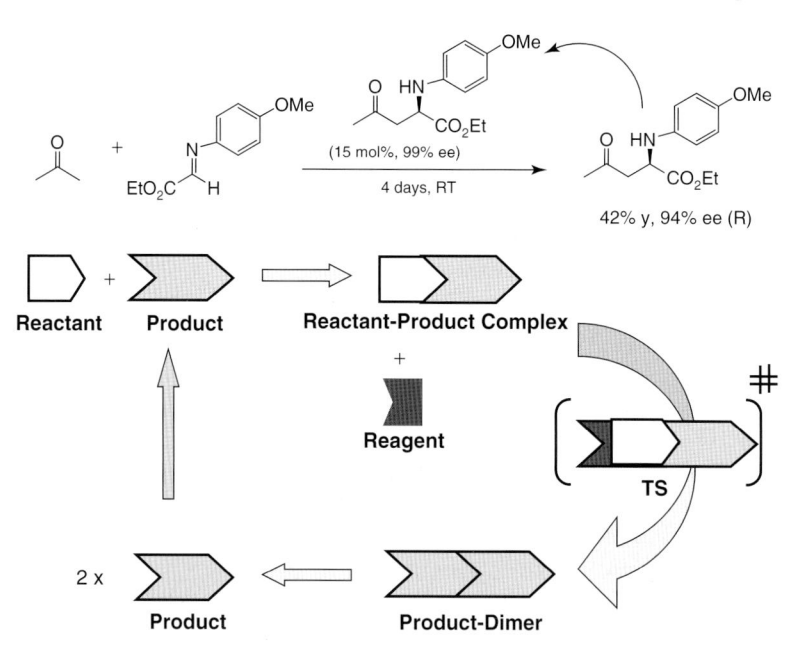

Figure 23.7 Asymmetric autocatalysis in an organocatalytic reaction. The proposed mechanism involves only a monomeric product autocatalyst as the active catalytic species.

supposed to form hydrogen bonded adducts with the reactant imine via specific "recognition sites," providing an explanation for the experimentally found chirality induction, which is similarly effective as with well-known external organocatalysts, like proline. Even without added catalyst, product with 9.4% ee at 31% yield was obtained after four days reaction time [31].

In principle, a conceivable enantioenrichment in certain amino acids might have led to a corresponding enantioenrichment in the sugars, because it has been shown first by Pizzarello and Weber [53] and later by Cordova [54] that amino acids might be efficient catalysts in the aldol reactions involved in producing carbohydrates. The connection of amino acid handedness and those of sugars has also been recently established by Nanita and Cooks [55]. Very recently, it was also suggested that asymmetric autocatalysis in the aldol reaction might play a role in gluconeogenesis via the formose reaction [56].

23.3
Autocatalysis and Self-replication

Self-replication appears to be an indispensable feature of the molecules of life (RNA and DNA) and necessary to sustain their homochirality–once achieved–against the relentless trend to racemization. Eigen and Schuster have theoretically studied self-replicating and autocatalytic systems [57]. Philp [58] and von Kiedrowski [59] have elegantly shown the connection of autocatalysis and self-replication. While von Kiedrowski reported studies on the replication and autocatalysis of oligomeric nucleotide sequences [59], his group also found that autocatalysis does not necessarily have to be self-accelerating – a property often attributed to autocatalytic or autoinductive processes – because the initial product of the autocatalytic step in the reaction of, for example, non-polymeric organic molecules can be considered a, for instance hydrogen-bonded, dimer of product template molecules (Figure 23.8), rather than isolated monomers themselves, as, for example, is assumed in the original Frank scheme [26]. This mechanism, which provides a reasonable mechanism of chiral induction in chiral autocatalysis, requires the presence of appropriate recognition sites (e.g., hydrogen bond donor/acceptor pairs) in the product (or substrate) molecules.

Such a reaction scheme, which involves subsequent dissociation of the dimer product to release the monomeric catalytic template molecules [58], appears more plausible in organo-autocatalytic reactions [52, 56] than schemes that require the regeneration of the catalyst or those that give only monomeric initial product. However, the mechanism is not nonlinear and therefore cannot give asymmetric amplification, because the ratio of formation rates of the enantiomers is proportional to the ratio of their concentrations: $d[R]/d[S] = [R]/[S]$. Very recently, Rebek, a pioneer of self-replicating simple organic non-nucleotide molecules, elegantly demonstrated an experimental example where the product of the autocatalytic step is even able to catalyze organocatalytically a different reaction in which the self-replicator is reproduced (Scheme 23.2) [60].

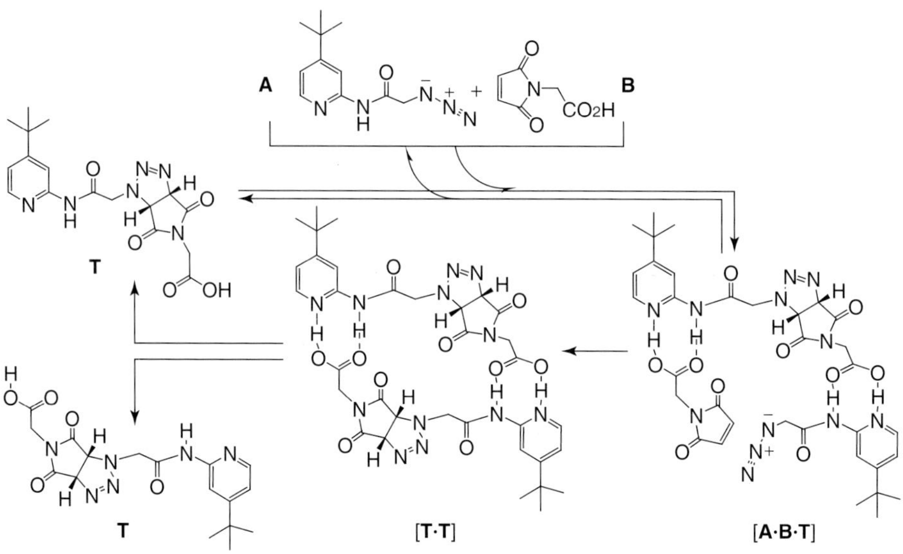

Figure 23.8 Example of a minimal organic self-replicator with the autocatalytic template T.

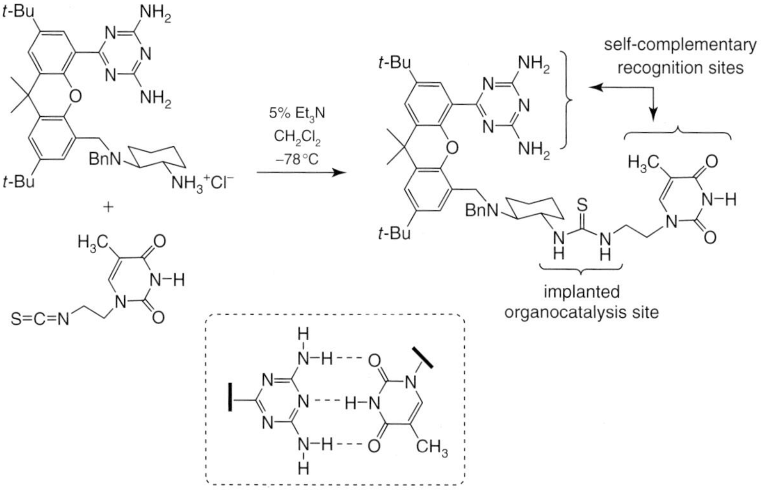

Scheme 23.2 Organocatalytic synthetic self-replicator of Rebek [60]; recognition sites are hydrogen bond donor/acceptor pairs.

23.4
Polymerization and Aggregation Models of Enantioenrichment

Würthner et al. [22] have demonstrated recently that in an ensemble of polymeric dyes homochiral strands can evolve from the initial aggregates even if the aggregation is fed on a near-racemic pool of monomeric precursors, employing a

23.5 Phase Equilibria

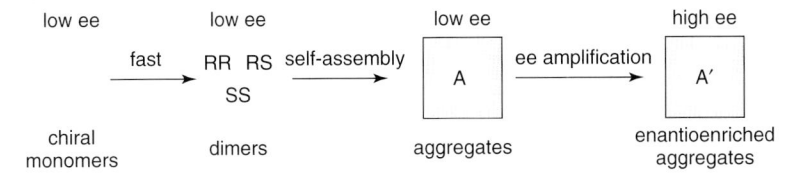

```
low ee              low ee              low ee                      high ee
         fast    RR  RS   self-assembly          ee amplification
    ──────────▶      SS        ──────────▶  [ A ]  ──────────────▶  [ A' ]
chiral              RS
monomers           dimers                 aggregates              enantioenriched
                                                                    aggregates
```

Figure 23.9 Würthner's polymerization model of enantioenrichment.

kinetic (autocatalytic) explanation for the "majority rules" effect of asymmetric amplification in supramolecular chemistry (Figure 23.9) [61]. Very recently, it was even shown that supramolecular aggregation on surfaces could result in drastic symmetry breaking and asymmetric amplification [62].

Lahav and coworkers have, for example, very recently and elegantly shown that homochiral oligopeptides can show spontaneous enantioenrichment during their formation reaction, due to the additional unexpected formation of racemic beta-sheets [19]. A further example is the theoretical recycling reaction model of Plasson et al. [30], which is actually a model for polymerization with epimerization and has been studied in depth by Brandenburg et al. [63].

A non-racemic non-equilibrium steady state is assumed to evolve from a racemic state far from equilibrium [64]. Notably, the "recycling" in this model does refer to the reactivation of the chiral monomers resulting from depolymerization, rather than to a "minor enantiomer recycling" that leads back to the prochiral reactant, discussed earlier [31]. In contradistinction to the original Frank mechanism, homochiral product dimers are assumed to be lower in energy than heterochiral dimers, but higher in energy than monomers [30]. The Plasson mechanism is, however, nevertheless a kinetic bifurcation mechanism (like Frank's), at least when only dimer species are involved. With longer chains, next-nearest neighbor interactions could be taken into account, which would render the polymerization and epimerization steps nonlinear with a reaction order slightly larger than one.

The enantioenrichment is here principally twofold: first, individual polymers might partially depolymerize to be rebuilt with a higher degree of enantiopurity; secondly, the proportion of homochiral polymers of the same sense of handedness can grow in an ensemble of polymeric strands of different lengths. This scenario might conceivably also apply to biopolymers. In this case, the origin of homochirality in biomolecules would not be due to symmetry breaking and true asymmetric amplification acting on a pool of near-racemic precursor molecules (like, e.g., amino acids), but would rather be the consequence of the autopoietic processes of life itself. It is conceivable that the formation of homochiral biopolymers and the development of the replication mechanism of these polymers had developed in sync.

23.5
Phase Equilibria

Enantioenrichment through asymmetric autocatalysis is based on nonlinear reaction kinetics under non-equilibrium thermodynamic conditions. Another

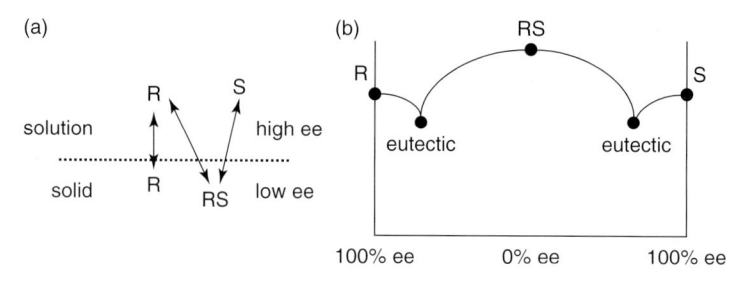

Figure 23.10 (a) Situation in solution; (b) an example binary melting diagram. RS denotes the racemic compound (marked by a dystectic point on the liquidus curve) and R and S are pure enantiomers.

possibility of phase-specific enantioenrichment through – thermodynamically, rather than kinetically – driven enantiospecific phase re-distribution processes has been put forward first by Hayashi [65] and later by Blackmond [66] and by Breslow [67]. The ternary system, composed of the two amino acid enantiomers and water, has a ternary phase diagram with eutectic points corresponding to a non-racemic distribution of the enantiomers in the solution and the solid phases when the amino acid forms a racemic compound, while those amino acids that form conglomerates have only racemic eutectics as thermodynamic fixed points, in accord with the Gibbs' phase rule [65, 66]. The racemic compound is here assumed to have a higher melting point than the enantiopure solids. Figure 23.10 depicts the situation in solution (a) with an exemplary binary melting diagram (b), RS denotes the racemic compound (marked by a dystectic point on the liquidus curve), and R and S are pure enantiomers. The ee value of, for example, serine in solution increased to the equilibrium value of 99.5% in a heterogeneous mixture of scalemic solid serine under its saturated solution, while the solid phase showed a compensating asymmetric depletion [66].

Soloshonok made the remarkable observation that a similar self-disproportionation of enantiomers with enantioenrichment in one of the two phases (accompanied by asymmetric depletion in the other phase) may also occur during chromatography or sublimation, re-crystallization, distillation, or other phase transitions [68]. Soloshonok explained this by the difference in stability of homochiral versus heterochiral aggregates. Suhm and coworkers found that the self-disproportionation behavior observed by Soloshonok may be shown only by compounds that form true racemates (i.e., racemic compounds), and when a certain critical ee value in the initial scalemic mixture is exceeded [69].

As most biochemical reactions take place in solution, such solution phase enantioenrichment is possibly relevant in the origin of homochirality scenarios. Moreover, about 90% of all chiral compounds form true racemates. A potential drawback could be the unexplained fate of the racemic or asymmetrically depleted solid amino acid phases.

Even properties of liquid crystal phases have been suggested earlier as possible cause for the biological homochirality [70].

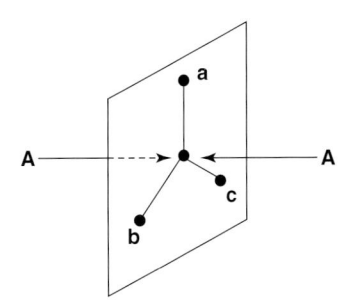

Figure 23.11 Achiral reactant A "sees" only one of the two enantiotopic faces and, therefore, reacts stereospecifically.

As an alternative, Cooks reported in 2007 that solid serine at 3% ee shows remarkable genuine asymmetric amplification to 69% ee on sublimation [71]. The mechanism of this amplification is still not clear.

23.6
Adsorption on Chiral Surfaces

Even achiral compounds or ions may crystallize in chiral space groups. Hence, apart from the above thermodynamic or kinetic schemes, a method of chiral resolution (separation of enantiomers) in the biochemical context is suggested by the possibility of enantiospecific adsorption of small chiral molecules like amino acids on chiral crystal surfaces of, for example, calcite, as a special case of chiral recognition [72]. Such mechanisms could probably have had a high chance of realization in prebiotic scenarios.

The enantiotopic faces of crystals have also recently been employed in such a "geometric" approach to absolute asymmetric synthesis in a stereoselective reduction reaction by Kuhn [73], acting on similar observations by Holland and Richardson made 20 years earlier [74]. An achiral reactant A "sees" only one of the two enantiotopic faces and reacts stereospecifically (Figure 23.11). Lahav further proposed that chiral additives possessing a greater affinity or structural resemblance to one of the enantiomers in their crystalline state may hamper kinetically the crystallization of that enantiomer, resulting in accumulation of the opposite enantiomer in the solid phase ("Lahav's rule of reversal") [75]. Acting on these predictions, Vlieg et al. demonstrated recently that this approach can be used to deracemize scalemic conglomerates of chiral compounds to enantiopurity in the solid state by grinding them under their saturated solution in the presence of a chiral additive, while the opposite enantiomer stays in solution [76].

23.7
Spontaneous Symmetry Breaking in Conglomerate Crystallizations

Pincock [77] and later Kondepudi [78] observed that a conglomerate forming inorganic salt ($NaClO_3$) spontaneously crystallizes with random enantiomeric excess from supersaturated solutions. Kondepudi found that vigorous stirring led to a fat-tailed (leptokurtic) histogram distribution of crystal enantiomeric excesses, while Pincock earlier observed a Gaussian distribution with a maximum likelihood for the racemic outcome [77, 78]. For an organic racemizing compound in unstirred supersaturated solutions, a similar observation was already made in the 1940s by Havinga [79, 80].

Ribó more recently pointed out that primary nucleation during crystallization from the supersaturated solution phase through cooperative "chiral recognition" interactions could result in sizable enantiomeric excesses in the crystalline phase [81]. Alternatively, secondary nucleation has been invoked as the cause of high cee (crystal enantiomeric excess) values in vigorously stirred supersaturated solutions under kinetic control (and when crystallization and dissolution are balanced), because secondary nucleation implies the autocatalytic nature of the crystallization process in the presence of a crystal surface, which results in self-amplification of enantiomeric excess in the solid phase [82]. In 2005, Viedma demonstrated in a legendary experiment that a racemic conglomerate mixture of an inorganic salt with achiral ions in the (merely saturated) solution can be fully deracemized when crystal abrasion through vigorous stirring is assisted by added glass beads [83]. Viedma explained his observation qualitatively by thermodynamically controlled dissolution of the smallest crystals (due to the Gibbs–Thomson rule) in conjunction with the kinetically controlled nonlinear autocatalytic crystal growth, which was already proposed earlier by Uwaha as the cause for the conglomerate deracemization [84]. Afterwards, Ribó [81] and, later, Blackmond [85] developed a conceptual expansion of the Viedma experiment to the deracemization of intrinsically chiral compounds that racemize rapidly in solution. In 2008, Vlieg, Blackmond, and associates reported a first stunning experimental realization of this idea: the stirred slurry of a racemic or nearly racemic chiral proteinogenic amino acid compound was fully deracemized to a chiral solid phase with 100% ee and with almost complete conversion rates (Figure 23.12) [86]. The compound racemized rapidly in the solution phase in the presence of a strong base at room temperature (with a half-life of about 10 min), while the deracemization process itself took weeks to complete. The increase in crystal ee value with time was exponential. Ostwald ripening, a thermodynamic mechanism of crystal growth, was employed by the authors to explain the deracemization process [86, 87]. The remarkable experimental process was later optimized further by Vlieg, Kaptein, Kellogg, and coworkers to allow complete deracemization in days or even hours, rather than weeks [87]. Soon after the first experimental reports, again Uwaha explained also these newer observations on the basis of a rate equation model that involves nonlinear kinetics of crystal growth, extending his earlier model [84] to account for solution phase racemization [88]. Uwaha proposed that, instead of monomeric molecules, subcritical clusters are incorporated into the growing bulk crystals. The connection of these fascinating results to the origin of homochirality is not at once apparent. Nevertheless, wind- or water-tossed sand or pebbles might conceivably have provided for the grinding in half-dried puddles, for example, at sea-shores.

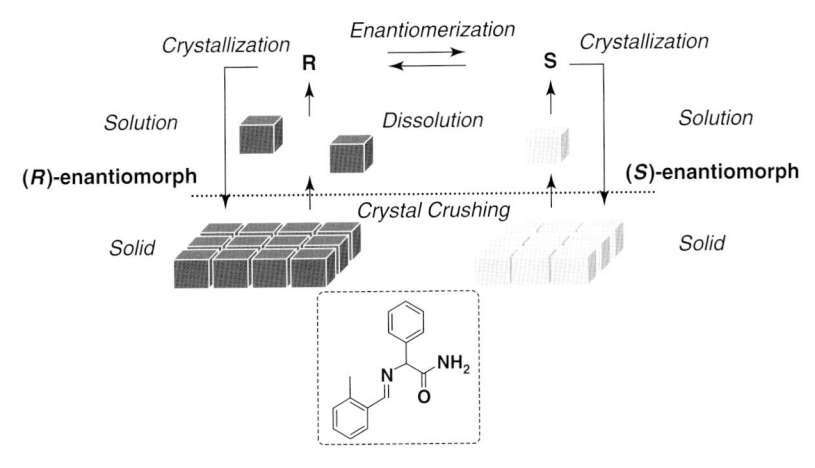

23.7 Spontaneous Symmetry Breaking in Conglomerate Crystallizations

Figure 23.12 Complete conversion of racemic crystal conglomerates into a single chiral state. Crystallization, crystal crushing, and dissolution form two competing process cycles for the R and S species, respectively – coupled by the enantiomerization reaction. The example shown is based on Reference [86]. The minor enantiomer is recycled through dissolution and enantiomerization.

Most recently, Tsogoeva, Mauksch, and coworkers reported that the slurry of the chiral product of a (reversible) Mannich reaction stirred under its saturated toluene solution can also be deracemized, and without the presence of a strong base (and without the presence of glass beads), combining conglomerate deracemization and asymmetric synthesis [89]. The Mannich product is here proposed to enantiomerize via the reactant imine – instead of directly; Mannich and retro-Mannich reaction steps are catalyzed by racemic or achiral organocatalysts. Functionalized amino acids can be formed by the Mannich-type reaction.

A solution phase excess of the solid phase's minor enantiomer was observed during the initial stages of the deracemization experiment [89, 90]. This surprising effect was explained in accord with Uwaha's autocatalytic crystal growth model, in which such an observation is implied [88]: nonlinear growth leads to faster deposition of the major enantiomer from solution [90]. Notably, the incorporation of clusters is not the only possibility for achieving a nonlinear growth rate: monomers could conceivably also aggregate and interact at neighboring sites at the crystal growth front.

Paradoxically, even at mildly elevated temperatures and in presence of a catalyst, the racemization half-life of the dissolved Mannich product was days in solution, while the total deracemization of the solid took only hours, which is a behavior opposite to the observations of Vlieg and associates, where direct and solution phase enantiomerization was faster, rather than slower, than deracemization [86, 87]. To explain this, a mechanism of autocatalytic enantiomerization at the crystal surface was invoked, a mechanism recently proposed theoretically by Saito et al. to explain deracemization of stirred conglomerate slurries [91]. The surface-assisted racemization in Saito's model, however, plays a dual role: first, the implied nonlinearity of the enantiomerization process (due to its dependence on the respective crystal surface areas of the enantiomorphic bulk crystals) is employed to explain the physical cause for the deracemization and, second, it provides a different mechanism for the chemical process of enantiomerization, independent from a solution phase racemization [90]. Asymmetric autocatalysis, a process known from homogeneous systems [44, 51], might therefore be extended conceptually to heterogeneous systems [90].

Slowly racemizing chiral compounds have not been considered before in concepts for complete conglomerate deracemization [81, 85–87]. Very recently, Bolm was able to employ a similar process in the deracemization of an aldol reaction product [92].

Viedma, Blackmond, and coworkers observed for the deracemization of a proteinogenic amino acid, applying the base-catalyzed direct enantiomerization method in heated slurries, that in absence of grinding the increase in crystal ee value was less than exponential [93]. They also observed the temperature dependence of the deracemization process and found that heating accelerates the deracemization process, which was explained by the temperature-induced activation of the solution-phase enantiomerization.

23.8
Symmetry Breaking in Reaction–Diffusion Models, Collision Kinetics, and Membrane Diffusion

Nonlinear kinetics, required to break the mirror symmetry spontaneously, are not only exhibited by chemical reaction kinetics, or in the autocatalytic physical process of crystal growth, but also by the combination of chemical reactions with, for example, the physical process of diffusion, which has a nonlinear time-dependence on the concentration gradients, as shown by Gayathri and Rao in a recent theoretical paper [94].

A further idea was proposed by Shinitzky, stating that the diffusion and, hence, permeability of phase boundaries or membranes for amino acids could differentiate between the enantiomers due to chiroselective hydration via ortho-H_2O [95].

Toxvaerd showed theoretically, by employing kinetics based on collision theory in molecular dynamics simulations, that a simple chiral carbohydrate, glyceraldehyde, may deracemize at high particle densities via reaction-step dependent activation barriers for the keto–enol tautomerization [96].

23.9
Concluding Remarks and Outlook

As we have seen, several competing theories are vying to explain the origin of biological homochirality, which appears so central to life in the forms we are familiar with. The race between these different approaches is far from decided, as more theories and observations are reported. While some theories for the endogenous origin of homochirality stress a thermodynamic origin of enantioenrichment in the solution phase, others put more weight on mirror symmetry breaking kinetic mechanisms of asymmetric amplification building up upon initial imbalances in the enantiomeric compositions in homogeneous ensembles of chiral molecules. Supramolecular aggregation and polymerization could also result in a homochiral world of biosystems. A further suggestion involves symmetry breaking crystallizations or the exploitation of chiral influences already present in Nature, for example, on the surface of chiral crystals. We have apparently not reached yet a level of understanding of prebiotic chemistry that allows us to decide between alternative explanations. Whatever the specific cause for the initial chiral imbalance, and for its amplification to homochirality, life itself is preserving the once developed dominance of one sense of handedness through the eternal process of reproduction through self-replication.

References

1. Losick, R., Watson, J.D., Baker, T.A., Bell, S., Gann, A., and Levine, M.W. (2008) *Molecular Biology of the Gene*, Pearson/Benjamin Cummings, San Francisco.
2. Meierhenrich, U. (2008) *Amino Acids and the Asymmetry of Life*, Springer, Berlin.
3. Lam, H., Oh, D.-C., Cava, F., Takacs, C.N., Clardy, J., de Pedro, M.A., and Waldor, M.K. (2009) D-amino acids govern stationary phase cell wall remodeling in bacteria. *Science*, **325**, 1552–1555.
4. Polak, E.H., Fombon, A.M., Tilquin, C., and Punter, P.H. (1989) Sensory evidence for olfactory receptors with opposite chiral selectivity. *Behav. Brain Res.*, **31**(3), 199–206.
5. Fahlbusch, K.-G., Hammerschmidt, F.-J., Panten, J., Pickenhagen, W., Schatkowski, D., Bauer, K., Garbe, D., and Surburg, H. (2002) Flavors and fragrances, *Ullmann's Encyclopedia of Industrial Chemistry*, Wiley-VCH Verlag GmbH, Weinheim.
6. Calvin, M. (1969) *Chemical Evolution*, Oxford University Press, London.
7. Siegel, J.S. (1998) The homochiral imperative of molecular evolution. *Chirality*, **10**, 24–27.
8. Mason, S. (1985) Chemical evolution: origin of biomolecular chirality. *Nature*, **314**, 400–401.
9. Blackmond, D.G. (2010) The origin of biological homochirality. *Cold Spring Harbor. Perspect. Biol.*, **2**, a002147.
10. Cronin, J.R. and Pizarello, S. (1997) Enantiometric excesses in meteoritic amino acids. *Science*, **275**, 951–955.
11. Bonner, W.A. (1994) Enantioselective autocatalysis, spontaneous resolution and the prebiotic generation of chirality. *Orig. Life Evol. Biosph.*, **24**, 63–78.
12. Bailey, J. (2001) Astronomical sources of circularly polarized light and the origin of homochirality. *Orig. Life Evol. Biosph.*, **31**, 167–183.
13. Brown, R.H. (1985) Amino acid dating. *Origins*, **12**(1), 8–25.
14. Florkin, M. (1969) in *Organic Geochemistry – Methods and Results* (eds G. Eglinton and M.T.J. Murphy), Springer-Verlag, New York and Amsterdam, pp. 498–520.
15. Bada, J. and Miller, S.L. (1985) The composition of the primitive atmosphere and the synthesis of organic compounds on the early Earth. *Workshop on the Early Earth: The Interval from Accretion to the Older Archean* (eds L.D. Ashwal and K. Burke) (SEE N85-33062 21-91), Lunar and Planetary Institution, Houston, pp. 8–10.

16. Miller, S.L. and Orgel, L.E. (1974) *The Origins of Life on Earth*, Prentice Hall Inc., Englewood Cliffs, NJ.
17. Wesendrup, R., Laerdahl, J.K., Compton, R.N., and Schwerdtfeger, P. (2003) Biomolecular homochirality and electroweak interactions. I. The Yamagata hypothesis. *J. Phys. Chem. A*, **107**, 6668–6673.
18. Tranter, G.E. (1987) Parity violation and the origins of biomolecular handedness. *Biosystems*, **20**, 37–48.
19. Rubinstein, I., Clodic, G., Bolbach, G., Weissbuch, I., and Lahav, M. (2008) Racemic beta-sheets as templates for the generation of homochiral (isotactic) peptides from aqueous solutions of (RS)-valine or –leucine N-carboxyanhydrides: relevance for biochirogenesis. *Chem. Eur. J.*, **14**, 10999–11009.
20. Beardsley, R.E. (1962) Amino acid cross resistance in Agrobacterium tumefaciens. *J. Bacteriol.*, **84**, 1237–1240.
21. Blackmond, D.G. (2004) Asymmetric autocatalysis and its implications for the origin of homochirality. *Proc. Natl. Acad. Sci. U.S.A.*, **101**, 5732–5736.
22. Lohr, A. and Würthner, F. (2008) Evolution of homochiral helical dye assemblies: involvement of autocatalysis in the "majority rules" effect. *Angew. Chem. Int. Ed.*, **120**(7), 1252–1256.
23. Rivera Islas, J., Lavabre, D., Grevy, J.M., Lamoneda, R.H., Cabrera, H.R., Micheau, J.C., and Buhse, T. (2005) Mirror symmetry breaking in the Soai reaction: a kinetic understanding. *Proc. Natl. Acad. Sci. U.S.A.*, **102**, 13743–13748.
24. Kitamura, M., Suga, S., Niwa, M., and Noyori, R. (1995) Self- and non-self recognition of asymmetric catalysts. Nonlinear effects in the amino alcohol-promoted enantioselective addition of dialkylzinc to aldehydes. *J. Am. Chem. Soc.*, **117**, 4832–4842.
25. Puchot, C., Samuel, O., Dunach, E., Zhao, S., Agami, C., and Kagan, H.B. (1986) Non-linear effects in asymmetric synthesis. *J. Am. Chem. Soc.*, **108**, 2353–2357.
26. Frank, C.F. (1953) On spontaneous asymmetric synthesis. *Biochim. Biophys. Acta*, **11**, 459–464.
27. Kondepudi, D.K. and Nelson, G.W. (1983) Chiral symmetry breaking in non-equilibrium systems. *Phys. Rev. Lett.*, **50**, 1023–1026.
28. Gutman, I., Todorovic, D., Vuckovic, M., and Jungwirth, P. (1992) Modelling spontaneous chiral stereoselection: the Frank mechanism with racemization. *J. Chem. Soc., Faraday Trans.*, **88**, 1123–1127.
29. Avetisov, V. and Goldanskii, V. (1996) Mirror symmetry breaking at the molecular level. *Proc. Natl. Acad. Sci. U.S.A.*, **93**, 11435–11442.
30. Plasson, R., Bersini, H., and Commeyras, A. (2004) Recycling Frank: spontaneous emergence of homochirality in noncatalytic systems. *Proc. Natl. Acad. Sci. U.S.A.*, **101**, 16733–16738.
31. Mauksch, M., Tsogoeva, S.B., Wei, S., and Martynova, I. (2007) Demonstration of spontaneous chiral symmetry breaking in asymmetric Mannich and aldol reactions. *Chirality*, **19**, 816–825.
32. Boiteux, A. and Hess, B. (1981) Design of glycolysis. *Philos. Trans. R. Soc. London Ser. B.*, **293**, 5–22.
33. Sugimoro, T., Hyuga, H., and Saito, Y. (2008) Fluctuation induced homochirality. *J. Phys. Soc. Jpn.*, **77**, 064606.
34. Ribo, J.M. and Hochberg, D. (2008) Stability of racemic and chiral steady states in open and closed chemical systems. *Phys. Lett. A*, **373**(1), 111–122.
35. Blackmond, D.G. (2009) Challenging the concept of "recycling" as a mechanism for the evolution of homochirality in chemical reactions. *Chirality*, **21**, 359–362.
36. Mauksch, M. and Tsogoeva, S.B. (2008) Spontaneous emergence of homochirality via coherently coupled antagonistic and reversible reaction cycles. *ChemPhysChem*, **9**, 2359–2372.
37. Nicolis, G. and Prigogine, I. (1981) Symmetry breaking and pattern selection in far-from-equilibrium systems. *Proc. Natl. Acad. Sci. U.S.A.*, **78**(2), 659–663.
38. Iwamoto, K. (2003) Spontaneous appearance of chirally asymmetric steady

states in a reaction model including Michaelis-Menten type catalytic reactions. *Phys. Chem. Chem. Phys.*, **5**, 3616–3621.

39. Qian, H. (2006) Open-system nonequilibrium steady state: statistical thermodynamics, fluctuations, and chemical oscillations. *J. Phys. Chem. B*, **110**, 15063–15074.

40. Huber, C. and Wächtershäuser, G. (1998) Peptides by activation of amino acids with CO on (Ni,Fe)S surfaces: implications for the origin of life. *Science*, **281**, 670–672.

41. Crusats, J., Hochberg, D., Moyano, A., and Ribo, J.M. (2009) Frank model and spontaneous emergence of chirality in closed systems. *ChemPhysChem*, **10**, 2123–2131.

42. Seelig, F.F. (1971) System-theoretic model for the spontaneous generation of optical antipodes in strongly asymmetric yield. *J. Theor. Biol.*, **31**, 355–361.

43. Decker, P. (1975) Evolution in bioids: hypercompetivity as a source of bistability and a possible role of metal complexes as prenucleoprotic mediators of molecular asymmetry. *Orig. Life*, **6**, 211–218.

44. Soai, K., Shibata, T., Morioka, H., and Choji, K. (1995) Asymmetric autocatalysis and amplification of enantiomeric excess of a chiral molecule. *Nature*, **378**, 767–768.

45. Schiaffino, L. and Ercolani, G. (2009) Amplification of chirality and enantioselectivity in the asymmetric autocatalytic Soai reaction. *ChemPhysChem*, **10**(14), 2508–2515.

46. Bailey, P.D. (1995) On the self-replication of chirality. *J. Chem. Soc., Chem. Commun.*, 1797–1798.

47. Blackmond, D.G., McMillan, C.R., Ramdeehul, S., Schorm, A., and Brown, J.M. (2001) Origins of asymmetric amplification in autocatalytic alkylzinc additions. *J. Am. Chem. Soc.*, **123**, 10103–10104.

48. Buono, F.G. and Blackmond, D.G. (2003) Kinetic evidence for a tetrameric transition state in the asymmetric autocatalytic alkylation of pyrimidyl aldehydes. *J. Am. Chem. Soc.*, **125**, 8978–8979.

49. Barabas, B., Caglioti, L., Micskei, K., and Palyi, G. (2009) Data-based stochastic approach to absolute asymmetric synthesis by autocatalysis. *Bull. Chem. Soc. Jpn.*, **82**, 1372–1376.

50. Mislow, K. (2003) Absolute asymmetric synthesis: a commentary. *Collect. Czech. Chem. Commun.*, **68**, 849–863.

51. Wynberg, H. (1989) Asymmetric autocatalysis: facts and fancy. *J. Macromol. Sci. A*, **26**(8), 1033–1041.

52. Mauksch, M., Tsogoeva, S.B., Martynova, I.M., and Wei, S. (2007) Evidence of asymmetric autocatalysis in organocatalytic reactions. *Angew. Chem. Int. Ed.*, **46**, 393–396.

53. Pizzarello, S. and Weber, A.L. (2004) Prebiotic amino acids as asymmetric catalysts. *Science*, **303**, 1151.

54. Cordova, A., Sunden, H., Xu, Y., Ibrahem, I., Zou, W., and Engqvist, M. (2006) Sugar-assisted kinetic resolution of amino acids and amplification of enantiomeric excess of organic molecules. *Chem. Eur. J.*, **12**, 5446–5451.

55. Nanita, S.C. and Cooks, R.G. (2006) Serine octamers: cluster formation, reactions, and implications for biomolecule homochirality. *Angew. Chem. Int. Ed.*, **45**, 554–569.

56. Mauksch, M., Wei, S.-W., Freund, M., Zamfir, A., and Tsogoeva, S.B. (2010) Spontaneous mirror symmetry breaking in the aldol reaction and its potential relevance in prebiotic chemistry. *Orig. Life Evol. Biosph.*, **40**(1), 79–91.

57. Eigen, M. and Schuster, P. (1978) The hypercycle. A principle of natural self-organization. Part B: the abstract hypercycle. *Naturwissenschaften*, **65**, 7–41.

58. Quayle, J.M., Slawin, A.M.Z., and Philp, D. (2002) A structurally-simple minimal self-replicating system. *Tetrahedron Lett.*, **43**, 7729–7733.

59. von Kiedrowski, G.A. (1986) Self-replicating hexadeoxynucleotide. *Angew. Chem. Int. Ed.*, **25**, 932–935.

60. Kamioka, S., Ajami, D., and Rebek, J. (2009) Synthetic autocatalysts show organocatalysis of other reactions. *Chem. Commun.*, 7324–7326.

61. Green, M.M., Park, J.-W., Sato, T., Teramoto, A., Lifson, S., Selinger, R.L.B., and Selinger, J.V. (1999) The macromolecular route to chiral amplification. *Angew. Chem. Int. Ed.*, **38**, 3138–3154.
62. Haq, S., Liu, N., Humblot, V., Jansen, A.P.J., and Raval, R. (2009) Drastic symmetry breaking in supramolecular organization of enantiomerically unbalanced monolayers at surfaces. *Nat. Chem.*, **1**, 409–414.
63. Brandenburg, A., Lehto, H.J., and Lehto, K.M. (2007) Homochirality in an early peptide world. *Astrobiology*, **7**(5), 725–732.
64. Plasson, R. and Bersini, H. (2009) Energetic and entropic analysis of mirror symmetry breaking processes in a recycled microreversible chemical system. *J. Phys. Chem. B*, **113**, 3477–3490.
65. Hayashi, Y., Matsuzawa, M., Yamaguchi, J., Yonehara, S., Matsumoto, Y., Shoji, M., Hashizume, D., and Koshino, H. (2006) Large nonlinear effect observed in the enantiomeric excess of proline in solution and that in the solid state. *Angew. Chem. Int. Ed.*, **45**, 4593–4597.
66. Klussmann, M., Iwamura, H., Mathew, S.P., Wells, D.H., Pandya, U., Alan Armstrong, A., and Blackmond, D.G. (2006) Thermodynamic control of asymmetric amplification in amino acid catalysis. *Nature*, **441**, 621–623.
67. Breslow, R. and Levine, M.S. (2006) Amplification of enantiomeric concentrations under credible prebiotic conditions. *Proc. Natl. Acad. Sci. U.S.A.*, **103**, 12979–12980.
68. Soloshonok, V.A. (2006) Remarkable amplification of the self-disproportionation of enantiomers on achiral-phase chromatography columns. *Angew. Chem. Int. Ed.*, **118**, 780–783.
69. Albrecht, M., Soloshonok, V.A., Schrader, L., Yasumoto, M., and Suhm, M.A. (2009) Chirality-dependent sublimation of α-(trifluoromethyl)-lactic acid: Relative vapor pressures of racemic, eutectic, and enantiomerically pure forms, and vibrational spectroscopy of isolated (S,S) and (S,R) dimers. *J. Fluor. Chem.*, **131**, 495–504.
70. Thiemann, W. and Teutsch, H. (1990) Possible amplification of enantiomer excess through structural properties of liquid crystal – a model for origin of optical activity in the biosphere. *Orig. Life. Evol. Biosph.*, **20**, 121–126.
71. Perry, R.H., Wu, C., Nefliu, M., and Cooks, R.G. (2007) Serine sublimes with spontaneous chiral amplification. *Chem. Commun.*, 1071–1073.
72. Hazen, R.M., Filley, T.R., and Goodfriend, G.A. (2001) Selective adsorption of L- and D-amino acids on calcite: implications for biochemical homochirality. *Proc. Natl. Acad. Sci. U.S.A.*, **98**, 5487–5490.
73. Kuhn, A. and Fischer, P. (2009) Absolute asymmetric reduction based on the relative orientation of achiral reactants. *Angew. Chem. Int. Ed.*, **48**, 6857–6860.
74. Chenchaiah, P.C., Holland, H.L., and Richardson, M.F. (1982) A new approach to the synthesis of chiral molecules from nonchiral reactants. Asymmetric induction by reaction at one surface of a single (nonchiral) crystal. *J. Chem. Soc., Chem. Commun.*, 436–437.
75. Addadi, L., Berkovitch-yellin, Z., Weissbuch, I., Lahav, M., and Leiserowitz, L. (1983) Morphology engineering of organic crystals with the assistance of "tailor-made" growth inhibitors. *Mol. Cryst. Liq. Cryst.*, **96**, 1–17.
76. Noorduin, W.L., van der Asdonk, P., Meekes, H., van Enckevort, W.J.P., Kaptein, B., Leeman, M., Kellogg, R.M., and Vlieg, E. (2009) Complete chiral resolution using additive-induced crystal size bifurcation during grinding. *Angew. Chem. Int. Ed.*, **48**, 3278–3280.
77. Pincock, R.E., Perkins, R.R., Ma, A.S., and Wilson, K.R. (1971) Probability distribution of enantiomorphous forms in spontaneous generation of optically active substances. *Science*, **174**, 1018–1020.
78. Kondepudi, D.K., Kaufman, R.J., and Singh, N. (1990) Chiral symmetry breaking in sodium chlorate crystallization. *Science*, **250**, 975–976.
79. Havinga, E. (1954) Spontaneous formation of optically active substances. *Biochim. Biophys. Acta*, **13**, 171–174.

80. Kostyanovsky, R.G., Kostyanovsky, V.R., Kadorkina, G.K., and Lyssenko, K.A. (2001) Wedekind-Fock-Havinga salt Me(Et)N$^+$(All)PhI−∗CHCl$_3$ as historically the first object for absolute asymmetric synthesis: spontaneous resolution, structure and absolute configuration. *Mendeleev Commun.*, **11**, 1−42.

81. Crusats, J., Veintemillas-Verdaguer, S., and Ribó, J.M. (2006) Homochirality as a consequence of thermodynamic equilibrium? *Chem. Eur. J.*, **12**, 7776−7781.

82. Asakura, K., Nagasaka, Y., Osanai, S., and Kondepudi, D.K. (2005) Kinetic model for chiral symmetry breaking transition in growth front of conglomerate crystal phase. *J. Phys. Chem. B*, **109**, 1586−1592.

83. Viedma, C. (2005) Chiral symmetry breaking during crystallization: complete chiral purity induced by nonlinear autocatalysis and recycling. *Phys. Rev. Lett.*, **94**(6), 065504.

84. Uwaha, M. (2004) A model for complete chiral crystallisation. *J. Phys. Soc. Jpn.*, **73**, 2601−2603.

85. Blackmond, D.G. (2007) "Chiral Amnesia" as a driving force for solid-phase homochirality. *Chem. Eur. J.*, **13**, 3290−3295.

86. Noorduin, W.L., Izumi, T., Millemaggi, A., Leeman, M., Meekes, H., van Enckevort, W.J.P., Kellog, R.M., Kaptein, B., Vlieg, E., and Blackmond, D.G. (2008) Emergence of a single solid chiral state from a nearly racemic amino acid derivative. *J. Am. Chem. Soc.*, **130**, 1158−1159.

87. Noorduin, W.L., Vlieg, E., Kellogg, R.M., and Kaptein, B. (2009) From Ostwald ripening to single chirality. *Angew. Chem. Int. Ed.*, **48**, 9600−9606.

88. Uwaha, M. (2008) Simple models for chirality conversion of crystals and molecules by grinding. *J. Phys. Soc. Jpn.*, **77**, 083802−083805.

89. Tsogoeva, S.B., Wei, S., Freund, M., and Mauksch, M. (2009) Generation of highly enantioenriched crystalline products in reversible asymmetric reactions with racemic or achiral catalysts. *Angew. Chem. Int. Ed.*, **48**, 590−594.

90. Wei, S., Mauksch, M., and Tsogoeva, S.B. (2009) Autocatalytic enantiomerisation at the crystal surface in deracemization of scalemic conglomerates. *Chem. Eur. J.*, **15**, 10255−10262.

91. Saito, Y. and Hyuga, H. (2008) Chiral crystal growth under grinding. *J. Phys. Soc. Jpn.*, **77**, 113001−113004.

92. Flock, A.M., Reucher, C.M.M., and Bolm, C. (2010) Enantioenrichment by iterative retro-aldol/aldol reaction catalyzed by an achiral or racemic base. *Chem. Eur. J.*, **16**, 3918−3921.

93. Viedma, C., Ortiz, J.E., de Torres, T., Izumi, T., and Blackmond, D.G. (2008) Evolution of solid phase homochirality for a proteinogenic amino acid. *J. Am. Chem. Soc.*, **130**, 15274−15275.

94. Gayathri, V.S. and Rao, M. (2007) Fluctuation-induced chiral symmetry breaking in autocatalytic reaction-diffusion systems. *Europhys. Lett.*, **80**, 28001−28006.

95. Scolnik, Y., Portnaya, I., Cogan, U., Tal, S., Haimovitz, R., Fridkin, M., Elitzur, A.C., Deamer, D.W., and Shinitzky, M. (2006) Subtle differences in structural transitions between poly-L- and poly-D amino acids of equal length in water. *Phys. Chem. Chem. Phys.*, **8**, 333−339.

96. Toxvaerd, S. (2009) Origin of homochirality in biosystems. *Int. J. Mol. Sci.*, **10**, 1290−1299.

Part VI
Conclusion: From Natural Facts to Chemical Fictions

24
Artifacts and Natural Substances Formed Spontaneously
Pierre Champy

> *M'enfin!*
> Gaston Lagaffe

> *In the supreme horror of that second I forgot what had horrified me, and the burst of black memory vanished in a chaos of echoing images.*
> Howard Phillips Lovecraft

> *Ceci n'est pas une pipe.*
> René Magritte

24.1
Introduction

A definition of natural product artifacts can be easily constructed on etymological foundations. The word "artifact (/artefact)" originates from Latin: its prefix "*art-*" arises from "*ars, artis*," a very general word for "way of being," a meaning that latter evolved to "ability acquired through study or practice." "*Art-*" thus embraced, in numerous variations, "human activities tending toward order." Complemented with the suffix "*-fact*" ("*facere*" – to do), it expressed the concept of craft, with positive ("*artifex*" – profession) or negative ("*artificium*" – cunning) aspects. "Artificial," "*artis factum*," thus in opposition with "natural," was taken up in medical English as "artifact" to designate a living tissue alteration provoked by scientific intervention [1]. "Artifact" now bears several meanings in the common language or within specific lexical fields. For the anthropologist, an artifact is "a human-made object that gives information about the culture of its creator and users," or "a product transformed by Man, distinguishable from another provoked by a natural phenomenon."[1] For any scientist, an artifact is an "undesired alteration in data, introduced by a technique and/or technology." In our context, "data" will be transposed to a molecular level, and considered through the natural products chemist's

---

1) Note that both definitions can apply to extraction artifacts.

Biomimetic Organic Synthesis, First Edition. Edited by Erwan Poupon and Bastien Nay.
© 2011 Wiley-VCH Verlag GmbH & Co. KGaA. Published 2011 by Wiley-VCH Verlag GmbH & Co. KGaA.

looking glass. After this primary focus, precision is needed. Indeed, both *"ars"* and *"factum"* shall prevent us from considering some natural products as artifacts:

- Inconstant metabolites (e.g., inducible metabolites or trace compounds arising from unspecific metabolism) can barely be called artifacts.
- Taxonomic mistakes prior to chemical study induce artifactual results, mainly at a chemotaxonomic level (Figure 24.1). Metabolites originating from symbiotic or pathogen microorganisms may also cause chemotaxonomical misinterpretations [2].[2)]
- Similarly, isolating a phthalate or describing the trifluoroacetic acid (TFA) salt of an alkaloid as a new natural compound is a virtual artifact only.[3)]

We will consider here an artifact as a natural product chemically altered while manipulated, with the exception of deliberate derivatization and degradation procedures. This molecule can be modified in its matrix or under purified form. The event leading to its formation can take place as soon as harvest, treatment, storage, and processing of the producer organism. It more frequently occurs during the extraction phase or isolation process. Purely analytical artifacts, though commonly observed, will not be dealt with in this chapter. An artifact is not always identified as such. Its true nature may be evidenced on the basis of several clues:

- appearance during isolation, generally noticed by TLC or HPLC-UV/MS; the precursor is sometimes monitored;
- absence when a new procedure is applied (e.g., solvent change);
- obtention of a racemic form or of several enantiomers when enzymatic, stereospecific biosynthesis is expected;
- an unexpected oxidation state for the molecule;
- a dubious structure regarding biosynthetic origin or a lack of match with chemotaxonomic standards;
- work with a notoriously unstable group of compounds;
- critical analysis of the protocol used (i.e., pH and redox conditions, light exposure, energy transfer, time spent in solution, etc.) or inappropriate storage and shelf-life.

In case of doubt, authors frequently try to investigate the artifactual nature of the compounds they isolate (modification of isolation protocol; deliberate conversion of presumed precursor; etc.). Such demonstrations will be presented, if available, for the examples displayed throughout this chapter.

In some interesting cases, a molecule can definitely look like an artifact but prove to be natural. Small molecules isolated from the bryozoan *Biflustra perfragilis* (Membraniporidae; methanol, chloromethane, dichloromethane, methanethiol,

2) Marine products are often subject to caution, and endophytic compounds obtained during the study of higher plants are now largely considered.

3) Several examples can be found in the recent literature.

"This is not an acetogenin of Annonaceae"

Ampelocissus sp., Vitaceae, root

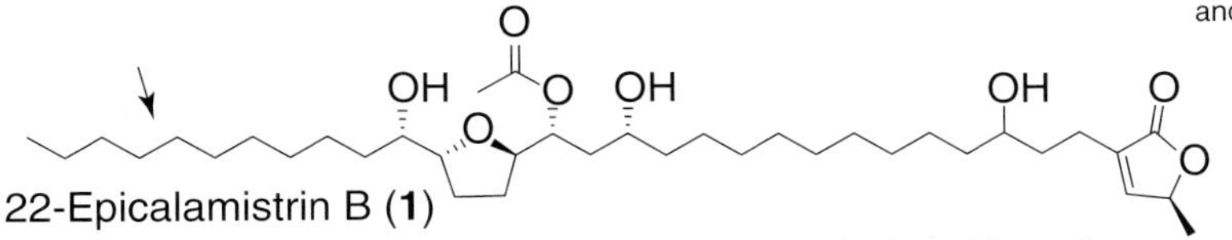

22-Epicalamistrin B (1)

"To our knowledge this is the first report of acetogenins occurring in the Vitaceae", state Pettit et al. Isolation of (**1**), highly interesting from a chemotaxonomical viewpoint, is indeed peculiar and requires confirmation, as Annonaceous acetogenins were thought to be exclusively produced in the unrelated Annonaceae taxa. The plant material was apparently poorly identified: Was the right species harvested? [*J. Nat. Prod.* 2008, **71**, 130-133]

In the 1990s, *Aristolochia fangchi* roots (Aristolochiaceae) were mistaken with those of *Stephania tetrandra* (Menispermaceae) because of near-homonymy in Chinese, and incorporated in weight loss dietary supplements.

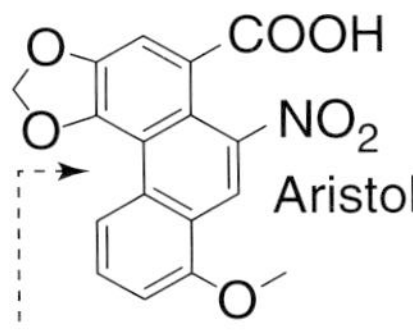

Aristolochic acid I (**2**)

highly nephrotoxic upon chronic exposure

Figure 24.1 Possible or proven botanical misidentifications [3, 4].

dimethyl sulfide, dimethyl disulfide) have been identified as genuine products rather than laboratory contaminants [5]. Nevertheless, many spontaneously self-assembled compounds may constitute borderline cases. Examples have been reviewed by Gravel and Poupon [6].

Artifacts of natural molecules are feared and abhorred by the common pharmacognosist. Although often perceived as nebulous entities, they are frequently encountered [7],[4],[5] echoing throughout the halls of literature: the keywords "artifact/artefact" yield about 2000 answers in a general search on the website of the journal *Phytochemistry*. An extensive review of published data is therefore not proposed here. However, relevant, generally recent examples were chosen, so as to give an overview of the generation of artifacts in regard to the reactivity of natural products. They will be presented according to their main causes (enzymatic, oxidative, photo- and thermally-induced processes, basic and acidic-catalysis, working solvents), and, secondarily, stigmata.

24.2
Glucosidases as Triggers for Formation of By-products

Some plant secondary metabolites are classified as phytoanticipins. When plant tissue in which they are present is disrupted they are bio-activated by the action of enzymes – generally β-glucosidases. These binary systems – two sets of components that when separated are relatively inert – provide plants with an immediate chemical defense against herbivores and pathogens [17]. Hydrolysis products can be viewed as artifacts (e.g., aldehydes produced by hydrolysis of cyanogenic glycosides and spontaneous release of HCN [18]). They often are reactive and can spontaneously rearrange, as in the case of ranunculine (3), a hemiterpenic glucoside from Ranunculaceae (*Ranunculus*, *Clematis* spp.). It yields protoanemonine (4), a reactive compound susceptible to nucleophilic attacks and responsible for contact dermatitis in humans (Scheme 24.1) [19].

Glucosinolates are inert thio-S-(β-D-glucopyranosyl)-(Z)-N-hydroximinosulfate esters with a variable side chain. They are found almost exclusively in Brassicaceae spp. Upon injury of plant tissues they are hydrolyzed by myrosinases (β-thioglucoside hydrolases) to yield unstable thiohydroxamate sulfonates that spontaneously rearrange to insecticidal isothiocyanates or to other compounds, depending mostly on pH conditions. Glucosinolates occur in many vegetables, and are of interest for human health: by-products and the conditions of their

---

4) Notably, R. Verpoorte and coworkers recently published a review about extraction artifacts due to commonly used solvents [7].
5) Notably, the whole history of culinary traditions is marked by chemical processing in search of palatability or taste [8], yielding "artifactual" molecules (e.g., Maillard reaction [9], fermentation of *Camellia sinensis* (Theaceae) leaves [10–12], and interactions of polyphenols during wine maturation [13]). Medical traditions include preparation methods aimed at equilibrating the risk/benefit ratio of herbal remedies (e.g., oxidation of anthrones glycosides of bark of *Rhamnus purshiana* (Rhamnaceae) [14]; hydrolysis of norditerpenic alkaloid esters of *Aconitum* spp. tuber and roots (Ranunculaceae) [15, 16]).

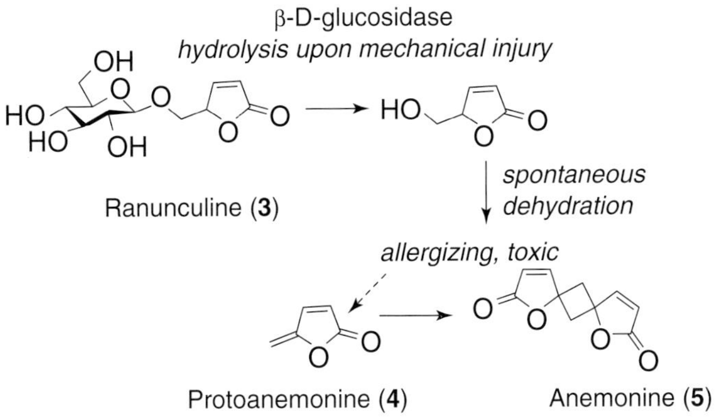

Scheme 24.1 Hemiterpenes from Ranunculaceae that cause irritation.

formation (including cooking conditions) have thus been extensively studied and reviewed (Scheme 24.2) [20].

24.3
Oxidation Processes

Spontaneous oxidative processes are common causes of artifact formation. The newly formed compounds gain stability, with extended conjugation. These artifacts frequently occur during the isolation process and can often be efficiently tracked down. Many marine products are involved, some appearing to be unstable in atmospheric rather than aqueous environment, especially when exposed to light. However, plant products of various biogenetic origins also yield examples, as in the aforementioned case of *Rhamnus* hydroxyanthracenes (see footnote 5). Oxidized positions are quite various, and it can be noticed that enzymatic intervention would seem likely in several cases. Therefore, authors sometimes appear to be very suspicious about artifactual origins, and discussions are frequent (Scheme 24.3).

24.3.1
Thiol Oxidation

The numerous oxidation and rearrangement products of 2-propenesulfenic acid (**9**) and of allicin (**10**), spontaneously formed from **9** after enzymatic hydrolysis of (+)-S-2-propenyl-L-cysteine-S-oxide (**8**, a phytoanticipin from garlic, *Allium sativum*, Asparagaceae, ex-Alliaceae; Scheme 24.4), have been reviewed extensively in an elegant and comprehensive article by Erick Block, who raises the following question in regard to this horde of organic sulfur compounds: "What is a natural product?" [22].[6]

---

6) Block and, more recently, Rose *et al.* also present compounds from other Alliaceae species. [22].

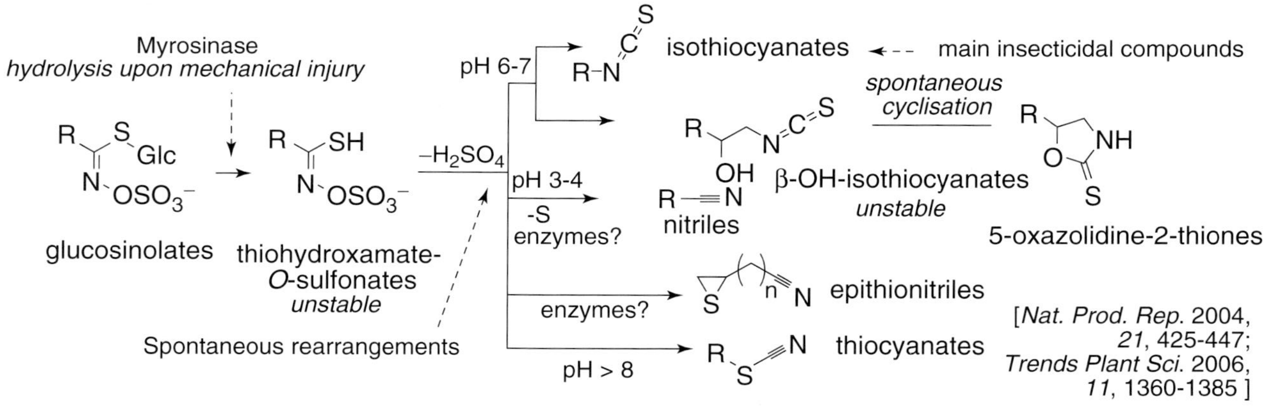

Scheme 24.2 Rearrangement products following hydrolysis of glucosinolates.

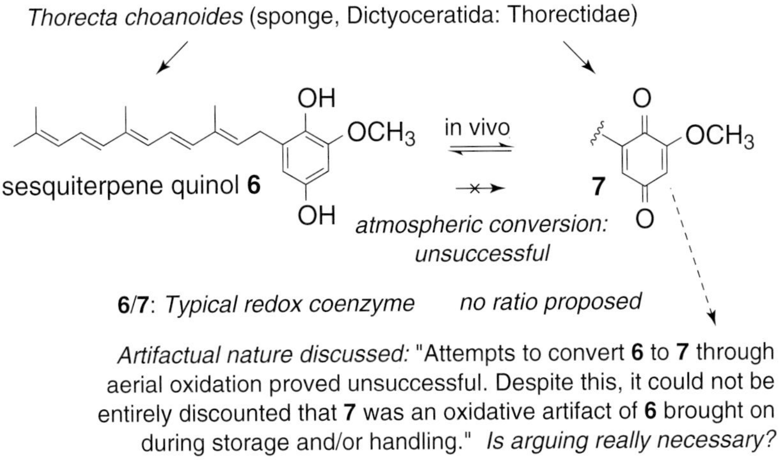

Thorecta choanoides (sponge, Dictyoceratida: Thorectidae)

sesquiterpene quinol **6** ⇌ **7** (in vivo)

atmospheric conversion: unsuccessful

6/7: *Typical redox coenzyme*   no ratio proposed

Artifactual nature discussed: "Attempts to convert **6** to **7** through aerial oxidation proved unsuccessful. Despite this, it could not be entirely discounted that **7** was an oxidative artifact of **6** brought on during storage and/or handling." *Is arguing really necessary?*

Scheme 24.3 Oxidized coenzyme Q-3 (**7**): the *redox status quandary* [21].

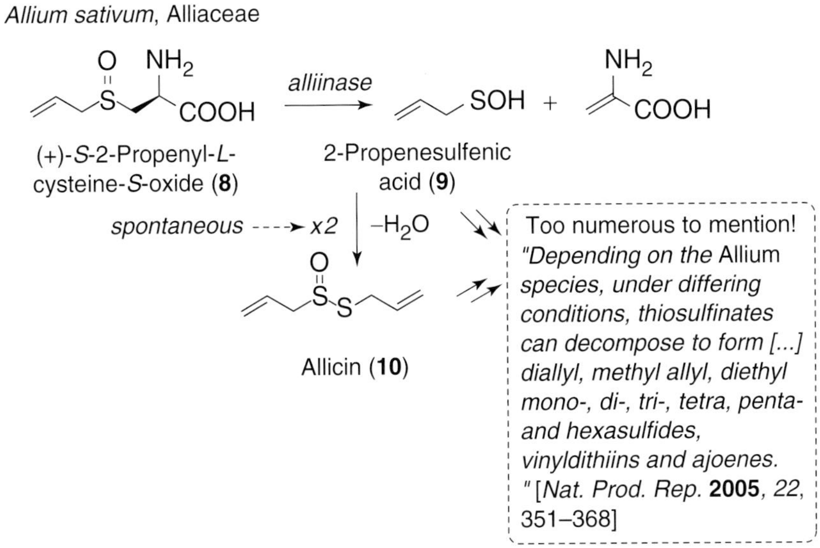

Allium sativum, Alliaceae

(+)-S-2-Propenyl-L-cysteine-S-oxide (**8**) —alliinase→ 2-Propenesulfenic acid (**9**) + NH$_2$-CH-COOH

spontaneous ×2, $-H_2O$ → Allicin (**10**)

Too numerous to mention! "Depending on the *Allium* species, under differing conditions, thiosulfinates can decompose to form [...] diallyl, methyl allyl, diethyl mono-, di-, tri-, tetra-, penta- and hexasulfides, vinyldithiins and ajoenes." [*Nat. Prod. Rep.* **2005**, *22*, 351–368]

Scheme 24.4 Allicin (**10**) as the first step in a long series of artifacts.

Unrelated species are similarly problematic; for example, oxidation and rearrangement of the volatile compounds in the pungent (and thus controversial) durian fruit, *Durio zibethinus*, Malvaceae (ex-Bombacaceae/Durionaceae) [23].

Dithiorhodysinin disulfide (**12**), a furanosesquiterpene isolated from the nudibranch *Ceratosoma brevicaudatum*, was reported to arise from a thiol monomer (**11**) upon exposure to air. Compound **11** was previously obtained from the sponge (and prey) *Dysidea avara*, but not **12**. It is hypothesized that **12** might originate from the nudibranch's diet, corresponding to an *in vivo* storage form rather than to an artifact (Scheme 24.5) [24].

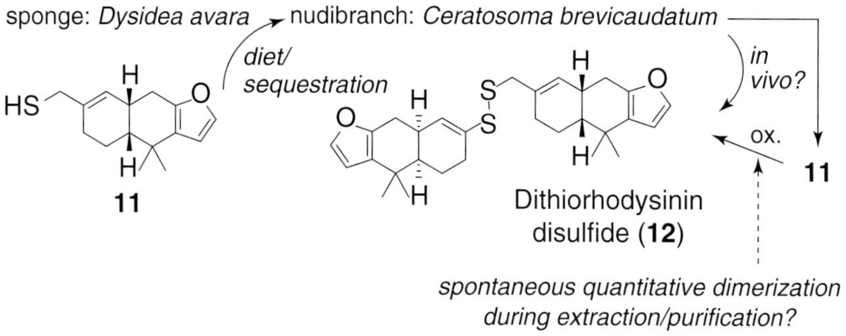

Scheme 24.5 Ambiguous fate of a marine thiol group.

Pyrroloiminoquinolines from a sponge of the *Latrunculia* genus (Latrunculiidae) yielded another example of such dimerization. Discorhabdin B (**13**) was shown to undergo degradation upon light exposure. Among formed products, discorhabdin W (**14**) was identified as a disulfide dimer of **13**; compound **14** converts, under reducing conditions, into the unstable monomer **15**, which in turn spontaneously cyclizes to form **13** (Scheme 24.6) [25].

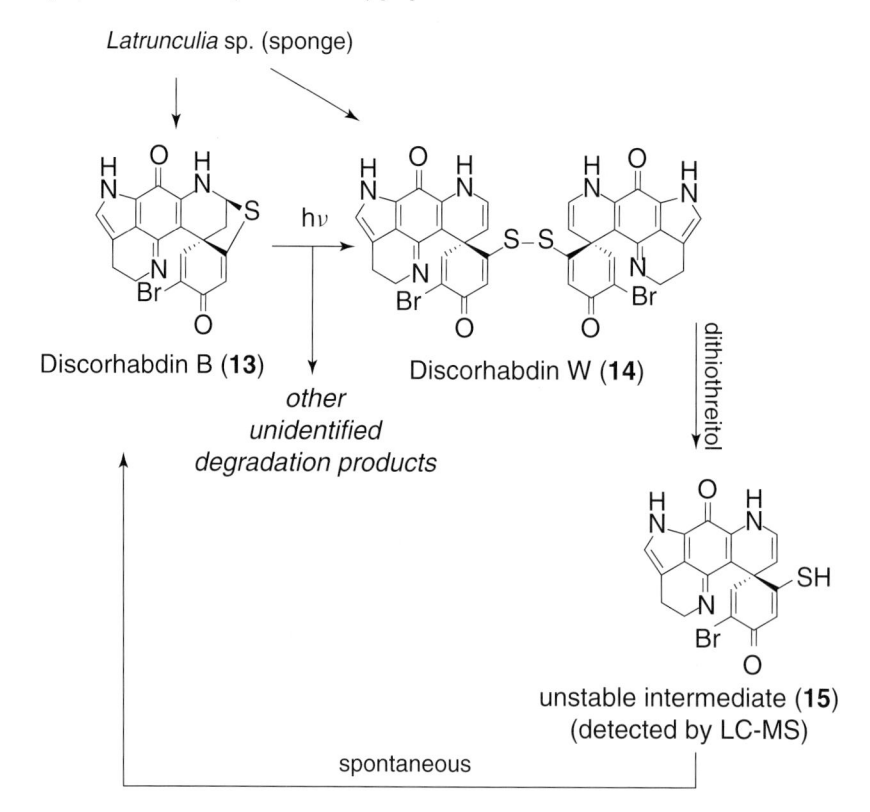

Scheme 24.6 Dimerization and recovery of discorhabdin B (**13**).

24.3.2
Oxidation Processes of Oxygenated Functions

Oxidation of conjugated oxygenated positions can occur within a crude extract, but are more frequently observed during isolation. A precursor is sometimes obtained along with the newly formed metabolite, as in the case of clathrins B (**16**) and C (**17**), from an Australian sponge (*Clathria* sp.). Here, the authors clearly identified the oxidation process of (**16**). Complete conjugation is remarkable for clathrin C (**17**) (Scheme 24.7) [26].

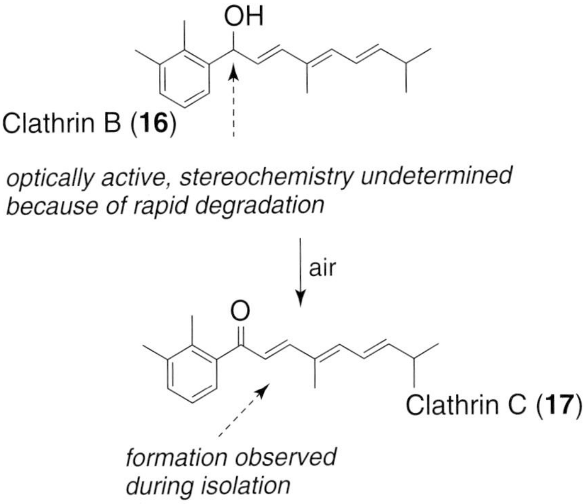

Clathrin B (**16**)

optically active, stereochemistry undetermined because of rapid degradation

↓ air

Clathrin C (**17**)

formation observed during isolation

Scheme 24.7 Clathrins from *Clathria* sp.

Oxidation of hydroperoxide-bearing carbons is also exemplified in the literature, with an obvious gain of stability for the artifactual derivatives. Among clerod-3,4-ene diterpenes found in *Aristolochia* spp. (Aristolochiaceae), two presenting a peroxide function at C2 (**18** and **20**) were isolated, accompanied by carbonyl analogs (**19** and **21**) (Scheme 24.8). Compounds **18** and **20** were shown to undergo rapid decomposition to the corresponding enones **19** and **21**, respectively, suggesting that 2-oxo-3-clerodenes could be artifacts [27]. This might indeed be the case for 2-oxopopulifolic acid (**22**) [28] and *rel*-(5S,8R,9S,10R)-2-oxo-*ent*-3-cleroden-15-oic acid (**23**) [29], the precursors of which were not obtained. However, apparent decomposition products of hydroperoxides cannot always be regarded as artifacts, especially when the stability of the hydroperoxide permits a full purification process and structural analysis (e.g., musambin A (**24**) from *Markhamia lutea* (Bignoniaceae), for which no degradation to musambin C (**25**) was observed [30]).

Lambertellols A (**26**) and B (**27**) were isolated from the filamentous fungus *Lambertella corni-maris* under conditions of co-culture with the pathogenic fungus *Molinia fructigena*, along with lambertellin (**30**). Isomerization between **26** and **27** was observed during structural studies. It occurred under mild conditions:

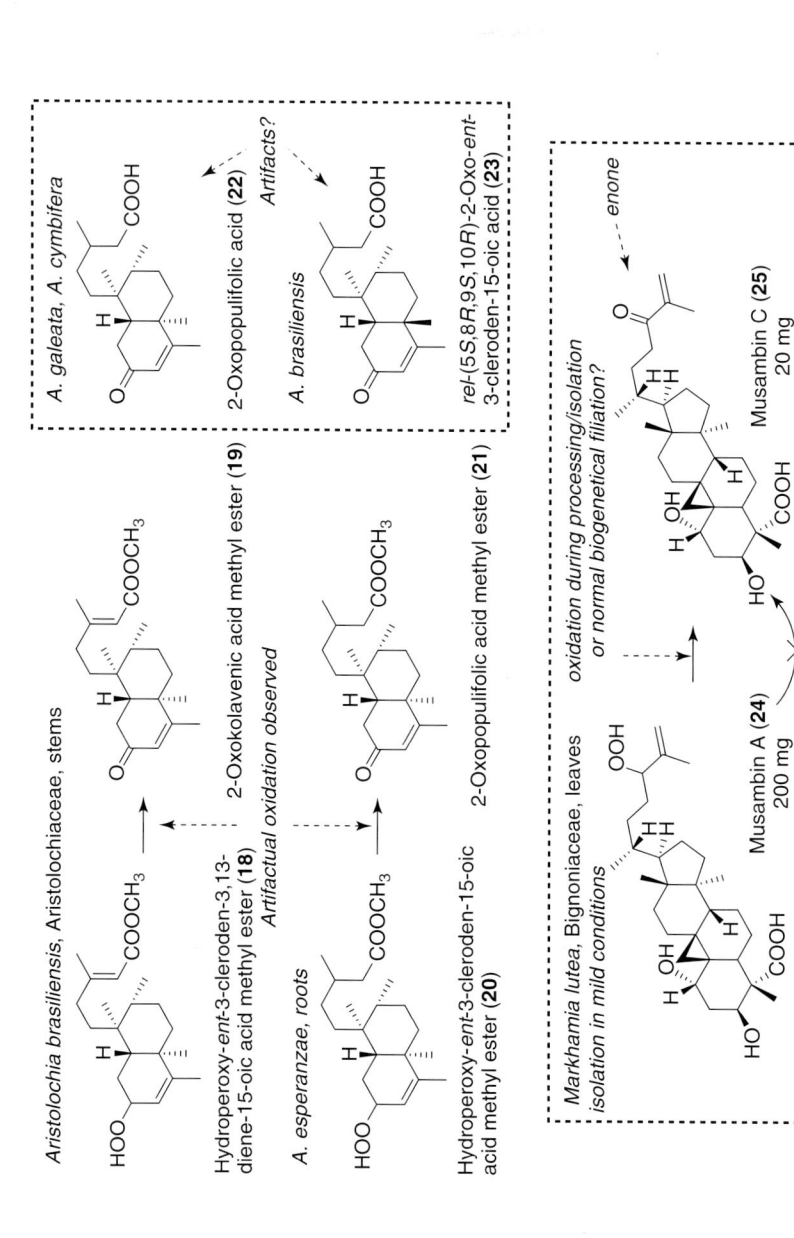

858 | *24 Artifacts and Natural Substances Formed Spontaneously*

Scheme 24.8 Possible artifacts from hydroperoxides.

chromatography on silica gel, standing in methanol or within fungus culture medium. Biosynthesis of 26 and 27 is induced in response to acidification of the medium. It is therefore likely that both molecules co-exist in "real-life" conditions. Nevertheless, conversion of 26 and 27 into 30 was observed under the same conditions as isomerization. To explain this interconversion, the authors proposed a retro-Michael type opening of the butenolide ring (26/27) to a carboxylic intermediate (28), followed by oxidation of its semiquinone moiety into a quinone (29). An intramolecular Michael-type ring closure at position C2 furnishes a pyrone, with a final oxidative aromatization to give 30. Interestingly, this artifact (30) is the active substance responsible for growth inhibition of M. fructigena during mycoparasitism by L. corni-maris (Scheme 24.9) [31].

24.3.3
Newly Oxygenated Products

Spontaneous oxygenations of unsaturated positions due to air exposure are frequently described, and are most likely produced by photocatalyzed free-radical mechanisms. Some compounds proposed as being artifacts may also appear to be enzymatically formed: an epoxy derivative (32) of radicinol (31; Scheme 24.10a), occurring in aging but not in fresh cultures of the phytopathogen *Alternaria chrysanthemi* (Deuteromycetes), is supposed to be due to oxidation by $^1O_2$ [32][7] but a change in the metabolic profile of the organism could also explain its appearance. Spontaneous oxygenation can also be quickly followed by dehydration. This pathway can be hypothesized for formation of ochrone B (34), isolated from *Coelogyne ochracea* (Orchidaceae), along with its 9,10-dihydro analog ochrone A (33). Despite a possible biosynthetic relationship, 34 proved to be an artifact due to aerial oxidation, as it was found in pure 33 after standing for several days (Scheme 24.10b) [33].

Dysidea herbacea (Aplysillidae), a common sponge of tropical seas, allowed isolation of the known furanosesquiterpene herbacin (35) and of a new hydroxylbutenolide 36, the relative stereochemistry of which was further confirmed by conversion of 35 in quantitative yield upon photochemical oxygenation. A complementary search for a putative one-ringed furanic biogenetic precursor of 36 (37) proved negative [34]. Interestingly, the closely related compound 38, previously isolated from *Dysidea pallescens*, thus appears to be artifactual (Scheme 24.11).

During isolation of N-methylwelwitindolinone C isonitrile (39) from the terrestrial cyanobacteria *Fischerella muscicola* and *F. major*, formation of three minor oxidized derivatives (42–44) was observed, and could be reproduced under UV irradiation. The authors proposed a pathway from 39, with photocatalyzed peroxidation leading to the hydroperoxide (40), the reduction of which yields 43. The isonitrile group in 43 slowly hydrolyses to give 44 during fractionation; 44 is readily obtained by photocatalysis from pure 43. Alternatively, from

---
7) The authors did not try to convert isolated 59 into 58.

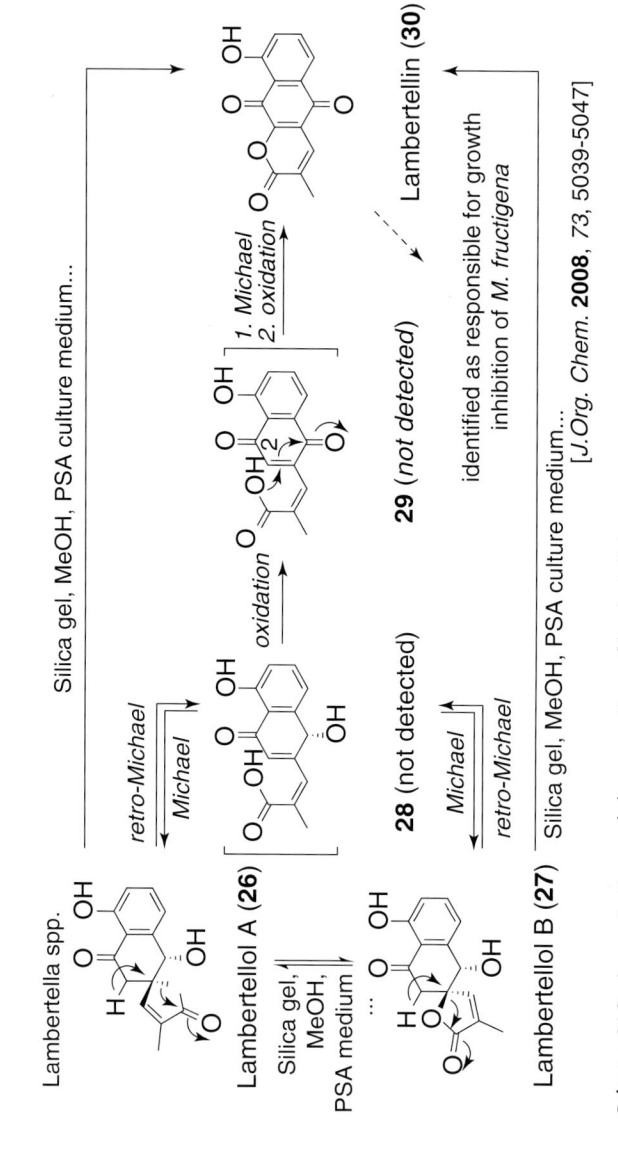

Scheme 24.9 Isomerization and decomposition of lambertellol.

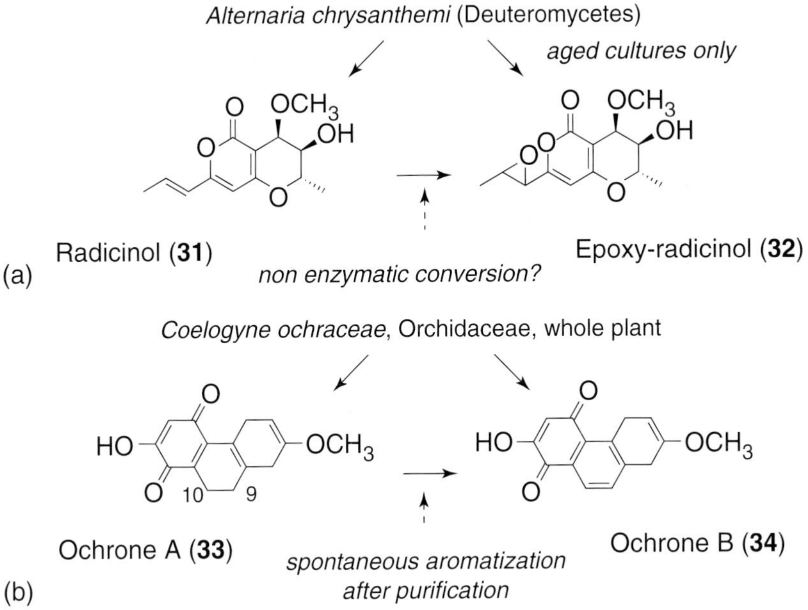

Scheme 24.10 Miscellaneous oxidation artifacts.

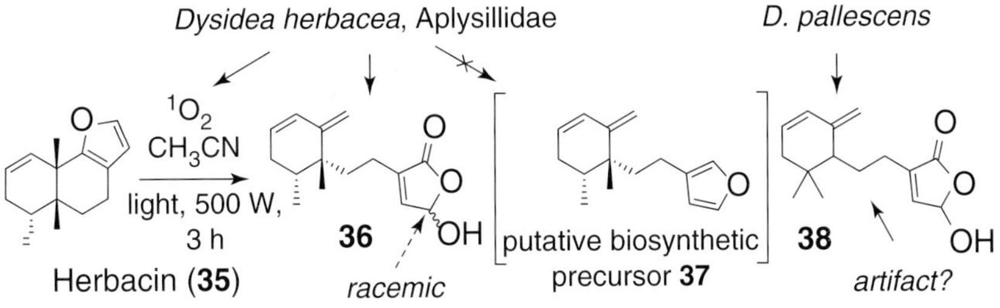

Scheme 24.11 Artifactual sesquiterpenic lactones from *Dysidea* spp.

intermediate **40**, an intramolecular epoxidation of the chloroalkene group would lead to the putative chloroepoxide **41**. Cyclization of the newly formed hydroxyl (positions C3 to C14), with concomitant opening of the epoxide ring (C14/C15), induces formation of a keto group at C15, with Cl as leaving group, to give **42**[8] (Scheme 24.12) [35].

Spontaneous oxygenation of the α-position of carbonyl functions is also stated to be easy by reaction with molecular oxygen from air under various conditions.

---

8) Alternatively, **42** could be formed by rearrangement of the epoxy intermediate (**41**) to a chloroketone prior to intramolecular cyclization (not shown). According to the authors, this latter route is much less likely, as chloroepoxides require relatively high temperatures to rearrange to chloroketones. Notably, during fractionation, the authors employed mild conditions.

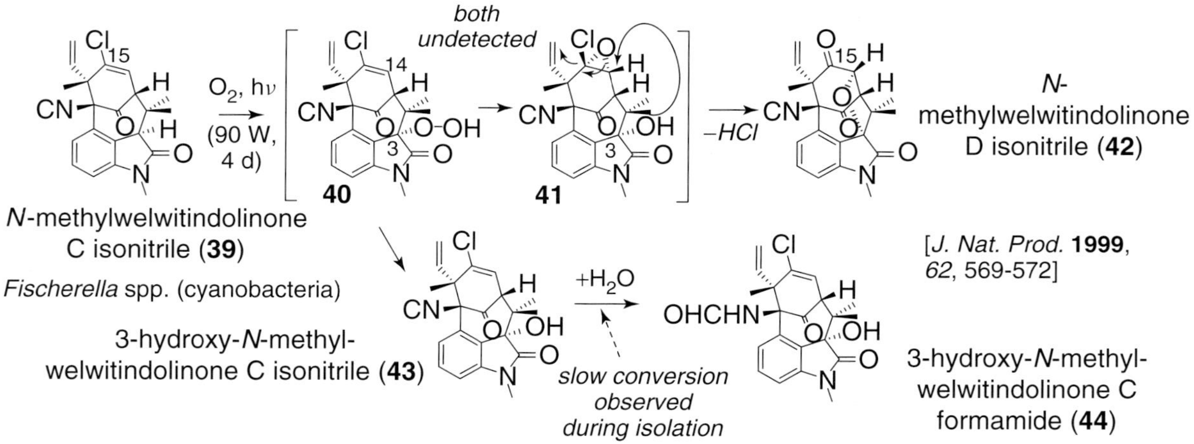

862 | *24 Artifacts and Natural Substances Formed Spontaneously*

Scheme 24.12 Pathways to oxidized welwitindolinones from *Fischerella* spp.

Re-investigation of the roots of *Pueraria mirifica* (Fabaceae), led to isolation of deoxymiroestrol (**45**), a potent phytoestrogen, along with the known hydroxylated analog miroestrol (**46**). The authors proposed miroestrol to be an artifact of oxidation with atmospheric O_2: indeed, they observed facile aerial conversion of deoxymiroestrol (**45**) into miroestrol (**46**) and isomiroestrol (**47**) under mild conditions (methanolic solution, room temperature, one week; Scheme 24.13). The rate of conversion was not given by the authors. Noticeably, **46** showed lower affinity for estrogen receptor than its precursor, challenging conservation of the drug [36].

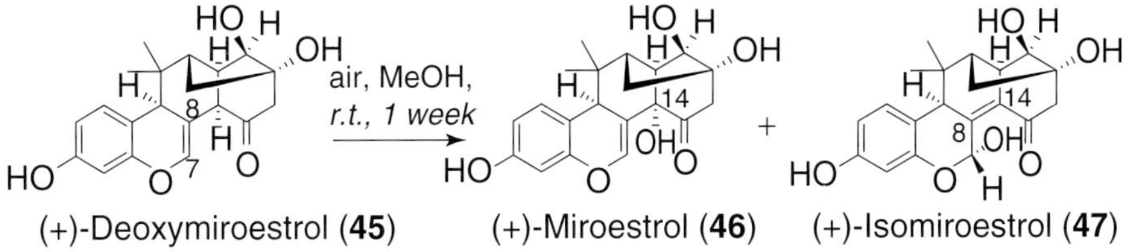

Scheme 24.13 Artifactual oxidation of deoxymiroestrol from *Pueraria mirifica*.

Stemona spp. (ex-*Roxburghia*, Stemonaceae, Dioscoreales) and related genera are a source of alkaloids characterized by a pyrrolo-azepine nucleus. Among lactonic representatives, tuberostemonine (**48**) and oxo-tuberostemonine (**49**) were described in the 1970s. With regard to **48**, **49** shows relocation of lactone ring from C11 to C1, while keeping the same configuration at C11, and displays a double bond at C9=C9a. The possibility that **49** might be an artifact formed by air oxidation of **48** was proposed, since it was also obtained from **48** by oxidation with mercuric acetate (Scheme 24.14) [37]. However, a biosynthetic mechanism for formation of **49**, with a sequential mechanism implicating double bond formation/lactone opening/closure from a putative hydroxylated precursor does not seem impossible. In this way, the semi-synthetic procedure applied for **48** would be biomimetic.

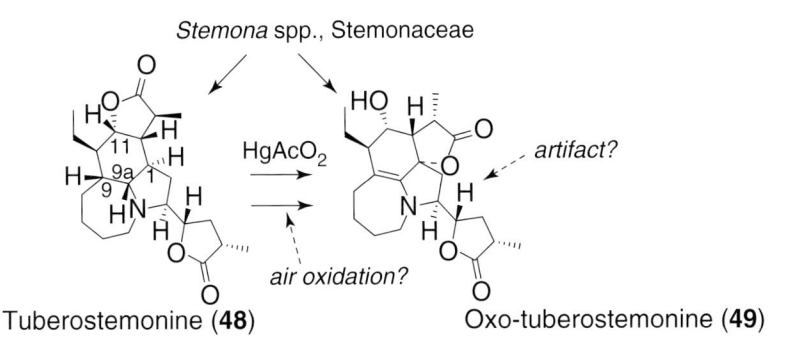

Scheme 24.14 Putative artifactual oxidation of a *Stemona* alkaloid.

24.3.4
Oxidative Coupling

Spontaneous dimerization of alkaloids is described as the result of nucleophilic attack or of oxidative coupling. The artifactual nature of bipowine (**51**) and bipowinone (**52**), two rare bis-aporphines from *Popowia pisocarpa* (Annonaceae) was investigated. If spontaneous oxidation of **51** to **52** was observed, dimerization of their putative precursors wilsonirine (**50**) and pancoridine (**53**), respectively, was not. However, the authors were able to obtain 7,7′-bisdehydronorglaucine (**56**) from dehydronorglaucin (**55**, from **54**) in solution at room temperature (Scheme 24.15).

Some bisacridones resulting from a C–C linkage between an aromatic moiety and a terpenic dihydrofuran (e.g., **57**–**59**) [38] or dihydropyran (**60**) [39] were obtained from Taiwanese *Citrus* spp. (Rutaceae). Being optically inactive, they were suggested to be either artifacts or non-enzymatically formed native compounds. Nevertheless, bisacridones with ether linkages between isoprene units and phenols were isolated from *Glycosmis* spp. (e.g., **61**, **62**) [40]. Extractions were conducted using acetone under long reflux conditions (30–40 h), while authors working with ethanol as an extraction solvent seem to obtain monomers only (e.g., **63**, **64**) [41]. A Norrish type I reaction of the solvent could be hypothesized as a source of free radicals,[9] but, notably, more "classical" artifacts were not obtained (see Section 24.9). However, it can be noticed that the species investigated appear to have intense oxidative metabolism, with frequent isolation of biscoumarins. A parallel can be drawn with bis-ether flavanolignans from milk thistle (*Silybum marianum*, Asteraceae): silybin (**65**) and isosilybin (**66**) are mixtures of *trans* diastereoisomers, which exemplify non-stereospecific biosynthetic radical coupling (Figure 24.2). It is thus difficult to conclude whether achiral bisacridones from Rutaceae are artifacts or chemomarkers of part of this family.

24.3.5
The *N*-Oxide and Oxoalkaloid Cases

Alkaloids of diverse classes are encountered in their *N*-oxide forms, in various taxa (isoquinolines from Magnoliales, indolomonoterpenes from Gentianales, pyrrolizidines from Asterales, various alkaloidic skeletons from Rutales, etc.), but can be considered collectively. The *N*-oxides are generally minor compounds in comparison to their counterparts. To our knowledge, there is no clade-specialization for *N*-oxygenation in higher plants. Such *N*-oxides can sometimes be artifacts due to air oxidation – especially in the case of poor storage conditions. If such compounds appear to be artifactual, most authors reckon their formation is biosynthetically deliberate, and might be linked to plant stress-conditions. Indeed, N-oxygenation is catalyzed by classical monooxygenases. Notably, mamalian phase I metabolites produced by liver CYPs sometimes are *N*-oxides [42]. The artifactual nature of a

---

9) Photochemical cleavage of ketone into two free radical intermediates.

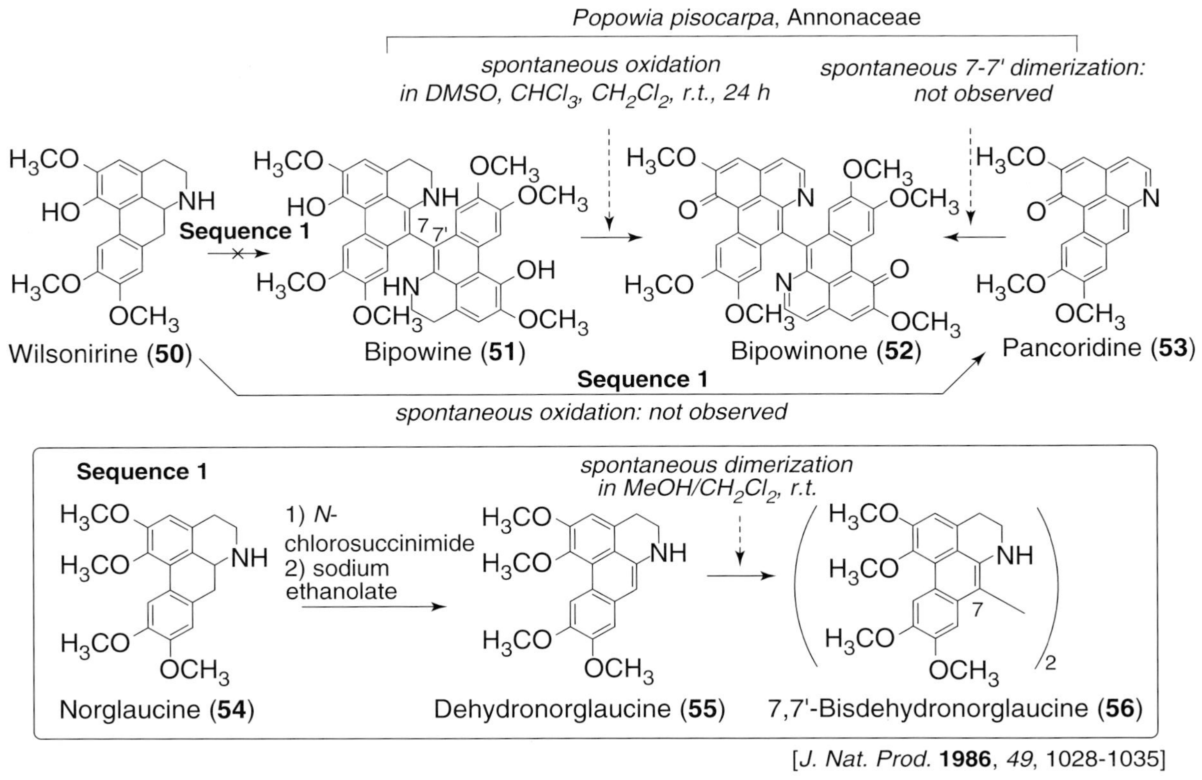

24.3 Oxidation Processes | 865

Scheme 24.15 Possible artifactual formation of bis-aporphines.

[*J. Nat. Prod.* **1986**, *49*, 1028-1035]

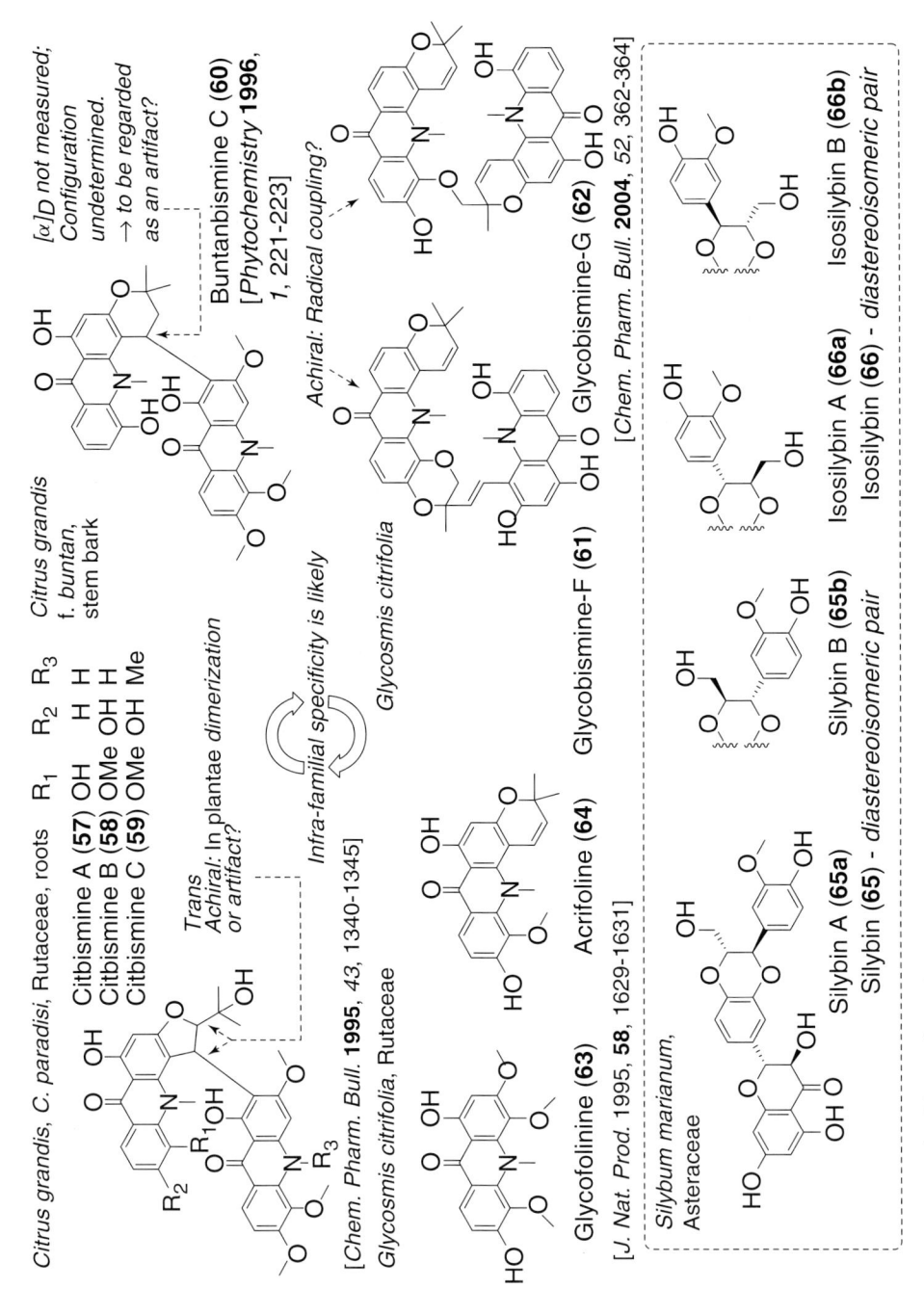

Figure 24.2 Achiral bisacridones from Rutaceae.

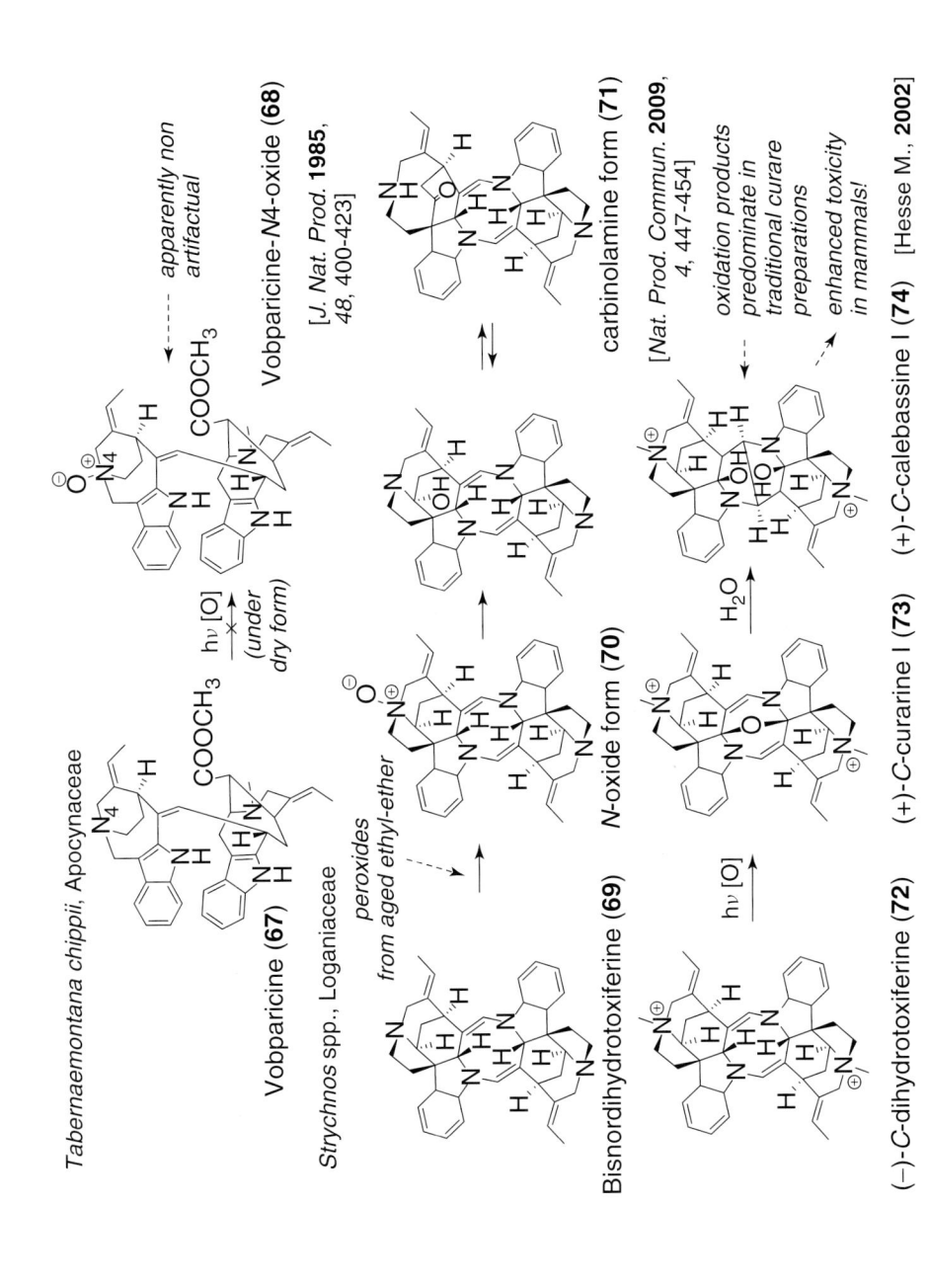

24.3 Oxidation Processes | 867

Tabernaemontana chippii, Apocynaceae

Vobparicine (**67**) — hν [O] (under dry form) → Vobparicine-*N*4-oxide (**68**)

apparently non artifactual

[*J. Nat. Prod.* **1985**, *48*, 400-423]

Strychnos spp., Loganiaceae

Bisnordihydrotoxiferine (**69**) ⇌ *N*-oxide form (**70**) ⇌ carbinolamine form (**71**)

peroxides from aged ethyl-ether

[*Nat. Prod. Commun.* **2009**, *4*, 447-454]

oxidation products predominate in traditional curare preparations

enhanced toxicity in mammals!

(−)-*C*-dihydrotoxiferine (**72**)   hν [O] →   (+)-*C*-curarine I (**73**)   H₂O →   (+)-*C*-calebassine I (**74**)   [Hesse M., **2002**]

Scheme 24.16 Oxidized indolomonoterpenes.

trace indolomonoterpenic N-oxide from *Tabernaemontana chippii* (Apocynaceae) root bark was discussed as follows:

> "Whether or not vobparicine-N4-oxide (**68**) is an artifact is not clear. The facts suggest that it may be a genuine product because, following column chromatography, **68** was isolated from a different fraction than vobparicine (**67**) and also fractions containing vobparicine never showed any accompanying spot of the N4-oxide"

(Scheme 24.16) [43]. However, it was shown that some N-oxidized indolomonoterpenes such as **70** were of artifactual origin, due to peroxides formed in diethyl ether after long exposure of the solvent to air and light [7].

Note that spontaneous oxidation of indolomonoterpenes can also take place at other positions, as observed for conopharyngine (**75**; *Tabernaemontana* spp. and other Apocynaceae) and its 3-hydroxy analog **77**, which, respectively, yield hydroxyindolenine derivatives **76** and **78** upon standing for a long time (Scheme 24.17) [44].

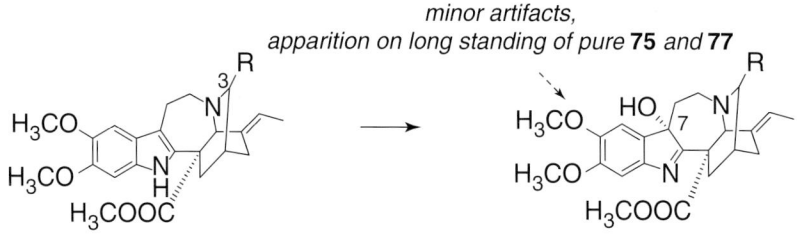

Conopharyngine (**75**): R=H
3R/S-Hydroxyconopharyngine (**77**): R=OH

Conopharyngine-hydroxyindolenine (**76**): R=H
3R/S-Hydroxyconopharyngine-hydroxyindolenine (**78**): R=OH

Scheme 24.17 Formation of hydroxyindolenine derivatives.

Oxoalkaloid derivatives raise similar questions about their true nature. Their artifactual formation can be very swift, as in the air oxidation of vasicoline (**79**) to vasicolinone (**80**) during fractionation or in its purified form (Scheme 24.18) [45].

Oxoaporphines and dioxoaporphines seem to appear spontaneously during long and uncareful storage of extracts or of isolated compounds, with an extension

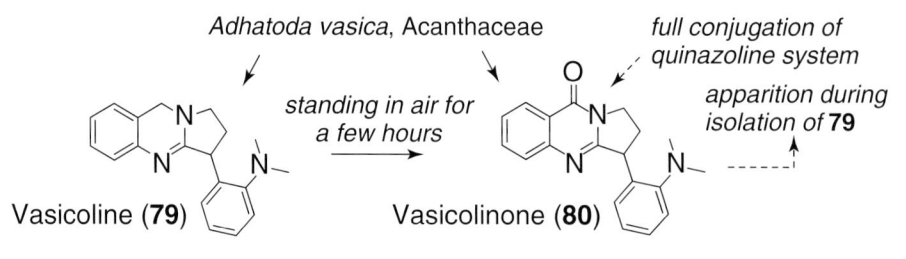

Vasicoline (**79**) → Vasicolinone (**80**)

Adhatoda vasica, Acanthaceae
standing in air for a few hours
full conjugation of quinazoline system
apparition during isolation of **79**

Scheme 24.18 Oxidation of vasicoline (**79**).

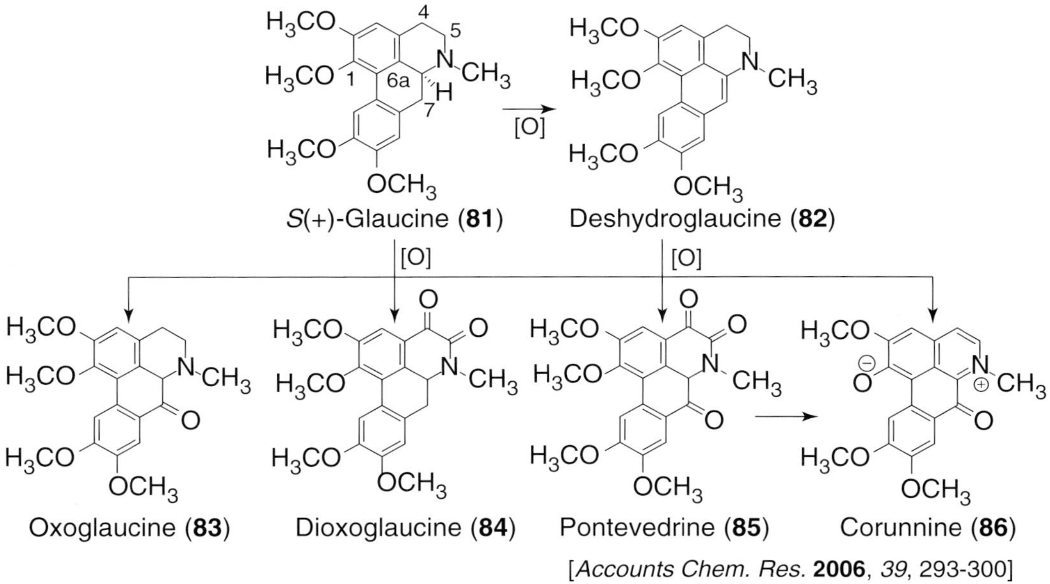

Scheme 24.19 *In plantae* oxidative pathways toward oxoaporphinoids.

of their conjugated domains. However, most of the derivatives obtained probably originate from a specific oxidation by light and singlet oxygen *prior to* isolation, as part of plant defense against pathogens, with some oxoalkaloids being proposed as phototoxic phytoalexins. For example, upon mechanical injury of Magnoliales or Ranunculales, the aporphine glaucine (**81**) is rapidly oxidized into oxoglaucine (**83**) (major product), ponteverdine (**85**) and other oxidized derivatives (**82, 84, 86**; Scheme 24.19) [46]. *In planta* spontaneous oxidation of enzymatic monooxygenation products can also be hypothesized.

It was noticed that dioxoglaucine (**84**) affords an iminium (i.e., 6,6a-dehydro-**84**) under UV irradiation. Similar examples of formation of iminium ions are proposed below.

24.4
Exposure to Light

As previously seen, light is a necessary condition for oxygenation by $^1O_2$ to give detectable artifacts. Oxidative light degradation is sometimes a quickly occurring phenomenon, as in the case of rotenoids [47].[10] Many different compounds have been already shown to undergo conversion and degradation by various photochemical reactions. Photochemical epimerization of stereogenic centers and isomerization of double bonds are also observed in several natural series.

24.4.1
Isomerization and Epimerization

A remarkable instance is that of the essential oil of aniseed fruit (*Pimpinella anisum* L., Apiaceae), which is used in aromatherapy and in the food industry. It contains 80–95% *(E)*-anethole (**87**), which is poorly toxic. However, hepatic toxicity of the *(Z)*-isomer (**88**) was observed in humans.[11] Both drug and essential oil must be stored away from light, as *(E)*-anethole isomerizes into its *(Z)*-isomer under UV irradiation [14]. Stilbenoids are also well known for undergoing light-induced isomerization. The olefinic photo-isomerization process, in the excited state, involves twisting (about the former double bond) of stilbene fragments relative to one another. The *(E)*-form is favored, as in resveratrol (**90**), in regard to *(Z)*-form (**89**). Combretastatin A4 (**91**), bearing a *(Z)*-configuration, is a cytotoxic derivative that inspired semi-synthetic derivatives undergoing clinical

10) This was studied with the neurotoxic pesticide rotenone, a prenylated isoflavonoid mainly produced by *Derris, Milletia, Tephrosia* spp. (Fabaceae). Rotenone is a champion of photochemical degradation: It undergoes *O*-demethylation, epimerization, epoxidation, hydroxylation, and dehydration [47].

11) The European Pharmacopoeia sets an upper limit of 0.5% (Z)-anethole in the essential oil, and the acceptable limit for dietary intake is 2.5 mg kg^{-1} According to Bruneton, a heavy drinker of aniseed flavored beverages is likely to ingest more than 250 mg total anethole a day! [14].

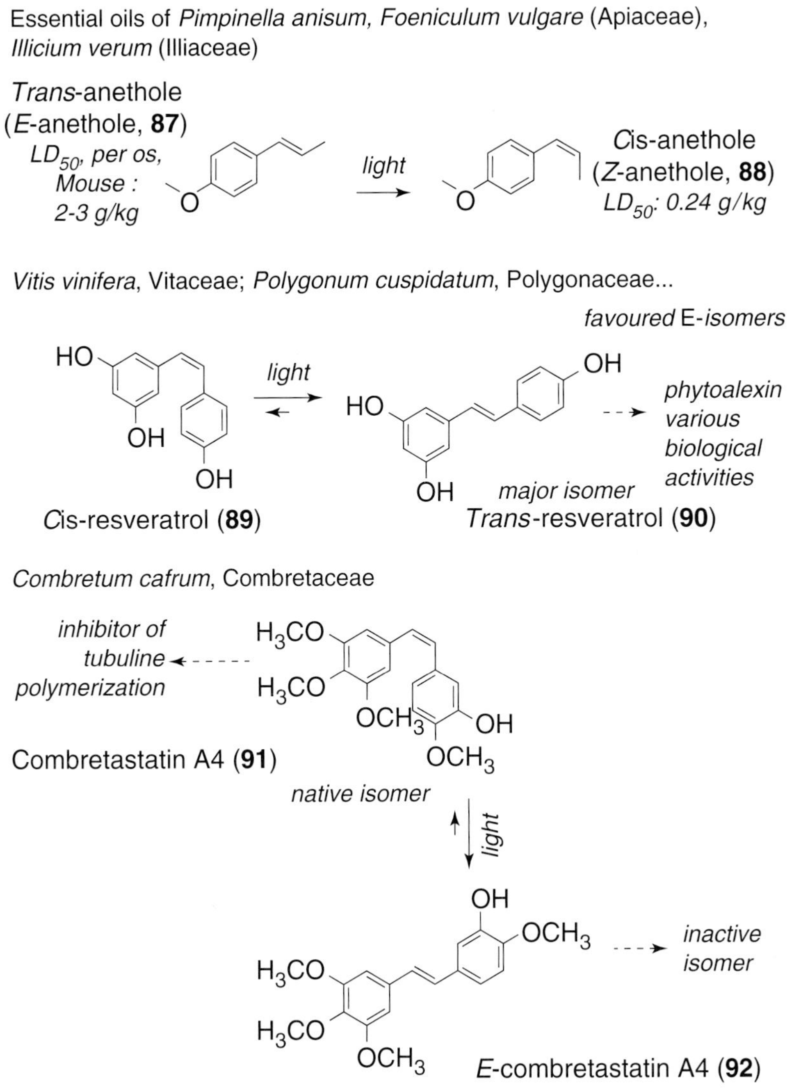

Essential oils of *Pimpinella anisum, Foeniculum vulgare* (Apiaceae), *Illicium verum* (Illiaceae)

Trans-anethole
(*E*-anethole, **87**)
LD_{50}, per os,
Mouse :
2-3 g/kg

— light →

Cis-anethole
(*Z*-anethole, **88**)
LD_{50}: 0.24 g/kg

Vitis vinifera, Vitaceae; *Polygonum cuspidatum*, Polygonaceae...

favoured E-isomers

Cis-resveratrol (**89**) ⇌ light ⇌ *Trans*-resveratrol (**90**) major isomer → phytoalexin various biological activities

Combretum cafrum, Combretaceae

inhibitor of tubuline polymerization ← ---- Combretastatin A4 (**91**) native isomer

↕ light

E-combretastatin A4 (**92**) ---→ inactive isomer

Scheme 24.20 Light-induced isomerization of anethole and stilbenoids

trials. It is readily isomerized to its inactive (*E*)-form **92** (Scheme 24.20) [48]. Many other cases exist in various chemical series, some being hypothetical but likely artifacts in regard to biosynthetic processes and known analogs (e.g., histrionicotoxins from South American frogs (*Dendrobates* spp.) with *trans*-enynes [49]).

Epimerization processes are also observed following photo-excitation. Oxidation of reserpine (**93**) under light exposure leads to the iminium 3-dehydroreserpine (**94**), and also yields the 3-(*S*) isomer isoreserpine (**95**). The reaction is favored in chloroformic solutions [7]. (+)-Milnamide A (**96**), a metabolite of marine sponges

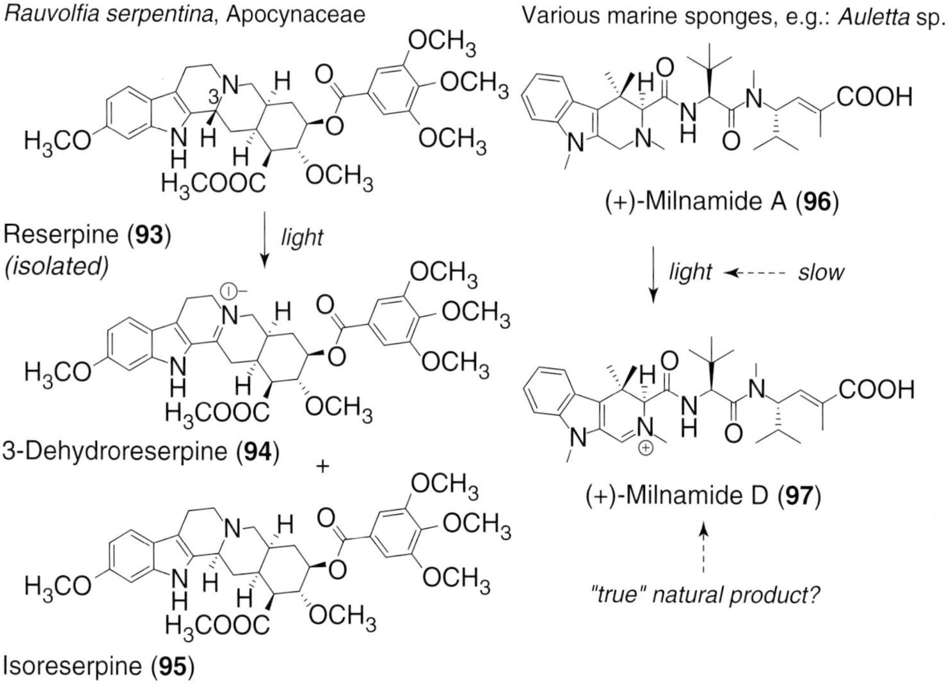

Scheme 24.21 Reserpine (**93**) and milnamide A (**96**): photoinduced artifacts.

(e.g., *Auletta* spp.), was shown to auto-oxidize to (+)-milnamide D (**97**), suggesting a non-biogenetic origin for this compound. The rate of oxidation was similarly accelerated in aged $CHCl_3$ and $CDCl_3$ solutions (Scheme 24.21) [50].

24.4.2
Rearrangements

Rearrangements due to UV exposure in the course of purification of plant metabolites can be exemplified with sesquiterpenes such as *ent*-bicyclogermacrene (**98**), which converts into *ent*-spathulenol (**99**) after peroxidation by atmospheric oxygen (Scheme 24.22) [51].

A complex example was provided by Majetich *et al.*, who studied the conversion of (+)-komaroviquinone (**102**) into (+)-komarovispirone (**103**). Both compounds had been previously isolated from *Dracocephalum komarovi* (Lamiaceae), a plant used in Uzbek traditional phytotherapy. It appears that **103** is rapidly formed upon light exposure of **102**. The authors proposed a radical mechanism (Scheme 24.23), and suggested formation from plant material [52], that is, a true artifactual nature for **103** [53].[12]

12) Note that the authors underline the analogy with light-induced rearrangement of (+)-verbenone into (+)-chrysantenone.

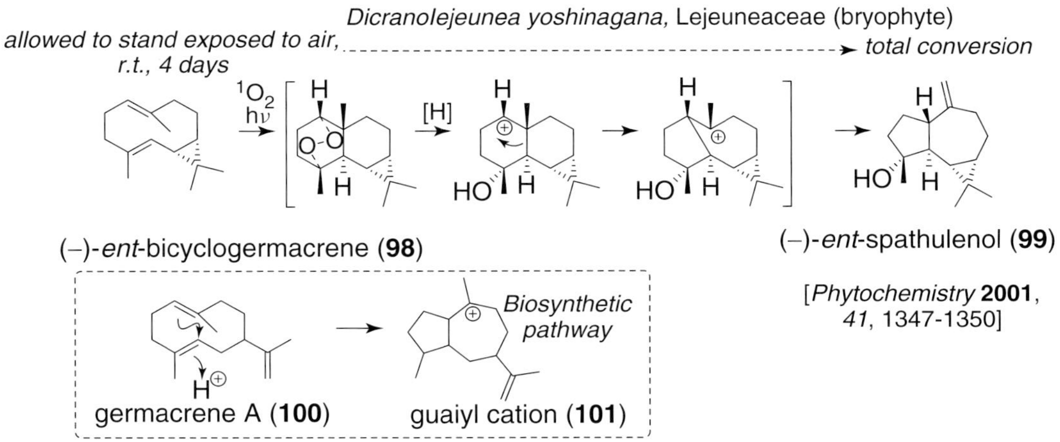

Scheme 24.22 Photoinduced oxidation and rearrangement of *ent*-bicyclogermacrene (**98**).

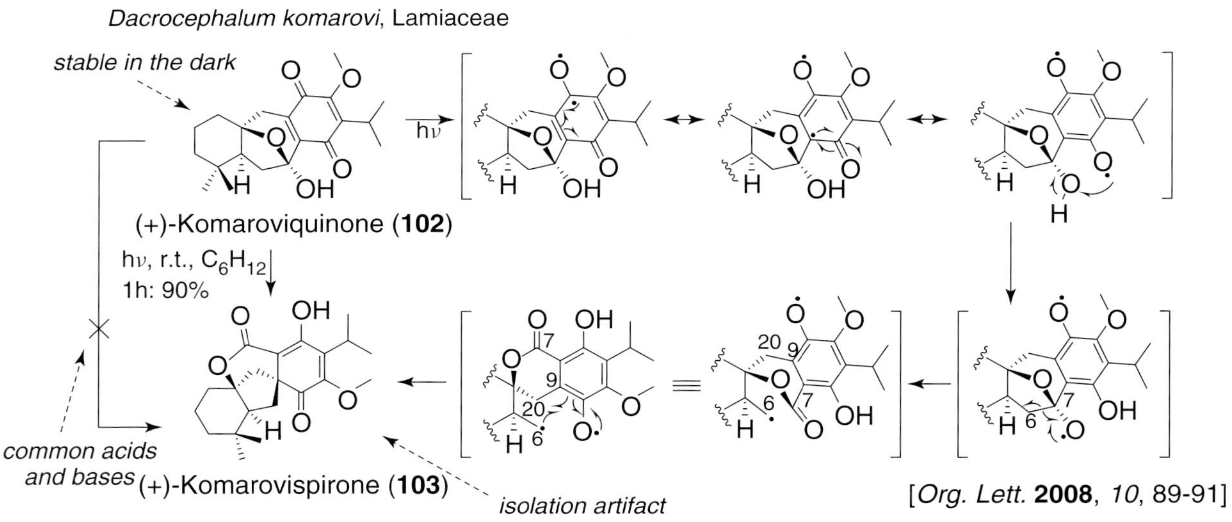

Scheme 24.23 Photoinduced rearrangement of komaroviquinone (**102**).

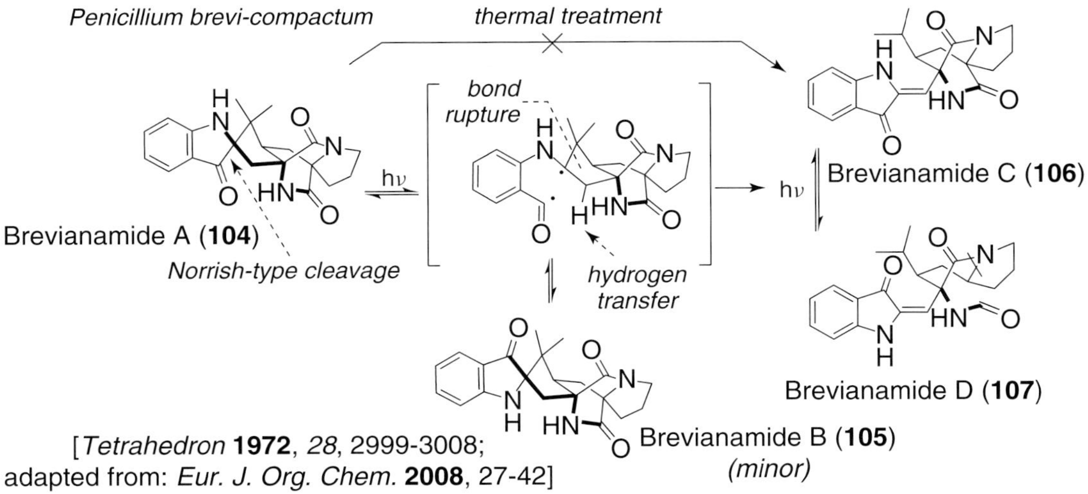

Scheme 24.24 Photoinduced rearrangements of brevianamide A (**104**).

Brevianamides A (**104**), B (**105**), C (**106**), and D (**107**) were isolated from *Penicillium brevi-compactum*. A thorough experimental showed that **105–107** are photochemical isomerization or cleavage products of **104**, probably resulting from a Norrish-type reaction (Scheme 24.24). These compounds were not isolated when the fungus was cultivated in the dark and when the purification procedure was conducted with low light [54].

24.4.3
Photocycloaddition and Photodimerization

Biomimetic synthesis literature offers delightful accounts of auto-assemblies under UV irradiation, the [2 + 2] photocycloaddition of α,β-unsaturated ketones or esters to alkenes or alkynes being largely used [6, 55]. Natural products obtained using this strategy do not necessarily correspond to artifactual compounds, but rather illustrate peculiarities in the field of biosynthesis. Reactions of α,β-unsaturated ketones can be induced by simple daylight absorption [55]. Lactones and pyrones are thus considered as "photoactive," and artifactual intra/homo/hetero-photocycloadditions of natural products often implicate such moieties. In addition, artifactual cyclizations of polyenes are reported. Generally, these reactions can take place without a solvent, but acetone can act as a sensitizer [55].

From the African shrub *Ethulia vernonioides* (Asteraceae), two terpenic 5-methylcoumarins, **108** and **109**, were isolated. Compound **109**, a hexacyclic compound, was evidenced to be an easily formed artifactual intramolecular cycloadduct of **108** (Scheme 24.25) [56].

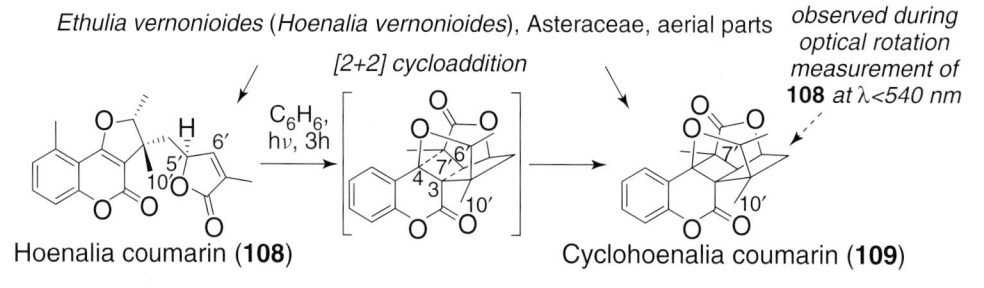

Scheme 24.25 Intramolecular [2 + 2] cycloaddition of "hoenalia coumarin" (**108**).

Numerous marine organisms have yielded cycloaddition products. Their artifactual nature outside of the marine environment is hard to discriminate. Sharma *et al.* explored the photoinduced auto-cyclization of polypropionates from shell-devoid molluscs of the order Sacoglossa, studying the influence of UV absorption by sea-water on the rate of formation of tridachiahydropyrone (**112**) and of phototridachiahydropyrone (**113**) from a putative precursor **110**. Note that **112** only is a natural product. Compound **113** was the photochemically preferred product in the

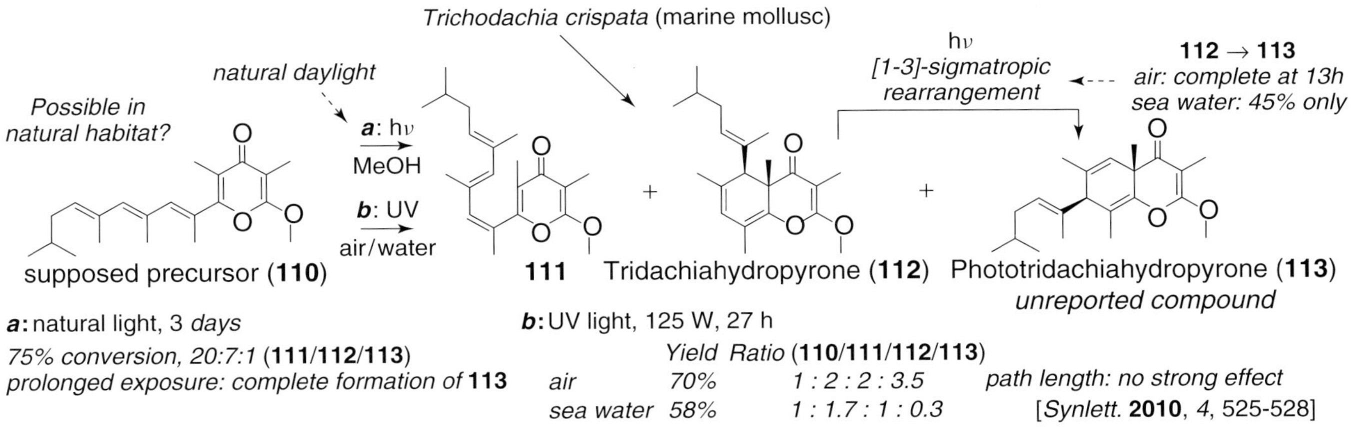

Possible in natural habitat?

supposed precursor (110) → [*a*: hν, MeOH; *b*: UV, air/water] → *Trichodachia crispata* (marine mollusc) → **111** + **Tridachiahydropyrone (112)** --[hν, [1-3]-sigmatropic rearrangement]--> **Phototridachiahydropyrone (113)** *unreported compound*

a: natural light, 3 days
75% conversion, 20:7:1 (**111/112/113**)
prolonged exposure: complete formation of **113**

b: UV light, 125 W, 27 h
Yield   Ratio (**110/111/112/113**)
air   70%   1 : 2 : 2 : 3.5
sea water 58%   1 : 1.7 : 1 : 0.3

112 → 113
air: complete at 13h
sea water: 45% only
path length: no strong effect
[*Synlett.* **2010**, *4*, 525-528]

Scheme 24.26 Influence of medium on photocatalyzed synthesis of marine polypropionates.

total synthesis of **112** from **111**, and has not been described in a natural source so far. Its formation appears to be partly inhibited in sea-water,[13] which is in keeping with its absorption spectrum (Scheme 24.26) [57].

From the bark of an *Aniba* sp. (Lauraceae), the styrylpyrone **114** was obtained, along with two dimers (**115, 116**) whose origin was uncertain, as Lauraceae spp. are prone to oxidative coupling. Found in much lower amounts than **114**, both **115** and **116** were shown to be photoinduced dimerization artifacts (Scheme 24.27) [58].[14]

Biyouyanagin A (**120**) and B (**121**) are minor compounds isolated from *Hypericum chinense* (Clusiaceae). Their total synthesis was performed by Nicolaou *et al.*, through [2 + 2] photocycloaddition, from *ent*-zingiberene (**118**) and hyperolactone C (**119**) under conditions compatible with extraction conditions (Scheme 24.28) [59]. Biyouyanagins **120** and **121** were not demonstrated to appear during purification, and are likely to form *in planta* without enzymatic intervention from readily available precursors. They do though appear as *potential* artifacts.

Finally, orinocin (**124**), a short pyrone isolated from light-exposed cultures of *Streptomyces orinoci*, was linked to the aureothin family of *Streptomyces* metabolites: indeed, an isomer of neoaureothin (**122**) was also obtained, and two relatively stable diastereoisomers (**123a,b**) were identified in light-working conditions only. Pure **122** was exposed to daylight, to partially afford **123a,b** by a $8\pi - 6\pi$ electrocyclic rearrangement cascade. This step was followed by photoinduced retro-[2 + 2]-cycloaddition, giving **124**. This entire artifactual sequence was termed *"polyene splicing"* (Scheme 24.29) [60].

The literature sometimes depicts the photochemical synthesis of natural cycloadducts with protocols that are far from real-life[15] conditions (monochromatic light, high intensity irradiation, sensitizers, etc.) [61]. However, in several cases, compounds are readily formed: determination of their genuine natural character requires specifically conducted studies, and is a matter of debate.

24.5
Heat and Pressure

Heat and pressure often provide the energy necessary for artifactual formation through various mechanisms, including oxidation processes.

---

13) Owing to the presence of organic compounds; distillated water had no major influence.
14) To our knowledge, this is not described from the well-studied *Piper methysticum* (kawa, Piperaceae).
15) Or *"real-lab"* – from an extractor's perspective.

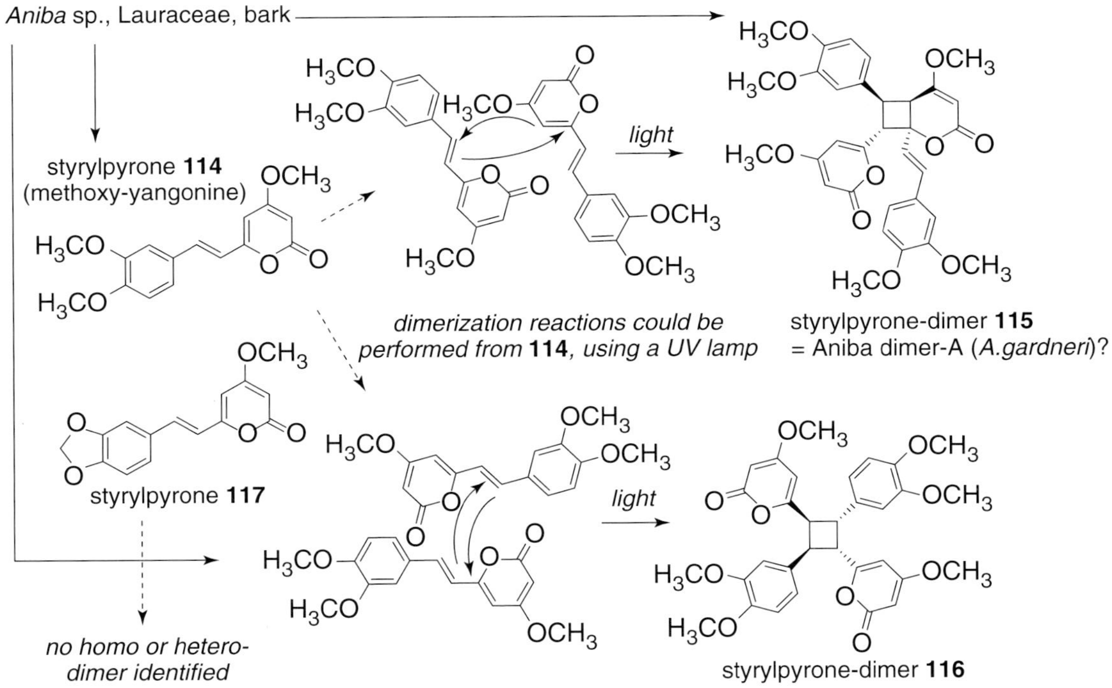

Scheme 24.27 Styrylpyrone photodimers.

To our knowledge, the authors did not express concern about the hypothetic artifactual nature for **149** *and* **150**.

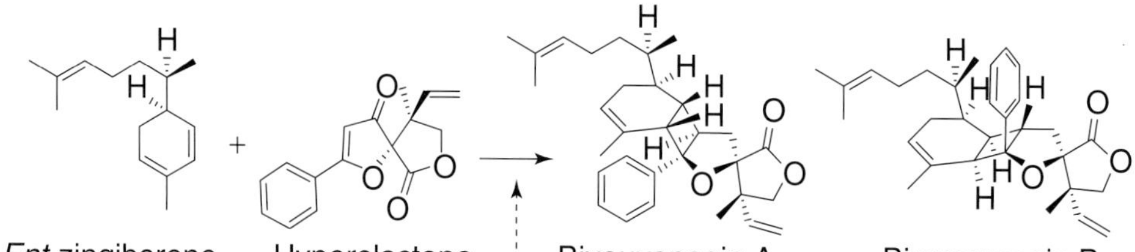

Ent-zingiberene (**118**) + Hyperolactone C (**119**) → Biyouyanagin A (**120**, *51%*) + Biyouyanagin B (**121**, *3%*)

CH_2Cl_2, hv, 320 nm, sentisizer (2'acetophenone), 5°C, 12 h

[*Chem. Eur. J.* **2010**, *16*, 7678-7682]

Scheme 24.28 Biomimetic total synthesis of biyouyanagins A (**120**) and B (**121**).

24.5.1
Epimerization and Isomerization In or Out of Solutions

Seeds of *Zizyphus vulgaris* var. *spinosus* (*Zizyphus jujuba* var. *spinosa*, Rhamnaceae) are traditionally used in China and in the Middle East as a sedative remedy. Cyclopeptides such as sanjoinine A (frangufoline, **125**) were identified as active principles, with GABA (γ-aminobutyric acid) agonist activity. Chinese *materia medica* describes roasting of this material to improve hypnotic activity. Compound **125** was shown to isomerize at its *exo*-amino acid (dimethyl-phenylalanine) to give sanjoinine Ah1 (**126**), its heat induced artifact, which exerts a more potent pharmacological activity *in vivo*, probably gaining affinity for GABA receptors (Scheme 24.30) [62].

Epimerization at C2 of the major catechins in a crude extract of green tea (*Camellia sinensis*) was investigated in aqueous solutions. The epimers of (−)-epicatechin and (−)-derivatives (e.g., **127**, **128**) can be observed under conditions of tea brewing (90 °C, 3 min, tea sample). After 30 min, about 40% conversion is observed (Scheme 24.31) [63]. Such epimerization processes are thus expected for related molecules, casting doubt on the stereochemistry of tannins or comparable compounds obtained at boiling temperature or using a Soxhlet apparatus.

24.5.2
Hydrodistillation

Calorific energy provided by hydrodistillation can be a cause of rearrangement of natural volatile compounds. Confirmation of artifactual nature is not easy, as gas chromatography conditions are likely to induce the same reactions. Hydrodistillation of the aerial parts of *Laggera tomentosa* (Asteraceae) was shown to yield (−)-chrysanthenone (**134**), (−)/(+)-filifolone (**132/135**) and *(Z)*-isogeranic acid (**133**), all likely to be artifacts. The main constituent of the oil, (+)-chrysanthenone

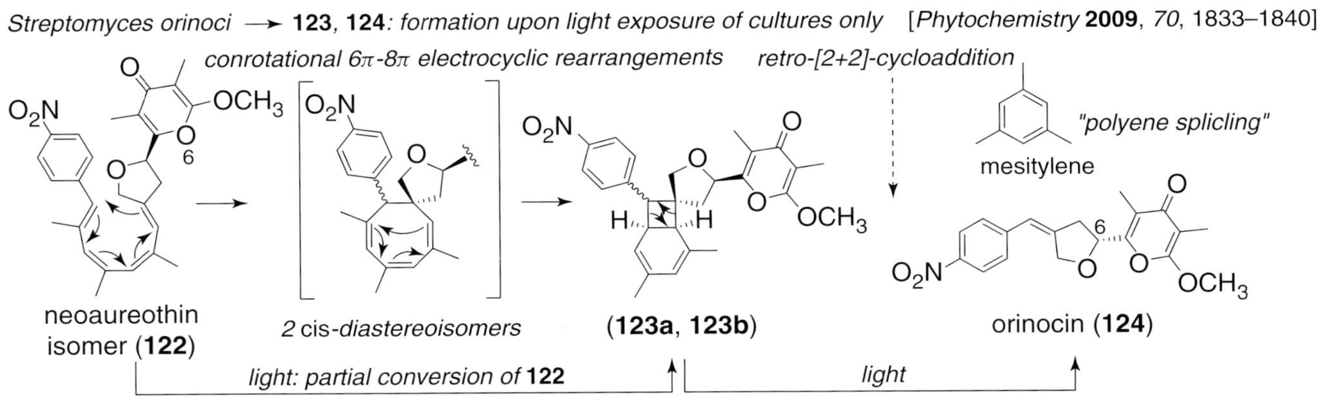

Streptomyces orinoci ⟶ **123, 124**: formation upon light exposure of cultures only [*Phytochemistry* **2009**, *70*, 1833–1840]

conrotational 6π-8π electrocyclic rearrangements   retro-[2+2]-cycloaddition

"polyene splicing"

neoaureothin isomer (**122**) → 2 *cis*-diastereoisomers (**123a, 123b**) → orinocin (**124**) + mesitylene

light: partial conversion of **122**   light

Scheme 24.29 Non enzymatic "polyene splicing."

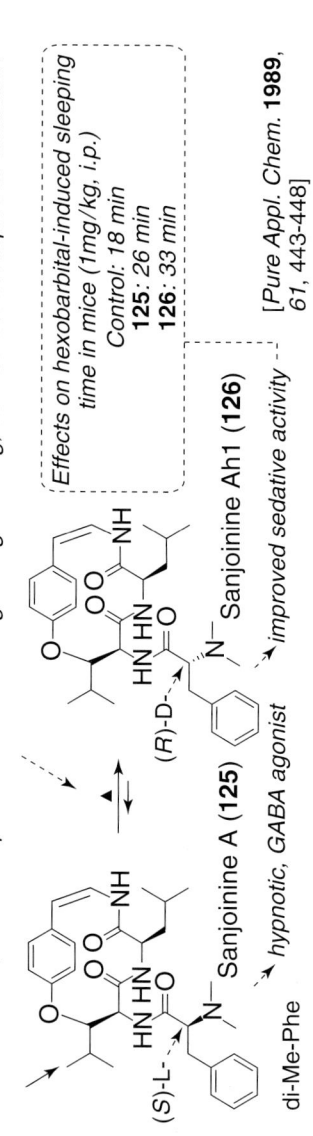

Scheme 24.30 Traditional roasting of *Zizyphus jujuba* seeds.

Zizyphus jujuba, Rhamnaceae, seeds → Sanjoinine A (**125**), di-Me-Phe, (S)-L-, hypnotic, GABA agonist

epimerization occurring during roasting, but not with hot aqueous extraction → Sanjoinine Ah1 (**126**), (R)-D-, improved sedative activity

Effects on hexobarbital-induced sleeping time in mice (1mg/kg, i.p.)
Control: 18 min
125: 26 min
126: 33 min

[*Pure Appl. Chem.* **1989**, *61*, 443-448]

Tea (*Camellia sinensis* aqueous leaves extract)   *Epimerization:* R = H: (+)-epigallocatechin **129**

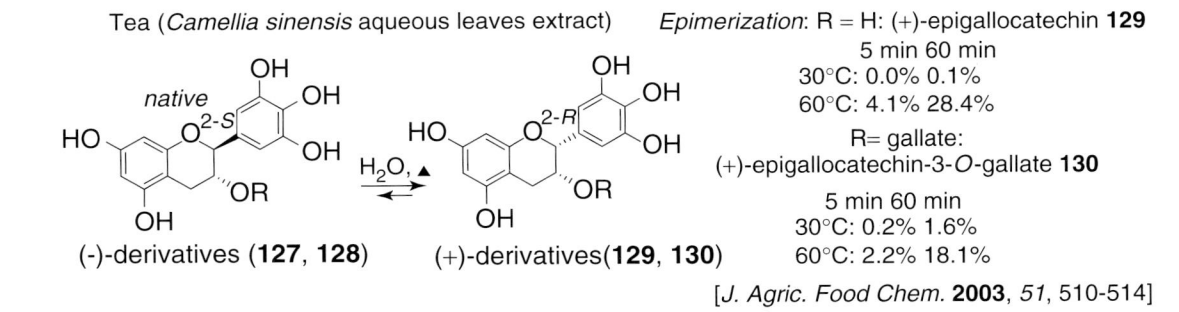

                                             5 min 60 min
                                     30 °C: 0.0% 0.1%
                                     60 °C: 4.1% 28.4%

R = gallate: (+)-epigallocatechin-3-*O*-gallate **130**
                                        5 min 60 min
                                  30 °C: 0.2% 1.6%
                                  60 °C: 2.2% 18.1%

(−)-derivatives (**127**, **128**) $\xrightarrow{H_2O, \triangle}$ (+)-derivatives (**129**, **130**)

[*J. Agric. Food Chem.* **2003**, *51*, 510-514]

Scheme 24.31 Epimerization of gallocatechins during tea brewing.

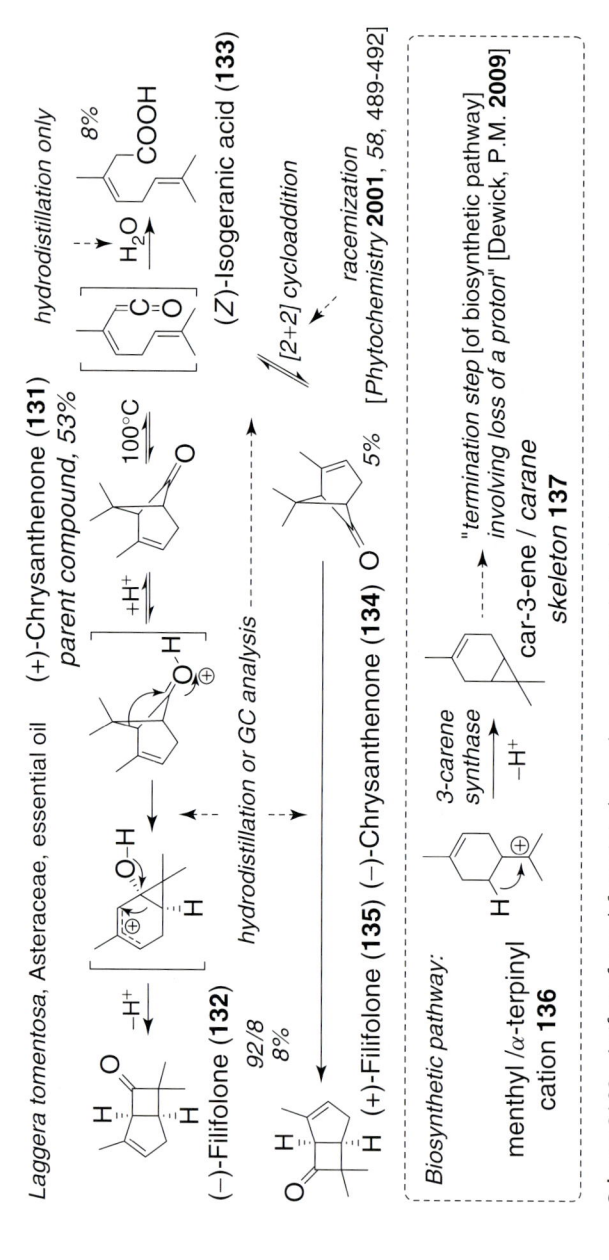

Scheme 24.32 Artifacts formed from (+)-chrysanthenone (**131**) during hydrodistillation.

(**131**) is believed to be their precursor, undergoing thermal and solvolytic reactions (Scheme 24.32) [64].

Even though not occurring in the gas phase, hydrolysis of glucosides was observed during hydrodistillation, and can be mentioned here: *(Z)*- and *(E)*-5-ethylidene-2(5*H*)-furanones (**142** and **143**) were obtained from foliage of *Halocarpus biformis* (Podocarpaceae) using this technique [65], but were absent from an hexane extract. Aqueous cold extraction yielded two potential glycosylated precursors (**138**, **139**). After treatment with a β-D-glucosidase, they gave the aglycones **140** and **141**, which easily dehydrated to **142** and **143** (Scheme 24.33) [66].[16] From *Osmunda japonica* (Osmundaceae, whole plant), (4*R*,5*S*)-osmundalactone (**145**) and **146** (diastereoisomer of **140** and **141**) were both identified as artifactual products of the acid hydrolysis of osmundaline (**144**) during extraction [67]. Hydrated and saturated derivatives **146** and **148** were later isolated, following a non-drastic protocol (Scheme 24.33) [68].[17] Notably, dehydration of **146** was not reported.

24.5.3
Decarboxylation Processes

The most famous example of heat-induced decarboxylated products is that of azulenes from Asteraceae, which has impacted the field of European phytotherapy: matricin (**149**), a sesquiterpenic lactone from *Chamaemelum* and *Chamomilla* spp., can be considered as an unstable prodrug, easily yielding the cyclooxygenase (COX) inhibitors **150** and **151** during infusion of the plant material [14]. Hydrodistillation induces similar reactions, which explain the blue color of some Asteraceae essential oils. This case is evocative of the very unstable pyrroloquinoline carboxylic pigment sanguinone A (**152**), isolated along with sanguinone B (**153**) from the fruiting bodies of the mushroom *Mycena sanguinolenta* (Marasmiaceae, Trichomatales); **153** was identified as a highly conjugated decarboxylation artifact from **152**, formed under milder conditions than **151** (Scheme 24.34) [69].

24.5.4
Supercritical CO_2

In a convincing chemical study of hop strobiles (*Humulus lupulus*, Cannabaceae) having undergone previous supercritical CO_2 extraction [70], the authors obtained the well known and abundant xanthohumol (**154**) along with new chalcone derivatives under racemic forms (**155–158**). These derivatives might be artifacts from **154** formed during pretreatment of the plant material, with probable oxidation to epoxy intermediates that are likely to be unstable under high pressure

---

16) See compound **3** (Scheme 24.1).
17) The role of glucosidases in the formation of these molecules appears plausible, and might be artifactual.

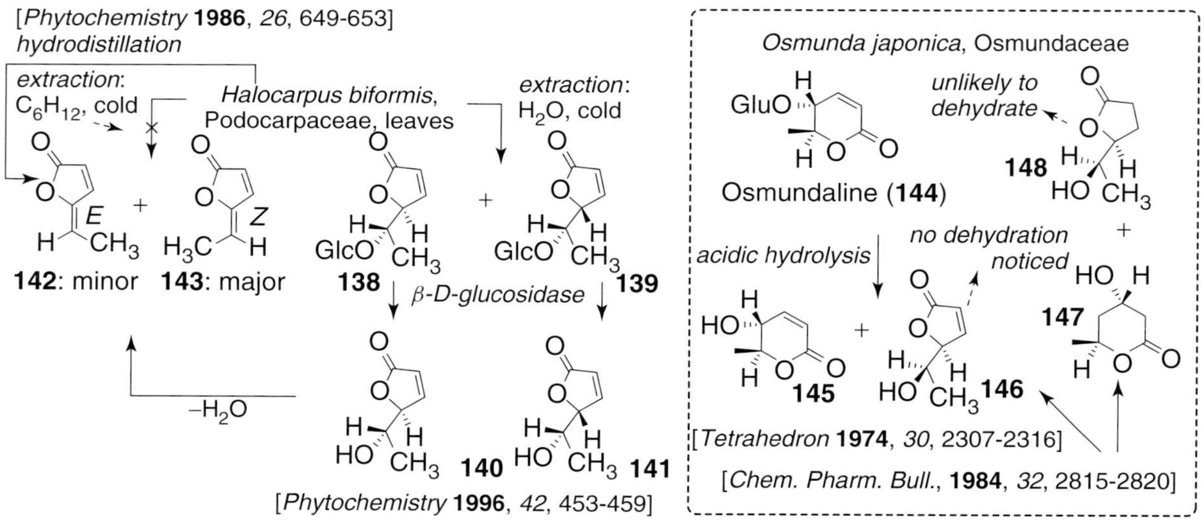

886 | *24 Artifacts and Natural Substances Formed Spontaneously*

Scheme 24.33 Artifactual lactones of *Halocarpus biformis* formed during hydrodistillation.

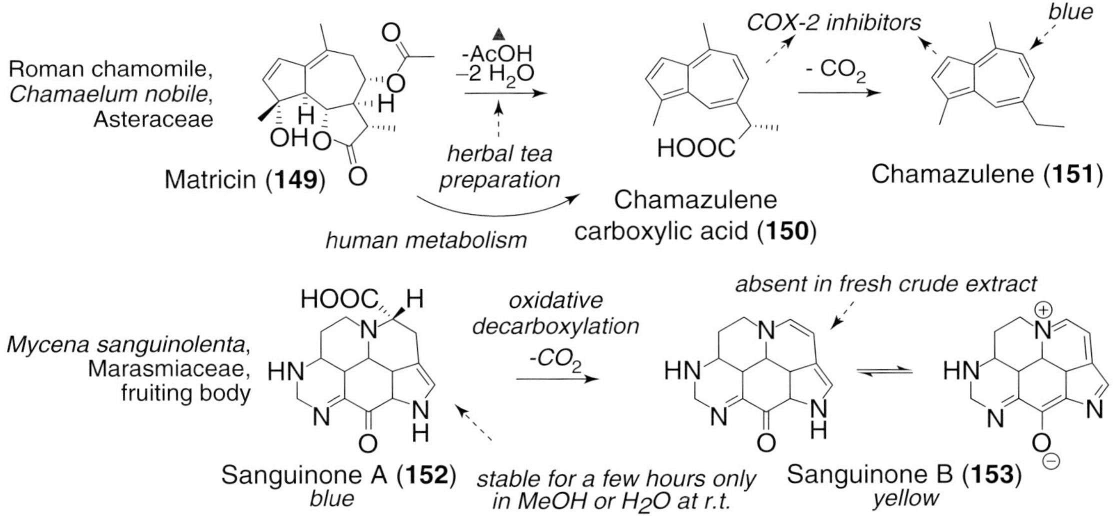

Scheme 24.34 Heat-induced degradation of matricin (**149**) and sanguinone A (**152**).

conditions. Interestingly, some of the compounds were identical to microbial bioconversion products of **154**. A cyclic derivative (**160**) of desmethylxanthohumol was also obtained, but not its parent compound (**159**). It is also noteworthy that (±)-prenylnaringenin (**161**), described as an isomerization artifact of **159** [71], was not retrieved (Scheme 24.35) [70].

24.6
Alkaline Media

24.6.1
Amination Processes

During alkaloid extraction and purification, alkalinization of plant material or of aqueous extracts is performed. Researchers eventually use diluted NaOH, but ammonia is much preferred, with subsequent artifacts identified throughout the literature. The newly formed compounds often have interesting and original structures. For example, from the bark of *Desmos dumosus* (Annonaceae), desmosine (**163**) was quantitatively obtained from 5-hydroxy-6,7-dimethoxyflavone (**162**). The artifactual nature of the newly formed compound, though obvious, was verified by authors by modifying their working conditions [72]. Similarly, from *Desmos dunalii*, dunaliine A (**165**) was obtained along with the chalcone **164**. The authors did not challenge the artifactual nature of **165**, but proposed a pathway from **164** involving a putative, secondary oxidation step (Scheme 24.36) [73].

The secoiridoid base gentianine (**167**), mainly isolated from Gentianaceae and Loganiaceae spp., similarly arises from treatment with ammonia [74]. After glucose hydrolysis of gentiopicroside (**166**) or swertiamarine (**168**),[18] nucleophilic addition of NH_3 on the newly-formed dialdehyde system, followed by dehydration, generates the pyridine ring of **167** [75].[19] Compound **167** was sometimes obtained with carbonates in the alkalinization steps, suggesting it could be a natural compound. Hesse considers that

> "it is conceivable that the action of Na_2CO_3 might bring about the release of NH_3 from certain plant materials"

[77]. However, **167** and derivatives were isolated in neutral conditions (e.g., Goodeniaceae [78]). A biosynthetic scheme is known, with early introduction of a nitrogen atom in the secoiridoid skeleton (Scheme 24.37) [75].

18) This reaction was observed with other iridoids.
19) Note the importance of spontaneous nucleophilic addition on an aldehyde (and of the Pictet–Spengler reaction) in the non-enzymatic synthesis of alkaloids from biological primary amines *ex vivo* (e.g., Maillard reactions in meat [20]) [76a] or *in vivo*. For example, the tetrahydro-isoquinolines (R/S)-salsolinols were demonstrated to arise from dopamine (a catecholaminergic neurotransmitter) and acetaldehyde (the metabolite of ethanol) in the brain of alcoholics, causing parkinsonism [76b]. Tetrahydro-β-carbolines are similarly formed from serotonin (5-hydroxytryptamin) [76c]. Such reactions can also implicate endogenous carboxylates [76].

24.6 Alkaline Media

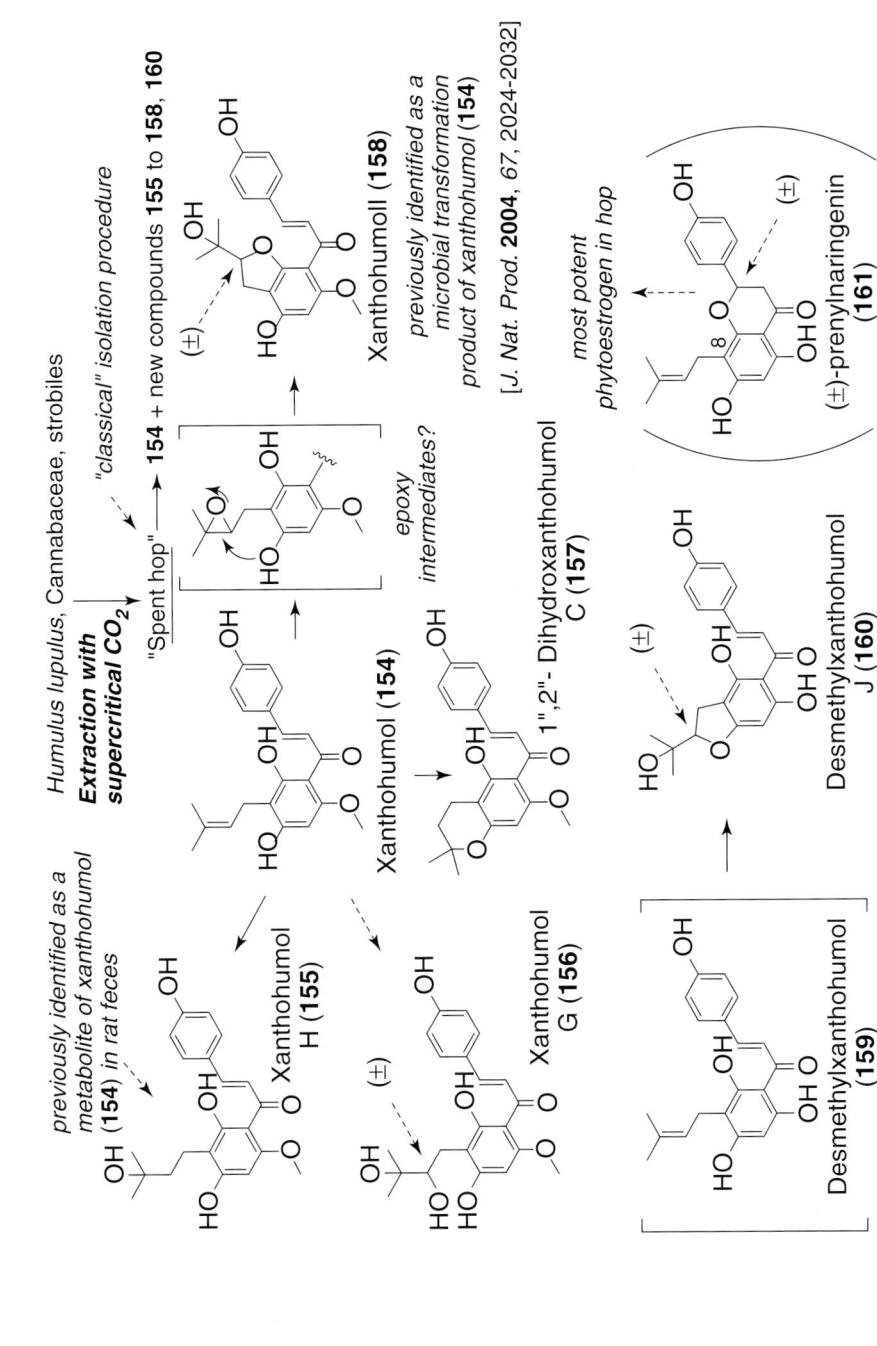

Scheme 24.35 Chalcones obtained from hops after supercritical CO_2 treatment.

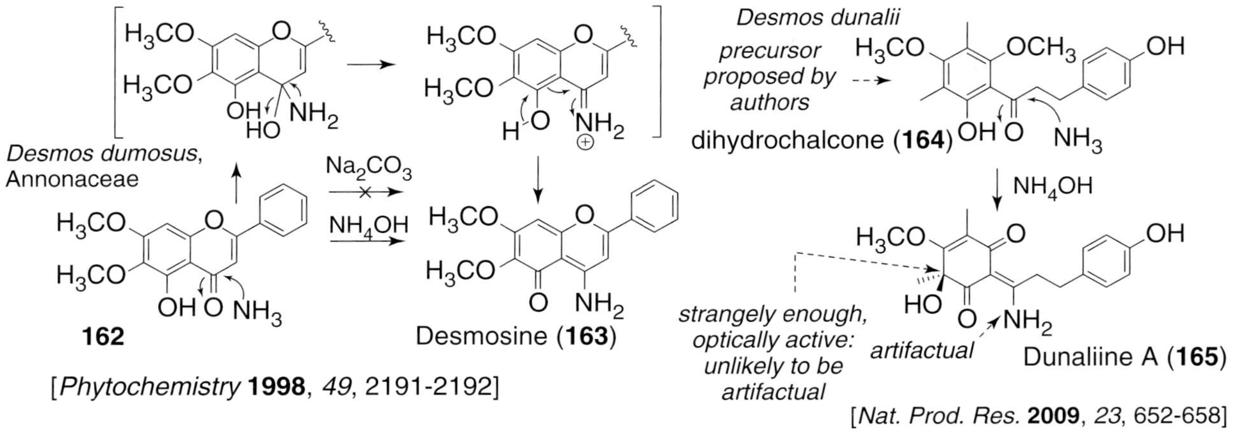

Scheme 24.36 Artifactual alkaloidic phenylchromane derivatives from *Desmos* spp.

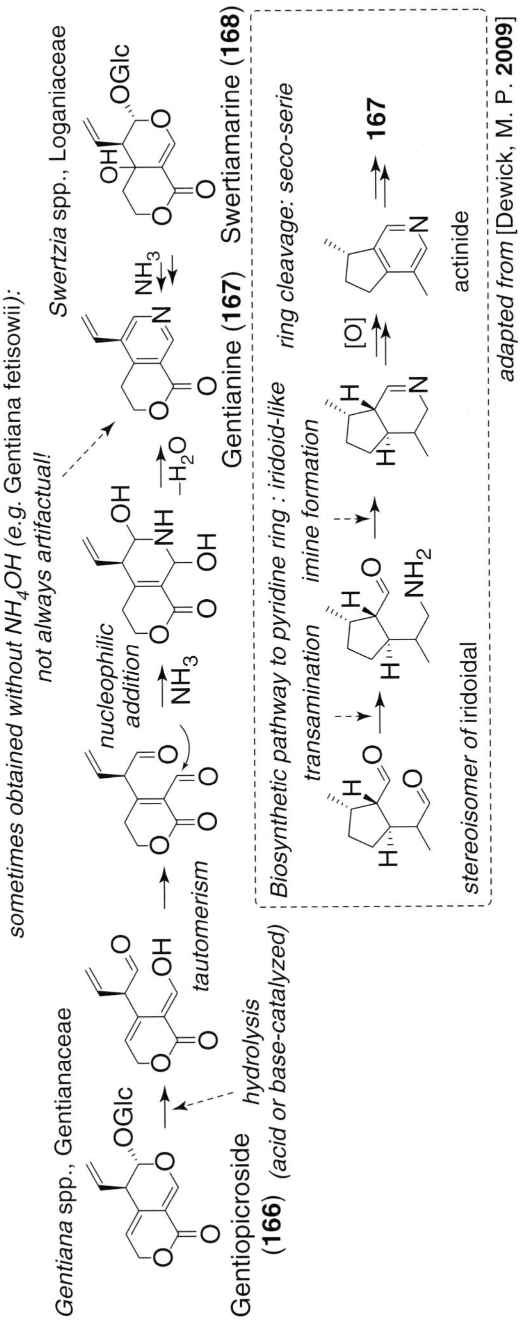

Scheme 24.37 Gentianine (**167**) as an artifactual alkaloidic compound.

24.6.2
Other Base-Catalyzed Reactions

Agujasterone C (**169**), an ecdysteroid bearing a hydroxyl-group at C11, was obtained from *Leuzea carthamoides* (Asteraceae) along with 5-deoxykaladasterone (**170**), an analog with a dienone system. As **170** was absent from the crude extracts of the plant, it was identified as an artifactual dehydration product of **169**, arising from alumina fractionation of the extract, as shown with pure **169**. This finding challenges the status of kaladasterone (**172**), isolated along with muristerone A (**171**) from seeds of various *Ipomea* species (Convolvulaceae) and proposed as being a genuine natural product (Scheme 24.38) [79].[20]

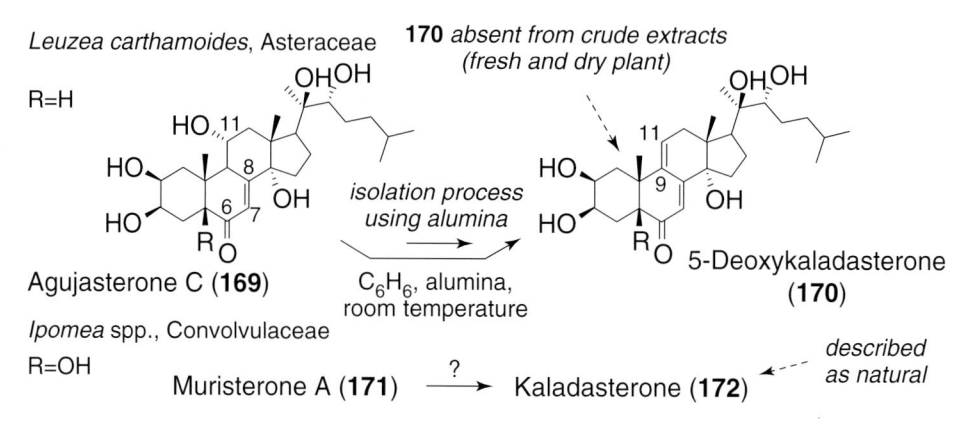

Scheme 24.38 Alumina-catalyzed dehydration of ecdysteroids.

From *Delphinium* spp. roots (Ranunculaceae), after basic treatment (NH$_4$OH) for alkaloid extraction, methyllycaconitine (**173**) by-products (**174**, **175**), resulting from pyrrolidine-2,5-dione aminolysis, were retrieved. Similarly, avhadaridine (**177**), originating from lycaconitine (**176**), was also obtained. Surprisingly, Shamma *et al.* isolated **177** from *Delphinium cashmirianum* in neutral conditions. They considered the compound as an artifact. Nevertheless, the artifactual nature of its methyl-ester counterpart, cashmiradelphine (**178**), was not determined: this product was apparently present in the crude ethanolic extract, but could be readily obtained from **176** in methanol at boiling temperature or at room temperature with basic (alumina) or acid catalysis (silica gel) (Scheme 24.39) [80].

Lounasmaa *et al.* have published a comprehensive review on the artifactual formation of a series of indolomonoterpenic alkaloids from *Rauvolfia* spp. (Apocynaceae) in the sarpagine series, implicating the Cannizzaro reaction[21] that might take place during alkaloid purification steps [81]. For example, O-acetylpreperakine (**183**) can be expected to originate from the ring-opened form (Z)-vellosimine

---

20) Note that the authentic nature of other Δ^{9-11}-ecdysteroids has repeatedly been questioned [79].

21) Redox disproportionation of non-enolizable aldehydes to carboxylic acids and alcohols, catalyzed by strong bases.

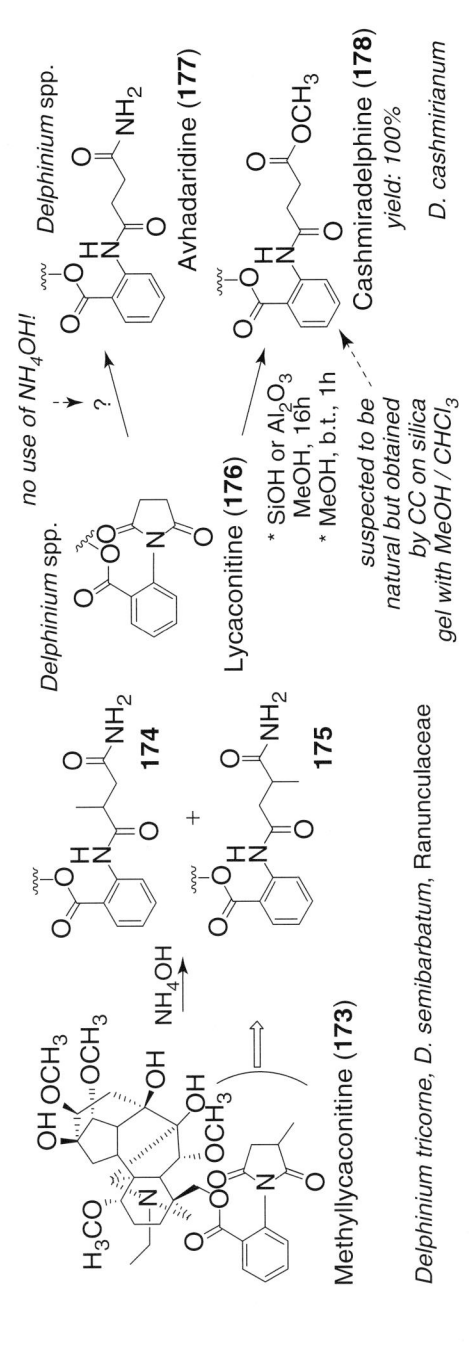

Scheme 24.39 By-products of imide-bearing aconitine-like alkaloids.

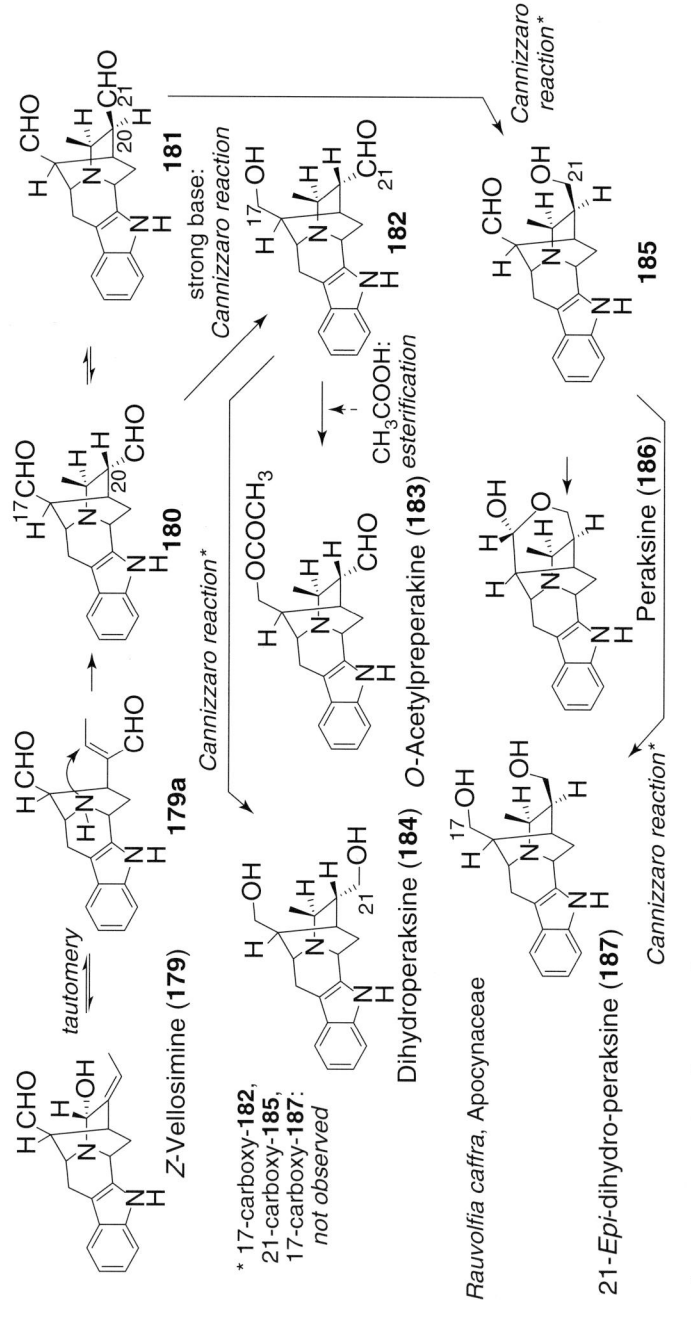

894 | *24 Artifacts and Natural Substances Formed Spontaneously*

Scheme 24.40 Artifacts from (Z)-vellomisine (**179**) via Cannizzaro reactions.

(**179a**). Note the acetylation of **182**, a product of the Cannizzaro reaction, during isolation procedures (Scheme 24.40).[22] In the ajmaline series, similar reactions were observed (Scheme 24.41).

The two series are in close relationship, via interconversions: under hydrolytic conditions (acidic or basic) the acetyl group of vomilenines (**188**, **190**) may be cleaved. In the case of *(Z)*-vomilenine (**190**) this leads to deacetyl-*(Z)*-vomilenine (**194**), which is in equilibrium with 16-epi-*(Z)*-vellosimine (**195**) and *(Z)*-vellosimine (**179**) (Scheme 24.42) [81].

24.7
Acidic Conditions during Purifications

Natural products nearly always encounter acidic catalysts in the course of isolation. This is obvious in the case of liquid/liquid extraction steps during purification of alkaloids, which require aqueous acidic media. However, acid-catalyzed reactions can be due to trace impurities in organic solvents. They also might take place on silica gel, as in the retro-Michael conversion of the instable alkaloid stephacidin B (**197**; *Aspergillus ochraceus*) into its monomer avrainvillamide (**196**) (Scheme 24.43) [82]. The various classes of acid-induced artifacts observed are summarized below, and in Section 24.8 on protic solvents.

24.7.1
Epimerization

Acidic conditions can promote epimerization. An illustration published recently for a *p*-hydroxybenzyl-substituted furanic ring in the stilbene tetramer kobophenol A (*Caragana sinica*, Fabaceae) challenges the conservation of medicinal extracts [83]. Similarly, radix of *Asarum heterotropoides* (Aristolochiaceae), a Chinese medicinal plant, contains furofuranic lignans, among which sesamin (**198**) and asarinin (**199**) are identified as main active principles. Variable amounts of these molecules are obtained using different extraction methods, and several reports indicate that **198** epimerizes to **199** under acidic conditions, with ring opening similar to that previously shown. Sesamin (**198**) is supposed to be more stable, as it bears two substituents in *exo* positions, against one in *exo* and the other in *endo* positions for **199**. A study of the epimerization process was published, showing an unexpected ratio of **198** and **199** (Scheme 24.44) [84].

The lignan cubebin (**200**), isolated from *Aristolochia lagesiana* and *Aristolochia pubescens* (Aristolochiaceae) provides a similar example. It was shown to be a mixture of **200/201** in a 3 : 2 proportion, suggesting the presence of two epimers (at C9 and/or C8) or a conformational equilibrium (Scheme 24.45). It is well known that lactols can undergo ring-opening and closing in solution: acid-catalyzed ring-opening of **200**, in $CDCl_3$ solution, leads to inversion of the stereogenic

22) Use of 10% CH_3COOH.

Scheme 24.41 Formation of raucaffrinoline (**193**) via the Cannizzaro reaction.

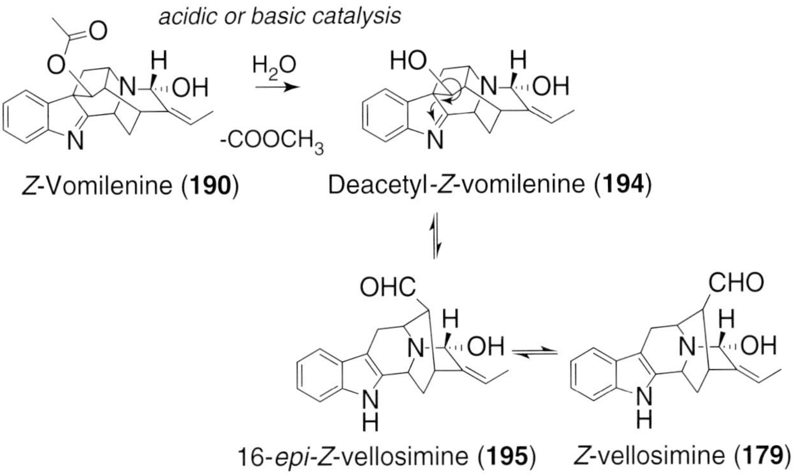

Scheme 24.42 Artifactual relationship between the ajmaline/sarpagine series.

center at C9. An intermediary aldehyde was detected by NMR under appropriate conditions (temperatures above 26 °C, long period of storage in CDCl$_3$ solution). However, it is not possible to establish whether both anomers are natural compounds [85].

24.7.2
Hydrolysis

Ester hydrolysis is sometimes observed in relatively mild conditions.[23] Bromophenols coupled to pyroglutamic acids **202** and **203** were isolated from the red alga *Rhodomela confervoides* (Rhodomelaceae). Compound **203** is a possible artifact since its conversion from **202** was observed on standing in aqueous methanol (Scheme 24.46). However, the methylation of **203** did not occur upon heating a methanolic solution either with or without silica gel at 45 °C for 48 h [87].

Artifactual hydrolysis in a heterosidic series generally occurs in strong acidic conditions. However, the quinolinic alkaloid **205**, from the marine cyanobacterium *Lyngbya majuscula*, is supposed to be an artifact from its glycosidic counterpart **204**, possibly catalyzed by silica gel (Scheme 24.47a) [88]. Extraction with hot water (or protic solvents) is suspected to induce hydrolysis of salicylic glucosides, as evidenced with *Populus deltoides* (Salicaceae) (Scheme 24.47b) [89]. Freeze-drying (i.e., low pressure) in external flasks, without temperature control, caused the same phenomenon in fresh Salicaceae leaves, due to thawing [90].[24]

23) Note also the occurrence of trans-acetylation reactions on vicinal diol systems, observed in various solvents (e.g., salvinorins D and E [86]).

24) In comparison to air-dried leaves, the content in phenolic glycosides (e.g., salicortin and 2′-cinnamoylsalicortin) dropped by about 50%, giving rise to the normally undetectable aglycon salicin.

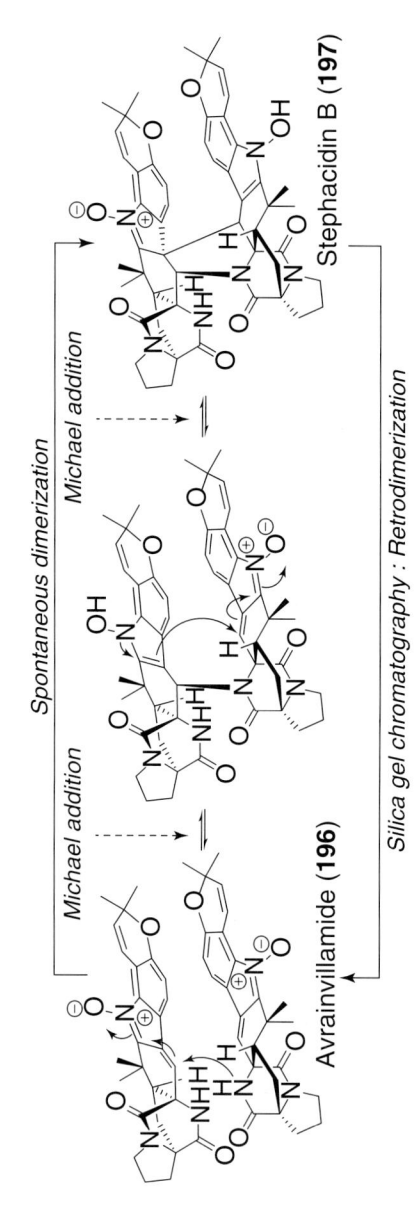

898 | *24 Artifacts and Natural Substances Formed Spontaneously*

Scheme 24.43 Stephacidin B (**197**) on silica gel.

more stable in silico *Which one is the artifact?*

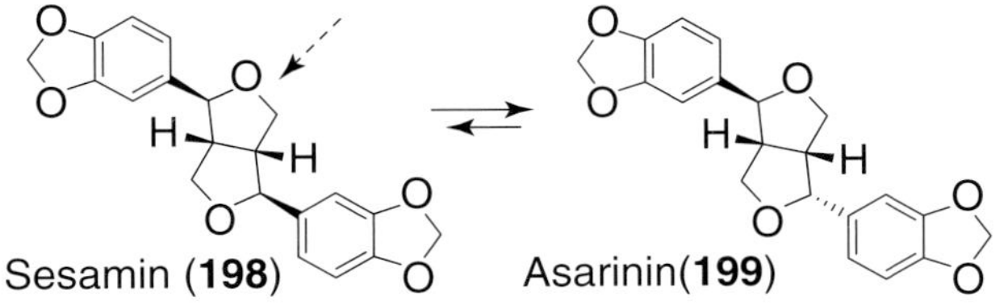

Sesamin (**198**) Asarinin(**199**)

	ratio **198**/**199**
Extraction by soaking in acetone, room temp.:	*1:1*
Vacuum distillation (10mmHg,140°C, 15min):	*1:2*
Steam distillation (110°C, 3h):	*1:2*
198 or **199** in EtOH/HCl (10%), boiling temp.	*4.5:5.5**

*(*equilibrium at 1.5h)*

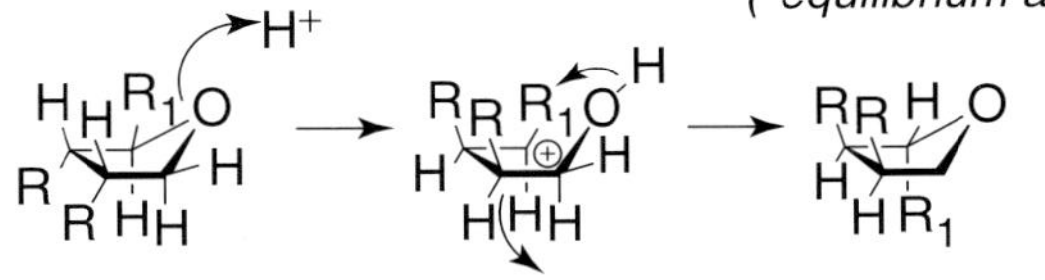

Asarum heterotropoides var. *mamdshuricum*, Aristolochiaceae R_1= methylenedioxyphenyl, R = furan moiety

Scheme 24.44 Epimerization of furofuranic lignans under various extraction conditions.

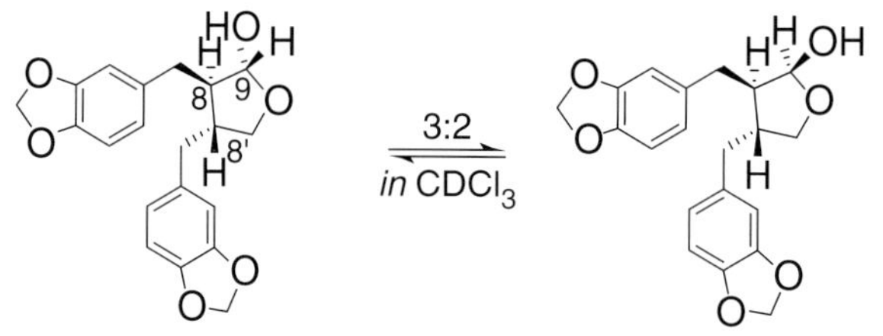

(8R,8'R,9R)-cubebin (**200**)   (8R,8'R,9S)-cubebin (**201**)

Scheme 24.45 Cubebin anomers from *Aristolochia* spp.

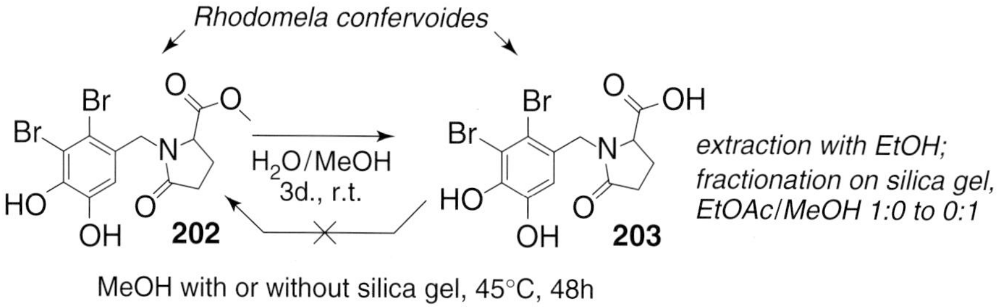

Scheme 24.46 Methyl-ester hydrolysis.

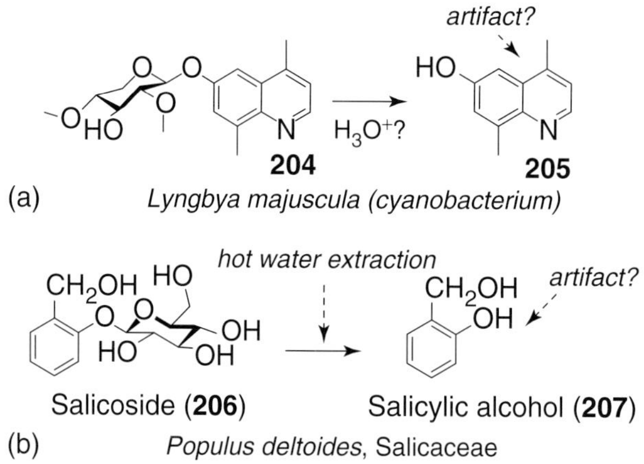

(a) *Lyngbya majuscula* (cyanobacterium)

(b) *Populus deltoides*, Salicaceae

Salicoside (**206**)   Salicylic alcohol (**207**)

Scheme 24.47 Hydrolysis of heterosides.

24.7.3
Other Acid-Catalyzed Reactions

Acid-catalyzed formation of iminiums from norrhoeadines (**208**) via acetal hydrolysis, followed by imine formation and dehydration, is particularly well known (Scheme 24.48) [91].

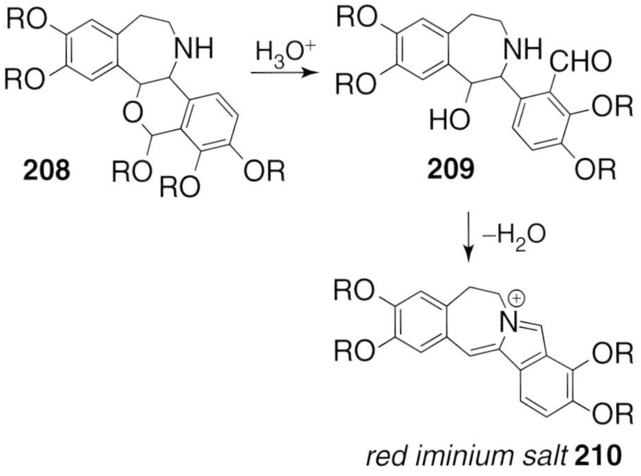

Scheme 24.48 Norrhoedanines in acidic conditions.

Several indoloquinoline-type alkaloids were found in the African shrub *Cryptolepis sanguinolenta* (Apocynaceae, ex-Asclepiadaceae), generating a debate on the origin of alkaloids cryptolepinone (**212**) and hydroxycryptolepine (**213**), which appear to be the keto and enol forms of the same compound [92]. Cryptolepinone (**212/213**), obtained from roots of the plant under acidic conditions, was not detected in neutral extracts. Forced formation of **212** from **211**, dissolved in organic solvents (MeOH or CH$_3$CN), was challenged in the absence or presence of acid (HCl 0.1 M), light, and/or air over 19 days. Increasing amounts of cryptolepinone (**212**) were detected, with the highest concentration (2.5%) in acid conditions under air exposure. Nevertheless, treatment of **211** with *m*-chloroperbenzoic acid (*m*-CPBA) gave a 16.5% yield of **212** (Scheme 24.49).

Acid-catalyzed formation of an acetal during silica chromatography was observed for laurencione (**215**), from the red alga *Laurencia spectabilis*. The molecule occurs as a mixture of two interconverting forms: (±)-2-hydroxy-2-methyldihydrofuran-3(2*H*)-one and 5-hydroxy-2,3-pentanedione. Chromatography on silica gel produces an artifactual dimeric spiroacetal **216** [93] previously isolated from *Laurencia pinnatifida* [94]. Regeneration of **215** is readily observed by acid hydrolysis (Scheme 24.50).

From Rutaceae spp., two groups recently isolated an acetonic acetal (**218**) similar to the prenylated coumarin phebalosin (**217**). Acidic hydrolysis of the epoxide group in **217** appears probable, and would lead to a vicinal diol (not isolated) likely to react with acetone. However, the authors claimed not to have used acetone as solvent, while considering [95] – or not – **218** to be an artifact! Notably, the accompanying murralongin (**219**) [96] – obviously a rearrangement product, as its formation was observed in the course of purification – could have a similar origin (Scheme 24.51).

Acid-catalyzed cyclizations of natural products are relatively frequent. Aposphaerin B (**221**), a minor achiral metabolite obtained from cultures of the fungus *Paraphaeosphaeria quadriseptata*, is thought to arise from cytosporone F (**220**) during extraction and chromatographic separation using methanol, by

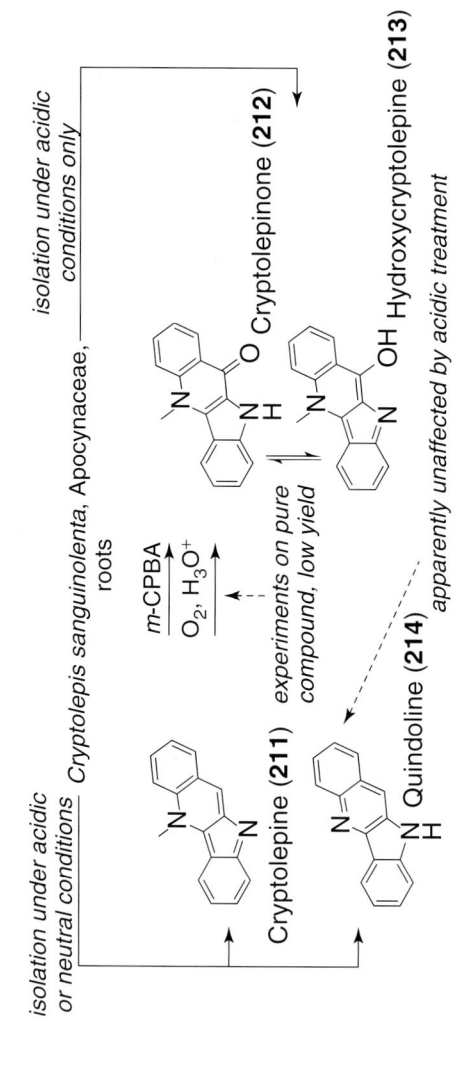

Scheme 24.49 Cryptolepine (**211**) and cryptolepinone (**212**) from *Cryptolepis sanguinolenta*.

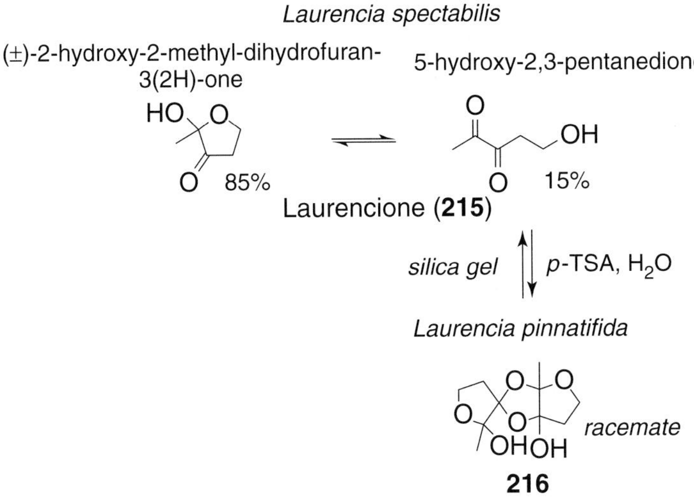

Laurencia spectabilis

(±)-2-hydroxy-2-methyl-dihydrofuran-3(2H)-one   5-hydroxy-2,3-pentanedione

85% ⇌ 15%

Laurencione (**215**)

silica gel ↕ *p*-TSA, H$_2$O

Laurencia pinnatifida

216 racemate

Scheme 24.50 Artifactual formation of a spiroketal from laurencione (**215**).

a Michael-type addition of a phenol to the enone moiety.[25] This was further suggested by the treatment of **220** with *p*-toluenesulfonic acid, resulting in the formation of **221** (Scheme 24.52a) [97]. Likewise, the minor chromane **223** from the Australian marine brown alga *Peritbalia caudata* (Sporochnaceae) might be an artifact from precursor **222**. Formation of **223** was described via acid- or UV-catalyzed cyclization of **222** by Blackman *et al.* [98].[26] However, latter attempts to induce this cyclization through exposure of **222** to silica gel at room temperature in the presence of sunlight, in common extraction solvents over prolonged periods, proved unsuccessful, rendering the issue of artifactual formation unclear (Scheme 24.52b) [99].

Finally, *Thalictrum* (Ranunculaceae) saponins (**224**) can be cited. Acidic hydrolysis of their sugar moieties also causes hydrolysis of their lactol rings, followed by a nucleophilic attack of a free vicinal hydroxy group to give a furanic ring. Dehydration would yield squarrofuric acid (**225**).[27] Similarly, acidic hydrolysis of thalicoside A (**226**) yields a furanic artifact (**227**) [100]. Notably, in this case, acidic treatment induces opening of the cyclopropane ring – a classical degradation process in the cycloartane series (Scheme 24.53).

24.8
Protic Solvents

Protic solvents are extensively used in the isolation of natural products, either for extraction (MeOH, EtOH, less frequently H$_2$O) or purification by normal-

---

25) Note that **220** and **221**, both ethyl esters at the C8 position, are possibly natural (no ethanol used during isolation).
26) See compound **157** (Scheme 24.35).
27) Surprisingly, a saturated analog is obtained.

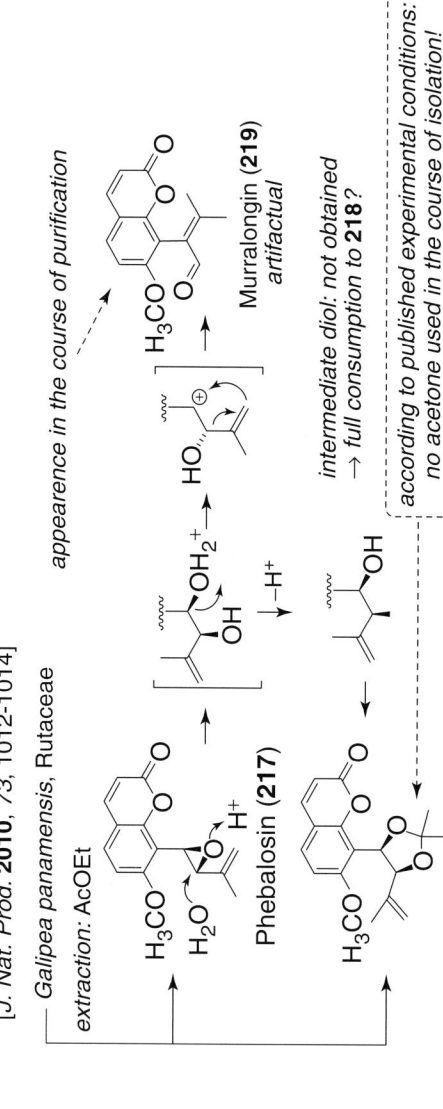

Scheme 24.51 Artifactual coumarins from Rutaceae spp.

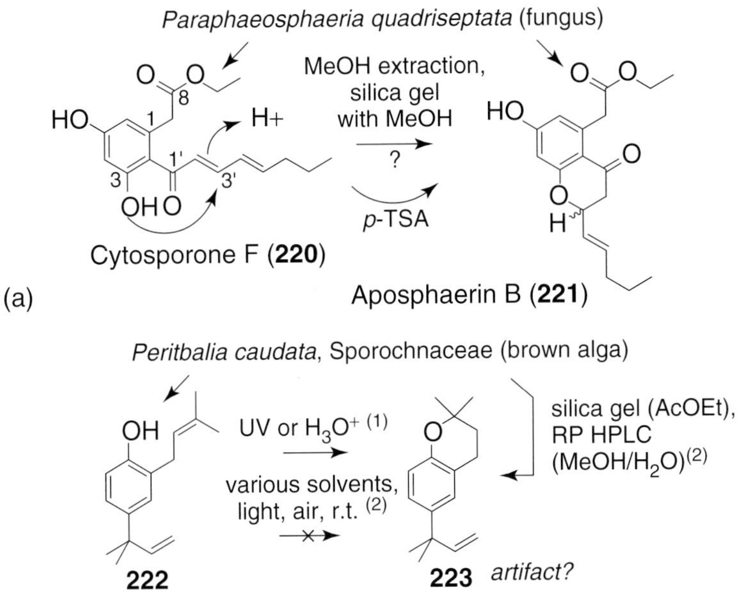

Paraphaeosphaeria quadriseptata (fungus)

(a) Cytosporone F (**220**) → Aposphaerin B (**221**)

(MeOH extraction, silica gel with MeOH; H+; p-TSA)

Peritbalia caudata, Sporochnaceae (brown alga)

(b) **222** → **223** artifact?

(UV or H$_3$O$^+$ (1); various solvents, light, air, r.t. (2); silica gel (AcOEt), RP HPLC (MeOH/H$_2$O) (2))

(1) [*Aust. J. Chem.* **1979**, *32*, 2783]; (2) [*J. Nat. Prod.* **1994**, *57*, 849-851]

Scheme 24.52 Acid-catalyzed intramolecular Michael additions.

or reversed-phase chromatography.[28] Alcohols easily react with several functions, especially under heating conditions or during long storage of wet extracts or raw material. Participation of neighboring groups is frequently possible. Reactions depicted below are often acid or base catalyzed.

24.8.1
Lactonic Compounds: Epimerization, Transesterification

Lactonic compounds are especially sensitive to protic solvents, under various conditions. For example, acetogenins of Annonaceae bear a S-γ-methyl butyrolactone, which is frequently α,β-unsaturated. Epimerization of this stereogenic center is easy in methanolic solutions containing traces of alkali.[29] Nevertheless, a series of oxopropyl-substituted lactonic "isoacetogenins" have been frequently obtained, as inseparable mixtures of *cis/trans* isomers. Duret *et al.* showed that these compounds occur by translactonization of natural acetogenins [101] bearing a hydroxyl group at the β-position relative to the unsaturated lactone [e.g., annonacin (**228**)], and proposed pathways involving acidic or basic catalysis [102]. However, silica gel is not a strong enough catalyst to induce this reaction, which could be observed upon thorough heating in alcoholic solution or in water (Scheme 24.54) [103].

---

28) Not to forget deuterated solvents in NMR experiments – in which case, artifact formation is easy to detect.

29) After extraction of isoquinolinic alkaloids – the other major class of interest in the family.

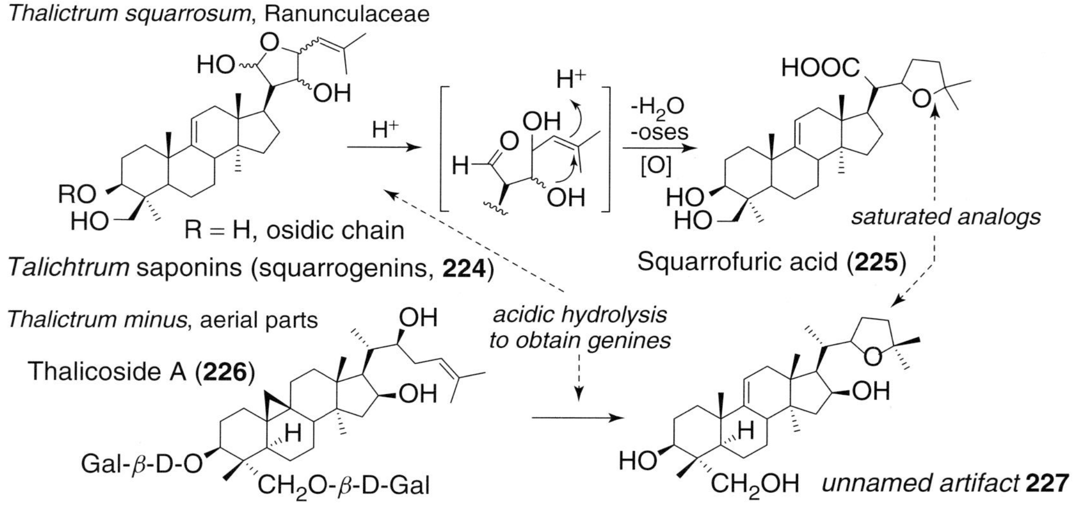

Scheme 24.53 Artifactual cyclization of *Thalictrum* saponins during acidic hydrolysis.

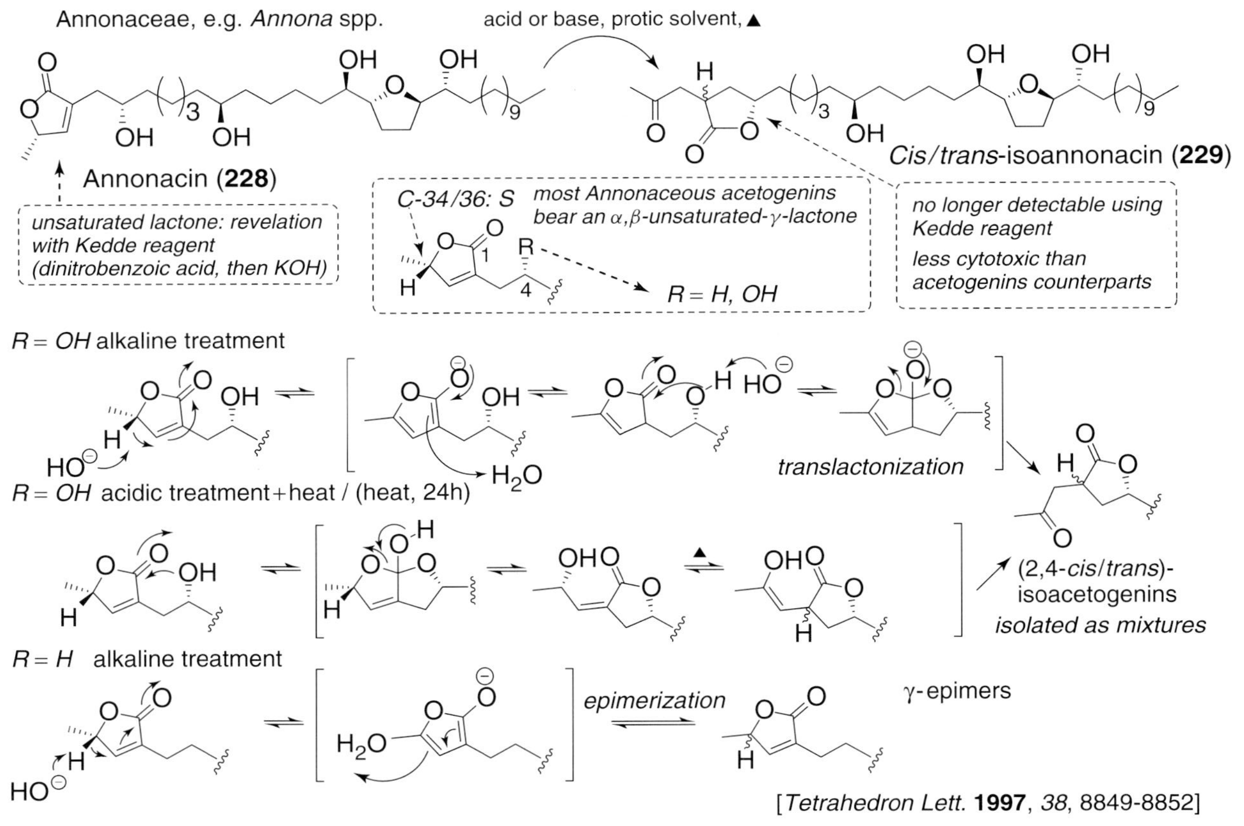

Scheme 24.54 Epimerization and translactonization of annonaceous acetogenins.

[*Tetrahedron Lett.* **1997**, *38*, 8849–8852]

Interestingly, Li et al. isolated isoannonacin (**229**) in its *trans*-form only, from seeds of *Annona muricata*, suggesting a possible enzymatic pathway *in planta* [104].

Complete lactone hydrolysis or alcoholysis can be induced under drastic conditions. However, partial conversion of lactones into esters can occur when traces of acid are present, as exemplified with lactonic eudesmanes **230–233**, the methyl ester derivatives of which (**234–237**) were observed in a methanolic extract prepared at boiling temperature, as well as during fractionation with silica gel. Their artifactual nature was challenged (Scheme 24.55) [105]. Surprisingly, such reactions are not frequently mentioned in the numerous studies on Asteraceae lactonic sesquiterpenes. Note that methanolysis of macrolactones has also been described [106].

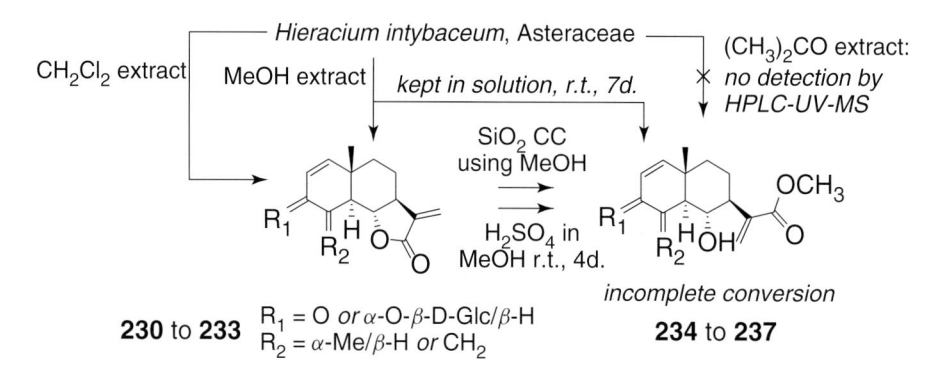

Scheme 24.55 Methanolysis of lactonic sesquiterpenes.

24.8.2
Esterification, Transesterification

Under weak acidic conditions, in alcoholic solvents, esterification or transesterification reactions are reported, especially during extraction at boiling temperature, as previously shown in the case of cyclic esters. Artifacts (e.g., **238–242**, methyl ester of **243**, **244**; Figure 24.3) arise in solubilized extracts upon long standing or, more frequently, during fractionation on silica gel. Traces of HCl from chloroform can also act as catalyst. Unexpected methyl and ethyl esters are thus quite commonly encountered, in various natural series. Adducts with butanol are scarce, as this solvent is less used (mostly for liquid/liquid partitions).

Noteworthy examples can also be found in the iridoid series, with asperuloside (**245**), acetyl-asperulosidic acid (**247**), and its carboxylic derivatives at C10 (**248**). The authors challenged the obvious artifactual nature of ethyl esters (**246**, **249**) of these molecules, avoiding ethanol and using methanol throughout their protocols. However, unclear results were obtained, evidencing the possible implication of solvent impurities. Nevertheless, silica was not the catalyst according to the protocols used (Scheme 24.56) [108].

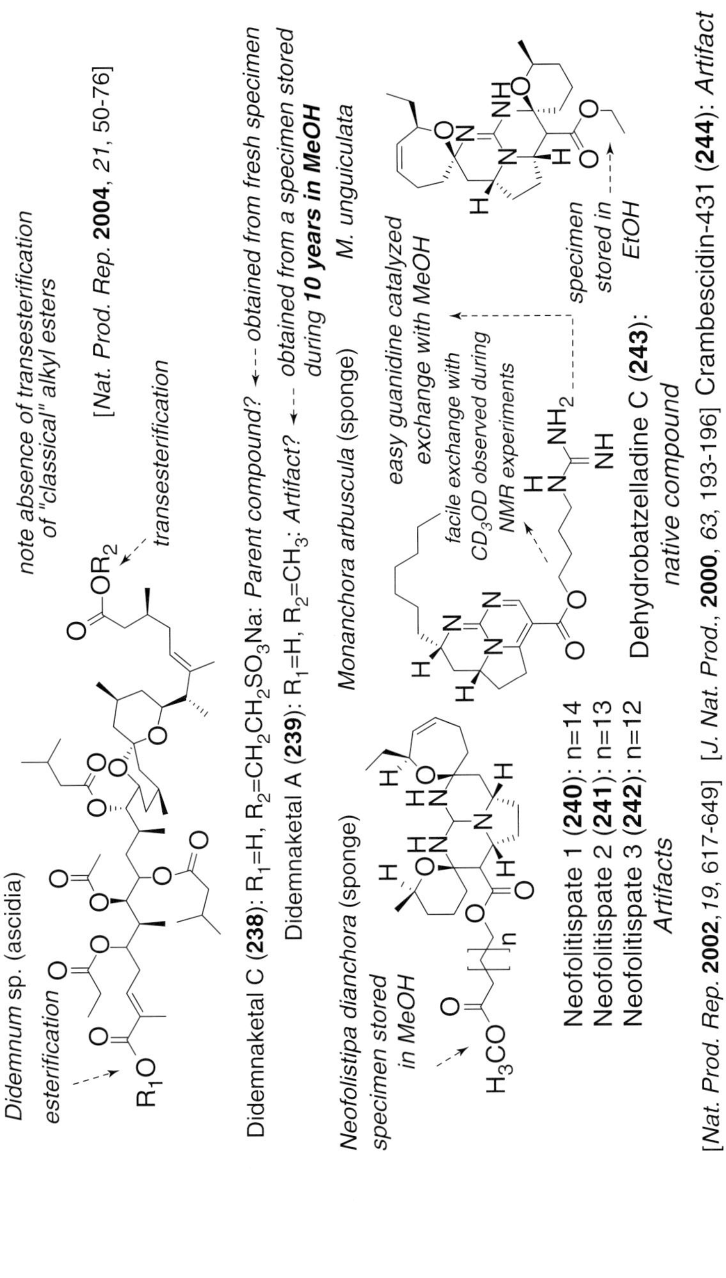

24.8 Protic Solvents

Didemnum sp. (ascidia)
esterification

note absence of transesterification of "classical" alkyl esters

[*Nat. Prod. Rep.* **2004**, *21*, 50-76]

transesterification

Didemnaketal C (**238**): R_1=H, R_2=$CH_2CH_2SO_3Na$: *Parent compound?* ← *obtained from fresh specimen*
Didemnaketal A (**239**): R_1=H, R_2=CH_3: *Artifact?* ← *obtained from a specimen stored during 10 years in MeOH*

Monanchora arbuscula (sponge)   *M. unguiculata*

easy guanidine catalyzed exchange with MeOH

facile exchange with CD_3OD observed during NMR experiments

specimen stored in EtOH

Dehydrobatzelladine C (**243**): native compound   Crambescidin-431 (**244**): *Artifact*

Neofolistipa dianchora (sponge)
specimen stored in MeOH

Neofolitispate 1 (**240**): n=14
Neofolitispate 2 (**241**): n=13
Neofolitispate 3 (**242**): n=12
Artifacts

[*Nat. Prod. Rep.* **2002**, *19*, 617-649]   [*J. Nat. Prod.*, **2000**, *63*, 193-196]

Figure 24.3 Miscellaneous esterification and transesterification artifacts [107].

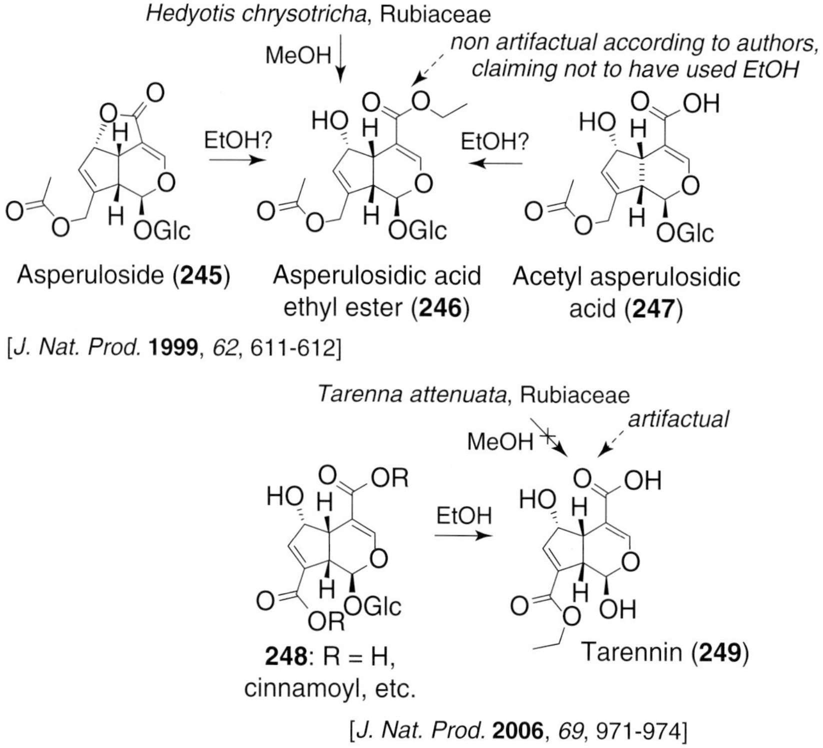

Scheme 24.56 Esterification and transesterification of Rubiaceae iridoids.

24.8.3
"Apparent Alkylations"

Substitutions of hydroxyl, phenol, or ether groups are frequent, under conditions similar to that of transesterifications, by S_N1 or S_N2 type reactions. These reactions preferentially occur on hemiacetals to form acetals. They are also reported when strongly inducing neighboring groups are present, or for resonance-stabilized intermediate carbocations. Ethanol or butanol "adducts" can easily be recognized as non-natural. Artifactual methylated compounds are less likely to be identified, and might therefore be largely underestimated: they are generally evidenced when chromatographic profiles of extracts prepared with different solvents are compared. Scheme 24.57 presents examples of artifactual methyl ethers (**251**, **253–255**) caused by the use of methanol [109]. The recently published isolation of xanthepinone (**256**) from a fungus of the *Rhizina* genus provides a nice example of methanolysis of an acetal finally yielding a rearranged methyl-acetalic artifact (**257**, Scheme 24.58) [110].

Nucleophilic substitution of a halogen can also constitute a mechanism for generating methoxy-bearing artifactual marine products. This is proposed for methoxydechlorochartelline A (**259**) in relation to chartelline A (**258**) (Scheme 24.59) [111].

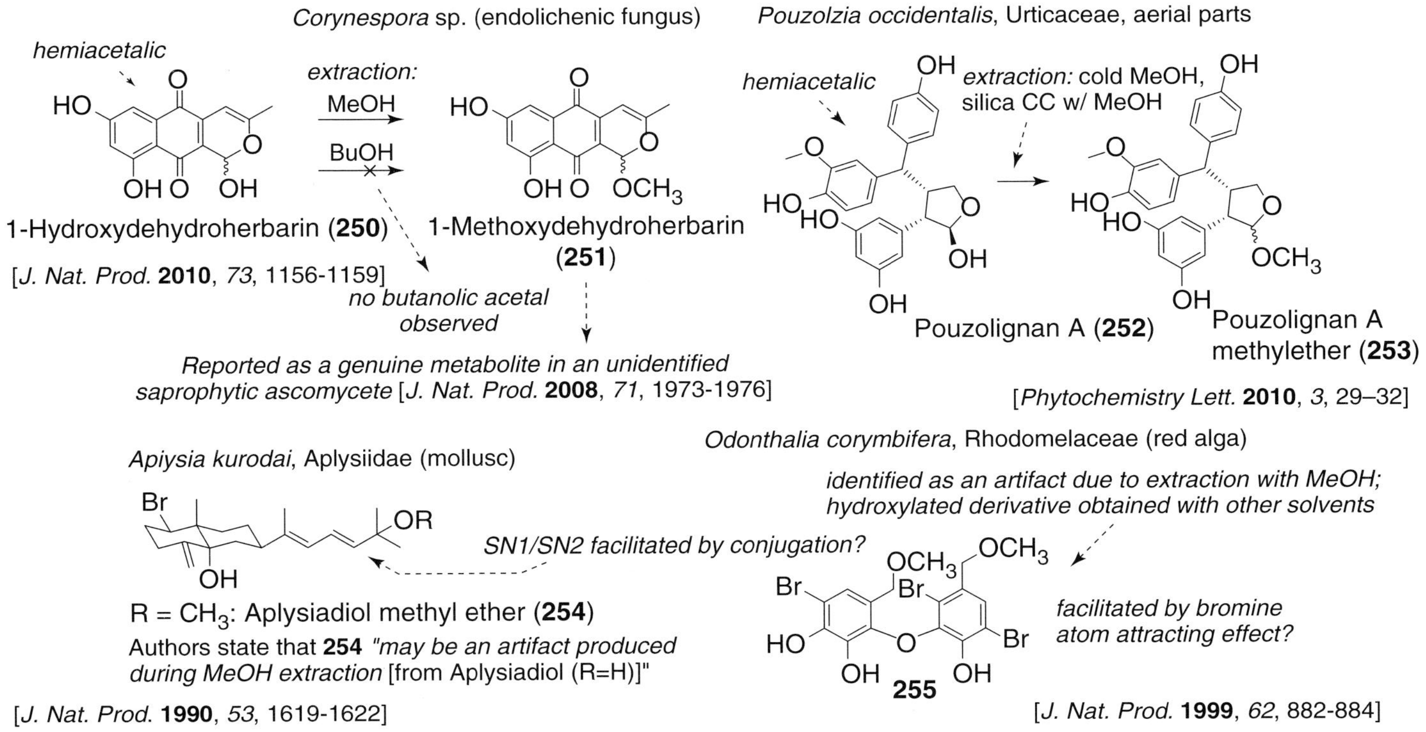

Scheme 24.57 Miscellaneous methylation artifacts.

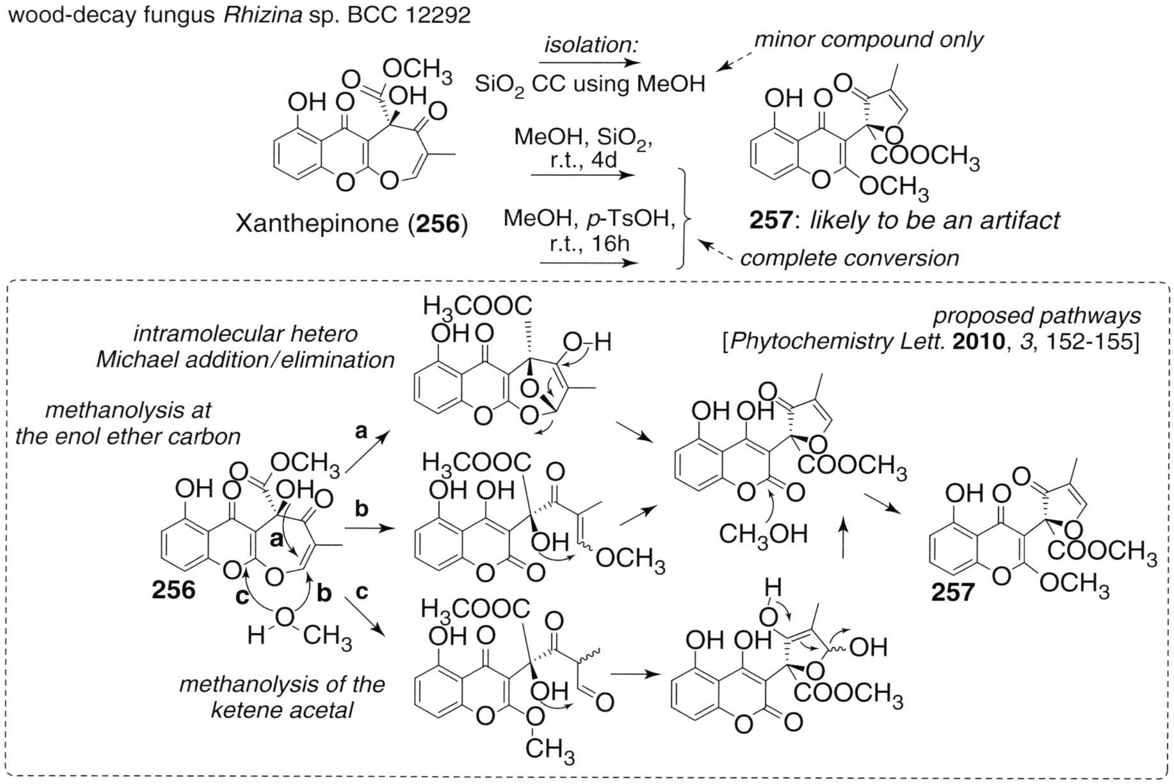

Scheme 24.58 Methanol induced rearrangement of xanthepinone (**256**).

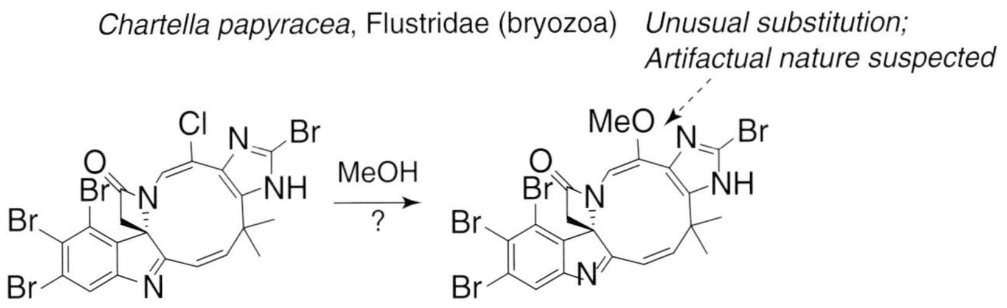

Chartella papyracea, Flustridae (bryozoa)   Unusual substitution;
Artifactual nature suspected

[Structures: Chartelline A (**258**) → (MeOH, ?) → Methoxydechlorochartelline A (**259**)]

Scheme 24.59 Methanol substitution of chartelline A (**258**).

Methoxylation artifacts are sometimes due to nucleophilic addition on conjugated systems, such as for puupehenone (**260**), which bears a quinone-methide system. Under fractionation of the chloromethylenic extract of a marine sponge (*Hyrtios* sp.) with methanol, **260** easily yielded 15α-methoxypuupehenol (**261**), thus becoming a trace compound [112]. Interestingly, **260** could be regenerated from **261** by filtration on silica gel (Scheme 24.60) [113].[30] Only a few reports depict such reactions, although they possibly occur frequently.

Ethylation artifacts – also found in various chemical series – are easily distinguished as ethyl-ether groups are rare in natural products. These artifacts are generally generated during extraction; ethanol is barely used in chromatographic procedures. However, traces of ethanol used as a stabilizer (e.g., in chloroform) can also yield such artifacts (Scheme 24.61) [114]. Such reactions also occur in the case of acetoxy acetals, the instability of which was noted by several authors (Scheme 24.62) [115].

Butanol adducts are described in the literature, such as 4-*O*-butylpaeoniflorin (**273**) from *Paeonia suffruticosa* (Paeoniaceae), formed from paeniflorin (**272**) during fractionation of a crude methanolic extract with butanol [116]. No evidence of a catalyst can be found in the protocol used, but the hemiacetalic position (C4) in *Paeonia* monoterpenes is very reactive. Indeed, the roots of *Paeonia lactiflora*, which are used in Chinese medicine, are sometimes traditionally fumigated with sulfur to preserve their white color (*white-peony root*) and to prevent bacterial growth. This processing method induces a fall in the amount of **272** and related hemiacetalic compounds to yield sulfated analogs at C4 (e.g., sodium paeoniflorin sulfonate **271**) (Scheme 24.63) [117].

24.8.4
Formation of Acetals

Formation of acetals from aldehydes can be observed under mild conditions: methanolic extraction of secoiridoids (e.g., secologanin **279**) easily yields dimethyl

---

30) Interestingly, **261** was less cytotoxic than **260** (this is quite understandable in regard to reactivity), while showing improved in vitro activity against *Plasmodium falciparum*.

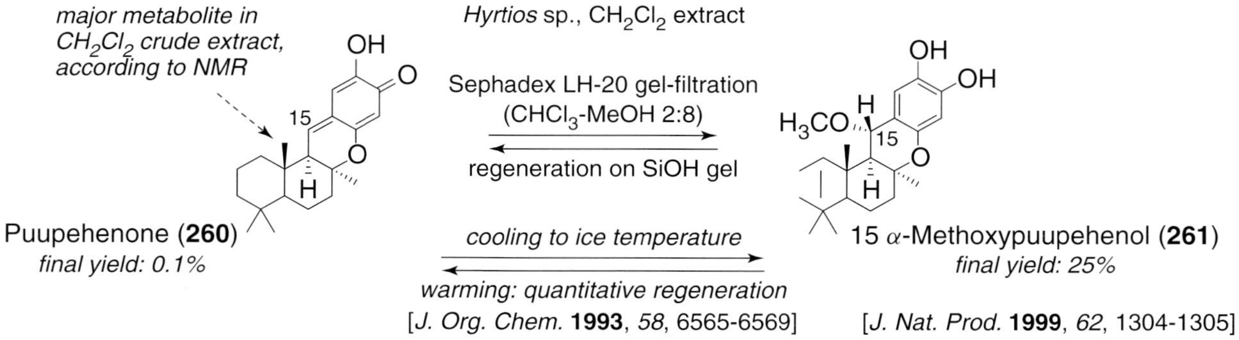

Puupehenone (260)
final yield: 0.1%
major metabolite in CH_2Cl_2 crude extract, according to NMR

Hyrtios sp., CH_2Cl_2 extract

Sephadex LH-20 gel-filtration ($CHCl_3$-MeOH 2:8)
regeneration on SiOH gel
[*J. Org. Chem.* **1993**, *58*, 6565-6569]

cooling to ice temperature
warming: quantitative regeneration

15 α-Methoxypuupehenol (261)
final yield: 25%
[*J. Nat. Prod.* **1999**, *62*, 1304-1305]

Scheme 24.60 Methanol adduct on puupehenone (**260**).

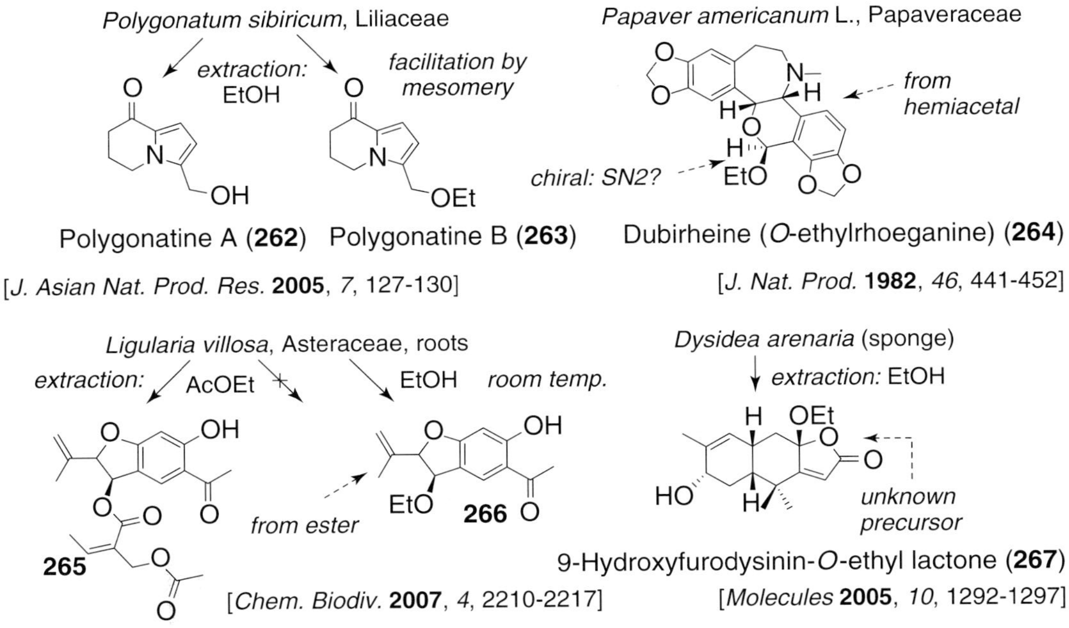

24.8 Protic Solvents | 915

Polygonatum sibiricum, Liliaceae

Papaver americanum L., Papaveraceae

Polygonatine A (**262**)   Polygonatine B (**263**)   Dubirheine (*O*-ethylrhoeganine) (**264**)

[*J. Asian Nat. Prod. Res.* **2005**, *7*, 127-130]

[*J. Nat. Prod.* **1982**, *46*, 441-452]

Ligularia villosa, Asteraceae, roots

Dysidea arenaria (sponge)

9-Hydroxyfurodysinin-*O*-ethyl lactone (**267**)

[*Chem. Biodiv.* **2007**, *4*, 2210-2217]

[*Molecules* **2005**, *10*, 1292-1297]

Scheme 24.61 Miscellaneous ethylation artifacts.

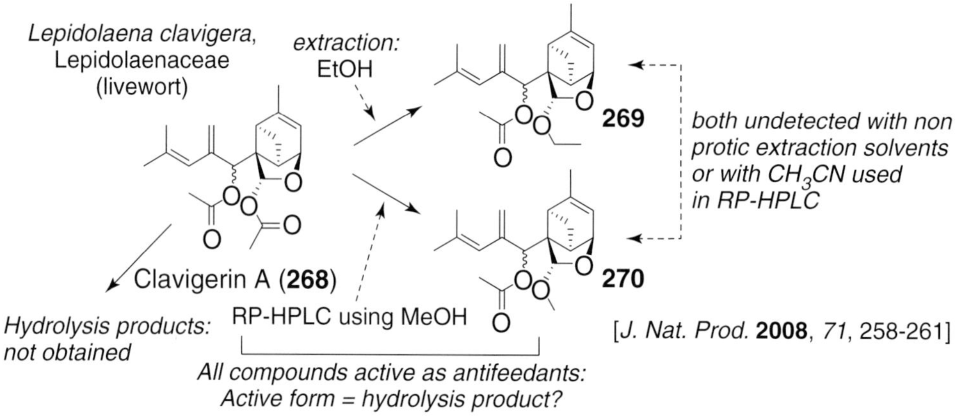

Scheme 24.62 Artifactual alcoholysis of acetoxy-acetalic sesquiterpenes.

acetal artifacts (**275**), as shown with *Lonicera korolkovii* [118] and *Lonicera japonica* (Caprifoliaceae) [119]. Conversion of secologanic acid (**274**) into **276** was also observed. Similarly, extraction of *L. japonica* aerial parts with water, followed by partition with *n*-butanol, yielded low amounts of secologanin dibutyl acetal (**277**) and butylsecologanic acid (**278**). Both molecules were obviously absent in crude extracts [120]. A complex artifact, korolkoside (**280**), showing both acetal and ester artifacts, was also isolated from *L. korolkovii* (Scheme 24.64) [118]. Surprisingly, it displays acetalization of the aldehydic group of a secologanin moiety by glucose of a second monomer (C4/C6 diol system). This feature, atypical for a natural product, was probably acid-catalyzed in anhydrous conditions during fractionation. However, the authors state that the methyl esters and dimeric nature of **280** are of natural origin, according to an unpublished study of *Lonicera morrowii* conducted *"without acidic conditions."*

In the isocedrene series, a peculiar acetal (**282**) was obtained from *Pleocarphus revolutus* (Asteraceae, aerial parts) after methanolic extraction, along with its dialdehydic counterpart **281** (Scheme 24.65). However, the artifactual nature of **282** was not proved [121]. 3,5-Dibromo-1-hydroxy-4-oxo-2,5-cyclohexadien-l-acetamide (**283**), from the sponges *Aplysina fistularis* forma *fulva* [122] and *Aplysina caissara* (Verongidae) [123], is also of interest: the presence of two bromine atoms in *ortho* positions to the ketone facilitates acetal formation during extraction with methanol, to give **284**.

24.9
Acetone-Derived Artifacts

Use of acetone in acidic conditions can quite easily give rise to acetals on vicinal diol systems, as previously exemplified (compound **218**, Scheme 24.51).

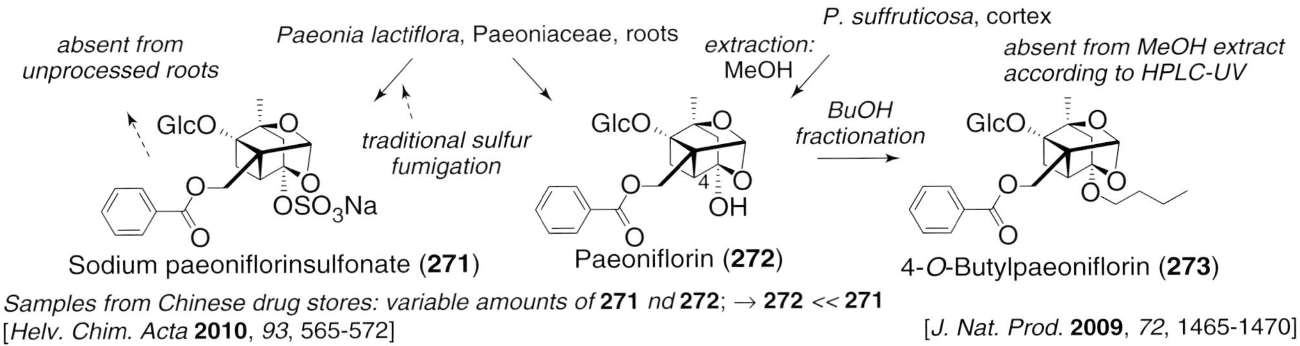

24.9 Acetone-Derived Artifacts

absent from unprocessed roots

Sodium paeoniflorinsulfonate (**271**)

Paeonia lactiflora, Paeoniaceae, roots

traditional sulfur fumigation

P. suffruticosa, cortex
absent from MeOH extract according to HPLC-UV

extraction: MeOH

Paeoniflorin (**272**)

BuOH fractionation

4-O-Butylpaeoniflorin (**273**)

[*J. Nat. Prod.* **2009**, *72*, 1465-1470]

Samples from Chinese drug stores: variable amounts of **271** nd **272**; → **272** << **271**
[*Helv. Chim. Acta* **2010**, *93*, 565-572]

Scheme 24.63 Nucleophilic substitutions on paeoniflorin (**272**).

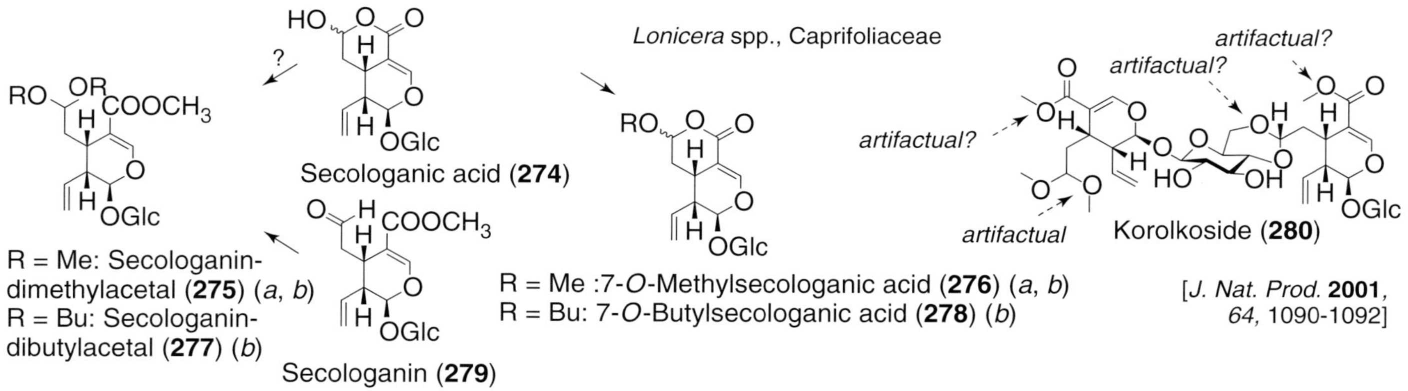

Scheme 24.64 Secoiridoid acetals from *Lonicera* spp.

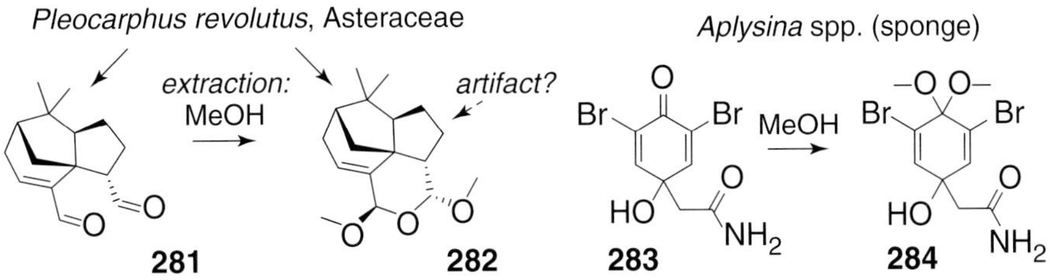

Scheme 24.65 Acetals from *Pleocarphus revolutus* and *Aplysina* spp.

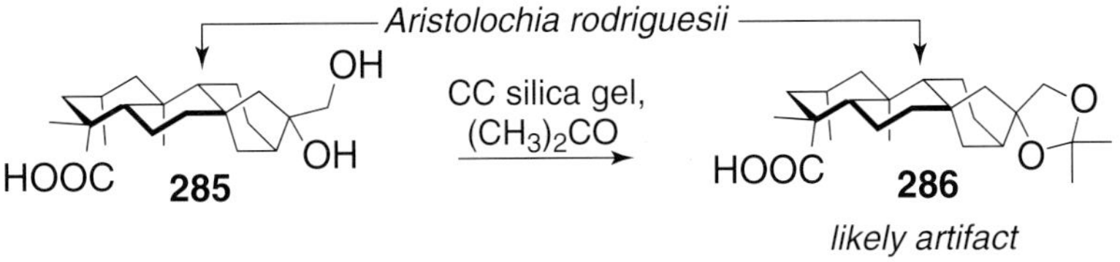

Scheme 24.66 An acetal caused by acetone under acidic conditions.

The kauranoid **286** from *Aristolochia rodriguesii* (Aristolochiaceae), an acetonide of *ent*-16β,17-dihydroxy-(−)-kauran-19-oic acid (**285**), is probably an artifact of purification using acetone on silica gel (Scheme 24.66) [29].

Acetone can also form specific artifacts with alkaloids: under acidic conditions, adducts on enamines are observed, for example, for indolomonoterpenes [1,2-dehydrobeninine (**287**) and its acetonide derivative **288** (Scheme 24.67) or for lysine-derived alkaloids [lycodine (**289**) and its artifact hydroxypropyllycodine (**290**)]. Hesse points out that the presence of an acetonyl side chain in an alkaloid does not dictate that the compound is an artifact [77], as in the case of (−)-pelletierine (**291**, *Punica granatum*, Punicaceae) [75].

24.10
Halogenated Solvents

Chloroform and methylene chloride are often used in natural products fractionation and analysis. Halogen-containing solvents are unstable. Chloroform, depending on purity and age, tends to decompose into HCl and phosgene ($COCl_2$) when exposed to light and oxygen. Phosgene easily yields artifacts, especially with amines [e.g., (−)-verbaskine (**292**) and ovihernangenine (**293**) (Figure 24.4)] [77], and is a source of free radicals. Ethyl chloroformate, which is also formed in aging chloroform containing ethanol used as a stabilizer to prevent the previous reaction, is also able to react with amines. These aspects have been reviewed [7, 124, 125]. Notably, methylene chloride also tends, to a lesser extent, to produce HCl.

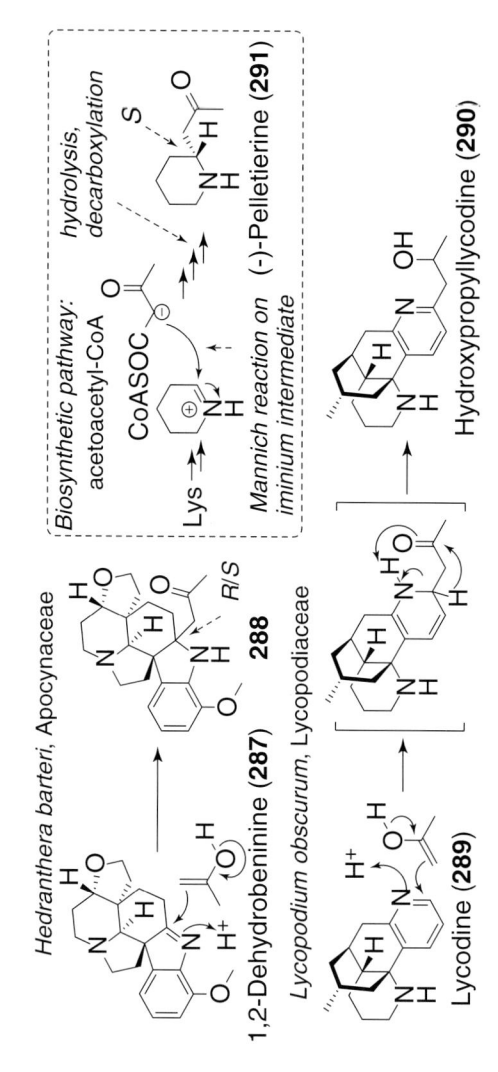

Scheme 24.67 Acetone adducts on alkaloids under acidic conditions.

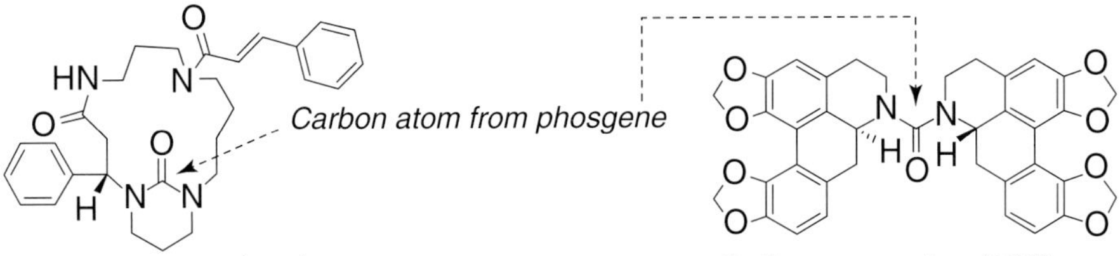

(−)-Verbaskine (292)
Verbascum nobile,
Scrophulariaceae

Ovihernangenine (293)
Hernandia nymphaeifolia,
Hernandiaceae

Figure 24.4 Phosgene adducts.

Some natural compounds can be halogenated in the course of isolation. Artifacts from terrestrial plants are easily recognizable: even though organochlorines and organobromines have been detected across all phyla, including higher plants, they are scarcely encountered as genuine molecules in plants [126]. Terpenoids exemplify several mechanisms of chlorination. The sesquiterpene madolin Q (**294**) from *Aristolochia heterophylla* is considered to be an artifact formed during extraction and separation with chloroform, by substitution with a $^{\bullet}CCl_3$ radical, according to authors [29]. A chlorinated triterpene glycoside **296** was obtained from *Cimicifuga racemosa* (Ranunculaceae) roots, along with its hydroxylated counterpart **295** (Scheme 24.68). Compound **296**, undetected in the crude methanol extract, apparently appeared during partition using chloroform and *n*-butanol. The authors proposed a reaction of HCl with the tertiary hydroxyl function at C25 and conducted a series of experiments in which a dilute chloroformic solution of **295** was subjected to HCl vapor: **295** partly converted into **296**, confirming it is an isolation artifact. Triterpenic saponosides from *Securidaca longepedunculata* (Polygalaceae) obtained by extraction with methylene chloride have two types of genine, namely, senegenine (**298**) and the chlorinated presenegenine (**297**) [127]. Artifactual chlorinated alkaloids are also observed (e.g., **299**) [128]; chloroform often reacts with quartenarized compounds by nucleophilic addition (e.g., **300**) [129].

Finally, cases of tertiary amines reacting with alkyl halides (dichloromethane, chloroform, and their impurities) during extraction to form quaternary ammonium salts were identified [e.g., macrosalhine (**301**, Scheme 24.69), and see Scheme 24.40]:

> "Caution is therefore needed when quaternary ammonium derivatives are isolated using these solvents" [81].

24.11
Protoberberines, a *"Cabinet de Curiosités"*

Protoberberines are quaternary isoquinolines encountered mostly in Magnoliales and Ranunculales (e.g., Berberidaceae, Papaveraceae, Ranunculaceae). The ease

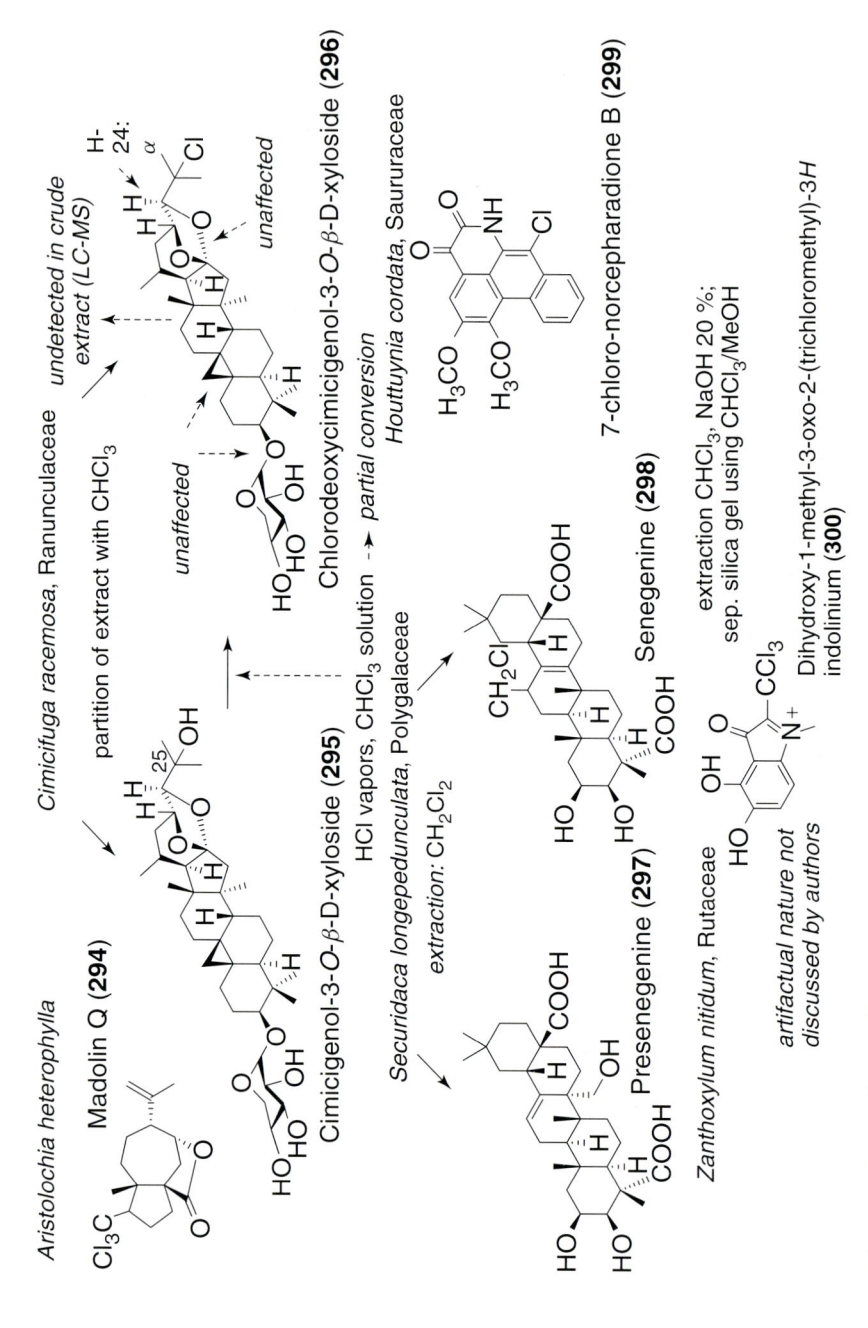

922 | *24 Artifacts and Natural Substances Formed Spontaneously*

Aristolochia heterophylla

Madolin Q (**294**)

Cimicifuga racemosa, Ranunculaceae

partition of extract with CHCl$_3$ → undetected in crude extract (LC-MS)

H-24: α

Cimicigenol-3-*O*-β-D-xyloside (**295**) → HCl vapors, CHCl$_3$ solution, partial conversion → Chlorodeoxycimicigenol-3-*O*-β-D-xyloside (**296**)

Securidaca longepedunculata, Polygalaceae
extraction: CH$_2$Cl$_2$

Presenegenine (**297**) → Senegenine (**298**)

Houttuynia cordata, Saururaceae

7-chloro-norcepharadione B (**299**)

Zanthoxylum nitidum, Rutaceae

artifactual nature not discussed by authors

extraction CHCl$_3$, NaOH 20 %; sep. silica gel using CHCl$_3$/MeOH

Dihydroxy-1-methyl-3-oxo-2-(trichloromethyl)-3*H* indolinium (**300**)

Scheme 24.68 Artifactual chlorinated terpenoids and alkaloids.

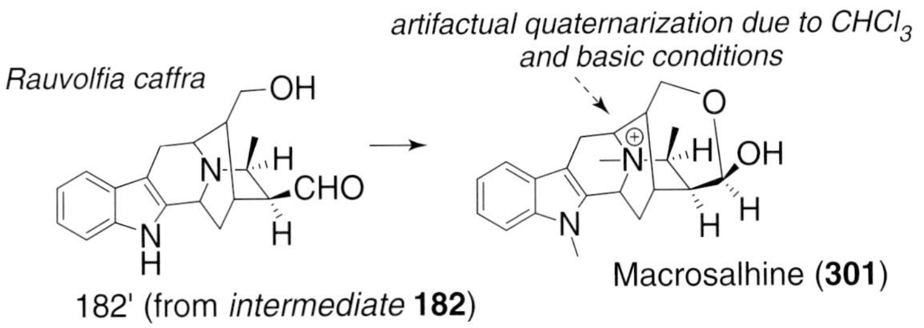

Scheme 24.69 Quaternization of the indolomonoterpenic alkaloid macrosalhine.

with which these alkaloids undergo artifactual transformations justifies their presentation as the concluding part of this chapter. The above-mentioned oxidation, dimerization, amination, addition of acetone, and chlorination reactions can be exemplified with the two major representatives of this class, berberine (**302**) and palmatine (**314**, Scheme 24.72 below), the reactivity and common artifacts of which have been reviewed [130, 131]. First, these native quaternary salts are easily converted into their base forms (**303** for berberine) by strong bases. Both diastereomers of bimolecular aminoacetals, formed by the condensation reaction between two molecules of the base form, were detected as minor products (**304** for berberine). They are immediately converted back into salts by traces of acid. Nevertheless, quaternary protoberberine salts are unstable in the presence of concentrated alkali: the alkaloid bases formed are subsequently transformed into 7,8-dihydro- or 8-keto-derivatives by disproportionation of the pyridine ring (**305**, **306**, Scheme 24.70).

Analogically, nucleophilic addition of alcohols is easy, inducing formation of 8-methoxy and 8-ethoxy-7,8-dihydroderivatives (**308** and **309**, respectively, from berberine), during the isolation or separation procedures in alkaline media, in the presence of methanol and ethanol. These unstable artifacts are easily converted back into their quaternary forms in an acidic environment. 8-Amino-derivatives (**307**) are formed with primary amines in methanol or ethanol (Scheme 24.71). When chloroform is used during extraction under alkaline conditions, 8-trichloromethyl adducts are also easily formed; trichloromethylberberine (**310**) and trichloromethylpalmatine (**315**) are, thus, frequently observed. The behavior of **310** on a silica gel column was studied: oxidized products **311**–**313** are obtained. A similar study of **315** led to isolation of **316**–**318** (Scheme 24.72) [7].

In a protic medium, 7,8-dihydroprotoberberine derivatives (e.g., **305**) are in equilibrium with their iminium forms (**305a**). The latter species are unstable and undergo rapid disproportionation to give a mixture of corresponding protoberberines (e.g., **302**) and tetrahydroprotoberberines (e.g., **319**). Notably, natural tetrahydroprotoberberines are natively *(S)*-enantiomers (Scheme 24.73) [75].

Finally, berberine (**302**) was shown to undergo facile nucleophilic attack by acetone during fractionation, yielding the heptacyclic karachine (**320**, Scheme 24.74) [77].

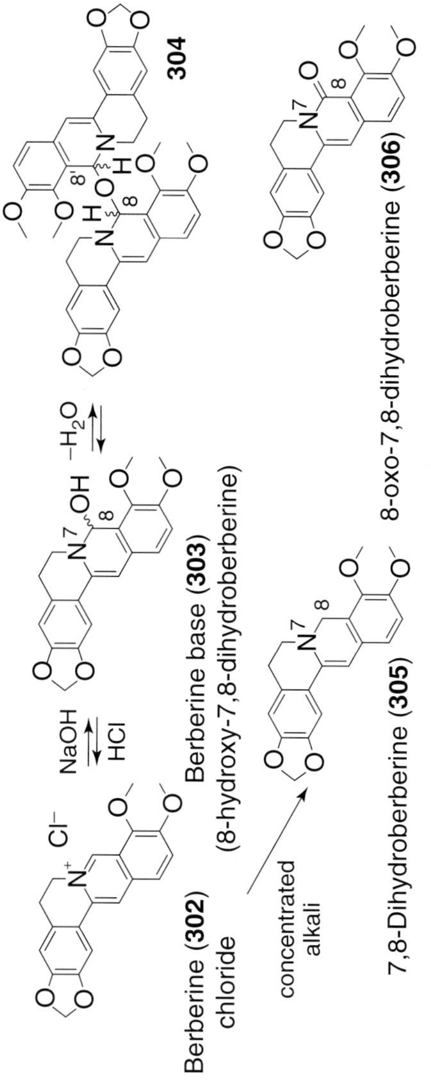

Scheme 24.70 Alkaline treatment of protoberberines.

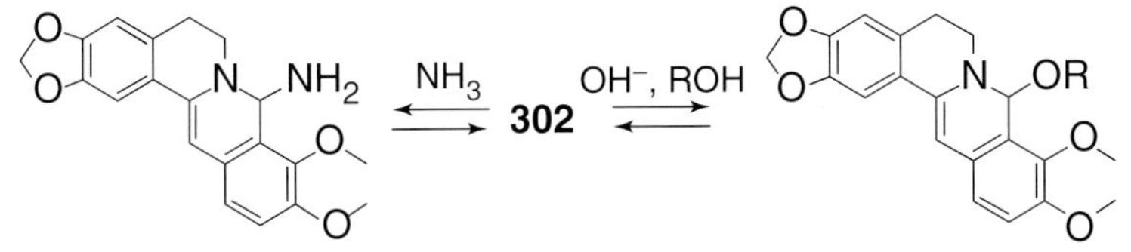

8-Amino-7,8-dihydroberberine (307)

Scheme 24.71 Nucleophilic additions on berberine in alkaline conditions.

R = Me: 8-Methoxy-7,8-dihydroberberine (**308**)
R = Et: 8-Ethoxy-7,8-dihydroberberine (**309**)

Similar reactions can be expected for chemo-equivalents. For example, nucleophilic attacks also occur on benzophenanthridines (Papaveraceae), with these artifacts being more stable due to lower pK. 6-Hydroxy-5,6-dihydrobenzophenanthridines also yield dimerization products [130]. This reactivity was exemplified by easy conversion of chelilutine (**321**; *Sanguinaria canadensis*, *Eschscholtzia californica*, Papaveraceae) into **322–324** under basic conditions (Scheme 24.75) [132].

It is also noteworthy that a parallel reactivity exists for dehydroaporphines (e.g., **55**), which are considered as key intermediates in the biosynthesis of 7-oxygenated and 7-alkylated aporphines. Their enamine group is nucleophilic enough to readily react during isolation. Obviously, such compounds should be studied or purified using adapted precautions.

24.12
Conclusion

The variety and often poor-description of favorable *"conditions + reactivity"* combinations leading to artifacts render the topic intricate. In most cases, formation of an artifact is due to uncontrolled elements, such as unidentified solvent impurities (carelessness and fumbling are rarely stated in "Material and Methods" sections). Some practices appear to be at risk for the phytochemist, with several solvents to be considered with caution, especially with thermal treatment.[31] Surprisingly, some artifact-prone protocols do not seem, in numerous reports, to yield any by-product. Nevertheless, discrepancies in the literature are frequent for identical or chemo-equivalent natural products, without any apparent clue as to why.

The nearly virtual fractionation of crude extracts allowed by development of hyphenated techniques is now a frequent starting point in phytochemical studies. It permits a reduction in duration and number of manipulation steps and, obviously, in formation of artifacts – while also facilitating their detection and

---

31) The use of Soxhlet apparatus only scarcely appears in recent literature, but long extractions at boiling temperature are frequently performed.

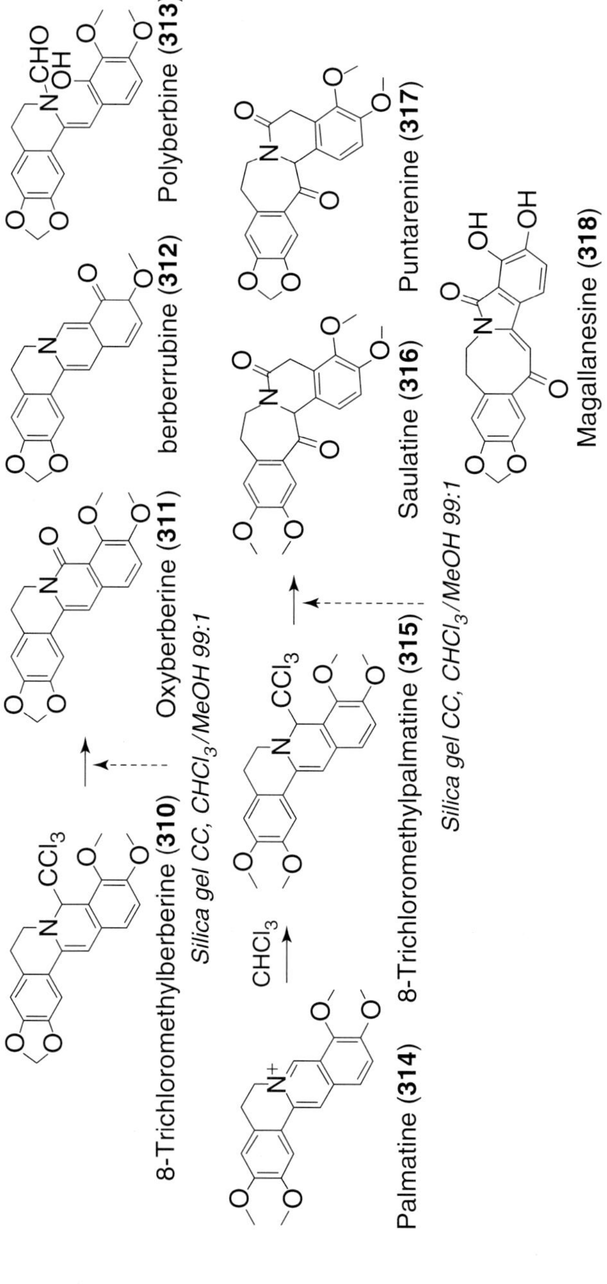

Scheme 24.72 Chloroformic adducts and their oxidation artifacts.

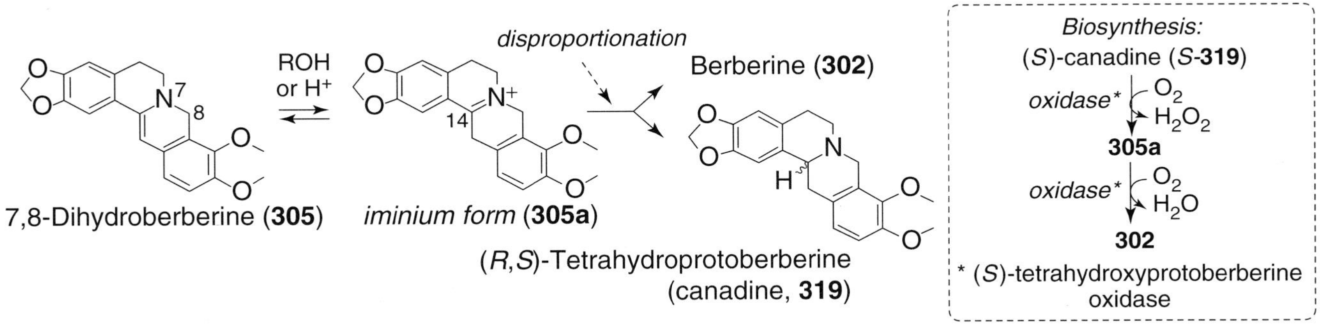

Scheme 24.73 Disproportionation of dihydroprotoberberines.

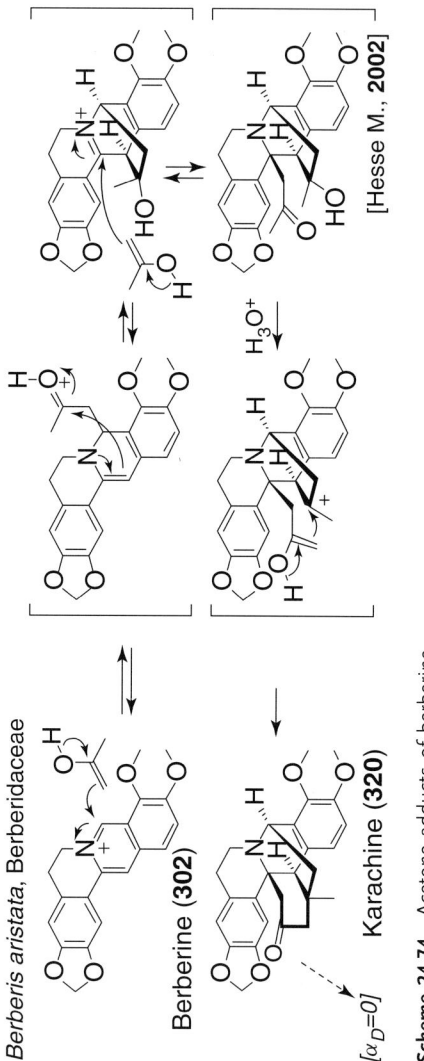

Scheme 24.74 Acetone adducts of berberine.

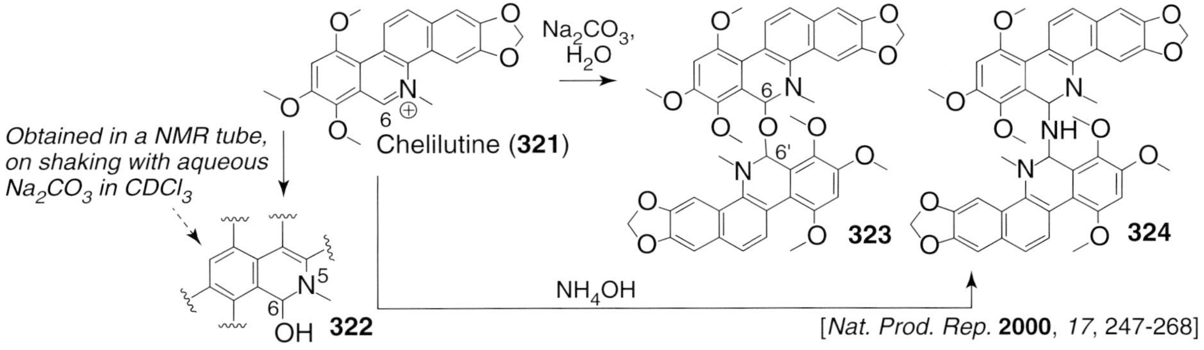

Scheme 24.75 Reactivity of benzophenanthridines.

structural identification. Recognizing the artifactual nature of an isolated natural product can be rather difficult. Measuring the ratio between biosynthesis and artifactual formation is yet another exciting challenge. Regarding this particular aspect, dereplication[32] can help in defining metabolic "priorities," especially in chemotaxonomic or ecological perspectives.[33]

The reactivity of natural products from various series sometimes affords artifacts of complex structures[34] under mild conditions. Parts of these examples are evocative of biogenetic processes, and are generally compatible with simple, "green" procedures. With several drawbacks (selectivity, stereocontrol, yield, etc.) but genuine spontaneity, the chemistry of artifacts can thus offer helpful pieces – readily possible and accessible synthesis steps – in the jigsaw of rationale and efficient biomimetic strategies.

References

1. Rey, A., Tomi, M., and Hordé, T. (2006) *Dictionnaire Historique de la Langue Française*, Le Robert-Sejer, Paris.
2. (a) Gunatilaka, A.A.L. (2006) *J. Nat. Prod.*, **69**, 509–526; (b) Strobel, G., Daisy, B., and Castillo, U. (2004) *J. Nat. Prod.*, **67**, 257–268.
3. Pettit, G.R., Mukku, V.J.R.V., and Cragg, G. (2008) *J. Nat. Prod.*, **71**, 130–133.
4. (a) Wu, K.M., Farrelly, J.G., and Upton, R. (2007) *Phytomedicine*, **14**, 273–279; (b) Heinrich, M., Chan, J., and Wanke, S. (2009) *J. Ethnopharmacol.*, **125**, 108–144.
5. Blackman, A.J., Davies, N.W., and Ralph, C.E. (1992) *Biochem. Syst. Ecol.*, **20**, 339–342.
6. Gravel, E. and Poupon, E. (2008) *Eur. J. Org. Chem.*, 27–42.
7. Maltese, F., Van der Kooy, F., and Verpoorte, R. (2009) *Nat. Prod. Commun.*, **4**, 447–454.
8. Barham, P., Skibsted, L.H., and Bredie, W.L.P. (2010) *Chem. Rev.*, **110**, 2313–2365.
9. Davidek, T., Clety, N., and Devaud, S. (2003) *J. Agric. Food Chem.*, **51**, 7259–7265.
10. Balentine, D.A., Wiseman, S.A., and Bouwens, L.C. (1997) *Crit. Rev. Food Sci. Nutr.*, **37**, 693–704.
11. Tanaka, T., Mine, C., and Watarumi, T. (2002) *J. Nat. Prod.*, **65**, 1582–1587.
12. Li, Y., Shibahara, Y., and Tanaka, T. (2010) *J. Nat. Prod.*, **73**, 33–39.
13. Quideau, S., Jourdes, M., and Lefeuvre, D. (2005) *Chem. Eur. J.*, **11**, 6503–6513.
14. Bruneton, J. (2009) *Pharmacognosie, Phytochimie, Plantes Médicinales*, 4th edn, Lavoisier, Paris.
15. Singhuber, J., Zhu, M., and Prinz, S. (2009) *J. Ethnopharmacol.*, **126**, 18–30.
16. Yue, H., Pi, Z.F., and Li, H.L. (2008) *Phytochem. Anal.*, **19**, 141–147.
17. Moran, A.V., Jørgensen, K., and Jørgensen, C. (2008) *Phytochemistry*, **69**, 1795–1813.
18. (a) Jones, D.A. (1998) *Phytochemistry*, **47**, 155–162; (b) Vetter, J. (2000) *Toxicon*, **38**, 11–36.
19. Bruneton, J. (2005) *Plantes Toxiques: Végétaux Dangereux Pour L'homme et les*

32) And constitution of a correctly stored extract library for later comparisons.
33) For example, compounds **26**, **30** (Scheme 24.9); **57**–**62** (Figure 24.2); **110**–**113** (Scheme 24.26).
34) Or key intermediates such as, for example, **194** (Scheme 24.42).

Animaux, 3rd edn, Éditions Tec et Doc, Paris.
20. (a) Holst, B. and Williamson, G. (2004) *Nat. Prod. Rep.*, **21**, 425–447; (b) Kliebenstein, D.J., Kroymann, J., and Mitchell-Olds, T. (2005) *Curr. Opin. Plant Biol.*, **8**, 264–271; (c) Grubb, C.D. and Abel, S. (2006) *Trends Plant Sci.*, **11**, 1360–1385.
21. Bonny, M.L. and Capo, R.J. (1994) *J. Nat. Prod.*, **47**, 539–540.
22. (a) Block, E. (1992) *Angew. Chem. Int. Ed. Engl.*, **31**, 1135–1178; (b) see also: Rose, P., Whiteman, M., and Moore, P.K. (2005) *Nat. Prod. Rep.*, **22**, 351–368; (c) Lanzotti, V. (2006) *J. Chromatogr. A*, **1112**, 3–22.
23. Voon, Y.Y., Abdul Hamida, N.S., and Rusul, G. (2007) *Postharvest Biol. Technol.*, **46**, 76–85.
24. Ksebati, M.B. and Schmitz, F.J. (1988) *J. Nat. Prod.*, **51**, 857–861.
25. Lang, G., Pinkert, A., and Blunt, W.G. (2005) *J. Nat. Prod.*, **68**, 1796–1798.
26. Capon, R.J., Miller, M., and Rooney, F. (2000) *J. Nat. Prod.*, **63**, 821–824.
27. Lopes, L.M.X., Bolzani, Vda.S., and Trevisan, L.M.V. (1987) *Phytochemistry*, **26**, 2781–2784.
28. Leitao, G.G., Kaplan, M.A.C., and Galeffi, C. (1992) *Phytochemistry*, **31**, 3277–3279.
29. Wu, T.-S., Damu, A.M., and Su, C.-R. (2004) *Nat. Prod. Rep.*, **21**, 594–624.
30. Lacroix, D., Prado, S., and Deville, A. (2009) *Phytochemistry*, **70**, 1239–1245.
31. Nomiya, M., Murakami, T., and Takada, N. (2008) *J. Org. Chem.*, **73**, 5039–5047.
32. Sheridan, H. and Canning, A.-M. (1999) *J. Nat. Prod.*, **62**, 1568–1569.
33. Bhaskar, M.U., Rao, L.J.M., and Rao, N.S.P. (1991) *J. Nat. Prod.*, **54**, 386–389.
34. Venkateswarlu, Y., Bibiani, M.A.F., and Reddy, M.V.R. (1994) *J. Nat. Prod.*, **57**, 827–828.
35. Jimenez, J.I., Huber, U., and Moore, R.E. (1999) *J. Nat. Prod.*, **62**, 569–572.
36. Chansakaow, S., Ishikawa, T., and Seki, H. (2000) *J. Nat. Prod.*, **63**, 173–175.
37. Pilli, R.A. and Ferreira de Oliveira, Mda.A. (2000) *Nat. Prod. Rep.*, **17**, 117–127.
38. (a) Takemura, Y., Matsushita, Y., and Nagareya, N. (1995) *Chem. Pharm. Bull.*, **43**, 1340–1345; (b) Ju-Ichi, M., Takemura, Y., and Nagareya, N. (1996) *Heterocycles*, **42**, 237–240.
39. Wu, T.-S., Huang, S.-C., and Wu, P.-L. (1996) *Phytochemistry*, **42**, 221–223.
40. Negi, N., Jinguji, Y., and Ushijima, K. (2004) *Chem. Pharm. Bull.*, **52**, 362–364.
41. (a) Ono, T., Ito, C., and Furukawa, H. (1995) *J. Nat. Prod.*, **58**, 1629–1631; (b) Michael, J.P. (1997) *Nat. Prod. Rep.*, **6**, 605–618.
42. Desrivot, J., Edlund, P.-O., and Svensson, R. (2007) *Toxicology*, **235**, 27–38.
43. Van Beek, T.A., Verpoorte, R., and Baerheim, A. (1985) *J. Nat. Prod.*, **48**, 400–423.
44. Hootele, C., Levy, R., and Kaisin, M. (1967) *Bull. Soc. Chim. Belg.*, **76**, 300.
45. Hesse, M. (1981) *Alkaloid Chemistry*, John Wiley & Sons, Ltd, Chichester.
46. Flors, C. and Nonell, S. (2006) *Acc. Chem. Res.*, **39**, 293–300.
47. (a) Cheng, H.M., Yomamoto, I., and Casida, J. (1972) *J. Agric. Food Chem.*, **20**, 850–855; (b) Cabras, P., Caboni, P., and Cabras, M. (2002) *J. Agric. Food Chem.*, **50**, 2576–2580; (c) Draper, W.M. (2002) *Analyst*, **127**, 1370–1374.
48. Waldeck, D.H. (1991) *Chem. Rev.*, **91**, 415–436.
49. Spande, T.F., Jain, P., and Garraffo, H.M. (1999) *J. Nat. Prod.*, **62**, 5–21.
50. Liu, C., Masuno, M.N., and MacMillan, J.B. (2004) *Angew. Chem. Int. Ed.*, **43**, 5951–5954.
51. Toyota, M., Koyama, H., and Asawaka, Y. (1996) *Phytochemistry*, **41**, 1347–1350.
52. Majetich, G. and Yu, J. (2008) *Org. Lett.*, **10**, 89–91.
53. Hurst, J.J. and Whitham, G.H. (1960) *J. Chem. Soc.*, 2864–2869.
54. Birch, A.J. and Russell, R.A. (1972) *Tetrahedron*, **28**, 2999–3008.
55. Hoffman, N. (2008) *Chem. Rev.*, **108**, 1052–1103.

56. Schuster, N., Christiansen, C., and Jakupovic, J. (1993) *Phytochemistry*, **34**, 1179–1181.
57. Sharma, P. and Moses, J.E. (2010) *Synlett*, 525–528. See references citations therein for further examples.
58. Rossi, M.H., Yoshida, M., and Maia, J.G.S. (1997) *Phytochemistry*, **45**, 1263–1269.
59. (a) Nicolaou, K.C., Sanchini, S., and Wu, T.R. (2010) *Chem. Eur. J.*, **16**, 7678–7682; (b) Tanaka, N., Okasaka, M., and Ishimaru, Y. (2005) *Org. Lett.*, **7**, 2997–2999; (c) Tanaka, N., Kashiwada, Y., and Kim, S.Y. (2009) *J. Nat. Prod.*, **72**, 1447–1452; (d) Nicolaou, K.C., Wu, T.R., and Sarlah, D. (2008) *J. Am. Chem. Soc.*, **130**, 11114–11121.
60. Busch, B. and Hertweck, C. (2009) *Phytochemistry*, **70**, 1833–1840.
61. Hoffman, N. (2008) *Chem. Rev.*, **108**, 1052–1103.
62. Han, B.H., Park, M.H., and Park, J.H. (1989) *Pure Appl. Chem.*, **61**, 443–448.
63. Suzuki, M., Sano, M., and Yoshida, R. (2003) *J. Agric. Food Chem.*, **51**, 510–514.
64. Asfaw, N., Storesund, H.J., and Skattebøl, L. (2001) *Phytochemistry*, **58**, 489–492.
65. Hayman, A.R., Perry, N.B., and Weavers, R.T. (1986) *Phytochemistry*, **26**, 649–653.
66. Perry, N.B., Benn, M.H., and Foster, L.M. (1996) *Phytochemistry*, **42**, 453–459.
67. Hollenbeak, K.H. and Kuehne, M.E. (1974) *Tetrahedron*, **30**, 2307–2316.
68. Numata, A., Hokimoto, K., and Takemura, T. (1984) *Chem. Pharm. Bull.*, **32**, 2815–2820.
69. Peters, S. and Spiteller, P. (2007) *J. Nat. Prod.*, **70**, 1274–1277.
70. Chadwick, L.R., Nikolic, D., and Burdette, J.E. (2004) *J. Nat. Prod.*, **67**, 2024–2032.
71. Stevens, J.F., Taylor, A.W., and Deinzer, M.L. (1999) *J. Chromatogr. A.*, **832**, 97–107.
72. Sulaiman, M., Martin, M.-T., and Paies, M. (1998) *Phytochemistry*, **49**, 2191–2192.
73. Awang, K., Abdullah, Z., and Mukhtar, M.R. (2009) *Nat. Prod. Res.*, **23**, 652–658.
74. Koch, M. (1965) Gentianine et swertziamarine de l' *Anthocleista procera* Leprieur ex-Bureau (Loganiacées). Université Paris dissertation.
75. Dewick, P.M. (2009) *Medicinal Natural Products: A Biosynthetic Approach*, 3rd edn, John Wiley & Sons, Ltd, Chichester.
76. (a) Herraiz, T. (2000) *J. Chromatogr. A*, **881**, 483–499; (b) Nagatsu, T. (1997) *Neurosci. Res.*, **29**, 99–111; (c) Fekkes, D., Tuiten, A., and Bom, I. (2001) *Neurosci. Lett.*, **303**, 145–148.
77. Hesse, M. (2002) *Alkaloids: Nature's Curse or Blessing?* John Wiley & Sons, Ltd, Chichester.
78. Ghisalberti, E.L. (2004) *Fitoterapia*, **75**, 429–446.
79. Szendrei, K., Varga, E., and Hadju, Z. (1988) *J. Nat. Prod.*, **51**, 993–995.
80. Shamma, M., Chinnasamy, P., and Miana, G.A. (1979) *J. Nat. Prod.*, **42**, 615–623.
81. Lounasmaa, M. and Hanhinen, P. (2004) *J. Nat. Prod.*, **63**, 1456–1460.
82. Escolano, C. (2005) *Angew. Chem. Int. Ed.*, **44**, 7670–7673.
83. Cheng, C., Liang, G., and Hu, C. (2008) *Molecules*, **13**, 938–942.
84. Li, C.-Y., Chow, T.J., and Wu, T.-S. (2005) *J. Nat. Prod.*, **68**, 1622–1624.
85. de Pascoli, I.C., Nascimento, I.R., and Lopes, L.M.X. (2006) *Phytochemistry*, **67**, 735–742.
86. Kutrzeba, L.M., Li, X.-C., and Ding, Y. (2010) *J. Nat. Prod.*, **73**, 707–708.
87. Zhao, J., Ma, M., and Wang, S. (2005) *J. Nat. Prod.*, **68**, 691–694.
88. Orjala, J. and Gerwick, W.H. (1997) *Phytochemistry*, **45**, 1087–1090.
89. Picard, S. and Chenault, J. (1994) *J. Nat. Prod.*, **57**, 808–810.
90. Orians, M. (1995) *J. Chem. Ecol.*, **21**, 1235–1243.
91. Montgomery, C.T., Cassels, B.K., and Shamma, M. (1982) *J. Nat. Prod.*, **46**, 441–452.
92. Fort, D.M., Litvak, J., and Chen, J.L. (1998) *J. Nat. Prod.*, **61**, 1528–1530; *Erratum* in: (1998) *J. Nat. Prod.*, **61**, 1577.

93. Aelterman, W., De Kimpe, N., and Kalinin, V. (1997) *J. Nat. Prod.*, **60**, 385–386.
94. De Kimpe, N., Georgieva, A., and Boeykens, M. (1995) *J. Org. Chem.*, **60**, 5262–5265.
95. Chia, Y., Chang, F., and Wang, J. (2008) *Molecules*, **13**, 122–128.
96. Arango, V., Robledo, S., and Séon-Méniel, B. (2010) *J. Nat. Prod.*, **73**, 1012–1014.
97. Paranagama, P.A., Wijeratne, E.M.K., and Gunatilaka, A.A.L. (2007) *J. Nat. Prod.*, **70**, 1939–1945.
98. Blackman, A.J., Dragar, C., and Wells, R.J. (1979) *Aust. J. Chem.*, **32**, 2783.
99. Rochefort, S.J. and Capon, R.J. (1994) *J. Nat. Prod.*, **57**, 849–851.
100. Khamidullina, E.A., Gromova, A.S., and Lutsky, V.I. (2006) *Nat. Prod. Rep.*, **23**, 117–129.
101. Bermejo, A., Figadére, B., and Zafra-Polo, M.-C. (2005) *Nat. Prod. Rep.*, **22**, 269–303.
102. Duret, P., Figadére, B., and Hocquemiller, R. (1997) *Tetrahedron Lett.*, **38**, 8849–8852.
103. Duval, R. (2003) Hémisynthése et évaluation biologique d'analogs de la squamocine, une acétogénine d'Annonaceae. Contribution à l'étude du mécanisme d'action. Université Paris-Sud 11 dissertation.
104. Li, D.-Y., Yu, J.-G., and Zhu, J.-X. (2001) *J. Asian Nat. Prod. Res.*, **3**, 267–276.
105. Grass, S., Zidorn, C., and Ellmerer, E.P. (2004) *Chem. Biodiv.*, **1**, 353–360.
106. (a) Luesch, H., Yoshida, W.Y., and Moore, R.E. (2002) *Tetrahedron*, **56**, 7959–7967; (b) Sone, H., Kondo, T., and Kiryu, M. (1995) *J. Org. Chem.*, **60**, 4774–4781.
107. (a) Braekman, J.C., Daloze, D., and Tavares, R. (2000) *J. Nat. Prod.*, **63**, 193–196; (b) Roberto, G. and Berlinck, S. (2002) *Nat. Prod. Rep.*, **19**, 617–649; (c) Faulkner, D.J., Newman, D.J., and Cragg, G.M. (2004) *Nat. Prod. Rep.*, **21**, 50–76.
108. (a) Peng, J.N., Feng, X.-Z., and Liang, X.-T. (1999) *J. Nat. Prod.*, **62**, 611–612; (b) Yang, X.-W., Ma, Y.-L., and He, H.-P. (2006) *J. Nat. Prod.*, **69**, 971–974.
109. (a) Okija, M., Yoshida, Y., and Okumura, M. (1990) *J. Nat. Prod.*, **53**, 1619–1622; (b) Kurihara, H., Mitani, T., and Kawabata, J. (1999) *J. Nat. Prod.*, **62**, 882–884; (c) Opatz, T., Kolshorn, H., and Thines, E. (2008) *J. Nat. Prod.*, **71**, 1973–1976; (d) Mohammed, M., Maxwell, A.R., and Ramsewak, R. (2010) *Phytochem. Lett.*, **3**, 29–32; (e) Wijeratne, E.M.K., Bashyal, B.P., and Gunatilaka, M.K. (2010) *J. Nat. Prod.*, **73**, 1156–1159.
110. Isaka, M., Sappan, M., and Auncharoen, P. (2010) *Phytochem. Lett.*, **3**, 152–155.
111. Anthoni, U., Bock, K., and Chevolot, L.C. (1987) *J. Org. Chem.*, **52**, 5638–5639.
112. Hamann, M.T. and Scheuer, P. (1993) *J. Org. Chem.*, **58**, 6565–6569.
113. Bourguet-Kondracki, M.L., Lacombe, F., and Guyot, M. (1999) *J. Nat. Prod.*, **62**, 1304–1305.
114. (a) Sun, L.-R., Li, X., and Wang, S.-X. (2005) *J. Asian Nat. Prod. Res.*, **7**, 127–130; (b) Piggott, A.M. and Karuso, P. (2005) *Molecules*, **10**, 1292–1297; (c) Kuroda, C., Kiuchi, K., and Torihata, A. (2007) *Chem. Biodiv.*, **4**, 2210–2217.
115. Perry, N.B., Burgess, E.J., and Foster, L.M. (2008) *J. Nat. Prod.*, **71**, 258–261.
116. Ha, D.T., Ngoc, T.M., and Lee, I.-S. (2009) *J. Nat. Prod.*, **72**, 1465–1470.
117. Cheng, Y., Cheng, P., and Zhang, H. (2010) *Helv. Chim. Acta.*, **93**, 565–572.
118. Kita, M., Kigoshi, H., and Uemura, D. (2001) *J. Nat. Prod.*, **64**, 1090–1092.
119. Kawai, H., Kuroyanagi, M., and Ueno, A. (1988) *Chem. Pharm. Bull.*, **36**, 3664–3666.
120. Tomassini, L., Cometa, M.F., and Nicoletti, M. (1995) *J. Nat. Prod.*, **58**, 1756–1758.
121. Zedro, C., Bohlmann, F., and Niemeyer, H.M. (1988) *J. Nat. Prod.*, **51**, 509–512.
122. Ciminiello, P., Costantino, V., and Fattorusso, E. (1994) *J. Nat. Prod.*, **65**, 705–712.

123. Saeki, B.M., Granato, A.C., and Berlinck, R.G.S. (2002) *J. Nat. Prod.*, **65**, 796–799.
124. Cone, E.J., Buchwald, W.F., and Darwin, W.D. (1982) *Drug Metabol. Dispos.*, **10**, 561–567.
125. Mauden, K.E., Will, S.M.R., and Lambert, W.E. (2007) *J. Chromatogr. B*, **848**, 384–390.
126. Gribble, G.W. (1999) *Chem. Soc. Rev.*, **28**, 335–346.
127. Neuwinger, H.D. (1996) *African Ethnobotany: Poisons and Drugs*, Chapman & Hall, Weinheim.
128. Guinaudeau, H., Lebéuf, M., and Cavé, A. (1994) *J. Nat. Prod.*, **57**, 1033–1135.
129. Hu, J., Wang, W.-D., and Shen, Y.-H. (2007) *Helv. Chim. Acta*, **90**, 720–722.
130. Marek, R., Sečkàrovà, P., and Hulovà, D. (2003) *J. Nat. Prod.*, **66**, 481–486.
131. Grycovà, L., Dostàl, J., and Marek, R. (2007) *Phytochemistry*, **68**, 150–175.
132. Bentley, K.W. (2000) *Nat. Prod. Rep.*, **17**, 247–268.

Index

a

abyssomicins 512–513
acalyphidin production 662
acetals formation 913–916
– by acetone under acidic conditions 919
– secoiridoid acetals from *Lonicera* spp 918
acetone-derived artifacts 916–919
acetoxy-acetalic sesquiterpenes, artifactual alcoholysis of 916
acetyl co-enzyme A (AcCoA) 594
acid-catalyzed reactions 900–903
– coumarins from Rutaceae spp 904
– cryptolepine from 902
– intramolecular Michael additions 905
– of natural products 901
– norrhoedanines in acidic conditions 901
– protic solvents 903–906
– spiroketal from laurencione 903
acrolein scenario 182–200
– *endo*-intramolecular Diels–Alder reaction 198–199
– keramaphidin skeleton conversion into ircinal/manzamine skeleton 200
– transannular hydride transfers 199–200
Actephila excelsa 769
actinorhodin 486
acyl carrier protein (ACP) 473
acyl transfer
– Scott's conditions for 475
– using glycoluril 478
acylphloroglucinols 434–436
Adler–Becker oxidation 727
Aeschynomene mimosifolia 747
Agelas wiedenmayeri 231
agelastatins 250–253
– Wardrop's synthesis 252
ageliferins 254–255
ajmaline/sarpagine series 897

(±)-akuammicine 102
aldol condensation 360
aldolization–crotonization process 286
aldotripiperideine 12
alkaline media 888–895
– amination processes 888–892
alkaloids, I–X
– arginine-derived: *see* in Chapter 1
– FR-901483: *see* in Chapter 2
– guanidinium alkaloids: chapter 7
– lysine-derived: *see* in Chapter 1
– manzamine alkaloids: Chapter 6
– non aminoacid derived alkaloids: Chapter 8
– ornithine-derived: *see* in Chapter 1
– peptide alkaloids:
– – aryl-peptide alkaloids: *see* in Chapter 9
– – azole-peptide alkaloids: *see* in Chapter 9
– – complex peptide alkaloids: *see* in Chapter 10
– – indole-oxidized peptide alkaloids: *see* in Chapter 10
– pyrrole-2-aminoimidazole: Chapter 7
– TAN-1251: *see* in Chapter 2
– tryptophan-derived
– – dioxopiperazine alkaloids: Chapter 4
– – indolemonoterpene alkaloids: Chapter 3
– – modified indole nucleus: Chapter 5
– tyrosine-derived: Chapter 2
3-alkylpiperidines 181–182, *see also* manzamine alkaloids, biomimetic synthesis
alkylpyridines with unusual linking patterns 194–195
– pyrinadine A, biomimetic synthesis 195
– pyrinodemin A, biomimetic synthesis of 194
3-Alkylpyridiniums 191
allicin 855
Alternaria solani 507

Biomimetic Organic Synthesis, First Edition. Edited by Erwan Poupon and Bastien Nay.
© 2011 Wiley-VCH Verlag GmbH & Co. KGaA. Published 2011 by Wiley-VCH Verlag GmbH & Co. KGaA.

amauromine, Danishefsky's total synthesis 126
amide coupling 385
amination processes 888–892
amino acids
– in primary metabolism 4
– from primary metabolism to alkaloid biosynthesis 5–6
– structure of 4
4-aminobutyraldehyde 9
aminopentadienals 190, 213–215
aminopentadienimines 189
ampelopsin D 710–711
ampelopsin F 713
amurensin H 715
anchinopeptolides 26
– synthesis 26–29
anethole 871
anhydrovinblastine series 110–113
anigorufone 768
Anigozanthos preissii 768
Annona muricata 908
annonaceous acetogenins 907
antheridic acid from gibberellins 408–410
anthracenoids 494–495
– biomimetic access to 494–495
– chrysophanol, Harris' biomimetic synthesis 495
– emodin, Harris' biomimetic synthesis 495
– nanaomycine, Yamaguchi's biomimetic synthesis 494
– Yamaguchi's access to 497
Anthromyces ramosus 699
anticancer vinblastine series 110–113
antirhine derivatives 99
Aplidium longithorax 777
alkylations, apparent 910–913
– chartelline A, methanol substitution of 913
– puupehenone, methanol adduct on 914
– xanthepinone, methanol induced rearrangement of 912
aquaticol 772
L-arginine 3
arginine, alkaloids derived from 18–30
Aristolochia heterophylla 921
Aristolochia lagesiana 895
Aristolochia pubescens 895
Aristolochia rodriguesii 919
Aristotelia alkaloids 93–96
(+)-aristoteline (16) 95
aromatic polyketides 485–499, see also anthracenoid derivatives; benzenoid derivatives; naphthalenoid derivatives; tetracyclic derivatives
– Collie and coworkers 485
– cyclization towards chrysophanol 487
– folding leading to 486
– pathways leading to 486
aromatic rings, biomimetic access to 471–499
Artemisia absinthium 407
artifacts and natural substances formation 849–930, see also hydrodistillation; hydrolysis, artifactual; oxidation processes
– acetals formation 913–916
– acetone-derived artifacts 916–919
– acid-catalyzed reactions 900–903
– acidic conditions during purifications 895–903
– ajmaline/sarpagine series 897
– alkylations 910–913
– alcoholysis 916
– base-catalyzed reactions 892–895
– botanical misidentifications 851
– chloroformic adducts 926
– decarboxylation processes 885
– ecdysteroids 892
– esterification, transesterification 908–910
– by exposure to light 870–878, see also light
– gentianine 891
– glucosidases 852–853
– halogenated solvents 919–921
– heat and pressure in 878–888
– – biyouyanagins A and B 880
– – matricin degradation 887
– – sanguinone A degradation 887
– – styrylpyrone photodimers 879
– hemiterpenes 853
– imide-bearing aconitine-like alkaloids 893
– methyl-ester hydrolysis 900
– miscellaneous methylation artifacts 911
– miscellaneous oxidation artifacts 861
– phenylchromane derivatives from *Desmos* spp 890
– protic solvents 903–906
– protoberberines 921–925
– raucaffrinoline 896
– Rubiaceae iridoids 910
– supercritical CO_2 885–888
– transesterification artifacts 909
– (Z)-vellomisine 894
aryl-containing peptide alkaloids 317–350
– cyclopeptides containing biaryls 344–345
Asarum heterotropoides 895
Asarum teitonense 769
asatone 772

Ascochyta rabiei 507
Aspergillus nidulans 755
Aspergillus ochraceus 135, 158
Aspergillus terreus 506, 755
Aspidosperma alkaloids 106
– fragmentation 106–109
– rearrangements of 106–109
aspochalasin Z 764
autocatalytic enantioselective reactions 825–833
– of asymmetric amplification 827
– density functional theory (DFT) calculations 829
– dissipative non-productive reaction cycle 828
– Kagan and Noyori models of 826
– reversible Frank-type mechanism 828
– Soai reaction 830
– – Blackmond–Brown model in 831
avrainvillamide 135–136
Axinella vaceletti 233
axinellamines A/B 257, 262–263
azaspiracids 291–293
– azaspiracid-1, 3D structure of 292
– FGHI rings of 292
– HI rings of, Forsyth's biomimetic approach to 293
– structure of 292
aziridinium mechanism 259
– for massadine formation 259
– for tetramer stylissadine A formation 259–261
azole-containing peptide alkaloids 321–336, *see also* lissoclinamides; thiangazole; thiostrepton
– biomimetic cyclodehydration 324
– enzymatic heterocycle formation 324
– GE2270A 334–336
– structural features 321–323
– synthesis of, completion 333

b

Baeyer–Villiger rearrangement 740
Baldwin's hypothesis development 195–200
– from cyclostellettamines to keramaphidin-type alkaloids 195–200
– pyridinium alkaloids and manzamine A-type alkaloids, linking 195–197
barakol, Harris' biomimetic synthesis of 493
barbaline 294
basidiolides 425
bastadin biosynthesis 342–343

benzenoids 487–492
– biomimetic access to 487–492
– masked hexaketides, Schmidt's condensation of 492
– pentacarbonyl derivative, Harris' biomimetic cyclizations of 492
– β-polyketones synthesis 488
– pyrones 489
– tetra-β-carbonyl compounds 489
– tri-β-carbonyl entities synthesis 489
– β-triketoacids 490
– zearalenone, Barrett's synthesis of 491
benzo[*a*]tetracenoid derivatives 498–499
benzo[*kl*]xanthenes 686–688
– manganese-mediated biomimetic synthesis 688
benzodiazepinones, hydrogenation 806
benzophenanthridines 929
benzosceptrin A 266
benzothiazines, hydrogenation 806
benzoxanthenone lignans 683–686
– biomimetic synthesis of 683–686
– carpanones 683
– mechanism 689–670
benzoxazines, hydrogenation 806–813
benzoxazinones, hydrogenation 806
berberine 923
betulin 423
biaryls, cyclopeptides containing 344–345
– biphenomycins A–C 344
bicyclo[2.2.2]diazaoctanes 126–141
– hetero-Diels–Alder formation of 127
– Sammes' model study 127
bicyclo[3.3.1]nonane-2,4,9-triones 448
bielschowskysin 416
BINOL-derivative, Brook-type-rearrangement 791
bio-inspired transfer hydrogenations 787–818
– benzodiazepinones 806
– benzothiazines 806
– benzoxazines 806–813
– benzoxazinones 806
– BINOL-phosphoric acid derivatives, synthesis 792
– Brønsted acid catalyzed transfer hydrogenation 788–799
– cascade sequences 814–817
– dehydrogenases as model 787–788
– diazepines 806
– enamides 798
– α-imino esters 796

bio-inspired transfer hydrogenations (contd.)
– indoles   805–806
– ketimines   793
– piperidines   813
– pyridines   813–814
– quinoxalines   806
– quinoxalinones   806
biological homochirality   823–841, see also autocatalytic enantioselective reactions
– chiral surfaces, adsorption on   837
– collision kinetics, symmetry breaking in   840
– conglomerate crystallizations, spontaneous symmetry breaking in   837–840
– enantioenrichment, polymerization and aggregation models of   834–835
– membrane diffusion, symmetry breaking in   840
– phase equilibria   835–837
– Plasson mechanism   835
– reaction–diffusion models, symmetry breaking in   840
– self-replication and   833–834
biphenomycins biaryls   344–345
bipinnatin J, photochemical rearrangement of   414
bisacridones from *Rutaceae*   866
bis(2-oxo-3-oxazolindinyl)phosphinic chloride (BOPCl)   133
bis-aporphines   864
bisorbicillinoids   772–773
bisorbicillinol   726
– Nicolaou and Pettus' synthesis   726
– from sorbicillin   726
bis-steroidal pyrazines   299–300
– biomimetic pseudo-combinatorial approach to   299
– unsymmetrical approaches to   300
biyouyanagins A and B   880
Black's cascade   612
Blackmond–Brown model   831
(±)-borreverine   173–175
– biosynthesis proposal by Koch et al.   173
botanical misidentifications   851
bouvardin synthesis   342
brazilide A   735
brevetoxin B   545
– Nakanishi's hypothesis   549
brevetoxins   580–582
(±)-brevianamides   155–158
– brevianamide A   875–876
– brevianamide B, biosynthesis
– – by Williams et al.   157, 159

brevianamides   118
– biosynthesis of   118
– – brevianamide B   133
– – brevianamide E   119
– – brevianamide F   119
– biosynthetic proposal for   128
– brevianamide F   118
– – Danishefsky's total synthesis of   120
– – Kametani's total synthesis of   120
Brønsted acid catalyzed transfer hydrogenation   788–799
bukittinggine   303
bukittinggine, Heathcock's synthesis of   303
Burgess cyclodehydration   325

c

caffeic acid phenethyl ester (CAPE)   687
caged xanthones   452
– *Garcinia* xanthones, biomimetic synthesis   455–464
– – non-biomimetic synthesis of   460–463
– – Theodorakis' unified approach to   459
Callyspongia species   186
calycanthines   168–171
– biomimetic synthesis
– – by Scott et al.   169
– – by Stoltz et al.   169–170
– biosynthesis proposal by Woodward and Robinson   169
camphene   397–399
camphorsulfonic acid (CSA)   627
camptothecin   93, 111, 152
– biosynthesis by Hutchinson et al.   153
– (±)-camptothecin   150–154
– synthesis by Winterfeldt et al.   153
Cane–Celmer–Westley hypothesis   540–541, 554
cararosinol C and D   713
carbinolamine formation   385
carpanone   683–685, 724
– Chapman's synthesis of   725
– related sequence for dehydrodiisoeugenol   725
Carpinus tschonoskii   663
caryolane, biomimetic studies in   402–404
caryophyllenes in sesquiterpene biosyntheses   401–402
cascade reactions, biomimetic   524–530
– hirsutellones   525–530
– tetronasin synthesis   524–525
cascade sequences
– hydrogenation   814–817

– – Brønsted acid catalyzed   815
– oxidative dearomatization   731–733
cassiarins A and B   284–285
– Yao's biomimetic synthesis of   286
cassigarol B   716
castalagin   666, 670
– pyrolytic degradations of   671
Castanea crenata   669
catechol oxidation   768–775
(±)-cedrene   731
celogentin C   357, 363–368
– A-ring Trp-Leu linkage origin   365
– bioactivity   363
– B-ring Trp-His linkage formation   365
– C–H activation–indolylation   367–368
– isolation   363
– retrobiosynthetic proposal for   364
– structures of   363
– synthetic approaches to   366
cephalostatins   294–298
– structures   295
Ceratosoma brevicaudatum   855
cermizine C   51–52
C-glycosidic ellagitannins   663–670
– oxidation of   669–670
– reactions at C1 positions   665–669
chaetoglobosin   761–763
Chamaecyparis obtusa   769, 774
Chartella papyracea   160
chartelline A, methanol substitution of   913
chartelline C
– biosynthesis proposal for   162
– (±)-Chartelline C   160–164
– endgame of synthesis   164
– synthetic route by Baran *et al.*   163
chebulagic acid synthesis   662–664
Chichibabin synthesis of pyridines   188
chimonanthines   168–171
chiral surfaces, adsorption on   837
chloptosin   369–374
– chloptosin pyrroloindole core synthesis   373
– retrobiosynthetic simplification of   370
– synthetic approaches to   372
chloroformic adducts   926
chloropeptin I   391
chloropeptin II/complestatin   391
chlorothricin   763–764
Chrossopetalum rhacoma   774
Claisen/Diels–Alder/Claisen reaction cascade   456–457
clathrins   857
clathrodins   227
– biogenetic hypothesis for   229–233

– post-clathrodin in P-2-AIs biogenesis   228
– pre-clathrodin in P-2-AIs biogenesis   228
clovane series, biomimetic studies in   402–404
clusianone, total synthesis of   448–451
– (±)-clusianone   438–439
– – double Michael reaction   438–439
– – Porco synthesis   438
– by Marazano and coworkers   451
– by Simpkins group   449
– through 'carbanions' differentiation   443–445
cochleamycins   514–521
– biosynthetic hypothesis for   518
– isolation   517
– Paquette's partial synthesis of   519
– Roush synthesis of   519
collision kinetics, symmetry breaking in   840
colombiasin A
– (–)-colombiasin A   412
– Nicolaou's synthesis of   412
– Rychnovsky's synthesis   413
communesins   168–171
– biomimetic synthesis by Stoltz *et al.*   169–170
– study by Funk *et al.*   171
complanadine A, total synthesis of   53–54
complex peptide alkaloids   357–392
complex terpenoids, biomimetic rearrangements of   397–428, see also monoterpene rearrangements
– miscellaneous diterpenes   417–420
conglomerate crystallizations, spontaneous symmetry breaking in   837–840
Cordia globifera   425
cordiachrome C   427
coriariin A   659
corilagin   653
Corynanthe alkaloids   95–99
Corynanthe skeleton into *Strychnos* skeleton   99–102
CP-225917   505–506
CP-263114   505–506
cryptolepine from   902
Cryptolepis sanguinolenta   901
CuI-mediated cyclization   347
curcuphenol   731–732
Cutleria multifida   616
(+)-cycloanchinopeptolide D   28
cyanophenyloxazolopiperidine   15
cyclic imine marine alkaloids   275–284

cyclic peptides containing aryl-alkyl ethers 336–339
cyclic peptides containing biaryl ethers 339–343
– bastadin biosynthesis 342–343
– bouvardin synthesis 342
– eurypamide B synthesis 341
– isodityrosine subunit 340
– – synthesis strategies 340
– K-13 synthesis 342
cyclooroidin 238
cyclopamine 422
cyclostellettamine alkaloids 215–217
– cyclostellettamine B 191
cyclostreptin 514
cytochalasin D 762

d

dalesconols 715
Danishefsky's synthesis 120
– of amauromine 126
– of spirotryprostatins 122–123
daphniglaucine A 294
daphnilactone A 304
Daphniphyllum alkaloids 298–305
– daphnilactone A 304
– Heathcock group in 300
– methyl homodaphniphyllate 304
– methyl homosecodaphniphyllate synthesis 301–302
– *proto*-daphniphylline biosynthesis 301
dearomatization, *see* oxidative dearomatization
debromodispacamides B, Al-Mourabit's biomimetic synthesis 236–237
debromodispacamides D, Al-Mourabit's biomimetic synthesis 236–238
decahydroquinoline alkaloids 285–288
– biosynthetic origin of 287
– *cis*-195A, Amat's biomimetic synthesis of 287
– *Dendrobates histrionicus* 286
– *Dendrobates pumilio* 285
decalin systems 506–509
– Diels–Alder reactions affording 506–509
decarboxylation processes 885
decarboxylative Grob-type anti-elimination 360
dehydrodiisoeugenol 725
dehydroellagitannins 648
dehydroellagitannins conversion into related ellagitannins 659–663
dehydrohexahydroxydiphenic acid esters synthesis from methyl gallate 652

dehydrohexahydroxydiphenoyl (DHHDP) ester 645
dehydropiperidines 332
Delphinium cashmirianum 892
density functional theory (DFT) 829, 832
deoxybrevianamide E, Kametani's total synthesis of 120
Desmos dumosus 888
1,8-diazabicyclo[5.4.0]undec-7-ene (DBU) 335
diazepines, hydrogenation 806
diazonamides 375–382
– aminal core formation 377
– biaryl coupling 377
– 2,4-bisoxazole core, biomimetic synthesis of 379
– bisoxazole ring system via oxidative dehydrative cyclization 379
– diazonamide A 357, 375
– α-hydroxyvaline formation 377
– indole–indole coupling 381–382
– late-stage aromatic chlorination 378–379
– oxidative annulation 379–380
– reductive aminal formation 380–381
– retrobiosynthetic analysis of 376
– sequential nucleophilic 1,2-addition, electrophilic aromatic substitution 380
dibromoagelaspongin 238–243
– biomimetic hypotheses for 241
– *rac*-dibromoagelaspongin
– – Al-Mourabit's approach to 244
– – Feldman's total synthesis of 242
dibromophakellin 243–247
– Nagasawa's enantioselective synthesis of 249
– Romo's enantioselective synthesis of 248
– synthesis from dihydrooroidin 233–234
dibromophakellstatin 243–247
– Feldman's synthesis of 245
– Lindel's synthesis of (−)-dibromophakellstatin 246
– Romo's synthesis of (+)-dibromophakellstatin 246
2,3-dichloro-5,6-dicyano-1,4-benzoquinone (DDQ) 158, 292
(+)-11,11-Dideoxyverticillin A 141, 144, 171–173
– dimerization synthesis strategy by Movassaghi *et al.* 172
– endgame of synthesis by Movassaghi *et al.* 173

Diels–Alder reaction/cycloaddition 124–131, 461, 506–524, 753–782
- [4 + 2] adducts derived from terpenoid dienes 780
- after reactive substrates formation by oxidation enzymes 767–779
- anigorufone 768
- aspochalasin Z 764
- bicyclo[2.2.2]diazaoctanes formation 127
- biomimetic Diels–Alder reaction 129
- biomimetic syntheses involving 506–524
- catechol oxidation 768–775
- chaetoglobosin 761–763
- chlorothricin 763–764
- cyclization of trienes 507
- cyclopentadiene formation 779
- cytochalasin D 762
- decalin systems 506–509
- equisetin 761–763
- indanomycin 764–766
- intramolecular, catalysis of 760
- kijanimicin 763–764
- lachanthocarpone 768
- longithorone 779
- lovastatin nonaketide synthase (LovB) 755–756
- macrophomate synthase 756–758
- nargenicin A 508–509
- in nature 754–760
- ortho-quinol formation 767
- ortho-quinone formation 767
- phenol oxidation 768–775
- prenyl side chain dehydrogenation, conjugated diene derived from 775–779
- solanapyrone synthase (SPS) 758–760
- solanapyrones 507–508
- spinosyn 766–767
- spiro systems 512–514
- superstolide A 509
- tetrahydroindane systems 509–512
- tetrocarcin A 763–764
- transannular Diels–Alder (TADA) reaction 508
diethyl azodicarboxylate (DEAD) 135
diffusion ordered spectroscopy (DOSY) 481
dihydrooroidin, dibromophakellin synthesis from 233–234
dihydropyridines
- biomimetic synthesis 188–189
- chemistry, Baldwin's hypothesis 186
- isomerization of 189
- Marazano biomimetic synthesis of 188

dihydropyridinium salts, biomimetic synthesis 188–189
diisopropyl-carbodiimide (DIC) mediated amide bond formation 152
dimerization 10–11, 173–174
- dimeric ellagitannin synthesis 658–659
- dimeric indolomonoterpene alkaloids 110–113
- intermolecular 724–727
- process towards isoglaucanic acid 504–505
- in pyrrolidine series 10
- in pyrroloindole peptide alkaloids 370
dimers (and oligomers) 184
dimethylallyl pyrophosphate (DMAPP) 129
dioxepandehydrothyrsiferol, epoxide-opening cascades in 546
dioxoaporphines 868
dioxopiperazine alkaloids 117–147, see also bicyclo[2.2.2]diazaoctanes; Prenylated indole alkaloids
- derived from tryptophan and amino acids 122–126
dioxopiperazines
- synthesis 131
- from tryptophan and proline 119–122
1,1-diphenyl-2-picrylhydrazyl (DPPH) 772
diptoindonesin D 714
dirigent protein 679–680
discorhabdins 154–155
- biosynthesis proposed by Munro et al. 155
- discorhabdin B 856
- (±)-discorhabdin C and E 154–155
- – synthesis by Heathcock et al. 156
dispacamide A, Al-Mourabit's synthesis of 238
dissipative non-productive reaction cycle 828
diterpene rearrangements 408–420
- antheridic acid biomimetic synthesis from gibberellins 408–410
- early chemical diversity in 408
- miscellaneous diterpenes 417–420
(−)-dityryptophenaline 143
double Michael reaction 438–439
Durio zibethinus 855
Dysidea herbacea 859
Dysidea pallescens 859

e

ecdysteroids, alumina-catalyzed dehydration of 892
ecteinascidin 743 (ET 743) 382–391
- biomimetic strategy 383–385
- biosynthesis 383–385

ecteinascidin 743 (ET 743) (*contd.*)
– bridge formation   389–390
– – Corey's strategy for   389
– pentacycle formation   385–389
– – Corey's approach to   386
– – Danishefsky's synthesis of   387
– – Fukuyama's work   387
– – Williams' synthesis of   386
– – Zhu's approach   387
– proposed biosynthesis for   384
– saframycin A biosynthesis   383
– synthetic approaches to   385
Elaeocarpus alkaloids, biomimetic syntheses of   19–22
electrophilic aromatic substitution   380
6π electrocyclizations, polyketides 598–612, see also tridachiahydropyrones
eleutherinol, Harris' biomimetic synthesis of   493
elisapterosin B   412
– Rychnovsky's synthesis   413
ellagitannins   639–640, 642–659
– synthesis with 3,6-(*R*)-HHDP group   651
– biosynthesis   640–642
– with $^1C_4$ glucopyranose cores   645–651
– corilagin   653
– decomposition   640
– dehydroellagitannins conversion into   659–663
– – acalyphidin production   662
– – benzyl-protected dehydrodigallic acid synthesis   660
– – castalagin   666
– – *C*-glycosidic ellagitannins   665
– – chebulagic acid synthesis   662–663
– – dehydrodigallic acid derivative synthesis   662
– – DHHDP esters reduction   659–662
– – DHHDP esters   663
– – dimeric ellagitannin coriariin A   661
– – mallotusinin production   662
– – pyranose-type ellagitannins   665
– – thiol compounds reaction   662–663
– – vescalagin   666
– dehydrohexahydroxydiphenic acid esters synthesis from methyl gallate   652
– 2,4-DHHDP esters   647–649
– dimeric ellagitannin synthesis   658–659
– epigallocatechin gallate oxidation during tea fermentation   650
– hexahydroxydiphenic acid, double esterification of   651–658
– 2,4-HHDP ester   647–649
– pedunculagin   658

– synthesis by biaryl coupling of galloyl esters   642–645
– tellimagrandin I   654
– Ullmann-type biaryl couplings   646
ellipticine   103
ellipticine-type alkaloids   102–105
elysiapyrones   624–626
emodin, Harris' biomimetic synthesis   495
– chrysophanol, Harris' biomimetic synthesis   495
enamides
– Brønsted acid catalyzed transfer hydrogenation of   788–799
– hydrogenation   798
enantioenrichment
– polymerization and aggregation models of   834–835
– Würthner's model   835
endiandric acids   612–618
– Nicolaou's biomimetic synthesis   617
endocyclic enamines   12–13, 39
7-endo epoxide ring opening   360
endo-intramolecular Diels–Alder reaction   198–199
enshuol   562
ent-17-epialantrypinone   142
ent-alantrypinone   118, 142
epidithiodioxopiperazines   141–146
– *ent*-alantrypinone synthesis   142
epimerization   895–897
– lactonic compounds   905–908
– light-induced   870–872
– – furofuranic lignans under   899
– – of gallocatechins during tea brewing   883
– – in or out of solutions   880
epinitraramine   37
epoxide-opening cascades in polycyclic polyethers   550–583
– bis-tetrahydrofurans synthesis via   556
– enshuol   562
– *ent*-abudinol B synthesis via   564
– enzymatic ester hydrolysis   555
– first-generation approach to   557
– glabrescol   561
– ladder polyethers synthesis   565–583
– omaezakianol synthesis via   563
– polyether ionophores synthesis   550–554
– – applications of   554–558
– – bis-tetrahydrofurans   553
– – 2,5-linked tetrahydrofurans   553
– second-generation approach to   557
– single-electron oxidation of homobenzylic ethers   555

– in squalene-derived polyethers synthesis 558–565
– third-generation approach to 557
epoxide-opening reactions
– Baldwin's rules in 538–539
– regioselectivity control in 539
epoxyquinols A–C 615
epoxysorbicillinol 741
epoxytwinol 615
equisetin 761–763
erinacine E, Nakada's biomimetic synthesis of 426
ervatamine alkaloids 102–105
ervitsine alkaloids 102–105
erythromycin 595
eurypamide B synthesis 341
eusynstyelamide A 26
exiguamines 737–738

f

fastigiatine, total synthesis of 52–53
fatty acid biosynthesis 594–597
fatty acid synthases (FASs) 473
ficuseptine 25–26
Fischerella muscicola 859
(−)-Fischerindole I 164–166
fissoldhimine 22–25
– biogenetically inspired heterodimerization toward 24
– biosynthetic hypotheses 23
– structures 23
flavin mononucleotide (FMN) 679
forbesione, biomimetic synthesis of 456
– Nicolaou approach to 458–459
– via Claisen/Diels–Alder/Claisen reaction cascade 456
FR182877 514–521
– acyclic system related to 517
– biosynthetic origin for 515
– large-scale synthesis of 516
– Sorensen's biomimetic synthesis of 515
FR-901483 compounds 61–86
– Ciufolini synthesis of 80–86
– Snider synthesis of 64–67
– – aldol step in 66
– Sorensen synthesis of 78–79
– synthesis via oxidative amidation chemistry 77–86
– total syntheses of 63–71
– Wardrop approach to 77
fredericamycins 743–744
frondosins 745
furanocembranoids, biomimetic relationships among 414–417
– bipinnatin J as precursor 415
furofuran lignans, biomimetic synthesis of 681–683
– enzyme-mediated 682–683
– metal-catalyzed approaches 682

g

Galbulimima alkaloids 271–275, 509
– Baldwin's biomimetic synthesis of 511
– Class I 272–273
– – Baldwin's biosynthetic hypothesis for 272, 274
– Class II 273–275
– – Movassaghi's biosynthetic hypothesis for 275–276
– Class III 273–275
– – Movassaghi's biosynthetic hypothesis for 275–276
galiellalactones 511–512
galloyl esters 642–645
gambogin synthesis by Nicolaou group 458–459
Garcinia forbesii 452
Garcinia hanburyi 452
Garcinia subelliptica 434
Garcinia xanthones, biomimetic synthesis 455–464
gardenamide 293–294
garsubellin A, total synthesis of 441–443
– by Danishefsky *et al.* 442
– by Shibasaki group 442
– by Simpkins group 450
GE2270A 334–336
– Nicolaou and Bach works 335
geissoschizine 101
(−)-Gelselegine 166–168
– biosynthesis proposal by Sakai *et al.* 167
Gelsemium elegans 166
gentianine 891
Geranium thunbergii 648
Gibbs' phase rule 836
Gibbs–Thomson rule 838
GKK1032 compounds
– biosynthetic origin of 528
– cyclization mechanism 528
– Oikawa's hypothesis for 529
glabrescol 561
gliotoxin 118, 145
– Kishi's total synthesis of 146
globiferin, biomimetic conversion into cordiachrome C 427
glucosidases, as by-products formation triggers 852–853
glucosinolates, hydrolysis of 854

glutaconaldehydes 213–215
glutacondialdehydes 190
glutamate dehydrogenase (GDH) 787
glutaraldehyde
– alkaloid skeletons from 13–15
– condensations of 14
glycosidation 347
grandione 768
griseorhodin A 743
guanidinium alkaloids 225–267
– biomimetic synthesis of 225–267
gymnodimines 282–284, 514
– Kishi's biomimetic approach to 283
– plausible biosynthetic origin of 283
– structure 283
gypsetin 118, 126–127
– synthesis of 127

h

halicyclamines 201–203
– Baldwin–Marazano concepts 207
– biomimetic models toward 205–208
– first generation approach to 206
– halicyclamine A, biomimetic synthesis 207
– second generation approach to 206
Halocarpus biformis 885
Haloxylon salicornicum 12
heliocides 770
hemibrevetoxins 580–582
hemiterpenes 853
Hericium erinaceum 424
hetero-Diels–Alder formation of bicyclo[2.2.2]diazaoctanes 127
heterosides hydrolysis 900
Heteroyohimbines 95–99
hexacyclinic acid 514–521
– biosynthetic origin for 515
hexafluoro-isopropanol (HFIP) 730
hexahydroxydiphenic acid (HHDP) double esterification of 651–658
himandravine 509
himastatin 357, 369–374
– himastatin pyrroloindole core synthesis 372–373
– synthetic approaches to 372
himbacine 272, 509
himbeline 509
Hirsutella nivea 525
hirsutellones 525–530
– 6,5,6-fused system of 525–530
– macrocycle of 525–530
– Nicolaou's total synthesis of 529
– structure of 527
(±)-hobartine 95

homo-Wagner–Meerwein transposition 734
hopeahainol A 708
hopeanol 708
Horner–Wadsworth–Emmons type olefinations 341
horse radish peroxidase (HRP) 679
Humulus lupulus, MPAPs from 435
Husson's strategy (modified Polonovski reaction) 188
Hutchinson's biosynthesis 152
hydrodistillation 880–885
– artifacts from (+)-chrysanthenone 884
– lactones of *Halocarpus biformis* formed during 886
– polyene splicing 881
– *Zizyphus jujuba* seeds 882
hydrolysis, artifactual 897–900
– cubebin anomers from *Aristolochia* spp 900
– of heterosides 900
– methyl-ester hydrolysis 900
– stephacidin B on silica gel 898
hydroperoxides, artifacts from 858
6-hydroxymusizin, Harris' biomimetic synthesis 493
hymenialdisines 247–250
hymenin 249
(−)-hyperforin, total synthesis 445–448
– catalytic asymmetric synthesis of 446
– *ent*-hyperforin 447
hyperguinone B 440–441
Hypericum chinense 878
Hypericum papuanum 434, 440
Hypericum perforatum 434

i

Iboga alkaloids 106
(±)-ialibinone A and B 440–441
imidazole, biomimetic conditions using 476
imide-bearing aconitine-like alkaloids 893
imines, Brønsted acid catalyzed transfer hydrogenation of 788–799
imino esters, Brønsted acid catalyzed transfer hydrogenation of 788–799
α-imino esters, hydrogenation 796
– organocatalytic asymmetric transfer 797
indanomycin 764–766
indole alkaloids 149–175, *see also* modified indole nucleus alkaloids
– indole nucleus conversion into first derivatives 150
indole–indole coupling 381–382
– Witkop-type photo-induced macrocyclization 382

indolemonoterpene alkaloids 91–113
– botanical distribution 91–93
– classification 91–93
indole-oxidized cyclopeptides 357–382,
 see also chloptosin; himastatin
– celogentin C 357, 363–368, see also
 individual entry
– indolyl–phenyl coupling 359
– NCS mediated oxidative coupling 368
– TMC-95A-D 357–363
indoles, hydrogenation 805–806
– asymmetric brønsted acid catalyzed
 805–806
indolomonoterpenes 867
indolomonoterpenic alkaloid macrosalhine,
 quaternization of 923
intramolecular Diels–Alder (IMDA)
 cycloaddition 138, 273, 506, 754
intramolecular Heck reaction 385
iodotrimethylsilane (TMSI) 126
ircinal A, biogenesis 210
ircinal alkaloids 200–201
– (4 + 2) cycloaddition strategy towards an
 ircinal model 203
isatisine A 174
islandicin 486
isoacetogenins 905
isoampelopsin D 711
isoanhydrovinblastine 112
(±)-isoborreverine 173–175
– biosynthesis proposal by Koch et al. 173
isocaryophyllene 404
isoglaucanic acid, dimerization process
 towards 504–505
isomerization, light-induced 870–872
– anethole 871
– in or out of solutions 880
– stilbenoids 871

j
jasminiflorine 174
juliprosine 25–26
juliprosopine 25–26
Juncus acutus 774

k
K-13 synthesis 342
Kametani's total synthesis 120
kapakahine A 391
Karenia brevis 545, 548
keramaphidin alkaloids 200–201
keramaphidin B, biomimetic total synthesis
 197–198
– Baldwin's hypothesis validation 198

– model studies 197
keramaphidin model, selective oxidation of
 205
ketimines, hydrogenation 793
– asymmetric biomimetic transfer 793
– enantioselective biomimetic transfer 793
– metal-free asymmetric transfer 793
ketosynthase (KS) 473
kijanimicin 763–764
Knoevenagel condensation 366–367, 483
komaroviquinone 874
kutzneride 371
kuwanons 776

l
lachanthocarpone 768
lactonic compounds 905–908
– annonaceous acetogenins 907
– epimerization 905–908
– methanolysis of 908
– *Thalictrum* saponins cyclization during
 acidic hydrolysis 906
– transesterification 905–908
ladder polyethers 545–550
– dioxepandehydrothyrsiferol 546
– Giner's proposal for biosynthesis 549
– Nakanishi's hypothesis 548
– structures 547
– synthesis 565–583
– – applications 580–583
– – 6-*endo* cyclization 567
– – fused polyether systems 567–580
– – iterative approaches 565–567
– – Jamison proposal 577
– – McDonald group 570–573
– – Murai's work 569
– – THP : THF selectivity in 578
– *trans-syn-trans* arrangement 545
Laggera tomentosa 880
Lahav's rule of reversal 837
lambertellol 860
lateriflorone biosynthesis 461
Laurencia pinnatifida 901
Laurencia spectabilis 901
Leuzea carthamoides 892
life's single chirality 823–841, see also
 biological homochirality
light, see also electrocyclizations
– in artifacts and natural substances
 formation 870–878
– – epimerization 870–872
– – isomerization 870–872
– – milnamide A 872
– – reserpine 872

light, see also electrocyclizations (contd.)
– photochemical reactions: see Chapter 16
– photocycloaddition   876–878
– photodimerization   876–878
– rearrangements by   872–876
– – brevianamide A   875–876
– – *ent*-bicyclogermacrene   873
– – komaroviquinone   874
lignans   677–691
– biomimetic synthesis of   681
– – benzo[*kl*]xanthenes   686–688
– – benzoxanthenone lignans   683–686
– – furofuran lignans   681–683
– – podophyllotoxins   681
– chemotypes of   678–679
Liquidambar formosana   663
lissoclinamides   326–328
– to heptapeptide   327
– entry into secondary metabolism   5–6
– metabolism toward alkaloids   6
Lobelia alkaloids   44
longithorone   779
– longithorone A   427
Lonicera japonica   916
Lonicera korolkoviii   916
Lonicera morrowii   916
– entry into secondary metabolism   5
lovastatin   507, 755–756
– lovastatin nonaketide synthase (LovB)   755–756
lupine alkaloids   31–34
– biomimetic conversion into oleane skeletons   423
– biomimetic synthesis of   32–33
– – oxidative deamination step   32
Lycopodium-like alkaloids   50–51
– *Lycopodium* alkaloids   44–54
lycoposerramine series   49–50
Lyngbya majuscula   897
lysine-derived alkaloids   3–54
– biomimetic synthesis of   30–42
– L-arginine   3
– L-lysine   3
– L-ornithine   3
lysine-derived reactive units   11–15
– oxidative degradation of free L-lysine   11–12
– Schöpf's pioneering works   13
– tetrahydropyridine   12

m
macquarimicins   514–521
– biosynthetic hypothesis for   518
– structure   517

– Tadano's biomimetic synthesis   521
– Tadano's model study   520
macrocyclic complex alkaloids   183
macrocyclization   340
macrolactamization   338, 347, 360, 373–374
macrophomate synthase (MPS)   754, 756–758
– catalytic mechanism of   756–757
– Michael–aldol route   756
macroxine   174
madangamine alkaloids   208–210
– madangamine C type alkaloids
– – biogenesis   209
– – biomimetic synthesis   210
maitotoxin   595
(+)-makomakine   95
Malbranchea aurantiaca   140
malbrancheamides   118, 136
– malbrancheamide B   136
– proposed biosynthesis of   138
– total syntheses of   138
mallotusinin production   662
malondialdehyde scenario   182–191, 200–203
– aminopentadienal connection   202
– halicyclamine connection   201–203
– keramaphidin/ircinal connection   200–201
malonic acid half-thioesters (MAHTs)   474
malonyl activation   475–477
malonyl acyl transferase (MAT)   474
malonyl half thioesters (MAHT)   479
– asymmetric and organocatalytic addition to nitroolefins   482
– Shair's catalytic aldol condensation with   479
– – asymmetric   480
Mannich bisannulation   385
manzamine A
– ABC-ring system synthesis of   204
– AB-ring system synthesis of   204
– biomimetic models toward   203–204
manzamine alkaloids synthesis   181–221, see also Baldwin's hypothesis development; pyridinium marine sponge alkaloids
– acrolein scenario   182–191, see also *individual entry*
– 3-alkylpyridiniums biosynthetic hypotheses based on pachychaline series   187
– aminopentadienals   190, 213–215
– Baldwin's hypothesis, dihydropyridine chemistry   186
– biomimetic C5 reactive units from Zincke reaction   189–191
– Chichibabin synthesis of pyridines   188

- cyclostellettamine A type   184–185
- from cyclostellettamines to keramaphidin and halicyclamine/haliclonamine alkaloids   218
- dimers (and oligomers)   184
- from fatty acids to long-chain aminoaldehydes and sarain alkaloids   215
- from fatty aldehydes precursors to simple 3-alkyl-pyridine alkaloids   182–187
- from ircinal and pro-ircinals to manzamine A alkaloids   218
- glutaconaldehydes   213–215, see also glutacondialdehydes
- glutacondialdehydes   190, see also glutaconaldehydes
- Husson's strategy (modified Polonovski reaction)   188
- ircinal pathway, spinal cord of manzamine metabolism   218
- macrocyclic complex alkaloids   183
- madangamine alkaloids   208–210
- malondialdehyde scenario   182–191, see also individual entry
- manzamine alkaloid chemistry, milestones in   185
- Marazano biomimetic synthesis of dihydropyridine   188
- Marazano modified hypothesis, pyridinium chemistry   186
- modified hypothesis testing in laboratory   203–208
- – biomimetic models toward halicyclamines   205–208
- – biomimetic models toward manzamine A   203–204
- – (4 + 2) cycloaddition strategy towards an ircinal model   203
- monomers   184
- nakadomarine A, biomimetic model of   210–211
- from pro-ircinals to madangamine alkaloids   218–219
- pyridine ring formation   186
- theonelladine A type   184–185
- total syntheses of   219–220
- towards a universal scenario   215–219
- xestospongins   184–185, 191–193
Marazano biomimetic synthesis of dihydropyridine   188
Marazano's hypothesis   201
marcfortine C   135
- total synthesis of   137
marcfortines   155–158

marine diterpenes biomimetic synthesis from *Pseudopterogorgia elisabethae*   410–414
- colombiasin A   412
marine polypropionates   877
marine pyrrole-2-aminoimidazole alkaloids, See pyrrole-2-aminoimidazole (P-2-AI) marine alkaloids
marine thiol group   856
Markhamia lutea   857
masked hexaketides, Schmidt's condensation of   492
massadine chloride   263–265
massadine   257, 263–265
- Baran's biogenetic hypothesis   264
- formation, intramolecular aziridinium mediated mechanism for   259
- Romo's biosynthesis proposal for   261
mauritiamine   253–254
meleagrine   174
meloscandonine   174
membrane diffusion, symmetry breaking in   840
meroterpenoids, biomimetic synthesis of   424–425
mersicarpine   174
metal-catalyzed cross coupling, Trp-Tyr biaryl bond formation by   361
methanolysis of lactonic sesquiterpenes   908
methoxymethyl (MOM)-protection   497
6-*O*-methylforbesione synthesis   458–459
methyl homodaphniphyllate   304
methyl homosecodaphniphyllate   301
methyllateriflorone synthesis   459–460
methyllateriflorone, total synthesis of   462
7-methylcycloocta-1,3,5-triene   618
Mg(II) salts, biomimetic conditions using catalytic   476
milnamide A   872
minfiensine   174
modified indole nucleus alkaloids   149–175, see also camptothecin; discorhabdins
- biomimetic synthesis of   149–175
- – monoterpenoid indole alkaloids   150
modified Julia coupling   360
Monascus ruber   506
mongolicumin A   686–687
monocyclic polyprenylated acylphloroglucinols (MPAPs)   434
- from *Humulus lupulus*   435
monomers   184
monoterpene rearrangements   397–401
- century since Wagner's structure of camphene   397–399
monoterpenoid indole alkaloids   150

Montmorillonite K-10 (MK-10)   165
Morus bombycis   775
Myrioneuron alkaloids   34–39

n

nakadomarine A, biomimetic model of   210–211
nakamuric acid, Baran's synthesis of   255
nanaomycine, Yamaguchi's biomimetic synthesis   494
naphthalenoid derivatives   492–494
– barakol, Harris' biomimetic synthesis of   493
– biomimetic access to   492–494
– eleutherinol, Harris' biomimetic synthesis of   493
– 6-hydroxymusizin, Harris' biomimetic synthesis   493
– naphthyl cyclization of β-hexaketones   493
– polyketides into, Yamaguchi's aromatic cyclization   494
nargenicin A   508–509
N-chlorosuccinimide (NCS) mediated oxidative coupling   368
Negishi-cross coupling   340
nemorosone, total synthesis through 'carbanions' differentiation   443–445
neocarzinostatin   595
neolignans   677
neopupukeananes   406
neoselaginellic acid   174
neosymbioimine   276–279
nepalensinol B   713
N-heterocycles, asymmetric organocatalytic reduction   800–814
– enantioselective hydride transfer   800
– enantioselective protonation   800
Nicotiana tabacum   10
nicotinamide adenine dinucleotide (NADH)   787
nitraramine, biomimetic synthesis of   35–37
nitraria alkaloids   14, 34–39
nitrophenyl pyrones   618–621
N-methylcytisine conversion into kuraramine   33–34
N-methyltriazolinedione (MTAD)   125
Nocardia argentinensis   508
nonadride series   504–506
– biomimetic studies in   504–506
– CP-225917   505–506
– CP-263114   505–506
– dimerization process towards isoglaucanic acid   504–505
– Sutherland's biomimetic studies   504

non-amino acid origin alkaloids, biomimetic synthesis   271–307, *see also* cyclic imine marine alkaloids; *Galbulimima* alkaloids
non-aromatic polycyclic polyketides   503–530, *see also* nonadride series
non-prenylated indole alkaloids   141–146, *see also* epidithiodioxopiperazines
non-ribosomal peptide synthesis (NRPS)   319–320, 346
norrhoedanines in acidic conditions   901
norzoanthamine   290
notoamide J synthesis   121
– Williams' biomimetic synthesis of   124
nucleophilic 1,2-addition   380

o

ocellapyrones   621–624
– ocellapyrone A, electrocyclic formation of   624
o-iodoxybenzoic acid (IBX)   726
– dimerization of 2,6-xylenol   726
– Pettus' oxidative dearomatization   726
okaramine N   118, 125
oligomeric ellagitannins   658
oligomers   695–718, *see also* resveratrol-based family of oligomers
– synthetic approaches to   695–718
olivacine alkaloids   102–105
omaezakianol synthesis   563
o-quinone dimerization   727
L-ornithine   3
ornithine alkaloids   18–30
– reactive units   9–11
– – 4-aminobutyraldehyde   9
oroidin   237–238
– Al-Mourabit's synthesis of   239
– Lindel's conversion into *rac*-cyclooroidin   240
orsellinic acid   486
ortho-quinone methide capture   385
Osmunda japonica   885
oxasqualenoids   542–544, 558–560
oxidation processes   853–870
– achiral bisacridones from *Rutaceae*   866
– allicin   855
– bis-aporphines   864
– dioxoaporphines   868
– discorhabdin B   856
– hydroperoxides, artifacts from   858
– indolomonoterpenes   867
– lambertellol   860
– marine thiol group   856
– newly oxygenated products   859–864
– *N*-oxide and oxoalkaloid cases   865–870

- oxidative coupling 865
- oxoaporphines 868–869
- of oxygenated functions 857–859
- pyrroloiminoquinolines 856
- *Tabernaemontana* spp. 868
- thiol oxidation 853–856
- vasicoline 868
- welwitindolinones from *Fischerella* spp 862
oxidative cyclization 347, 440
oxidative dearomatization 723–747
- Adler–Becker oxidation, Singh's application of 727
- Canesi's 735
- Danishefsky's 730
- Diels–Alder dimerization 725
- Feldman's 734
- Gaunt's 732
- Heathcock's synthesis of styelsamine B 732
- initial intermediate 723–724
- intermolecular dimerizations 724–727
- intramolecular cascade sequences 731–733
- intramolecular cycloadditions 729–731
- Liao's 729
- Majetich's 737
- Morrow's 728
- Nakatsuka's 740
- Njardarson's 730
- Pettus' 731, 735
- phenol oxidative cascades 741–747
- Porco's 734
- Quideau's 739
- rearrangements 733–737
- Rodríguez's 733
- Rogić's 740
- Sarpong's 729
- sequences 723–724
- sequential reactions initiated by 723–747
- sequential ring rupture
- - and contraction 737–739
- - and expansion 739
- Sigman's enantioselective 728
- Sorensen's 731
- Stoltz' 729
- successive intermolecular reactions 727–729, 741
- successive intramolecular reactions 741
- successive tautomerizations 733–737
- Takeya's *o*-quinone dimerization 727
- Tejera's 740
- Trauner's 736, 738
- Wood's 730
- Yamamura's 728

oxidative diversification 415
oxindole fragment, stereocontrolled oxidation of 361–362
oxindoles synthesis 139
oxoaporphines 868–869
oxysceptrins 254–255
- Baran's synthesis of 255

p

P-2-AIs simple dimers, biomimetic synthesis 253–255
- ageliferins 254–255
- mauritiamine 253–254
- oxysceptrins 254–255
- sceptrins 254–255
Pachychalina species 186
Paeonia lactiflora 913
Paeonia suffruticosa 913
paeoniflorin 917
palau'amine 255–257, 265
- Al-Mourabit's biogenetic proposal for 260
- axinellamine A 257
- axinellamine B 257
- massadine 257
- synthesis
- - first proposal based on Diels–Alder key step 257
- - Kinnel's biogenetic proposal for 258
- - Scheuer biogenetic proposal for 258
paliurine F synthesis 339
pallavicinolide A 417, 420
pallidol 702, 713
paraherquamide A, biosynthesis of 130
paraherquamides 155–158
Paraphaeosphaeria quadriseptata 901
paucifloral F 711
pedunculagin 658
pelletierine
- based metabolism 42–54
- biomimetic synthesis 43–44
Penicillium brevicompactum 117
Penicillium glaucum 504
Penicillium islandicum 486
Penicillium purpurogenum 504
penifulvins 405–406
pentacarbonyl derivative, Harris' biomimetic cyclizations of 492
pentacycle formation 385–389, *see also under* Ecteinascidin 743 (ET 743)
pentacyclization 305–306
peptide alkaloids 317–318, *see also* aryl-containing peptide alkaloids; azole-containing peptide alkaloids
- aryl-alkyl ether peptide alkaloids 337

peptide alkaloids (*contd.*)
- - ring-closing strategies in   338
- - ring formations in biosynthesis of   337
- biosynthesis, key features   319–321
- covalent folding of peptide chains into   321
- cyclic peptides containing biaryl ethers   339–343
- cyclized by aryl side chains oxidation   336–350
peptide fragment coupling   347
perovskone   420, 770
phalarine   174
phalloidin   391
phenol oxidation   768–775
phenol oxidative cascades   741–747
- additional natural compounds arising   746
- Hertweck's   743
- Pettus'   742, 745
- Porco's   746
- Shen's   744
- Steglich's   743
- Zhao's   743
phenols, oxidative amidation of   71–77
- Honda oxidative cyclization   75
- Knapp iodocyclization   72
- oxidative spirocyclization   72
- stereoselective cyclization of   73
- Wardrop oxidative cyclization   76
phenoxonium species   727
phenyl iodine diacetate (PIDA)   725
phenyl iodine(bis)trifluoroacetate (PIFA)   728
phlegmariurine series   49–50
phloracetophenone   486
phosgene adducts   921
photochemical reactions, *see also* light, electrocyclizations
photocycloaddition   876–878
- hoenalia coumarin   876
- marine polypropionates   877
photodimerization   876–878
Phyllanthus emblica   645
Pictet Spengler cyclization   121, 385
pinnatoxins   279–282
- (−)-pinnatoxin A, Kishi's biomimetic synthesis   281
- pinnatoxin A, biosynthetic origin of   281
L-pipecolic acid   6
pipecolic acids   15–18
- biomimetic access to   15–18
- biosynthesis   15–16
- - by photocatalysis   18
- containing secondary metabolites   16
- importance   15–16

- Rossen's biomimetic synthesis of   17
- Yamada's biomimetic access to   17
piperidines, hydrogenation   813
Plakortis angulospiculatus   509
p-nitrophenyl pyrones   619
podophyllotoxins   681–682
polyamine alkaloids   7–8
- polyamine backbones in   7
polycyclic polyethers, *see also* ladder polyethers; polyether ionophores; squalene
- biosynthesis   539–550
- epoxide-opening cascades in   550–583, *see also individual entry*
- structure   539–550
polycyclic polyprenylated acylphloroglucinols (PPAPs)   433–452
- biomimetic synthesis of   436–441
- biosynthesis of   434–436
- classification of   434
- from MPAPs   437
- non-biomimetic synthesis of   441–451
- - Garsubellin A   441–443
- synthesis via oxidative cyclization reactions   440
- Type A PPAPs   439–440
- - via an intramolecular Michael addition   440
polyene/polyene splicing   607, 878, 881
polyepoxide opening, polyether natural products synthesis via   537–584
- synthetic considerations, Baldwin's rules   538–539
- - 4-*exo-trig* reactions   538
- - 5-*endo-trig* reactions   538
polyether ionophores   539–542
- Cane–Celmer–Westley hypothesis   540
- *endo* cyclizations   540
- structures of   541
polyketide assembly mimics/polyketide synthases (PKSs)   472–485, *see also* aromatic polyketides
- C–C connection mechanism in   473
- - addition–decarboxylation for   481
- structure   473
- Type-a mimics   475–478
- - acyl transfer, Scott's conditions for   475
- - catalytic Mg(II)salts, biomimetic conditions using   476
- - imidazole, biomimetic conditions using   476
- - malonyl activation   475–477
- - thioesters, biomimetic conditions using   476
- - without malonyl activation   477–478

- Type-b mimics 478–479
- – Coltart's aldol addition with non-activated thioester 482
- – enolate formation before nucleophilic addition 480
- – Fagnou's metal-free decarboxylative condensation 480
- – malonyl activation 479–482
- – without malonyl activation 482–483
- Type-c mimics 483–485
- – Barbas III asymmetric and organocatalytic addition of thioesters 483
- – Birch reduction–ozonolysis reaction 485
- – List's condensation of MAHO in 484
- – reaction mimic with MAHT 484
polyketides (PK) 284–293, 485–499, 503–530, 591–632
- aromatic polyketides 485–499
- biological electrocyclizations 628–631
- biomimetic analysis 597–598
- biosynthetic origin proposed by Morita 285
- Black's electrocyclization cascade hypothesis 617
- cassiarins A and B 284–285
- decalin systems 506-509
- electrocyclic reactions 592
- 6π electrocyclizations, 598–612, see also individual entry
- electrocyclization reactions toward 591–632
- elysiapyrones 624–625
- endiandric acids A–G 615
- enzyme catalysis 628–631
- epoxyquinols A–C 615
- epoxytwinol 615
- fatty acid biosynthesis 594–597
- general biosynthesis 596
- nitrophenyl pyrones 618–621
- nonadrides 504–506
- non aromatic polyketides 503–530
- p-nitrophenyl pyrones 619
- shimalactones 625–628
- structure 285
- 8π systems and black 8π–6π electrocyclic cascade 612–628
- torreyanic acid 614
polyolefin cyclization 421
polyprenylated phloroglucinols 433–464, see also polycyclic polyprenylated acylphloroglucinols (PPAPs)
polyprenylated xanthones 452–464
- biosynthesis of 454–455
- – CGX motif via cascade of nucleophilic attacks 455

- – CGX motif via Claisen/Diels–Alder reaction cascade 455
- caged xanthones 452
- Diels–Alder cycloaddition 461
- Wessely/Diels–Alder strategy 461
Pomerantz-Fritsch reaction 385
Popowia pisocarpa 865
Populus deltoides 897
Porco synthesis of clusianone 438
potassium hexamethyldisilazide (KHMDS) 525
prenyl side chain dehydrogenation 775–779
prenyl-9-borabicyclo[3.3.1]nonane (prenyl-9-BBN) 119
prenylated indole alkaloids 117–141
- notoamide J synthesis 121
presilphiperfolanol 404
pretetramide, Harris' biomimetic synthesis of 496
pro-ircinal alkaloids 218–219
- to madangamine alkaloids 218–219
- to nakadomarine alkaloids 219
L-proline 6 see also pyrrole-2-aminoimidazoles
proline, dioxopiperazines derived from 119–122
protoberberines 921–925
- acetone adducts of 928
- alkaline treatment of 924
- – nucleophilic additions on 925
- dihydroprotoberberines 927
- indolomonoterpenic alkaloid macrosalhine, quaternization of 923
proto-daphniphylline 301
- pentacyclization of 305
przewalskin A 743
Pseudomonas fluorescens 608
pseudopelletierine 43–44
Pseudopterogorgia bipinnata 414
Pseudopterogorgia elisabethae, marine diterpenes biomimetic synthesis from 410–414
Pseudopterogorgia kallos 416
pseudorubrenoic acid A 611
pteriatoxins 279–282
Pueraria mirifica 863
(−)-pumiliotoxin C, Amat's biomimetic synthesis of 287
Pummerer oxidative cyclization 240
Punica granatum 919
purifications, acidic conditions during 895–903
- epimerization 895–897
putrescine *N*-methyltransferase (PMT) 5

puupehenone, methanol adduct on 914
pyranose-type ellagitannins 665–667
pyridine alkaloids 215–217
– Chichibabin synthesis of 188
– hydrogenation 813–814
pyridinium chemistry, Marazano modified hypothesis 186
pyridinium marine sponge alkaloids, biomimetic synthesis 191–195, see also xestospongins
– 3-alkylpyridiniums 191
– alkylpyridines with unusual linking patterns 194–195
– cyclostellettamine B 191
– upenamides, synthetic approaches to 193
– Zincke-type pyridine ring-opening 193–194
pyridinium salts 181–182, see also manzamine alkaloids synthesis
pyrinadine A, biomimetic synthesis 195
pyrinodemin A, biomimetic synthesis of 194
pyrones
– Harris' biomimetic access to 489
– as masked tetraketide 490
pyrrole-2-aminoimidazole (P-2-AI) marine alkaloids 225–267, see also clathrodins
– Al-Mourabit's retro-biogenetic proposal for 232
– biomimetic synthesis of 225–267
– George Büchi's work 233–234
– new challenging P-2-AI synthetic targets and perspectives 266–267
– P-2-AI biosynthesis, common chemical pathway for 256–257
– P-2-AI linear monomers, biomimetic synthesis 237–238
– P-2-AI polycyclic monomers, biomimetic synthesis 234–253, see also cyclized monomers
– P-2-AIs simple dimers, biomimetic synthesis 253–255
– synthetic achievements 261–265
– tautomerism in building blocks of 229
pyrrolizidine ring, biomimetic access to 18–19
pyrroloiminoquinolines 856
pyrroloindole-based peptide alkaloids 369
– dimeric 370

q

quadrangularin A 701, 710
Quercus robur 669
quinolines, asymmetric organocatalytic reduction 800–805

– asymmetric biomimetic transfer 798
– Brønsted acid catalyzed transfer 801
– organocatalytic asymmetric transfer 798
– 2,3-substituted quinolines 803
– 3-substituted quinolines 804
– 4-substituted quinolines 804
quinolinic acid 194
quinoxalines, hydrogenation 806
quinoxalinones, hydrogenation 806

r

rameswaralide 417–419
raucaffrinoline via Cannizzaro reaction 896
reaction–diffusion models, symmetry breaking in 840
red tides 545
reductive aminal formation 380–381
reserpine 872
resveratrol-based family of oligomers 695–718, see also oligomers
– biosynthetic approaches 697–705
– davidiol A from 704
– indane-containing members of 711
– palladium-based reactions 706
– quadrangularin A 701
– stepwise synthetic approaches 705–717
– – work toward single targets within 705–709
– synthetic approaches to 695–718
– universal, controlled synthesis approach 709–717
– ε-viniferin from 698–700
rhazinilam 93
Rhodomela confervoides 897
ribosomal peptide synthesis (RPS) 319–320
ritterazines 294–298
– structures 295
Robinson-Gabriel cyclodehydration 325
Rubiaceae iridoids 910
rubifolide conversion into coralloidolides A, B, C, and E 416
rufescidride 686–687
Ru-mediated S_NAr-cyclization 340

s

Saccharopolyspora spinosa 521
saframycin A biosynthesis, gene cluster-based proposal for 383
Salvia leucantha 427
Salvia prionitis 769
salvileucalins A and B 428
Sammes' model study of cycloaddition 127
sanguiin H5, synthesis of 645
sanjoinine G1 synthesis 339

sarains
- biomimetic models of, side branch of manzamine tree   211–213
- biomimetic synthesis   212
- - first sarain A model   213
- - second sarain A model   213
- sarain A-type alkaloids
- - biogenesis   212
sceptrins   254–255
secologanin   150
- derived indolomonoterpene alkaloids   95–109
- derived quinoline alkaloids   109–110
Securidaca longepedunculata   921
Securiflustra securifrons   162
Sedum alkaloids   44
self-replication   833–834
senepodine G   51–52
serratezomine A   47
serratinine   47
- into lycoposerramine B   47–49
- into serratezomine A   48
sesquiterpene rearrangements   401–408
- caryophyllenes in   401–402
- miscellaneous sesquiterpene rearrangements   406–408
shimalactones   625–628
shoreaphenol   708
silphinane series, oxidative rearrangements in   405–406
silphinyl mesylate   405
Silybum marianum   865
silydianin   754
siomycin A   331
SNF4435 C and D   618–621
- Baldwin's approach   620, 623
- Parker's approach   622
- Trauner's approach   620
Soai reaction   830–831
sodium dodecyl sulfate polyacrylamide gel electrophoresis (SDS-PAGE)   759
solanapyrone synthase (SPS)   758–760
- *endo/exo*-selectivities   759
solanapyrones   507–508
solasodine   294
Sophora flavescens   33
sophoradiol, biomimetic synthesis of   421
sorbicillin   726
spinosyns   766–767
- biomimetic TADA reactions toward   521–524
- biosynthesis of   522
- Roush's total synthesis of   523

spiro systems, Diels–Alder reactions   512–514
- abyssomicin C   512–513
- gymnodimine   514
spirolactam formation   73
spirotryprostatin A, Danishefsky's synthesis   123
spirotryprostatin B   118
- Danishefsky's synthesis   122
($\pm$)-sporidesmin A, total synthesis of   145
spontaneous phenol-aldehyde cyclization   385
squalene, polyethers derived from   542–545, 558–565
'stabilized' iodoxybenzoic acid (SIBX)   739, 769
Strecker reaction   385
Stemona spp.   863
Stenus comma   39–42
stenusine   39–42
- natural versus biomimetic   41
- putative biosynthetic pathway   40
- stereochemical particulars   40
- structure   40
stephacidins
- biosynthesis proposal for   160
- stephacidin A   134–136
- - conversion to stephacidin B   136
- - improved biomimetic synthesis of   135
- stephacidin B   136
- (+)-Stephacidin A   158–160
- - biosynthesis through notoamide S   136
- - total synthesis by Baran *et al.*   161
- (−)-Stephacidin B   158–160
- - synthesis by Baran *et al.*   162
- - biosynthesis through notoamide S   136
stilbene synthase   697
stilbenoids   871
strellidimine   113
Streptomyces antibioticus   764
Streptomyces coelicolor   486
Streptomyces fradie   496
Streptomyces longisporoflavus   524
Streptomyces orinoci   878
strictosidine alkaloids   95–99
strictosidine   92, 150
strychnine   103
strychnochromine   174
styelsamine B   732
- Heathcock's synthesis of   732
stylissadine A formation
- aziridinium mechanism for   259–261
- from massadine, Baran and Köck's proposal   261

stylissazole C 266
styrylpyrone photodimers 879
supercritical CO_2 treatment 885–888
– chalcones obtained from 889
superstolide A 509
– Roush's total synthesis of 510
Suzuki-Miyaura coupling 345, 360, 564
symbioimine 276–279
– biomimetic synthesis of 277
– – Snider's approach 278
– – Thomson's approach 278
– biosynthetic origin of 278
– Chruma's contribution to 279

t

Tabernaemontana spp. 868
TAN-1251 compounds 61–86
– Ciufolini synthesis of 80–86
– Honda synthesis of 79–83
– – aldol cyclization 85
– Snider synthesis of 68–71
– – solvent effects in 69
– synthesis via oxidative amidation chemistry 77–86
– total syntheses of 63–71
– Wardrop approach to 77
tangutorine 37–39
tannins 639–672, *see also* ellagitannins
– condensed tannins 639
– hydrolyzable tannins 639
tautomerism in building blocks of P-2-AI monomer clathrodin 229
Teichaxinella morchella 231
tellimagrandin I 643–644
terengganesine B 174
terpene precursors alkaloids 293–305, *see also Daphniphyllum* alkaloids
– barbaline 294
– cephalostatins 294–298
– daphniglaucine A 294
– gardenamide 293–294
– ritterazines 294–298
– solasodine 294
terrecyclene 405
tetracyclic derivatives 495–499
– anthracenoids, Yamaguchi's access to 497
– benzo[a]tetracenoid derivatives 498–499
– biomimetic access to 495–499
– – tetracenoid derivatives 495–496
– pretetramide, Harris' biomimetic synthesis of 496
– tetrangomycin, Krohn's synthesis of 498
– tetraphenoid derivatives 496–498
– (–)-urdamicynone 497

tetrahydroanabasine chemistry 12–13
tetrahydrofuran ring 548
tetrahydroindane systems 509–512
– Diels–Alder reactions affording 509–512
– galiellalactones 511–512
– spiculoic acid A 509
– superstolide A, Roush's total synthesis of 510
tetrahydropyran ring 548
tetrahydropyridine 12
tetrangomycin, Krohn's synthesis of 498
tetrapetalone C 746
2,2,6,6-tetramethylpiperidine (TMP) 415
tetrocarcin A 763–764
tetronasin
– biosynthetic origin of 524
– Ley's formal synthesis 526
– synthesis 524–525
– Yoshii's total synthesis of 527
thallium trinitrate (TTN) mediated cyclization 341
Thapsia garganica 424
theonelladine alkaloids 215–217
theozymes 427
thiangazole 324–326
– to pentapeptide, hypothetical biomimetic simplification of 325
– strategic disconnections in total syntheses 325
thioesters
– biomimetic conditions using 476
– condensation between 476
– malonyl thioesters, self-condensation of 477
thiol compounds reaction 662–663
thiol oxidation 853–856
thiostrepton 328–334
TMC-95A-D 357–363
– (Z)-enamide side-chain 360
– late-stage stereoselective (Z)-enamide formation 362–363
– 3-methyl-2-oxopentanoic side-chain T origin 360
– retrobiosynthetic analysis of 359
– synthetic approaches to 360
topaquinone 32–33
Torreya grandis 768
torreyanic acid 610–612, 614
Townsend–McDonald hypothesis 550
trachyopsane A, biomimetic synthesis of 406
transamination 385
transannular Diels–Alder (TADA) reaction 508
transannular hydride transfers 199–200

transesterification artifacts   909
transesterification, lactonic compounds   905–908
transtaganolides   425
tridachiahydropyrones   599–603
– biomimetic analysis of   600
– biomimetic synthesis of   602
tridachione family   603–608
– 9,10-deoxytridachione   604–606
– oxytridachiahydropyrone   603
– polyene   607
– pseudorubrenoic acid A   608–610
trienes, cyclization of   507
trimethylsilyl trifluoromethanesulfonate (TMSOTf)   564
triquinane series, biomimetic studies in   404–405
tris(dimethylamino)sulfonium difluorotrimethylsilicate (TASF)   735
triterpene rearrangements   420–424
tropinone chemistry   29–30
Trp-Tyr biaryl bond formation by metal-catalyzed cross coupling   361
tryptamine   150
tryptophan alkaloids   91–113, see also indolemonoterpene alkaloids
– biomimetic synthesis of   117–147, see also bicyclo[2.2.2]diazaoctanes; non-prenylated indole alkaloids; prenylated indole alkaloids
– – dioxopiperazines derived from   119–122
– – tryprostatin B, biomimetic total synthesis of   121
TTN-oxidative coupling   340
tyrosine alkaloids   61–86

u

Ugi four component reaction   385
Ullmann-coupling   340–341, 646
upenamides, synthetic approaches to   193
(−)-urdamicynone, Yamaguchi's synthesis   497
usambarine   93

v

vancomycin   345–350
– biaryl-ether formation during biosynthesis of   347
– Evans' synthesis of   349
– Nicolaou's synthesis of   348
– structure of   346
vasicoline   868

vellomisine   894
Veratrum californicum   421
(+)-versicolamide B   118
versicolamides, asymmetric synthesis   140
vescalagin   666, 670
– pyrolytic degradations of   671
vincadifformine   109
vincamine   109
vincorine   174
vincoside alkaloids   95–99
ε-viniferin
– biogenetic explorations using   703
– davidiol A from   704
– from resveratrol   698–700
VM55599   129, 155–158
– biosynthesis of   130
– (−)-VM55599, asymmetric total synthesis of   132
– Williams' biomimetic total synthesis of   130
Vorbrüggen condensation   72

w

Wagner–Meerwein rearrangement   398–400, 410
Wardrop oxidative cyclization   76
welwitindolinones
– from *Fischerella* spp, oxidation   862
– welwitindolinone A
– – biosynthesis proposal for   165–166
– – synthesis by Baran *et al.*   165
– – (+)-Welwitindolinone A   164–166
Wieland–Gümlich aldehyde   101
Williams' biomimetic synthesis
– of notoamide J   124
– of VM55599   130
(+)-WIN 64821 synthesis   143
Winterfeldt-Witkop cyclization   153
Witkop-type photo-induced macrocyclization   382
Woodward–Hoffmann rules   616
Würthner's polymerization model   835

x

xanthepinone, methanol induced rearrangement of   912
xanthones   433–464, see also polyprenylated xanthones
xestospongins   191–193, 215–217
– biomimetic synthesis by the Baldwin group   193
– xestospongins A   192

y

yohimbine 95–99
yunnaneic acid H 686–687
Yuzuriha 298

z

zamamidine C, retrobiosynthesis 220
zearalenone, Barrett's synthesis of 491
Zincke reaction 189–191
– pyridine ring-opening 193–194
Zizyphus jujuba 882
zoanthamine alkaloids 288–291
– biosynthetic origin proposed by Uemura 289
– cyclization substrate synthesis 290
– Kobayashi's biomimetic approach to 291
– proposed biosynthetic route for 290